The Clinical Neurophysiology Primer

THE CLINICAL NEUROPHYSIOLOGY PRIMER

Edited by

ANDREW S. BLUM, MD, PhD

Comprehensive Epilepsy Program
Department of Neurology
Rhode Island Hospital, Brown Medical School
Providence, RI

SEWARD B. RUTKOVE, MD

Department of Neurology
Beth Israel Deaconess Medical Center
Boston, MA

HUMANA PRESS ✻ TOTOWA, NEW JERSEY

999 Riverview Drive, Suite 208
Totowa, New Jersey 07512

humanapress.com

Due diligence has been taken by the publishers, editors, and authors of this book to assure the accuracy of the information published and to describe generally accepted practices. The contributors herein have carefully checked to ensure that the drug selections and dosages set forth in this text are accurate and in accord with the standards accepted at the time of publication. Notwithstanding, as new research, changes in government regulations, and knowledge from clinical experience relating to drug therapy and drug reactions constantly occurs, the reader is advised to check the product information provided by the manufacturer of each drug for any change in dosages or for additional warnings and contraindications. This is of utmost importance when the recommended drug herein is a new or infrequently used drug. It is the responsibility of the treating physician to determine dosages and treatment strategies for individual patients. Further it is the responsibility of the health care provider to ascertain the Food and Drug Administration status of each drug or device used in their clinical practice. The publisher, editors, and authors are not responsible for errors or omissions or for any consequences from the application of the information presented in this book and make no warranty, express or implied, with respect to the contents in this publication.

Cover design by Karen Schulz

Cover illustration: *(Foreground)* Needle EMG, positive sharp wave (Fig. 4, Chapter 14; *see* complete caption and discussion on pp. 233–234). *(Background)* Epileptiform abnormalities, three-Hertz EEG of generalized spike-and-slow wave activity (Fig. 10, Chapter 8; *see* complete caption on p. 114 and discussion on p. 112).

For additional copies, pricing for bulk purchases, and/or information about other Humana titles, contact Humana at the above address or at any of the following numbers: Tel.: 973-256-1699; Fax: 973-256-8341; E-mail: orders@humanapr.com or visit our website at www.humanapress.com

The opinions expressed herein are the views of the authors and may not necessarily reflect the official policy of the National Institute on Drug Abuse or any other parts of the US Department of Health and Human Services. The US Government does not endorse or favor any specific commercial product or company. Trade, proprietary, or company names appearing in this publication are used only because they are considered essential in the context of the studies reported herein.

This publication is printed on acid-free paper. ∞

ANSI Z39.48-1984 (American National Standards Institute) Permanence of Paper for Printed Library Materials.

Printed in the United States of America. 10 9 8 7 6 5 4 3 2 1

eISBN 978-1-59745-271-7

Library of Congress Cataloging-in-Publication Data

Essentials of clinical neurophysiology / edited by Andrew S. Blum,
Seward B. Rutkove.
p. ; cm.
Includes bibliographical references and index.
ISBN 0-89603-996-X (alk. paper)
1. Electroencephalography. 2. Electromyography. 3. Neurophysiology. 4. Nervous system--Diseases--Diagnosis. I. Blum, Andrew S. II. Rutkove, Seward B.
[DNLM: 1. Nervous System Physiology. 2. Electroencephalography. 3. Electromyography. 4. Electrophysiology. 5. Nervous System Diseases--diagnosis. WL 102 E776 2007]
RC386.6.E43E87 2007
616.8'047547--dc22

2006014475

PREFACE

With the growth of combined clinical neurophysiology fellowship training programs and their corresponding "pan-physiology" board examinations, there has been an increased need for educational materials that span the range of clinical neurophysiology topics. *The Clinical Neurophysiology Primer* aims to meet this need by providing a broad and intentionally basic treatment of the most central topics within clinical neurophysiology.

The Clinical Neurophysiology Primer initially took shape within the clinical neurophysiology sections at Beth Israel Deaconess Medical Center and Rhode Island Hospital, as an outgrowth of their fellowships' didactic lecture series. Faculty and trainees at these and affiliated teaching hospitals participate in a series of lectures over the course of the academic year designed to acquaint trainees with the elements of clinical neurophysiology, supplementing their clinical experiences. We hope that this primer will prove valuable to others as a companion book intended for clinical neurophysiology fellows and neurology residents, to be used in conjunction with such a program of lectures.

The Clinical Neurophysiology Primer is divided into four parts. The first addresses background topics integral to, and shared by, all the disciplines within clinical neurophysiology. These treat such topics as basic electronics and the neural basis for the central and peripheral electrical potentials that we study in the laboratory. Part II addresses the most central topics pertinent to the application and analysis of electroencephalography. Part III tackles similar key topics pivotal to understanding neuromuscular disease pathophysiology and correlates found with nerve conduction studies and electromyography. The last part covers topics in related fields of clinical neurophysiology: autonomic testing, evoked potentials, sleep studies, and their applications. The primer is multiauthored. Many of the contributing authors are faculty, or were trainees, at our fellowship programs. Inevitably other contributors also joined the effort. Each chapter has appended references or bibliographies that provide the reader with additional sources of information to expand upon the introductory materials covered here. Chapter lengths also vary considerably in size, in part related to the breadth of the material incorporated. Finally, each chapter ends with a set of questions and answers to aid trainees in gauging their mastery of the materials.

We hope this primer will fulfill its intended role as a starting point for fellows engaged in clinical neurophysiology training, for those pursuing more focused training in areas within clinical neurophysiology, and for neurology residents aiming to acquire a basic understanding of these disciplines.

***Andrew S. Blum,* MD, PhD**
***Seward B. Rutkove,* MD**

CONTENTS

IV Autonomic Testing, Evoked Potentials, and Sleep

Contributors

JUAN A. ACOSTA, MD • *Department of Neurology, Beth Israel Deaconess Medical Center, Harvard Medical School, Boston, MA*

NITHI S. ANAND, MD, MPH • *Department of Neurology, University of Massachusetts Medical Center, Worcester, MA*

JACOB R. BERGER, MD • *Comprehensive Epilepsy Program, Department of Neurology, Rhode Island Hospital, Brown Medical School, Providence, RI*

ANN M. BERGIN, MB, MRCP • *Division of Epilepsy and Clinical Neurophysiology, Department of Neurology, Children's Hospital, Harvard Medical School, Boston, MA*

ANDREW S. BLUM, MD, PHD • *Comprehensive Epilepsy Program, Department of Neurology, Rhode Island Hospital, Brown Medical School, Providence, RI*

BLAISE F. D. BOURGEOIS, MD • *Division of Epilepsy and Clinical Neurophysiology, Department of Neurology, Children's Hospital, Harvard Medical School, Boston, MA*

EDWARD B. BROMFIELD, MD • *Department of Neurology, Brigham and Women's Hospital, Harvard Medical School, Boston, MA*

JAMES B. CARESS, MD • *Department of Neurology, Wake Forest University School of Medicine, Winston Salem, NC*

RICHARD L. CERVONE, MD • *Comprehensive Epilepsy Program, Department of Neurology, Rhode Island Hospital, Brown Medical School, Providence, RI*

DAVID CHAD, MD • *Department of Neurology,University of Massachusetts Medical Center, Worcester, MA*

BERNARD S. CHANG, MD • *Comprehensive Epilepsy Center, Department of Neurology, Beth Israel Deaconess Medical Center, Harvard Medical School, Boston, MA*

EDWARD M. DONNELLY, MD • *Comprehensive Epilepsy Program, Department of Neurology, Rhode Island Hospital, Brown Medical School, Providence, RI*

FRANK W. DRISLANE, MD • *Comprehensive Epilepsy Center, Department of Neurology, Beth Israel Deaconess Medical Center, Harvard Medical School, Boston, MA*

BARBARA A. DWORETZKY, MD • *Department of Neurology, Brigham and Women's Hospital, Harvard Medical School, Boston, MA*

GREGORY J. ESPER, MD • *Department of Neurology, Emory University School of Medicine, Atlanta, GA*

ROY FREEMAN, MD • *Department of Neurology, Beth Israel Deaconess Medical Center, Harvard Medical School, Boston, MA*

MASON C. GASPER, DO, MPH • *Comprehensive Epilepsy Program, Department of Neurology, Rhode Island Hospital, Brown Medical School, Providence, RI*

JAMES M. GILCHRIST, MD • *Department of Neurology, Rhode Island Hospital, Brown Medical School, Providence, RI*

GREGORY L. HOLMES, MD • *Section of Neurology, Neuroscience Center at Dartmouth, Dartmouth Medical School, Lebanon, NH*

DEVANAND JILLAPALLI, MD • *Department of Neurology, SUNY Upstate Medical University, Syracuse, NY*

PETER B. KANG, MD • *Department of Neurology, Children's Hospital, Boston, MA*

ROUSTEM KHAZIPOV, MD, PHD • *Institut de Neurobiologie de la Méditerranée, Marseilles, France*

MASAHITO KOBAYASHI, MD • *Department of Neurosurgery, Keio University School of Medicine, Tokyo, Japan*

MILIND J. KOTHARI, MD • *Department of Neurology, Penn State Hershey Medical Center, Hershey, PA*

JEAN K. MATHESON, MD • *Department of Neurology, Beth Israel Deaconess Medical Center, Harvard Medical School, Boston, MA*

FREDERICK K. NAHM, MD, PHD • *Yale New Haven Health Systems, Greenwich, CT*

ANDREJA PACKARD, MD, PHD • *Department of Neurology, Boston University Medical Center, Boston, MA*

ALVARO PASCUAL-LEONE, MD, PHD • *Department of Neurology, Beth Israel Deaconess Medical Center, Harvard Medical School, Boston, MA*

ELIZABETH M. RAYNOR, MD • *Department of Neurology, Beth Israel Deaconess Medical Center, Harvard Medical School, Boston, MA*

JAMES J. RIVIELLO, JR., MD • *Division of Epilepsy and Clinical Neurophysiology, Department of Neurology, Children's Hospital, Harvard Medical School, Boston, MA*

SEWARD B. RUTKOVE, MD • *Department of Neurology, Beth Israel Deaconess Medical Center, Harvard Medical School, Boston, MA*

GEORGE SACHS, MD, PHD • *Department of Neurology, Rhode Island Hospital, Brown University, Providence, RI*

DONALD L. SCHOMER, MD, PHD • *Comprehensive Epilepsy Center, Department of Neurology, Beth Israel Deaconess Medical Center, Harvard Medical School, Boston, MA*

KEVIN R. SCOTT, MD • *Department of Neurology, Penn State Hershey Medical Center, Hershey, PA*

JEREMY M. SHEFNER, MD, PHD • *Department of Neurology, SUNY Upstate Medical University, Syracuse, NY*

CHRISTOPHER M. SINCLAIR, MD • *Comprehensive Epilepsy Program, Department of Neurology, Rhode Island Hospital, Brown Medical School, Providence, RI*

RANDIP SINGH, MD • *Department of Neurology, Beth Israel Deaconess Medical Center, Harvard Medical School, Boston, MA*

ANNABEL K. WANG, MD • *Department of Neurology, Mount Sinai School of Medicine, New York, NY*

NANON E. WINSLOW, REEGT • *Department of Neurology, Brigham and Women's Hospital, Harvard Medical School, Boston, MA*

I
Basic Considerations

1
Basic Electronics in Clinical Neurophysiology

Christopher M. Sinclair, Mason C. Gasper, and Andrew S. Blum

Summary

A basic understanding of simple electronics is vital for the student of clinical neurophysiology to better understand how we begin to analyze neurobiological systems. The elements of basic circuits have relevant and tangible application to the way in which we model the behavior of neural systems in the laboratory. This chapter helps to define and assemble these varied circuit elements for the student. This base of understanding is then used to illustrate how simple electronic circuits can filter and amplify biological data. The composition and behavior of commonly used electrodes are discussed, as are the varied montages we use to record and/or display the measured data, as in an EEG. Attention is devoted to digital signal analysis because modern clinical neurophysiology increasingly relies on digital sampling for ease of data analysis and storage. Lastly, electrical safety issues are considered, particularly as they apply to the clinical neurophysiology arena.

Key Words: Amplifier; circuit element; digital conversion; electrical safety; electrode; electronic filter; montage.

1. GENERAL PRINCIPLES

An understanding of the nature of electricity and the behavior of charged particles begins with one fundamental principle—like charges repel and opposite charges attract. If a collection of charges, whether positive or negative, are unevenly distributed, there is an inherent drive for those charges to redistribute to achieve electrical neutrality. This drive may be considered the electrical potential.

The MKS (meter-kilogram-second) unit of energy (E) is the joule (J). One joule is defined as the energy required to accelerate a 1-kg mass by 1 m/s^2 over a distance of 1 m. The unit of charge (Q) is the coulomb (C). One coulomb is defined as 6.24×10^{18} individual units of charge, where a single electron carries one unit of negative charge. Separated charges (that have not achieved electrical neutrality) are a form of stored or potential energy, and this energy will be expended as the charge separation is neutralized. In the MKS system, 1 J of energy is needed to separate 1 C of charge against an electrical potential of 1 V. Stated more concisely, 1 J = 1 V × 1 C.

2. CURRENT

The flow of electrons in response to an existing or applied electrical potential, or *voltage*, is known as *current*. Current (I) is simply some quantity of charge (Q) moving in some quantity of time (t). Mathematically, this is expressed as:

$$I = Q/t$$

From: *The Clinical Neurophysiology Primer*
Edited by: A. S. Blum and S. B. Rutkove © Humana Press Inc., Totowa, NJ

where I is the current in amperes (A), Q is the quantity of charge in coulombs, and t is the time in seconds required for the transfer of charge. The current must travel through a medium that consists of other particles, and this medium may interfere with the efficient flow of charge; it presents *resistance* (R) to that flow. Thus, the current is not only affected by the applied potential but also by the amount of resistance in the conducting medium. Various media conduct electricity with variable efficiency. Metals conduct very well because of their abundant free electrons and, thus, are termed *conductors*. Conversely, materials that lack free electrons to facilitate the flow of charge resist this flow, and are known as *insulators*. Although the flow of electricity is achieved through the movement of electrons, current is conventionally described to flow from the positive pole to the negative pole. Thus, the direction of current refers to the movement of positive rather than negative charge. Current may consist of other forms of charge apart from electrons. Current may also be conveyed by ions (regardless of charge polarity) in a tissue or solution, as is the case in the conduction of muscle or nerve potentials.

3. CIRCUIT ELEMENTS

3.1. Resistors

Under everyday conditions, current meets with some resistance to flow, much as friction opposes the movement of an object over a surface. Some energy or force is expended in overcoming this resistance. The voltage (or potential) difference across a given resistance is known as the voltage "drop," and the relationship between these parameters and the resultant current is given by Ohm's law:

$$V = IR$$

The unit of resistance is the ohm (Ω), which is defined as the resistance (R) that will dissipate 1 J of energy when a current of 1 A flows through it for a period of 1 s. Practically speaking, resistors are made from materials that do not easily allow the free movement of electrons, such as carbon. Very high resistance materials that are the most restrictive toward the movement of electrons, such as air, rubber, or glass, make the best insulators. The greater the distance that current must traverse through a resistive material, the more resistance to flow there will be. It is, thus, useful to alter the length of a resistive material to vary the current flow. As given by Ohm's law, resistance and current vary inversely with one another ($R = V/I$). Therefore, a reduction in the length of a resistive medium by half will lead to a doubling of the current. The *potentiometer* (voltmeter) uses this principle by providing a way to vary the length of a resistor (and thereby vary the current flow) to advantage.

Resistance in the acquisition of a biological test, such as an EEG, does not only derive from the material of the wiring in use. Resistance derives from any material through which current must pass. For example, resistive elements in the EEG include not just the electrode wiring but also the scalp–electrode interface and the internal circuitry of the machine. Resistance is provided by anything that lies between the positively charged terminal of a circuit (the *cathode*) and the negatively charged terminal (the *anode*). If the resistance is infinitely large, then the current becomes infinitely small (or ceases). This produces a circuit that is "open." Circuit breakers act in this way to ensure the safety of an electrical system. If resistance is reduced to a miniscule value, this permits a relatively large current, and is deemed a "short circuit." Any resistance between the anode and cathode that allows current to flow but is neither infinitely large nor extremely small is a "closed circuit."

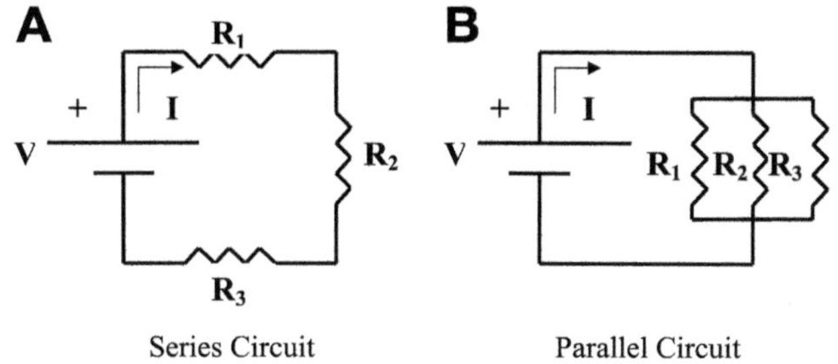

Fig. 1. This figure contrasts the organization of a series circuit (**A**) and a parallel circuit (**B**). In a series circuit, equal current must flow through each resistor in turn. Therefore, the resistors function as a voltage divider. The resistance, R_{comb}, is given by $R_1 + R_2 + R_3$. The parallel circuit functions as a current divider, with equal voltage across each resistor. The combined resistance is given by $1/R_{comb} = 1/R_1 + 1/R_2 + 1/R_3$.

Each element in a circuit contributes its own resistance. If multiple resistive elements exist in a succession along a circuit, they are said to be in *series* (Fig. 1A). If they are configured to allow current to travel in multiple alternate paths, they are said to be in *parallel* (Fig. 1B). Because the series configuration fractionates the total *voltage* across each of the resistive elements, it is also known as a *voltage divider*. Addition of these resistive elements creates a resistor of greater length that is equivalent to the sum of all the component resistances. Therefore, the equivalent resistance (R_{eq}) for a series circuit may be obtained by summing the individual resistances in the circuit as such:

$$R_{eq} = R_1 + R_2 + R_3$$

By contrast, a parallel circuit will allow *current* to fractionate and travel any of a number of paths, and, therefore, is known as a *current divider*. The several routes that the current may travel effectively reduces the total resistance to flow to less than that of any of the component resistances in the circuit. This is represented by the following relationship:

$$1/R_{eq} = 1/R_1 + 1/R_2 + 1/R_3$$

In considering a complete circuit, there are two other applicable laws. Kirchoff's current law states that the sum of current flowing into and out of any circuit node must be zero. Kirchoff's voltage law states that the sum of all voltage steps (voltage sources and drops) around a complete circuit must be zero.

3.2. Capacitors

A *capacitor* is a device that permits the storage of charge. It consists of two parallel conducting plates closely apposed to one another but separated by a small distance and an interposed insulating material, the *dielectric*. The gap between the plates provides a large resistance to the flow of current from plate to plate. As such, when a potential is applied across a circuit containing a capacitor, positive charge will accumulate on the positive plate, attracting negative charge to the opposite plate. Current flows between the plates via the circuit without charge actually crossing the dielectric gap between the plates. The accumulation of separated charge creates a potential difference across the plates that eventually balances the potential applied across the circuit, and current flow then ceases. Several factors affect the magnitude of charge, or *capacitance*, that may be stored by a capacitor. This is proportional to the size of the plates of the capacitor, inversely proportional to the distance between those plates, and is affected by the dielectric material between the plates. The MKS unit for capacitance is the farad (F). A farad will store 1 C of charge on the plates of a capacitor with an applied potential difference of 1 V. This is mathematically expressed as:

$$C = Q/V$$

where C is the capacitance in farads, Q is the charge in coulombs, and V is the voltage in volts across the plates. In practice, most circuits use capacitance on the order of microfarads or picofarads.

If you differentiate both sides of the above capacitance equation with respect to time and rearrange the result, you obtain the following relation:

$$I = C \times \mathrm{d}V/\mathrm{d}t$$

or current is equal to capacitance multiplied by the change in voltage with respect to time. Thus, if the voltage is unchanging ($\mathrm{d}V/\mathrm{d}t = 0$), then current flow becomes zero. This is the case with a *direct current* (DC) circuit, wherein current flows directly between the anode and cathode with an invariant voltage. Once the potential difference between the plates of the capacitor has equaled that applied constant voltage, current flow ceases. Conversely, a continually varying potential will be able to maintain current flow across a circuit that includes such a capacitive element. This is the effect produced by *alternating current* (AC) that, as the name implies, is constantly oscillating between two alternating poles. (AC will be described in more detail later.) Thus, a capacitor will *pass* AC flow but will block DC flow. This impeding effect of the capacitor is known as *capacitive reactance* and is defined as follows:

$$X_C = 1/(2\pi f C)$$

where X_C is in ohms, f is the frequency of the current in hertz, and C is in farads. One can see that as the frequency of the current approaches zero (as in DC), the capacitive reactance (resistance to flow) becomes infinitely large.

Capacitance is crucial to any system that can maintain separated charge and, thereby, store potential energy for use in doing work. The lipid bilayer membrane of nerve tissue is a superb capacitor, which both permits and restricts the flow of ionic currents. It is these intermittent fluctuations in biological currents that ultimately produce the potentials of interest in clinical neurophysiology, such as in EEG. However, other sources of biological capacitance can also interfere with these signals, such as the capacitive resistance in the cerebrospinal fluid, skull, and scalp. As the equation for X_C predicts, these will affect differing neuronal frequencies to different degrees. For example, 3 Hz activities through 2 μF of capacitance will have an $X_C = 1/[2 \cdot 3.14 \cdot 3\ \mathrm{Hz} \cdot (2 \times 10^{-6})\ \mathrm{F}] = 26.5\ \mathrm{k}\Omega$, which is much larger than the 4.4 kΩ reactance seen by 18-Hz beta frequencies. This illustrates how much more capacitive reactance there is to low frequencies vs higher frequencies with scalp recordings.

Multiple capacitors in a circuit interact in a manner that is opposite to the behavior of resistors. When arranged in *parallel*, there is an additive effect as such:

$$C_{eq} = C_1 + C_2 + C_3$$

and when arranged in *series*, the equivalent capacitance is less than any of the individual values, as such:

$$1/C_{eq} = 1/C_1 + 1/C_2 + 1/C_3$$

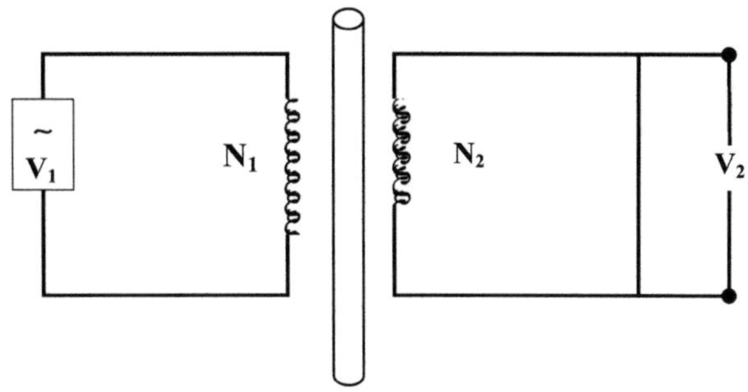

Fig. 2. This figure illustrates a transformer, which is based on the principle of induction. An alternating current (AC) with voltage, V_1, is applied to an inducer, represented by the coil with N_1 turns. Another coil with N_2 turns shares the same rod. AC flowing through the coil induces a magnetic field that then induces a reciprocal electrical field (voltage) in the second coil. The ratio of coil loops determines the change in voltage in the second circuit; fewer turns leads to a proportionately reduced voltage in the second circuit.

3.3. *Inductors*

An inductor consists of a continuous coil of wire called a solenoid. Current flowing in this coil generates a magnetic field whose axis passes through the coil (with directionality dictated by the right hand rule). Because of the equivalence of electricity and magnetism (i.e., Maxwell's equations), this magnetic field can *induce* an electromotive force (emf, ε) in a nearby conductor, if the magnetic field is variable over time. The magnetic field can vary if the current flow in the coil varies. The relationship of this emf to the current is:

$$\varepsilon = -\mathrm{L} \times \mathrm{d}I/\mathrm{d}t,$$

where L is a constant called the *inductance* of the device. The negative sign in the equation indicates that the changing current (dI) induces an emf that opposes that change. The unit of inductance (L) is the henry (H). The inductance of a solenoid is proportional to the number of turns in the coil.

A changing current (i.e., AC) passing through a coil will generate a changing magnetic field that passes through its core. If a second coil of wire is wrapped around a nearby section of this core, the changing magnetic field will generate a reciprocal emf and current in the second coil. One can tap this feature to step voltage from one value to another, as in a *transformer* (Fig. 2). Because inductance (L) depends on the number of turns (N) in the coils, if the number of turns in the first coil (N1) is greater than in the second coil (N2), then the inductance will decrease in the second coil. From the above equation, if L decreases, then dI/dt will increase proportionately. The induced emf (or voltage) in circuit two will decrease in proportion to the drop in inductance. Therefore, voltage varies directly with L and current varies inversely with L, whereas the total energy (power) in the system is conserved. As current steps up, voltage steps down. These vary according to the ratio L1/L2, which is directly related to N1/N2.

Inductance is similar to resistance in that it poses an impediment to the motion of charge generated by another source. For example, an AC source, with its associated emf, providing a current through a circuit with an inductor, will be opposed by the emf generated by that inductor. The inductor's emf is, in effect, subtracted from that of the circuit to determine the net potential. This property is known as the *inductive reactance* (X_L), which is:

$$X_L = 2\pi f\mathrm{L}$$

where X_L is in ohms and the frequency (f) is in Hz.

4. POWER

Energy is simply charge moving across some potential energy gradient ($E = QV$). *Power* is the rate of transfer of this energy, or mathematically:

$$P = E/t$$

A useful permutation of power for use in electrical circuits is as follows:

$$P = Q \cdot V/t = (Q/t) \cdot V = IV$$

where I is the current and V is the voltage. The SI unit of power is the watt (W), which is equivalent to 1 J/s (energy per time).

Recall the previous transformer discussion. As the voltage climbs, the current drops proportionally, and the product of these (the power) will remain constant. Of course, this is an ideal, and a transformer in the real world will lose something in the transfer (albeit not much). Their typical efficiency is on the order of 90 to 99%. The *intensity* of power (β) is often represented as a ratio with a second power level on a normalized, logarithmic scale with units in decibels. The decibel ratio of two power levels is:

$$\beta = 10_{\log}(\mathrm{P2/P1})$$

5. ALTERNATING CURRENT

AC has very useful properties, particularly in circuits involving capacitors and inductors. In the previous treatment of inductors, we saw that a current generated in a coil around a magnetic material could induce a magnetic flux that, in turn, would result in an emf (voltage) across the circuit. An AC generator operates on a similar principle with a minor difference. That is, a magnetic field across a rotating wire will cause a changing field in that wire that will, in turn, induce an alternating emf and current in the wire.

The wire is part of a circuit that must be rotated by some external energy source. Practical examples of this are wind, falling water at a hydroelectric plant, or burning coal. The resulting rotation will produce a sinusoidal flow of AC with a characteristic amplitude and frequency.

As is apparent from the sinusoidal nature of AC, the average current over any given complete cycle is zero. The quantity of current delivered, however, relates to the amplitude of the sine wave. Because AC is only at its maximum amplitude for an instant, it will not produce the same heating effect as an equivalent DC, nor will it produce an equivalent current. It will instead produce an effect that is similar to a DC of lesser quantity. The effective current in an AC circuit across a resistor is given by the *root mean square* (rms) value. This can be shown to be $I_{rms} = I_m/\sqrt{2} = 0.707 I_m$, where I_m is the maximum amplitude and I_{rms} is the effective equivalent to DC. The direction of this current makes no difference in the power of the system because $P = IV$, which is equivalent to I^2R.

6. IMPEDANCE

Impedance (Z) is the term used for the combined effects of resistance along with capacitive and inductive reactance in an RC circuit (a circuit that includes a resistor and capacitor in series) passing AC current:

$$Z = [R^2 + (X_C - X_L)^2]^{1/2}$$

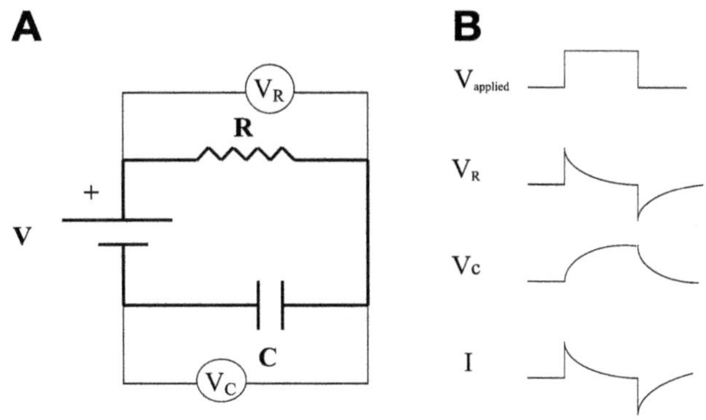

Fig. 3. This figure depicts a simple resistance/capacitance (RC) circuit. In **(A)**, V_R is the voltage drop across the resistor and V_C is the voltage drop across the capacitor. In **(B)**, the behavior of such a circuit in response to an applied square wave pulse ($V_{applied}$) is illustrated. The voltage across the capacitor, V_C, gradually builds as charge accrues on the capacitor. This process follows a logarithmic function and has a time course shown in V_C. Eventually, V_C opposes the flow of current in the circuit. When the applied voltage is zero at the end of the pulse, the capacitor discharges in a reciprocal fashion. By contrast, V_R is maximal at the onset of the applied pulse, but as the capacitor charges and opposes the source voltage, current flow decreases and then stops; V_R then reaches zero because current, I, becomes 0 ($V = IR$). Note that V_R and I behave similarly because they are directly proportional to one another. In this way, V_C behaves as a high-frequency (low-pass) filter, whereas V_R behaves as a low-frequency (high-pass) filter.

The inductive reactance is subtracted from the capacitive reactance because they have opposite phase. In an AC circuit, Ohm's law takes the form $V = IZ$, where Z is the term for resistance in this type of circuit. This is analogous to Ohm's law as applied to DC circuits ($V = IR$).

7. TIME CONSTANTS

With this understanding of basic circuit elements, we can now examine how simple circuits behave and permit basic electronic filtering of waveform data. An RC circuit is shown in Fig. 3A. When voltage is applied to the circuit, current flows across the resistor and begins to accumulate on the capacitor. As the capacitor becomes fully charged, it accrues a voltage that opposes further flow of current through the circuit. If the power source is turned off, the capacitor discharges in the opposite direction of current flow as it charged.

The charging (and discharging) behavior of a capacitor over time is exponential. Its kinetics are described using a time constant, τ, which is that time required for the capacitor to reach approx 63% of its charge. This is 1 – 1/e, where e is the base of the natural logarithm (~2.718). This time is independent of the applied voltage, but rather depends on the resistor and capacitor combination. In an RC circuit, the time constant can be calculated as:

$$\tau = R \times C$$

A larger resistor permits less current to flow to fill the capacitor, thereby prolonging the time constant of an RC circuit. Similarly, a larger capacitor takes longer to charge, thus, prolonging τ.

8. FILTERS

Let us return to the RC circuit and apply a voltage square wave pulse (Fig. 3B). At the outset, the voltage change is seen across the resistor, but there is a lag in the appearance of voltage across the capacitor. At steady state (with no current flow after the capacitor is fully charged), there is no measurable voltage drop across the resistor and there is maximal voltage across the capacitor. The sum of the voltages across these elements always equals the input voltage. Thus, the voltage output across the resistor is very sensitive to sudden changes in input voltage (high frequencies) but insensitive to relatively unchanging voltage (low frequencies); the opposite is true of the capacitor. Therefore, these circuit elements form the basis of high- and low-frequency electronic filters of variable input waveforms (as we record in EMG and EEG). That is, the resistive element serves as a low-frequency filter, and the capacitive element serves as a high-frequency filter.

The low-frequency (or high-pass) filter is helpful in EEG, for instance, in blunting slow DC potentials that are of lesser interest. A shorter time constant makes for a more stringent low-frequency filter (higher cutoff frequency). The relationship between τ and the low-frequency filter is given by:

$$F_{cutoff} = 1/(2\pi\tau) \approx 0.16/\tau,$$

where F_{cutoff} is that frequency above which greater than 70% of the input amplitudes will pass.

The high-frequency (low-pass) filter is based on the capacitive element and is useful in EEG, to attenuate undesired frequencies that may stem from muscle activity near the scalp leads. The cutoff frequency is defined similarly as in the low-frequency filter. The combination of the low- and high-frequency filters defines the operative bandwidth in use. This is the range of input frequencies that will be allowed through for further analysis.

With such RC circuit-based filters (e.g., as in older analog EEG machines), input data is not filtered in an all-or-none fashion, but rather there is a roll-off in the restriction of input frequencies above and below the low- and high-frequency filter settings (Fig. 4). By combining such circuits, one can obtain filters that are more specific, such as the 60-Hz "notch" filter, which more dramatically blunts 60-Hz inputs, a common source of artifact in typical recording environments because of ambient electrical noise. This discussion notwithstanding, modern digital EEGs filter input data using different methodologies than described, with much steeper frequency response characteristics than available with simple RC circuits. It is important to note that overly stringent filtering can distort the output data, for instance, making waveforms seem less sharp than in reality. This can become clinically relevant to the interpretation of the data, for instance, in the recognition of subtle notched morphologies on EEG.

9. AMPLIFIERS

Amplifiers are electronic devices that serve to multiply an input signal by a constant. This amplification factor is called gain and is related to the ratio, V_{out}/V_{in}. It is common to express gain in decibels as $20 \times \log_{10}(V_{out}/V_{in})$. The dynamic range of an amplifier refers to the voltage range over which the amplifier behaves linearly. The sensitivity control on an EEG machine helps to modify the dynamic range of the amplifier. Sensitivity is expressed as microvolts per millimeter and refers to the size of the deflection on the paper or screen that represents this voltage. Typical sensitivity settings for EEG are 7 μV/mm. Increasing the amplifier gain requires lowering the sensitivity; they are inversely related. EEG amplifiers have multiple circuit elements that include voltage regulators, filters, and calibration circuits, among other elements.

The heart of the EEG machine is the differential amplifier. The difference between the input voltages from two electrodes relative to a reference electrode (ideally close to the

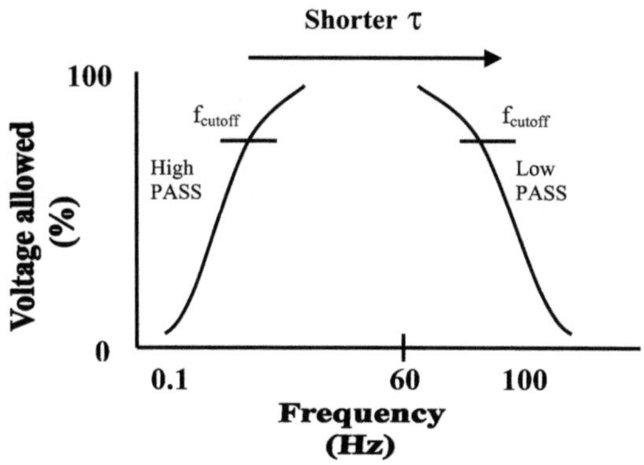

Fig. 4. Frequency response of filters. This figure shows the percentage of input voltage that is allowed as a function of frequency in relation to applied filters based on analog resistance/capacitance (RC) circuits with various time constants, τ. Such filters exhibit a "roll off" in their attenuation of input frequencies. As τ shortens, the curves for both the high- and low-pass filters are shifted toward higher frequencies. The cutoff frequency is inversely related to τ. For a given filter circuit, it is the frequency above which approx 70% ($0.16/\tau$) of the input amplitudes will pass through for analysis. The notch filter is designed to specifically filter out 60-Hz inputs, because these are frequently artifactual in origin.

recording leads) is amplified and serves as output. This method serves to subtract common artifactual noise that may be contaminating both input electrodes. One example is ambient 60-Hz noise from the local recording environment. This subtraction of common noise is called common mode rejection. The capacity of an amplifier to perform common mode rejection is described by the common mode rejection ratio, which is equal to the common signal voltage divided by the nonamplified output voltage.

The common mode rejection ratio for many amplifiers in modern EEG devices is 10,000. In a differential amplifier, by convention, if input 1 is negative with respect to input 2, then the pen deflection is upward. If input 1 is positive with respect to input 2, then the deflection is downward.

10. ELECTRODES

The above principles can now be applied to the acquisition of neurophysiological data. This begins with the electrode and the interface between the subject and the electrode. Electrodes connect the patient to the circuits of the neurophysiological recorder. They serve to detect and conduct electrical potentials from the patient to the machine. They are metallic, and an electrolyte paste is used to help conduct current and reduce movement artifacts. Electrodes may be nonreversible (polarized) or reversible (nonpolarized). Polarized electrodes are prone to develop significant capacitance, and this may interfere with the faithful transmission of underlying biological signals (the electrode behaves like a low-frequency filter). Reversible electrodes, such as those of silver chloride, are preferred for common neurophysiological applications. Polarization is avoided because the chloride ion is common to both the electrode and the electrolyte. Other metals can be used, such as gold or platinum, but may be costly.

Because the electrode–electrolyte interface resembles a simple RC circuit, it is important for the electrode impedance to be as low as possible, typically less than 5 kΩ, for scalp electrodes.

Large electrode impedances foster large artifact potentials caused by even small local currents induced by ambient electric fields (Ohm's law: $V = IR$, $V = IZ$). This leads to troublesome electronic noise. Preparing the site (e.g., with alcohol) to remove oils helps to lower the impedance to desired levels. Because inputs from neighboring contacts are compared using differential amplifiers in EEG, their various input impedances must be comparable. If there is an impedance mismatch, the signal fidelity will be degraded, because common sources of electronic artifact will not be efficiently subtracted away. Higher impedances favor lower than true amplitudes and loss of lower frequencies. Excessively low impedances (e.g., a "salt bridge" caused by a smeared electrode gel between contacts) can also cause erroneously low amplitudes. Calibration maneuvers using square wave inputs and biocalibration signals help to assay the fidelity of the recording setup.

Many other types of electrodes exist. Subdermal needle electrodes have generally a smaller area of contact, hence, higher input impedances, and, therefore, more susceptibility to noise. Sphenoidal leads are long, thin leads made of stainless steel or platinum placed to position the tip lateral to the foramen ovale. They may be more sensitive to potentials from the anterior temporal lobe than surface electrodes; this view has been, however, disputed. Stereotactically implanted depth electrodes made of stainless steel or platinum may also be used to detect activity from deeper contacts, such as the amygdala, hippocampus, or cingulum, among others. These leads have low impedance and can remain indwelling for weeks, if needed. They circumvent other problems of surface recording, such as muscle artifact and filtering effects of the dura and scalp. Subdural strips and grids made up of stainless steel or platinum discs embedded in a sheet of plastic can be surgically placed to cover the cortex. These permit not only the recording of brain activity, but also the mapping of eloquent cortex (e.g., language cortex) by stimulation of the underlying cortex with various testing paradigms.

11. EEG MONTAGES

In the recording of the EEG, electrodes are typically placed on the scalp using the 10-20 system (Fig. 5). In this standardized method, contacts are named by their location (frontopolar, frontal, central, parietal, temporal, occipital, and auricular). They are also numbered with odd numbers over the left hemisphere, even numbers over the right, and z referring to the midline. The particular sequence in which the EEG data is displayed is called the montage. Montages may be bipolar or referential. Bipolar montages involve a comparison of voltages recorded from (usually adjacent) active electrodes in a chain-like fashion, (e.g., front-to-back or side-to-side). By contrast, referential recordings involve a comparison of each electrode to an (ideally) inactive electrode. Examples of reference sites include the ear, the mastoid, the vertex, or an average of many active leads ("average reference").

Bipolar montages visually map the peak of voltage negativity over the scalp owing to the property of phase reversal that emerges from the serial comparisons between adjacent electrodes along a chain. As one ascends and descends the underlying voltage "hill" with each bipolar comparison along a chain, the sign of the comparison flips from positive to negative (Fig. 6). Referential recordings do not exhibit phase reversals, but may show a truer picture of the relative amplitudes of voltage at each electrode. Negativity at the "active" lead is defined as an upward deflection in the display. In addition to displaying the channels based on comparisons among the electrodes, other channels are often used to permit comonitoring of cardiac rhythm, eye movement, respiration, and EMG activity as needed for the study at hand.

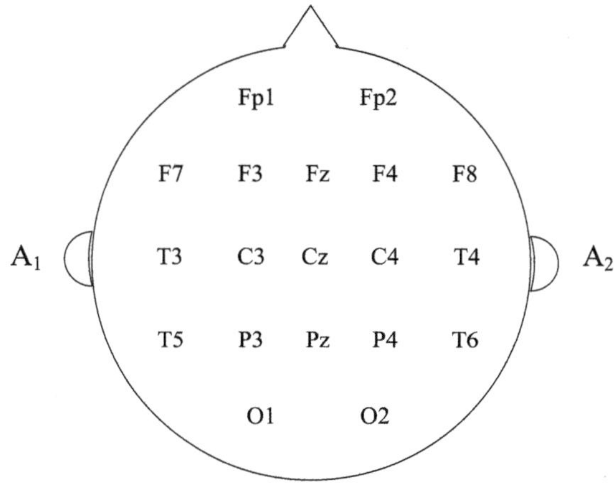

Fig. 5. The International 10-20 system of electrode placement. This figure depicts an overhead view of the most commonly used array of standardized electrodes in EEG. Odd numbers refer to the left, even numbers to the right. Electrodes are named according to location: F, frontal; C, central; P, parietal; T, temporal; Fp, frontopolar; O, occipital; and A, auricular. Sphenoidal leads (Sp) and supplemental temporal (T1, T2) leads are not shown.

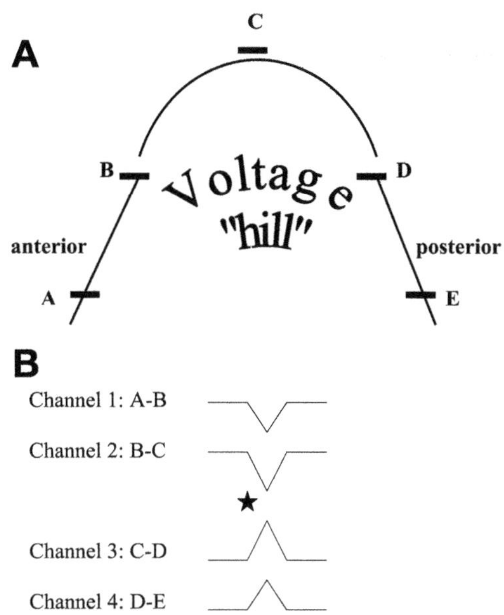

Fig. 6. Bipolar EEG recordings and phase reversal. A hypothetical region of scalp negativity is illustrated topologically, as a voltage "hill" **(A)**. The higher the hill, the greater the negative field potential at that point along the anterior–posterior axis, with electrode A most anterior, and electrode E most posterior. The hill peaks at C. In **(B)**, the waveforms derived from bipolar comparisons between adjacent electrodes are illustrated, as in a bipolar recording. By convention, if contact 1 is more negative than contact 2, the deflection is upward. Because these comparisons are performed serially along the anterior–posterior axis, the difference between adjacent channels flips polarity as one traverses the peak of the hill. The flip in polarity leads to the highly visible phase reversal on a bipolar tracing, denoted by the star in (b). The shared contact of the phase reversal maps the voltage peak, electrode C in this case.

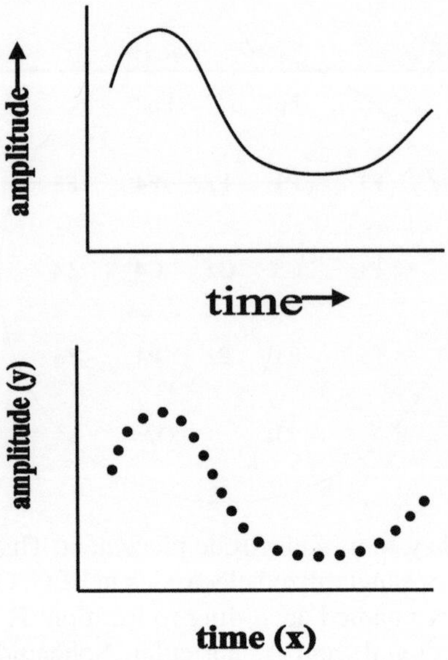

Fig. 7. Analog vs digital. The upper curve depicts an analog signal with variable amplitude over time. Digitization relies on using discrete intervals along both the *x*- and *y*-axes and making sampling measurements of the analog signal at discrete time intervals and assigning discrete amplitude values. The ability of a digitized output signal to faithfully resemble its analog input depends on the fineness of these discrete steps (resolution) along these axes. Sampling at greater than twice the Nyquist frequency ensures an adequate degree of digital sampling to minimize aliasing.

12. DIGITAL SIGNAL ANALYSIS

As elsewhere in our lives, the digital revolution has become firmly entrenched within the arena of clinical neurophysiology. Increasingly, our neurophysiological recording systems and our displays of recorded waveforms are based on computers. This affords numerous advantages, such as the reformatting of EEG data into different montages, refiltering data, and/or applying spike and seizure detection systems to accrued data sets. Storage and sharing of data is simpler; no longer are rooms dedicated to the storage of reams of paper tracings.

Computers require that data be digitized. However, the input waveforms that we strive to faithfully record occur in an analog universe. Analog refers to continuous data with respect to time and with respect to value. In EEG, the scalp voltages we record are generated continuously over time and with a continuous range of amplitudes. Digital, on the other hand, refers to the intermittent sampling of data. Measurements are sampled at discrete moments and amplitudes are assigned to nearest discrete values. The degree of fidelity with which the digitally sampled data resembles the original analog data depends on how finely it is sampled, that is, how many discrete steps there are along the time (*x*) axis and the amplitude (*y*) axis (Fig. 7).

Analog signals are first amplified, then filtered, then subjected to digital conversion. The analog-to-digital converter (ADC) is at the core of the generation of digital signals. The ADC samples the continuous data and assigns a binary code to represent the voltage at that moment in time. Many modern recorders have a sampling rate of 256-Hz per channel. The total throughput refers to the number of channels times the sampling rate. For a 21-channel EEG, this leads to a throughput of 5276 samples/s. One can have a sampling system race through all of these in sequence, or parallel samplers can analyze each channel singly.

It is important to have an adequate sampling rate to avoid undue distortion of the input waveforms. The Nyquist frequency is the highest input frequency component of interest to be recorded. To avoid distortion, the sampling rate should be greater than twice the Nyquist frequency. If one samples at less than twice the Nyquist frequency, *aliasing* will occur. Aliasing refers to the artifactual creation of low frequencies in ADC, by undersampling high-frequency inputs. The alias frequency is mathematically defined as the sampling frequency minus the Nyquist frequency.

In addition, another means to deter such ADC artifact is to apply a high-frequency filter to the input data to cut off undesirable high-frequency elements.

The dynamic range of a recording system refers to the range of measurable values, from highest to lowest; in EEG, this refers to the recordable voltage range. The dynamic range is the $\log_{10}(V_h/V_l)$. The resolution of a digital recording system depends on the range of recordable values divided by the number of digitally assignable increments. This, in turn, depends on the number of available digital bits in the recording system. With n-bit data, one has 2^n possible values; for a 12-bit EEG recording system, this is $2^{12} = 4096$ available voltage increments. This can be regarded as plus or minus 2048 units, 0 to 2047. The available dynamic range available is $\log_{10}(2047/1) = 2.8$. Obviously, more bits lead to finer resolution.

In an analog system, values at the extremes may be clipped because of inadequacies in the recording or display capacity. Generally, digital systems afford better dynamic range, and much less risk of loss of information at the extremes of the recordable range. Although digital recording systems typically have very good resolution and dynamic range, our ability to display this data depends on the digital resolution of our monitors. When the monitor's resolution becomes limiting, we can improve our viewing resolution by restricting our analysis to a subset of the data.

As can be seen, digitization of neurophysiological data affords numerous advantages to previous analog methods. For instance, digitization has improved our EEG analysis by permitting switching the viewing montage, permitting the adjustment of filters and sensitivity settings *post hoc*, improving the dynamic range and preservation of data at the extremes of the recording range, lowering the cost for data storage and data sharing, and facilitating the recall of previous data for comparison. In the arena of long-term video EEG monitoring, digitization has been a tremendous boon. It is now simpler to integrate the video and EEG data stream, easier to locate events of interest, and simpler to store the data over time. Computerized methods now exist to apply search techniques to automate the detection of spikes and seizures within a digital data set. Algorithms exist and are being perfected to allow for frequency analysis of data (Fourier transformation—spectral analysis), current source mapping, and real-time seizure detection and even seizure prediction technology that may one day revolutionize the treatment of epilepsy.

13. ELECTRICAL SAFETY

Electrodiagnostic procedures are extremely safe, however, injuries can occur and may include pain, dermal burns, seizure, and ventricular fibrillation. A current of 1 mA at 60 Hz applied to dry skin is the threshold for pain sensation. Lethal current injury causing ventricular fibrillation exceeds 100 to 300 mA. Be aware of low-resistance pathways between external current sources and the heart (e.g., indwelling cardiac catheters), because currents less than 100 mA can cause serious cardiac arrhythmias under such circumstances. Following a few simple rules will minimize the risk of injury while recording an EEG or EMG.

High-risk populations among those undergoing EEG evaluation include neonates and patients with intravascular catheters or instruments. Lower risk populations include non-neonates without implanted medical instrumentation and bystanders who are near instrumentation, possibly in contact with an electrical device.

The overriding safety issue in recording an EEG is the risk of exposing the patient to current. The amount of current is the most important predictor of electrical injury. A secondary concern, especially with older EEG equipment in which small sparks are generated when the power switch is activated, is prevention of flammable gas explosion.

To reduce the risk of patient exposure to current, proper grounding must be ensured. The electrical circuit in EEG consists of a current provided by the "hot" contact in a wall outlet, which travels through the EEG machine and returns to the outlet's neutral contact. Excess current can accrue from various sources that escape this circuit (i.e., a short circuit within the instrument that directs current to the covering of the EEG machine). This potentially dangerous leaked current can fortunately be shunted to the ground contact, which is the path of lowest resistance. However, should ground failure occur, the leak current may flow preferentially through a low-resistance pathway, which may include the patient.

In intensive care units or surgical operating rooms, multiple machines may often be attached to a patient and may lead to the use of multiple grounds. A *ground loop* may be formed if each of these several grounds has a slightly different potential, allowing excess current to flow from the higher resistant ground to the lower resistant ground. This may be hazardous if this ground loop includes a nearby person or the patient. A simple rule to avoid this hazard is to connect the patient to only one *common ground* that serves all of the devices in use.

Basic requirements for proper grounding include:

1. EEG equipment should always be connected to a three-pronged wall receptacle, never a two-pronged plug. Three-pronged plugs include a black "hot" connector, a white neutral connector, and a green ground connector. The ground is absent in two-pronged plugs. The black connector has AC current at 110 V. The neutral connector is the reference for the "hot" line but is not necessarily at 0 V. The ground is the building ground connection.
2. A working fuse within the machine to break the circuit, should high current travel through the ground, alleviating a potential safety risk to the patient.
3. Adequate power outlets for patient-related equipment, typically marked by a green dot, indicating a higher standard of safety.
4. Regular evaluation and testing of equipment. Equipment and plug grounds must be periodically tested with an ohmmeter, which should show a resistance of less than 0.1 ohms when assessing the drop in current from cover to ground connector.
5. Ensure that the ground plug is firmly in place in the socket.
6. Connect all devices into the same outlet using a grounding bar. This will ensure a common low resistance ground pathway for all stray currents. In the case of multiple devices in use, placing a ground on the patient may actually increase the danger of electrical shock. Because another purpose of the ground in EEG and EMG is to reduce ambient 60-Hz interference, avoidance of multiple grounds will actually improve the quality of the recording.

Excessive current that requires shunting arises typically because of a short circuit within the equipment or a leakage current. Leakage currents from stray conductive and stray inductive sources mainly arise from the power cable. The cable contains three wires covered by insulation, which creates a type of capacitor in which currents can flow between wires (from "hot" to the others). Longer cords create greater capacitance, with a larger potential for current leak (up to 70 μA) into the instrument.

Unwanted inductive current is a less significant source of current leak than cable capacitance. Inductive currents are generated by the magnetic field surrounding a wire that creates current in other wires. This may arise when wires are coiled. Steps to avoid generation of excessive currents are:

1. Power cords are rated in capacitance per foot and selected by instrument manufacturers to ensure minimal leakage.
2. Extension cords should never be used. A leakage current produced by adding a 6-foot power extension cord will range from a minimum of 7 mA to 60 mA.
3. Keep wires from excessive coiling.

SUGGESTED READING

American EEG Society. Guidelines in EEG. J Clin Neurophysiol 1986;3:1–147.
American EEG Society. Guidelines for standard electrode position nomenclature. J Clin Neurophysiol 1991;8:200–202.
Hughes JR. EEG in Clinical Practice. Butterworth-Heinemann, Boston, MA, 1994.
Laks MM, Berson A. Electrical safety measures for physicians using electromedical equipment. Sem Neurol 1995;15:311–316.
Litt B, Fisher RS. EEG engineering principles. In: Current Practice of Clinical Electroencephalography, 2nd ed. (Daly DD, Pedley TA, eds.) Raven, New York, NY, 1990.
Misulis KE. Basic electronics for clinical neurophysiology. J Clin Neurophysiol 1989;6:41–74.
Misulis KE, Head TC. Essentials of Clinical Neurophysiology, 3rd edition. Butterworth-Heinemann, Boston, MA, 2003.
Niedermeyer E, Lopes da Silva FH, eds. Electroencephalography: Basic Principles, Clinical Applications and Related Fields, 3rd ed. Williams and Wilkins, Baltimore, MD, 1993.
Seaba P. Electrical safety. Am J EEG Tech 1980;20:1–13.

REVIEW QUESTIONS

1. In a simple circuit with a single resistor, if the applied voltage is 10 V, and the resistance is 5 Ω, what is the current?
2. In a series circuit with two resistive elements, R_1 and R_2, what is the derived combined resistance in the circuit?
3. In a parallel circuit with two resistors, R_1 and R_2, what is the resultant combined resistance in the circuit?
4. How does the resistance to current flow attributable to a capacitor (capacitive reactance) relate to the frequency of the AC?
5. In a transformer with a coil (N1) of 20 turns and an applied voltage of 100 V, what will be the resultant voltage in its associated secondary circuit if its coil (N2) has only 5 turns?
6. What is impedance?
7. In an RC circuit, what is its derived time constant? What is a time constant?
8. Which element in an RC circuit behaves as a low-pass (high-frequency) filter?
9. What is the Nyquist frequency? If the Nyquist frequency for EEG is considered to be 100 Hz, what should the sampling frequency be to avoid aliasing?
10. What is a ground loop?

REVIEW ANSWERS

1. Using Ohm's law, $V = IR$, the derived current is 2 A.
2. The combined resistance in a series circuit is $R_{comb} = R_1 + R_2$.
3. This is a current divider. The resultant combined resistance is $1/R_{comb} = 1/R_1 + 1/R_2$.
4. Capacitive reactance is inversely related to the applied frequency of AC.

5. In a transformer, the ratio of the coil number defines the step-down or step-up relationships. In this instance, N1/N2 = 20/5 = 4. Therefore, the induced voltage will be 1/4 of the original voltage, or 100 V/4 = 25 V.
6. Impedance is the combined resistances to current flow in a circuit passing AC. It may include not only resistance from a resistor, but also capacitive and inductive reactances.
7. The time constant, τ, of a circuit with resistance R and capacitance C is defined by: $\tau = R \times C$. The time constant, τ, refers to the time needed for a capacitor in an RC circuit to reach 63% (1 – 1/e) of its full charge.
8. The voltage reading across the capacitor in an RC circuit behaves as the high-frequency filter. Adjustment of this element may serve to help filter out higher-frequency artifact, such as EMG activity in the recording of an EEG.
9. The Nyquist frequency is the highest input frequency that is deemed desirable to detect. If the Nyquist frequency is 100 Hz, then the sampling frequency should be at least 200 Hz to avoid aliasing.
10. A ground loop occurs when multiple grounds are applied to a subject because of various electrical devices, such as in an intensive care unit setting. If these various grounds are not identical in voltage, an electrical pathway can exist that includes the patient, through which stray currents may travel, thus, endangering the patient. A common ground can avoid this hazard.

2
Basic Neurophysiology and the Cortical Basis of EEG

Gregory L. Holmes and Roustem Khazipov

Summary

Clinical neurophysiological studies are based on the recording of both spontaneous electrical activity, as with the EEG, or with stimulated responses, such as evoked potentials. The electrical signaling within these neuronal circuits is responsible for both the spontaneous and evoked electrical activity that is routinely measured in the clinical neurophysiology laboratory. This chapter will review some of the basic concepts of neuronal signaling that are important to the understanding of clinical neurophysiology.

Key Words: Action potential; dipole; ion channel; membrane potential; neurotransmission; paroxysmal depolarization shift; postsynaptic potential.

1. INTRODUCTION

The human brain contains more than 100 billion neurons, the majority of which have the ability to influence many other neurons. It is through this communication process, termed signaling, that electrical activity is generated, resulting in the human EEG.

The cells of the nervous system can be divided into two major categories: neurons and neuroglial cells. Although neurons come in many shapes and sizes, the major components of most neurons are the dendrites, which receive information; the cell body, which processes and integrates the information; and the axon, which conducts signals to other brain regions. Neuroglia, often referred to as glia, do not participate directly in electrical signaling, although they are critical in maintaining support of the neurons. The three major categories of glial cells are the astrocytes, which maintain the correct metabolic milieu for neuronal signaling; oligodendrocytes, which myelinate neurons; and microglia, which serve as the brain's macrophages and assist in brain recovery from injury.

Neurons are organized into neuronal ensembles of circuits that process specific kinds of information. Neurons that carry information into the circuit are termed afferent neurons, whereas neurons signaling information away from the circuit are referred to as efferent neurons. Nerve cells that only participate in the local aspects of a circuit are called interneurons. Processing circuits are combined to form systems that server broader function, such as memory, vision, or hearing.

Clinical neurophysiological studies are based on the recording of both spontaneous electrical activity, as with the EEG, or with stimulated responses, such as evoked potentials. The electrical signaling within these neuronal circuits is responsible for both the spontaneous and evoked electrical activity that is routinely measured in the clinical neurophysiology laboratory.

From: *The Clinical Neurophysiology Primer*
Edited by: A. S. Blum and S. B. Rutkove © Humana Press Inc., Totowa, NJ

This chapter will review some of the basic concepts of neuronal signaling that are important to the understanding of clinical neurophysiology.

2. BASIS OF BRAIN ELECTRICAL ACTIVITY

2.1. *Membrane Polarity*

All neurons and glia have lipid bilayer membranes separating the delicate internal machinery of the cell from the external environment. The neuronal membrane is an excellent insulator and separates different concentration of ions inside the cell from those outside the cell. The activity of ion channels is fundamental to signaling in the nervous system. The movement of ions that carry electrical charge through ion channels results in voltage changes across the membrane.

Electrical potentials are generated across the membranes of neurons because there are differences in the concentration of specific ions across the membrane, and the membrane is selectively permeable to ion flow. Movement of ions across membranes occurs through ion channels, which are proteins that transverse the neuronal membrane and allow certain ions to cross in the direction of their concentration gradient. Whereas Na^+ and Cl^- are more concentrated outside the cell, K^+ and organic anions, consisting of amino acids and proteins, are more concentrated inside the cell. Na^+ and Cl^- ions, therefore, tend to flow into the cell along this concentration gradient, whereas K^+ ions tend to flow outward. Because of their size, large organic anions are unable to move out of the intracellular compartment.

Ion flow is not only determined by ionic concentration gradients, but also is dependent on the selective permeability of ion channels (often referred to as the channel's conductance), as well as electrical forces that arise from the membrane potential. Ultimately, in the resting state, ion flow ceases when concentration dependent forces are offset by opposing electrical forces based on the voltage across the membrane, the resting membrane potential. Because of the selective permeability of ion channels and the unequal distribution of anions and cations inside and outside of the neuron, there arises a potential difference across the neuronal membrane. The charge separation gives rise to a difference of electrical potential, or voltage, across the membrane, called the membrane potential. The membrane potential, V_m, is defined as

$$V_m = V_{in} - V_{out}$$

where V_{in} is the potential on the inside of the cell and V_{out} the potential on the outside. At rest, the membrane potential is called the resting membrane potential. The outside of the cell is by convention defined as zero and the resting membrane potential is, therefore, equal to the voltage inside the cell. In neurons, the usual range is –60 mV to –70 mV.

Ions are, thus, subject to two forces driving them across the membrane: a chemical driving force that depends on the concentration gradient across the membrane and an electrical driving force that depends on the electrical potential across the membrane. Ions flow from high-concentration areas to low-concentration areas (chemical driving force); ions flow to areas of opposite charge, like charges repel, unlike charges attract (electrical driving force). The net electrochemical driving force is determined by the electrical potential difference across the membrane and the concentration gradient of the ions selective for the channel.

To illustrate these physiological features, the flow of K^+ ions will be considered (Fig. 1). Because K^+ ions are highly concentrated inside the cell, K^+ ions tend to diffuse from inside to outside the cell down their chemical concentration gradient. Because of this ionic flow, the

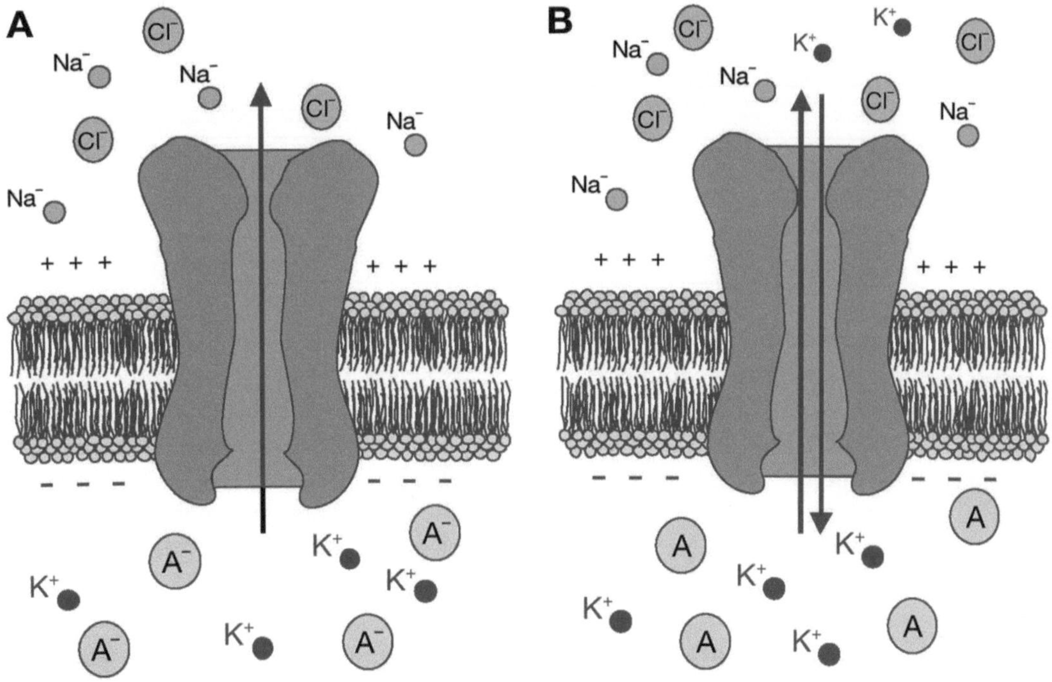

Fig. 1. K^+ channel. In neurons, K^+ has a higher concentration inside neurons than outside **(A)**. Because of the concentration differences, K^+ diffuses from inside the cell to the outside. With K^+ outflow, the inside of the cell becomes even more negative, because the K^+ ion is carrying a positive charge. At some point, a steady state is reached, in which the electrical and chemical driving forces are equal and opposite, and there is a balance between K^+ entering and leaving the cell **(B)**. Modified from Kandel et al., 2000, with permission.

outside of the membrane accumulates a slight positive charge and the inside a slight negative charge. As K^+ ions accumulate outside the cell, there grows a counteracting electrostatic force opposing further K^+ efflux, because the positive charges repel each other. Once K^+ diffusion has proceeded to a certain point, a potential develops across the membrane at which the electrical force driving K^+ ions into the cell exactly balances the chemical force driving K^+ ions out of the cell, that is, the outward movement of K^+ (driven by its concentration gradient) is equal to the inward movement of K^+ (driven by the electrical potential difference across the membrane). This potential is called the potassium equilibrium potential, E_K.

The equilibrium potential for any ion X can be calculated from the Nernst equation:

$$E_X = \mathrm{R}T/\mathrm{zF} \times \ln([\mathrm{X}]_o/[\mathrm{X}]_i)$$

where R is the gas constant, T the temperature, z the valence of the ion, F the Faraday constant, and [X] the concentrations of the ion inside (i) and outside (o) of the cell. Because RT/F equals 25 mV at 25°C, z is +1 for K^+, and the concentrations of K^+ outside and inside the axon are 20 mM and 400 mM, respectively, the Nernst equation for K^+ in the squid axon can be rewritten as:

$$E_K = \frac{58\ \mathrm{mV}}{1}\log\frac{[20]}{[400]} \approx -75\ \mathrm{mV}$$

Table 1
Extracellular and Intracellular Ion Concentrations in Squid Axon

Ion	*Intracellular (mM)*	*Extracellular (mM)*
Potassium	400	20
Sodium	50	440
Chloride	50	560
Calcium	0.00001	2

Na^+ is more common outside the cell than inside, therefore, it tends to flow into the cell down its chemical concentration gradient. There is also an electrical driving force that drives Na^+ into the cell by virtue of the negative electrical potential difference across the membrane. The equilibrium potential of +55 for Na^+ is not reached at resting conditions because there are so many more open K^+ channels than Na^+ channels. The relatively larger conductance of K^+ compared with that of Na^+ at resting conditions dictates the resting membrane potential of the cell. Thus, despite the large chemical and electrical forces driving Na^+ into the cell, the influx of Na^+ is small compared with K^+ if there are many open channels.

The resting membrane potential, V_m, is not equal to either E_K or E_{Na}, but lies between them. As a rule, if V_m is determined by two or more ions, the influence of each ion is determined not only by the concentration of the ion inside and outside the cell but also by the relative permeability of the membrane to each ion. The Goldman equation describes voltage when more than one ion is active:

$$V = -58 \log \frac{P_k[K]_i + P_{Na}[Na]_i + P_{Cl}[Cl]_o}{P_k[K]_o + P_{Na}[Na]_o + P_{Cl}[Cl]_i}$$

where V is the voltage and P is the permeability of the membrane to each ion. Often, conductance, G, is used in lieu of permeability, P, in biological systems.

The Goldman equation is an extension of the Nernst equation that considers the relative permeabilities of the ions involved. Table 1 provides the extracellular and intracellular ion concentrations that apply in common physiological circumstances.

Because ion leaks could eventually result in a run down of Na^+ and K^+ gradients, the resting membrane potential would eventually be altered. The Na^+–K^+ pump, which moves Na^+ and K^+ ions against their net electrochemical gradients, extrudes Na^+ from the cell while bringing K^+ into the cell. The energy to run this pump comes from the hydrolysis of ATP. At the resting membrane potential, the cell is not in equilibrium but, rather, in a steady state. The continuous passive influx of Na^+ and efflux of K^+ ions is counterbalanced by the Na^+–K^+ pump.

3. CHANNEL GATING

Thus far, we have considered selectively permeable ion channels in the resting state. There are also gated channels that exist in several configurations. The term gating is used to describe the transition of a channel between these different states. Most gated channels are closed when the membrane is at rest. Each ion channel has at least one open state and one or two closed states. However, many ion channels have two or more conformational states that are relatively stable.

Three regulatory mechanisms determine whether a channel remains open. In voltage-gated channels, the voltage across the membrane determines whether a conformational change in the membrane occurs that may open the channel. In ligand-mediated channels, the ligand binds to the channel, either at an extracellular site, as with neurotransmitters, such as glutamate or γ-aminobutyric acid (GABA), or at an intracellular site, as with certain cytoplasmic compounds, such as Ca^{2+} and nucleotides. Ligand binding causes a conformational change that causes the channel to open. Finally, ligands can also activate cellular signaling cascades that can covalently modify a channel through phosphorylation (*see* the discussion of metabotropic receptors in Subheading 7.). This covalent modification also influences channel behavior.

4. ACTION POTENTIAL

For a neuron to transmit information, it must generate an electrical current, termed an action potential. Development of the action potential requires an electrical or chemical stimulus, which alters ion flow into the cell.

The electrical current that flows into and out of the cell is carried by ions, both positively charged (cations) and negatively charged (anions). The direction of current flow is conventionally defined as the direction of net movement of positive charge. Cations move in the direction of the electrical current, whereas anions move in the opposite direction. Depending on the exact nature of transmembrane ionic flow, the charge separation across the resting membrane is disturbed, altering the polarity of the membrane. A reduction of charge separation results in a less negative membrane potential and is termed depolarization, whereas an increase in charge separation leading to a more negative membrane potential is called hyperpolarization.

When the membrane potential reaches a threshold (~–55 to –60 mV), the voltage-gated Na^+ channels open rapidly. The influx of Na^+ makes the interior of the cell more positive than before. This sudden marked increase in depolarization causes still more voltage-gated Na^+ channels to open, resulting in further acceleration of the depolarization. This positive feed forward cycle initiates an action potential and is responsible for its all-or-none character. Once initiated, the action potential behaves independently of its stimulus. Because the Na^+ conductance becomes so transiently large relative to other ionic conductances, the membrane potential approaches the equilibrium potential for Na^+ (+60 mV) during an action potential. Depolarization during the action potential is very large, but very brief, lasting only 1 ms. These features of the action potential allow neuronal signaling with high fidelity at a very high rate (up to hundreds of action potentials per second). Termination of the action potential is caused by rapid inactivation of Na^+ channels and delayed opening of voltage-gated K^+ channels. The delayed increase in K^+ efflux, combined with a decrease in Na^+ influx, produces a net efflux of positive charge from the cell, which continues until the cell has repolarized. Indeed, after an action potential, there is a brief hyperpolarization, during which time the cell is more refractory to immediate depolarization, before reestablishing the resting membrane voltage. Figure 2 demonstrates the sequential opening of voltage-gated Na^+ and K^+ channels during the action potential.

During the action potential, Na^+ channels undergo transitions among three different states: resting, activated, or inactivated. On depolarization, the channel goes from the resting (closed) state to the activated (open) state (Fig. 3). If the depolarization is brief, the channels go directly back to the resting state on repolarization. If the depolarization is

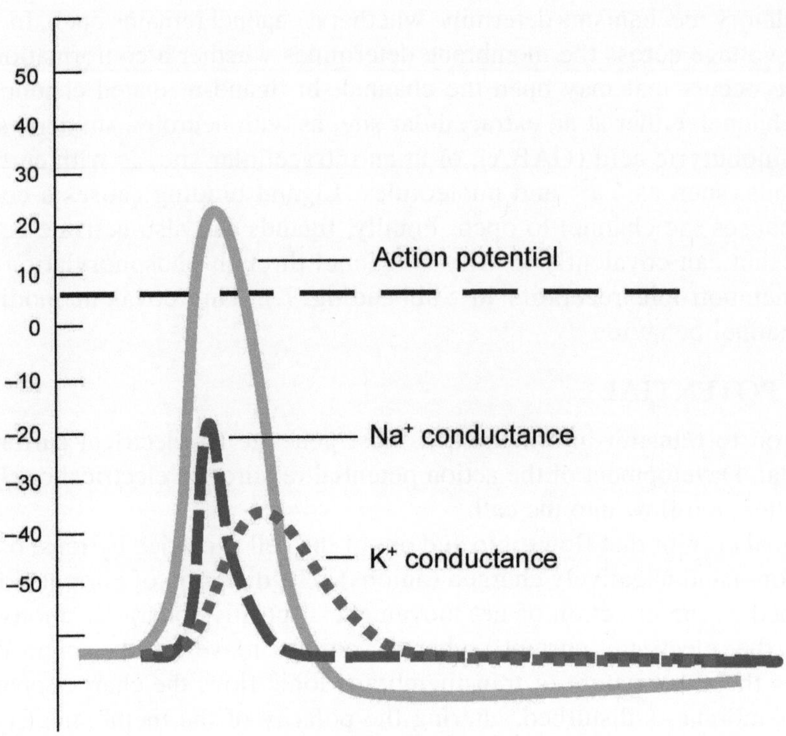

Fig. 2. Openings of channels during the action potential. The influx of Na^+ makes the interior of the cell more positive than before, increasing the degree of depolarization, which causes still more voltage-gated Na^+ channels to open, resulting in further acceleration of the depolarization. The positive feedback cycle initiates the action potential and is responsible for its all-or-none character. However, with the depolarization there is a greater electrical driving force on the K^+ ions and K^+ flow outward. The increase in K^+ efflux combined with a decrease in Na^+ influx results in an efflux of positive charge from the cell, which continues until the cell has repolarized to its resting membrane potential. Modified from Kandel et al., 2000 with permission.

maintained, the channels go from the open state to the inactivated closed state. Once the channel is inactivated, it cannot be opened by further depolarization. Channel inactivation can be reversed only by repolarization of the membrane to its negative resting potential, which allows the channel to switch from the inactivated to the resting state. Each Na^+ channel has two kinds of gates that must operate in conjunction for the channel to conduct Na^+ ions. An activation gate closes when the membrane is at its negative resting potential and is rapidly opened by depolarization; an inactivation gate is open at the resting potential and closes slowly in response to depolarization. The channel conducts ions only for the brief period during a depolarization when both gates are open. Repolarization reverses the two processes; closing the activation gate rapidly and opening the inactivation gate more slowly. After the channel has returned to the resting state, it can again be reactivated by further depolarization.

After an action potential, the Na^+ channels are inactivated and the K^+ channels are activated for a brief period of time. These transitory events make it more difficult for another action potential to be generated quickly. This refractory period limits the number of action

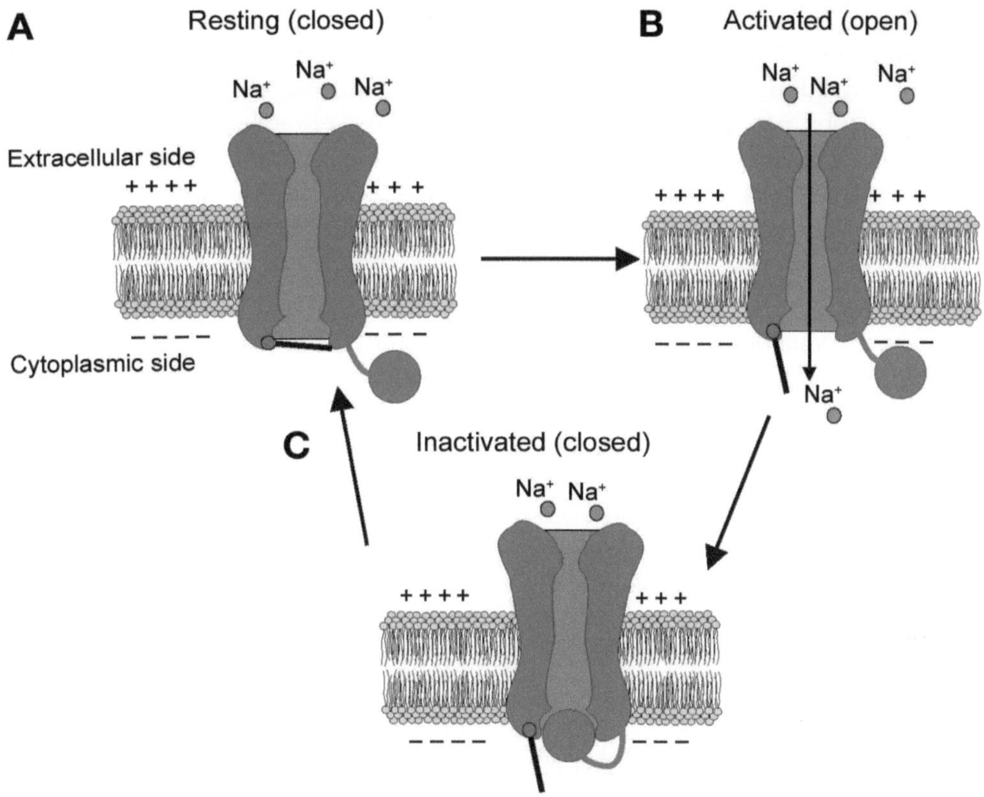

Fig. 3. Voltage-gated Na^+ channel. In the resting condition **(A)** the activation gate (black bar) is closed and the inactivation gate (ball and chain) is open. No Na^+ flows because of the closed activation gate. With depolarization of the membrane, there is a conformational change of the channel and the activation gate opens **(B)**. Na^+ flow then occurs. This is followed by inactivation by closure of the inactivation gate **(C)**, prohibiting the further flow of Na^+ ions. With repolarization of the membrane, the inactivation gate opens and the activation gate closes and the channel is ready for another cycle (A). Modified from Kandel et al., 2000, with permission.

potentials that a given nerve cell can produce per unit time. This phenomenon also explains why action potentials do not reverberate up and down the neuronal membrane.

Very small depolarizations do not trigger an action potential because they do not open a large enough number of Na^+ channels and increase the driving force on K^+ ions, favoring repolarization, as the depolarized membrane potential is further from E_K. Action potential threshold occurs when the inward Na^+ current just exceeds K^+ outflow. The net inward current produces an active depolarization, which initiates further Na^+ channel opening and generation of the action potential.

Extracellular electrodes can detect action potentials from individual neurons only if the size of the electrode is comparable to the size of the cell (tens of micrometers) and if the electrode is very close to the cell soma, where the action potential is generated. The amplitude of the extracellularly recorded action potential is small, on the order of tens of microvolts, and the duration is less than a millisecond. With conventional EEG electrodes, the action potentials from individual neurons are too small to be detected. However, when many neurons fire action potentials simultaneously, which can occur, for instance, in epileptic patients, their summated action potentials can be detected in EEG recordings as a "population spike."

5. TRANSMISSION OF ACTION POTENTIALS

The action potential can traverse long axonal distances without loss of amplitude despite the fact that neuronal membranes have relatively poor conducting properties. During the generation of the action potential, there is some passive flow of current downstream from the action potential. The passive current flow depolarizes the membrane potential in adjacent regions of the axon, opening Na^+ channels. The local depolarization results in another action potential, which then spreads again in a continuing cycle until the end of the axon is reached.

6. SIGNALING CHANGES

When the postsynaptic membrane is stimulated through either electrical stimulation via gap junctions or at chemical synapses, there is a change in membrane potential. The change in membrane potential is typically not instantaneous because the membrane has both a resistive and capacitive component. Neurons have three passive electrical properties that are important to electrical signaling: the resting membrane resistance, the membrane capacitance, and the intracellular axial resistance along the axons and dendrites. These membrane properties are important in determining whether an action potential will be generated.

All cell membranes have a resistance. The input resistance, R_{in}, of the cell determines how much the cell will depolarize in response to a steady current. The magnitude of the depolarization, ΔV is given by Ohm's law:

$$\Delta V = I \times R_{in}$$

Of two neurons receiving identical synaptic current inputs, the cell with the higher input resistance will have a greater change in membrane voltage. Input resistance depends on both the density of resting ion channels in the membrane and the size of the cell. The larger the neuron, the greater will be its membrane surface area and the lower the input resistance, because there will be more resting channels to conduct.

Membranes also act as capacitors. A capacitor consists of two conducting plates separated by an insulating layer. The fundamental property of a capacitor is its ability to store charges of opposite sign: positive charge on one plate, negative charge on the other. Voltage across a capacitor is proportional to the charge stored on the capacitor:

$$V = Q/C$$

where Q is the charge in coulombs and C is the capacitance in farads. To alter the voltage, charge must either be added or removed from the capacitor.

$$\Delta V = \Delta Q/C$$

The change in charge (ΔQ) is the result of the flow of current across the capacitor (I_c). Because current is the flow of charge per unit time ($I_c = \Delta Q/\Delta t$), the change in voltage across a capacitor can be calculated as a function of current and the time of current flow (Δt):

$$\Delta V = I_c \times \Delta t/C$$

The magnitude of the change in voltage across a capacitor in response to a current pulse depends on the duration of the current, as time is required to deposit and remove charge on the plates of the capacitor.

Capacitance is directly proportional to the area of the plates of the capacitor. The larger the area of a capacitor, the more charge it will store for a given potential difference. Because all biological membranes are composed of lipid bilayers with similar insulating properties, membrane capacitance increases with the size of the cell. More charge, and, therefore, current, is required to produce the same change in membrane potential in a larger neuron than in a smaller one. The capacitance of the membrane has the effect of reducing the rate at which the membrane potential changes in response to a current pulse. If the membrane had only resistive properties, a step pulse of outward current passed across it would change the membrane potential instantaneously. Biological membranes have both capacitive and resistive properties in parallel.

7. CELL-TO-CELL COMMUNICATION

Through either neurotransmitter release at chemical synapses or current flow through gap junctions, ligand-gated or voltage-gated channels open and elicit postsynaptic potentials. Postsynaptic potentials alter the probability that an action potential will be produced in the postsynaptic cell. If there is depolarization of the membrane, the potential is termed an excitatory postsynaptic potentials (EPSP), whereas, if there is hyperpolarization, the potential is called an inhibitory postsynaptic potential (IPSP). EPSPs bring the membrane potential closer to threshold for action potential generation, whereas IPSPs keep the membrane potential more negative than the threshold potential. In chemical synapses, whether the event is an EPSP or IPSP depends on the neurotransmitter released and the type of postsynaptic receptor activated. In the cerebral cortex, approx 90% of neurons (principal neurons) synthesize and release glutamate, the principal CNS excitatory neurotransmitter. The remaining neuronal populations (interneurons) release GABA, the principal inhibitory neurotransmitter of the cortex.

Direct electrical transmission from one cell to another occurs through gap junctions. Gap junctions consist of hexameric complexes formed by the close juxtaposition of pores consisting of proteins called connexons that span the neuronal membrane. The pore of a gap junction is larger than the pores of voltage-gated ion channels and can, therefore, transfer larger substances between cells, including intracellular metabolites. Electrical transmission across gap junctions occurs rapidly because passive current flow across the gap junction is virtually instantaneous. Gap junctions seem to have an important role in the synchronization of neuron firing, particularly in interneuronal networks.

Chemical synapses have a wider spacing between cells, termed the synaptic cleft, and operate through release of neurotransmitter stored in vesicles. The neurotransmitter diffuses from the presynaptic membrane to the postsynaptic membrane. Neurotransmitter release occurs when an action potential reaches the presynaptic terminal and initiates the opening of voltage-gated Ca^{2+} channels. This permits a rapid influx of Ca^{2+} into the presynaptic terminal. Elevation of intracellular Ca^{2+} causes synaptic vesicles to fuse with the plasma membrane of the presynaptic neuron. There are a number of calcium-binding proteins that participate in the cascade of events that lead to neurotransmitter release. The fusion of the vesicular and neuronal membranes allows release of the neurotransmitter. Figure 4 illustrates the process of neurotransmitter release.

Excitatory synaptic transmission is mediated by glutamate acting on target postsynaptic neurons via three types of ionotropic receptors, named after their selective agonists (α-amino-3-hydroxy-5-methylisoxazole-4-proprionic acid [AMPA], kainic acid [KA], and *N*-methyl-D-aspartate [NMDA]). Although all types of glutamate receptors respond to glutamate, they have individual characteristics. The AMPA receptor responds rapidly to glutamate with opening of

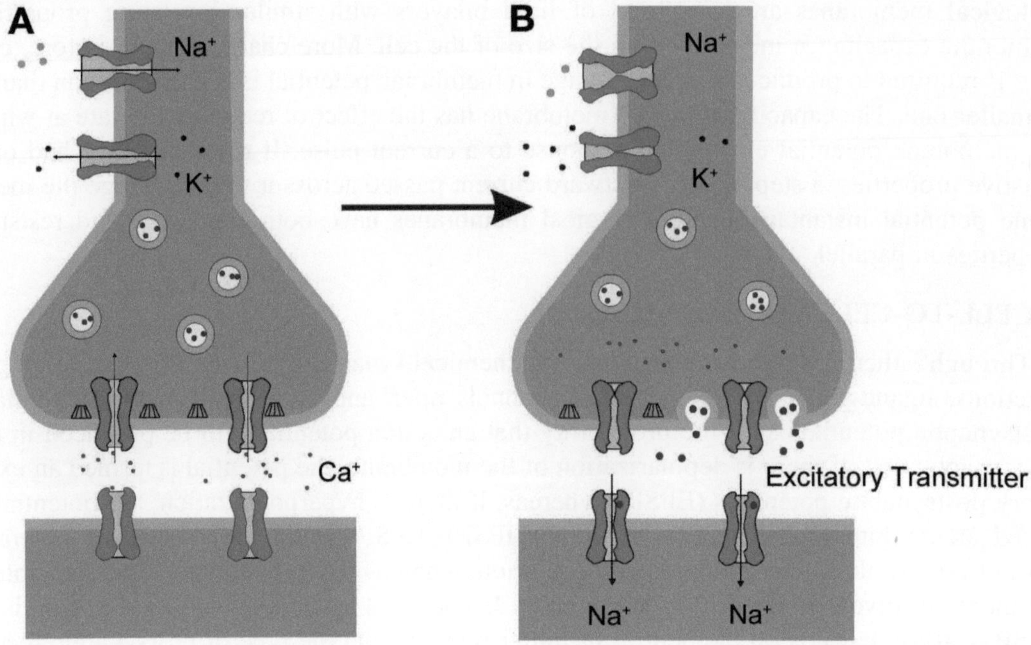

Fig. 4. Synaptic neurotransmission. An action potential travels down the axon until it reaches the synapse. The depolarization causes voltage-gated Ca^{2+} channels to open. The influx of Ca^{2+} results in high concentrations of Ca^{2+} near the active zone **(A)**. This triggers fusion of vesicles with neurotransmitter to the presynaptic cell membrane and emptying of the vesicles into the synaptic cleft **(B)**. The neurotransmitter crosses to the postsynaptic membrane and results in depolarization of the membrane if it is an excitatory neurotransmitter. With glutamate release, there is binding of the ligand to postsynaptic receptors (AMPA, KA, or NMDA) with subsequent inflow of Na^+ ions. Modified from Kandel et al., 2000 with permission.

channels permeable to Na^+ and K^+, resulting in depolarization. One synapse contains tens of AMPA receptors on the postsynaptic membrane, and current summation leads to an EPSP of approx 1 mV. As a result, simultaneous activation of several excitatory synapses is necessary to sufficiently activate a postsynaptic neuron to action potential threshold. Kainate receptors are similar to AMPA receptors but have slower kinetics. The third type of glutamate ionotropic receptor—the NMDA receptor—may not directly participate in information processing but seems to play a critical role in synaptic plasticity. The NMDA channel has characteristics of both a ligand-activated and voltage-sensitive channel. At resting potential, Mg^{2+} sits in the channel, blocking the flow of ions. Only with depolarization of the membrane is Mg^{2+} displaced and Na^+ and Ca^{2+} ions able to cross the channel. The high permeability of the NMDA receptor to Ca^{2+} underlies its role in synaptic plasticity, such as long-term potentiation of synaptic strength, which is hypothesized to participate in learning and memory.

GABA is the principal inhibitory transmitter of the brain. Inhibitory synapses made by interneurons and using GABA as their transmitter use two types of receptors, $GABA_A$ and $GABA_B$ receptors. $GABA_A$ receptors are ligand-gated ion channels, whereas $GABA_B$ receptors are metabotropic receptors (*see* next paragraph). $GABA_A$ receptors are inhibitory because their associated ion channels are permeable to Cl^-. Because the reversal potential for Cl- is more negative than the threshold for neuronal firing, Cl^- flow hinders action potential generation. Activation of $GABA_B$ receptors results in opening of K^+ channels that also inhibits the postsynaptic cell. In the spinal cord, GABA and glycine act as inhibitory transmitters.

Metabotropic receptors differ from ionotropic receptors in that they affect ion channels via the activation of G proteins. All metabotropic receptors are part of a superfamily of G protein-coupled receptors. In certain cases, the activation of G proteins by metabotropic receptors allows binding of activated G protein subunits directly to ion channels (this mechanism, for instance, operates in $GABA_B$ receptor-mediated opening of K^+ channels). In other cases, metabotropic receptors can be coupled to ionic channels via second messengers, such as cAMP or cGMP. Metabotropic receptors can also couple to intracellular effector enzymes. Activation of protein kinases can then phosphorylate ion channels or other proteins closely associated with ion channels, thereby altering channel function.

There are five biogenic amine neurotransmitters: three catecholamines, norepinephrine, epinephrine, and dopamine; and histamine and serotonin. Dopamine plays a role in control of body movements, and norepinephrine and serotonin in the modulation of sleep and wakefulness. The role of epinephrine and serotonin is less clear.

Acetylcholine is concentrated in two major CNS loci, the forebrain nuclear complex and the cholinergic nuclei of the brainstem tegmentum. Although not yet fully understood, it is known that acetylcholine is in involved in pain and chemosensory pathways as well as memory. Interestingly, autosomal dominant nocturnal frontal lobe epilepsy has been linked to chromosome 20q13.2 and a mutation in the gene *CHRNA4* encoding the α_4-subunit of a neuronal nicotinic acetylcholine receptor. Nicotinic cholinoreceptors are enriched in interneurons and deficiency in the excitation of these inhibitory cells possibly underlies the enhanced excitability thought to be operant in epilepsy.

8. PHYSIOLOGICAL BASIS OF THE EEG

The EEG represents a set of field potentials recorded by multiple electrodes on the surface of the scalp. The electrical activity of the EEG is an attenuated measure of the extracellular current flow from the summated activity of many neurons. The surface EEG predominately reflects the activity of cortical neurons close to the EEG electrodes. Deeper structures, such as the hippocampus, thalamus, or brainstem, do not contribute directly to the surface EEG. However, transmission of electrical impulses from distant sites has substantial effects on the surface EEG. For example, thalamocortical connections are critical in the synchronization of electrical activity, such as sleep spindles. Oscillatory EEG patterns arise because of pacemaker cells, in which membrane voltage fluctuates spontaneously, or because of the reciprocal interaction of excitatory and inhibitory neurons in circuit loops. The human EEG shows activity over the range of 1 to 30 Hz, with amplitudes in the range of 20 to 300 μV.

The waveforms recorded by the surface electrodes depend on the orientation and distance of the electrical source with respect to the recording electrode. To understand how the EEG is obtained, it is useful to examine a single pyramidal neuron situated in layer 5 of the cortex, with its apical dendritic arbor above, although it is clear that EEG activity derives from thousands of such neurons functioning within networks.

Figure 5 shows such a single neuron with current flowing into the dendrite at the site of generation of an EPSP, creating a current sink. Current flow must complete a loop and, therefore, this generates a source somewhere along the dendrites or cell body. The size of the voltage change created by the EPSP is predicted by Ohm's law, $V = IR$. The R_m (membrane resistance)

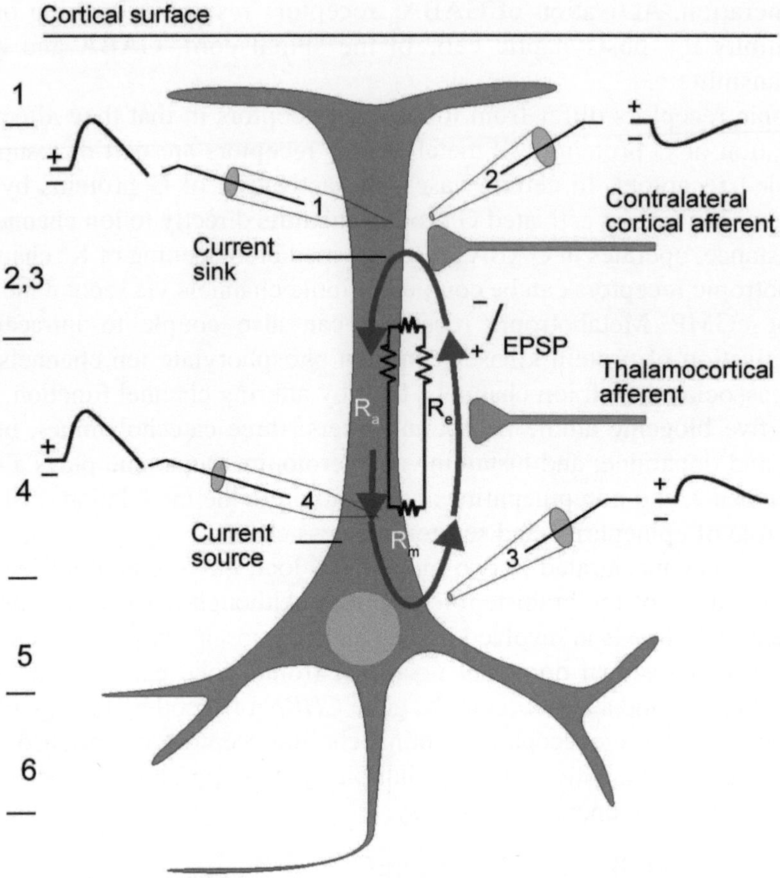

Fig. 5. Current flow in a cortical neuron. *See* text. Modified from Kandel et al., 2000 with permission.

is much larger than the extracellular fluid and, therefore, the corresponding voltage recorded by an intracellular electrode is larger, and opposite in polarity to an extracellular electrode positioned near the current sink. At the site of generation of an EPSP, the extracellular electrode detects current (positive ions) flowing away from the electrode into the cytoplasm as a negative voltage change, whereas the intracellular electrode detects a positive change in voltage caused by the influx of Na^+ ions. An extracellular electrode near the source has an opposite deflection. The direction of voltage change is determined by location in regards to the sink and source. In Fig. 5, note the differences in pen deflection depending on whether the extracellular electrode is near the source or sink.

If instead of an apical (layer 2/3) EPSP, we examine the instance of a basal IPSP, close to the soma of such a pyramidal neuron, the current flow loops and derived extracellular and intracellular recordings turn out to be identical to those described above for the superficial EPSP. In both cases (a superficial EPSP or a basal IPSP), a dipole is created with separation of charge oriented vertically in the cortex, with extracellular negativity in more superficial laminae, and extracellular positivity in deeper laminae.

Now consider how these extracellular field potentials will behave when recorded from the scalp. It is fortunate that the cortex is organized as a sheet just under the scalp, with an

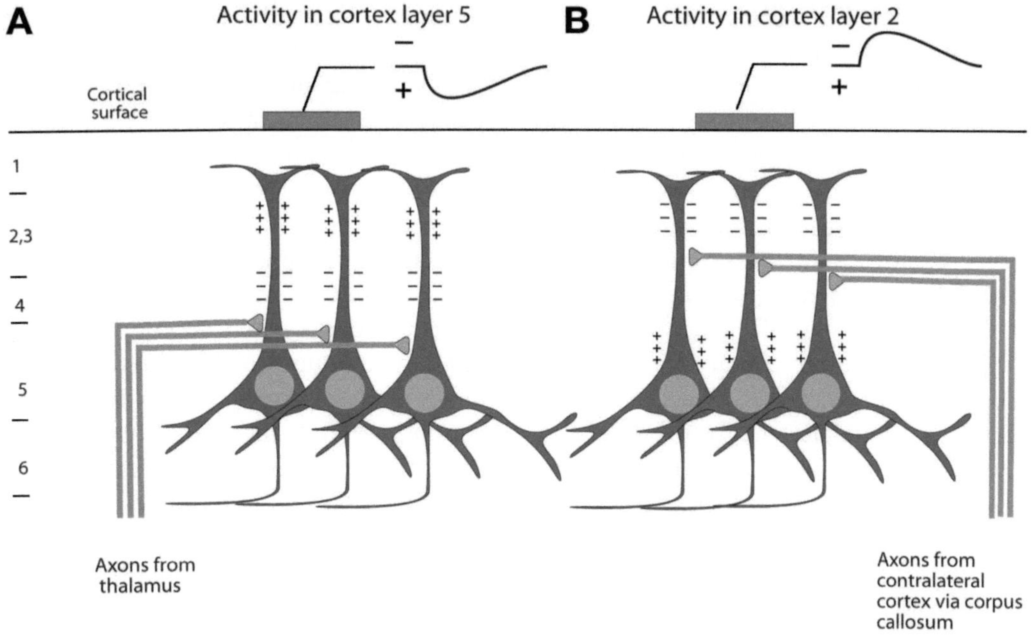

Fig. 6. Polarity of the surface EEG depends on the location of the synaptic activity within the cortex. *See* text. Modified from Kandel et al., 2000 with permission.

intrinsic columnar organization. It is these pivotal features that afford our ability to obtain a useful EEG signal. The highly organized columns of pyramidal neurons are arranged just so, with cell bodies in deeper laminae, and dendritic arbors extending upward to laminae that are more superficial. Excitatory afferent fibers innervate the superficial layer 2/3 dendrites, whereas inhibitory contacts favor the deeper cell bodies below. When such EPSPs are generated because of the coordinated excitation of numerous afferent inputs into layer 2/3 dendrites, broad regions of the cortex coordinately generate transient dipoles that lead to measurable extracellular voltage negativity, as recorded by scalp electrodes.

Figure 6 illustrates these anatomic and physiological properties. In Fig. 6, there are afferent inputs into either the apical dendrites (A) or cell body (B). In both cases, the afferent stimuli lead to depolarization (sink) with current flow into the cell body. This results in negativity extracellularly. The current flow in A results in a source in the apical dendrites, whereas, in B, the source is located at the soma. The examples thereby lead to two vertically oriented dipoles of opposing polarity. Surface EEG electrodes will detect the extracellular electrical fields generated closer to the cortical surface (i.e., the superficial laminae), and there will be less influence from activity occurring at the cell body. Therefore, the deflection of the pen is opposite in the two conditions. Again, as stated above, a deep IPSP will generate a similar scalp electrical field as will a superficial EPSP—both lead to scalp negativity.

Because of these geometric reasons, only vertically oriented dipoles are detectable with scalp electrodes. Thus, only those portions of the underlying cortical sheet that are parallel to the scalp are detectable with EEG electrodes. Portions of the cortex that run down the sulci, oriented orthogonal to the scalp, generate radially oriented dipoles, which are not well seen by scalp EEG technique. Magnetoencephalography may be able to detect such radial dipoles with much greater facility, however.

It follows that EEG detects the summed extracellular electrical field potentials from a swath of underlying cortex. It has been estimated that each EEG electrode "sees" the summed activity of roughly 6 cm^2 of underlying cortex. We obtain cogent signals because there is a significant amount of synchrony underlying the behavior of thousands of cortical neurons. This synchrony may be physiological, as seen in the alpha rhythm over the posterior channels. However, when the cortex becomes excessively synchronized, we may detect pathological EEG morphologies, termed spikes and sharp waves.

Such epileptiform features represent the summed activity of numerous rapidly firing neurons, which have been depolarized to threshold in a coordinated and excessively synchronized fashion. A wave of cortical excitation, termed the paroxysmal depolarization shift (PDS) is thought to be responsible for this broad and synchronous cortical excitation. The PDS is sufficiently strong to bring a large collection of neurons synchronously to threshold, generating a rapid and synchronized bursting of action potentials that "ride" atop the more sustained PDS potential. It is thought that PDS waves are followed by strong waves of hyperpolarization, perhaps recruited by inhibitory interneurons in the involved circuits. These pathological waves of cortical excitation and subsequent inhibition affecting broad regions of excessively synchronized cortex are thought to underlie the spikes, sharp waves, and sharp and slow wave complexes we recognize as pathological correlates in routine EEG interpretation.

REVIEW QUESTIONS

1. What is responsible for voltage potentials across neuronal membranes?
2. What forces govern ionic flow through neuronal channels?
3. Which ion is most important for determining the resting membrane potential? Why?
4. What are the two main mechanisms that permit gating of channel opening?
5. What is the main channel that, in a feed-forward fashion, leads to the rapid depolarization intrinsic to the action potential? Which ion facilitates repolarization?
6. As neuronal size increases, does depolarization require more or less current flow across the membrane?
7. What are the two types of subthreshold changes to membrane voltage that occur secondary to afferent input?
8. What are the three main types of glutamate receptors? How are they defined and how do they differ?
9. What are the main geometric and physiological properties of the cortex that permit the recording of the EEG signal?
10. What is the term for the wave of cortical depolarization that can support rapid and excessively synchronized (pathological) spiking of large collections of neurons?

REVIEW ANSWERS

1. Different concentrations of ionic species across the neuronal membrane and differential permeability to ionic flow across the membrane are the determinants of the membrane potential. The Goldmann equation represents this relationship mathematically.
2. Two forces govern ionic flow across channels, a chemical driving force defined by concentration gradients, and an electrical driving force determined by the membrane potential.
3. The resting membrane potential is principally influenced by the reversal potential for K^+. It has the highest permeability at resting conditions and, therefore, contributes the most in the Goldmann equation to the determination of the resting membrane potential.
4. Ligand-gated and voltage-gated channels represent two mechanisms for controlling the opening of gated channels.

5. The voltage-gated Na^+ channel is the principal ionic conductance that underlies the upstroke of the action potential. The K^+ ion facilitates repolarization.
6. As neuronal size grows, more current is required to achieve a similar level of depolarization, because the larger neuron has larger membrane surface area, which leads to higher capacitance.
7. The two main types of subthreshold stimuli are EPSPs and IPSPs.
8. AMPA, KA, and NMDA receptors are three distinct subtypes of glutamate receptors. AMPA and KA receptors are ligand-gated ionic channels, permeable to Na^+ and K^+. KA receptors have slower kinetics and differential binding properties with specific agonists. NMDA receptors have both ionotropic and metabotropic properties and are important in synaptic plasticity. NMDA receptors also affects Ca^{2+} translocation across the membrane.
9. The two main features of the cortex that permit EEG acquisition are its sheet-like organization parallel to the scalp (in large part), as well as its columnar organization, which leads to the generation of orthogonal electrical dipoles in the cortex, detectable by scalp EEG electrodes.
10. The paroxysmal depolarization shift is the term for the pathological wave of cortical excitation that leads to rapid and synchronized regional spiking detectable by EEG as epileptiform activity.

SUGGESTED READING

Armstrong CM, Bezanilla F. Inactivation of the sodium channel II. Gating current experiments. J Gen Physiol 1977;70:567–590.

Catterall WA. Structure and function of voltage-sensitive ion channels. Science 1988;242:50–61.

Friedman LK, Sperber EF, Moshe SL, Bennett MV, Zukin RS. Developmental regulation of glutamate and GABAA receptor gene expression in rat hippocampus following kainite-induced status epilepticus. Dev Neurosci 1997;19:529–542.

Furshpan EJ, Potter DD. Transmission at the giant motor synapses of the crayfish. J Physiol 1959;145:289–325.

Goldman DE. Potential, impedance, and rectification in membranes. J Gen Physiol 1943;27:37–60.

Hodgkin AL, Huxley AF. A quantitative description of membrane current and its application to conduction and excitation in nerve. J Physiol 1952;117:500–544.

Hodgkin AL, Katz B. The effect of sodium ions on the electrical activity of the giant axon of the squid. J Physiol 1949;108:37–77.

Jessell TM, Kandel ER. Synaptic transmission: a bi-directional and a self-modifiable form of cell-cell communication. Cell 1993;72(Suppl):1–30.

Kandel ER, Schwartz JH, Jessell TM. Principles of Neural Science. 4th ed. McGraw-Hill, New York, NY, 2000.

Unwin N. Neurotransmitter action: opening of ligand-gated ion channels. Cell 1993;72(Suppl):31–41.

Woodhull AM. Ionic blockage of sodium channels in nerve. J Gen Physiol 1973;61:687–708.

3
Ions, Membrane Potentials, and Myelin

Devanand Jillapalli and Jeremy M. Shefner

Summary

The neuron is uniquely suited for the transmission of electrical impulses. The neuronal membrane itself allows for charge separation; depending on the permeability of the membrane to a given type of ion, that ion will distribute across the membrane, producing a resting membrane potential, described by the Nernst equation. However, via the sodium–potassium (Na–K) pump, an active electrochemical gradient is maintained across the cell membrane, the magnitude of which can be calculated by knowing the relative concentrations of all the relevant ions, both in the intracellular fluid (ICF) and extracellular fluid (ECF), via the Goldman–Hodgkin–Katz equation. The development of action potentials are dependent on the presence of voltage-gated sodium channels, which open when the membrane itself is partially depolarized through mechanical, electrical, or chemical means. The initiation of an action potential creates a spreading area of voltage change, causing additional nearby channels to open, ultimately leading to the propagation of the action potential down the entire length of the axon. Myelin dramatically speeds the process of neuronal depolarization by producing salutatory conduction. Together, with the complex set of processes at the neuromuscular junction, neural transmission is effectively achieved.

Key Words: Action potential; ion; myelin; resting membrane potential; sodium–potassium pump; voltage gated.

1. INTRODUCTION

In the clinical neurophysiology laboratory, a variety of electrical signals are recorded from subjects. Understanding the underlying neurophysiology governing these electrical signals is very important if one is to correctly identify, record, and interpret the signals. In this chapter, we will first review the role of the semipermeable cell membrane; intracellular fluid (ICF) and extracellular fluid (ECF) compartments; and the concentration, charge, and permeability of the important ions in the generation of potential difference across the cell membrane. Later, we will see how local changes in the resting membrane potential of excitable cells can initiate propagation of the current to form the basis of communication from nerve to nerve and from nerve to muscle. This chapter repeats some of the information in chapter 2; however, additional information relevant to peripheral nerves is also included.

2. CELL MEMBRANE, ICF, AND ECF

The cell membrane separates the ICF from the ECF, and helps to maintain critical differences in ionic concentrations between the two fluid spaces. Of the electrically important ion species, sodium concentration is approximately 12 times greater in the ECF than in

From: *The Clinical Neurophysiology Primer*
Edited by: A. S. Blum and S. B. Rutkove © Humana Press Inc., Totowa, NJ

the ICF, whereas chloride concentration is approx 25 times greater in the ECF than the ICF. Potassium, however, is approx 35 times more abundant in the ICF than in the ECF. The ICF also contains proteins, organic phosphates, nucleic acids, and so on, which at physiological pH have a net negative charge. Because the cell membrane is a bilipid layer, it is quite permeable to lipid soluble substances, while acting as an effective barrier against many water-soluble particles. However, the cell membrane also contains specialized protein molecules, which act to selectively transport or allow passage of certain water-soluble particles. It is through these transport and channel proteins that movement of water-soluble ions and molecules occurs in and out of the cell. The mechanisms by which this movement occurs are *passive transport* and *active transport*.

2.1. Passive Transport

Passive transport is defined by movement that does not require the expenditure of energy. Two main forces contribute to passive movement of particles across the membraneconcentration gradient and electrical charge. All other factors being equal, particles will tend to move in such a manner that their concentration is equal throughout a given space. Thus, if a membrane is permeable to a certain particle, and that particle is more highly concentrated on one side of the membrane than the other side, diffusion forces will cause a net movement of particles toward the side with the lower concentration. Physiologically, this movement occurs through the transport proteins in the membrane. Thus, all other factors being equal, sodium ions will move from the ECF into the ICF, because there is a higher concentration of these ions in the ECF. Similarly, potassium ions tend to diffuse out of the cell, because their concentration is much higher in the ICF. If diffusion is the only force determining particle movement, the amount of particle movement will be determined by the magnitude of the concentration gradient, and by the extent to which the membrane is permeable to that particle. Along the cell membrane, permeability is determined by properties of the transport proteins, with selectivity based on the electrical and mechanical (diameter and shape) properties of the transport protein and the particle to be transported. Not all water-soluble ions and molecules move simply down the concentration gradient unaided. Some ions and molecules require a carrier to shuttle them across the cell membrane down the concentration gradient, referred to as *facilitated diffusion*.

The other major force that determines passive movement of particles across the nerve membrane is electrical charge. The particles of interest to us are all charged; potassium, sodium, and calcium are present as positively charged ions, whereas chloride carries a negative charge. Therefore, if particles move across the membrane according to concentration gradients, they carry with them either a net positivity or negativity. Because like charges repel each other, a buildup of either positive or negative charge will tend to inhibit the further transport of like charges.

2.2. Active Transport

In contrast to passive diffusion, active transport of substances across the cell membrane requires the expenditure of energy, usually in the form of high-energy phosphate bonds, such as ATP. Active transport involves the transport of ions in a direction not determined by concentration or electrical gradients. For example, positively charged sodium ions passively tend to move into the cell along both electrical and concentration gradients. However, active transport supports the movement of such ions in the opposite direction, against the passive forces. Both sodium and potassium ions are transported actively across the membrane, via a single mechanism called the sodium–potassium (Na–K) pump.

2.3. *The Na–K Pump*

The Na–K pump is present in all living cells. It is a complex protein with two subunitsthe α- and β-subunits. Because the role of Na–K pump is to move ions against a concentration gradient, the sodium receptor sites for this pump are located inside the cell membrane (where the sodium concentration is lower), whereas the potassium receptor sites on the pump are located outside the cell membrane (where the potassium concentration is lower). There are three sodium-binding sites and two potassium-binding sites in any Na–K pump. Thus, for every three sodium ions sent out of the cell, two potassium ions are brought inside.

3. THE RESTING MEMBRANE POTENTIAL

We have seen in the preceding paragraphs that the net diffusion of ions occurs down the concentration and electrical gradients. If a membrane were completely and equally permeable to all particles, diffusion would be allowed to proceed unchecked, and the various concentration differences across the cell membrane would eventually become equalized. However, in biological systems, both active transport and selective membrane permeability prevent such equalization. The resting membrane potential for a cell membrane is a product of the gradients that exist for single ion species, in combination with energy-requiring processes, such as active transport.

3.1. *The Nernst Potential*

In a hypothetical situation, let us assume that the cell membrane is permeable to only one single ion, which then tends to diffuse through the cell membrane along its gradient. With ion movement, there will be an accumulation of charge; this charge will tend to oppose further diffusion of like-charged particles with the force of opposition increasing as more ions travel across the membrane. At some point, the diffusion force responsible for ion movement along the concentration gradient will exactly equal the electrical opposing force. The potential difference at which this equalization of forces occurs is called the *Nernst potential* for that ion. For singly charged ions at 37C, the formula to calculate this potential (in millivolts) is: $61 \times \log_{10}$ (concentration of the ion outside divided by the concentration of the ion inside). Because the concentration of different ions is different across the cell membrane, one can see that the Nerst potential will be different for each ion. Using this formula, we can calculate the potential difference across the cell membrane for potassium and sodium ions as examples. For potassium, the Nernst potential is approximately –94 mV, with the negativity inside the cell membrane, whereas it is +61 mV for sodium. The effect that each ion species will have in determining the global resting membrane potential for a membrane is determined by the membrane permeability to than ion. Thus, if a membrane is completely impermeable to an ion, it will have no effect on the membrane potential of the cell, no matter what its Nernst potential might be.

3.2. *Goldman–Hodgkin–Katz Equation*

This equation objectifies the point made in the preceding paragraph, and suggests that the overall resting membrane potential for a biological membrane is a combination of the Nernst potentials for all permeable ion species inhabiting the perimembrane area, scaled by their respective permeabilities. At rest, the membrane is most permeable to chloride and, hence, the overall resting membrane potential is closest to the Nernst potential for chloride, at approximately –60 mV.

3.3. The Role of the Na–K Pump

The Goldman-Hogkin-Katz equation makes no inference regarding equilibrium. It simply states what the transmembrane potential will be given to the membrane conductances and relative concentrations of different ions at a given point in time. As an equilibrium state does not exist, without active intervention the concentrations of various ions will change, according to either their concentration or diffusion gradient. The concentrations discussed above are kept constant, however, but the activity of an energy requiring Na-K pump. The Na–K pump takes out three sodium ions for every two potassium ions it brings inside the cell, leaving a net *deficit* of positive charges (or net *excess* of negative charges) inside the cell; this also contributes to the electrical difference or gradient across the cell membrane, with the negativity inside the membrane and positivity outside.

In summary, the resting membrane potential of cells is largely determined by the passive diffusion of the ions through channel proteins in the cell membrane, according to both concentration and electrical gradients. The membrane is most permeable to chloride, so this ion contributes most to the resting membrane potential as a whole. In addition, the electrogenic Na–K pump, on a smaller but sustained scale, contributes to this resting membrane potential. As a result, the resting membrane potential in large nerve fiber is approx –60 mV, with the negativity inside the cell membrane. The resting membrane potential provides the substrate on which neural communication is based but by itself is not responsible for signal transmission. Information transfer requires the presence of a regenerative process, the action potential, which can transmit a signal over long distances without decay.

4. MEMBRANE ACTION POTENTIALS

Nerve and muscle action potentials occur as a result of sudden changes in permeability of specific ion channels, in response to local perturbations in the voltage of membrane potential. The channels involved are known to be *voltage dependent*; that is, permeability changes are dependent on changes in membrane potentials. Although voltage-dependent ion channels depend on changes in membrane potential to initiate permeability changes, once those changes occur, they revert back to baseline permeability levels quickly and irrespective of membrane potential. Thus, although an action potential needs a membrane potential change for its initiation, its termination occurs because of processes intrinsic to the channels themselves. Action potentials are, therefore, extremely brief, usually in the order of approx 1 ms. Action potentials are always all-or-none responses, because the voltage-dependent changes discussed either occur completely or not at all. To understand these action potentials, we will discuss what happens during depolarization and repolarization.

The process of action potential generation was first studied in *unmyelinated* axons of invertebrates. Such axons have both voltage-gated potassium and sodium channels. If a local area of membrane is briefly depolarized (i.e., rendered more positive) by approx 15 mV, a conformational change occurs in both ion channels, such that both become briefly much more permeable. Because of the high concentration of extracellular sodium, as well as the negativity of the intracellular space, sodium will rush into the cell through the open sodium channels. However, after approx 1 ms, the sodium channels revert to their resting, nearly closed, state. The effect of sodium influx is to render the local area of intracellular space

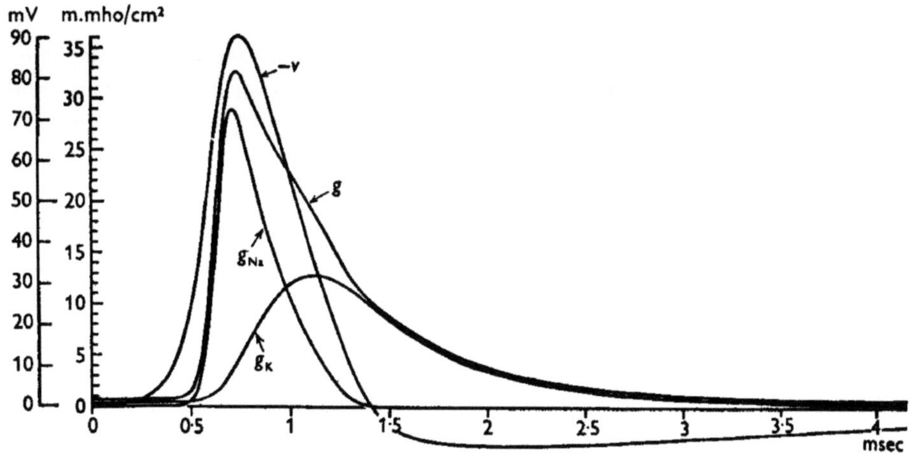

Fig. 1. The conductance of the sodium ions (g_{Na}) and potassium ions (g_K) in the action potential (Hodgkin–Huxley model, 1952), illustrating the rapid opening and closing of the sodium channels and, in contrast, the slower opening and delayed closure of the potassium channels.

briefly depolarized. In contrast to sodium channels, potassium channels open more slowly and close more slowly. Therefore, relatively long after the sodium channels have reverted to their baseline closed state, potassium channels remain open, allowing positively charged ions to leave and restoring the baseline negativity within the axon. These changes in the sodium and potassium conductance during an action potential are illustrated in Fig. 1. In *myelinated* axons, however, changes in potassium conductance do not seem to contribute to the repolarization process.

Local changes in myelinated or unmyelinated axons are unimportant in themselves, unless a mechanism exists for them to propagate along the axon. This mechanism is discussed next.

5. PROPAGATION OF THE ACTION POTENTIAL

The consequence of an action potential occurring in a local area of membrane is to depolarize that area. Local current spreading from that area depolarizes the surrounding resting membrane to an extent greater than necessary to cause neighboring voltage-gated sodium and potassium channels to open, resulting in the changes in sodium and potassium conductance that then produce the action potentials in these neighboring areas. This wave of depolarization occurs in a continuous fashion in unmyelinated fibers. The rate at which action potentials traverse an unmyelinated axon is determined by both the total amounts of ions traversing the membrane (ion flux), as well as the resistance to current flow within the axon. Both of these factors are a function of axon diameter; the larger the axon, the more ions flow into it, and the easier it is for current to flow along it. The relationship between fiber size and conduction velocity in nonmyelinated axons is nonlinear, which implies that *very large unmyelinated axons* would be necessary to provide conduction velocities required for normal vertebrate function. Furthermore, such large unmyelinated axons are not only impractical but also would be inefficient in signal transmission because of the large transmembrane capacitance and low transmembrane resistance and the need for higher energy expenditure. Myelinated axons, however, provide the correct combination of efficiency in signal conduction and energy conservation, as discussed next.

5.1. Myelin and Salutatory Conduction

The myelin sheath is a specialized structure existing within Schwann cells; it is a highly effective insulating material, wrapping around the axon in concentric spirals. Schwann cells are present every 1 to 2 mm along the axon. At the junction of contiguous Schwann cells are the *nodes of Ranvier*, an area that is devoid of myelin. In contrast to unmyelinated axons, sodium and potassium channels are not evenly dispersed along the axon surface. Sodium channels are found in the highest concentration at the nodes of Ranvier and are not usually found in between these nodes under the myelin. In contrast, potassium channels are found underneath the myelin sheath. Depolarization in myelinated axons, therefore, can occur only at nodal regions, where sodium channels reside. Once a nodal area has been depolarized, a potential difference exists between it and the next contiguous nodal area, resulting in longitudinal flow of current. The difference between longitudinal current flow in the unmyelinated axons and myelinated axons is that, in the latter, instead of spreading to the immediate contiguous membrane, spread occurs from one node to the next node. However, because these nodes are separated by 1 to 2 mm, the action potential transmission seems to jump from one node to the next in a discontinuous fashion, termed *saltatory conduction.* Myelin allows this form of conduction by reducing membrane capacitance and increasing transmembrane resistance, so that the stimulus decay is not significant by the time it reaches the next node. Saltatory conduction significantly increases speed of action potential propagation (~70 m/s in large myelinated fibers). Myelinated axons, thus, combine efficiency in signal conduction and energy conservation.

SUGGESTED READING

Armstrong CM. Voltage-dependent ion channels and their gating. Physiol Rev 1992;72:S5–S12.

Hodgkin AL. Huxley AF. A quantitative description of membrane current and its application to conduction and excitation in nerve. J Physiol 1952;117:500–544.

Hubbard JI. Microphysiology of vertebrate neuromuscular transmission. Physiol Rev 1973;53:674–715.

Preston DC, Shapiro BE. Electromyography and neuromuscular disorders. Butterworth-Heinemann, Boston, MA, 1998.

Waxman SG, Bennett MVL. Relative conduction velocities of small myelinated and non-myelinated fibers in the central nervous system. Nat New Bio 1972;238:217–219.

REVIEW QUESTIONS (CHOOSE THE BEST ANSWER FROM THE ALTERNATIVES)

1. During depolarization of the cell membrane, the movement of sodium ions from the ECF into the ICF uses which of the following physiological mechanisms?
   A. Active transport.
   B. Active transport through voltage-gated sodium channel.
   C. Facilitated diffusion.
   D. Passive transport through voltage-gated sodium channel.
2. The potential difference across the cell membrane on the *inside* of the cell is:
   A. Positive at the peak of sodium influx during an action potential.
   B. Negative at the peak of sodium influx during an action potential.
   C. Positive during the resting state of the membrane potential.
   D. Independent of the action potential.
   E. None of the above.
3. The role of the Na–K pump is:
   A. Passive transport of either sodium or potassium ions.
   B. Passive transport of sodium and potassium ions.

C. Maintenance of the resting membrane potential by active transport of sodium and potassium ions.
D. Maintenance of the resting membrane potential by passive transport of sodium and potassium ions.
E. Relies on the transformation of ADP to AMP for energy.

4. The resting membrane potential of a neuron is approximately:
   A. +60 μV.
   B. –30 mV.
   C. –60 μV.
   D. –90 mV.
   E. None of the above.
5. The closure of the sodium channels during an action potential is caused by:
   A. Changes in the membrane voltage.
   B. Changes in the potassium conductance.
   C. Spontaneous closure independent of membrane voltage changes.
   D. Changes in calcium conductance.
   E. Local electrical currents adjacent to the channels.
6. The process of repolarization in a nerve fiber is brought about by which of the following:
   A. Increased potassium conductance in the myelinated fiber.
   B. Decreased sodium conductance in the unmyelinated fiber.
   C. Increased potassium conductance in the unmyelinated fiber.
   D. Increased calcium conductance in the myelinated fiber.
   E. A and C.
7. Which of the following properties in a nerve fiber results in a faster conduction velocity?
   A. High transmembrane capacitance, low transmembrane resistance, and large fiber diameter.
   B. Low transmembrane capacitance, high transmembrane resistance, and large fiber diameter.
   C. Low transmembrane capacitance, low transmembrane resistance, and small fiber diameter.
   D. High transmembrane capacitance, high transmembrane resistance, and small fiber diameter.
8. Which of the following statements concerning chloride ions and channels is most accurate?
   A. They have no role in establishing the resting membrane potential; only potassium and sodium contribute.
   B. At rest, they distribute passively across the cell membrane.
   C. The opening of chloride channels helps in repolarization of the neuronal membrane.
   D. Chloride is necessary for the action of the Na–K pump.
9. The membrane permeability of which of the following ions contribute the most in the genesis of the resting membrane potential?
   A. Sodium and potassium.
   B. Potassium and calcium.
   C. Chloride and potassium.
   D. Sodium and chloride.
   E. Sodium, chloride, and potassium contribute equally.
10. Which of the following statements concerning saltatory conduction is false?
    A. It occurs only in myelinated nerves.
    B. It relies on high internodal resistance and low capacitance.
    C. It requires the opening of potassium channels in the internodal membrane.
    D. It may fail if internodal myelin thickness is reduced.
    E. The greater the internodal distance, the faster the conduction velocity will be, up to the point that conduction fails.

REVIEW ANSWERS

1. The correct answer is D. Although it is true that sodium ions move passively into the cell along the concentration gradient during depolarization, such movement occurs through the voltage-gated sodium channels after they are first opened.

2. The correct answer is A. The membrane potential is negative on the inside of the cell membrane at rest. However, at the peak of the sodium influx during an action potential, there is reversal of polarity, resulting in positivity on the inside of the membrane.
3. The correct answer is C. The Na–K pump actively transports three sodium ions for every two potassium ions against their respective concentration gradient. This unequal transport of positive charges helps maintain the potential difference across the cell membrane in a sustained manner.
4. The correct answer is E. None the above. The correct resting membrane for a neuron is approx –60 mV; for a muscle fiber it is closer to –90 mV.
5. The correct answer is C. Sodium channels open during an action potential because of changes in the membrane potential. However, the closure of the sodium channels occurs spontaneously because of processes intrinsic to the channels and independent of membrane voltage changes.
6. The correct answer is C. In unmyelinated nerve fibers, increased potassium conductance plays an important role in the process of repolarization. However, in myelinated axons, changes in potassium conductance do not seem to significantly contribute to the repolarization process.
7. The correct answer is B. Low transmembrane capacitance, high transmembrane resistance, and large fiber diameter is the correct choice.
8. The correct answer is B. At rest, chloride is distributed passively across the cell membrane. In fact, chloride channels remain open at rest, helping to stabilize the membrane potential.
9. The correct answer is C. The membrane is most permeable to potassium and chloride ions, so that these ions contribute the most in the genesis of the resting membrane potential.
10. The correct answer is C. Opening of potassium channels in the internodal membrane is not required for saltatory conduction to occur. All of the other statements are true.

4

Introduction to Volume Conduction

Seward B. Rutkove

Summary

The term "volume conduction" refers to the complex effects of measuring electrical potentials a distance from their source generators. Near-field potentials refer to those recorded in relative close proximity to the detector, whereas far-field potentials refer to those recorded at a considerable distance, as is most commonly the case in evoked potentials. A relative straightforward model of volume conduction can be worked through to assist in better understanding how volume conduction effects can impact the shape of a recorded neuronal potential. In fact, all motor and sensory nerve conduction waveforms are substantially impacted by volume conductive effects. The recording setup of sensory studies (i.e., whether they are bipolar or referential) also impacts the size and morphology of the recorded signals. In addition, the compound motor action potential actually represents a composite of both near- and far-field activity. The morphology of both spontaneous discharges and the motor unit potentials themselves evaluated during needle EMG are, in part, caused by the complex effects of volume conduction.

Key Words: Bipolar; far-field; fibrillation potential; near-field; positive sharp wave; source generator; referential; volume conduction.

1. WHAT IS VOLUME CONDUCTION? WHY IS IT RELEVANT?

Volume conduction is the term used to describe the effects of recording electrical potentials at a distance from their source generator. In other words, the recording electrodes are not in direct contact with the nerve or muscle; there is a medium of some sort separating the two. In truth, volume conduction plays a role in almost all clinical neurophysiological recordings, both central and peripheral, because recording electrodes are never placed in direct contact with the nerve cells generating the signal. There are some occasional examples in which volume conduction plays only a minor role because the amount of nonexcitable tissue between the depolarizing neuron and the recording electrodes is quite small. One example of this would be near-nerve nerve conduction studies, in which a needle electrode is placed in very close vicinity to the nerve of interest. Even so, electrical signals can take short cuts through ionic media and, therefore, distant activity can be detected.

Two terms are worth being familiar with immediately: far-field potentials and near-field potentials, and volume conduction theory plays into both. A *near-field potential* is one that is obtained with the recording electrode in relatively close proximity to the source of the signal (the generator), although, again, not in direct contact with it. Bipolar recording arrangements as discussed below in Section 3 are generally used to study exclude these potentials. Examples would include standard needle EMG, digital sensory conduction studies, motor studies, and

From: The *Clinical Neurophysiology Primer*
Edited by: A. S. Blum and S. B. Rutkove © Humana Press Inc., Totowa, NJ

standard electroencephalography. *Far-field potentials*, in contrast, are generated at a distance from the recording electrodes. These potentials are most critical in somatosensory evoked potentials (SSEPs), in which electrodes on the scalp can pick up electrical activity occurring in the brainstem. Referential recordings (again, discussed in Section 3, below) detect these potentials more easily. However, far-field potentials also play a role in other aspects of clinical neurophysiology, and even standard motor studies can be "contaminated" by far-field effects, as described below in Section 4.2.

To become more familiar with the effects of volume conduction, it is helpful to first review several examples. In the sections that follow, we will then deal with the underlying theory and come back to an explanation regarding why the observed phenomena occur.

Example 1: A motor response recorded from the motor point of abductor pollicis brevis produces a negative first phase; however, a motor response recorded off the motor point will produce an initial positive phase followed by a negative phase.

Example 2: An antidromic sensory response recorded from the sural or radial nerve usually has a small positive phase before the main negative phase; antidromic sensory responses from the ulnar and median nerves, on the other hand, have a major negative phase without the preceding positive phase.

Example 3: Fibrillation potentials and positive sharp waves, which are morphologically quite "dissimilar," are both generated by the single muscle fibers.

Example 4: The P9 potential observed on upper extremity SSEPs can be observed simultaneously at multiple recording sites.

2. WORKING MODEL OF VOLUME CONDUCTION

There are two basic models that have been used to explain the observed phenomena of volume conduction. The first involves the use of a waveform moving down a nerve sitting in a medium. The second relies upon the concepts of what is called "solid-angle geometry." Although both models are relatively straightforward, because our goal here is merely to provide a basic understanding of volume conduction, we will omit the discussion on the solid-angle geometry model, which can be obtained in the suggested reading.

Figure 1 demonstrates the depolarization of an axon moving from left to right and what will be observed depending on the exact location of the recording electrodes, both in terms of distance from the nerve (Levels 1, 2, and 3) and location along the nerve relative to the depolarizing wave front (Ovals A, B, and C). For purposes of discussion, it is easier to envision the waveform remaining stationery and the electrode moving from left to right across the diagram. In A, the waves of depolarization are initially moving *away from* the current source as they fan out in the medium, slightly against the direction of electrode movement (most obvious toward the extreme left side of A). By convention, this will produce a downward deflection on the oscilloscope. As we move through A, however, the current lines begin to move in the same direction as the electrode movement and the tracing on the oscilloscope begins to move upward. At the "O" line (the equipotential point), there is no net current flow across the electrode, and the oscilloscope tracing returns to baseline. In B, current flow is now toward the sink, in the same direction as the moving recording electrode, producing further upward movement of the tracing producing the rising peak of the main negative spike. In C, the current abruptly switches direction as we cross the point of maximal depolarization. Current again now moves in the opposite direction of the electrode and the oscilloscope shows a downward movement (but it now has to start at the apex of the negative peak, not at baseline, so it remains in negative territory). After crossing the equipotential line again, we

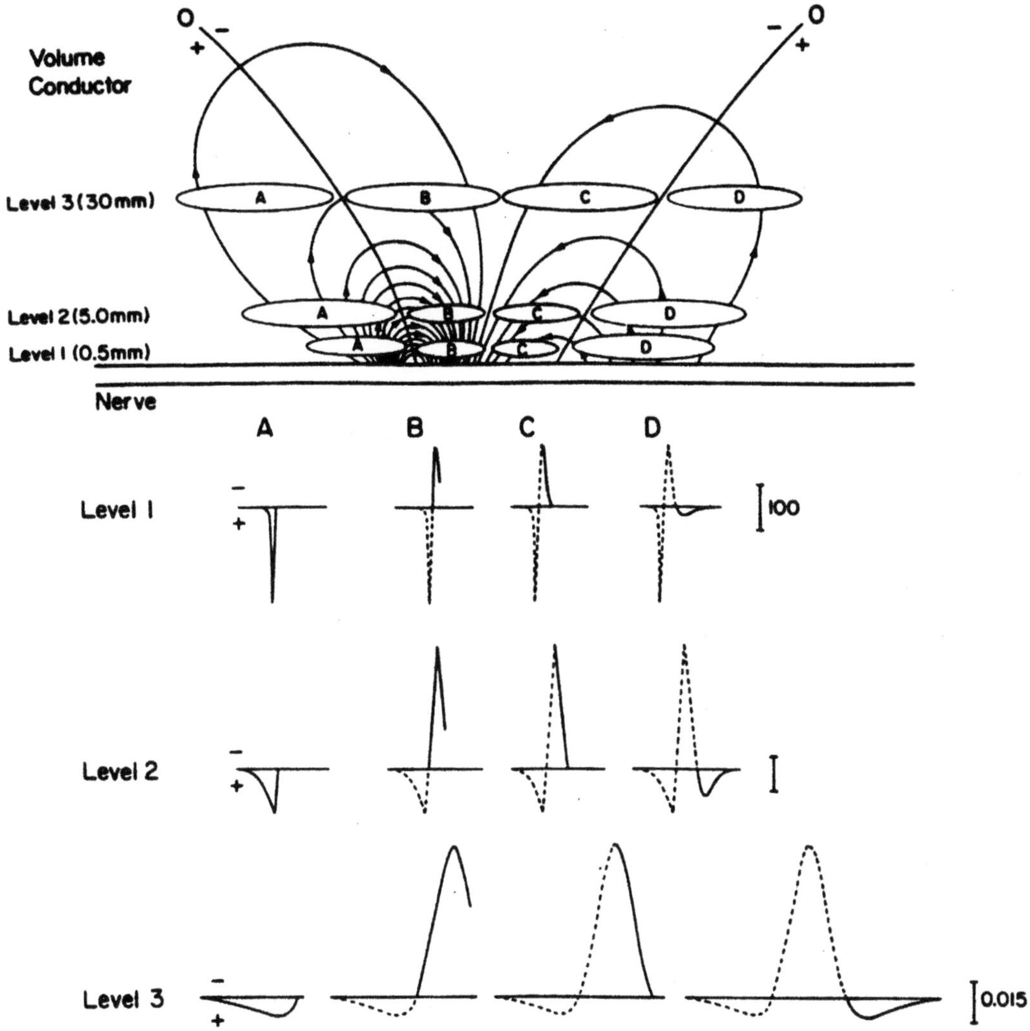

Fig. 1. A model of the effects of volume conduction on a recorded neuron or muscle potential. For discussion purposes, it is simpler to imagine the depolarization frozen in time and the electrode slowly being moved through positions A to D. Levels 1, 2, and 3 simply represent what would be seen depending on the proximity of the recording electrode to the nerve. Focusing on a single level (e.g., level 2) is simplest. By convention, if current lines are traveling opposite the direction of electrode movement, this will produce a downward deflection on the oscilloscope; if traveling in the same direction, it will produce a upward deflection. In A, the lines of current fanning out beyond the electrode source are initially traveling opposite the electrode movement, producing a downward (positive deflection), but then gradually start to travel in the same direction as the electrode, returning the deflection back to baseline, as the electrode "arrives at" the "zero" or equipotential line. In B, the current continues to travel in the same direction as the electrode movement and the oscilloscope traces an upward (negative) potential. As the electrode crosses the current sink (where the actually depolarization of the neuron is occurring), the lines of current abruptly change direction and the oscilloscope potential begins to move downward in C. This downward movement continues past the equipotential line and into D. However, within D, the fanning out of current lines gradually begins to move in the same direction as the electrode, producing an upward deflection on the oscilloscope and returning the tracing to baseline. Note that the potential is asymmetric, reflecting the fact that neuronal repolarization is a slower process with less dense current lines.

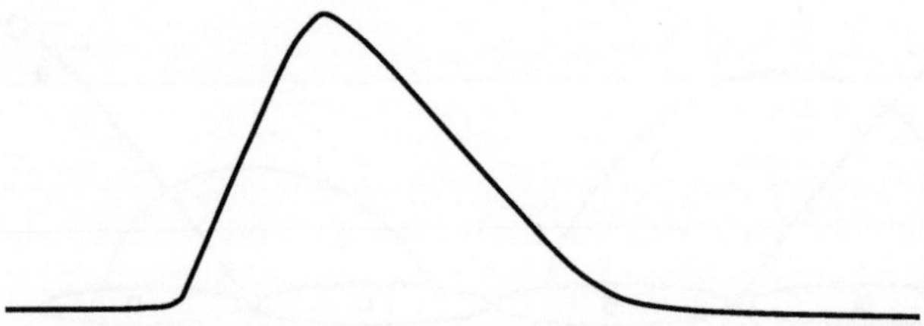

Fig. 2. A monophasic potential produced by a neuronal recording in the complete absence of volume conduction. Because there is no fanning out of current lines observed in Fig. 1, the initial and final positivities do not occur, leaving only the negative, slightly asymmetric monophasic negative potential.

enter D, where the current continues to flow in the direction opposite electrode movement initially, and the oscilloscope traces out the trailing positive peak. However, at some point in D, the fanning out current lines begin to move in relatively the same direction as the electrode, and the tracing moves up and returns to zero as the current reaches undetectable levels. It is also worth noting that the repolarization of the axon is a slower process than the depolarization (note how the lines of current flow are more densely packed on the left side than the right side of the diagram), leading to a sharper initial positive peak as compared with the final positive peak. When closer to the axon (the different levels), the peaks appear sharper than when further away from it.

It is worth considering what a potential would look like in the absence of a volume conductor. Using the same model, the major difference would be that no initial positivity nor final positivity would be identified, because the current would not fan out from the source generator. This would leave a monophasic negative spike (Fig. 2). This spike would have some asymmetry, however, given the fact that repolarization is slower than depolarization. In summary, the middle negative phase of a recorded potential will be present with or without volume conduction, but the end positivities are purely a consequence of volume conduction.

3. TWO OTHER CONCEPTS RELEVANT TO VOLUME CONDUCTION THEORY

All recordings in clinical neurophysiology are differential recordings, in that the electrical activity under one electrode (the "active" electrode or E1) is compared with that under the other electrode (the "reference" or "inactive" electrode, called E2). The final waveform obtained is a combination of what the E1 and the E2 electrodes are sensing; in mathematical terms, this would be described as E1 – E2. Hence, whatever E2 is "seeing" needs to be subtracted from E1.

Two terms are used to describe the two types of differential recordings: "referential" and "bipolar." Referential recordings refers to E2 being placed on a relatively electrically inactive structure, such as the earlobe in EEG recordings, or a bony prominence in EMG recordings. Bipolar recordings are those in which both E1 and E2 are being placed in relative proximity to electrical tissue, such as an antidromic digital sensory response stimulating the wrist and recording from the finger. For better or worse, although we like to think of E2 as electrically silent in referential recordings, in fact, far-field potentials can be recorded under this electrode, which can contribute substantially to the recorded waveforms. This issue will be discussed in further detail below in Section 4.2.

"Standing waves" (also called "virtual dipoles") may be produced when a depolarization of a nerve traverses from one body segment to another, for example, from arm to shoulder or hand to finger. The mechanism of this is complex, but likely is caused by several different effects, including a change in the anatomic uniformity of the volume conductor, extracellular resistivity, a change in the anatomic orientation of the propagating potential, or the abrupt termination of excitable tissue. Although this plays a role in both nerve conduction studies and all evoked potentials, the issue is most relevant in SSEPs.

4. VOLUME CONDUCTION AND ITS ROLE IN SPECIFIC TESTS

4.1. Sensory Nerve Conduction Studies

Let us begin our discussion by looking at a standard antidromic median sensory study recording from digit 3, a form of "bipolar study," in that both E1 and E2 are in proximity to the potential generators (the digital nerves of the middle finger). Figure 3A shows the standard recorded response. Note that there is an initial negativity and a final positivity. In Fig. 3B, the reference electrode is moved to digit 5, essentially changing the biopolar arrangement to a referential recording arrangement, where E2 is picking up no electrical activity. An initial positivity now appears on the recorded waveform, because of the approaching wave front, as described in Fig. 1. The reason that it was not visible with both electrodes on digit 3 is that both were sensing it simultaneously and, hence, it was subtracted out by the differential recording. With the reference electrode on digit 5, only the active electrode on digit 3 senses it and, thus, it appears. Note also that the amplitude of the negative peak is slightly higher. In Fig. 3C, the contribution of E2 is clarified by keeping it as the reference electrode and placing E1 on digit 5. What is seen is essentially an upside-down version of that seen in Fig. 3B, with an initial negativity and slightly lower amplitude at the now "upside down" negative peak. The amplitude is slightly reduced because the digit 3 electrode is very distal and the underlying electrical potential is smaller. The initial negativity seen here normally counterbalances the positivity of E1 (seen in Fig. 3B) in the bipolar recording (A), producing the flat baseline leading up to the major negativity. In Fig. 3D, the waveforms seen in Fig. 3B and 3C are digitally added, to produce approximately what is observed in Fig. 3A.

Thus, antidromic median and ulnar bipolar sensory studies typically do not demonstrate an initial positivity because they are truly bipolar and the approaching wave front is sensed simultaneously by both electrodes and is cancelled out. However, in standard antidromic radial sensory studies (recording from the snuff box) and antidromic sural sensory studies (recording from the ankle) a small positivity is typically identified. This likely reflects the fact that E1 and E2 are seeing slightly different environments and that the potentials are not truly bipolar with incomplete cancellation of the approaching wave front.

4.2. Motor Studies

In motor studies, there is generally no initial positivity on the compound motor action potential, similar to that observed on bipolar digital sensory studies. However, the reason why this occurs is different. Rather than the wave front being sensed by both electrodes and canceling, in this situation, the depolarization of the muscle fibers is actually originating directly beneath the active electrode, E1, and, hence, there is *no* approaching wave front. This can be demonstrated by moving E1 slightly off the motor point, which will then create a small positivity in front of the major negative phase.

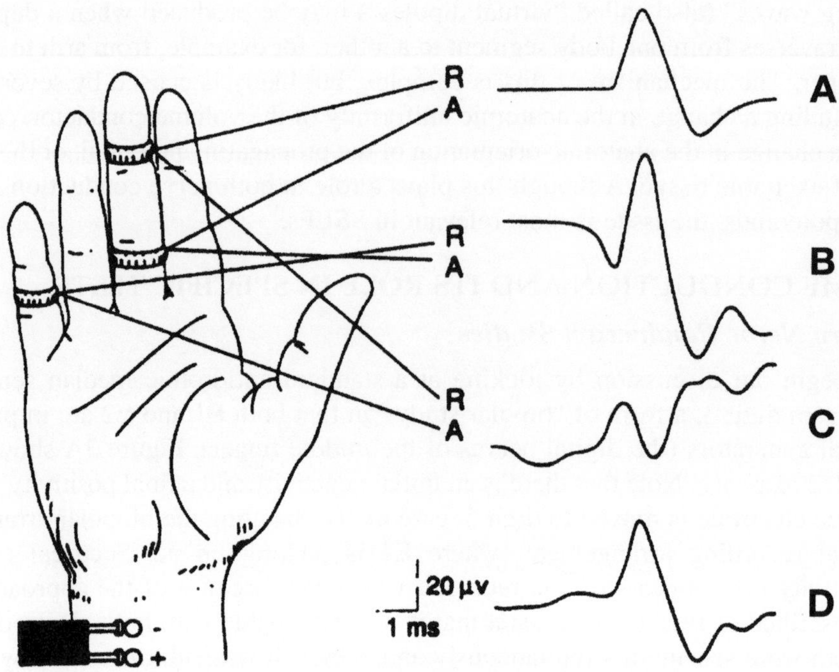

Fig. 3. Median digit 3 sensory studies. Comparisons are made between the standard bipolar arrangement (A) and other arrangements to better understand how the standard potential is generated. In A, the active electrode "A" and the reference electrode "R" are both placed on digit 3 to obtain the standard bipolar recording. In B, the reference electrode is moved to digit 5 to produce a true "referential recording." Note how an initial positivity is observed (because of the approaching wavefront) and the peak-to-peak amplitude is higher than the baseline-to-peak observed in A. In C, the reference electrode is maintained on digit 3, but the active electrode "A" is moved away. This reveals what is being recorded by the reference electrode alone: a small initial negativity and larger second positivity. In D, the responses from B and C are digitally added, giving approximately what is observed in A.

Regardless, the negative peak amplitude that is being measured is still very much a product of what the two electrodes are sensing. It is appealing to think that the active electrode is sensing the depolarization of the muscle and the inactive is seeing a flat line. This is actually pretty much the case for standard median and peroneal motor conduction studies, in which only a small subset of muscles on the hand or foot are depolarizing. However, this is definitely not true in ulnar and tibial motor studies, in which the majority of the muscles in the hand and foot are all depolarizing, essentially "electrifying" the entire structure. This was demonstrated by Kincaid et al., in a study in which the authors evaluated what the reference electrode was actually seeing in different parts of the hand (Figs. 4 and 5). In these studies, the active electrode was placed on different parts of the hand and the reference electrode was moved to the opposite side of the body so that E2 could be viewed as a true reference. As can be seen, nearly the entire hand is "electrified" when the ulnar nerve is stimulated (Fig. 4) and volume-conducted waveforms can be picked up all over the hand (even on the tips of the fingers); hence, there really is no inactive site for this muscle. When stimulating the median nerve, however, virtually no activity is picked up in any location, except in the region of the thenar eminence (Fig. 5). Similar results are also found by evaluating the tibial and peroneal nerves in the feet; the tibial nerve analogous to the ulnar and the peroneal nerve is analogous to the median.

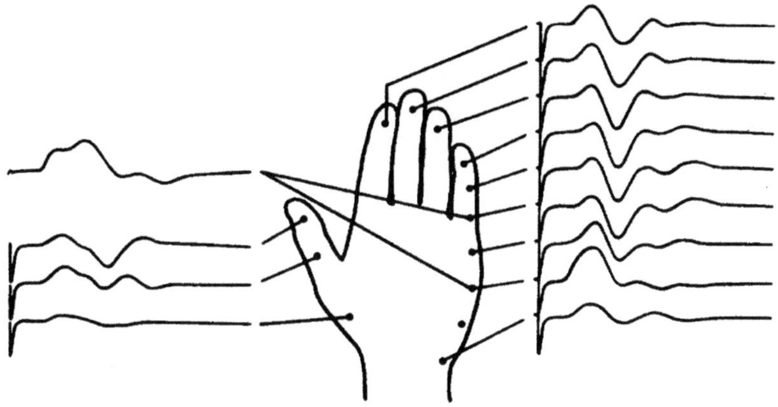

Fig. 4. Volume-conducted potentials recorded from various parts of the hand with stimulation of the *ulnar* nerve. Note how potentials can be recorded from nearly the entire hand, even the tips of the fingers. From Kincaid et al., 1993 with permission.

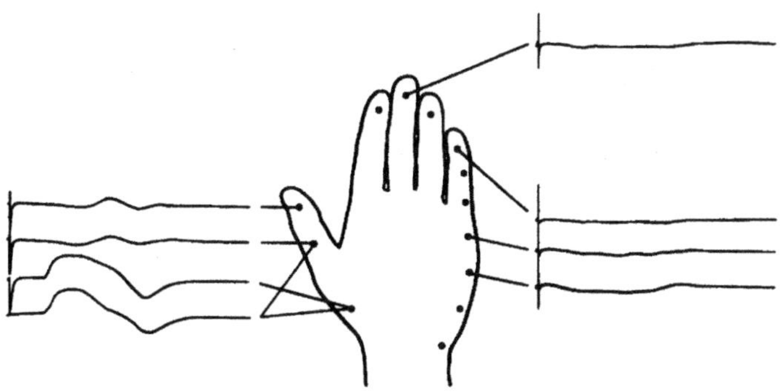

Fig. 5. Volume conducted potentials recorded from the hand with stimulation of the *median* nerve. Note how potentials can only be recorded in the vicinity of the thenar eminence, where the median-innervated muscles reside. From Kincaid et al., 1993 with permission.

Hence, when performing a standard ulnar motor study, the ultimate configuration of the motor response is a composite of what E1 and E2 are sensing. The large positivity that is sensed over the fifth metacarpal–phalangeal joint is subtracted from the response detected directly over the hypothenar muscles. Subtracting this positivity will cause it to become a negativity, leading to the characteristic "double-humped" ulnar motor potential (Fig. 6). The same exercise for the median nerve, on the other hand, produces essentially no change in the amplitude/configuration of the motor response because the E2 activity is negligible.

4.3. EMG

4.3.1. Fibrillation Potentials and Positive Sharp Waves

Fibrillation potentials and positive sharp waves are both produced by the electrical depolarization of single muscle fibers. However, they have dramatically different morphologies, as illustrated in Figs. 3 and 4, Chapter 14. They appear so dissimilar because of volume conduction effects. Fibrillation potentials are thought to represent the depolarization of a single muscle fiber observed from a short distance by the needle electrode. Hence, the reason for its triphasic nature is very similar to a sensory nerve conduction study, with an approaching

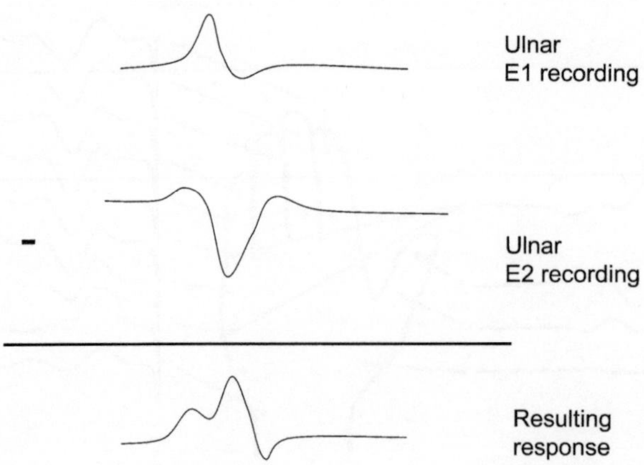

Fig. 6. Approximate digital addition of the ulnar motor response recorded from the hypothenar eminence and that recorded from the fifth MCP joint, leading to the typical "double-humped" ulnar motor response.

wave front causing the initial positivity, the major negative spike being induced by the depolarization traveling underneath the electrode, and the trailing positivity caused by the departing wavefront. This is in contrast with a positive sharp wave, in which it is proposed that needle is in actual contact with the fiber being recorded. The mechanical deformation of the myocyte membrane by the needle is thought to partially depolarize the cell. Hence, the approaching wave front produces an initial positivity and then the negativity begins to develop, but essentially aborts as it comes into contact with the depolarized area of muscle membrane. Although these are only models and the real phenomenon is likely more complex, they are helpful in understanding how two different waveforms can be produced by a single generator.

4.3.2. End-Plate Spikes

These potentials are similar to fibrillation potentials in that they represent the depolarization of a single muscle fiber. However, they are different in that needle movement at the end plate induces the firing of the muscle fiber and, hence, the action potential is generated immediately beneath the needle. Accordingly, unlike fibrillation potentials, they are usually biphasic—with an initial negativity followed by a positivity. The fact that these potentials are generated immediately beneath the needle is similar to a motor conduction study, in which the depolarization of the muscle fibers begins under the surface electrode.

4.3.3. Motor Unit Potentials

When in close proximity to the muscle fibers of a single motor unit, the motor unit potential will have higher amplitude and, when more distant, the amplitude will decrease and the general sharpness of this waveform will decrease. The duration tends to be more immune to these volume conduction and tissue filtering effects, with it being relatively consistent regardless of the distance the needle electrode is placed from the motor unit potential being recorded. Hence, the single most useful parameter when evaluating a motor unit is its duration.

4.4. Somatosensory Evoked Potentials

Somatosensory potentials are usually recorded in a bipolar fashion with both recording electrodes placed near the generator source (e.g., spinal cord or scalp). This approach measures near-field potentials. However, by widely spacing the recording electrodes, early waveforms in the form of far-field potentials can also be observed. Although initially thought to be due to a specific neuronal generator of some form, such as ganglia or individual neurons, it was recognized by Jun Kimura and others that these waveforms were actually being caused by a waveform traveling from one body segment into another, the so-called standing waves or virtual dipoles noted above in Section 3. These waveforms have also been called "junctional" or "boundary" potentials. The most prominent and most carefully studied of these is the P9 potential obtained with stimulation of the median nerve. It is thought that this is caused by the intersection of the arm with the trunk. Similar "virtual dipoles" can be demonstrated in standard digital sensory nerve conduction studies, if performed in a referential fashion; however, these are usually irrelevant when performed in a standard bipolar manner.

SUGGESTED READING

Dumitru D, Delisa JA. AAEM Minimonograph #10: Volume Conduction. Muscle Nerve 1991;14:605–624.

Frith RW, Benstead TJ, Daube JR. Stationary waves recorded at the shoulder after median nerve stimulation. Neurology 1986;36:1458–1464.

Kimura J, Kimura A, Ishida T, et al. What determines the latency and amplitude of stationary peaks in far-field recordings? Ann Neurol 1986;19(5):479–486.

Kincaid JC, Brashear A, Markand ON. The influence of the reference electrode on CMAP configuration. Muscle Nerve 1993;16(4):392–396.

Lagerlund TD. Volume conduction. In: Daube J, ed. Clinical Neurophysiology. Oxford University Press, New York, NY, 2002, pp. 28–36.

REVIEW QUESTIONS

1. What causes the initial positivity of a normal median compound motor action potential when the active electrode is *not* placed over the motor point of the abductor pollicis brevis muscle?
   A. Depolarization of distant muscles.
   B. A retreating wavefront.
   C. An approaching wavefront.
   D. A stationary wavefront.
   E. None of the above.
2. End-plate spikes lack an initial positivity because:
   A. The depolarization initiates immediately beneath the needle.
   B. They are not generated by muscle fibers.
   C. The axon initiates the depolarization.
   D. The wavefront is forced to abruptly stop by the presence of the needle.
   E. The depolarization involves only the proximal portion of the muscle fiber.
3. In the volume conduction model described, which of the following is true?
   A. The depolarizing wavefront produces current lines that fan out away from the current sink.
   B. There is symmetry between the depolarizing and repolarizing regions of the nerve.
   C. Inward current produces a negative deflection on the oscilloscope regardless of the location of the electrodes.
   D. The rise time of the observed potential is independent of distance from the axon.
   E. Electrons are moving in the direction opposite current flow.
4. Which of the following statements regarding positive sharp waves and fibrillation potentials is true?
   A. Positive waves represent an action potential traveling down a large muscle fiber; fibrillation potentials represent an action potential traveling down a small muscle fiber.
   B. Fibrillation potentials are recorded when the needle is in contact with a muscle fiber.
   C. Fibrillation potentials are induced by movement of the needle against the muscle fiber.
   D. Positive sharp waves result from an aborted muscle fiber depolarization.
   E. Fibrillation potentials initiate immediately beneath the recording needle electrode.

5. Which of the following SSEP peaks is caused by a virtual dipole?
   A. P9.
   B. P22.
   C. N13.
   D. N5.
   E. N20.
6. An initial positivity can be observed on median antidromic digit 2 sensory recordings by doing which of the following?
   A. Moving the active electrode to digit 5.
   B. Moving the reference electrode to digit 5.
   C. Moving both electrodes to digit 5.
   D. Moving the recording electrodes more closely together.
   E. B and D are both correct.
7. Which of the following statements concerning the reference electrode in a motor recording is most accurate?
   A. It is electrically inactive.
   B. It detects electrical activity from nondepolarizing muscles.
   C. Although electrical active, it usually does not contribute substantially to the recorded waveform with median nerve studies.
   D. Although electrical active, it usually does not contribute substantially to the recorded waveform with tibial nerve studies.
   E. Placing it over a bone ensures that no distant muscle activity is detected.
8. In the absence of a volume conductor, a triphasic wave (e.g., a fibrillation potential) would appear as which of the following?
   A. A biphasic wave with a prominent initial positive peak.
   B. A triphasic wave.
   C. A negative monophasic wave.
   D. A biphasic wave with a prominent trailing positive peak.
   E. An inverted triphasic wave.
9. Which of the following is most likely to have an initial positivity under normal recording conditions?
   A. An antidromic digit 3 median sensory response.
   B. An antidromic digit 5 ulnar sensory response.
   C. An antidromic sural response recorded at the ankle.
   D. A tibial motor response.
   E. A peroneal motor response.
10. An example of a far-field potential is:
    A. A median motor response with the recording electrode over abductor pollicis brevis.
    B. A sural sensory response recorded at the ankle.
    C. Ulnar motor activity detected at the tip of digit 5.
    D. The N14 peak, generated by the medial lemniscus, on upper extremity SSEPs detected on scalp recordings.
    E. C and D are both correct.

REVIEW ANSWERS

1. The correct answer is C. The approaching wavefront causes the initial positivity. Although the depolarization of distant muscles can also cause this phenomenon, this is unlikely to be the case here. A retreating wavefront contributes to the final positivity. Stationary wave fronts are generally not observed in median motor responses.
2. The correct answer is A. In most circumstances, end-plate spikes are generated by the needle tip directly causing a depolarization of the end plate, with the depolarization then traveling away

from the point of contact with the muscle. Hence, an initial negativity is followed by a final positivity. The axon does not initiate the depolarization. The abrupt ending of a muscle fiber depolarization caused by the presence of a needle causes a positive sharp wave, not end-plate spikes.

3. The correct answer is A. The fanning out of the current lines through the medium is essentially what causes volume conduction-related effects. There is *asymmetry* between the depolarizing and repolarizing regions of the nerve because repolarization is a slower process than depolarization. The inward current may produce a negative or positive deflection depending on where the recording electrode is relative to the depolarization. The rise time of the potential increases the further the electrode moves from the nerve. Electrons do not move freely through solutionions do.
4. The correct answer is D. Positive sharp waves result when a muscle fiber depolarization abruptly aborts as it travels down the muscle fiber, likely because the needle is in contact with the muscle fiber. Although the amplitude of fibrillation potentials and positive sharp waves is related to muscle fiber size, at least to some extent, size does not result in different morphologies of the recorded waveform. Fibrillation potentials are generally thought to be identified when the needle is not in contact with the muscle fiber. Fibrillation potentials can initiate anywhere relative to the needle electrode.
5. The correct answer is P9. This one is thought to be cause by the change in conduction as the potential crosses from the arm into the trunk.
6. The correct answer is B. By moving the reference electrode to digit 5, the approaching wavefront on the median nerve, which causes the initial positivity, will only be seen by the active electrode and, hence, will appear. Normally, it would be detected by both electrodes and, therefore, cancelled out. Moving the electrodes closer together would decrease the amplitude of the recorded response but would do little else.
7. The correct answer is C. In median and peroneal motor recordings, the reference electrode generally detects very little electrical activity. The reference electrode is "active" in that it is helping to produce the differential recording and, if a muscle is not depolarizing, there is no electrical activity to record. In tibial recordings, the reference electrode detects considerable volume-conducted electrical activity. Although we would like to think that placing the electrode over the bone ensures that no distant muscle activity will be detected, this is not true.
8. The correct answer is C. Without volume conduction, the arriving and departing waveforms would not be detected and the triphasic wave would become a monophasic, somewhat asymmetric negative peak, as shown in Fig. 2. The asymmetry is because the rate of repolarization is slower than the rate of depolarization. None of the other answers is valid.
9. The correct answer is C. Sural and radial sensory responses can be thought of more as "referential" than as "bipolar" because both electrodes are not oriented identically in relation to the nerve, allowing the initial positivity to appear. In standard digit 3 and digit 5 recordings, there is no positivity because they are bipolar recordings and the approaching waveform is seen simultaneously by both electrodes and cancelled out. Motor responses usually do not have initial positivities because the active electrode is placed immediately above the motor point, where the depolarization of the muscle fibers begins. Hence, there is no approaching waveform to cause the positivity.
10. The correct answer is E. In both cases, the electrical activity is being recorded at a distance rather than in the immediate proximity of the generator. In the other two examples, the recording electrodes are placed as close to the generator as possible.

II

Electroencephalography

5
The Normal EEG in an Adult

Donald L. Schomer

Summary

When an experienced electroencephalographer sits down to review an EEG, whether obtained on a pen/ink-based analog machine or from the cathode ray tube screen of a digital device, a number of mental integrations take place seamlessly. This chapter addresses the "normal" EEG observed in people older than 18 yr of age. Topics to be covered include the normal waking background rhythm (alpha rhythm); beta activity; mu, theta, and lambda waves; activation effects on the EEG; and features of normal sleep.

Key Words: Awake and sleep EEG; normal EEG; routine EEG.

1. INTRODUCTION

When an experienced electroencephalographer sits down to review an EEG, whether obtained on a pen/ink-based analog machine or off of a cathode ray tube screen from a digital device, a number of mental integrations take place seamlessly. This chapter examines the "normal" EEG observed in people older than 18 yr of age.

2. RECORDING PRINCIPALS

EEG recording electrodes are glued onto the scalp in an orderly fashion according to an agreed on measured placement, referred to as either the International 10-20 or 10-10 system *(1)*. The electrodes are plugged into a head-box, which allows the technician to record in either a bipolar or a referential fashion. The former is a system in which adjacent electrodes are connected to a differential amplifier. The latter is a system in which each electrode is connected to a differential amplifier and compared with a common electrode. The differential amplifier has two inputs (G1 and G2), and amplifies the difference in voltage at the two input sites. The output is then charted onto a graph in the case of an analog-based EEG machine, so that the difference in voltage is graphed against time. In the case of a computer-based system, the inputs are digitized and the amplifier registers the voltage differences and stores it as a digital signal for display on a cathode ray tube. In either case, the viewed information is essentially the same.

There are certain basic rules that are followed by the technician to ensure high quality and reproducible results. There are strict guidelines related to the placement of the electrodes so that the same electrodes end up in the same spots, regardless of which technician applies them. The electrodes need to be tested for impedance and maintained below 5000 Ω. The electrodes, when recording in a bipolar montage, are connected in straight lines going from

From: *The Clinical Neurophysiology Primer*
Edited by: A. S. Blum and S. B. Rutkove © Humana Press Inc., Totowa, NJ

front to back and from left to right. Displays of the output tend to follow similar rules, and the left side of the head is usually displayed above the right side.

There is also a polarity convention that all manufacturers of EEG equipment follow. Originally, when all of the device outputs were graphic pens, the convention was that if the G1 input was more negative than the G2 input, the pen deflection was upward. It follows that either a negative event at G1 or a positive event at G2 would have led to an upward deflection. A positive event at G1 or a negative event at G2 leads to a downward deflection. These rules still apply to digital recording devices, but there are obviously no pen deflections, simply movement of the signal off of the baseline. By connecting electrodes in linear arrays, one can graph fields and electrical polarity of given potentials over time *(2)*.

3. THE AWAKE AND RESTING STATE

After the electrodes are attached, the montage selected and the machine calibrated, the technician is ready to record. Recordings are done for between 20 min (minimum) and several hours in selected circumstances. The patient is recorded initially in the awake resting state, with eyes open, and recorded again with eyes closed. Unless there is a specific reason to the contrary, part of the recording session should be performed with the patient hyperventilating and receiving intermittent photic stimulation. The patient may also be allowed to fall asleep and, therefore, EEG can be recorded during sleep induction and sequencing into the various stages of sleep.

4. THE ALPHA RHYTHM

When electroencephalographers sit down to review a normal study, they are usually drawn first to the prominent background alpha rhythm. As with all repeating rhythms, it is most important to note the frequency and the location of the activity, its amplitude, and its reactivity. The alpha rhythm oscillation is between 8 and 13 Hz and is most prominent over the more posterior aspects of the head. Its amplitude or voltage in a bipolar (P4–O2) derivation is anywhere from 15 to 65 μV. This is reactive or responsive to mental activity and to eye opening or closure. The rhythm is partially or completely blocked by mental activity or by eye opening *(3)*. Likewise, it is enhanced by relaxation and by eye closure (Fig. 1).

For healthy adults, there is a bell-shaped distribution curve for the alpha frequency, centering around 10.0 Hz. The frequency is significantly affected by cerebral blood flow and may vary by up to 2 Hz in any individual, based on flow changes. It normally does not vary by more than 1 Hz during the course of the record and shows very little change during long time spans. Although it has been reported that the alpha rhythm may slow by 1 Hz every 10 yr after the age of 50 yr, in more recent studies that controlled for subtle disease states, such changes did not appear. In a few healthy subjects (<2%), no apparent alpha rhythm can be seen. In a few more subjects (<7%), a very low-voltage alpha rhythm is observed. Because recordings are bipolar, one can improve on alpha detections by increasing the inter-electrode distances. When this is done, the finding of an absent alpha rhythm is actually quite low. The frequency may also vary slightly in relationship to the menstrual cycle, where it is faster by up to 0.3 Hz during the follicular phase. This variation is probably below the level that can be visually recognized.

The alpha rhythm is most prominent over the posterior aspects of the head. In some individuals, it is overwhelmingly occipital in its distribution. In approximately one-third of healthy adults, the alpha rhythm is widely distributed and incorporates parietal and posterior temporal regions.

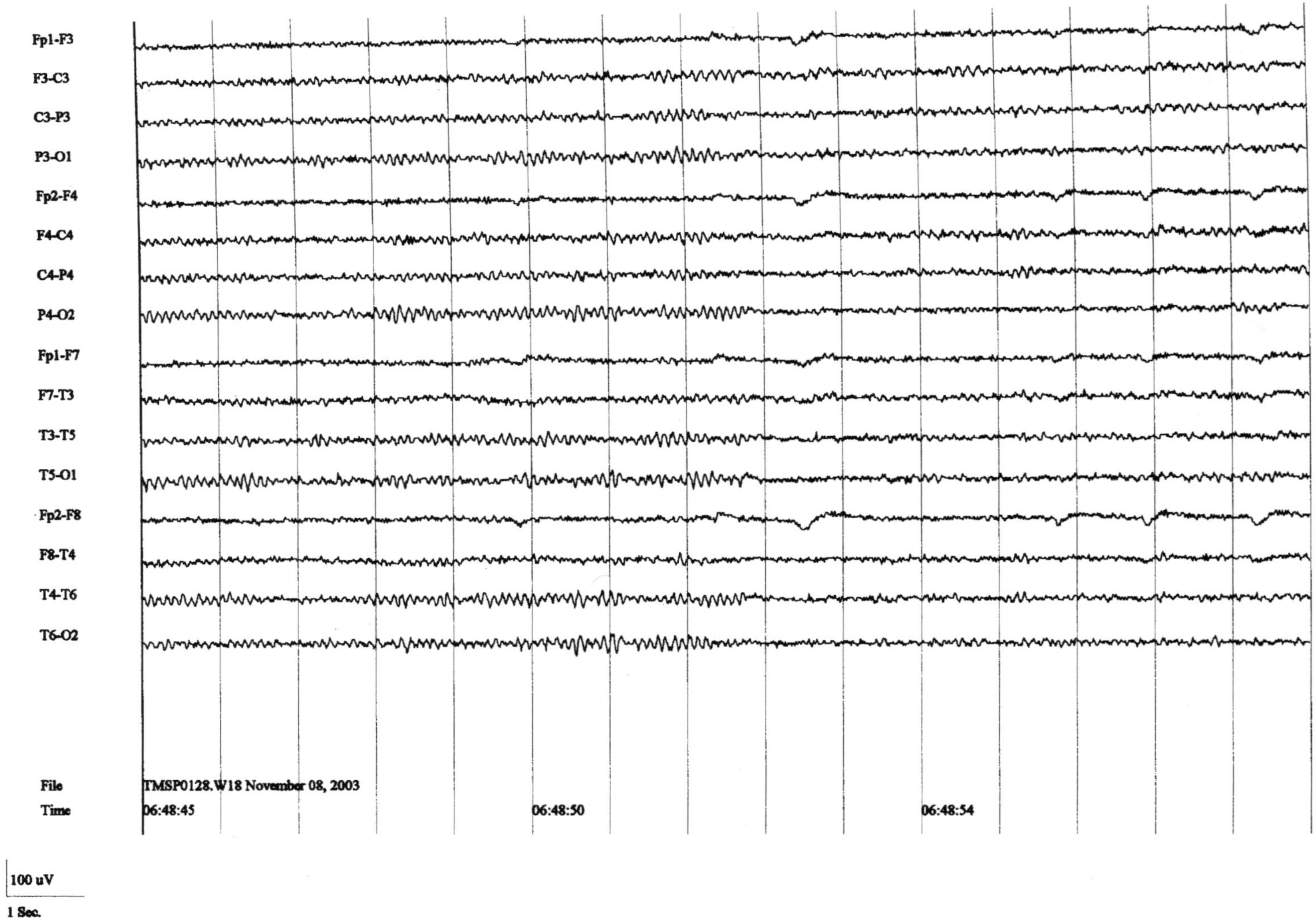

Fig. 1. This recording shows the presence of a normal alpha rhythm present over the posterior head regions. Note how the alpha rhythm attenuates just before a spontaneous eye opening marked by the eye movement artifact noted over the frontal electrodes, approximately midway through the recording.

There may be a side-to-side amplitude asymmetry, noted in approx 60% of people, in which the right side tends to be of slightly higher amplitude. The left side may be more predominant in left handed individuals. The difference in amplitude between the two hemispheres is not more than 50%. If it is more than that, it is almost always related to an identifiable abnormality. The skull can act as an amplitude attenuator and, therefore, as one ages and as one's skull thickens, the amplitude of the alpha rhythm tends to diminish.

The state of the eyes is important to the alpha rhythm. The alpha rhythm is the rhythm of the awake, resting, eyes-closed individual. In most people, the alpha rhythm will be blocked or significantly attenuated when the eyes are open (Fig. 1). Similarly, mental effort, such as doing simple math, likewise blocks alpha rhythm in approximately three-fourths of healthy subjects.

5. BETA ACTIVITY

Beta activity (>13 Hz) is defined by three relatively distinct frequency bands: 18 to 25 Hz activity, which is the most frequently encountered; 14 to 17 Hz activity, which is less common; and the still rarer, greater than 25 Hz activity. The first two frequencies are seen commonly over the frontal regions and become more prominent as the subject gets drowsy. These two beta rhythms are usually of low voltage (<25 μV). It can be markedly increased by the use of some drugs, most notably the benzodiazepines and barbiturates. However, in those circumstances, it is felt that the rhythm is a provoked cortical rhythm and not a normal one. A good review of beta activity is in the textbook on EEG by Niedermeyer *(4)*.

6. MU RHYTHM

The mu rhythm is a centrally located rhythm with a frequency of 8 to 10 Hz. It is thought to be the resting rhythm of the pre- and post-Rolandic cortex. It is more commonly noted in the younger adult population, in whom it is detected in up to 20% of people younger than 30 yr of age. The likelihood of finding it decreases with increasing age. It is reactive to motor programming. It is seen during times of relaxation and can be affected by the state of the eyes. When seen, the technician often asks the patient to make a fist contralateral to the side in which the rhythm has been observed. This movement or even the thought of this movement blocks the rhythm. The mu rhythm may appear quite asymmetrically, although that is not of any pathological significance, except for the possibility that there may a skull defect on the side of this rhythm (breach effect) (Fig. 2).

7. THETA RHYTHM

Theta activity is defined as activity between 4 and 7 Hz. Although theta activity is often a sign of disease or of sleep onset, it may also be seen as part of a normal awake EEG. Although some intermittent low-voltage theta activity is seen over the frontal–central regions in healthy people while resting and awake, this is usually not a well-developed nor regular rhythm. Under a circumstance in which the subject is performing some moderately difficult mental task, such as spelling or mathematics, one can occasionally see a well-developed theta rhythm in the frontal midline region *(3)*. A small amount of left temporal theta activity during wakefulness is also expected as a normal factor in aging, starting around the age of 50 yr *(3)*.

8. LAMBDA WAVE

Occasionally, ones sees occipital positive sharp slow waves in the record of a patient who is lying quietly in bed awake. These waves may be quite asymmetric and raise the possibility

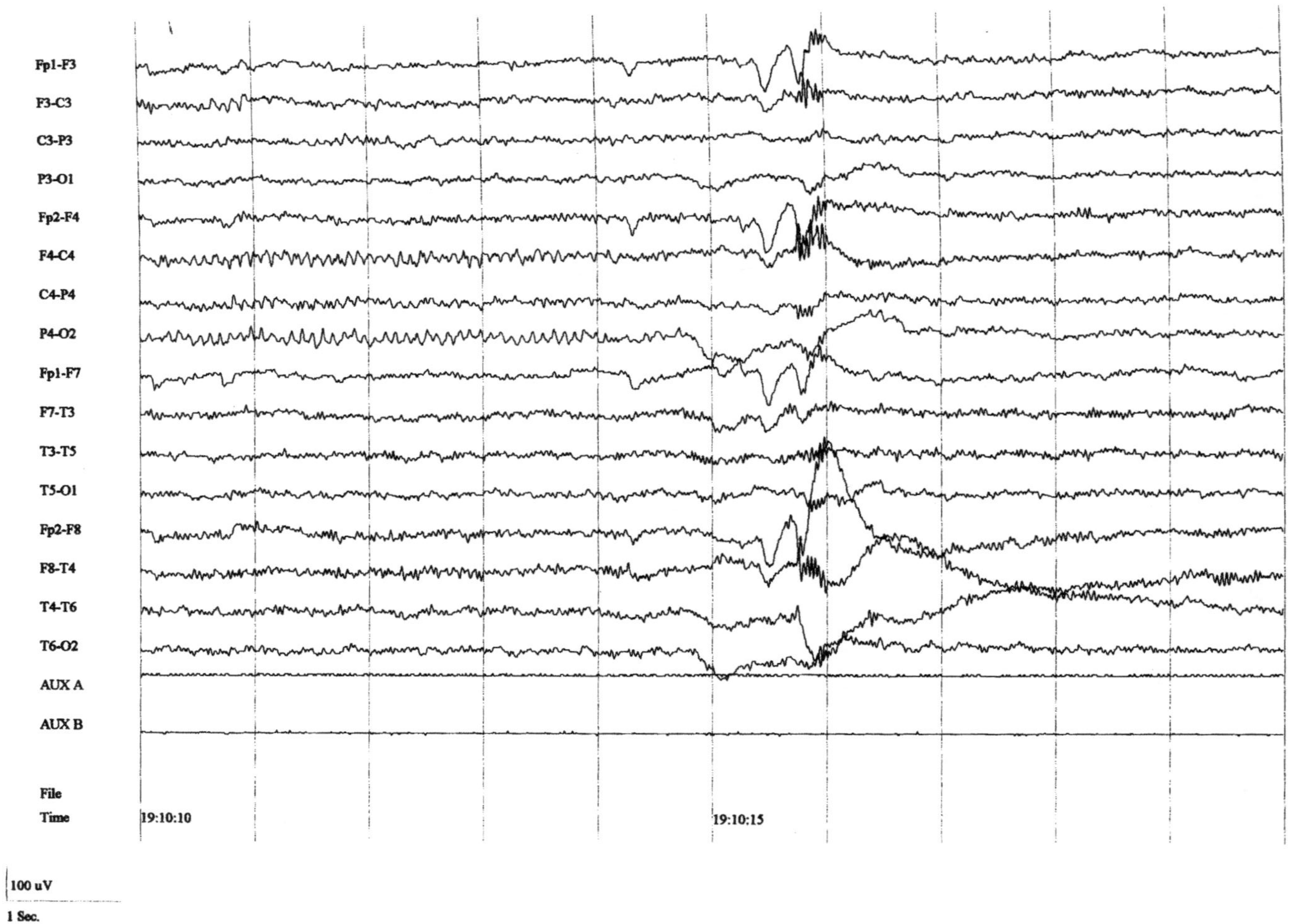

Fig. 2. This tracing shows the presence of a sharp-appearing 11-Hz rhythm present over the right central region that abruptly stops associated with an eye movement. This rhythm also blocks with movement of the right hand and is classic for the mu rhythm.

of a focal abnormality. The normal lambda wave is identified by doing a few simple maneuvers. The wave is felt to represent occipital lobe activity in a person who is actively reading or scanning the room. Often the subjects have their eyes open and are looking carefully at the ceiling tiles. The technician will have the patient reproduce the maneuver that they felt caused the waves to appear in the first place. They will then have the patient close their eyes, which will block the normal lambda wave but usually not have an effect on the abnormal activity. They will then have them open their eyes and look at a plain piece of paper, which also blocks the normal lambda wave but will not effect the epileptiform discharges, which are most often mistaken for a lambda wave *(3)*.

9. RESPONSE TO HYPERVENTILATION

Hyperventilation is a method of "activating" the EEG. It may bring out focal or generalized slowing in cases of structural disease or of more diffuse encephalopathic disorders. It also can bring out interictal epileptic discharges or trigger more overt symptomatic seizures. It should not be performed in the very elderly patient or in someone suspected of having any intracranial mass lesions or a recent transient ischemic event/stroke. In the adult population, it is normal to hyperventilate the patient for 3 to 5 min. The technician usually tries to, at minimum, double the base breathing rate, and to have the subject exhale more deeply than usual. This act causes the subject to exhale excessive amounts of CO_2 and become hypocapnic. The hypocapnia causes mild cerebral vessel vasoconstriction and, hence, mild cerebral hypoxia. The hypoxia and the hypocapnia together potentially can produce changes that may signal a disorder. One additional variable that must be considered is the blood glucose level at the time of this test. If the response is one of significant generalized slow wave activity, the technician often will give the patient a glass of orange juice and wait 10 to 15 min and repeat the exercise. If the slowing is still present, it takes on significance as a potential abnormality. The normal response in a healthy adult is to have no slow wave or epileptiform activity brought out by this maneuver. It is normal to sometimes see a patient exhibit greater alpha activity as a result of hyperventilating than they had otherwise shown.

10. RESPONSE TO INTERMITTENT PHOTIC STIMULATION

Similar to hyperventilation, intermittent photic stimulation is performed to "activate" the EEG. Photic stimulation is performed by using a commercial stroboscopic stimulator placed approx 1 m from the patient's eyes. They are asked to keep their eyes closed and look straight ahead while the ambient room lighting is turned down. The test is performed by alternating flashes varying from 1 to 35 Hz and lasting for 10 s and interrupted by 10 to 30 s with no stimulation. The variables that need to be controlled for are the distance from the subject to the strobe, the luminance of the strobe, direction of gaze, and level of consciousness. The most common abnormal activation is to produce epileptic activity in relationship to the photic stimulation. In healthy people, one may see any level of "driving." This phenomena appears in the occipital leads and is the result of a "flash" visual evoked response. Therefore, the background rhythm gets linked to the timing of the photic stimulator (Fig. 3). The first response appears shortly after the stimulator goes on (<100 ms) and stops when the stimulator shuts off. It is more likely to occur around the baseline background frequency (±2–4 Hz). One may also *see* what is referred to as a photomyogenic response as a normal variant. In this condition, widespread muscle twitching appears, which is timed to the stimulator. This reaction is felt to

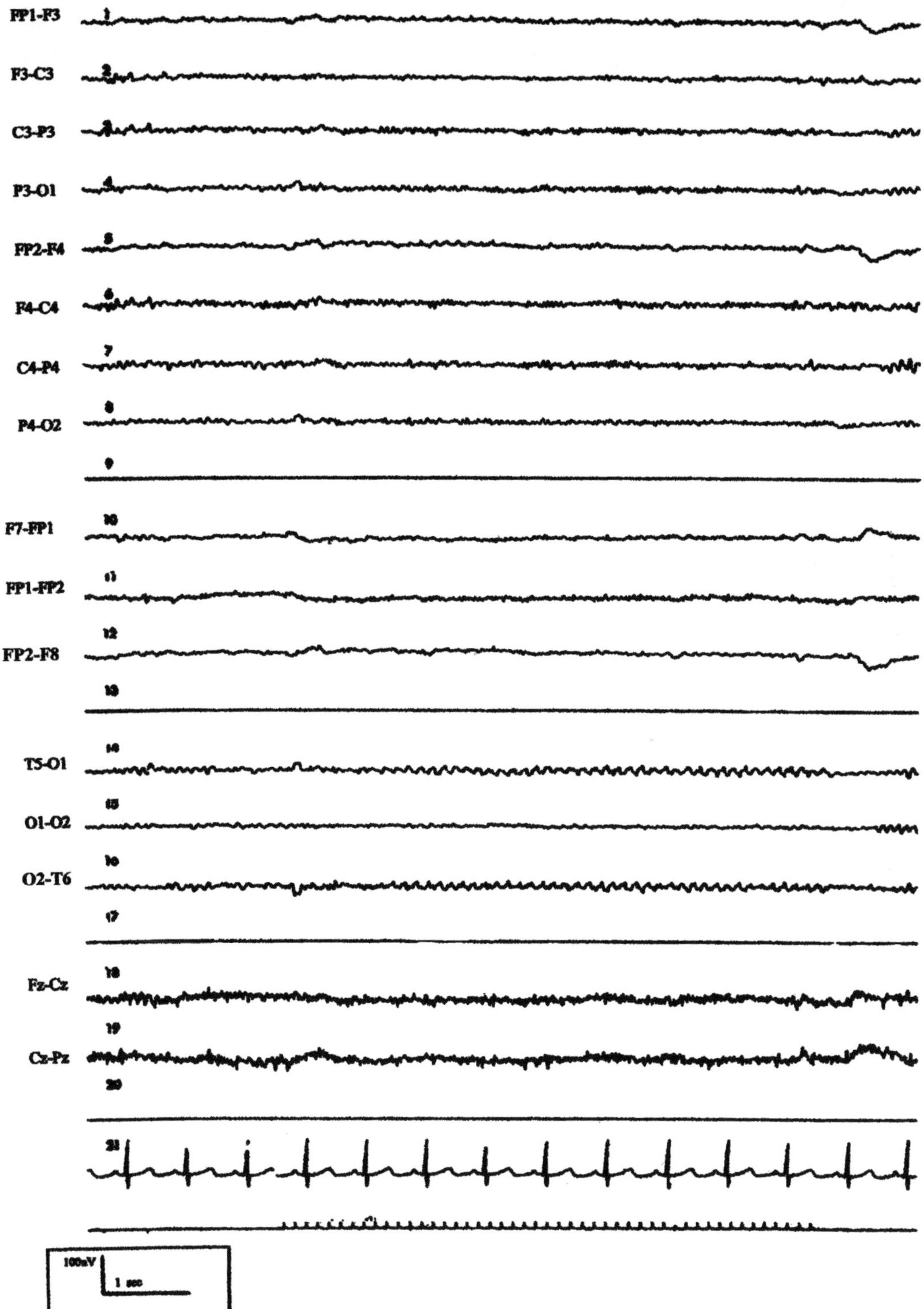

Fig. 3. Midway through this 20-s portion of a routine EEG is the 10-s period of intermittent photic stimulation noted by the artifact on the last channel. The frequency of the photic stimulator is eight flashes per second. There is a prominent evoked response noted over the occipital leads at the same frequency. This is a normal reaction.

represent a heightened brainstem-mediated reflex reaction to the photic stimulator. Additional enhanced levels of photic driving reactions are recognized and defined in a review by Waltz et al. *(5)*.

11. SLEEP

Many patients will come to an EEG lab and immediately fall asleep. Others may come with that in mind and never attain even a minimum level of drowsiness. For an electroencephalographer, sleep is useful to observe for several reasons. Similar to hyperventilation and photic stimulation, it may bring out latent abnormalities. There are some disorders, such as juvenile myoclonic epilepsy, in which waking up from sleep may be the only time one sees the markers for this disorder. In many EEG laboratories, the technician tries to get at least some sleep on every study. In other labs, the staff are asked to have the subject stay awake for significant portions of the night before the test to try to assure that the patient will fall asleep easily.

Four stages of non-rapid eye movement sleep are described next. Seldom does a subject get beyond Stage III during routine testing, but all stages are noted as they happen in a healthy person. Chiappa and Santamaria's book on drowsiness is an excellent dissertation on the subject *(6)*.

12. DROWSINESS OR STAGE I

The first suggestion that a patient is going to sleep is that there is a sudden drop in the voltage of their background alpha rhythm, followed by intermittent theta activity noted over the more posterior head regions (Fig. 4). This presleep or twilight state is usually followed by early Stage II. However, patients may alternate back and forth from awake resting to drowsiness for extended periods of time. In the healthy elderly patient, they may skip this state completely and go into Stage III or IV. Other healthy elderly patients may spend most of their sleep time in drowsiness and Stage II of sleep and have very little of the later stages noted.

Vertex sharp waves may begin to be seen in deeper stages of drowsiness, Stage Ib sleep. Vertex sharp waves are centered around the CZ electrode, with maximum voltage noted in the adjacent electrodes (C3 and C4). The wave may have complex morphology and consist of several phases. The initial phase is usually a surface negative biphasic sharp wave followed by a high-voltage surface negative slow wave, lasting up to 400 ms in duration.

13. STAGE II

This stage's onset is identified by the appearance of spindles. They are frontal–centrally predominant waves that occur as a cluster lasting for one to several seconds in duration. The spindle frequency is between 11 and 15 Hz and, in healthy adults, they are bilateral and synchronous in their appearance, with an amplitude up to 30 μV. Spindles may appear by themselves or following a vertex wave. In that latter situation, the wave is called a "K complex" (Fig. 5). Vertex waves are increasingly prevalent in Stage II sleep, and often persist in deeper stages as well. During the progression into deeper aspects of Stage II, the background rhythm continues to slow into the slower theta ranges, and high voltage generalized and frontally predominant delta slow waves are noted (<4 Hz).

During Stage II of sleep, occipital predominant high-voltage sharp slow waves are occasionally noted. These waves look like the lambda waves described in Section 8. These are referred to as either lambdoid waves or posterior occipital sharp transients of sleep. The physiological basis for posterior occipital sharp transients of sleep is probably very similar to

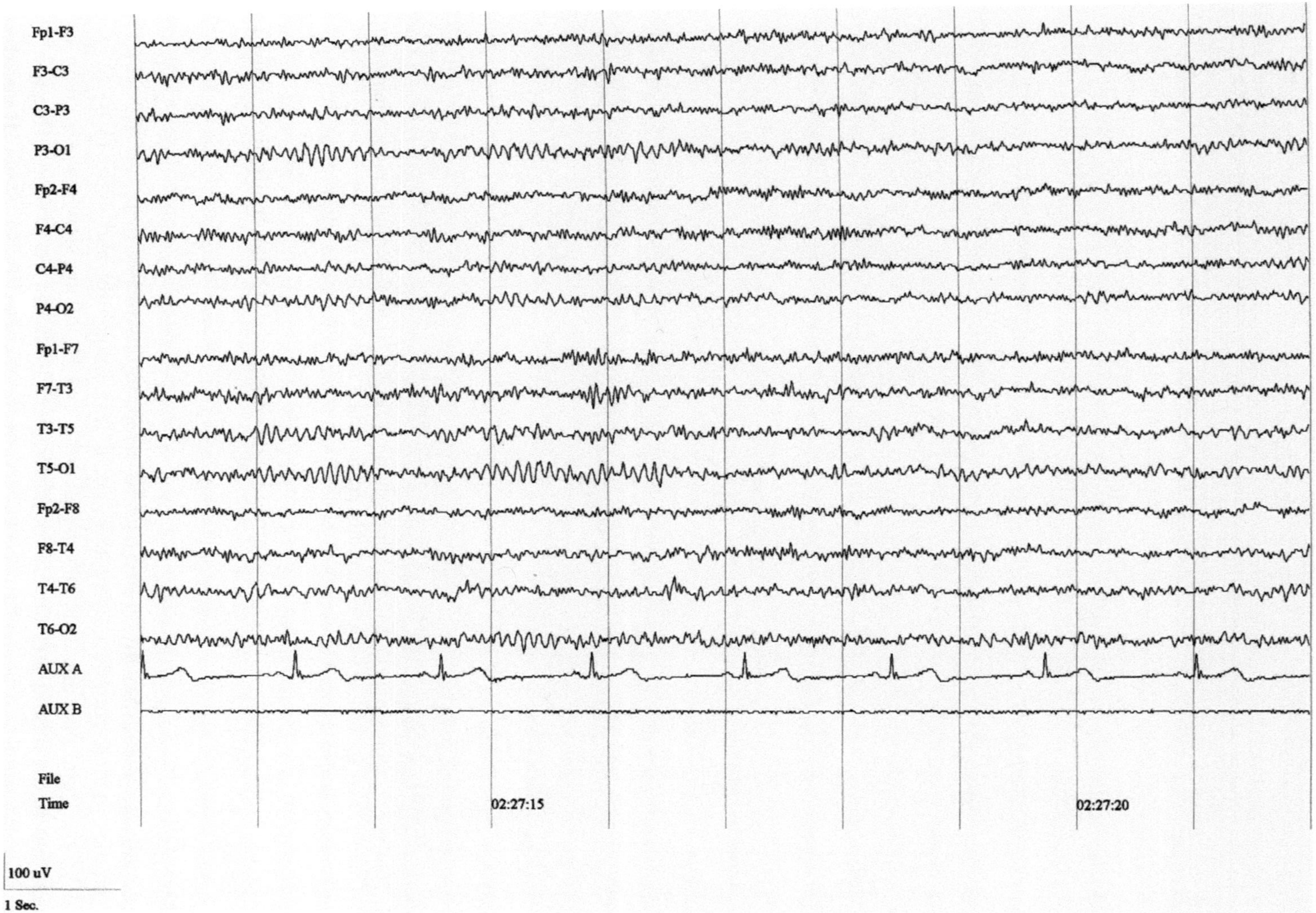

Fig. 4. Approximately half-way through this record, the alpha rhythm gradually fades and is replaced by slower theta frequency activity. There are no eye movement artifacts and this pattern evolved into Stage II sleep. Therefore, this tracing shows the transition from wakefulness to sleep, i.e., drowsiness.

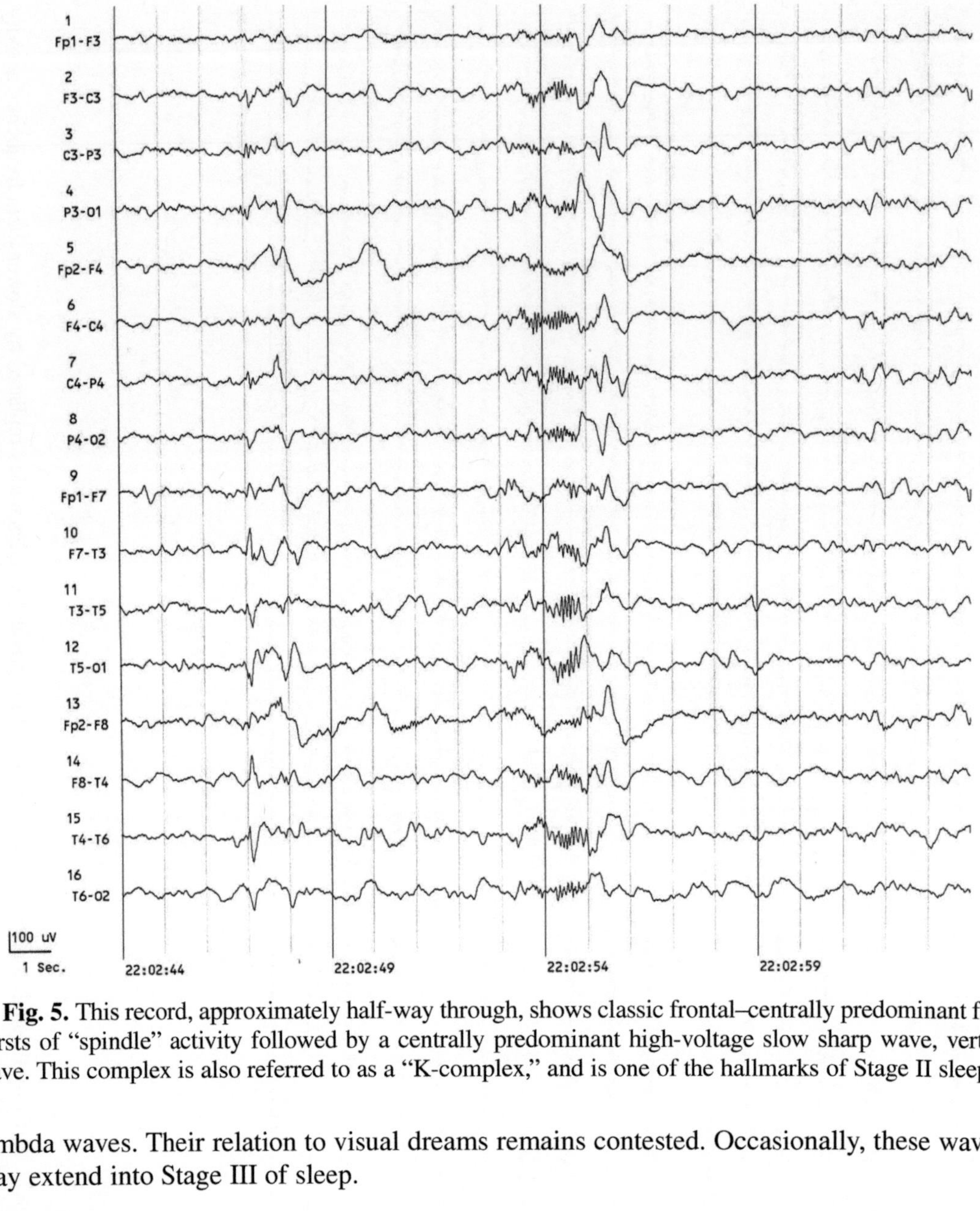

Fig. 5. This record, approximately half-way through, shows classic frontal–centrally predominant fast bursts of "spindle" activity followed by a centrally predominant high-voltage slow sharp wave, vertex wave. This complex is also referred to as a "K-complex," and is one of the hallmarks of Stage II sleep.

lambda waves. Their relation to visual dreams remains contested. Occasionally, these waves may extend into Stage III of sleep.

14. STAGE III

Occasionally a patient may go on to Stage III sleep. In this phase, there are still vertex waves, spindles, and K-complexes, but they occur less frequently and seem to be overwhelmed by the more prominent delta wave activity (<4 Hz). This slow wave activity is widespread but has a frontal and central predominance. It should be the most frequently noted rhythm and should be present for 20 to 50% of the record (Fig. 6). Stage III and the later Stage IV of sleep are most often associated with increases in the interictal discharges of temporal lobe epilepsy.

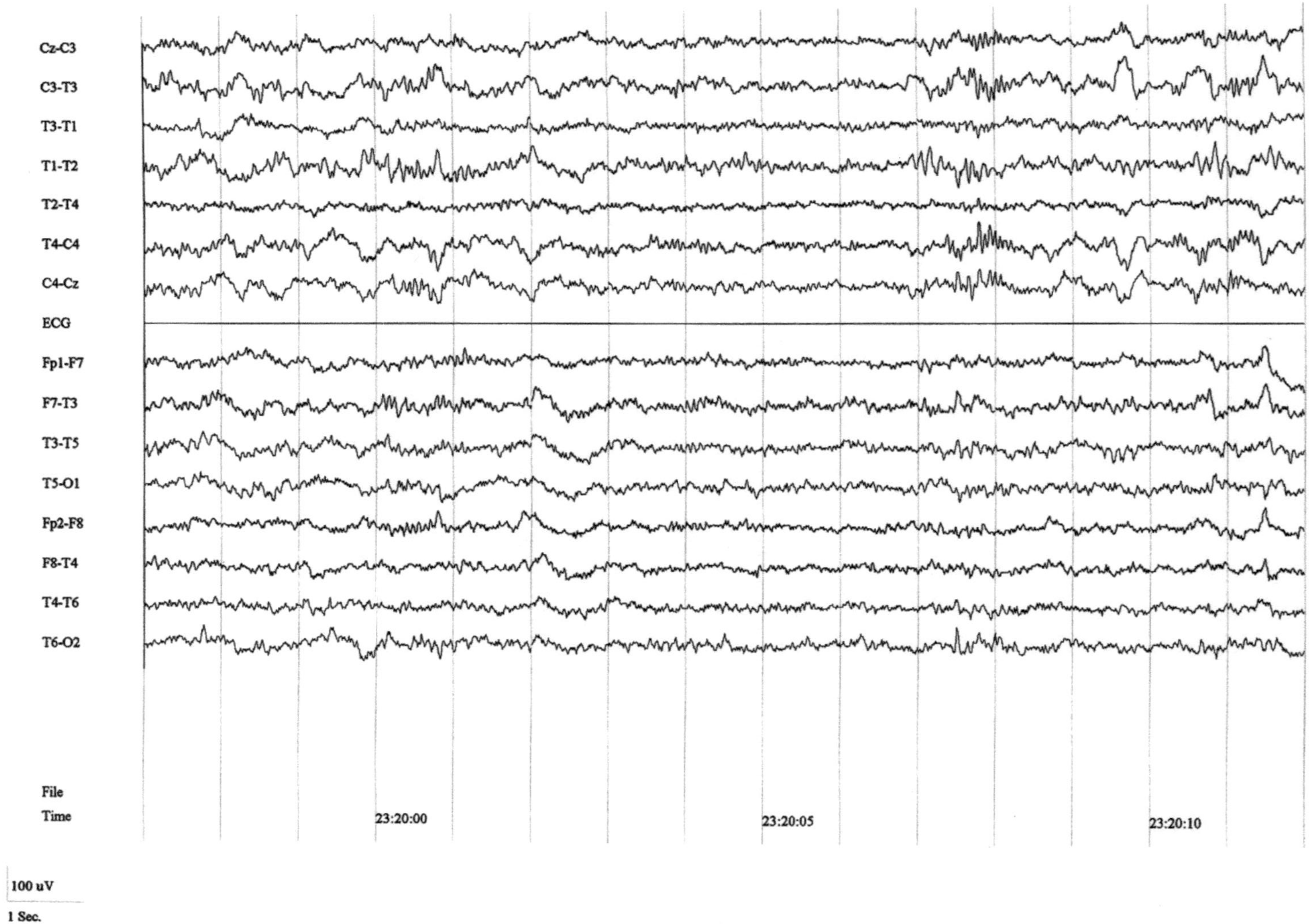

Fig. 6. As a subject progresses into Stage III sleep, as shown here, the background slows into the delta frequency range and is present approx 40% of the time.

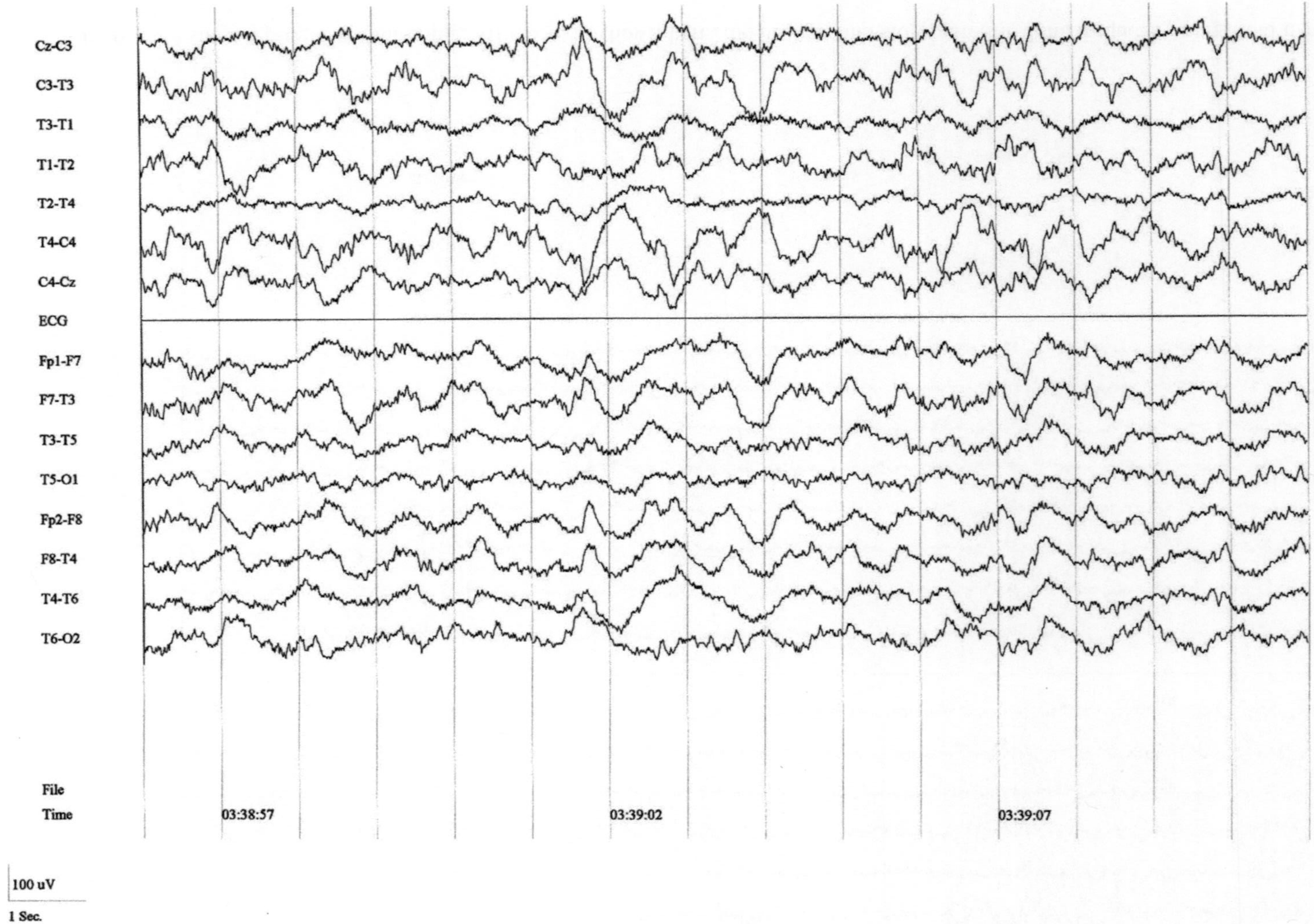

Fig. 7. When the background is more than 50% delta activity, the subject is considered to be in Stage IV of sleep, as noted on this portion of the record.

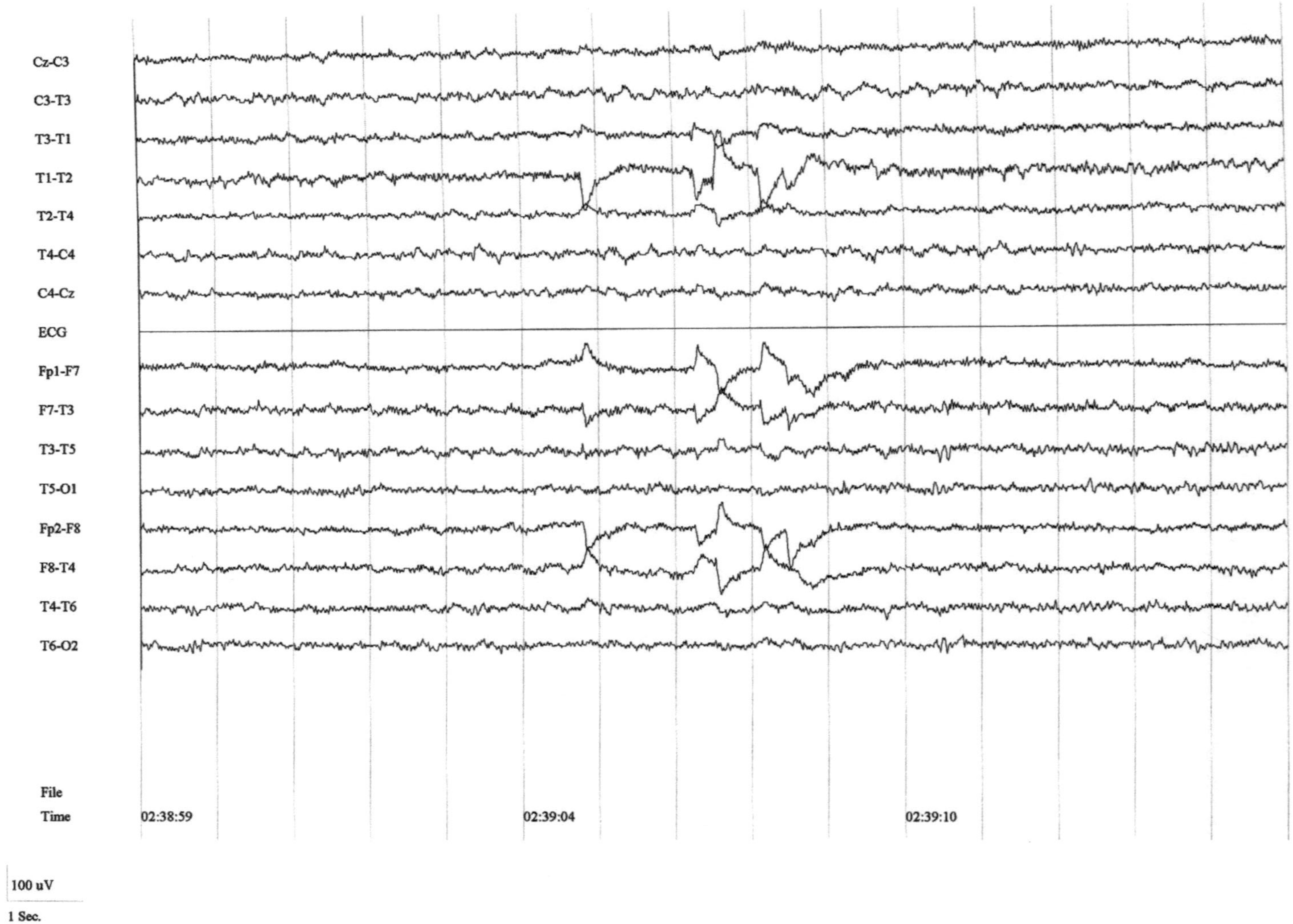

Fig. 8. Rapid eye movement sleep follows Stage IV in the sequencing of sleep. As noted here, the background is a low-voltage fast rhythm, as is seen in the wakeful state, but is also associated with lateral eye movements with a frequency of approx 1 Hz. These eye movements are seen over the lateral temporal leads, where they are out of phase with each other from side to side.

15. STAGE IV

Stage IV also has rudimentary vertex waves, spindles, and K-complexes, but shows further slowing and a more persistent delta-frequency background. In this stage, most (>50%) of the time of the recording is associated with continuously appearing delta activity (Fig. 7). This stage is only rarely achieved in the routine EEG recording.

16. REM

In sleep laboratories, the REM stage is divided into tonic and phasic components based on the simultaneous recording of other physiological parameters not routinely monitored during EEG acquisitions *(7)*. The other recordings include respiration effort and pulse oximetry, and chin and leg EMG activity. From a strictly EEG perspective, the background rhythm returns to what seems to be a drowsy or wakeful state, with low-voltage theta and alpha frequency activity, and with large, slow lateral eye movements (Fig. 8). This stage of sleep is also rarely encountered in routine EEG. When it is, perhaps ideally, the time from onset of drowsiness to the time of onset of REM should be measured. If it is short (<10 min), then the subject might consider a formal sleep lab referral to rule out narcolepsy.

REVIEW QUESTIONS

1. What system ensures consistent EEG recording technique? What is the upper limit of electrode impedance that is tolerated in EEG acquisition?
2. An upward deflection on an EEG conventionally indicates a negative or positive event?
3. What is the lower limit of normal for the waking posterior background rhythm in an adult patient? What are its characteristics?
4. What defines an abnormal degree of asymmetry in alpha amplitude?
5. What medications lead to excess beta activity? Can beta activity be a normal finding?
6. What maneuver may demonstrate reactivity of the mu rhythm?
7. What patients should not be asked to perform hyperventilation?
8. Are photic driving responses normal? What is the most characteristic frequency of photic driving?
9. What is the defining EEG feature of Stage II sleep?
10. What differentiates Stage III from Stage IV sleep?

REVIEW ANSWERS

1. The International 10-20 system ensures that electrodes are applied in a systematic fashion by the technologist to each patient. Electrode impedances are kept at 5000 Ω or less to minimize electrode-related artifact.
2. An upward deflection reflects a surface negative event.
3. The lower limit of normal for the adult waking posterior rhythm (the alpha rhythm) is 8 Hz. It is most prominently seen in the posterior channels and exhibits physiologic reactivity, specifically attenuating with eye opening or other mental tasks.
4. The alpha amplitude is said to be abnormally asymmetric if there is a greater than 50% discrepancy between the hemispheres. Particularly in right-handed people, the right side is usually of higher amplitude.
5. Benzodiazepines and barbiturates are most often associated with excessive beta activity. Beta activity can be a normal finding, and may be state dependent, because it is more often noted in drowsy tracings, especially in frontal channels.
6. Planning a movement with the contralateral hand or executing such a movement (e.g., make a fist) will attenuate the mu rhythm.

7. Hyperventilation should be avoided in the very elderly, in those suspected of an intracranial mass lesion, or in those with recent cerebral ischemia.
8. Photic driving responses are a normal phenomenon. However, their absence is common and is not pathological. When present, they should be fairly symmetrically evident. They are most common at close to the patient's alpha frequency.
9. Spindles are the defining element of Stage II sleep. Vertex waves are also abundant, as are K-complexes. Vertex waves may be seen in isolation in deeply drowsy (Stage Ib) subjects.
10. Stage III sleep entails delta activity involving 20 to 50% of the tracing, whereas Stage IV sleep exhibits greater than 50% delta activity.

REFERENCES

1. Nuwer MR. Recording electrode site nomenclature. J Clin Neurophysiol 1987;4:121–133.
2. Niedermeyer E. The EEG signal: polarity and field determination. In: Niedermeyer E, Lopes da Silva F, eds. Electroencephalography: Basic principles, Clinical Applications and Related Fields, 4th ed. Williams and Wilkins, Philadelphia, PA, 1999, pp. 143–148.
3. Kellaway P. Orderly approach to visual analysis: elements of the normal EEG and their characteristics in children and adults. In: Ebersole JS, Pedley TA, eds. Current Practice of Clinical Electroencephalography, 3rd ed. Lippincott Williams and Wilkins, Philadelphia, PA, 2003, pp. 100–159.
4. Niedermeyer E. The normal EEG of the waking adult. In: Niedermeyer E, Lopes da Silva F, eds. Electroencephalography: Basic principles, Clinical Applications and Related Fields, 4th ed. Williams and Wilkins, Philadelphia, PA, 1999, pp. 149–173.
5. Waltz S, Christen H-J, Doose H. The different patterns of the photoparoxysmal response: a genetic study. Electroencephal Clin Neurophysiol 1992;83:138–145.
6. Santamaria J, Chiappa KE. The EEG of Drowsiness. Demos Publications, New York, NY, 1987.
7. Niedermeyer E. Sleep and EEG. In: Niedermeyer E, Lopes da Silva F, eds. Electroencephalography: Basic principles, Clinical Applications and Related Fields, 4th ed. Williams and Wilkins, Philadelphia, PA, 1999, pp. 174–188.

6
Activation of the EEG

Barbara A. Dworetzky, Edward B. Bromfield, and Nanon E. Winslow

Summary

Various procedures are commonly used in the recording of the EEG in an effort to increase the diagnostic yield of the test. Common methods, such as hyperventilation (HV), photic stimulation, and sleep deprivation, referred to collectively as activation techniques, are traditionally used toward this end. Less common techniques, such as withdrawal of antiepileptic medications, use of specific triggers reported by the patient, and other idiosyncratic methods can be tried as well. This chapter will review methods of activation of the EEG, including HV, photic stimulation, and sleep deprivation. Historical background, physiological mechanisms, standard techniques, and clinical significance will also be reviewed.

Key Words: Hyperventilation; photic stimulation; photic driving; photoparoxysmal response; reflex epilepsy; sleep deprivation.

1. INTRODUCTION

Even in patients with a definite diagnosis of epilepsy, the first EEG will be normal 50% of the time. It is a common and accepted practice to arrange for additional EEGs if the first one is nondiagnostic, because three studies increase the sensitivity of detecting an abnormality to approx 90%. The EEG technologist is trained to use certain techniques to increase the likelihood that an abnormality will emerge during the 20- to 30-min sampling of brain activity that is obtained during a routine EEG. Common methods, such as hyperventilation (HV), photic stimulation, and sleep deprivation, referred to collectively as activation techniques, are traditionally used if an EEG would otherwise be interpreted as normal or "nonspecific." Less common techniques, such as withdrawal of antiepileptic medications, use of specific triggers reported by the patient, and other idiosyncratic methods can be tried as well. This chapter reviews methods of activation of the EEG, including HV, photic stimulation, and sleep deprivation. Historical background, physiological mechanisms, standard techniques, and clinical significance will also be reviewed.

2. HYPERVENTILATION

2.1. *Background*

Activation of seizures by HV was first reported in 1924, even before the discovery of the EEG. This technique became widely used in the diagnosis of absence seizures. HV responses can vary widely depending on age of patient and the amount of individual effort put forth. Common HV

From: *The Clinical Neurophysiology Primer*
Edited by: A. S. Blum and S. B. Rutkove © Humana Press Inc., Totowa, NJ

A

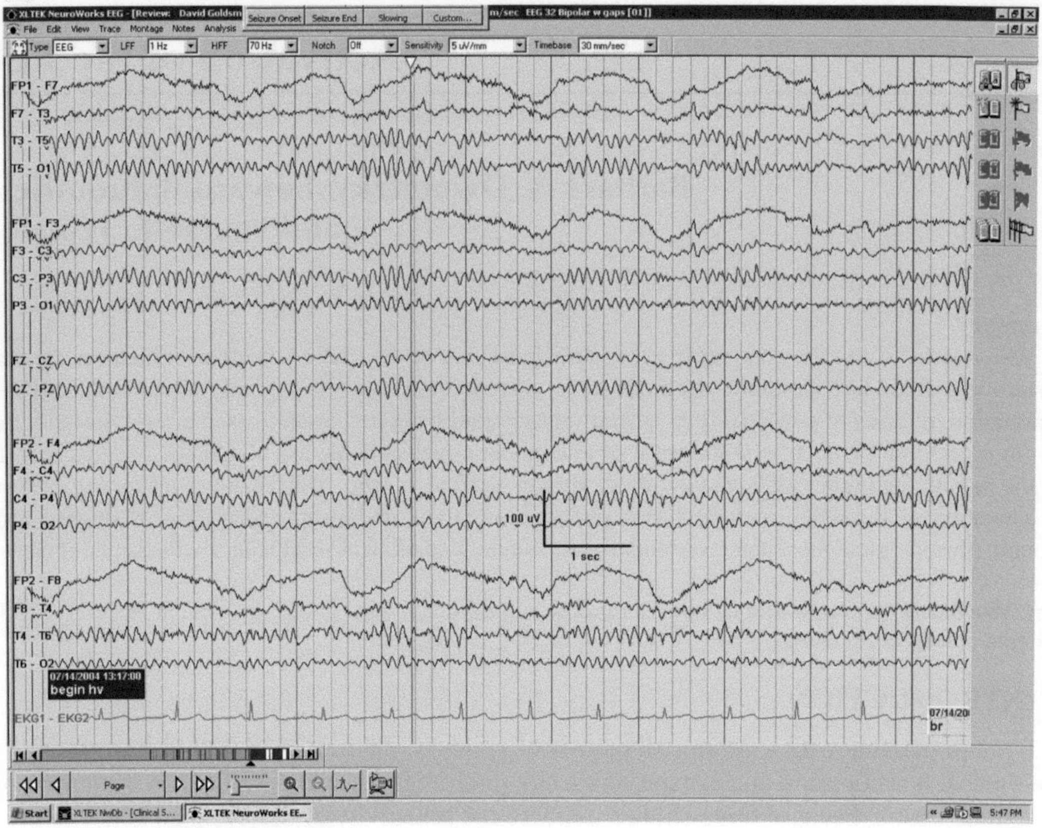

Fig. 1. (A) Start hyperventilation (HV) in an adult. *(Continued)*

responses in adults include no EEG change or mild slowing temporally, frontally, or diffusely. In younger individuals, in contrast, responses can be particularly dramatic, with extremely high voltage synchronous delta waves in bursts or runs. This can be further exaggerated if it has been many hours since the person has eaten a meal (relative hypoglycemia) (Fig. 1). Unless there are definitive spikes embedded within the synchronous delta activity or the record fails to return to baseline within 1 to 2 min of verified completion of over-breathing, the response should be interpreted as negative, because there is large variation in the normal response. Consistent focal features, whether epileptiform or not, are interpreted as abnormal.

2.2. Mechanism

With HV, there is a rise in PaO_2 and a drop in $PaCO_2$. To compensate for the resultant hypocapnia, the cerebral blood vessels constrict. The mechanism of the EEG response to HV is not yet understood, but several theories are reviewed by Takahashi. These include inadequate compensatory vasoconstriction, cerebral hypoperfusion as a result of vasoconstriction, increased neuronal excitability from respiratory alkalosis, synchronous activity of the thalamocortical projections that are enhanced by hypocapnia, and decreased activity of the mesencephalic reticular formation. The more dramatic changes noted in children suggest that immature autoregulation may explain the HV response, or that HV slowing is independent of cerebral blood flow. Whatever the true mechanism, HV is an accepted, standard technique for activation of the EEG.

B

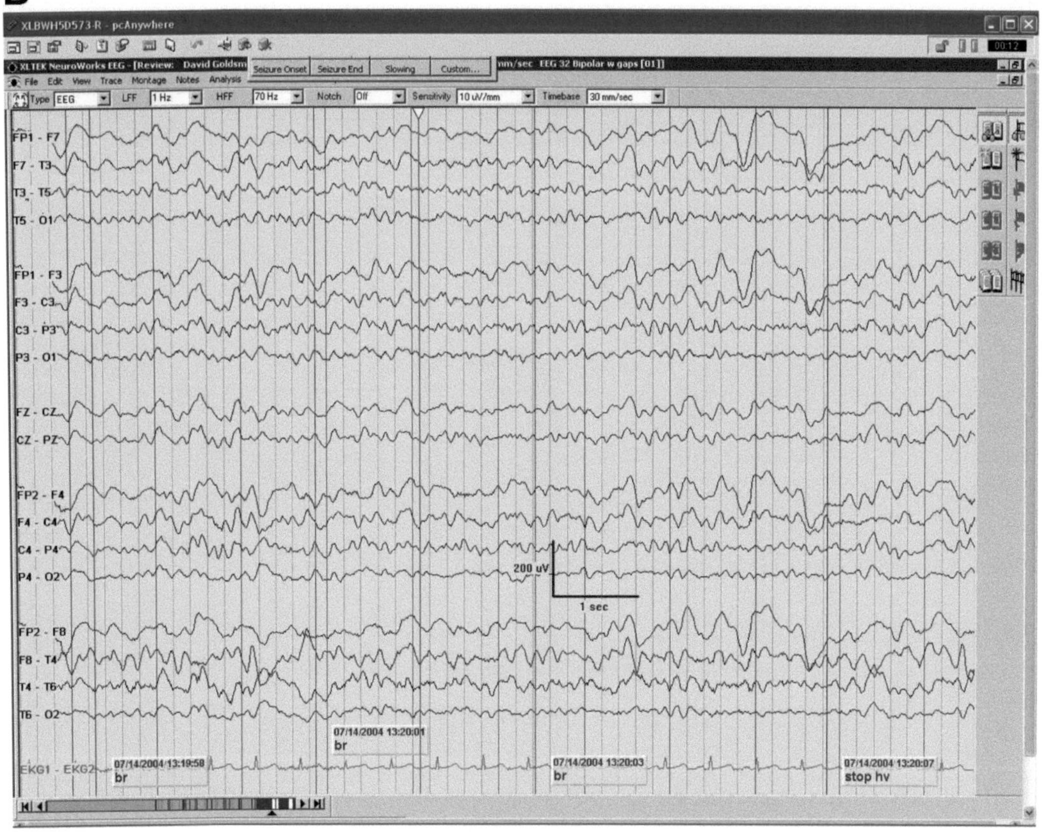

Fig. 1. (**B**) End HV 3 min later with marked bilateral slowing.

2.3. *Technique*

When there are no epileptiform discharges uncovered during a routine EEG, the HV procedure is explained to the patient. Those with severe cardiac or pulmonary disease, uncontrolled hypertension, or a recent vascular event, such as myocardial infarction, stroke, or transient ischemic attack, should not be exposed to this procedure, because hypocapnia and alkalosis may cause vasospasm or decrease cerebral perfusion. To begin the procedure, the technologist instructs the patient to breathe deeply and rapidly for 3 min. Patients are usually lying flat during a routine EEG, although it has been noted that the effects are enhanced by an upright posture, possibly because of the relative cerebral hypoperfusion. Patients should be told that they might experience symptoms of lightheadedness and tingling, particularly around the mouth and fingertips, although they can be reassured that this is reversible. As mentioned, the time since the last meal should be documented, because low glucose may enhance the response. The EEG is usually recorded on a bipolar montage with the patient's eyes closed, and should be run on the same montage, at least 1 min before starting the hyperventilating and continued for up to 3 min afterward, documenting the effort that the patient puts forth.

Syncope may lead to prolonged post-HV slowing, and patients with the rare vascular disease known as Moyamoya can have a delayed "re-buildup" referring to high-voltage diffuse slowing, which can occur even after HV is completed. Longer duration of HV (up to 6 min)

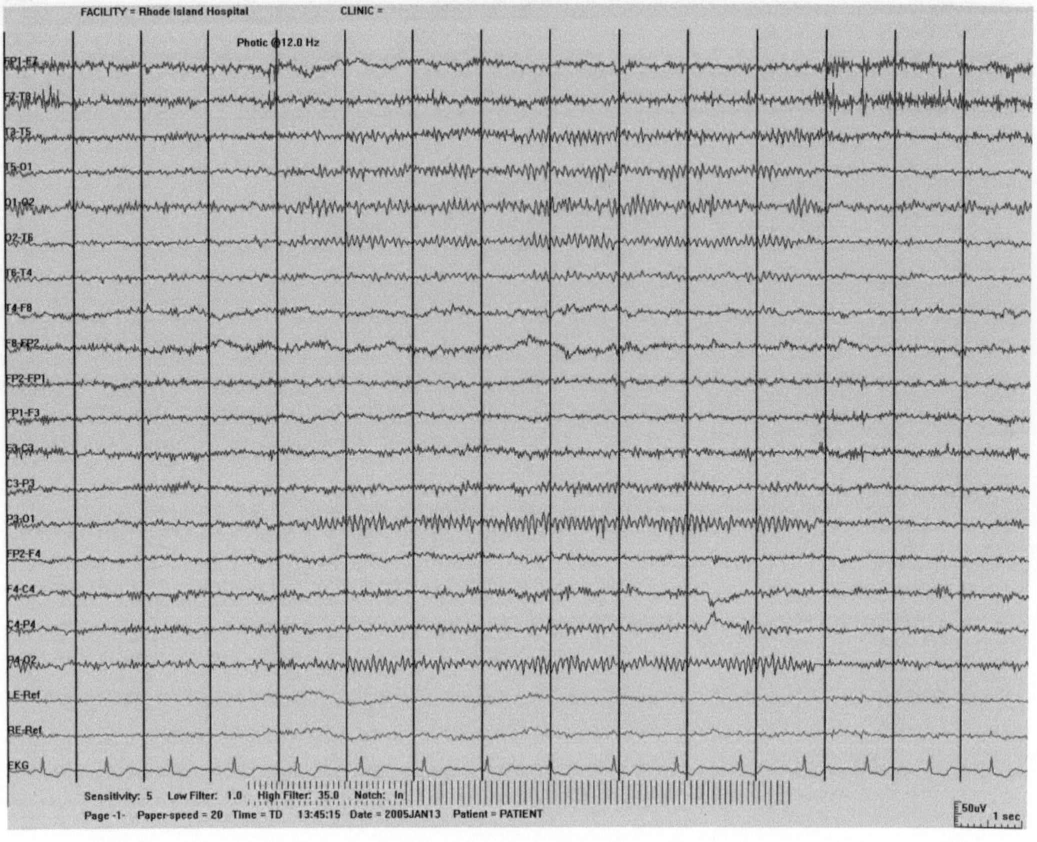

Fig. 2. (**A**) Symmetric photic driving with 12-Hz stimulation frequency. *(Continued)*

was shown by Adams and Luders to have a higher yield than 6 h of continuous EEG monitoring, justifying continuation of this technique when no epileptiform discharges are noted at 3 min, particularly in patients with absence seizures. Occasionally, patients become mildly confused and have difficulty discontinuing the procedure and may require gentle reminders to stop. If this is ineffective, rebreathing CO_2 with a paper bag or oxygen mask placed over the mouth can be useful. For the pediatric population, it is helpful to have children blow at a pinwheel, although for the very youngest, HV may be captured during the all too common periods of sobbing in the EEG laboratory. HV should be discontinued if repetitive discharges or seizures are elicited.

2.4. Clinical Significance

HV can enhance subtle but significant abnormalities, rendering insignificant ones unchanged or diminished. Additionally, it can be used as an induction technique for nonepileptic seizures. Previous studies have shown that, in patients with partial seizures and a normal baseline EEG, focal interictal discharges were elicited in 6 to 9% of patients using HV, whereas, in children with absence epilepsy, generalized spike and wave activity appeared in 80% of cases. A recent study by Holmes et al. however, suggested that even in known epilepsy patients, seizures were elicited by HV in less than 0.5%.

B

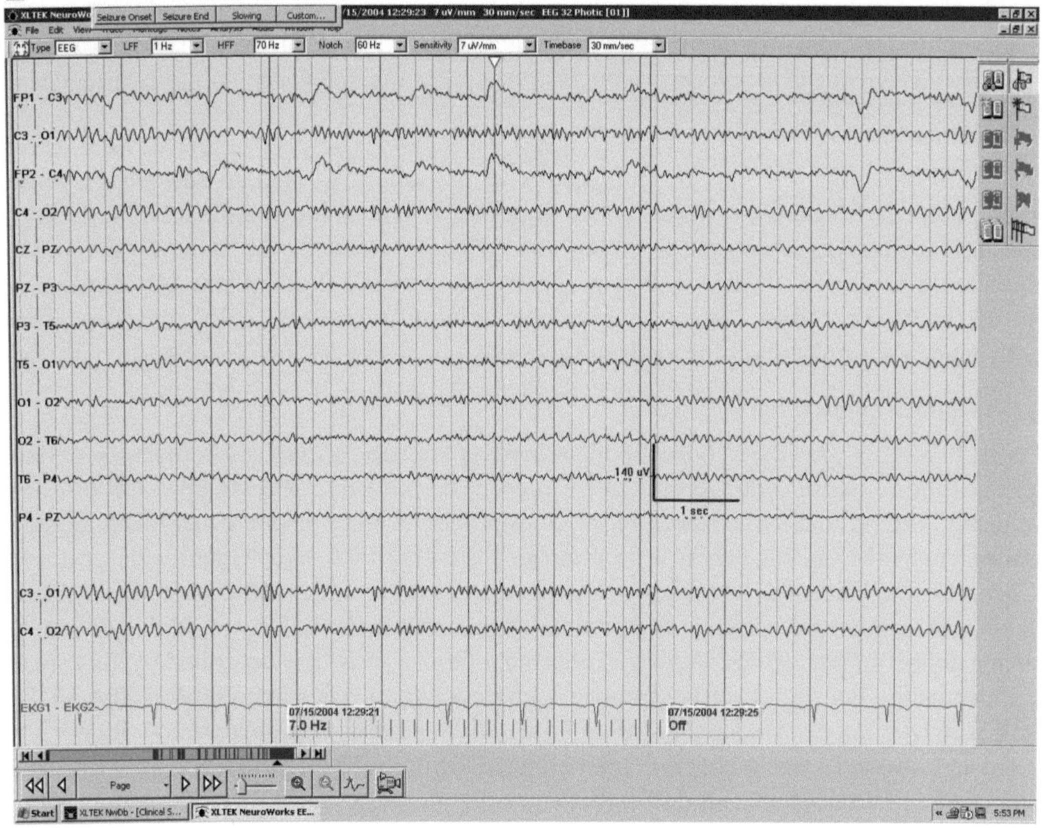

Fig. 2. (B) Photic driving "harmonic" response, intermittent photic stimulation at 7-Hz stimulation, 14-Hz driving.

3. PHOTIC STIMULATION

3.1. Background

Photic stimulation is performed in most EEG laboratories for nearly all patients referred for routine EEG. The earliest roots of this activation procedure may be traced back at least to ancient Greece, where descriptions of the potter's wheel to screen people who might have seizures were documented. Bright lights were noted to cause epileptic seizures at the turn of the 20th century. Intermittent photic stimulation (IPS) with a constant light source was used by 1934, and activation of epileptic discharges by using a strobe technique was soon reported. Clinical description of young patients' producing pleasurable feelings by waving their hands in front of a source of light was subsequently reported.

3.2. Mechanism

Physiologically, the photic response is a repetitive visual potential evoked by flash, as the response occurs in a specific time relationship to the light stimulus. Photic driving refers to this time-locked activity generated by visual cortex and seen at the back of the head (Fig. 2A). Certain stimulus frequencies more readily elicit responses, usually those between 4 and 30 Hz. Analysis at slower frequencies reveals that the actual response is of positive polarity and follows the stimulus by approx 100 ms. There is a wide range in amplitude of normal responses. When they are of particularly high amplitude, one might recognize a "harmonic," or multiple, of the presented stimulus frequency. For example, a double harmonic response occurs at twice the stimulus frequency (Fig. 2B) and a half harmonic occurs at half the stimulus frequency.

3.3. Technique

Most EEG machines are preprogrammed with the ability to present successive stimulus trains at particular frequencies. Standard techniques for photic stimulation include presenting the strobe light stimuli at a measured distance of 20 to 30 cm with the patient's eyes closed. Trains should begin at 1 or 3 Hz, and be presented for 4 to 10 s, with at least 4 s between each train, up to 30 Hz. Subjects may have trouble tolerating these visually presented stimuli directly, and often find it helpful to close their eyes.

3.4. Clinical Significance

Photic driving is considered normal unless there is dramatic voltage asymmetry or unilateral absence of the response (not just caused by varying harmonics), or if epileptiform discharges are elicited. The presence of a dramatic, high-amplitude driving response may indicate hyperautonomic responsiveness, such as is seen in alcohol withdrawal, hyperthyroidism, or migrainous disorders. Prominent driving, however, is not considered abnormal, even in the absence of such clinical correlates. Repetitive contractions of the frontalis muscle synchronized to the light flash at a delay of 50 to 60 ms, known as photomyoclonus, was first described by Gastaut, and this phenomenon does not clearly correlate with epilepsy or other neurological disease. Such muscle activity can be differentiated from cortical discharges by its suppression with eye opening, and anterior rather than posterior or generalized location. These twitches are always exactly time-locked and do not outlast the stimulus.

The phenomenon of photomyoclonus, similar to an exaggerated driving response, is normal, although, again, perhaps more associated with hyperautonomic states. Photoparoxysmal response (PPR), formerly known as photoconvulsive response, occurs when IPS generate bilaterally synchronous and usually generalized epileptiform discharges, which may outlast the stimulus by several seconds, or indeed even cause a seizure (Fig. 3). Photic-induced seizures are most commonly myoclonic although absence or tonic–clonic seizures can occur. Discharges are often generalized but more pronounced in frontal leads, and are more likely induced at middle stimulus frequencies (15–20 Hz) or with stimuli presented on eye closure while the patient is awake and alert. The presence of PPR is more likely to indicate idiopathic generalized epilepsy rather than partial epilepsy. This response can be a genetic trait, although it is rarely seen in patients without any history of seizures. Photic stimulation should be discontinued in the presence of polyspike and wave discharges.

Artifacts identified during photic stimulation can mimic PPR, particularly the electroretinal response (generated by retinal ganglion cells), and photoelectric effect (from photochemical response of the silver electrodes). Occipital spikes can be elicited with photic stimulation in progressive myoclonus epilepsy, such as that seen in infantile ceroid lipofuscinosis.

4. SLEEP DEPRIVATION

4.1. Background and Clinical Significance

The association of epileptic seizures and sleep has been recognized since at least the 19th century. Gowers, as cited by Chokroverty, found that 21% of patients had seizures exclusively during the night, especially during transitions into and out of sleep, as well as 1 to 2 h after

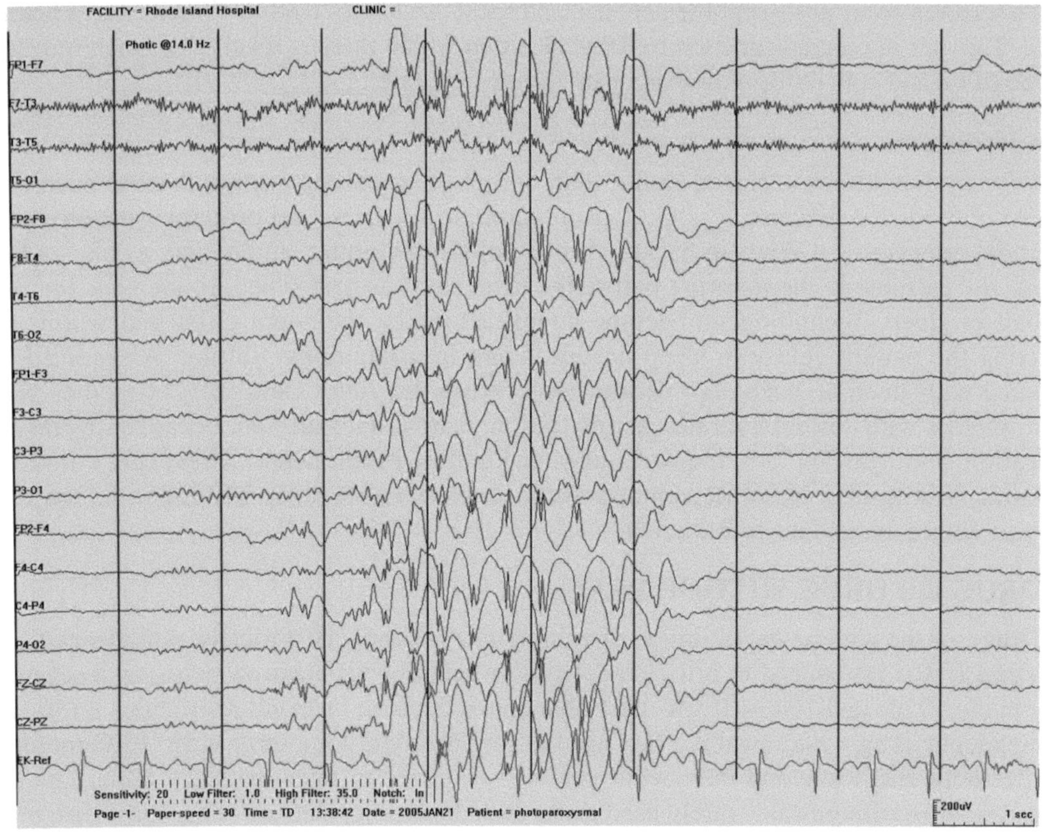

Fig. 3. Photoconvulsive response in an 11-yr-old girl with spells of unresponsiveness, aggravated by bright lights.

awakening. With the introduction of the EEG, epileptic discharges during sleep were discovered. An increase in discharges during sleep has been a longstanding observation among patients with grand mal seizures. Sleep deprivation has been demonstrated to increase the yield of epileptiform findings even in the absence of sleep production, and may be more activating than sedated sleep; a more recent study of Degen et al., however, did not confirm significant differences in activation of epileptiform activity in patients with and without complete sleep deprivation. Even more recent studies of Malow et al. have shown that slow wave sleep is most likely to include spikes, whereas rapid eye movement (REM) sleep is the least likely to include discharges, even less than waking. Klein et al. found that sleep was more activating than HV for patients with definite interictal epileptiform discharges. Glick's recent review of the effects of sleep deprivation on EEG concluded that there is a prominent activating effect of light sleep on the EEG, but no evidence of an overall increase in discharge rate after sleep deprivation in waking or other sleep states. Regarding specificity of sleep recording, activity that is "sharp appearing" but seen only during wakefulness and not in sleep is less likely to be "epileptiform" than discharges that continue or increase during sleep.

4.2. Mechanism

Neuronal synchronization is the underlying basis for epileptiform discharges and ictal events. Non-REM sleep, mediated by thalamocortical input, is a time of increased synchronization when seizures occur in susceptible patients, and focal discharges tend to demonstrate a broader field. This lies in contradistinction to REM sleep, in which there is relative desynchronization of the EEG, and epileptiform discharges or seizures are rare.

4.3. *Technique*

In many laboratories, partial rather than complete sleep deprivation is used because it is less onerous for patients, less likely to precipitate seizures, and probably just as useful. We tell patients to cut down to half of their usual sleep amount, and to stay awake and not drink any caffeine in the morning before the EEG study is performed, to increase the likelihood of sleep. Technologists become adept at making the room quiet and warm, and coaxing the patient gently to sleep despite being in a laboratory setting. A second EEG ordered with sleep or sleep deprivation can increase the yield. Generally, conscious sedation, such as with chloral hydrate administration, is no longer offered, because special certification and staffing are required and make this problematic. However, conscious sedation is still used in some pediatric facilities in which a sleep EEG may be the only way to obtain an artifact-free tracing.

5. DRUG OR DRUG WITHDRAWAL

Drugs, or the withdrawal of drugs, can precipitate seizures. Historically, pentylenetetrazol (Metrazol) was introduced to bring on convulsive seizures for electroconvulsive shock therapy in the 1930s and 1940s. It was later abandoned because of many restrictions on its use. Antiepileptic drug withdrawal is often used during inpatient long-term video EEG monitoring for both diagnostic and therapeutic reasons. Contrary to many patients' understanding, however, drug withdrawal is rarely used in the EEG laboratory outside of the long-term monitoring setting, because the likelihood of seeing interictal discharges, especially focal discharges, is not greatly affected by most antiepileptic drugs (and the risk of seizure is high). In the long-term monitoring setting, concerns raised that medication withdrawal would change the nature of the typical seizures and negate the usefulness for localization has dissipated with studies indicating that seizures may more likely generalize, but clinical seizure semiology and electrographic onset are not significantly altered.

6. OTHER METHODS

Additional idiosyncratic activation methods may be justified in patients with "reflex epilepsy" associated with specific stimuli or activities, such as reading or listening to specific music. Patients with psychogenic nonepileptic seizures may perceive similar relationships, and appropriate methods may also induce these spells and establish the correct diagnosis.

The importance of making a definitive diagnosis of epilepsy cannot be overstated. Although extended recording on either an ambulatory or inpatient basis may have the highest yield, this testing is not appropriate for patients with infrequent spells or those whose interictal discharges can be identified using the simpler and less expensive techniques outlined here. The more intensive techniques, however, are preferable to a "therapeutic trial," which may lead to prolonged, inappropriate use of potentially toxic medications, not to mention the associated stigma and activity restrictions. In some circumstances, it may be useful to proceed directly to ambulatory EEG after an initial study that includes HV, photic stimulation, and sleep and sleep deprivation is negative, rather than repeating the routine EEG.

SUGGESTED READING

Adams D, Luders H. Hyperventilation and 6-hour EEG recording in evaluation of absence seizures. Neurology 1981;31:1175–1177.

Ajmone-Marsan C. Pentylenetetrazol: historical notes and comments on its electroencephalographic activation properties. In: Luders H, Noachtar S, ed. Epileptic Seizures: Pathophysiology and Clinical Semiology. Churchill Livingstone, Philadelphia, PA, 2000, pp. 563–569.

Benbadis SR Johnson K, Anthony K, Caines G, et al. Induction of psychogenic nonepileptic seizures without placebo. Neurology 2000;55(12):1904–1905.

Chokroverty S, Quinto C. Sleep and epilepsy. In: Chokroverty S, ed. Sleep Disorders Medicine, 2nd ed. Butterworth Heinemann, Woburn, MA, 1999, pp. 697–727.

Degen R, Degen HE, Reker M. Sleep EEG with or without sleep deprivation? Does sleep deprivation activate more epileptic activity in patients suffering from different types of epilepsy? Eur Neurol 1987;26(1):51–59.

Doose H, Waltz S. Photosensitivity: genetics and clinical significance. Neuropediatrics 1993;24:249–255.

Drury I. Activation of seizures by hyperventilation. In: Luders H, Noachtar S, ed. Epileptic Seizures: Pathophysiology and Clinical Semiology. Churchill Livingstone, Philadelphia, PA, 2000, pp. 575–579.

Gastaut H. Effects des stimulations physiques sur l'EEG de l' homme. Electroencephalogr Clin Neurophysiol Supp. 1949;2:69–82.

Glick T. The sleep-deprived electroencephalogram: evidence and practice. Arch Neurol 2002; 59(8):1235–1239.

Holmes M, Dewaraja A, Vanhatalo S. Does hyperventilation elicit epileptic seizures? Epilepsia 2004;45(6):618–620.

Klein K, Knake S, Hamer HM, Ziegler A, Oertel WH, Rosenow F. Sleep but not hyperventilation increases the sensitivity of the EEG in patients with temporal lobe epilepsy. Epilepsy Res 2003;56(1):43–49.

Malow B, Selwa L, Ross D, Aldrich M. Lateralizing value of interictal spikes on overnight sleep-EEG studies in temporal lobe epilepsy. Epilepsia 1999;40(11):1587–1592.

Misulis K. Essentials of Clinical Neurophysiology. Butterworth-Heinemann, Woburn, MA, 1993.

Patel V, Maulsby R. How hyperventilation alters the electroencephalogram: a review of controversial viewpoints emphasizing neurophysiological mechanisms. J Clin Neurophysiol 1987;4:101–120.

Rosenow F, Luders H. Hearing-induced seizures. In: Luders H, Noachtar S, eds. Epileptic Seizures: Pathophysiology and Clinical Semiology. Churchill Livingstone, Philadelphia, PA, 2000, pp. 580–584.

Rosenow F, Luders H. Startle-induced seizures. In: Luders H, Noachtar S, eds. Epileptic Seizures: Pathophysiology and Clinical Semiology. Churchill Livingstone, Philadelphia, PA, 2000, pp. 585–592.

Takahashi T. Activation Methods. In: Niedermeyer E, Lopes Da Silva F, eds. Electroencephalography: Basic Principles, Clinical Applications, and Related Fields, 4th ed. Lippincott Williams & Wilkins, Philadelphia, PA, 1999, pp. 261–284.

Wolf P. Activation of seizures by reading and praxis. In: Luders H, Noachtar S, eds. Epileptic Seizures: Pathophysiology and Clinical Semiology. Churchill Livingstone, Philadelphia, PA, 2000, pp. 609–614.

REVIEW QUESTIONS

1. What is the value of HV to the EEG?
2. In what way does hypoglycemia influence the effect of HV on the EEG?
3. Who should not undergo HV?
4. What is the physiological basis for photic driving response?
5. When are photic driving responses abnormal?
6. What are PPRs and what do they signify?

7. What is photomyoclonus? Is it considered pathological?
8. What is the photoelectric (photocell) effect?
9. What additional activation methods can be used in the inpatient setting to improve the yield of monitoring?
10. Which sleep stage is least apt to activate epileptiform features?

REVIEW ANSWERS

1. HV is helpful in that it may accentuate focal or epileptiform abnormalities in the EEG. It has an especially high yield in the evaluation of patients with 3-Hz spike wave abnormalities associated with absence epilepsy.
2. Hypoglycemia tends to accentuate the effect of HV on the EEG.
3. Patients with recent or unstable CNS or cardiac ischemia, cardiopulmonary compromise, or uncontrolled hypertension should not undergo HV.
4. Photic driving responses are derived from visual evoked potential responses in the visual cortex.
5. Photic driving responses are abnormal if there are dramatic hemispheric voltage asymmetries or frankly unilateral responses. The affected hemisphere does not generate a response. Otherwise, photic driving responses are considered to be normal (epileptiform activities are also abnormal when elicited, but these are termed PPRs). Some investigators think that prominent photic driving responses, including at high stimulus frequencies, may be a marker of hyperautonomic states, but this phenomenon is insufficiently specific to be deemed pathological.
6. PPRs are denoted by epileptiform discharges generated in the context of IPS, sometimes continuing after the completion of the photic train. They signify an underlying epileptic process that is photically sensitive. They are usually encountered in association with generalized epilepsies, although they may also be seen in focal occipital epilepsies.
7. Photomyoclonus refers to entrained contraction of the frontalis muscle in the context of IPS. It is considered a normal phenomenon.
8. The photoelectric effect refers to an electrode-derived artifact caused by a photochemical reaction from the silver electrode in response to light stimulation itself. It may be improved (and verified) by shielding the involved lead from the direct illumination of the strobe source.
9. Other methods of EEG activation during inpatient long-term monitoring include sleep deprivation, anticonvulsant withdrawal, and occasionally exposure to stimuli in the context of reflex epilepsies (e.g., reading or writing epilepsies).
10. REM sleep is not conducive to epileptiform activity because it is a more desynchronized pattern. Non-REM stages are much more productive of epileptiform features.

7
Normal Variant EEG Patterns

Richard L. Cervone and Andrew S. Blum

Summary

The object of this chapter is to familiarize the reader with a number of commonly encountered normal variants of brain-derived EEG activity. The term "normal variant pattern" refers to those rhythms or waveforms that have features reminiscent of either interictal or ictal EEG abnormalities. However, these patterns have been found in a substantial proportion of tracings from healthy subjects and, therefore, are not currently thought to represent pathological entities. It is, therefore, vital that such patterns be appropriately recognized by the EEG reader as normal variants and not erroneously confused for pathological patterns. This chapter addresses four main categories of variant EEG activity:

1. Rhythmic patterns.
2. Epileptiform patterns.
3. Lambda and lambdoids.
4. Age-related variants.

EEG artifacts derived from sources other than brain-derived activity will not be reviewed.

Key Words: Benign; EEG; epileptiform; rhythm; variant.

1. RHYTHMIC VARIANT PATTERNS

There are six main types of rhythmic variant EEG patterns:

1. Alpha variant.
2. Mu rhythm.
3. Rhythmic temporal theta burst of drowsiness ("psychomotor variant").
4. Subclinical rhythmic electrographic (theta) discharges in adults (SREDA).
5. Midline theta rhythm.
6. Frontal arousal rhythm (FAR).

1.1. *Alpha Variant*

This pattern was described first by Goodwin in 1947. There are two types of alpha variants, "slow" and "fast." The slow (subharmonic) alpha variant appears as an abrupt rhythm usually at half the frequency of the patient's more typical waking background rhythm, and often of greater voltage (Fig. 1). The fast (harmonic) alpha variant may appear as a notched or bifurcated form of the patient's usual waking background rhythm, so that a superimposed harmonic rhythm of twice the alpha frequency occurs. Alpha variants are blocked with eye opening and exhibit a posterior predominance, just as with normal alpha rhythms. Alpha variants vary in their prevalence within a subject's tracing, alternating with periods of normal-appearing alpha

From: *The Clinical Neurophysiology Primer*
Edited by: A. S. Blum and S. B. Rutkove © Humana Press Inc., Totowa, NJ

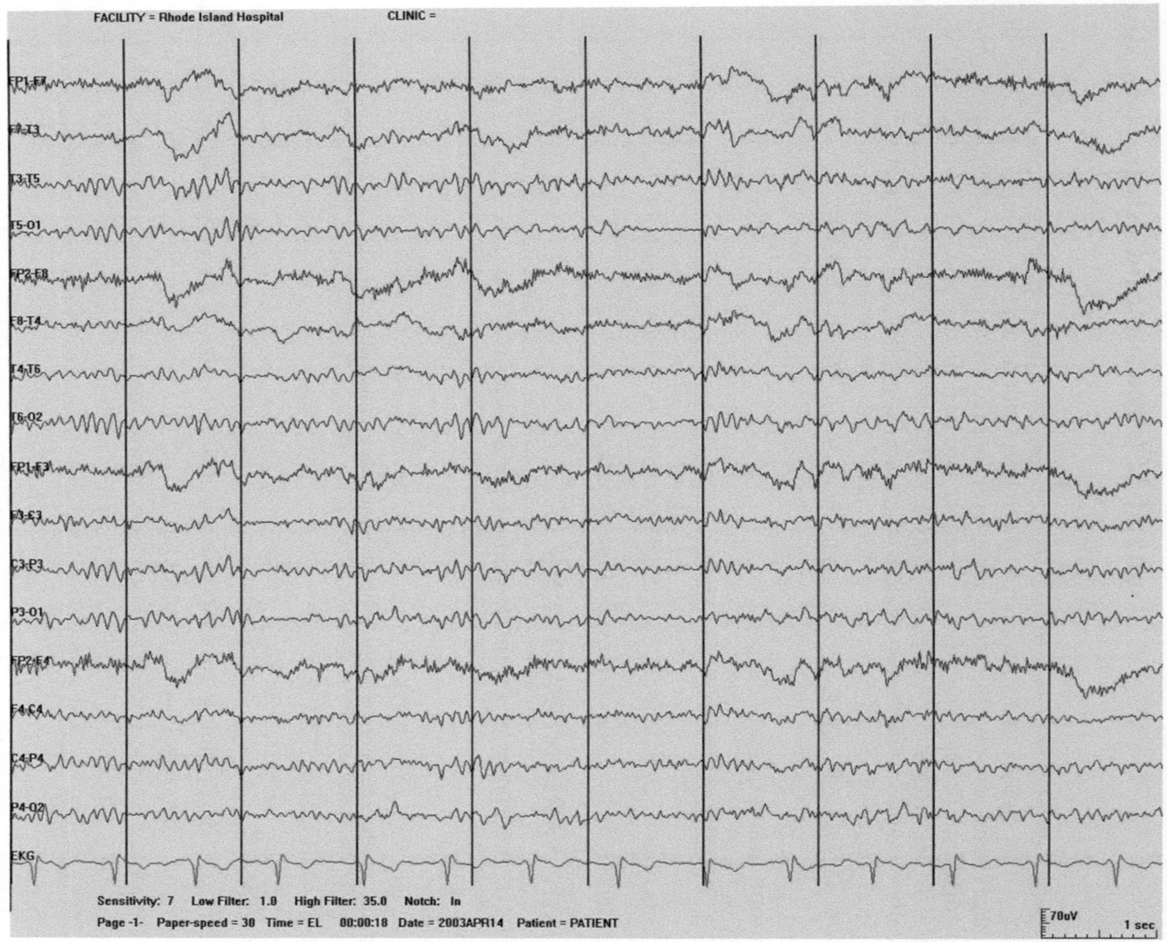

Fig. 1. Slow alpha. This subject is 46 yr of age. Note how the posterior background rhythm shifts from a bilateral 11-Hz rhythm earlier in the figure to a somewhat notched biposterior 5- to 6-Hz rhythm late in the figure.

activity. Alpha variants have been regarded as a physiological variation of the more familiar posterior background alpha activity, and do not predict increased convulsive tendency.

1.2. Mu Rhythm

Numerous terms have been applied to the mu rhythm, including arcade, comb, and wicket rhythms, owing to its morphology. Individual waveforms have an arciform morphology. This occurs in waking over the central regions, especially the C3, Cz, and C4 contacts (Fig. 2). It is closely associated with the sensorimotor cortex, hence the term "mu," for motor. Mu exhibits a frequency in the alpha range, typically at 9 to 11 Hz. Niedermeyer observed this rhythm in approx 14% of adolescents' EEG tracings, and less often in younger children and the elderly. Similar to the alpha activity of the occipital cortex, it exhibits physiological reactivity. It attenuates with contralateral limb movement or just planned movement of the contralateral limb. With direct cortical recording methods, a 20-Hz beta activity may be observed from the sensorimotor cortex, with similar reactivity. Thus, the scalp-recorded mu is likely a subharmonic of this underlying rhythm.

Mu is usually observed bilaterally with shifting predominance; it may, however, be asymmetrical and asynchronous. Mu activity is augmented in the setting of a focal skull breach. This could, for instance, explain some instances of highly lateralized mu rhythms. Exclusively lateralized mu should raise a suspicion of an abnormality in the hemisphere lacking mu activity. Sometimes, focal mu activity in the setting of a bony defect of the skull may be so sharp and of higher voltage as to falsely mimic an epileptogenic focus.

1.3. Rhythmic Temporal Theta Bursts of Drowsiness ("Psychomotor Variant")

Gibbs et al. called this pattern the "psychomotor variant" because it was thought to represent a temporal lobe or psychomotor seizure. This concept has been more recently discarded because this pattern is observed in asymptomatic healthy individuals and exhibits poor correlation with patients with true temporal lobe or psychomotor seizures. This pattern has also been called "rhythmic mid-temporal discharges" describing its character, location, and frequency.

This pattern may be present in waking or early drowsiness and usually in tracings of adults and adolescents. It wanes with deepening sleep. As its name implies, this particular pattern is found in the mid-temporal head regions, but can spread parasagittally. It is comprised of 5- to 7-Hz rhythms in bursts or trains lasting often longer than 10 s and sometimes beyond a minute (Fig. 3). This variant rhythm can exhibit variable morphologies, but is often sharply contoured. It is usually monomorphic; it does not evolve significantly in frequency or amplitude, as occurs in most ictal patterns. Rhythmic mid-temporal discharges can occur bilaterally or independently with shifting hemispheric predominance. This pattern is uncommon; its incidence is approx 0.5%, according to Gibbs et al.

1.4. Subclinical Rhythmic Electrographic Discharges in Adults

This variant pattern involves sharply contoured 5- to 7-Hz activities with a wide distribution, mainly over temporo-parietal derivations. It is usually bilateral, but can be asymmetrically disposed. SREDA can appear as repetitive monophasic sharp waves or as a single discharge followed seconds later by sharp waves that gradually accelerate to form a sustained, rhythmic train of theta activity. This may last from 20 s to several minutes, usually 40 to 80 s. Because of its duration and evolution, SREDA can easily be misinterpreted as an ictal pattern, even by experienced readers. Nevertheless, SREDA has not been shown to have any consistent

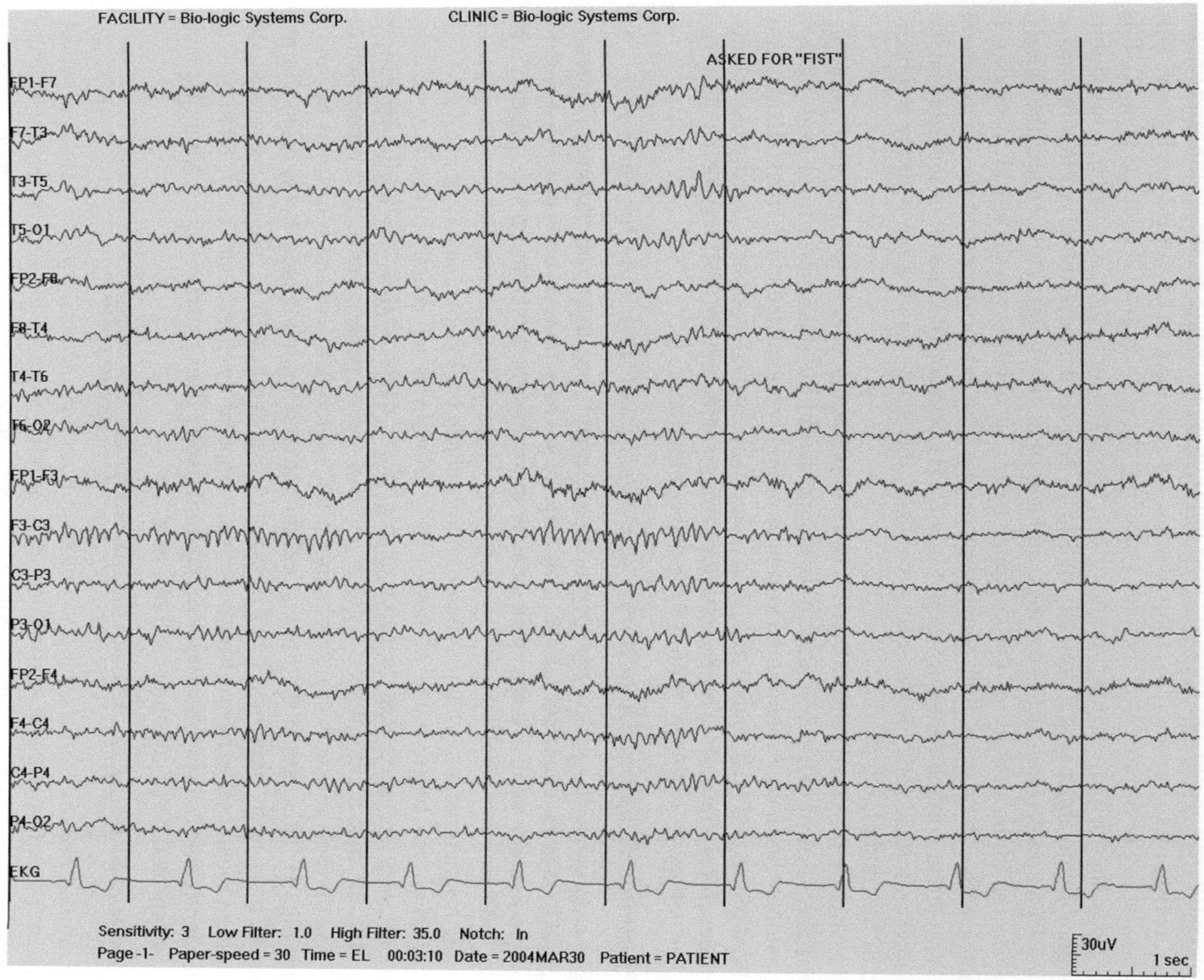

Fig. 2. Mu. This subject is 27 yr old. Note the prominent mu rhythm over the C3 contact. This blocks efficiently when the subject is asked to make a fist with the contralateral hand, as indicated.

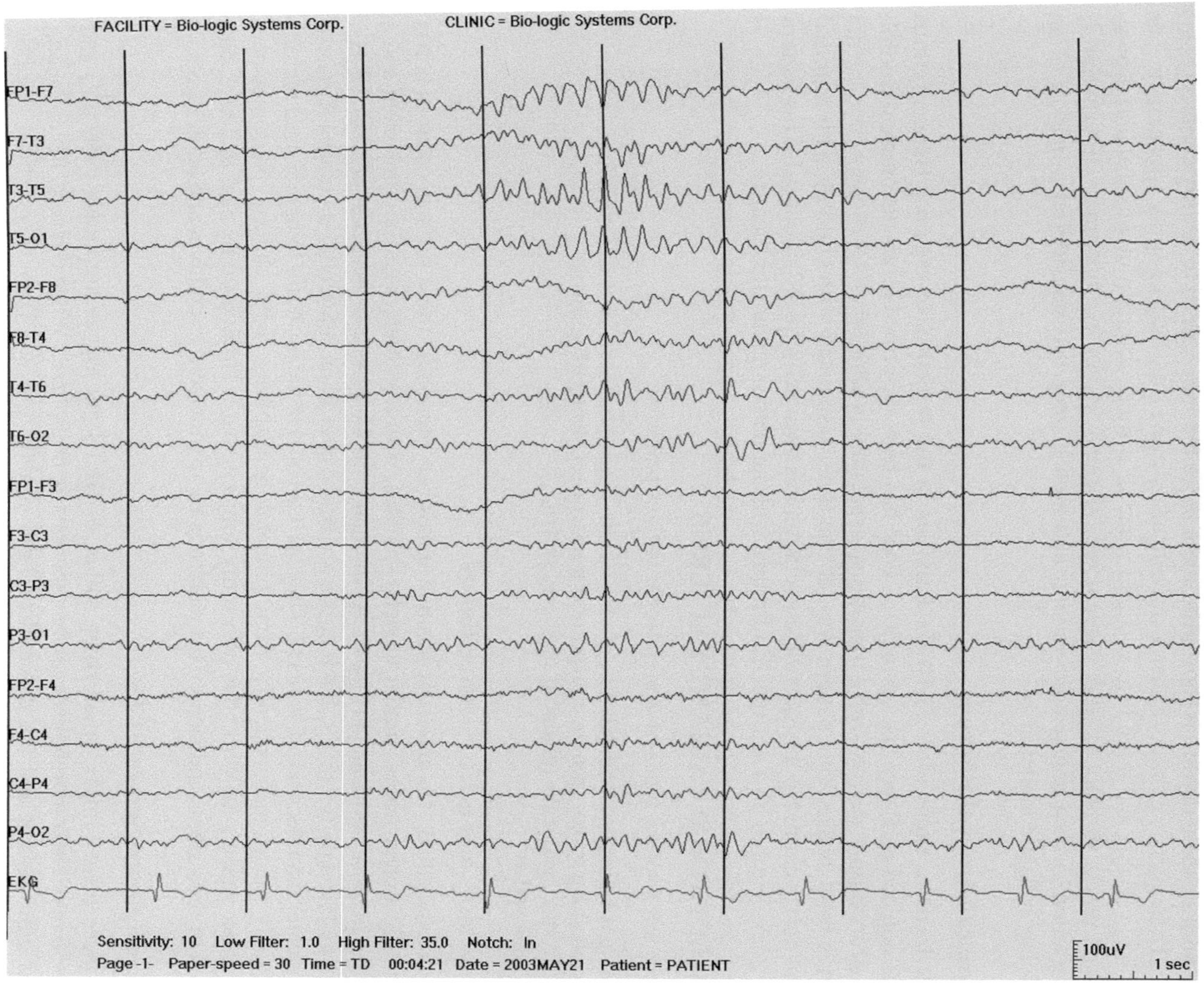

Fig. 3. Psychomotor variant. The subject is 20 yr of age. Note the run of monomorphic sharply contoured theta activity lasting 2 to 3 s over the left temporal channels.

correlation with seizures. SREDA is more typically seen in older adults, and more common at rest, drowsiness, or during hyperventilation.

1.5. Midline Theta Rhythm

The midline theta rhythm is most prominent at Cz but may spread to nearby contacts. This 5- to 7-Hz frequency exhibits either a smooth, arc-shaped (mu-like) or spiky appearance. The duration is variable and it tends to wax and wane. It is more common in wakeful and drowsy states and reacts variably to limb movements, alerting, and/or eye opening. This rhythm is now regarded to be a nonspecific variant, although it once was considered a marker of an underlying epileptic tendency.

1.6. Frontal Arousal Rhythm

FAR involves trains of 7- to 10-Hz activities in the frontal head regions. These rhythms may be notched and may last up to 20 s. FAR has been described as an uncommon rhythm that appears during sleep-to-wake transitions, especially in children. FAR disappears once the subject is fully awake. This pattern was, at one time, associated with children with minimal cerebral dysfunction, but this specific association has been subsequently doubted. This pattern is still considered to be a nonspecific finding without pathological significance.

2. EPILEPTIFORM VARIANT PATTERNS

There are 4 major types of epileptiform variant patterns:

1. 14- and 6-Hz positive bursts.
2. Small sharp spikes (benign epileptiform transients of sleep [BETS]).
3. 6-Hz spike and wave (phantom spike and wave).
4. Wicket spikes.

2.1. Fourteen- and 6-Hz Positive Bursts

Previously called "14- and 6-Hz positive spikes" or "ctenoids," these variants occur as bursts of rhythmic arched waves, similar to sleep spindles, with a smooth negative component and a spike-like positive component (Fig. 4). As the name implies, these trains occur at 14 and 6 Hz. The bursts last only 0.5 to 1 s. Such bursts are best captured on referential montages (because of the greater interelectrode distances). They are maximal at the posterior temporal head regions and usually occur independently from bilateral hemispheres with shifting predominance. This variant appears in 10 to 58% of healthy subjects, but is influenced by age, montage, and duration of drowsiness and sleep. This pattern is more prevalent in children and adolescents.

2.2. Small Sharp Spikes/BETS

As these names imply, small sharp spikes or BETS are low in amplitude (~50 μV) and brief (~50 ms). Their morphology can be monophasic or diphasic. When diphasic, the ascending limb is quite abrupt and the descending limb slightly less so. They may exhibit a subtle following slow wave. BETS are isolated and sporadic. They appear during drowsiness and light sleep in adults. They are usually unilateral but can appear independently (and rarely synchronously) from bilateral regions (Fig. 5). On a transverse montage, their field often illustrates a *transverse oblique dipole* (opposite polarities across the opposing hemispheres), an atypical finding in bona fide epileptiform discharges. Other distinguishing features

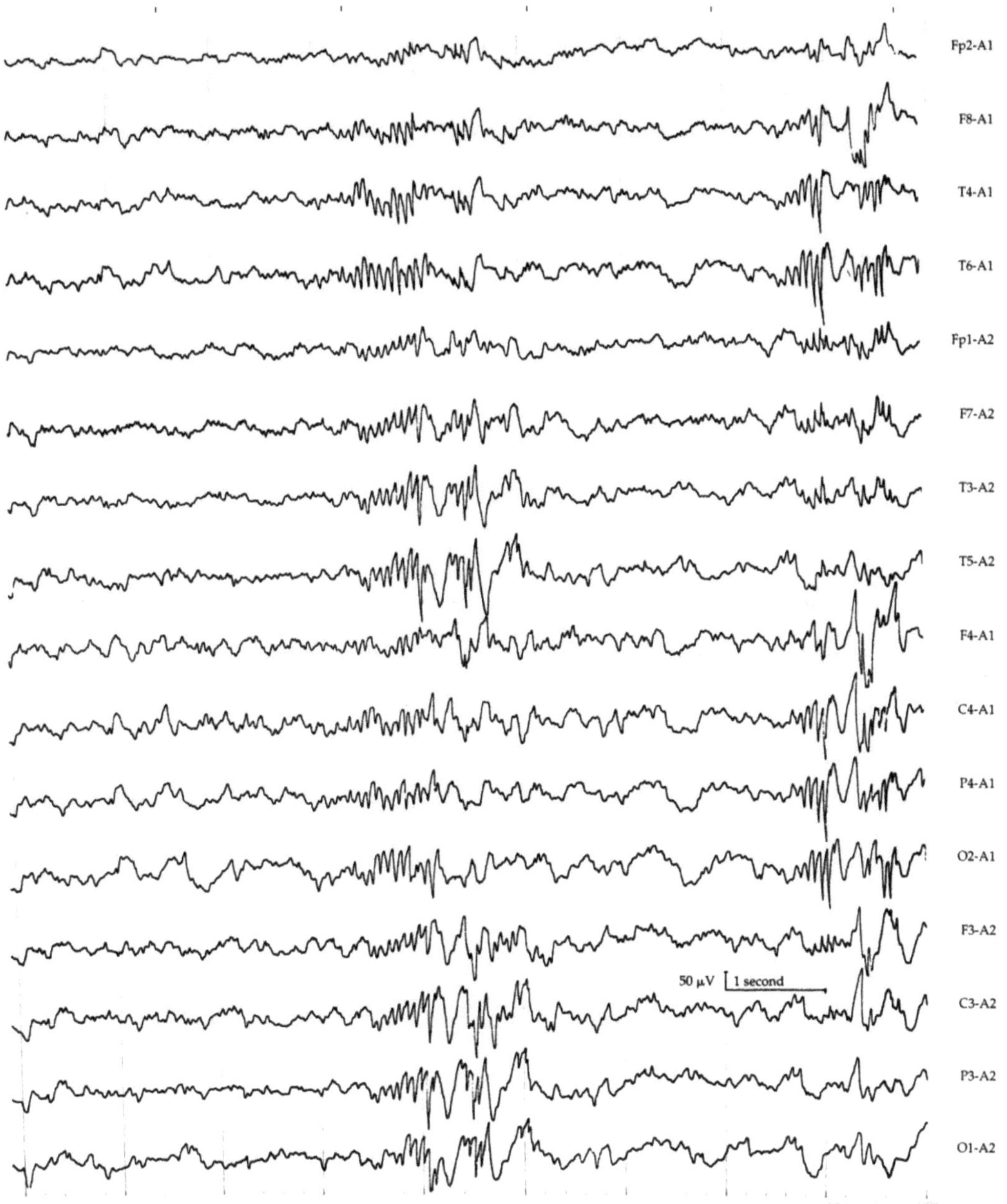

Fig. 4. Fourteen- and 6-Hz positive bursts. Note the bursts of spiky morphologies with surface positivity, here seen bilaterally with shifting laterality. They are best detected with long distance referential montages. Reprinted from Goldensohn et al., 1999 with permission.

between BETS and epileptiform activity are that BETS do not run in trains, distort the background, or coexist with rhythmic slowing, and BETS diminish with deepening sleep, whereas epileptiform discharges worsen with deeper sleep stages. White et al. reported the incidence of BETS to be comparable in healthy subjects (24%) as in symptomatic patients (20%). Thus, BETS seem unrelated to the diagnosis of epilepsy.

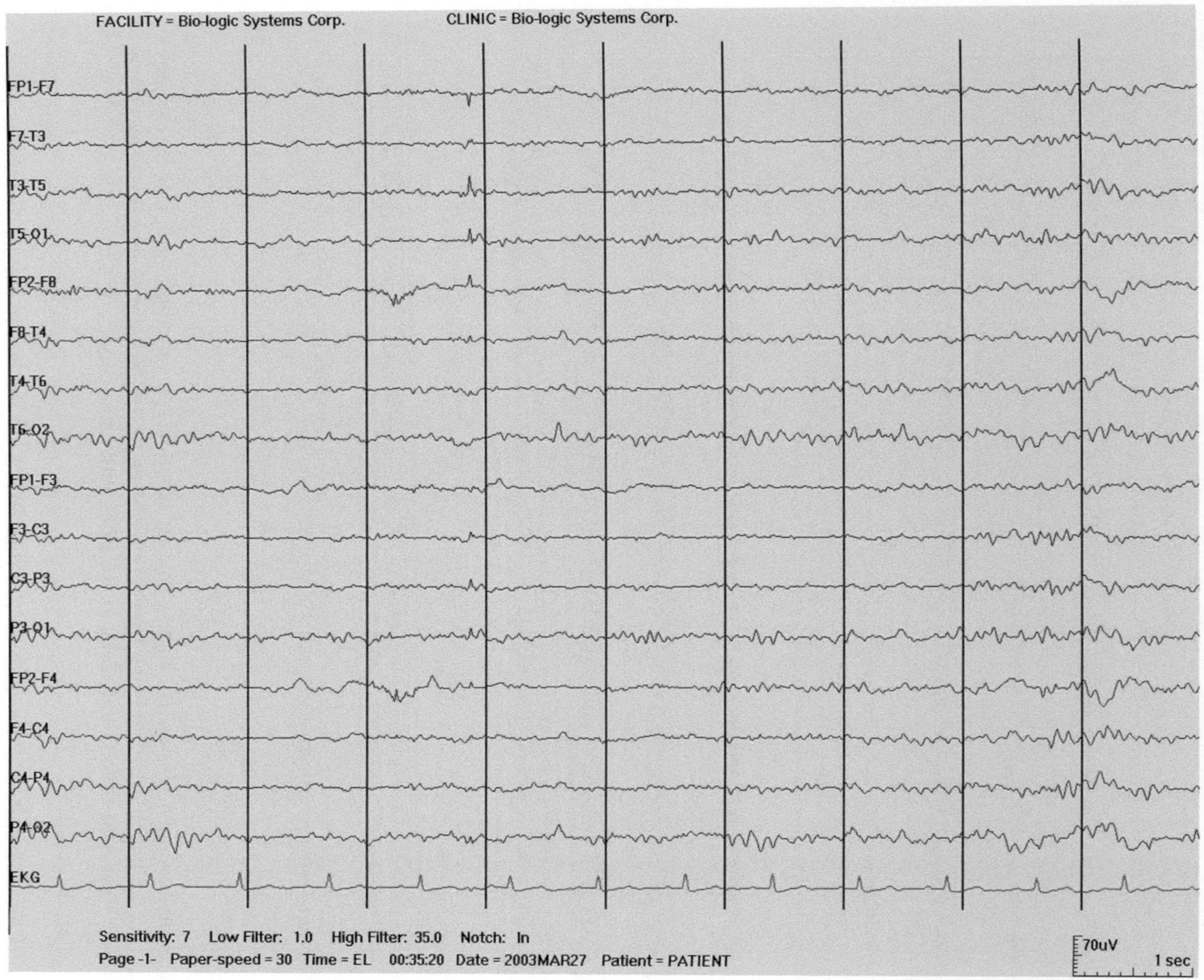

Fig. 5. Benign epileptiform transients of sleep (BETS). This tracing is from a 45-yr-old subject. Note the diminutive and very sharp diphasic transient seen broadly across the left temporal channels with spread to the left parasagittal chain of electrodes in the context of early sleep.

2.3. Six-Hertz Spike and Wave ("Phantom" Spike and Wave)

These rhythms have a frequency ranging from 5- to 7-Hz. They occur in brief bursts lasting 1 to 2 s, rarely up to 3 to 4 s. The spike component can be difficult to recognize because it is not only very brief but also of very low amplitude and, thus, has a fleeting quality. This subtle characteristic has given rise to the pithy term, "phantom" spike and wave. By contrast, its slow-wave component is broader in duration, higher in amplitude, and more widespread in distribution.

This pattern appears in waking or drowsiness in adolescents and adults. However, it is absent from slow-wave sleep. Silverman reported its overall incidence to be 2.5%. It usually has a diffuse, bilaterally synchronous distribution. At times, it is asymmetric or more regional.

Phantom spike and wave can appear quite similar to the 6-Hz positive spike burst. Infrequently, a transition may be seen on the same subject's EEG between these two types of variants. This pattern is thought by most to represent a benign finding. However, its morphology may be easily confused with an epileptiform pattern. Its failure to persist into slow wave sleep and its monomorphic quality permit its distinction from bona fide epileptiform discharges.

2.4. Wicket Spikes

This variant pattern appears as single spike-like waves or as intermittent trains of arc-like monophasic waves at 6- to 11-Hz (Fig. 6). Amplitudes range from 60 to 200 μV. Wicket spikes commonly appear from temporal channels and can be bilateral and synchronous or with shifting predominance. They occur mainly in drowsiness and light sleep in adults older than 30 yr of age.

When wicket spikes occur in isolation, they may be mistaken for an epileptiform discharge. Several features help differentiate isolated wicket spikes from pathological spikes. A similar morphology of the isolated wicket spike to those in a later train or cluster argues for the variant pattern and against an epileptiform discharge. The absence of a following slow-wave argues for the variant and against an epileptiform discharge. An unchanged background also argues more for the variant and against an epileptiform event.

3. LAMBDA AND LAMBDOIDS

3.1. Lambda

Lambda waves are sharp monophasic or biphasic waveforms that resemble the Greek letter lambda. They have a duration of 160 to 250 ms, an amplitude of 20 to 50 μV, and usually appear over bi-occipital leads, although occasionally may be unilateral (Fig. 7). Lambda waves depend on rapid saccadic eye movements with eyes open. This variant pattern is usually generated when the patient scans a complex patterned design in a well-illuminated room, for instance, dotted ceiling tiles in the laboratory. They block by staring at a featureless white surface or by closing the eyes. They are much more common in children 2 to 15 yr of age than in adults.

Slow lambda of youth, also known as shut-eye waves and posterior slow wave transients associated with eye movements, are associated with eye blinks in children. They appear over occipital channels as single, broad monophasic or diphasic, mainly surface negative waves. These morphologies last 200 to 400 ms and have an amplitude of 100 to 200 μV. They may be asymmetric. They are usually found in children younger than 10 yr of age.

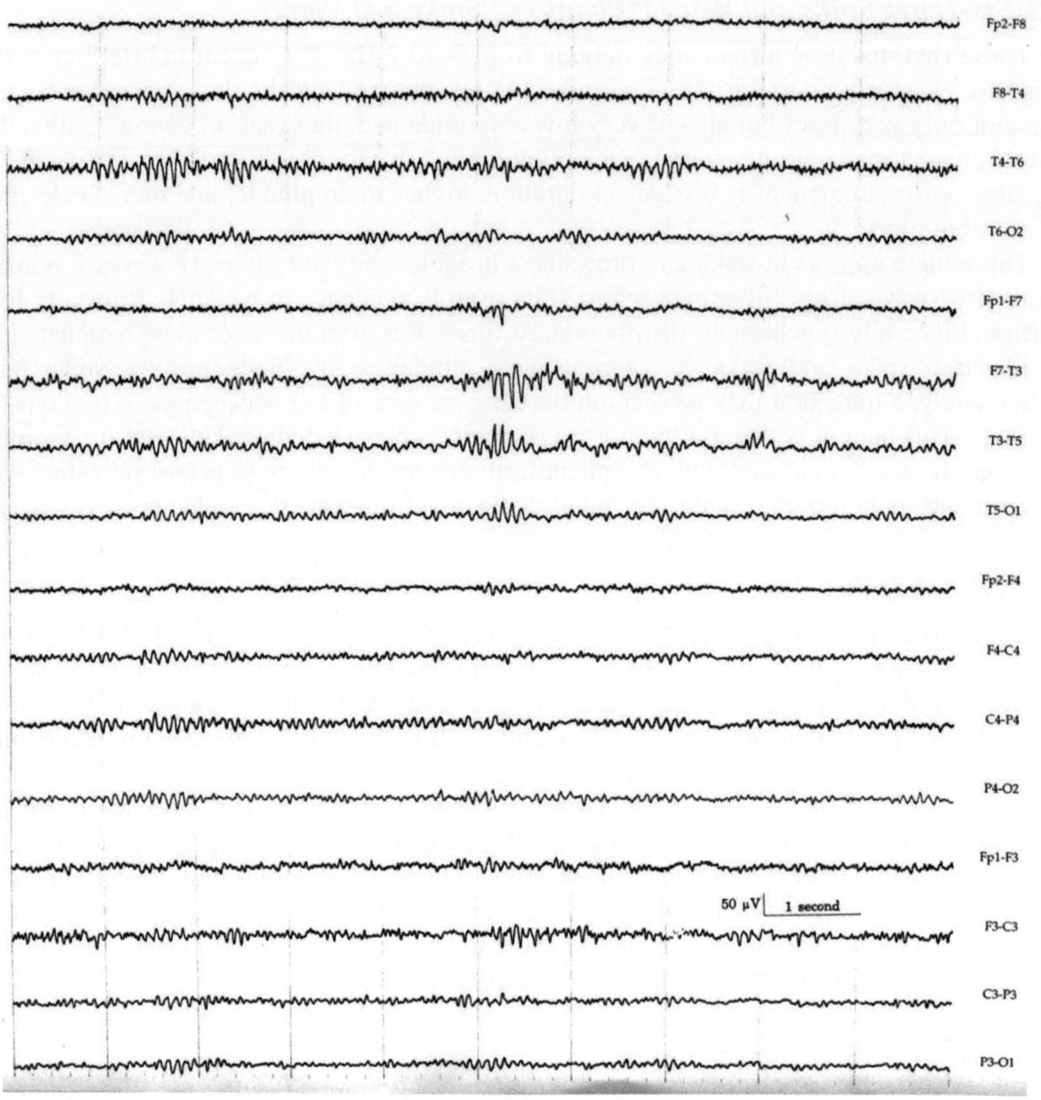

Fig. 6. Wicket spikes. Wicket spikes are often seen over temporal channels in drowsy or sleep recordings in adults. Reprinted from Goldensohn et al., 1999 with permission.

3.2. Lambdoids

Lambdoids, also called positive occipital sharp transients of sleep (POSTS) have a check-mark (biphasic) morphology with initial surface positivity and often appear in trains up to 4 to 5 Hz (Fig. 8). POSTS are usually synchronous but can be asymmetric in size. There are most commonly seen between 15 and 35 yr of age, and usually in light sleep. They may appear before the alpha rhythm completely evaporates in drowsiness. POSTS are a commonly encountered variant pattern on routine EEGs.

Slow lambdoids of youth, also known as cone-shaped waves or O-waves, are high voltage, diphasic slow transients seen over the occipital contacts and frequently with the occipital delta activity in deeper sleep states (Fig. 9). As the name implies, they are cone-shaped. They can be seen up to 5 yr of age.

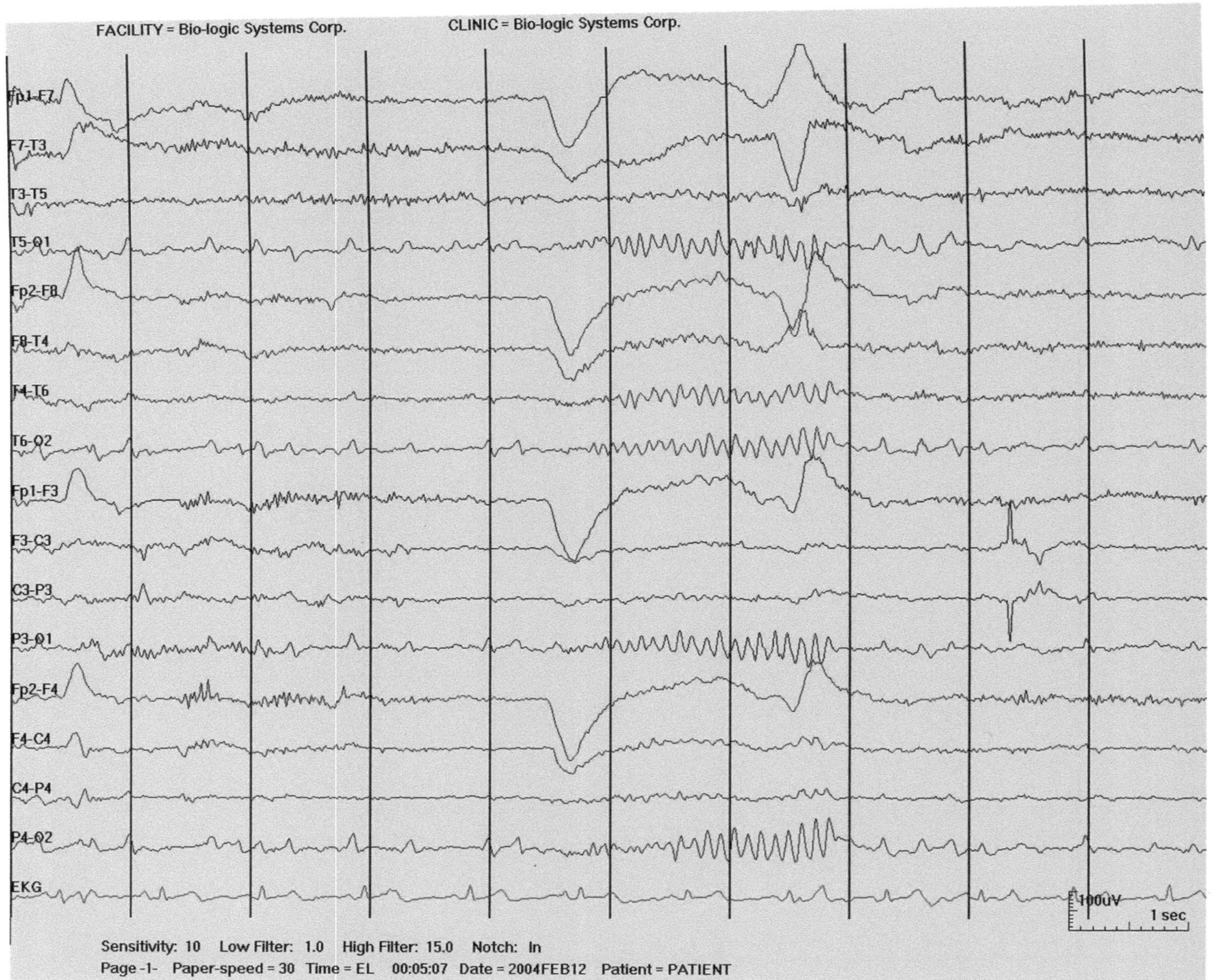

Fig. 7. Lambda. This is from a 42-yr-old subject. Note the bilateral triangular-shaped morphologies over the occipital derivations evident with eyes open in waking.

A

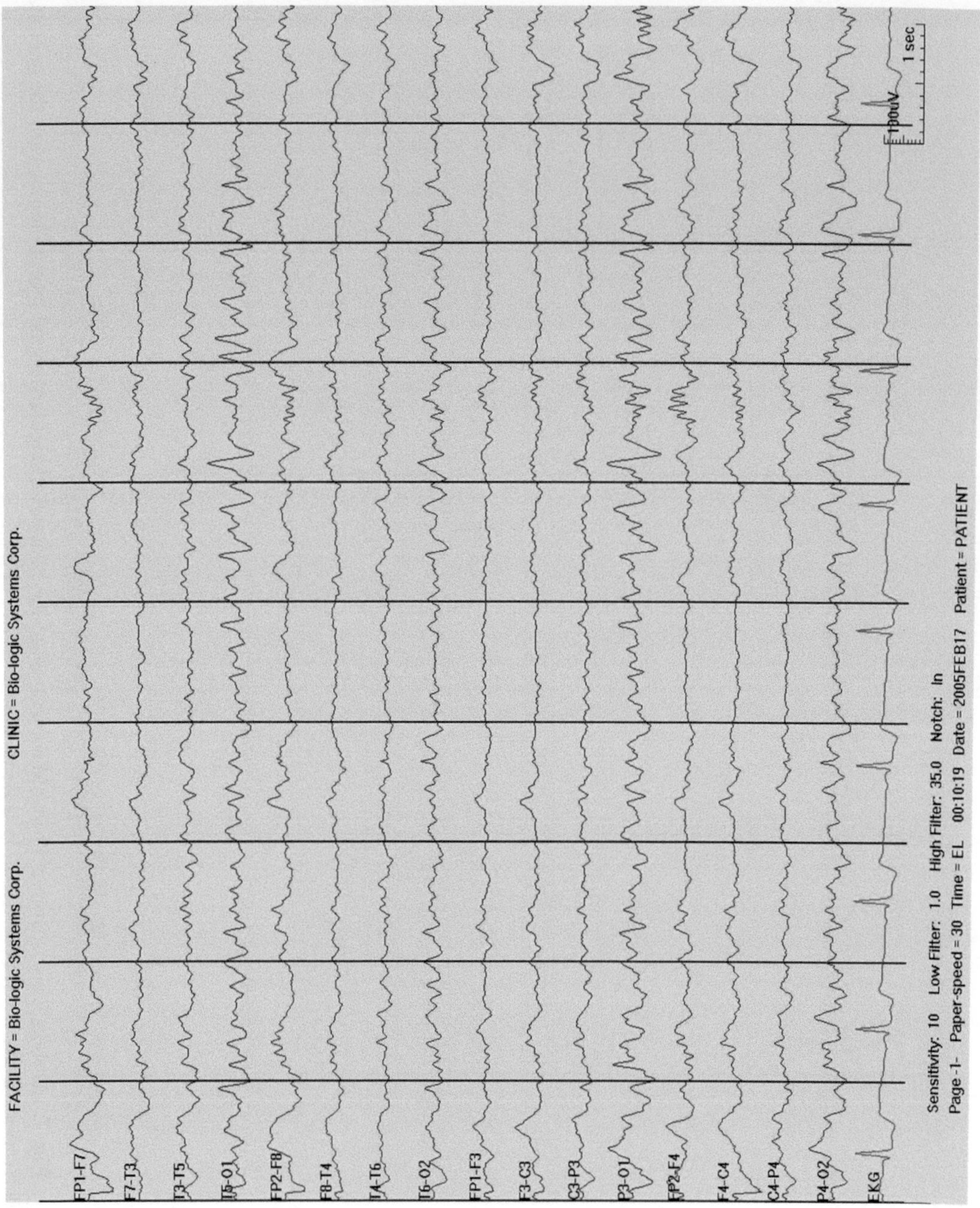

FACILITY = Bio-logic Systems Corp.
CLINIC = Bio-logic Systems Corp.
FP1-F7
F7-T3
T3-T5
T5-O1
FP2-F8
F8-T4
T4-T6
T6-O2
FP1-F3
F3-C3
C3-P3
P3-O1
FP2-F4
F4-C4
C4-P4
P4-O2
EKG
100uV
1 sec
Sensitivity: 10 Low Filter: 1.0 High Filter: 35.0 Notch: In
Page -1- Paper-speed = 30 Time = EL 00:10:19 Date = 2005FEB17 Patient = PATIENT

B

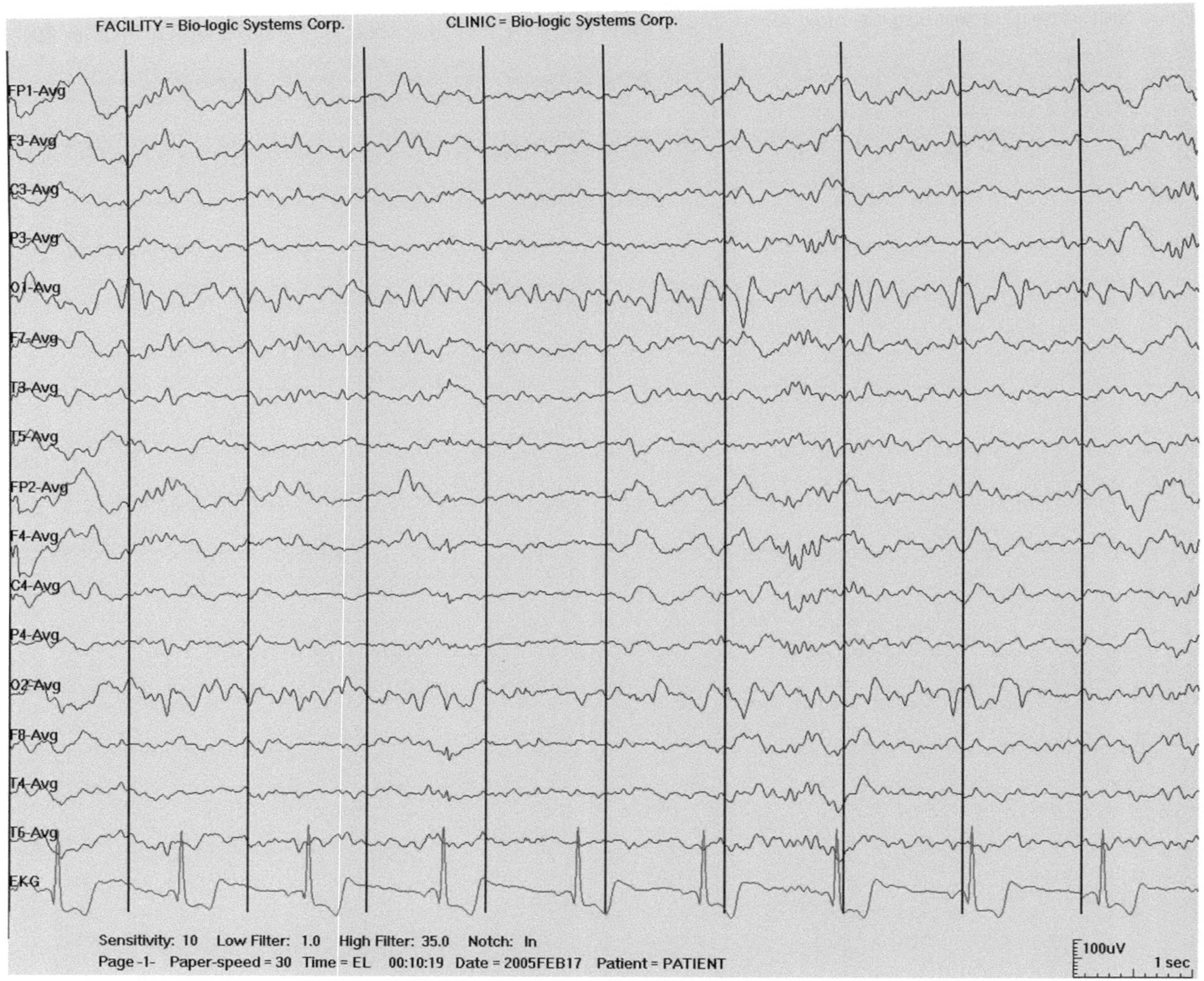

Fig. 8. Positive occipital sharp transients of sleep (POSTS). These derive from a drowsy 26-yr-old subject. (**A**) Illustrates POSTS occurring in a biposterior train at 2 to 4 Hz. In this anterior–posterior bipolar montage, POSTS have a check-mark appearance. The initial deflection is surface positive. This is better demonstrated in (**B**), which illustrates the same fragment

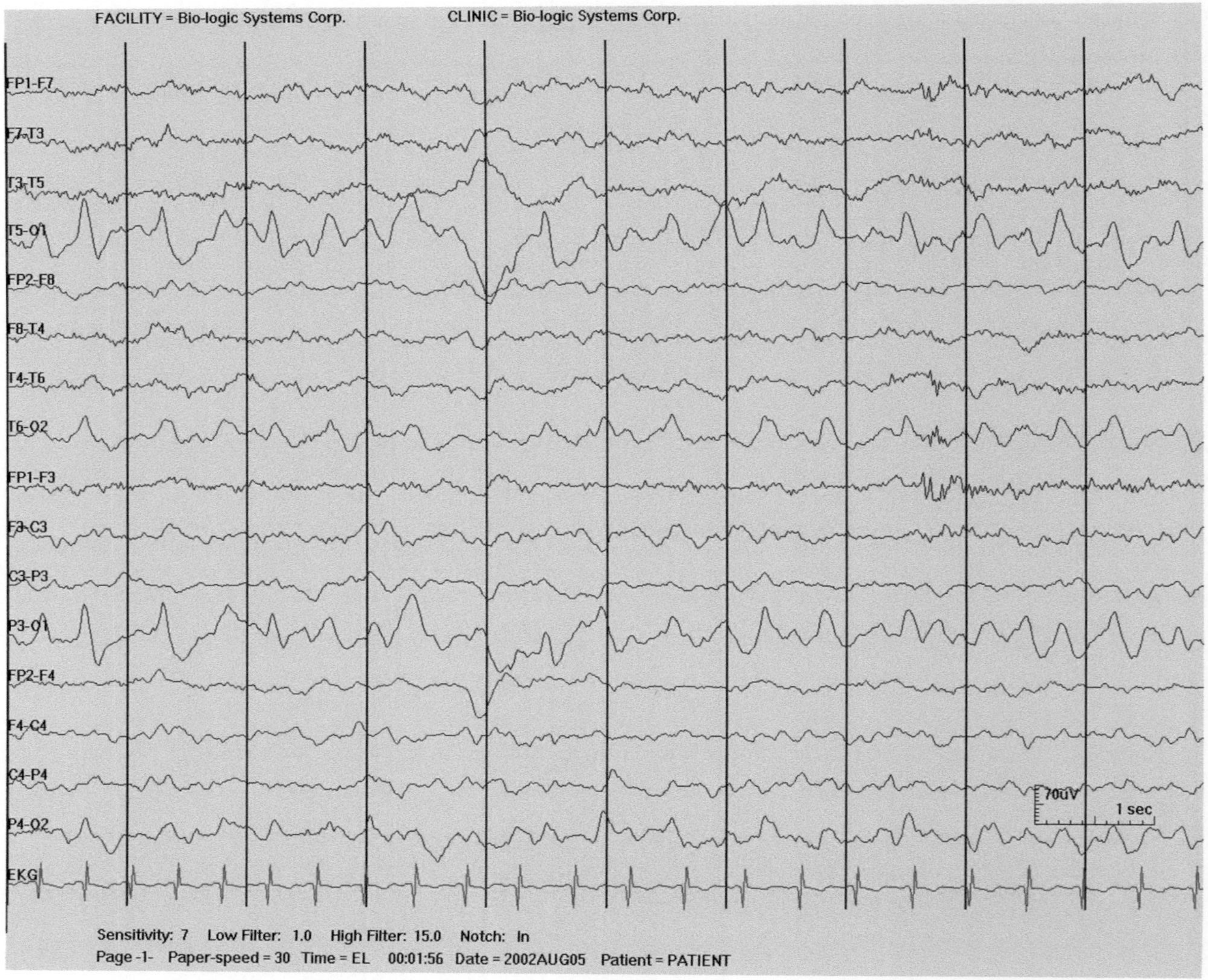

Fig. 9. Cone-shaped (O) waves. This is from a 3-mo-old child during sleep. Note the frequent high amplitude cone-shaped morphologies evident over the bi-occipital contacts in this tracing.

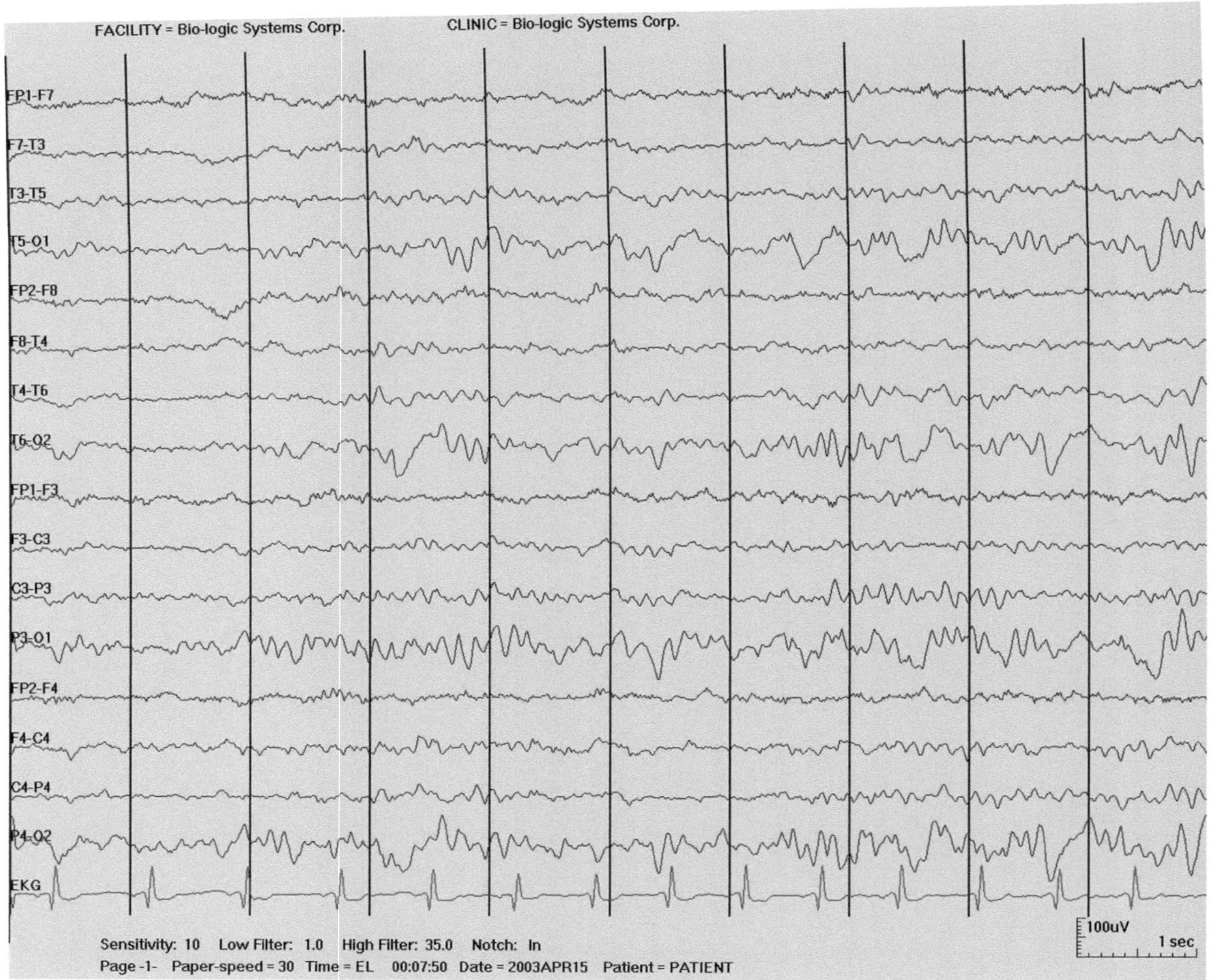

Fig. 10. Posterior slow waves of youth (youth waves). This is from the tracing of an 11-yr-old girl. Note the complex waveforms seen over the biposterior contacts. They illustrate the fused waveforms involving a delta waveform with a superimposed or "fused" alpha rhythm on top.

4. AGE-DEPENDENT VARIANTS

Posterior slow waves of youth, also called youth waves, posterior fused transients, and sail waves, are triangular, 2- to 4-Hz waveforms that coexist with other waking background rhythms (Fig. 10). They commonly form a complex that consists of the slow delta component with superimposed and following faster rhythms of the waking background rhythm, hence the term "fused." They are best seen over the posterior head regions in waking. These waves are occasionally asymmetric. These waveforms behave just like normal waking background rhythms in that they block with eye opening, increase with hyperventilation, and disappear in drowsiness. Youth waves usually appear between the ages of 8 to 20 yr, maximally in early adolescence.

Hyperventilation-induced slowing refers to a build-up of diffuse theta and delta activities in response to hyperventilation for several minutes during the EEG. This build-up may be abrupt or gradual and is often rhythmic and of high amplitude (up to 500 μV); it should resolve within 1 min after cessation of the patient's hyperventilation effort. This slowing is more prominent posteriorly in children and more anteriorly in young adults. This phenomenon is highly age dependent. It is extremely common in children and adolescents, becoming less common in young adults and thought by many to be uncommon after the age of 30 yr. As such, many laboratories regard this finding as a nonspecific abnormality in tracings of those older than 30 yr of age. Focal slowing or epileptiform activity evoked during hyperventilation is, however, pathological in nature.

Temporal theta in the elderly is a commonly encountered age-dependent pattern. This is a 4- to 5-Hz activity involving temporal channels and is thought by some to represent a subharmonic of the 8- to 10-Hz background rhythm common in the elderly. However, this rhythm is distinct from the alpha rhythm in that it persists with eye opening and even into drowsiness and light sleep. Hyperventilation augments this pattern's voltage and persistence. Temporal theta in the elderly usually occurs as very brief fragments lasting only two to three waves, rarely more than seven. Some have observed them to be more prevalent over the left hemisphere. Their significance remains controversial. Some regard these waveforms as normal variants but others have suggested that they may signify underlying vascular insufficiency, possibly subclinical in nature. Low-voltage intermittent temporal theta activity in asymptomatic elderly individuals has been estimated to occur with an incidence of approx 35%.

SUGGESTED READING

Arenas AM, Brenner RP, Reynolds CF III. Temporal slowing in the elderly revisited. Am J EEG Technol 1986;26:105–114.

Gibbs FA, Rich CL, Gibbs EL. Psychomotor variant type of seizure discharge. Neurology (Minneap) 1963;13:991–998.

Goldensohn ES, Legatt AD, Koszer S, Wolf SM. Goldensohn's EEG Interpretation: Problems of Overreading and Underreading, 2nd ed. Futura Publishing, Armonk, NY, 1999.

Goodwin, JE. The significance of alpha variants in the EEG, and their relationship to an epileptiform syndrome. Am J Psychiatry 1947;104:369–379.

Netchine S, Lairy GC. The EEG and psychology of the child. In: Handbook of Electroencephalography and Clinical Neurophysiology, Vol. 6: The Normal EEG Throughout Life. Part B: The Evolution of the EEG From Birth to Adulthood (Lairy GC, ed). Elsevier Scientific, Amsterdam, Holland, 1975, pp. 69–104

Niedermeyer E, Lopes da Silva FH, eds. Electroencephalography: Basic Principles, Clinical Applicatons and Related Fields, 3rd ed. Williams and Wilkins, Baltimore, MD, 1993.

Pedley TA. EEG patterns that mimic epileptiform discharges but have no association with seizures. In: Current Clinical Neurophysiology: Update on EEG and Evoked Potentials, (Henry CE, ed.). Elsevier/North Holland, New York, NY, 1980, pp. 307–336.

Silverman D. Phantom spike-waves and the fourteen and six per second positive spike pattern: a consideration of their relationship. Electroencephalogr Clin Neurophysiol 1967;22: 207–213.

Westmoreland BF. Normal and benign EEG patterns. Am J EEG Technol 1982;22:3–31.

Westmoreland BF. Chapter 8: Benign EEG variants and patterns of uncertain clinical significance. In: Current Principles and Practice of Clinical Electroencephalography, 2nd ed., (Daly DD, Pedley TA, eds.). Lippincott-Raven, Philadelphia, PA and New York, NY, 1997, pp. 243–251.

White JC, Langston JW, Pedley TA. Benign epileptiform transients of sleep: clarification of the small sharp spike controversy. Neurology 1977;27:1061–1068.

REVIEW QUESTIONS

1. Alpha variant may:
   A. Be a subharmonic (slow) frequency of normal alpha.
   B. Be a harmonic (fast) frequency of normal alpha.
   C. Block with eye opening.
   D. Be sporadic, intermingling with normal alpha or may predominate.
   E. All the above.
2. Mu has features that may include:
   A. Predominantly found at the C3 and/or C4 contacts.
   B. An arciform appearance.
   C. Blocking with contralateral upper extremity use or planned use.
   D. More common in adults vs children.
   E. All of the above.
3. Which one of the following statements concerning the psychomotor variant pattern is false?
   A. It is also known as rhythmic temporal theta bursts of drowsiness.
   B. It is typically seen in patients with temporal lobe ("Psychomotor") epilepsy.
   C. It usually exhibits a 5- to 7-Hz frequency.
   D. It usually occurs in restful wakefulness or in drowsiness.
   E. None of the above is false.
4. The 14- and 6-Hz positive bursts (14- and 6-Hz positive spikes or ctenoids) are best described as:
   A. More common in children and adolescents, declining in prevalence with advancing age.
   B. Maximal over the posterior temporal regions, especially on referential montage.
   C. Similar in morphology to sleep spindles with a sharp, spike-like positive component and a smooth, rounded negative phase.
   D. Exhibiting a brief duration of 0.5 to 1 s.
   E. All of the above.
5. BETS (small sharp spikes) typically:
   A. Have an amplitude of 50 μV and a duration of 50 ms.
   B. May be monophasic or diphasic with an abrupt ascent and a steep descent.
   C. May include a diminutive following slow wave.
   D. Appear in isolation, not in trains.
   E. All of the above.
6. Which of the following statements regarding phantom spike and wave are true?
   A. It persists into slow-wave sleep.
   B. This pattern often presents in lengthy trains up to 40 to 80 s.
   C. It resembles the 6-Hz positive spike burst variant, and both may coexist in the same tracing.
   D. The pattern is fleeting or "phantom-like," in that it may be easily blocked by eye opening.
   E. All of the above.

7. Lambda waves can be:
   A. Seen only with the eyes open.
   B. Seen when the patient scans a complex pattern and block when looking at a featureless object.
   C. Seen in the posterior (occipital) head regions.
   D. More common in children rather than adults.
   E. All of the above.
8. POSTS are:
   A. Usually seen in light sleep and typically just before the alpha rhythm disappears.
   B. Seen over a wide age range, including children to the elderly.
   C. Demonstrate a check-mark morphology.
   D. Usually synchronous over the occipital regions, but may show asymmetric amplitudes.
   E. All of the above.
9. Posterior slow waves of youth (youth waves):
   A. Block with eye opening.
   B. Accompany the posterior alpha rhythm.
   C. Typically increase with hyperventilation and decrease in drowsiness.
   D. Are triangular-contoured, with a frequency of 2 to 4 Hz.
   E. All of the above.
10. Temporal theta in the elderly:
    A. Is usually at 4 to 5 Hz .
    B. Typically occurs in brief runs of two to three waves, rarely greater than six to seven waves.
    C. Persists despite eye opening and despite drowsiness or light sleep.
    D. Is more common over the left hemisphere and may represent asymptomatic or subclinical cerebrovascular insufficiency in the elderly.
    E. All of the above.

REVIEW ANSWERS

1. The correct answer is E.
2. The correct answer is E.
3. The correct answer is B.
4. The correct answer is E.
5. The correct answer is E.
6. The correct answer is C.
7. The correct answer is E.
8. The correct answer is E.
9. The correct answer is E.
10. The correct answer is E.

8
Epileptiform Abnormalities

Bernard S. Chang and Frank W. Drislane

Summary

Despite the advances in neuroimaging technology in recent decades, the EEG is still the cornerstone of diagnostic testing for patients with epilepsy, and it remains the best real-time assessment of cerebral physiological function available. For patients with suspected seizure disorders, the routine EEG is used by clinicians to identify the presence of interictal epileptiform discharges that serve as a marker for epilepsy, and more prolonged ambulatory or inpatient EEG monitoring is used to record complete seizures. This chapter reviews interictal epileptiform abnormalities, ictal patterns observed in association with seizures, and periodic epileptiform discharge patterns.

Key Words: Epileptiform; ictal; interictal; periodic; polyspike; sharp wave; spike.

1. INTRODUCTION

Despite the advances in neuroimaging technology in recent decades, the EEG is still the cornerstone of diagnostic testing for patients with epilepsy, and it remains the best real-time assessment of cerebral physiological function available. For patients with suspected seizure disorders, the routine EEG is used by clinicians to identify the presence of interictal epileptiform discharges that serve as a marker for epilepsy, and more-prolonged ambulatory or inpatient EEG monitoring is used to record complete seizures. This chapter reviews interictal epileptiform abnormalities, ictal patterns observed in association with seizures, and periodic epileptiform discharge patterns.

2. INTERICTAL EPILEPTIFORM ABNORMALITIES

2.1. Spikes and Sharp Waves

2.1.1. Description

Spikes and sharp waves are sharply contoured waveforms that are distinct from the EEG background and usually have a negative polarity (Fig. 1). They can be of any voltage and can occur either singly or in repetitive runs with varying frequencies. They can be focal or generalized in distribution. Spikes, by definition, have a duration shorter than 70 ms, whereas sharp waves have a duration between 70 ms and 200 ms. These discharges classically have an asymmetric appearance, with the initial deflection characterized by a sharper slope than the return to baseline. They may be observed as isolated waveforms or they can be followed by a slow wave. Although similar deflections with a positive polarity are sometimes referred to as "positive spikes," nearly all epileptiform discharges of clinical interest are of negative polarity.

From: *The Clinical Neurophysiology Primer*
Edited by: A. S. Blum and S. B. Rutkove © Humana Press Inc., Totowa, NJ

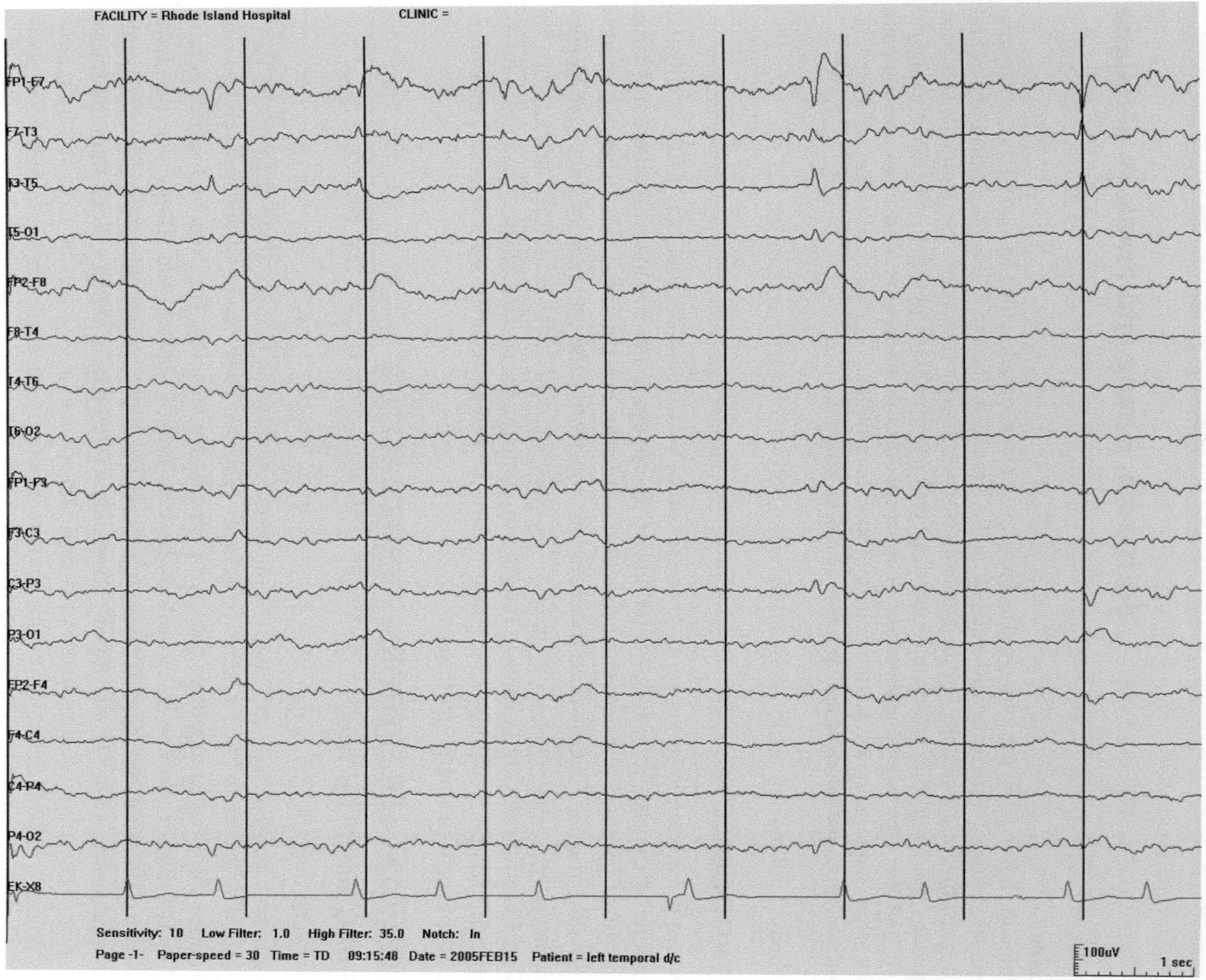

Fig. 1. Sharp waves. Note the run of sharp wave discharges evident over the anterior left temporal channels, with phase reversal at F7.

2.1.2. Physiological Basis and Significance

A spike is thought to be generated by the synchronous depolarization of thousands of cortical neurons located within an area of at least 6 cm^2. Sharp waves are thought by some to result from the synchronous activation of either a smaller pool of neurons or a pool of neurons further from the recording electrode, such as below the cortical surface. However, most electroencephalographers consider spikes and sharp waves to have the same physiological significance: they are interictal epileptiform discharges that serve as markers for epileptogenesis, either focal or generalized, and are suggestive of an underlying propensity toward seizures. They are observed only rarely in healthy individuals.

2.1.3. The Spike-and-Slow-Wave Complex

This term refers to the occurrence of a spike followed immediately by a slow wave, which can be of varying frequency and amplitude, and is usually distinct from the underlying background (Fig. 2). Sharp-and-slow wave complexes include a sharp wave as the initial waveform, rather than a spike. The following slow wave in these discharges may represent inhibition and subsequent hyperpolarization of cortical neurons that follows the initial synchronous depolarization. Spike-and-slow wave complexes are strongly suggestive of an underlying epileptic disorder.

2.1.4. Polyspikes

These are discharges characterized by multiple spikes observed in rapid succession, typically at frequencies of 10 Hz or faster (Fig. 3). They may be followed by a slow wave. Polyspikes may be observed in many generalized seizure disorders, particularly those in which myoclonus is a feature, and at times may occur in temporal association with a clinically noted myoclonic jerk.

2.1.5. Differential Diagnosis

Because of the clinical and physiological significance of spikes and sharp waves, it is important to recognize the features that distinguish them from other similar waveforms. Many of these are also discussed in Chapter 7, on normal EEG variants.

2.1.6. Vertex Waves

Vertex waves are broad, sharply contoured waveforms of negative polarity that are typically central in location and observed mostly during Stage II sleep (Fig. 4). Individual vertex waves can appear more prominently over one hemisphere, as long as they are not persistently unilateral. Because sleep can be associated with an exacerbation of epileptiform activity, distinguishing central spikes from vertex waves may be difficult in certain cases. However, a vertex wave usually has a symmetric morphology, whereas the initial deflection of a spike is usually steeper than the return to baseline.

2.1.7. Wicket Spikes

Wicket spikes are sharp negative-polarity waveforms that can be of high amplitude and are typically observed in the temporal regions of older adults (Fig. 5). Most commonly, they arise out of a high-amplitude background that is already somewhat sharp in morphology. They are of uncertain clinical significance but do not seem to suggest an underlying epileptic disorder.

2.1.8. Small Sharp Spikes, or Benign Epileptiform Transients of Sleep

Small sharp spikes, or benign epileptiform transients of sleep (BETS), are low-amplitude, short-duration waveforms that are symmetrically biphasic. As their name suggests, they are

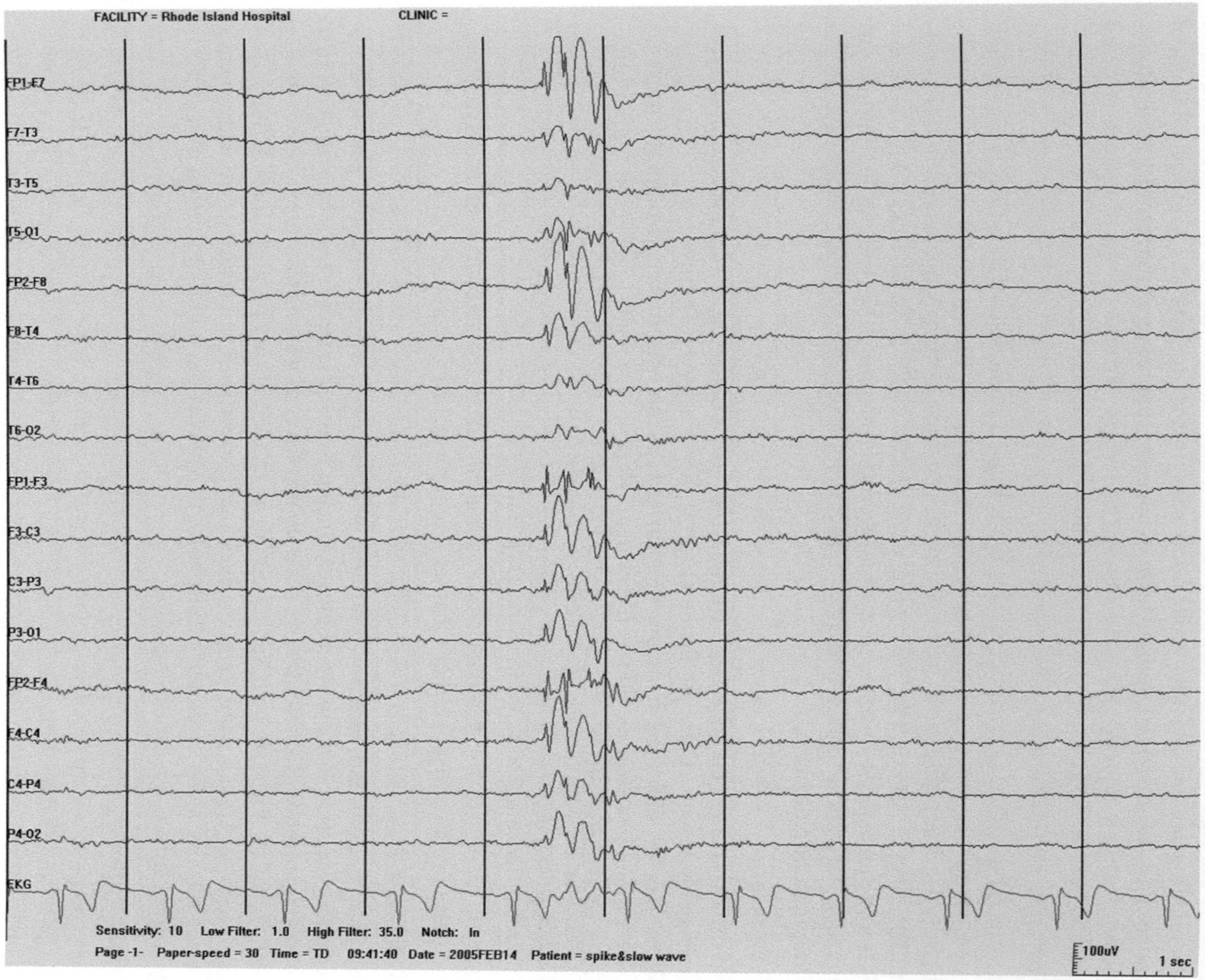

Fig. 2. Spike and slow wave. Note the characteristic spike followed by a slow wave, in this instance distributed bi-anteriorly at close to 4 Hz. Spikes may sometimes seem to be "buried" by the slow wave component, giving rise to a notched appearance, as seen in portions of this burst.

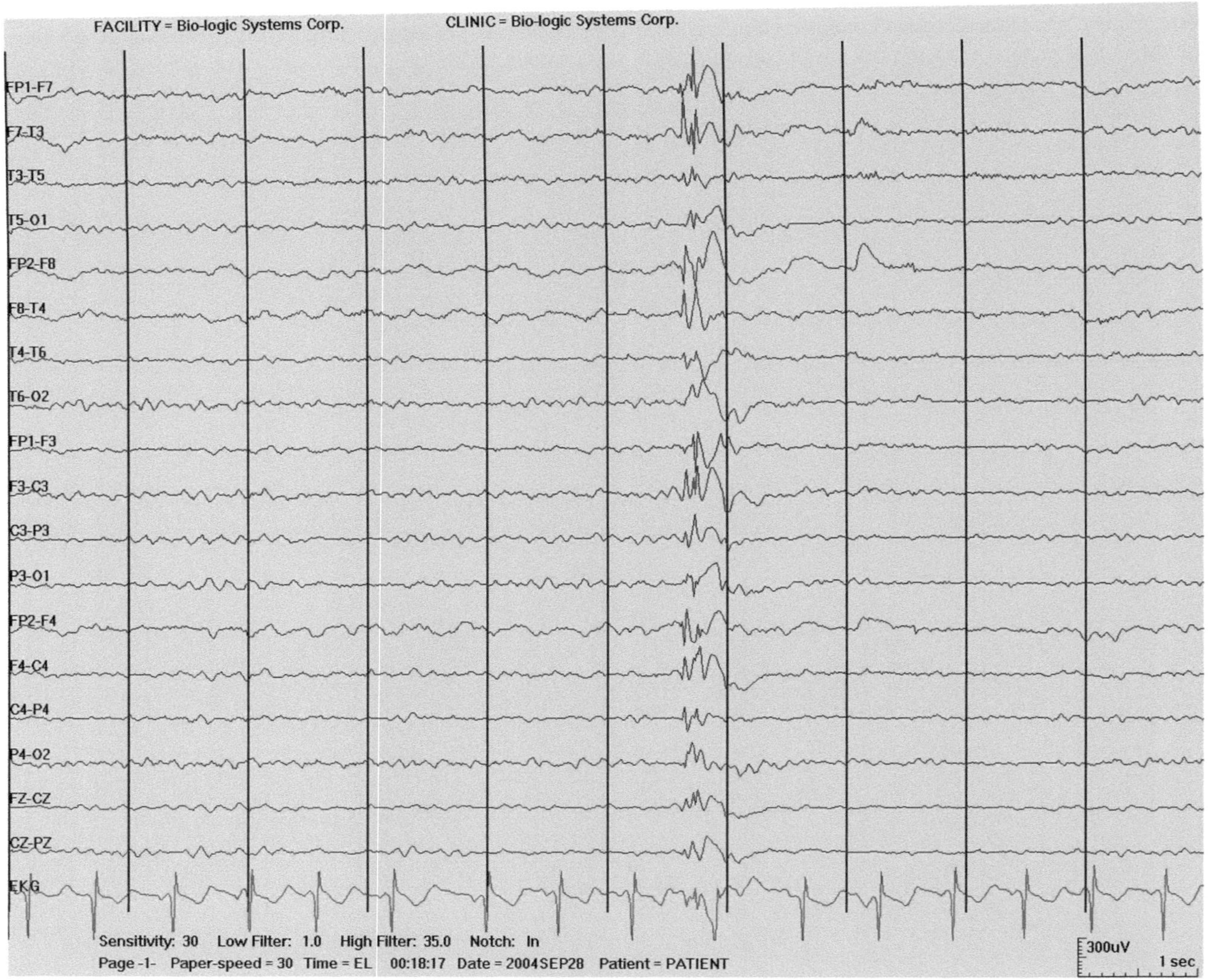

Fig. 3. Polyspike and slow wave. A bi-anterior burst of polyspikes appear, followed by a prominent slow wave. These are more commonly observed in generalized epilepsies, but are not unique to such.

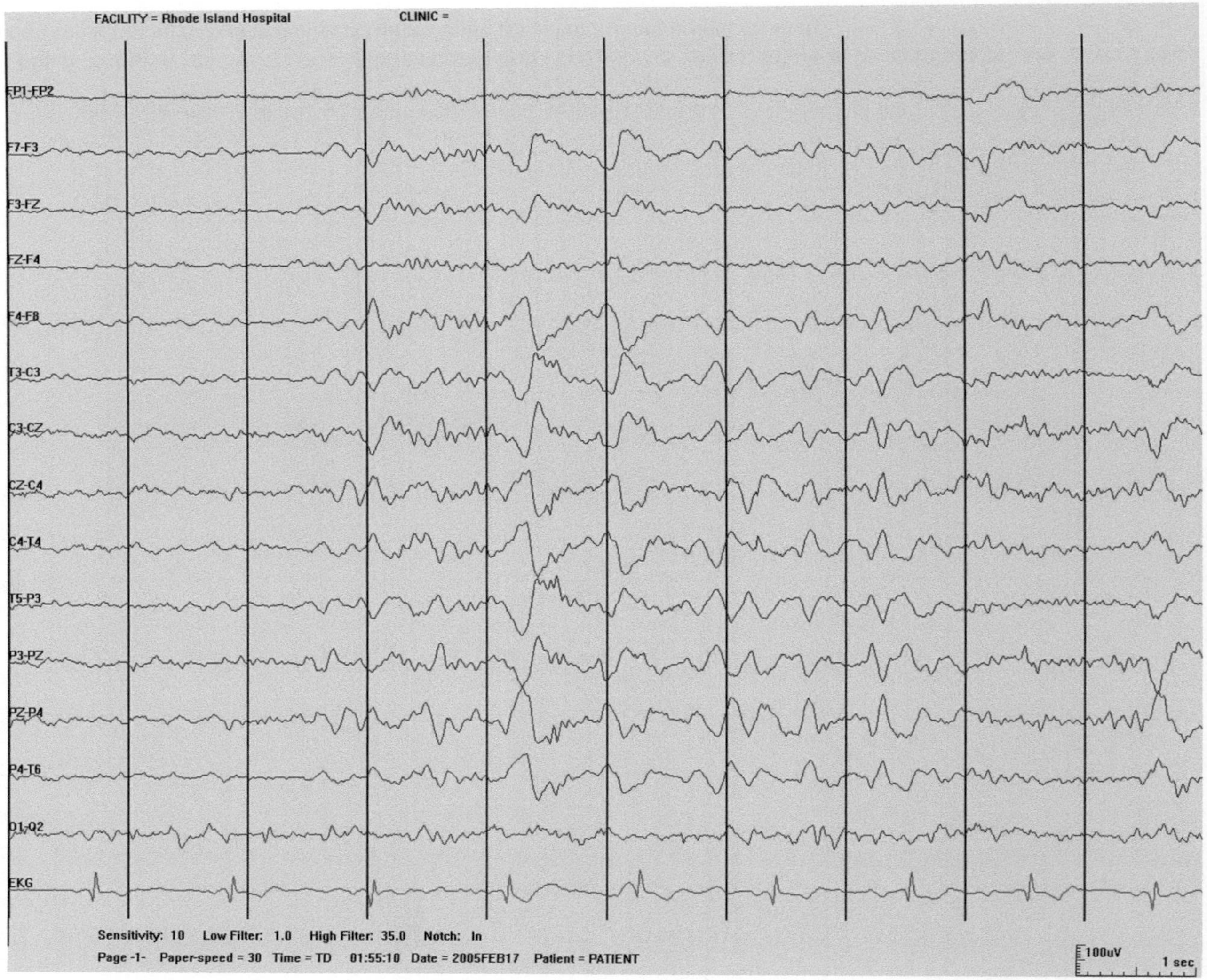

Fig. 4. Waveforms with epileptiform appearance: vertex waves. Vertex waves are typically central in location and are observed mostly during Stage II sleep. Here, vertex waves appear in a clustered fashion, with phase reversal about the vertex channels, as illustrated in this transverse montage.

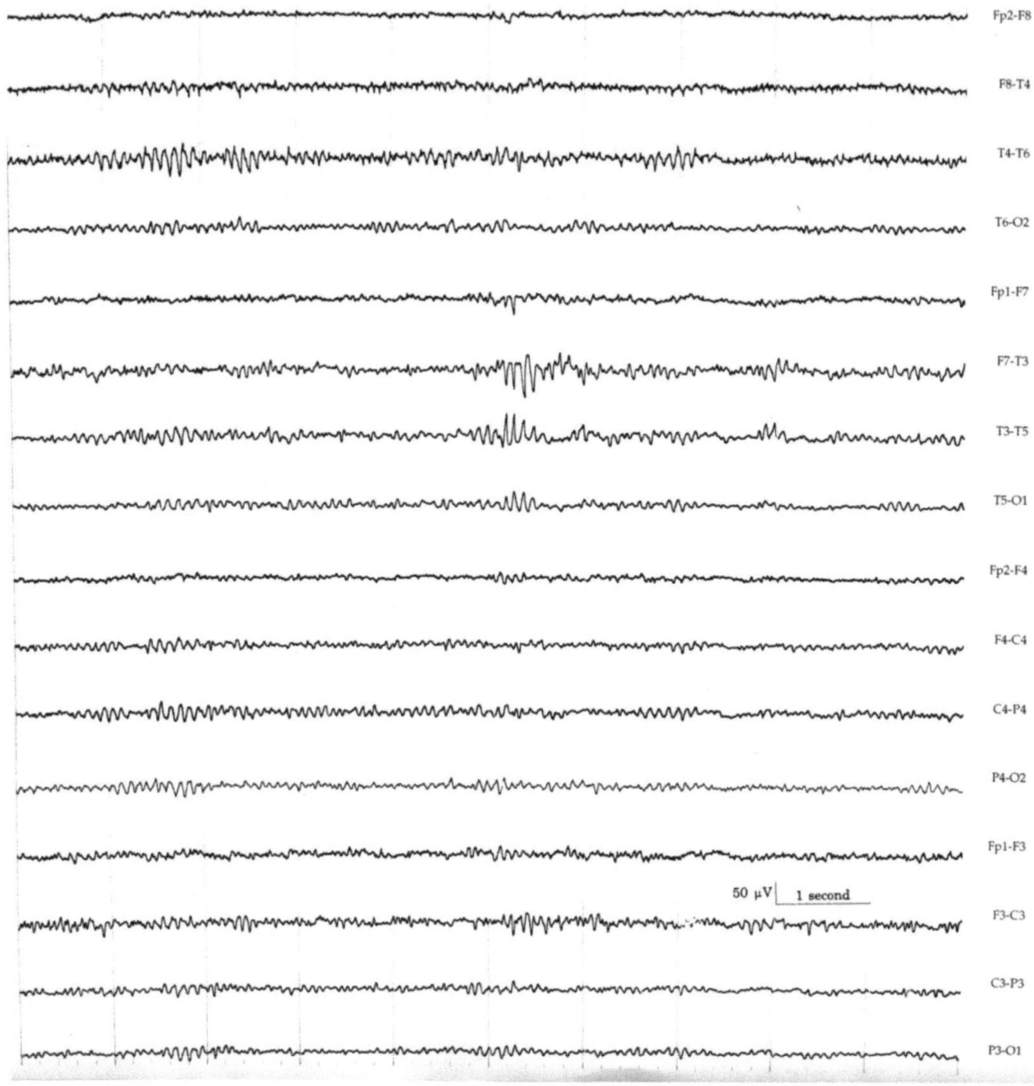

Fig. 5. Waveforms with epileptiform appearance: wicket spikes. Wicket spikes are of uncertain clinical significance but are typically observed in temporal channels in older adults. From Goldensohn et al., 1999 with permission.

observed during sleep, and are most common over the temporal regions, although they can be more widespread, unilateral or bilateral, and can have a horizontal dipole distribution. They are usually a normal EEG variant of no known clinical significance.

2.1.9. Rhythmic Midtemporal Theta Bursts, or Psychomotor Variant

Rhythmic midtemporal theta bursts, or psychomotor variant, are sharp in morphology and occur classically at a 6-Hz frequency. A characteristic feature is the notched appearance of the waveforms (Fig. 6). They most commonly occur in the temporal regions of young adults during sleep. They are a normal EEG variant of no clinical significance.

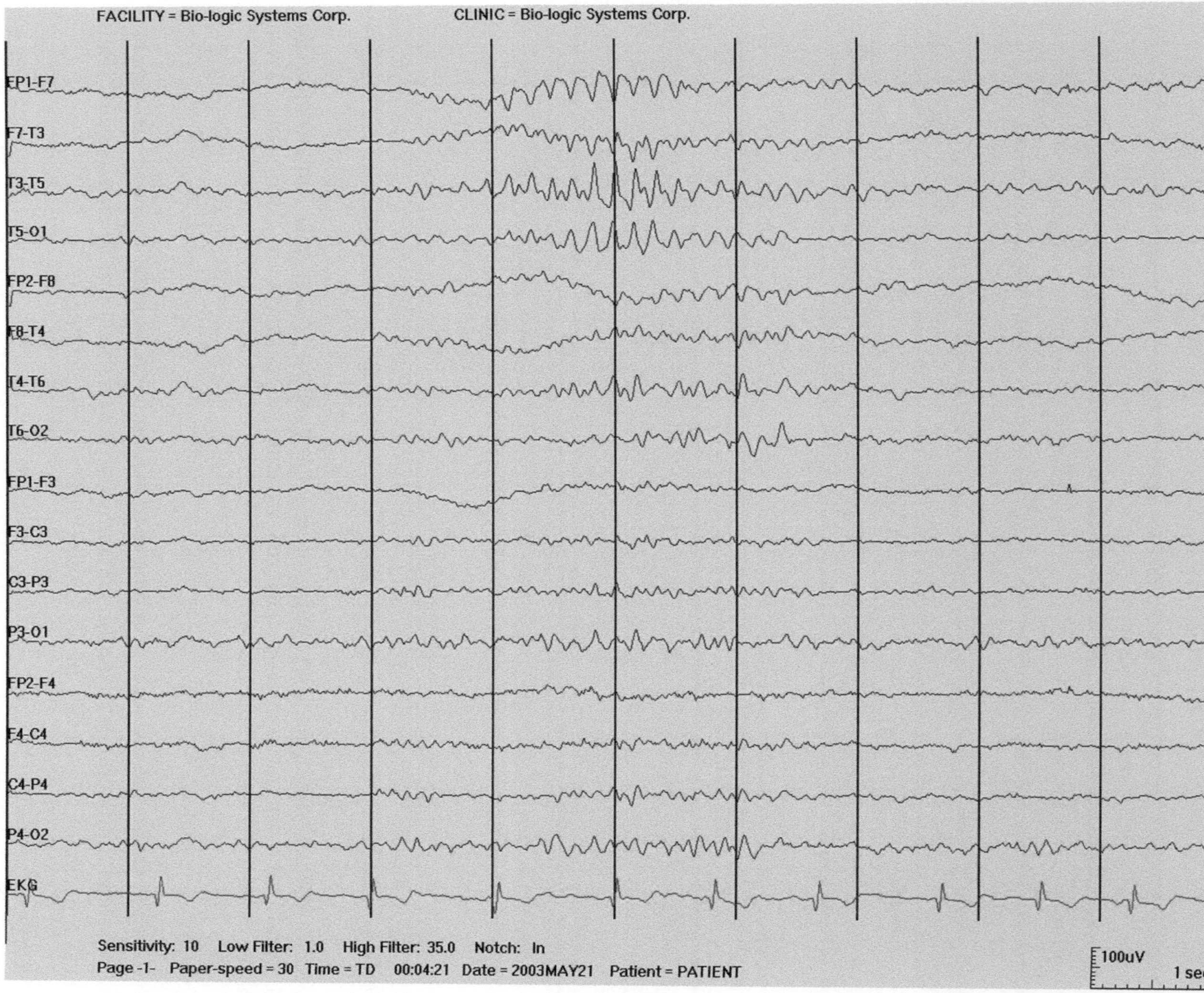

Fig. 6. Waveforms with epileptiform appearance: rhythmic midtemporal theta bursts. This pattern is also known as the "psychomotor variant." Note the run of unchanging 5-Hz activity over the left temporal channels lasting 3 s in this tracing. The patient is drowsy, as is commonly the case for this pattern. This pattern is no longer considered pathological.

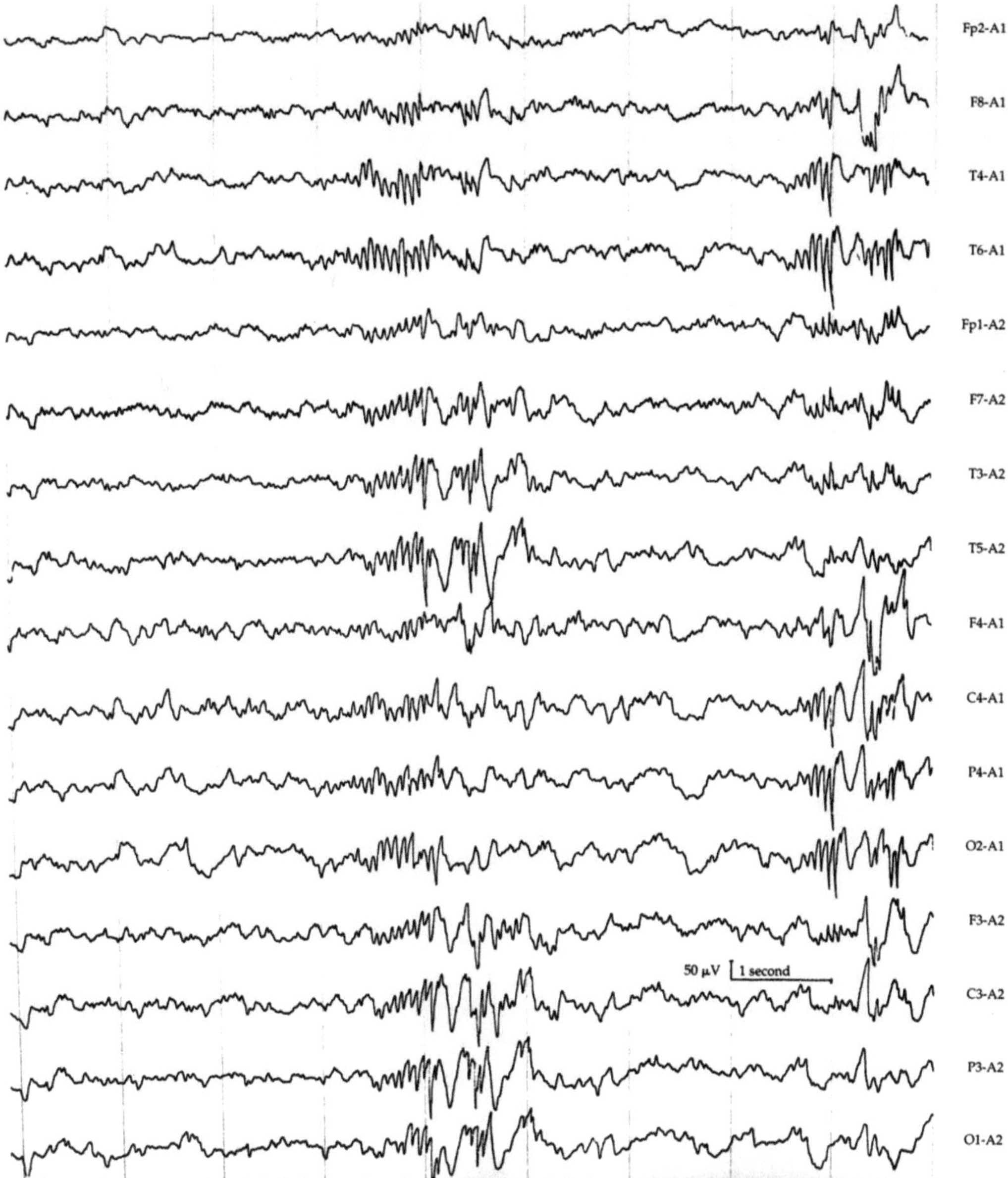

Fig. 7. Waveforms with epileptiform appearance: 14-and-6 Hz positive bursts. Fourteen- and 6-Hz positive bursts are seen with shifting laterality in this tracing. They are downward deflections, very sharply contoured, broadly distributed, and best seen with long-distance referential recording technique. They are a normal variant pattern. From Goldensohn et al., 1999 with permission.

2.1.10. Fourteen- and 6-Hz Positive Bursts

Fourteen- and 6-Hz positive bursts are distinguished from epileptiform spikes by their positive polarity and their characteristic mixture of 14-Hz and 6-Hz frequencies (Fig. 7). They usually occur in the recordings of adolescents and young adults during sleep, and are best observed using montages with long interelectrode distances or ear references. Although, in the past, electroencephalographers have associated 14- and 6-Hz positive bursts with a variety of neurological abnormalities, they are now thought to be a normal variant.

2.1.11. Occipital Spikes of Blindness

Occipital spikes of blindness can be observed in patients with acquired visual loss in childhood. They are epileptiform in appearance and often quite narrow but are not thought to suggest an underlying seizure disorder.

2.1.12. Six-Hertz Spike-and-Wave, or "Phantom" Spike-and-Wave

Six-Hertz spike-and-wave, or "phantom" spike-and-wave, is a discharge characterized by repetitive spike-and-slow wave complexes occurring at a 6-Hz frequency, with the spike being of very low amplitude compared with the following slow wave. It is typically observed in a generalized distribution but may be posteriorly predominant. Six-Hertz spike-and-wave is not thought to be associated with clinical seizures.

2.1.13. Sharply Contoured Artifact

Finally, sharply contoured artifact can also be confused for true spikes and sharp waves. ECG artifact is frequently sharp in morphology and is identified by a clear temporal association with the QRS complex on the cardiac rhythm strip. Muscle artifact can be quite sharp but is usually of an extremely fast frequency that distinguishes it from cerebrally generated activity. Electrode dysfunction or "pop" artifact can also be initially confused for epileptiform activity. However, this artifact characteristically lacks the appropriate field expected in a brain-derived discharge. It is also often much steeper at the outset than bona fide epileptiform discharges, with an exponential decay.

2.2. Epileptiform Discharges by Location

Spikes and sharp waves from several brain regions deserve particular mention because of their special significance or the frequency with which they are observed.

2.2.1. Temporal Spikes

The majority of partial seizures in adults stem from the temporal lobes, and, thus, temporal spikes and sharp waves are the most frequently observed focal interictal discharges (Fig. 1). There is a large literature on the significance of interictal temporal discharges in helping to determine the laterality of temporal lobe seizure onset, and synchronous bitemporal discharges or even unilateral discharges from the side contralateral to ictal onset can be observed, presumably because the two mesial temporal lobes are so intimately connected. Extra electrodes are frequently used to provide more coverage of temporal areas: "true anterior temporal" scalp electrodes are placed by some laboratories (referred to as T7 and T8 in American Clinical Neurophysiology Society nomenclature), whereas many use sphenoidal needle electrodes, which are inserted several centimeters deep to the surface below the zygomatic arch bilaterally and record from near the mesial temporal lobes.

2.2.2. Centrotemporal (or "Rolandic") Spikes

These are characteristic of the childhood syndrome of benign epilepsy with centrotemporal spikes (Rolandic epilepsy) and are observed over the centrotemporal region either unilaterally or bilaterally (Fig. 8). These are typically high-amplitude biphasic or, less often, polyphasic spikes or sharp waves, with a following slow wave. In patients with this syndrome, these discharges can occur quite frequently, particularly during sleep. Often the

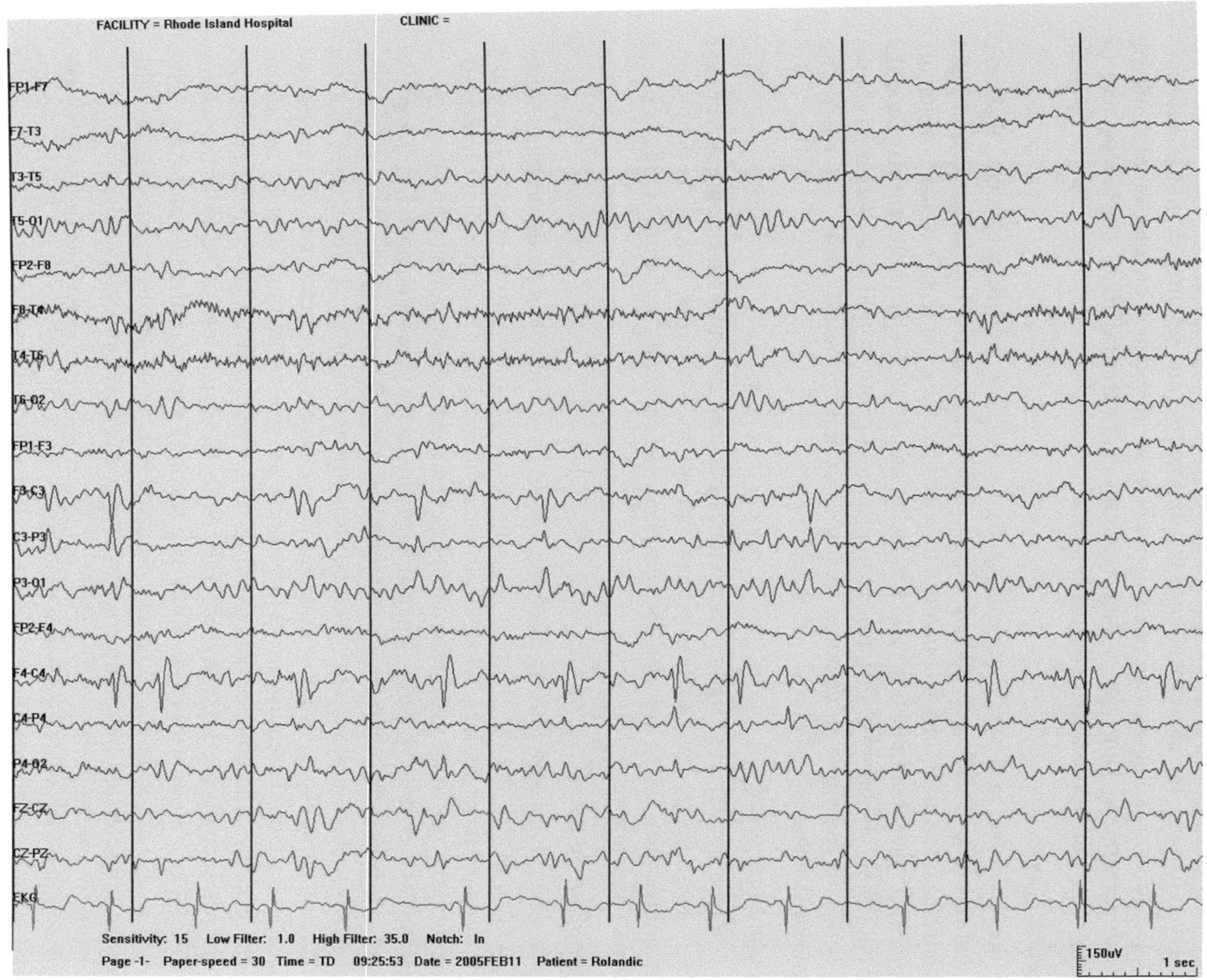

Fig. 8. Centrotemporal ("Rolandic") spikes. This bipolar tracing is from a 5-yr-old boy with a history of spells of convulsive activity involving the left arm. Note the frequent high-amplitude sharp and slow waves appearing independently from bicentral channels with some spread to involve the temporal channels as well, better seen on the right. These discharges are often sleep activated.

distribution suggests a horizontal dipole (i.e., a phase reversal in a referential recording or two simultaneous phase reversals of opposite polarity in a bipolar recording), although this is not invariable (Fig. 9). The clinical syndrome consists of simple partial seizures of the face, sometimes affecting speech or swallowing functions, and occasional generalized tonic–clonic (GTC) seizures. The syndrome is considered benign, in that underlying structural lesions are rarely observed (despite the clear focality of the discharges) and seizures do not usually persist beyond childhood. Rolandic epilepsy is thought to be inherited in an autosomal dominant fashion with incomplete penetrance, such that the characteristic EEG discharges can be observed in some children without clinical seizures.

2.2.3. Occipital Spikes

These are observed in a syndrome analogous to, but less well-defined than, Rolandic epilepsy, called benign childhood epilepsy with occipital paroxysms. The occipital discharges in this syndrome attenuate with eye opening. Importantly, these patients' seizures, which are characterized by visual hallucinations and headache, are not typically photosensitive, and the EEG does not show an abnormal response to intermittent photic stimulation. As noted above, occipital spikes in patients with acquired blindness do not necessarily imply an underlying tendency toward seizures.

2.2.4. Frontal Sharp Waves

These can be a normal EEG finding in neonates between 28 and 42 wk conceptional age and, in this age group, do not necessarily imply an epileptic disorder, as long as there is no persistent unilaterality. They are also termed *encoches frontales*, because of their classic "check-mark" appearance.

2.2.5. Generalized Spikes

In adults, generalized epileptiform discharges often have a bifrontal predominance, sometimes to such an extent that the sharp or spike morphology is only truly observed frontally and not in more posterior regions. Most electroencephalographers consider these bifrontally predominant discharges to be manifestations of generalized epileptogenesis, however.

Generalized spikes are typically suggestive of an underlying generalized epilepsy. As noted above, a bifrontal predominance is common. In most idiopathic generalized epilepsies, spike-and-slow wave complexes are observed, sometimes repeating at frequencies that are characteristic of particular epilepsy syndromes. For example, interictal runs of generalized spike-and-slow wave discharges recurring three times per second are most strongly associated with childhood or juvenile absence epilepsy (Fig. 10), whereas faster generalized discharges (four to six per second) are typically observed with the syndrome of juvenile myoclonic epilepsy (Fig. 11). "Slow" (less than three per second) generalized spike-and-slow wave discharges (Fig. 12) are an interictal pattern observed in Lennox–Gastaut syndrome, a childhood disorder characterized by multiple seizure types and developmental delay.

Occasionally in patients with partial epilepsy, a physiologically focal epileptiform discharge can spread so rapidly to involve the entire brain that, on visual analysis of the EEG, it seems generalized in onset, a phenomenon called secondary bilateral synchrony.

2.3. Activation Procedures and Precipitants

The two commonly used activation procedures during a routine EEG, hyperventilation and intermittent photic stimulation, can increase the chance of detecting interictal epileptiform

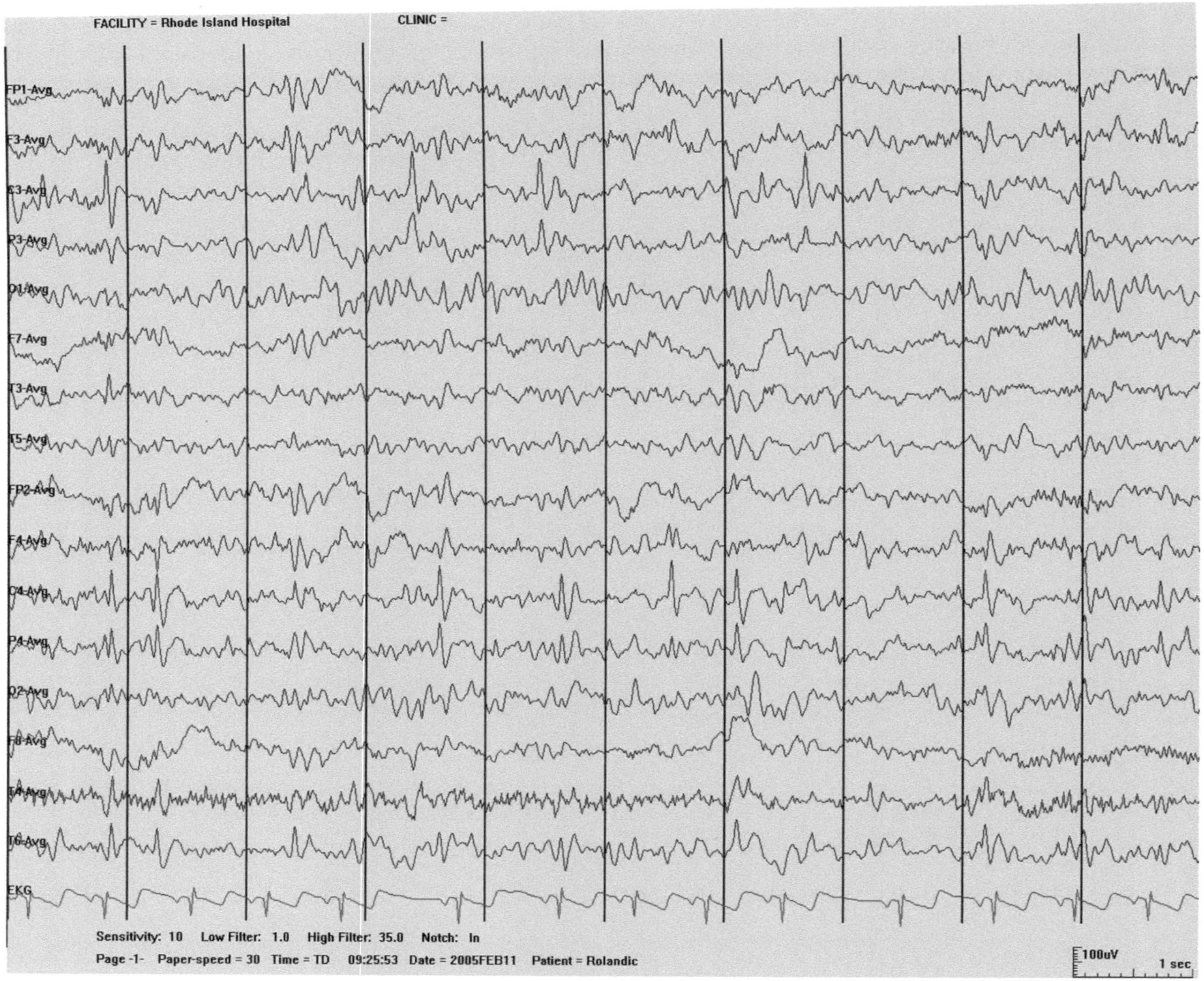

Fig. 9. Centrotemporal spikes. This referential tracing from the same patient as in Fig. 8 illustrates the horizontal dipole associated with centrotemporal discharges. Maximal negativity is seen from the peri-Rolandic channels, with the positive pole of the dipole appearing over anterior channels.

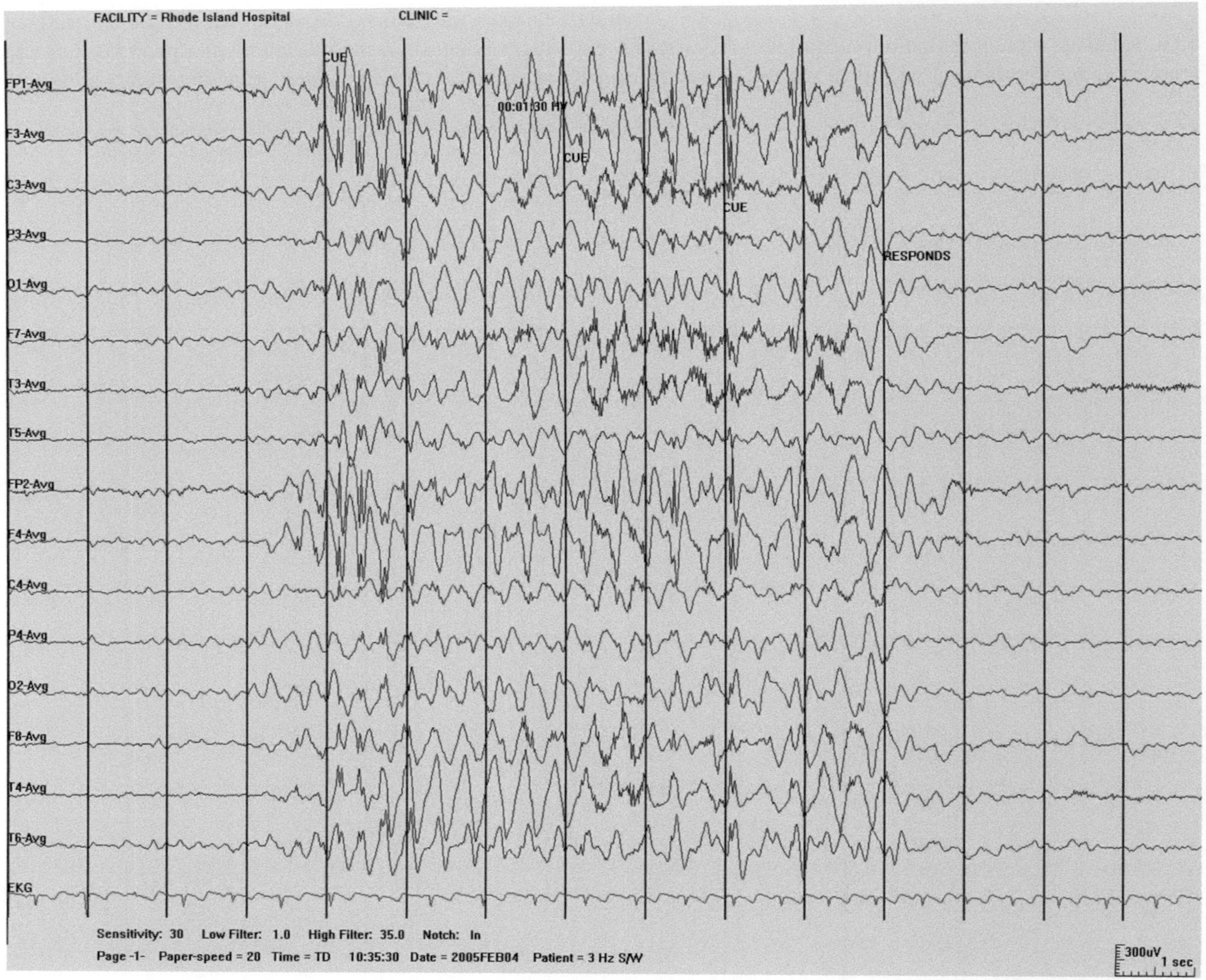

Fig. 10. Three-Hertz generalized spike-and-slow wave activity. This tracing derives from a 12-yr-old girl with staring spells and a positive family history. Note that this 8-s burst of bilaterally synchronous 3-Hz spike and wave activity arises during hyperventilation. The subject does not respond to the cue until the cessation of the spike and wave activity. The burst is maximal over bi-anterior channels.

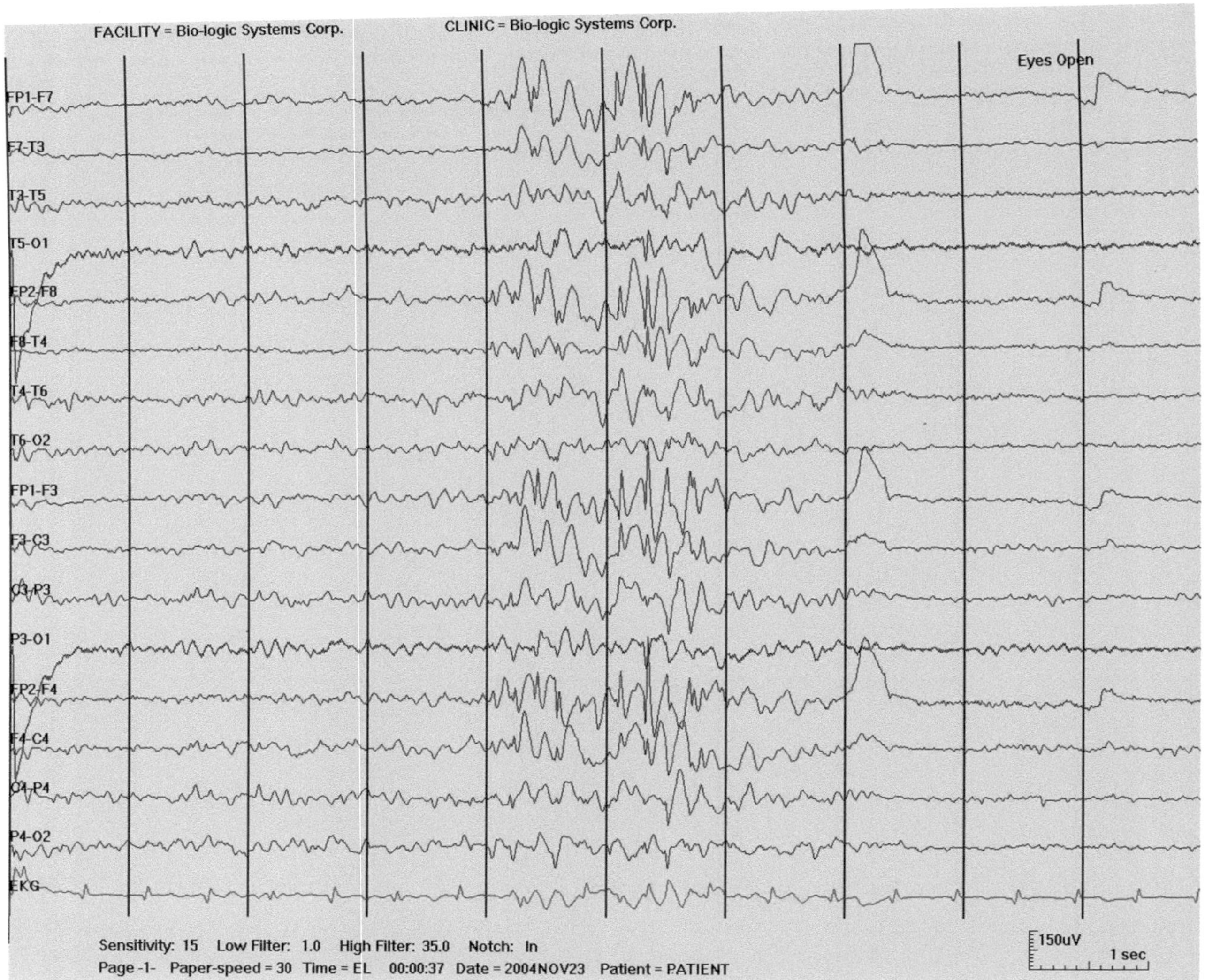

Fig. 11. Fast spike-and-wave activity. This tracing is from a 17-yr-old girl with juvenile myoclonic epilepsy. Note the synchronous, bi-anterior spike and wave activity at 4 to 5 Hz.

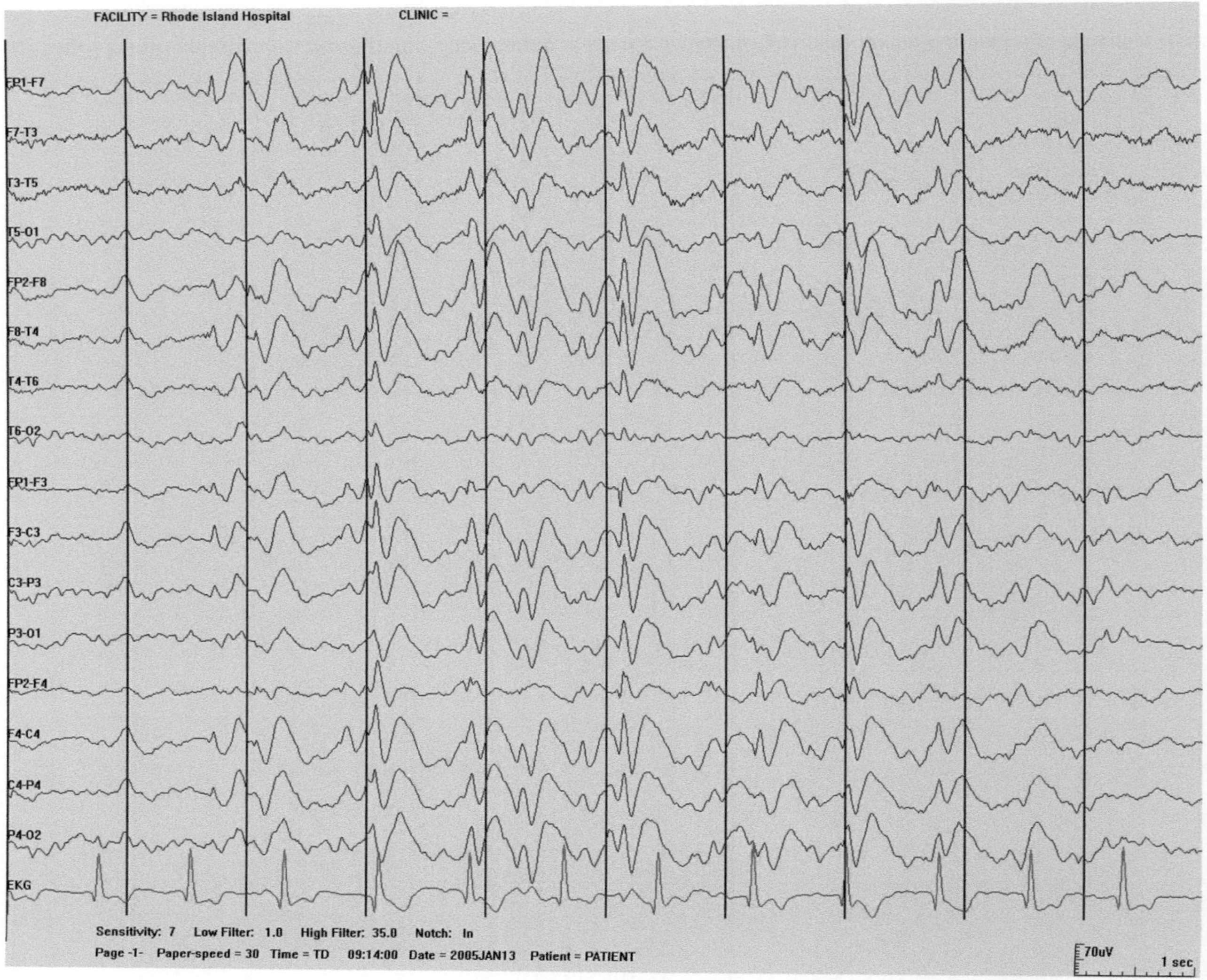

Fig. 12. "Slow" spike-and-slow wave activity. This tracing is from a 36-yr-old man with Lennox–Gastaut Syndrome (LGS). In this case, there is a run of 2- to 2½-Hz synchronous bi-anteriorly predominant sharp and slow wave activity lasting 7 s. Such activity, when prolonged, may appear during atypical absence seizures common to patients with LGS.

discharges in susceptible patients. Hyperventilation can bring out both focal and generalized discharges, although it is most useful in eliciting the three-per-second spike-and-slow wave pattern observed with absence seizures. Intermittent photic stimulation usually triggers generalized discharges in patients with photosensitive generalized epilepsies.

Drowsiness and sleep are also precipitants of epileptiform discharges. For this reason, routine EEGs in patients with suspected seizure disorders are often performed after sleep deprivation, to maximize the chances of recording an adequate period of sleep. The various stages of non-REM sleep, in particular, seem to be associated with increased epileptiform activity. Some patients may only have epileptiform discharges during sleep.

2.4. Clinical Interpretation of Epileptiform Abnormalities

The identification and characterization of interictal epileptiform discharges are useful clinically in several ways. First, their mere presence can help to confirm the diagnosis in a patient in whom a seizure disorder is suspected. However, their limitation in this regard must be understood: only approx 50% of patients with known epilepsy have interictal discharges on their first routine EEG. This number increases to 70% or greater with repeat studies. However, patients with mesial temporal lobe seizures may not have visible discharges if only standard scalp electrodes are used, and frontal lobe seizure patients also have a low incidence of interictal discharges. Elderly epilepsy patients have a lower incidence of interictal discharges than do younger patients. Finally, approx 1% of patients who do not have clinical seizures may have epileptiform discharges on their EEGs. In summary, then, although the finding of interictal epileptiform discharges is fairly specific for the diagnosis of epilepsy, a single EEG is not an exquisitely sensitive test for epilepsy, particularly in certain situations or populations.

Interictal discharges can be critical, however, in the characterization of a patient's epilepsy syndrome. For some patients, it is not possible on clinical grounds alone to determine whether seizures are partial or generalized in onset, and some syndromes have characteristic interictal EEG discharges that form part of their definition.

Thirdly, a clinician may wish to assess the frequency of interictal discharges in a known epilepsy patient, if there is a question of worsening seizures or frequent subclinical seizures, for example (surprisingly, however, frequent discharges do not necessarily correlate with frequent seizures in many cases). In addition, patients with epilepsy who have unexplained cognitive dysfunction or sleep disturbances may sometimes have very frequent interictal discharges during sleep, although the causal relationship between these discharges and their symptoms is not completely understood.

3. ICTAL PATTERNS

3.1. Introduction

Ictal patterns are those that are observed on EEG during clinical seizures. Although, intuitively, one might expect that a seizure should appear electrographically to be a monotonous repetition of multiple interictal epileptiform discharges in a row, this is not usually the case. Instead, the patterns during both focal- and generalized-onset seizures may have their own unique morphology and evolution, and, in some cases, a good correlation can be made between clinical seizure phenomenology and ictal EEG pattern.

3.2. Generalized Ictal Patterns

3.2.1. The Evolving Discharge of a GTC Seizure

The onset of a GTC seizure on EEG is typically characterized by the rapid (>10 per second) repetition of generalized spikes and polyspikes. These spikes usually increase in amplitude and decrease in frequency during the first 10 to 20 s of a GTC seizure, corresponding to the tonic phase observed clinically. As the clonic phase ensues, bursts of generalized high-amplitude spike-and-slow wave or polyspike discharges may be observed coinciding temporally with the clonic jerks of the extremities, with a low-voltage suppressed background observed in between the jerks. A gradual tapering of the frequency of the bursts is usually noted as the clonic jerks slow toward the end of the clinical seizure. Postictally, the background is suppressed, followed eventually by widespread delta-frequency slowing. It should be noted that, in most GTC seizures, much of the scalp EEG tracing from onset of tonic stiffening through the phase of clonic jerking is usually obscured by widespread muscle artifact. The actual epileptiform discharges may only partially be observed through the artifact, if at all.

3.2.2. Three-Per-Second Spike-and-Slow Wave Discharges

The classic EEG during an absence seizure demonstrates generalized spike-and-slow wave discharges repeating at a frequency of three per second (Fig. 10). These seizures may last only for a few seconds at a time and may occur tens or hundreds of times a day. Patients usually stare unresponsively during them, and may have minor automatisms. As described earlier, briefer interictal runs of the three-per-second pattern may be observed at other times, without any concomitant interruption in the patient's awareness.

3.2.3. Slow Spike-and-Wave Discharges

Prolonged runs of generalized spike-and-slow wave discharges repeating at a "slow" frequency (less than three per second; Fig. 12) are the pattern usually observed in association with atypical absence seizures, one of the seizure types observed in Lennox–Gastaut syndrome. During these seizures, patients stare but may also have complex automatisms. As mentioned before, patients with Lennox–Gastaut syndrome usually have briefer bursts of slow spike-and-wave discharges as an interictal pattern as well, without accompanying clinical manifestations.

3.2.4. Polyspike Discharges

Polyspikes (Fig. 3) are the cardinal feature of myoclonic seizures, such as those observed in the syndrome of juvenile myoclonic epilepsy. However, their presence is not universal, and repetitive polyspike discharges can be observed as an ictal pattern in association with other generalized seizure types as well. Sometimes the polyspike discharges on EEG coincide temporally with the frequency of clinical myoclonic jerks.

3.2.5. Electrodecremental Pattern

This ictal pattern is observed primarily with three seizure types: infantile spasms, tonic seizures, and atonic seizures. The EEG demonstrates a sudden widespread drop in amplitude, resulting in a low-voltage fast rhythm that persists for at least a few seconds. This decrement may be preceded by a high-voltage generalized slow wave. Infantile spasms, which are characterized by repeated episodes of flexion of the trunk and arms in jackknife-like movements, are usually associated with mental retardation, either cryptogenic or from a known etiology, such as tuberous sclerosis. In these cases, the interictal EEG may demonstrate hypsarrhythmia, an extremely high-voltage slow chaotic background with multifocal spikes. Tonic and atonic seizures can also manifest on EEG with an electrodecremental pattern, and are most commonly observed in Lennox–Gastaut syndrome.

3.3. *Focal Ictal Patterns*

Ictal EEG patterns observed in association with partial-onset seizures come in many forms. As a general rule, it is the *rhythmicity* and *evolution* of an EEG pattern that suggest to an electroencephalographer that a focal pattern of activity is ictal in nature, more than any other features. Here, we discuss briefly some of the commonly recognized focal ictal patterns. It should be noted that with all partial-onset seizures, secondary generalization to a tonic–clonic seizure is usually associated with an evolution of the ictal EEG pattern from localized activity to one typical for a GTC seizure, as described in Section 3.2.1.

3.3.1. Rhythmic Theta

Focal seizures, including many of temporal lobe origin, often begin with a focal area of rhythmic theta activity, which is often not sharp in morphology initially (Fig. 13). The theta activity may be quite localized in appearance, but will typically undergo an evolution in which the amplitude becomes higher, the frequency may become faster or slower, and the morphology may become sharper. In brief seizures, this may be the entire ictal pattern, with the rhythmic activity either ending abruptly or, more commonly, gradually waning with decreased amplitude and slower frequency. In longer seizures, this pattern may spread to adjacent channels as the clinical symptoms progress. Occasionally, localized rhythmic artifact confined to a small number of electrodes may be difficult to distinguish from an ictal pattern. In these cases, it is useful to remember that electrographic seizures typically display an evolving pattern, whereas rhythmic artifact may appear monotonous, and that rhythmic activity localized to a single electrode is more likely to be artifact than a seizure, except possibly in neonatal recordings.

3.3.2. Rhythmic Beta

Another electrographic pattern of focal seizure onset involves the appearance of rhythmic beta activity (Fig. 14). This can evolve to a pattern of slower frequencies and sharper morphologies. Although rhythmic beta activity is uncommonly observed in scalp ictal recordings, it is frequently observed in ictal recordings obtained with intracranial electrodes, because the removal of the filtering effect of the skull allows higher frequencies to be observed more readily.

3.3.3. Repetitive Spikes and Sharp Waves

At times focal seizures begin with repetitive, rhythmic spikes and sharp waves that appear similar to the patient's interictal epileptiform discharges. In these cases, what distinguishes a period of EEG activity as representing an ictal pattern rather than just frequent interictal discharges is the prominent rhythmicity of the discharges, their increased frequency compared with interictal discharges, and the overall evolution of their appearance and resolution.

A particular pattern of rhythmic *triangular sharp waves* has been described at the onset of some focal seizures; these sharp waves may be quite different from the patient's usual interictal discharges. This pattern of seizure onset is commonly observed in ictal recordings from the hippocampus using intracranial electrodes.

4. PERIODIC EPILEPTIFORM DISCHARGES

4.1. *Introduction*

Periodic discharges, as their name implies, are those that recur at regular intervals. In general, the term is used to refer to discharges that appear on the order of every second to every

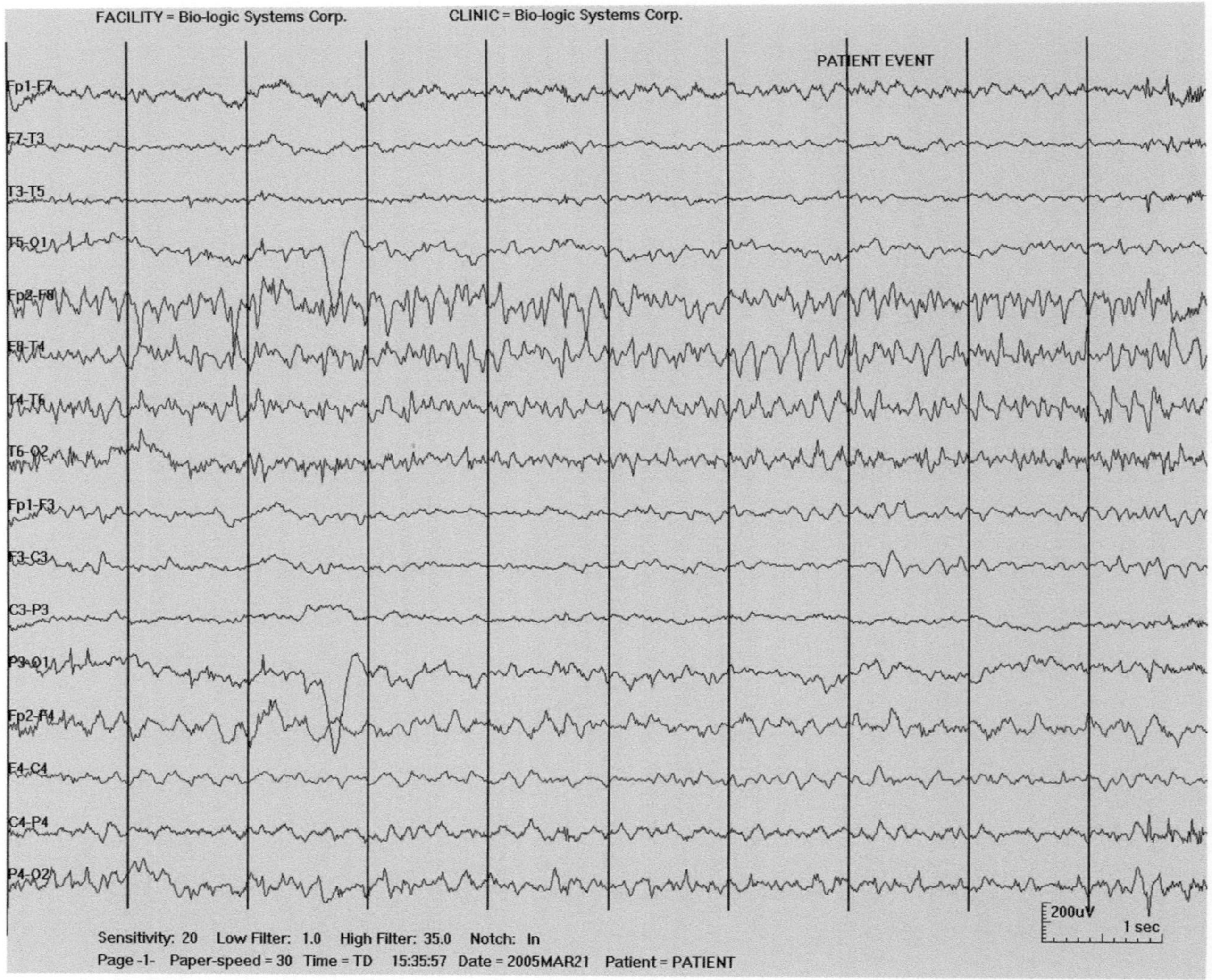

Fig. 13. Rhythmic theta. This tracing is from a 54-yr-old woman with refractory right temporal lobe epilepsy. This depicts an evolving electrographic seizure involving rhythmic and sharply contoured theta activity over right temporal channels, acquired during a complex partial seizure recorded during long-term monitoring.

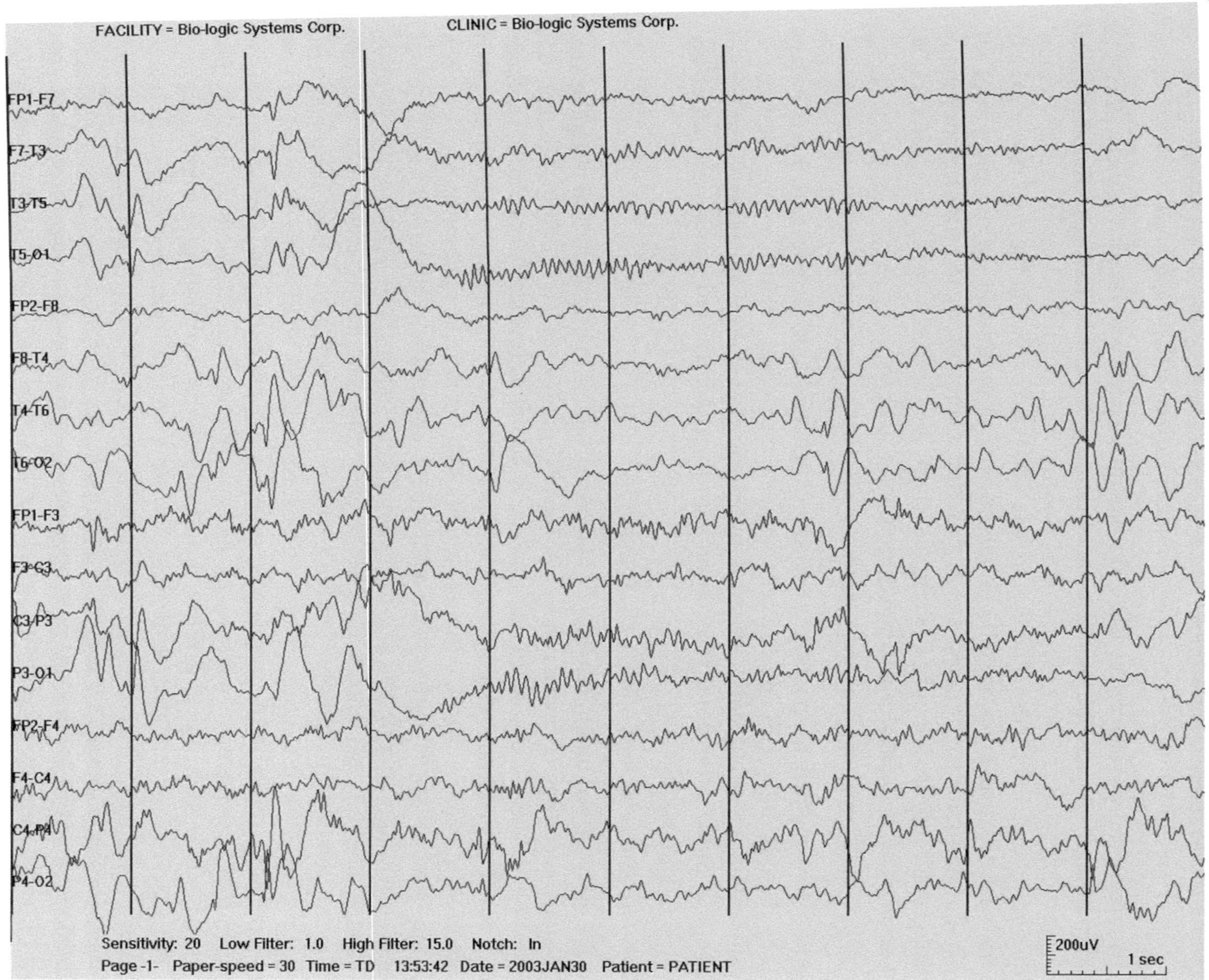

Fig. 14. Rhythmic beta. This tracing is from a 12-mo-old girl with infantile spasms. This particular page illustrates a sudden burst of left hemispheric rhythmic beta activity at 13 to 14 Hz for 6 s, associated with sudden dystonic arm movements. The burst of rhythmic beta activity begins with a sudden high-amplitude sharp wave discharge, maximal at T3.

several seconds, but not fast enough to be considered ongoing seizure activity. In some instances, periodic discharges on EEG are so characteristic of particular clinical diagnoses as to be nearly pathognomonic, whereas, in other cases, periodic discharges may be the nonspecific result of a variety of cerebral insults.

4.2. Patterns of Periodic Discharges

4.2.1. Periodic Generalized Sharp Waves With One-Per-Second Frequency

This pattern is the characteristic EEG finding in Creutzfeldt–Jakob Disease (CJD), a progressive spongiform encephalopathy caused by prions that leads to dementia and myoclonus, among other features (Fig. 15). More than 90% of all CJD patients will demonstrate this finding on EEG by the time the disease is advanced, but in the first few weeks or months the periodic discharges may be unilateral, not as prominent, or not present at all. They typically occur on a background that is slow and disorganized, and such an encephalopathic background may be the only EEG sign early in the disease course. New variant CJD does not seem to be associated with these periodic discharges.

4.2.2. Periodic Generalized Sharp Complexes Occurring Every 5 to 20 s

Subacute sclerosing panencephalitis, a late sequela of childhood measles infection, is characterized by these periodic EEG discharges that are separated by relatively long intervals. A noteworthy feature is that clinically observed myoclonic jerks, which are a prominent sign of the disease, can occur in temporal association with these generalized complexes, which are usually of very high voltage and occur on a slow, disorganized background.

4.2.3. Periodic Lateralized Epileptiform Discharges

These are aptly named unilateral sharply contoured waveforms that recur at a regular frequency, usually every one to few seconds. The waveform appearance is commonly that of a broad, high-voltage sharp wave, sharp-and-slow wave complex, or sharply contoured triphasic wave (Fig. 16). The intervening background on the side of the discharges is abnormal, often slow, and of low voltage. Most commonly, periodic lateralized epileptiform discharges (PLEDs) are observed in the setting of an acute or subacute structural lesion of the cerebral hemisphere, often large and destructive in nature. Thus, stroke (both ischemic and hemorrhagic), tumor, abscess, and encephalitis are among the common etiologies. In general, PLEDs are not thought to represent ongoing ictal activity, although a high percentage of patients with PLEDs have seizures at other times. Sometimes PLEDs are observed in late status epilepticus or during a recovery phase after status epilepticus, and less commonly a pattern of PLEDs may increase in frequency and evolve into clear ictal activity. PLEDs can persist for days to weeks after a cerebral insult, although their amplitude and frequency usually decrease during that period of time. Periodic epileptiform discharges occurring independently over the two cerebral hemispheres have been termed bilateral independent PLEDs; their presence sometimes suggests CNS infection.

The occurrence of PLEDs that are predominantly over the temporal regions suggests the diagnosis of herpes simplex encephalitis, a hemorrhagic infection caused by herpes simplex virus that has a predilection for the orbitofrontal and temporal regions and can rapidly lead to encephalopathy, seizures, coma, and death. Because of the fulminant nature of this disorder, the appearance of any periodic discharges over the temporal regions should immediately raise suspicion for this diagnosis in the appropriate clinical setting.

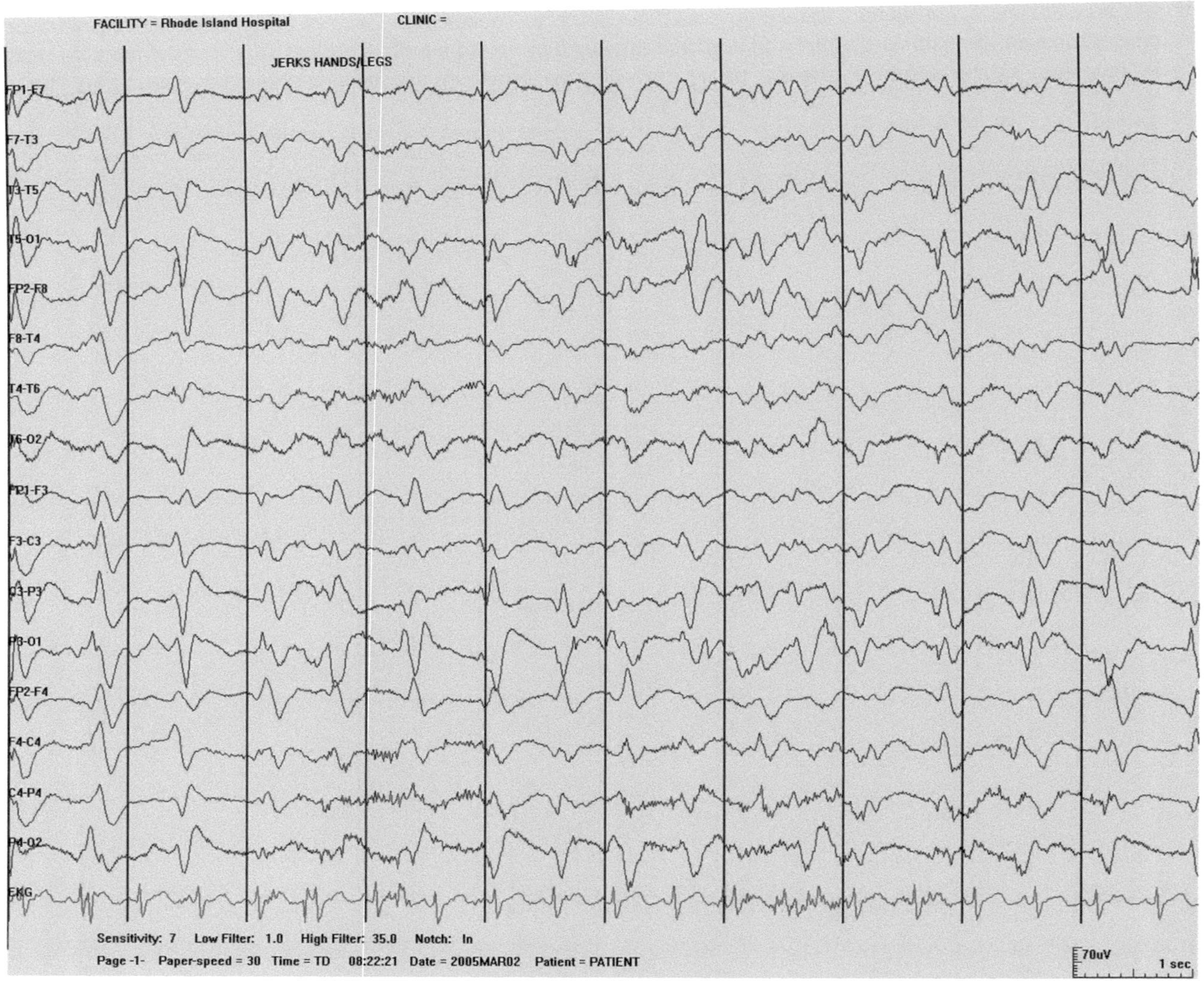

Fig. 15. Periodic generalized sharp waves in Creutzfeldt–Jakob disease. This tracing derives from a 68-yr-old man with a rapidly progressive dementia and myoclonus. The EEG demonstrates continuous, high-amplitude, periodic sharp waves at 1 to 1.5 Hz that are bilateral, although maximal over the left hemisphere. The technologist noted occasional clinical myoclonic jerking of the limbs during this acquisition.

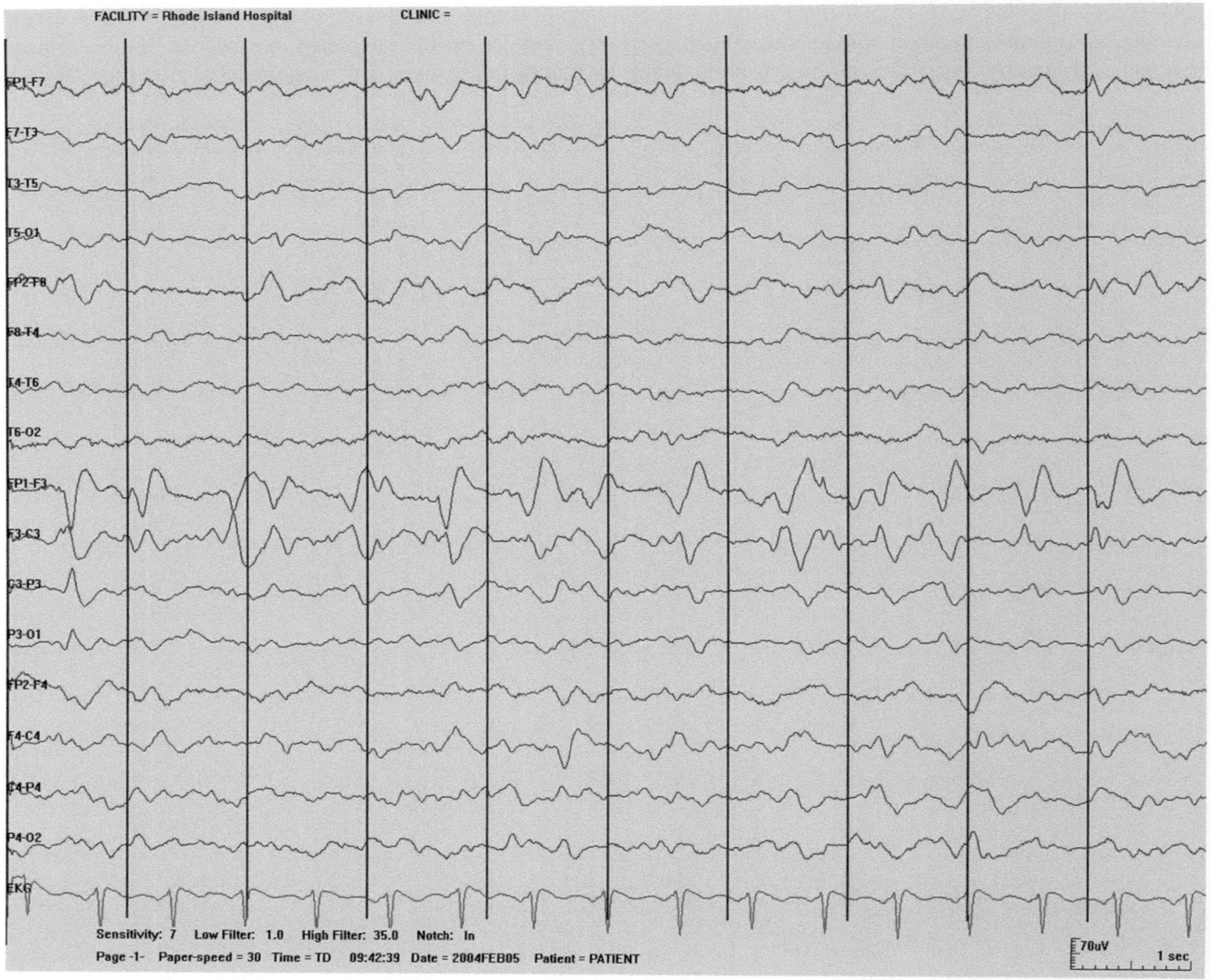

Fig. 16. Periodic lateralized epileptiform discharges (PLEDs). PLEDs are regularly recurring sharply contoured waveforms seen unilaterally. This sample is from a 66-yr-old man after resection of a parafalcine meningioma complicated by intracranial hemorrhage. There are fairly monotonous, recurrent sharp wave discharges from the left frontal region at 1.4 Hz.

SUGGESTED READING

Ebersole JS, Pedley TA. Current Practice of Clinical Electroencephalography. 3rd ed. Lippincott Williams & Wilkins, Philadelphia, PA, 2002.

Goldensohn ES, Legatt AD, Koszer S, Wolf SM. Goldensohn's EEG Interpretation: Problems of Overreading and Underreading. 2nd ed. Futura Publishing Company, New York, NY, 1999.

Niedermeyer E, Lopes da Silva F. Electroencephalography: Basic Principles, Clinical Applications, and Related Fields. 5th ed. Lippincott Williams and Wilkins, Philadelphia, PA, 2004.

Tyner FS, Knott JR, Mayer WB Jr. Fundamentals of EEG Technology: Basic Concepts and Methods. Lippincott Williams and Wilkins, Baltimore, MD, 1983.

REVIEW QUESTIONS

1. What are the definitions of a spike and a sharp wave?
2. What is the presumed basis for the slow wave after a spike or sharp wave?
3. When are polyspike discharges most commonly encountered?
4. Are occipital spikes of blindness indicative of an epileptic tendency?
5. What are the characteristic features of Rolandic spikes?
6. What are common frequencies of generalized spike and wave discharges and in what setting are they observed?
7. When are electrodecremental patterns encountered?
8. What are the features of sharp waves in CJD?
9. What are the main features of an EEG pattern that suggest an ictal event?
10. When are PLEDs encountered?

REVIEW ANSWERS

1. A spike is defined to be less than 70 ms in duration. Sharp waves have a duration between 70 and 200 ms.
2. The following slow wave is thought to derive from the synchronous repolarization of a collection of cortical neurons associated with a previous wave of cortical excitation.
3. Polyspikes are most frequently encountered in generalized epilepsies and in myoclonic epilepsies.
4. No, occipital spikes of blindness are observed with patients with acquired vision loss early in life. They are not thought to be predictive of epilepsy.
5. Rolandic spikes are high amplitude, biphasic spikes or sharp waves, often with a following slow wave, with surface negativity maximal involving the centrotemporal channels, and often exhibiting a horizontal dipole. They are very sleep activated.
6. Generalized spike and wave discharges are sometimes encountered at approx 3 Hz in childhood absence epilepsy, at 4 to 6 Hz in juvenile myoclonic epilepsy, and at less than 3 Hz in Lennox–Gastaut syndrome.
7. Electrodecremental patterns are observed in infantile spasms, tonic, and atonic seizure types. A generalized slow wave may occasionally precede the sudden, low-voltage, fast activity in this pattern.
8. CJD is frequently associated with periodic sharp waves at approx 1 Hz. This pattern may be unilateral early in the course of the disease, but later becomes generalized.
9. Rhythmicity and evolution of pattern are pivotal features that allow recognition of an electrographic seizure.
10. PLEDs are usually encountered in acute or subacute cerebral injuries. Etiologies may include stroke, tumor, abscess, and encephalitis, among others. Temporal lobe PLEDs raise a concern of herpes encephalitis. PLEDs may also be observed in the setting of recent status epilepticus.

9

Focal and Generalized Slowing, Coma, and Brain Death

Edward M. Donnelly and Andrew S. Blum

Summary

This chapter addresses the related topics of focal and generalized slowing, coma, and brain death. These EEG abnormalities are encountered in a wide range of clinical situations of variable severity. Focal and generalized slowing are both common and highly nonspecific findings in the EEG laboratory. Despite their lack of etiological specificity, EEG slowing and related patterns often bear important implications for both the location of CNS abnormalities and/or the prognosis for neurological recovery.

Key Words: Burst suppression; continuous; delta; diffuse; electrocerebral inactivity; focal; intermittent; theta; triphasic wave.

1. FOCAL SLOWING

Focal slowing in the EEG suggests an underlying abnormality but is of nonspecific etiology. It may reflect structural (i.e., tumor or infarct) or functional (i.e., postictal or migraine) abnormalities. There exist two spectra of severity, one pertaining to frequency, with slower rhythms representing more severe lesions, and one pertaining to persistence, with continuous slowing a more significant abnormality than intermittent slowing. An interhemispheric frequency difference of less than 1 Hz is not considered significant.

Continuous slowing in the EEG, whether focal or generalized, tends to take the appearance of either rhythmic monomorphic or arrhythmic polymorphic waveforms. These patterns often have differing significance. Continuous focal arrhythmic polymorphic slowing (Fig. 1) usually suggests some type of structural lesion in the underlying subcortical white matter. Abscesses, ischemic strokes, tumors, contusions, and so on, all may produce this pattern. The mechanism for this form of slowing may reflect disordered intracortical connectivity. Even transient functional disturbances, such as migraine and the postictal state, can be responsible. This illustrates the value of follow-up EEGs looking for evolution or resolution of any focal slowing that may be present. Rarely, focal slowing (and even focal seizures) can suggest toxic–metabolic disturbances, especially in hypoglycemic and hyperglycemic states. Continuous focal rhythmic monomorphic slowing, by contrast, is more commonly associated with underlying gray matter lesions. Note that focal cortical lesions are more likely to produce focal voltage attenuation or epileptiform abnormalities rather than focal slowing. Focal rhythmic monomorphic slow activity can also be intermittent. This is a less common pattern. The manner of appearance should be considered. Recurrent bursts of paroxysmal focal slowing may raise

From: *The Clinical Neurophysiology Primer*
Edited by: A. S. Blum and S. B. Rutkove © Humana Press Inc., Totowa, NJ

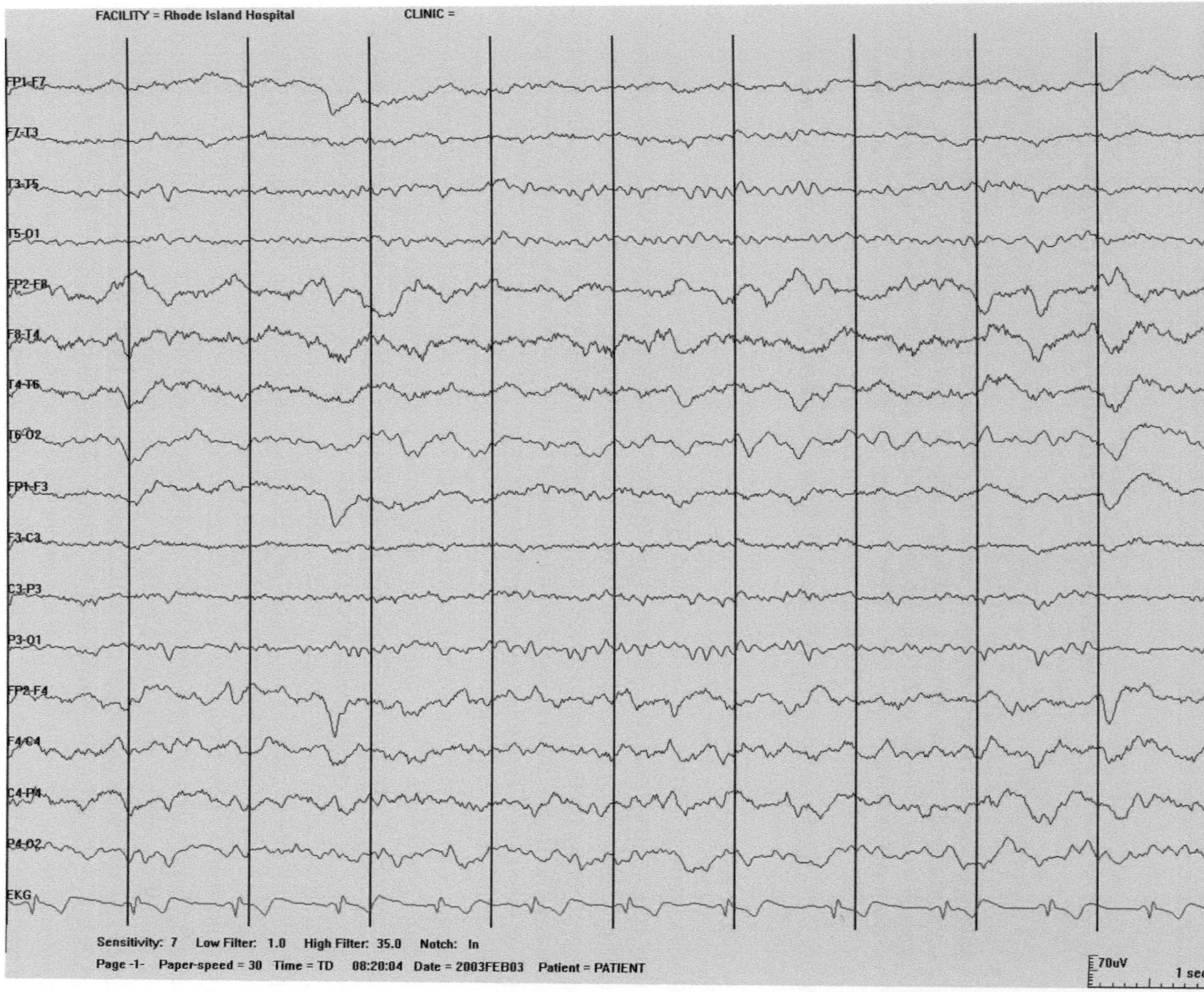

Fig. 1. Continuous focal arrhythmic polymorphic slowing. This EEG is from a 36-yr-old man with a right frontal brain tumor. Note the continuous polymorphic theta and delta activities involving the right hemispheric channels with relatively preserved normal rhythms over the left.

a suspicion for an underlying epileptogenic focus. A seizure focus is also suggested when the focal slowing shows exceptional rhythmicity or frequency evolution, that is, rhythmic slow waves that speed up and slow down. The location of focal intermittent rhythmic slowing may be significant. For example, if observed in the temporal regions, that is, temporal intermittent rhythmic delta, an underlying epileptogenic focus is more likely.

2. GENERALIZED SLOWING

In discussing generalized slowing of the EEG, several qualifiers must be mentioned. Is the slowing intermittent or continuous? Is it rhythmic–monomorphic or arrhythmic–polymorphic? In what context does it occur? For example, a buildup of generalized slowing during hyperventilation is a normal finding in children, adolescents, and young adults. Finally, some special examples will be considered.

Intermittent rhythmic delta activity (IRDA) tends to be monomorphic and is a commonly observed EEG abnormality. It is usually diffuse, bisynchronous, monomorphic, and reactive to eye opening. Hyperventilation may activate the pattern and sleep may attenuate it. Commonly, there is bianterior predominance to the slowing, hence, the term frontal IRDA (FIRDA) (Fig. 2). Note that children and adolescents often show a biposterior predominant IRDA and, thus, the term occipital IRDA activity has been applied.

IRDA is thought to be a projected rhythm and may reflect diffuse gray matter dysfunction, either cortical or subcortical. Acute or subacute disturbances are more likely to produce this pattern than chronic encephalopathies. FIRDA suggests a changing or evolving underlying disturbance~~an~~ encephalopathy that is either worsening or improving. Toxic–metabolic encephalopathies or electrolyte disturbances are typical underlying etiologies. Rarely, this pattern may accompany a postictal state. Eye blink or glossokinetic artifact should be excluded because they are common FIRDA imitators.

Continuous generalized slowing (Fig. 3) is an extremely common pattern, distinct from the intermittent rhythmic pattern described in the preceding paragraphs, although they may frequently appear together within the same tracing. Continuous slow patterns may refer solely to slowing of the posterior waking background rhythm. This observation usually implies a type of diffuse encephalopathy. In adults, a commonly used lower limit of normal for the waking background rhythm is 8 Hz. Varying degrees of background slowing may be encountered, including delta and theta frequencies. The degree of slowing of the posterior waking background rhythm is thought to correlate with the degree of clinical cerebral disturbance. As the encephalopathy deepens, other features may accrue in addition to progressive slowing of the posterior background rhythm. These include slowing of anterior rhythms; the normal frontal beta activity may slow to reveal varying degrees of frontal alpha or theta frequencies. In addition, the overall rhythmicity of the tracing wanes in conjunction with progressive deepening of encephalopathic states. In more marked encephalopathies, the entire tracing may become dominated by polymorphic slow forms, particularly delta activities, with much less of the reactivity and organization observed in the normal tracing. Owing to its fidelity as a surrogate marker of current CNS function, serial EEGs may be valuable to monitor the course of an acute or subacute encephalopathy.

Despite the helpful correlation between the degree of EEG slowing and the degree of cerebral dysfunction, there is no specificity to the observation of continuous slowing. It may be observed equally in static encephalopathies or in those of acute or subacute natures. Continuous diffuse slowing may arise in the setting of any diffuse CNS insult, including head trauma, hypoxic–ischemic injury, toxic or metabolic derangement, diffuse CNS infectious or

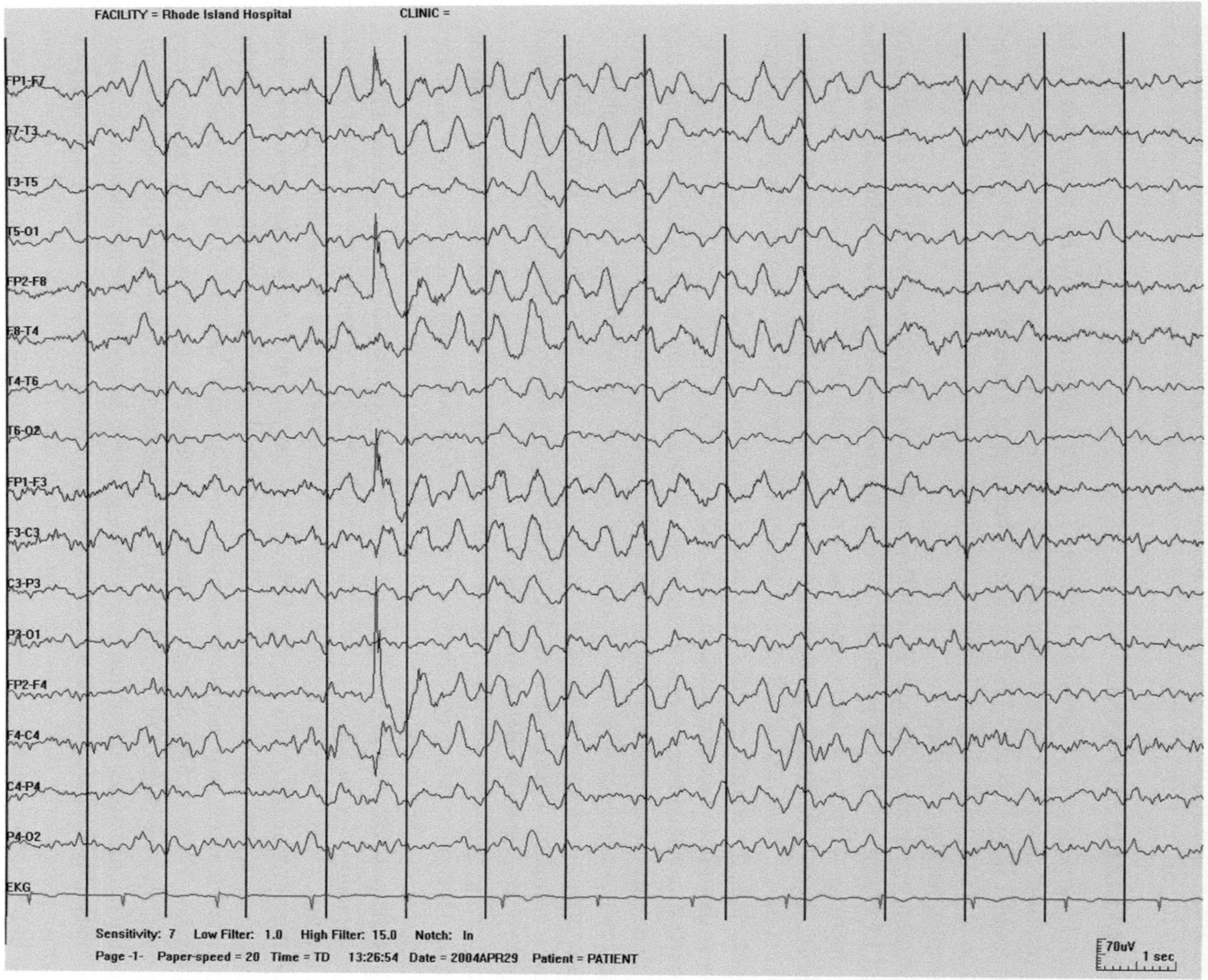

Fig. 2. Frontal intermittent rhythmic delta activity. This tracing is from an 83-yr-old woman with dementia, normal pressure hydrocephalus, and syncope. Note the bursts of rhythmic delta activity with bi-anterior predominance. There is also slowing of posterior background rhythms.

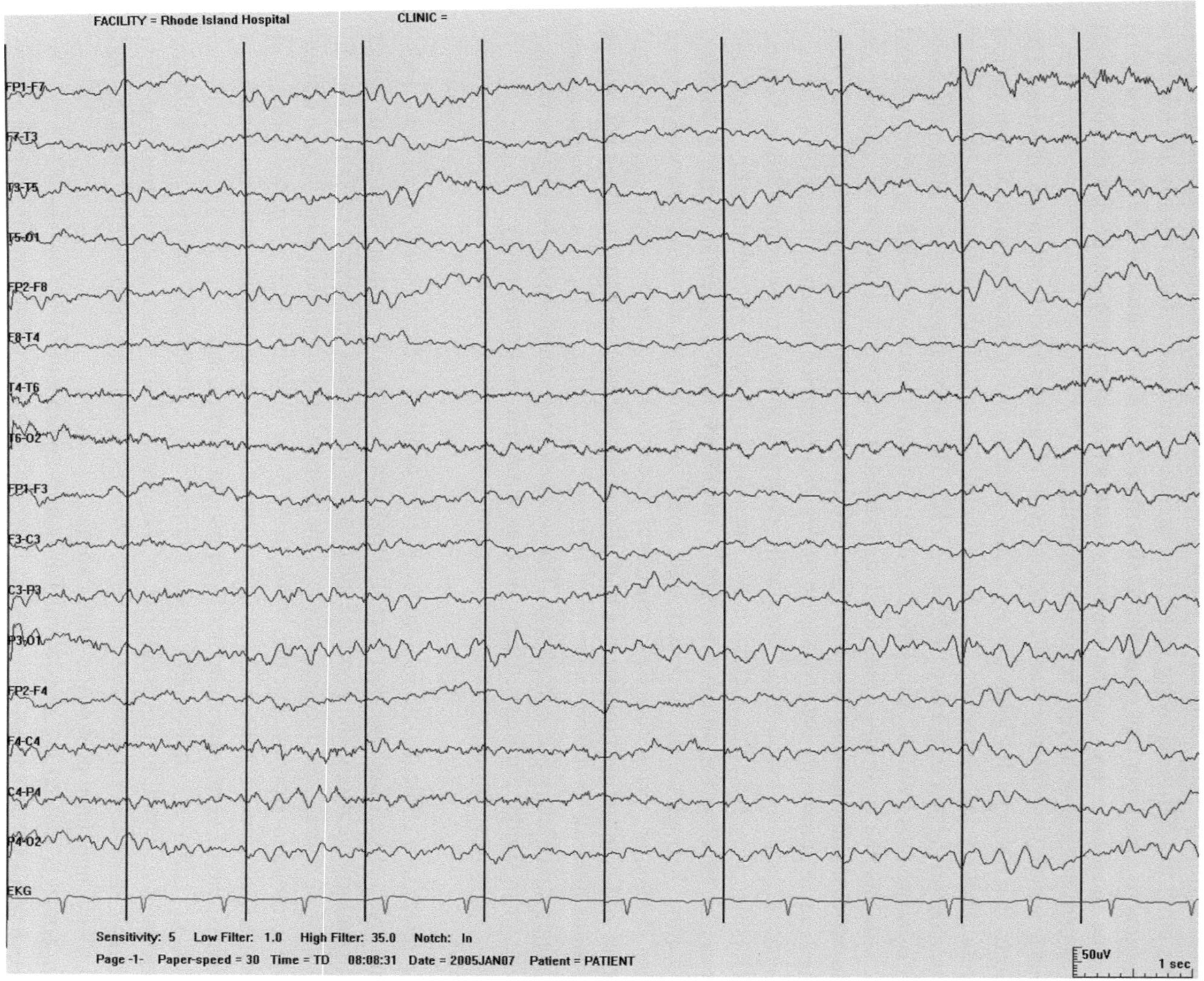

Fig. 3. Continuous generalized slowing. This is from a 73-yr-old man with a several year history of memory loss and recently increased confusion. Note the slightly irregular, continuous, 5- to 6-Hz activity evident biposteriorly as well as more diffusely in this tracing.

neoplastic processes, dementing illnesses, and even in multifocal conditions, such as multifocal, bihemispheric vascular insults. Indeed, because EEG offers relatively poor spatial resolution, as vascular events accrue in the CNS, the tracing may lose its focal/multifocal quality and may appear diffusely slow. Likewise, focal abnormalities may have less distinctive EEG signatures amid the diffuse slowing caused by an encephalopathy; focal details are lost. It is also important to remember that generalized slowing is a normal feature of the drowsy or sleep tracing. One must take stock of the patient's state when interpreting whether the observed slowing is pathological or merely reflective of state.

Triphasic waves (Fig. 4) represent a special type of generalized continuous slowing. The key features that distinguish triphasic waves from other forms of slowing include their typical triphasic morphology and a phase lag. The waves themselves are usually medium- to high-voltage slow waves occurring at a frequency of 1.5 to 2.5 Hz. They typically occur in a bilaterally symmetric, bisynchronous fashion. Although they may wax and wane somewhat in amplitude and frequency during the recording, they tend to exhibit a somewhat monotonous appearance. Triphasic waves usually show a phase lag of 25 to 140 ms across the anterior–posterior axis. This phase lag is more commonly observed in an anterior-to-posterior direction than vice versa.

Triphasic waves suggest a toxic–metabolic encephalopathy, most commonly a hepatic encephalopathy. However, this pattern is not specific for hepatic encephalopathy. Triphasic waves can also be observed in other metabolic disorders, such as uremia, hyperthyroidism, hypercalcemia, hypoglycemia, hyponatremia, and lithium intoxication. Alzheimer's disease and other dementias; prion diseases; structural pathologies, such as stroke and subdural hematoma; and cerebral carcinomatosis can also demonstrate this pattern. Triphasic waves may be quite difficult to differentiate from triphasic-appearing epileptiform morphologies, blunted sharp and slow wave complexes. This is even more problematic because both may equally occur in similar clinical settings, such as in uremic encephalopathy.

3. COMA

Coma refers to a clinical state in which a person exhibits a decreased level of consciousness with eyes closed and no purposeful responses to applied stimuli. Just as in milder encephalopathic conditions, the depth of the coma is paralleled by helpful EEG findings. In lighter forms of coma, the EEG may show some responsivity to stimuli with higher voltage and more prominent slowing. As the coma deepens, a blocking type response ensues, in which stimuli produce a voltage drop and attenuation of background activity. Finally, in deeper coma, the EEG becomes unreactive to patient stimulation.

Causes of coma are many and may include toxic–metabolic or hypoxic–ischemic encephalopathies as well as supratentorial or infratentorial structural pathologies. The EEG in coma may show several possible patterns, some of which can help identify etiology and some of which may have prognostic implications.

When coma is caused by nonconvulsive status epilepticus (Fig. 5), the EEG can be extremely helpful because it not only quickly establishes etiology, but may also permit assessment of subsequent anticonvulsant treatment efficacy. In no other instance is the specificity of the EEG in coma higher than in nonconvulsive status epilepticus. Keep in mind, however, that although the EEG may identify the cause of coma as an epileptic encephalopathy, there may be an as yet unidentified disturbance acting as a precipitant, for instance, hypoxic–ischemic encephalopathy, uremia, stroke, and so on. The EEG provides an immediate assessment of treatment efficacy in abolishing the epileptiform activity.

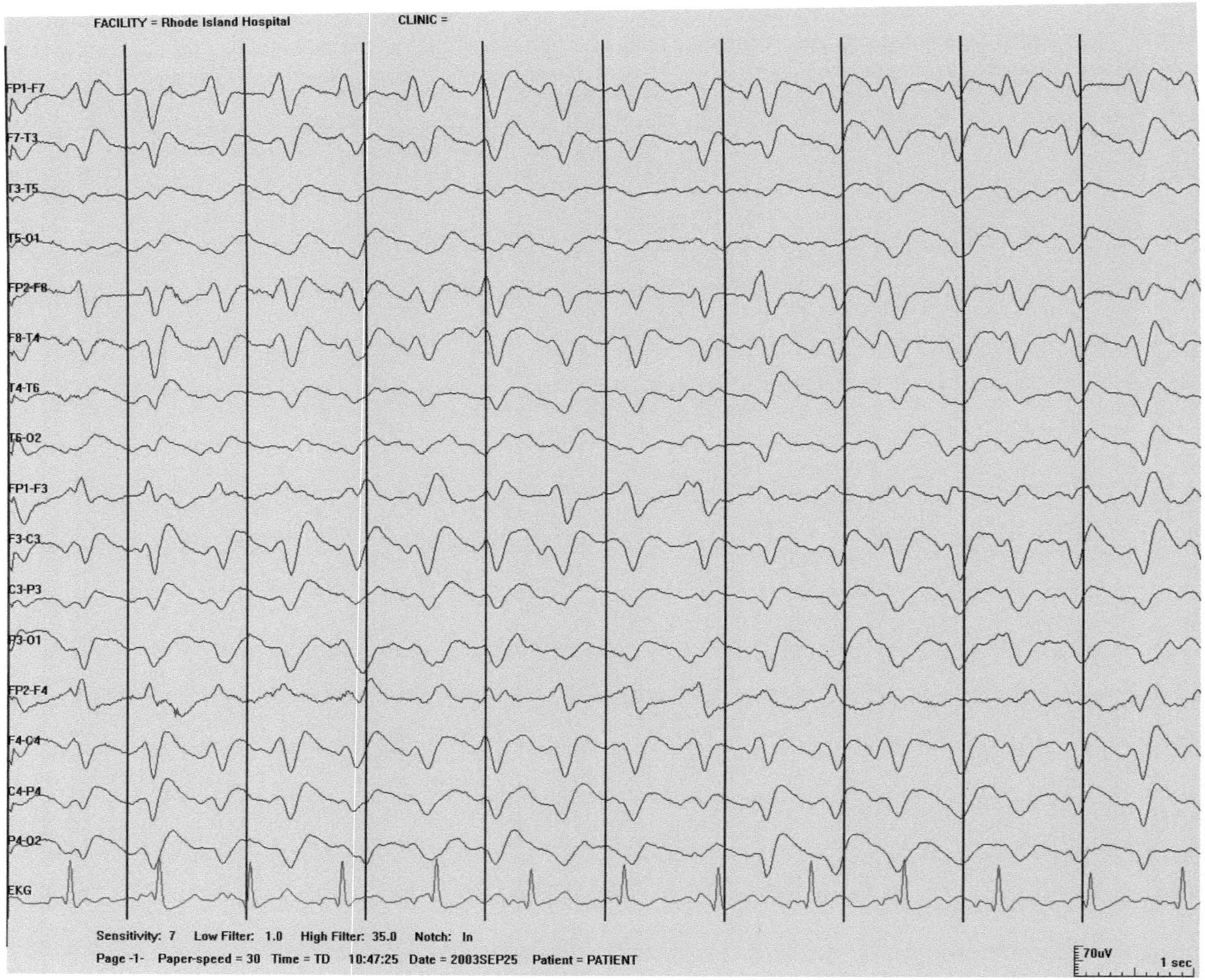

Fig. 4. Triphasic waves. This is derived from the EEG of a 46-yr-old man with hepatic encephalopathy. Note the bi-anteriorly predominant waveforms with triphasic morphology. There are no clear phase reversals or embedded sharp elements, and a subtle phase lag is evident along the anterior–posterior axis of the tracing.

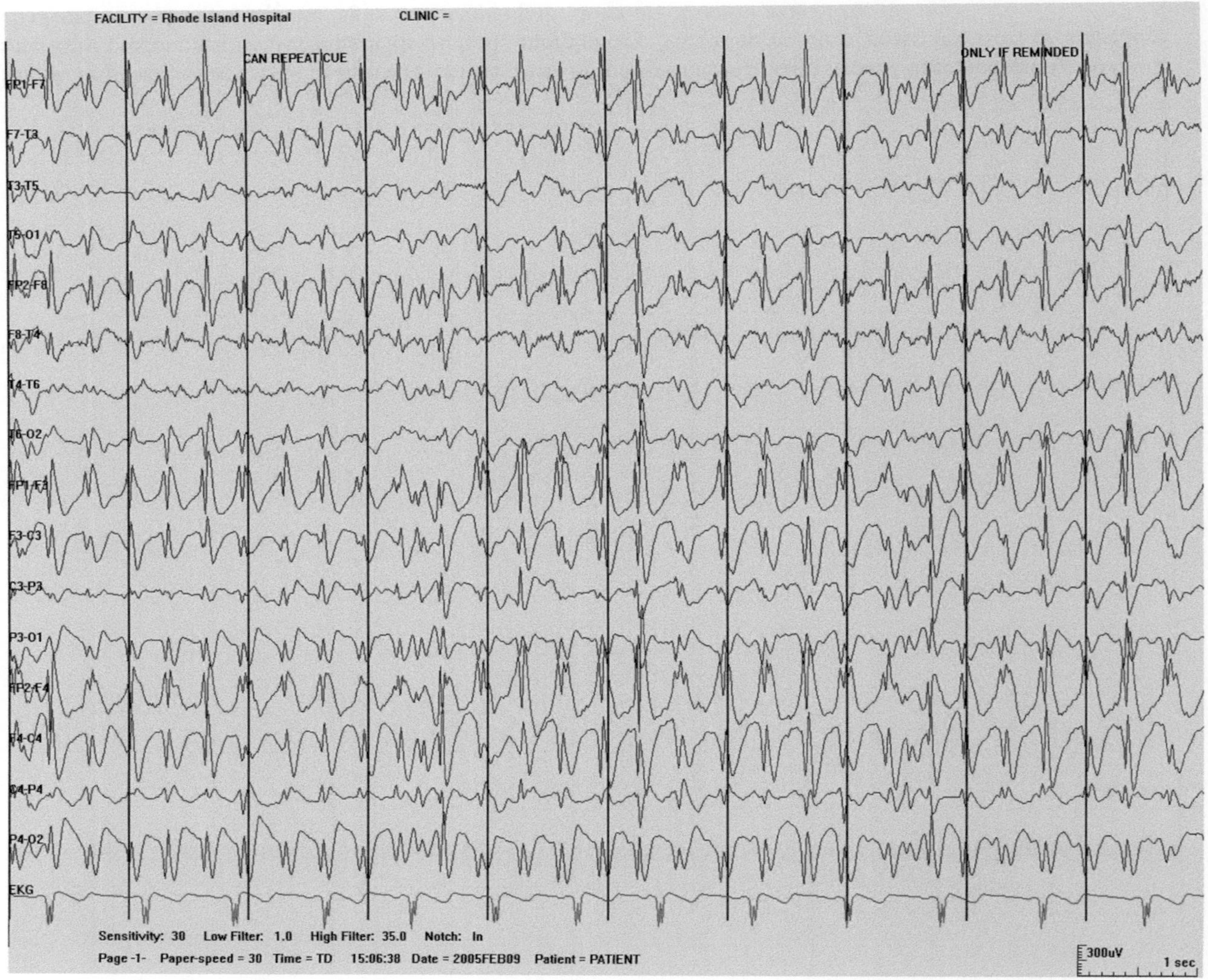

Fig. 5. Nonconvulsive status epilepticus. This tracing is from a 13-yr-old boy with absence epilepsy and a new prolonged confusional state. Note the incessant, bisynchronous, very high amplitude 3-Hz spike-and-wave pattern. The patient was quite confused, but could repeat if reminded to do so. This indicates absence status epilepticus.

Focal slowing in the EEG of a comatose patient may suggest a structural cause, such as a supratentorial structural lesion. Such lesions often produce coma in the setting of various cerebral herniation syndromes via mechanical compression of pontomesencephalic tegmental zones important in "alerting" the cortex and permitting wakefulness.

Generalized burst suppression (Fig. 6) is another common EEG pattern observed in coma. The bursts occur in a quasi-periodic fashion and may contain admixed sharp and/or spike and slow waves. Myoclonic jerks can accompany the bursting. Asynchronous bursting may reflect disordered interhemispheric cortical connectivity. Asymmetric burst voltage often signifies asymmetric cortical injury and/or raises the suspicion of a breach effect or an overlying fluid collection. The quasi-periodic bursts and suppressive intervals vary in duration with the depth of the coma. As the coma deepens, the bursts of activity become shorter and more infrequent and the suppressive intervals widen. This pattern reflects an exceptionally profound level of depressed consciousness. It is often observed during induction of general anesthesia. It is also the desired EEG pattern during administration of barbiturate therapy for refractory status epilepticus or to help control increased intracranial pressure after traumatic brain injury. Burst suppression suggests a poor prognosis, depending on the etiology. In the setting of a toxin- or medication-induced coma, the prognosis may be far better than in hypoxic–ischemic injury or trauma, in which burst suppression patterns may suggest a poor outcome.

Monotonous monorhythmic patterns can also be observed in the EEG of a comatose patient. Persistent, diffuse 8- to 12-Hz activity in a comatose patient is known as alpha coma (Fig. 7). This pattern, at first glance, may resemble normal background activity. However, the 8- to 12-Hz activity appears diffusely, not over posterior head regions, as in the normal waking background rhythm. Additionally, the pattern is completely unreactive to exogenous stimuli. Typical precipitants of alpha coma include brainstem lesions and hypoxic–ischemic mechanisms. It may also be observed as an ante mortem pattern as the patient progresses from burst suppression to electrocerebral inactivity (ECI). Alpha coma is, thus, thought to imply a very poor prognosis, particularly in the setting of anoxic injury. However, rare case reports have shown neurological recovery from this EEG pattern. Beta coma, theta coma, and delta coma are less common unreactive monomorphic EEG patterns whose prognostic significance is less clear.

Comatose patients can exhibit an EEG that seems to show features of normal sleep. Spindles, vertex waves, and K-complexes can be observed with cyclic variability. The EEG is distinguishable from normal sleep, however, because the patient is unarousable and the EEG does not react to applied stimuli. This pattern is sometimes referred to as spindle coma. These features associated with the EEG of sleep may disappear as the coma deepens.

Cheyne–Stokes respirations in a comatose patient may have a specific EEG correlate. An alternating pattern consisting of low-voltage irregular periods followed by higher-voltage slowing mirrors the respiratory rhythm changes. The cyclic alternating pattern may represent the effects of a cortical release phenomenon on the pacemaker function of the brainstem arousal system.

4. BRAIN DEATH

Brain death is a clinical diagnosis made when there is no evidence of brainstem function on successive neurological exams. Protocols for declaring brain death vary among institutions and according to the age of the patient. The EEG is one of several tests that can help confirm the diagnosis. Complete absence of brain-derived rhythms, ECI (Fig. 8), can help confirm the clinical diagnosis of brain death. Electrocardiogram and respirator derived artifacts are often all that

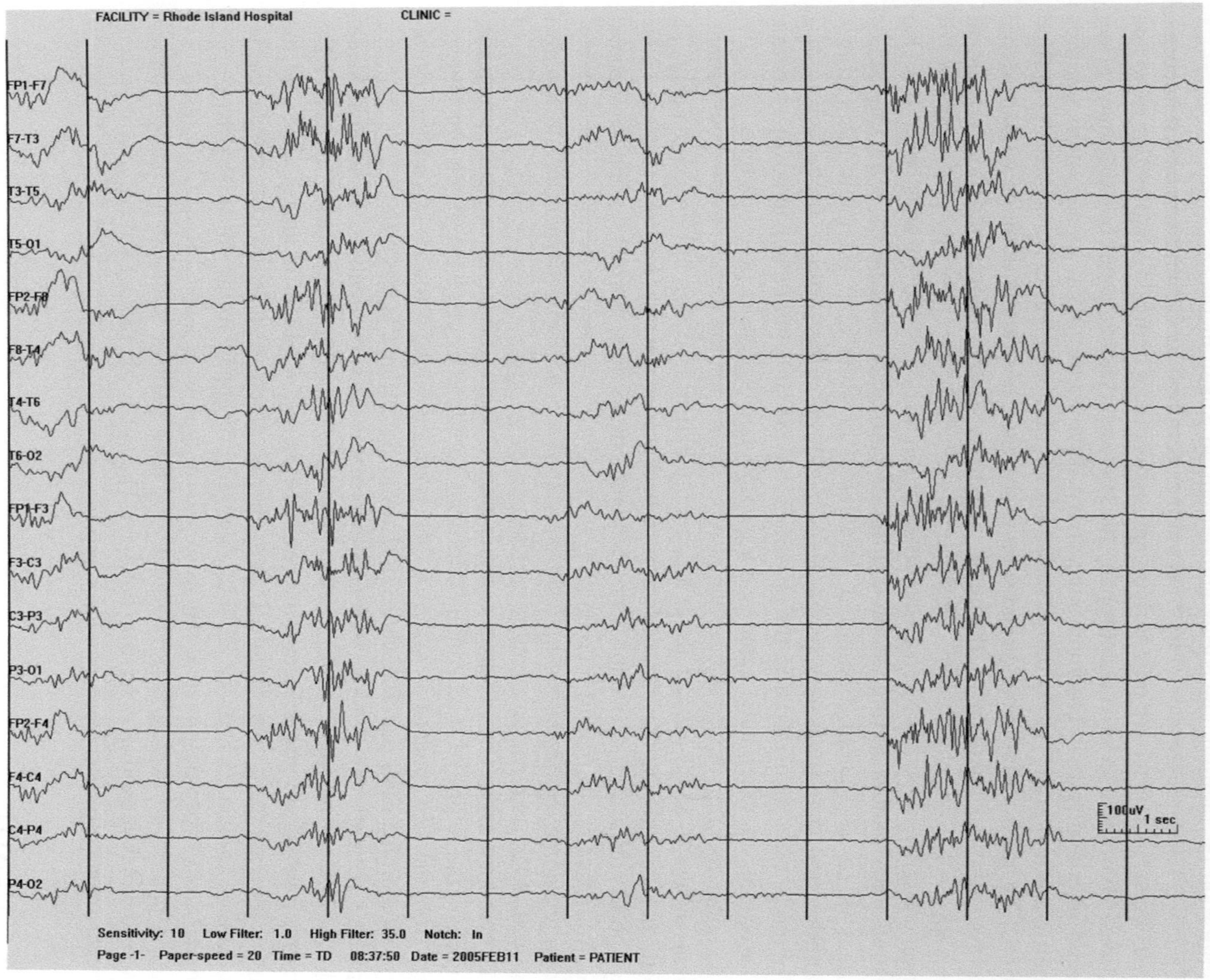

Fig. 6. Burst suppression. This 49-yr-old woman was recorded while undergoing induction with general anesthesia for vascular surgery. Bursts of bilaterally synchronous, higher amplitude mixed frequencies occur lasting 1 to 2 s, punctuated by periods of relative attenuation lasting 3 to 4 s.

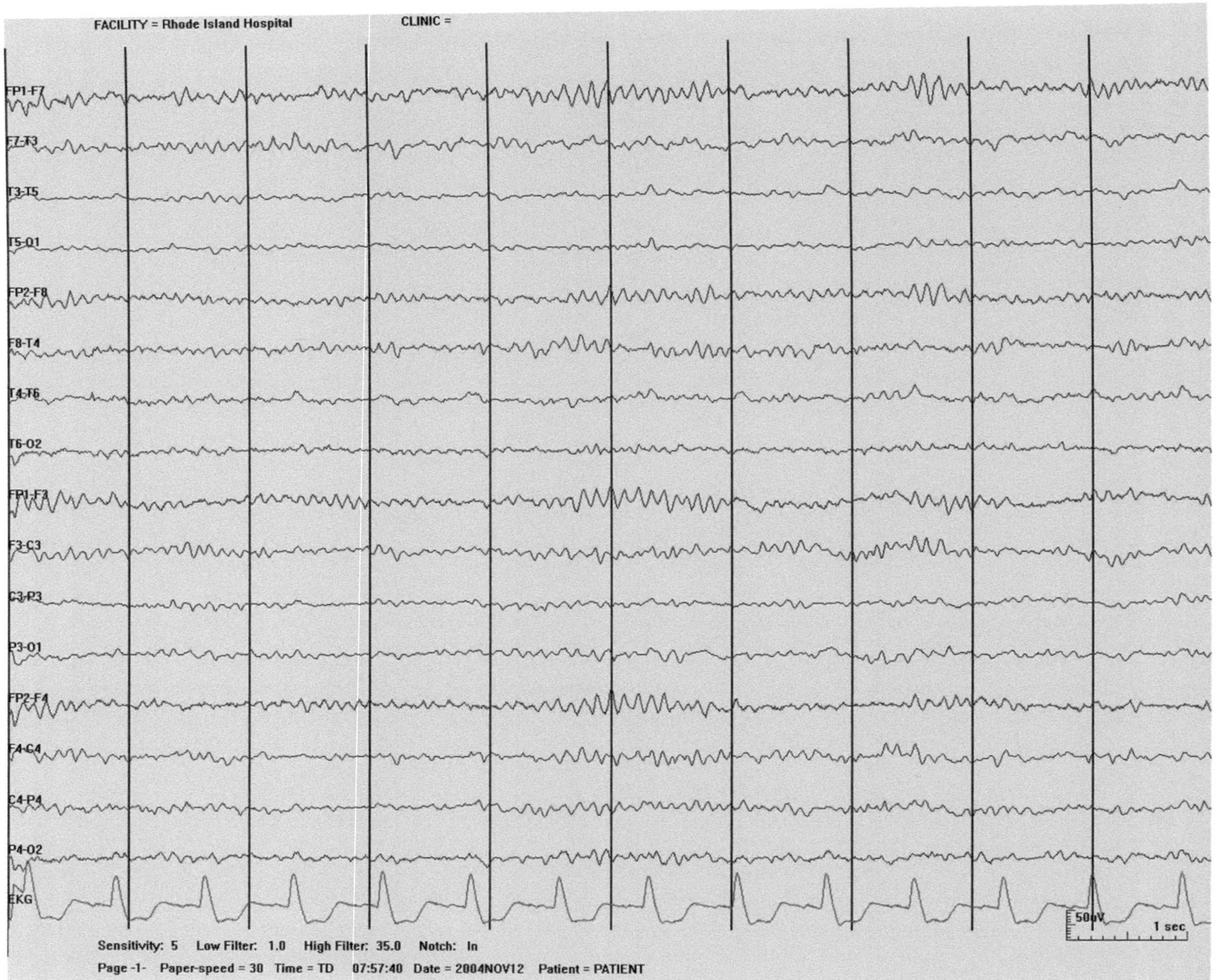

Fig. 7. Alpha coma. This recording is of a 59-yr-old man on a respirator in the medical intensive care unit, partly sedated, in coma. He was unreactive to noxious stimulation. Note the diffuse and fairly continuous 9- to 9.5-Hz activities, with reversal of the usual anterior–posterior voltage gradient. This activity failed to vary with attempts to arouse the patient.

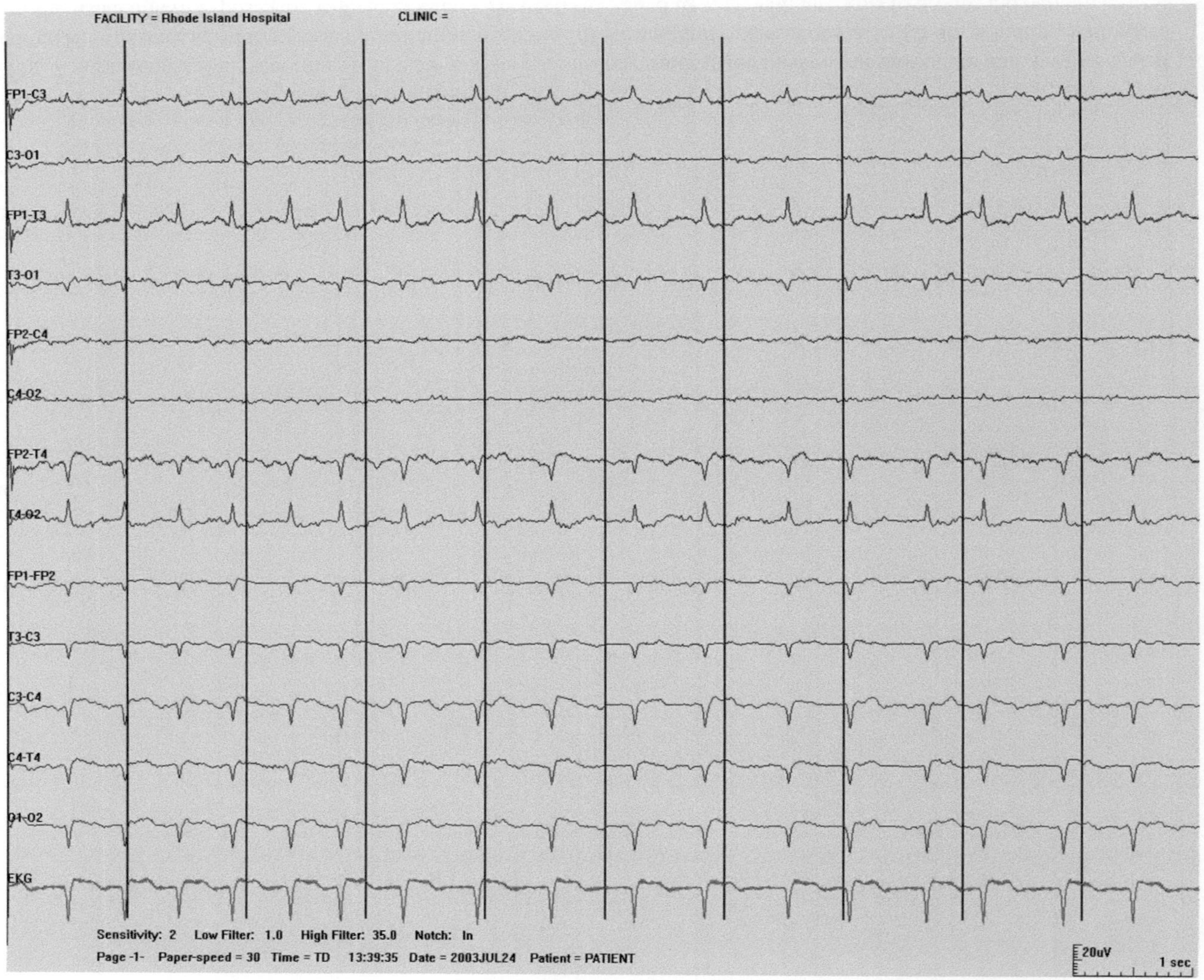

Fig. 8. Electrocerebral inactivity. This record is from a 66-yr-old man after a cardiopulmonary arrest. Note that the sensitivity is at 2 μV/mm and "double distance" electrode comparisons are in use. The record is dominated by amplified ECG-derived artifact. No convincing brain-derived rhythms are seen.

Table 1
EEG Criteria for Electrocerebral Inactivity

1.	A minimum of eight scalp electrodes should be used
2.	Interelectrode impedances should be less than 10,000 Ω but more than 100 Ω
3.	The integrity of the entire recording system must be verified
4.	Interelectrode distances should be at least 10 cm
5.	The sensitivity should be at least 2 μV/mm for at least 30 min of recording
6.	Appropriate filter settings should be used
7.	Additional monitoring techniques should be used when necessary
8.	There should be no EEG reactivity to afferent stimulation
9.	The recording should be made by a qualified technician
10.	A repeat EEG should be performed if there is doubt regarding the presence of electrocerebral inactivity

are observed in ECI. Several technical requirements must be met to ensure the validity of the finding. The American Electroencephalographic Society has published technical criteria that must be met before an EEG can be considered to fulfill ECI (Table 1).

One technicality deserving special mention is the diagnosis of brain death in infants and children. Persistence of the EEG ECI pattern must be documented in these age groups. For infants younger than 2 mo of age, two EEGs showing ECI must be obtained, separated by 48 h. For infants between 2 and 12 mo of age, the two EEGs showing ECI must be separated by 24 h. Of course, administration of sedative–hypnotic medications, for instance, barbiturates and benzodiazepines, negates the ability of the EEG to speak to brain death because the observed findings could be attributed to reversible, medication-induced effects. Variable requirements exist for the duration of time necessary since the last administration of sedative medication before the EEG can support the diagnosis of brain death. Other potential confounders of ECI must be excluded, such as hypothermia, hypotension, and severe electrolyte and glucose abnormalities, among others.

SUGGESTED READING

Bennett DR, Hughes JR, Korein J, Merlis JK, Suter C. Atlas of Electroencephalography in Coma and Cerebral Death: EEG at the Bedside or in the Intensive Care Unit. Raven, New York, NY, 1976.

Daly DD, Pedley TA. Current Practice of Clinical Electroencephalography, 2nd ed. Raven, New York, NY, 1990.

Lüers HO, Noachtar S. Atlas and Classification of Electroencephalography, 1st ed. WB Saunders, Philadelphia, PA, 2000.

Niedermeyer E. Electroencephalography: Basic Principles, Clinical Applications and Related Fields, 5th ed. Williams and Wilkins, Baltimore, MD, 2004.

Spehlman R. EEG Primer, 2nd ed. Elsevier, Amsterdam, Holland, 1991.

REVIEW QUESTIONS

1. What are the two main characteristics of slowing? What is their significance?
2. What is the significance of focal, arrhythmic polymorphic slowing?
3. What is the significance of focal, rhythmic monomorphic slowing?
4. What is IRDA and when is it encountered?
5. What are triphasic waves and when are they most commonly encountered?
6. How does the reactivity of an EEG in coma speak to the depth of coma?

7. How is the EEG beneficial in epileptic stupor, that is, nonconvulsive status epilepticus?
8. What do burst suppression patterns signify?
9. When is alpha coma encountered and what is its significance?
10. What are some technical criteria for recording ECI?

REVIEW ANSWERS

1. The frequency of the observed slowing and its persistence are two important characteristics. In general, the slower the recorded rhythms, the more severe the lesion. Additionally, in general, the more persistent the slowing, the more severe the process.
2. Focal, arrhythmic polymorphic slowing usually implies a focal subcortical white matter lesion. However, it is nonspecific to etiology.
3. Focal, rhythmic monomorphic slowing usually suggests a lesion of the underlying gray matter. However, when such a pattern becomes notably repetitively paroxysmal and burst-like, one should also entertain the possibility of a focal epileptic process.
4. IRDA usually signifies an acute or subacute process leading to diffuse cortical or subcortical gray matter dysfunction. It is usually bianteriorly predominant, that is, FIRDA, but may be posteriorly maximal in childhood and adolescence, that is, occipital IRDA.
5. Triphasic waves are broad waveforms with characteristic triphasic morphology. They are usually diffusely represented and bilaterally synchronous, at 1.5 to 2.5 Hz. Often, they exhibit a helpful phase lag across the anterior–posterior axis. They mainly occur in metabolic encephalopathies, usually in hepatic insufficiency, but are not exclusive to hepatic disease states.
6. As comatose states become deeper, the EEG becomes progressively less reactive to environmental stimuli. Lighter comas may still show some reactivity, even when this is clinically unapparent.
7. The EEG is critical to making the diagnosis of nonconvulsive status epilepticus. It is also invaluable in assessing the progress of therapeutic interventions in this setting.
8. Burst suppression is observed in many comatose states. As the periods of voltage suppression become longer, the coma becomes deeper. This pattern may be iatrogenic as in general anesthesia or in barbiturate coma for status epilepticus, among others. Thus, it may have a good prognosis in intoxications, but may carry a poor prognosis in other etiologies, for instance, hypoxic–ischemic insults.
9. Alpha coma is denoted by widespread alpha activity that is unreactive to applied stimuli. It implies a poor prognosis in most instances, although when associated with medication related coma, recovery may occur.
10. The criteria for ECI include a minimum of eight leads, 10-cm interelectrode distances or greater, sensitivity set to 2 μV/mm for at least 30 min of recording, among others (Table 1). In children, a repeat tracing may be necessary.

10
Normal Pediatric EEG

Ann M. Bergin and Blaise F. D. Bourgeois

Summary

This chapter provides an overview of the developmental changes in the pediatric EEG from early prematurity through adolescence. The early development of recognizable sleep–wake cycles in the premature infant, the changes that occur around term and the maturation of the EEG in the awake, drowsy, and sleep states are described. Changes in response to routine activation procedures are also described. Figures are provided throughout the text. Discussion of common variant patterns observed in childhood is also provided.

Key Words: Child; development; EEG; normal variant; sleep–wake cycles.

1. INTRODUCTION

From the first detectable EEG tracings in the extremely premature infant, to the mature EEG of the 18-yr-old subject, the pediatric EEG undergoes enormous changes in parallel with the great increase in size and complexity of the brain and its connections. Characteristic patterns of activity are observed at various stages during normal development. These patterns emerge, wax, and wane during a broadly predictable time span. Their persistence beyond the expected period may be an indicator of dysmaturity or injury. Recognition of the normal developmental progression and deviation from normal patterns is essential for identifying, understanding, and predicting recovery from injury.

2. NEONATAL PERIOD

The patterns observed in the neonatal EEG and the significance attributed to them depends on the conceptional maturity of the infant. Therefore, to evaluate the neonatal EEG, the reader must know the conceptional age of the infant (duration in weeks since last menstrual period/beginning of pregnancy), in addition to the postnatal age. Patterns observed also depend on the infant's state of arousal, and this should also be noted. The EEG patterns are evaluated in light of these conditions.

2.1. *Premature Infants*

2.1.1. Less Than 29 Wk

The most striking feature of the early premature EEG is the discontinuity of activity. Bursts of high-voltage, predominantly delta activity, mixed with other frequencies and sharp waves, are interspersed with periods of low-voltage quiescent recordings. Interburst intervals

From: *The Clinical Neurophysiology Primer*
Edited by: A. S. Blum and S. B. Rutkove © Humana Press Inc., Totowa, NJ

may be prolonged, up to 90 s or more, whereas active bursts are generally shorter, but may last up to 1 min. This pattern is described as trace discontinu (Fig. 1). At this early stage, there is virtually complete synchrony between the hemispheres. The EEG is invariant with clinical sleep or wakefulness, or with stimulation. The bursts of activity are predominant in the parasagittal and occipital areas, with relative inactivity in the temporal areas. Specific gestational features observed at this stage may include monomorphic occipital theta or delta activity, occurring in bursts lasting up to 5 or 6 s, sharp theta in the occiput of premature infants (STOP) (Fig. 2), and delta brushes (beta–delta complexes) initially confined to the Rolandic and occipital areas.

2.1.2. *Twenty-Nine to 32 Wk*

At this age, the EEG remains for the most part discontinuous, although the average duration of the interburst interval decreases (maximum ~60 s). Previously synchronous bursts of activity are now commonly asynchronous. This will persist, especially in quiet sleep (QS), until term and briefly beyond, during trace alterant. Brief periods of continuous activity correlate with active sleep (AS; rapid eye movement [REM] sleep) and irregular respirations, the earliest phase of differentiation of sleep–wake cycles. Specific gestational features include delta brushes that are more prominent in Rolandic, occipital, and parietal areas (Fig. 3). Bursts of higher voltage, sharply contoured theta activity become more common in the midtemporal regions (temporal theta bursts).

2.1.3. *Thirty-Three to 36 Wk*

At this stage, cyclic state changes between wakefulness, AS and QS become more easily defined, and the relationship between these electrical states and the attendant respiratory (irregular respiration in waking and AS; regular in QS) and musculoskeletal (clonic chin movements in QS; REM in AS) patterns become more consistent. The EEG background is a more continuous mix of theta and delta activity in waking and AS, with continuing cycling through periods of trace discontinu. The infant now responds to stimulation with diffuse attenuation of background activity (Fig. 4). Temporal theta bursts give way briefly (33–34 wk) to temporal alpha bursts before largely disappearing by 34 to 35 wk. Delta brushes persist and become more clearly associated with QS (Fig. 5). Sharp waves occur predominantly in central and temporal areas. Positive temporal sharp waves may occur singly or in runs, unilaterally or synchronously in both hemispheres. Toward 36 to 37 wk, frontal sharp waves (encoches frontales; Fig. 6), sometimes associated with rhythmic frontal delta activity (Fig. 7), become more prominent, and sharp waves in other areas wane in frequency.

2.2. Term Infants

Sleep–wake cycles are now fully developed. The waking record is a continuous mixture of frequencies. Infants at this stage move directly from waking into AS, characterized by two patterns of activity (mixed-frequency AS [Fig. 8], and low-voltage irregular sleep [Fig. 9]), both associated with irregular respiratory pattern and REM. QS is associated with two patterns on EEG, trace alternant (Fig. 10), which persists to 46 wk conceptional age, and with high-voltage slow pattern (Fig. 11), both of which are associated with regular respiratory pattern and absence of eye movements. Frontal and temporal sharp transients persist, observed predominantly in QS, but wane by 44 wk conceptional age. Interhemispheric synchrony is complete in waking and virtually complete in sleep.

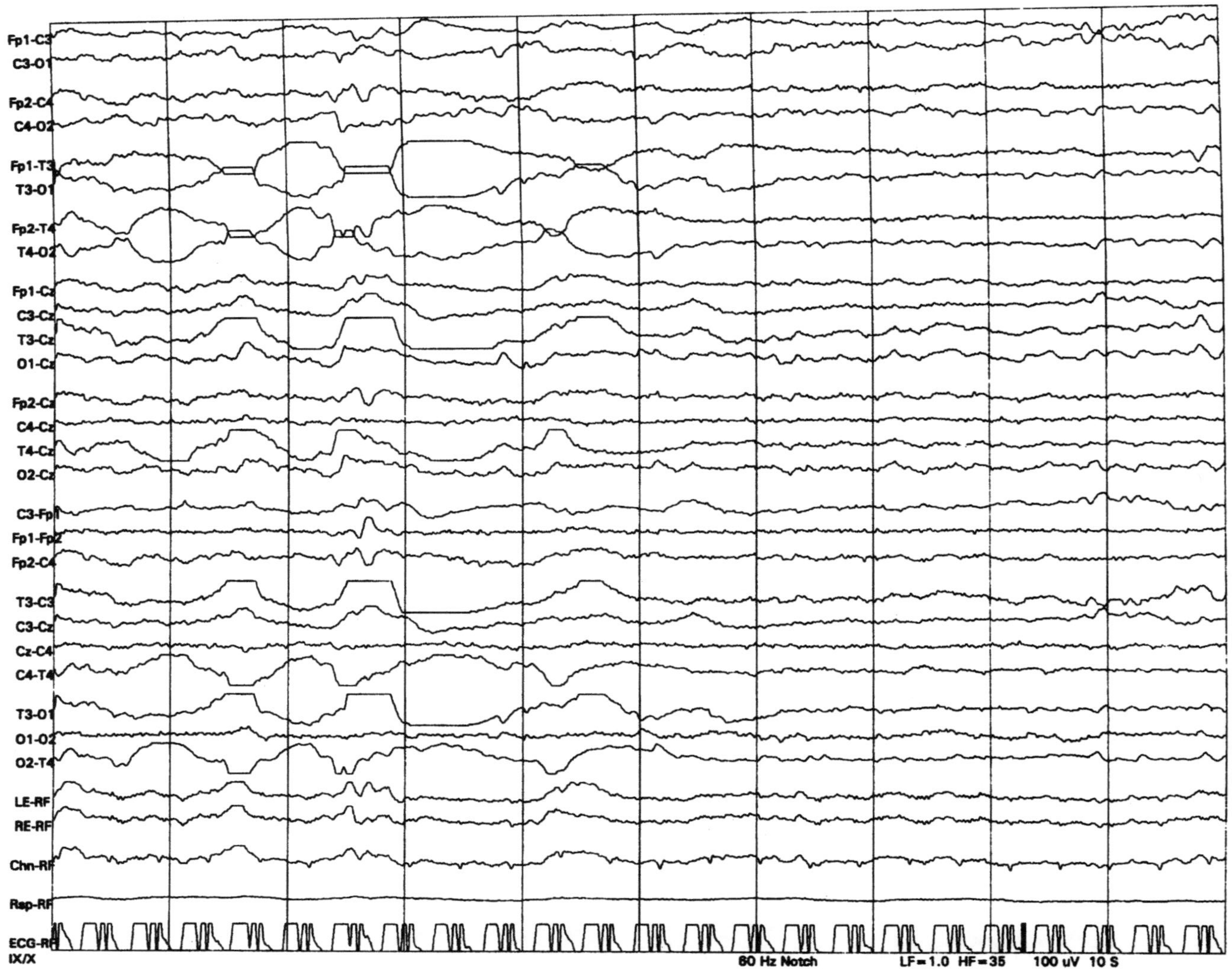

Fig. 1. Twenty-six weeks gestation. Trace discontinu. LE-RF, left eyelid to reference; RE-RF, right eyelid to reference; Chn-RF, chin to reference; LF, low frequency filter; HF, high frequency filter; Rsp, respiration.

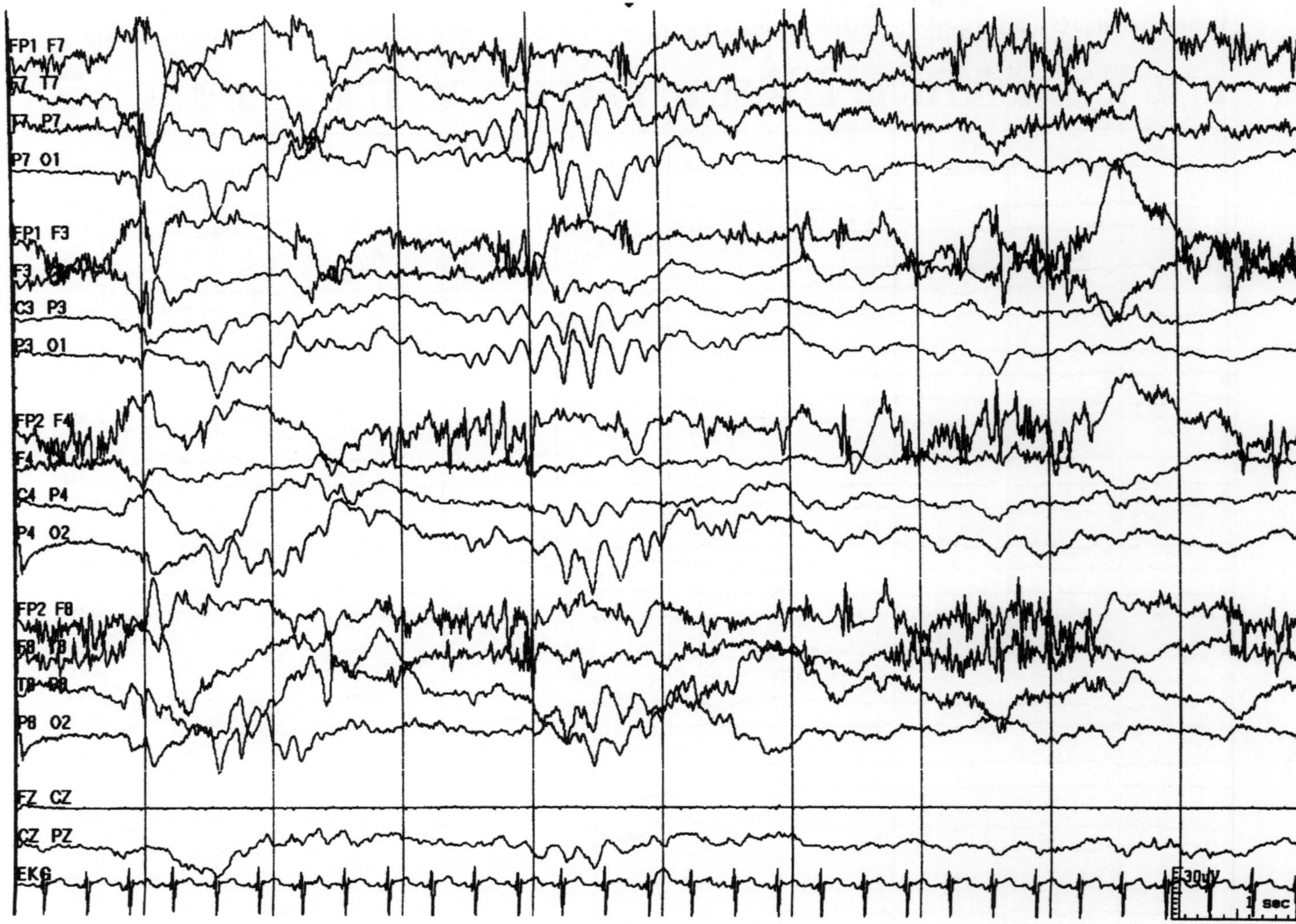

Fig. 2. Twenty-seven weeks gestation. Sharp theta in the occiput of premature infants (STOP).

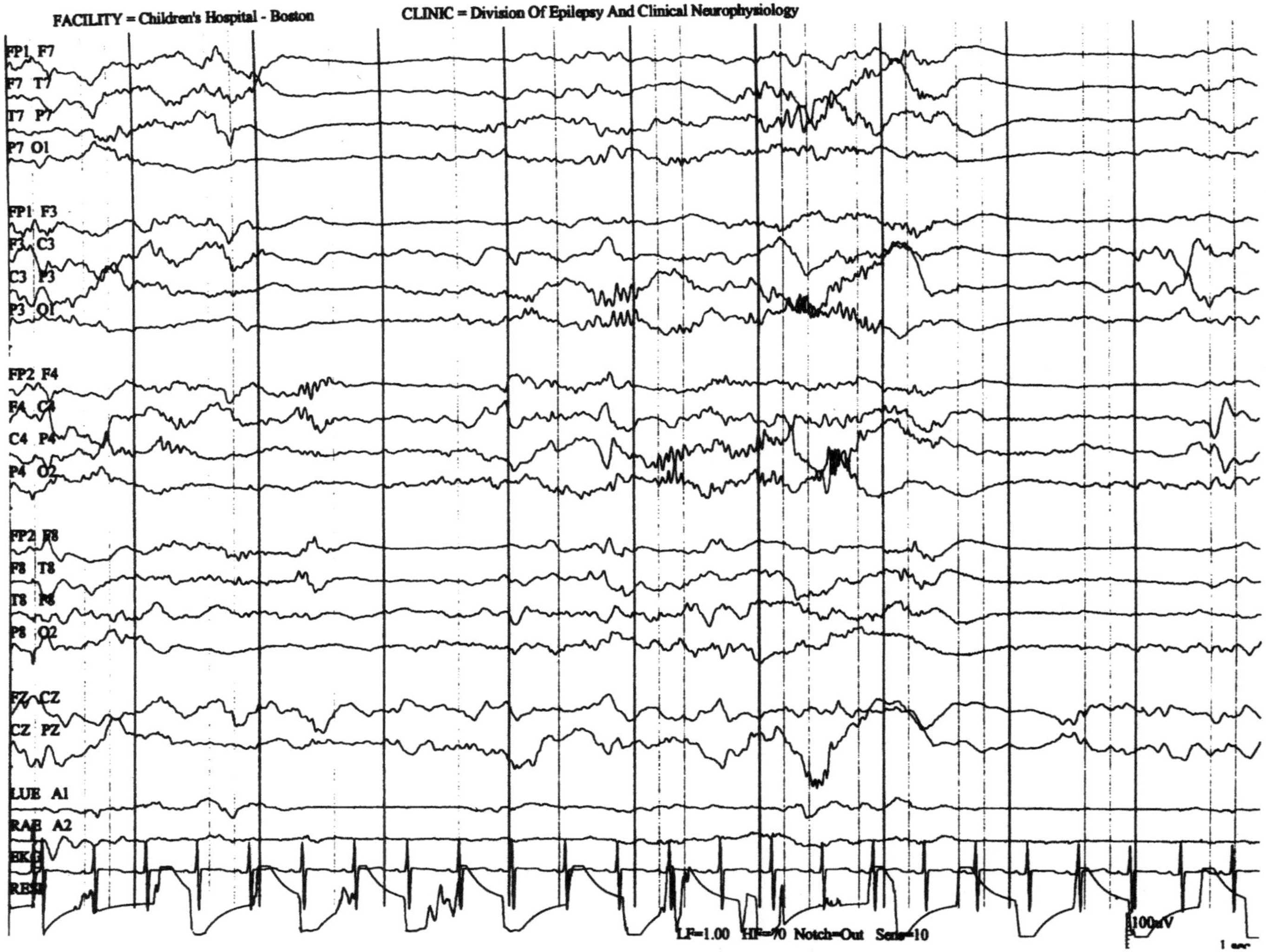

Fig. 3. Thirty weeks gestation. Delta brushes, prominent in the parietal areas. LUE, left under eye; RAE, right above eye; RESP, respiration.

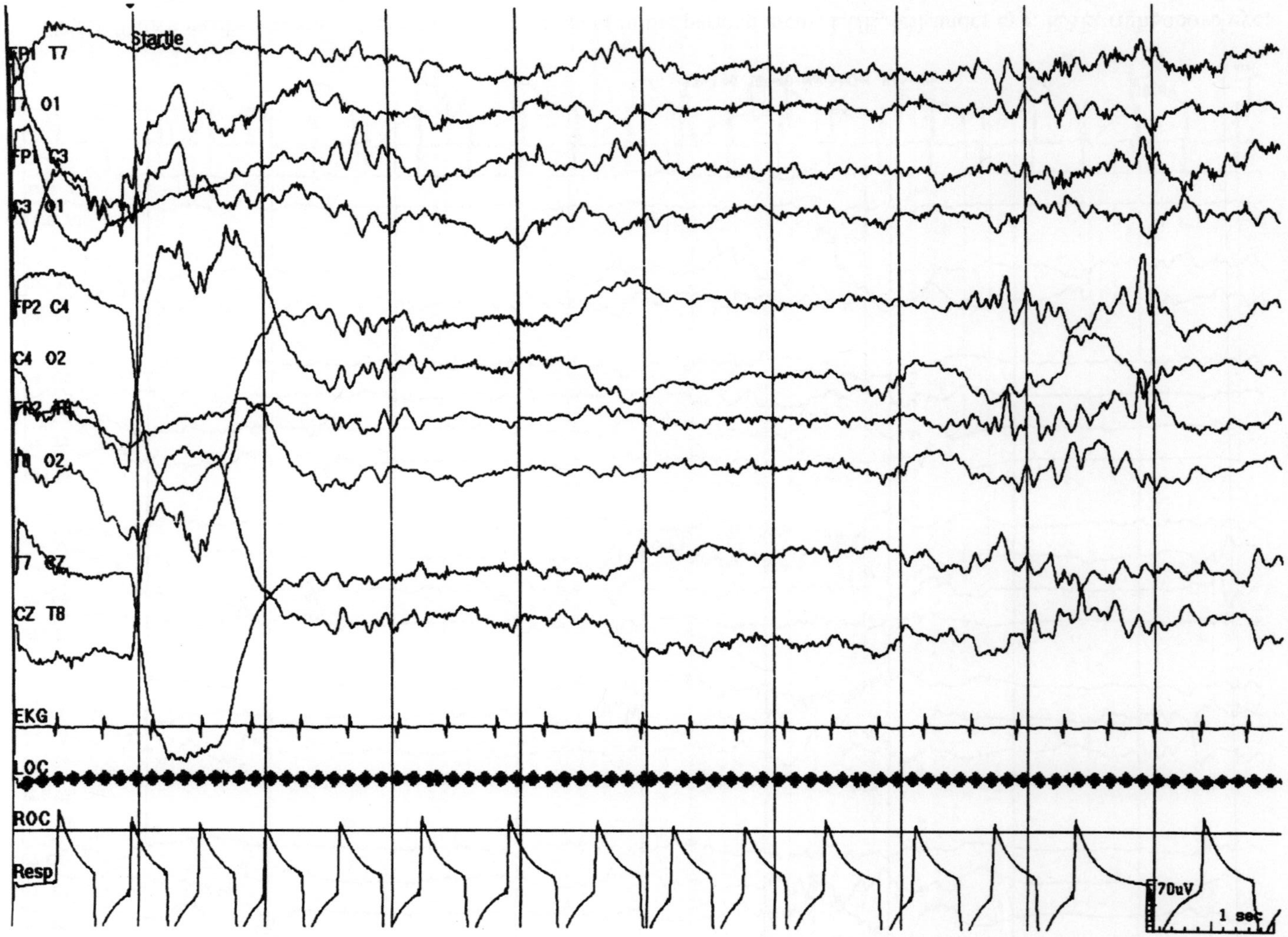

Fig. 4. Thirty-four weeks gestation. Startle, producing arousal with diffuse attenuation of background. LOC, left outer canthus; ROC, right outer canthus; Resp, respiration.

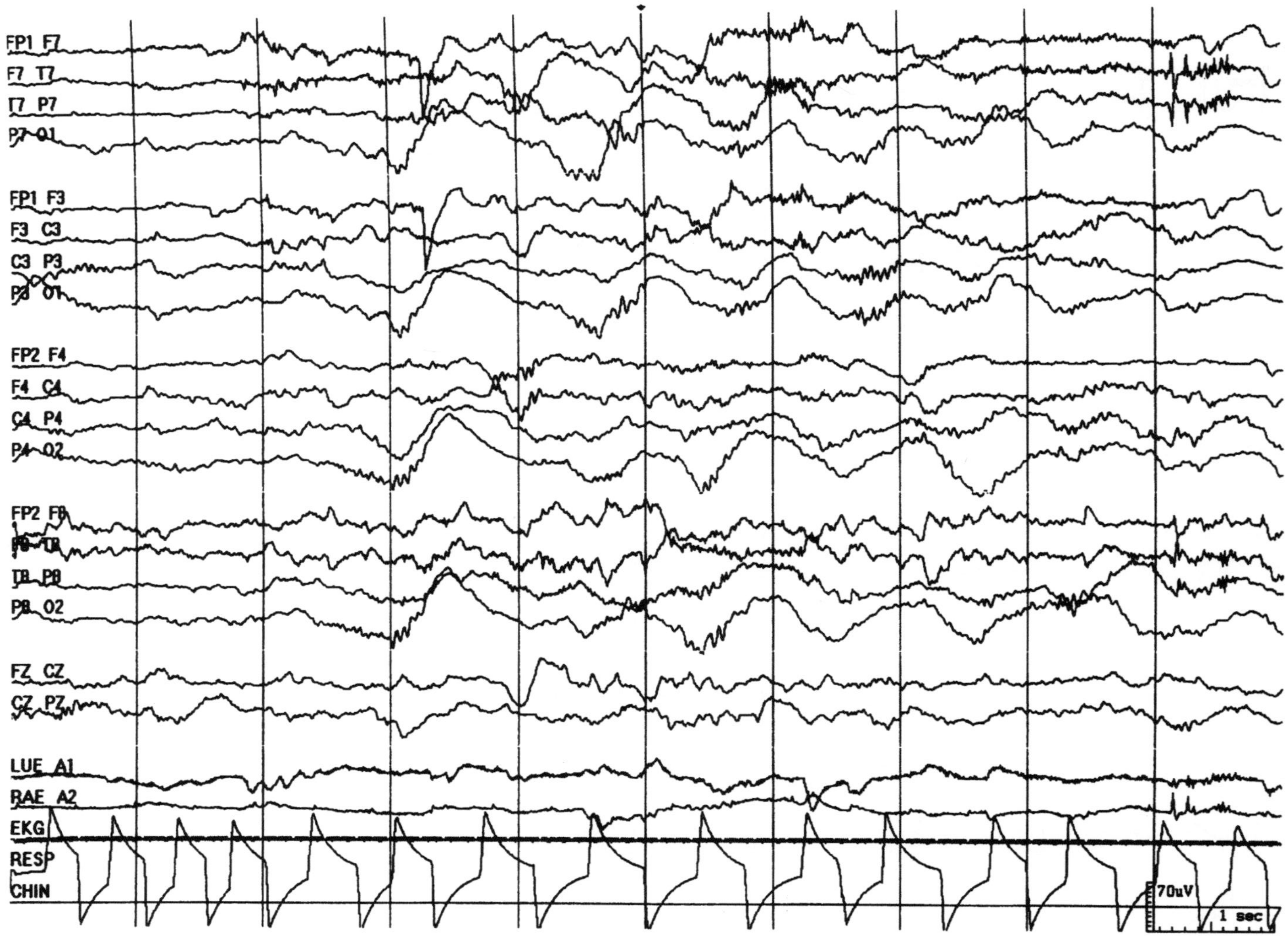

Fig. 5. Thirty-four weeks gestation. Occipital delta and delta brushes. RESP, respiration; LUE, left under eye; RAE, right above eye.

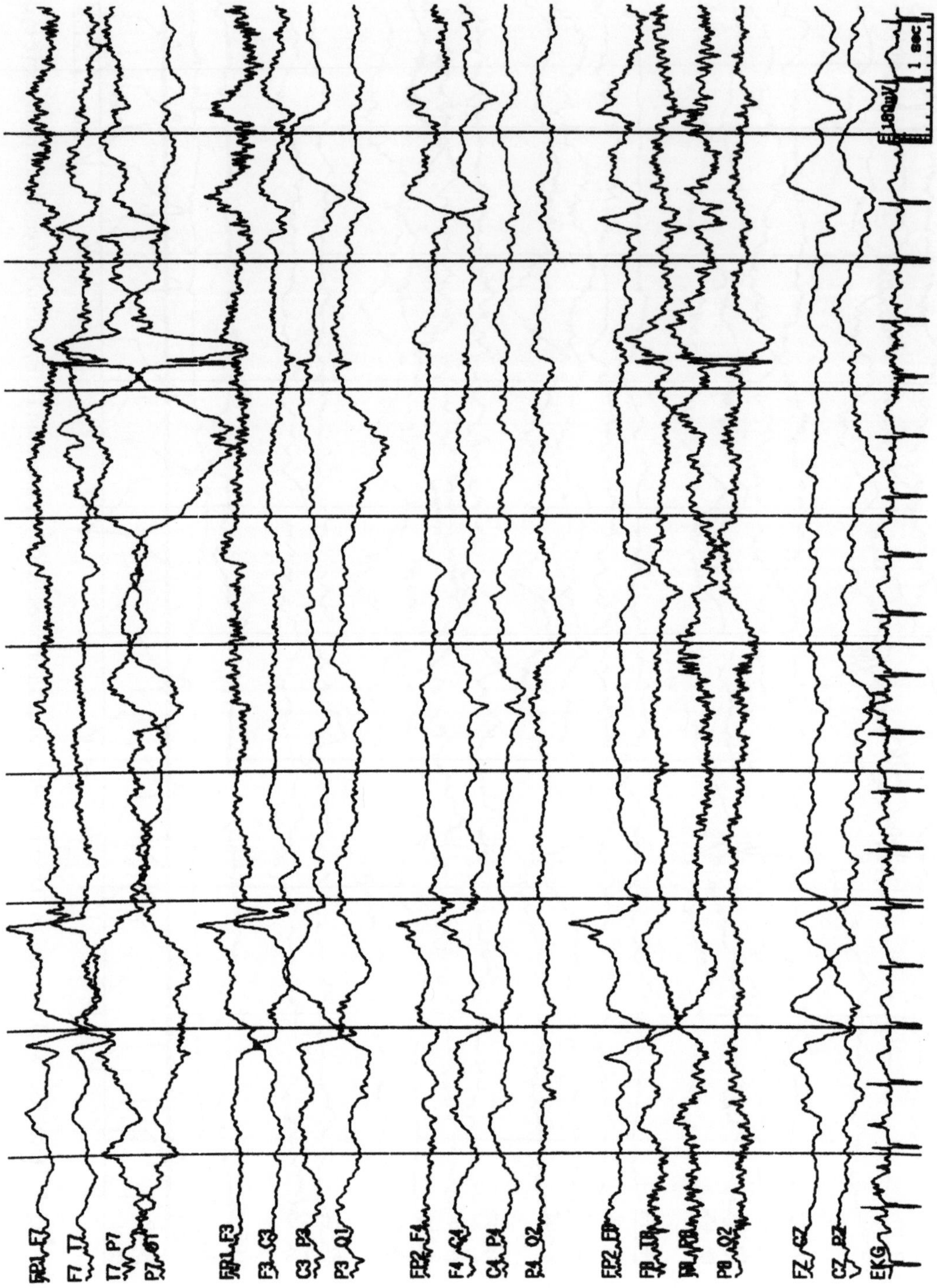

Fig. 6. Thirty-six weeks gestation. Frontal sharp waves (encoches frontales).

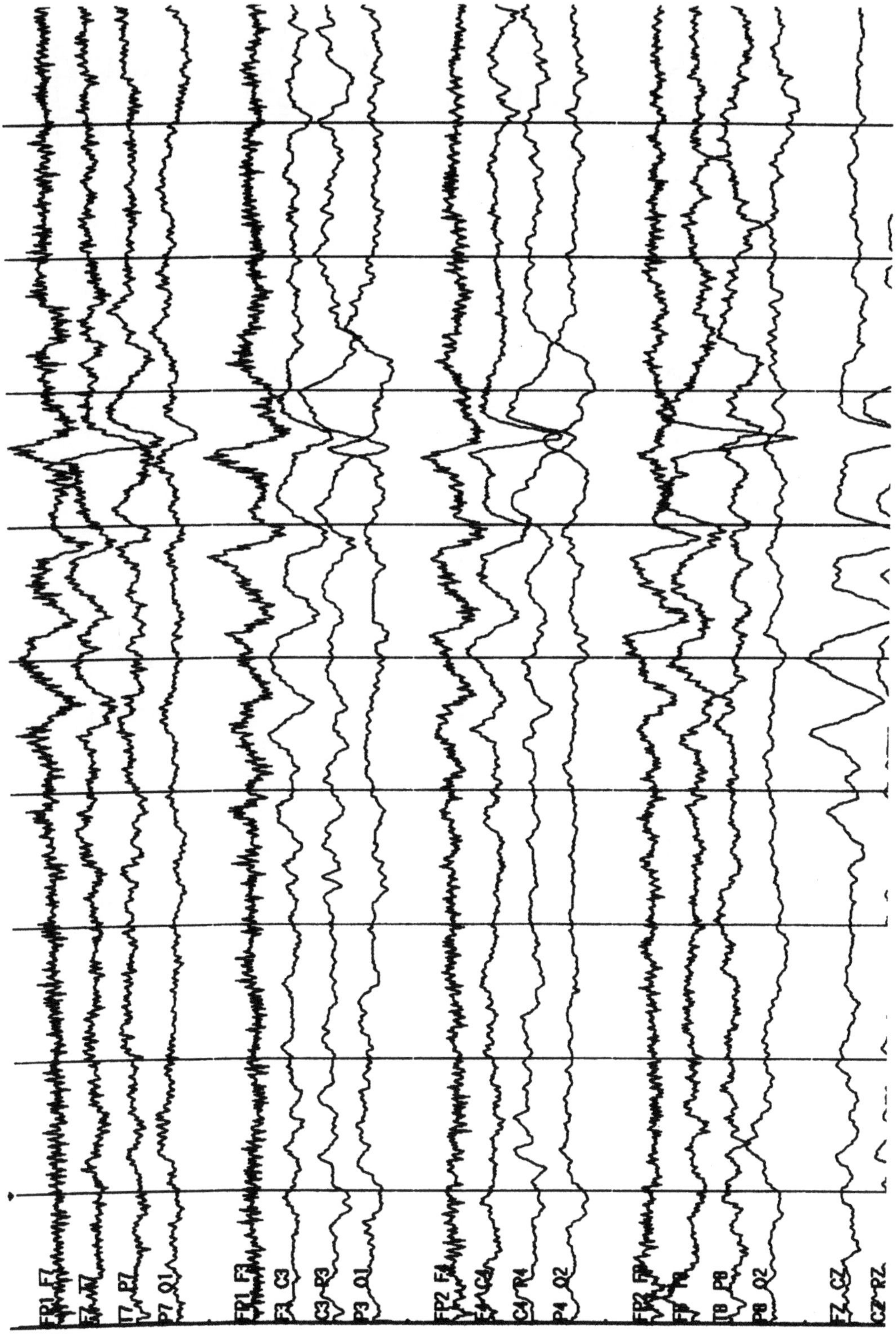

Fig. 7. Thirty-six weeks gestation. Frontal sharp waves and rhythmic frontal delta.

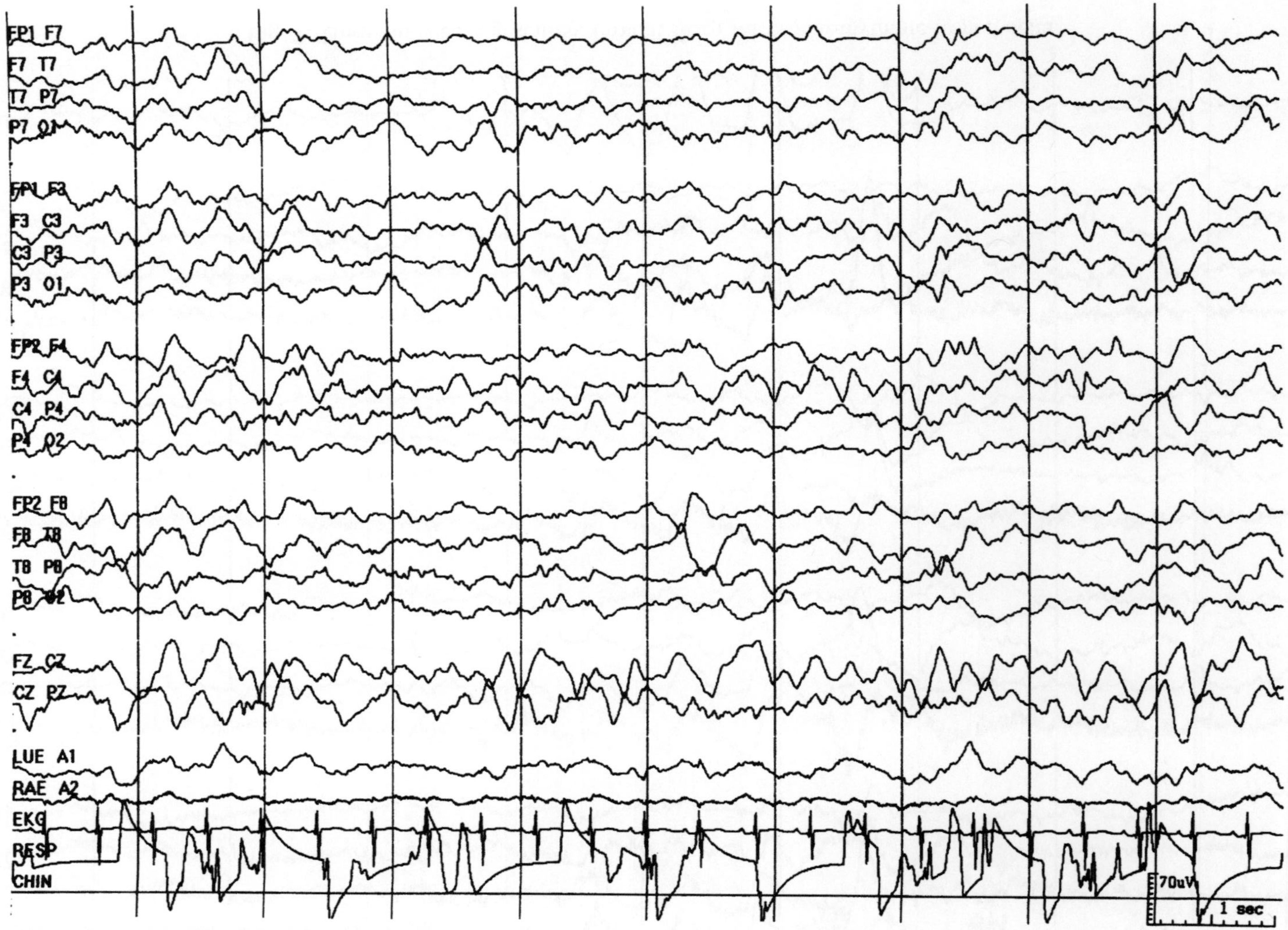

Fig. 8. Term. Active sleep, mixed-frequency pattern (note irregular respirations). RESP, respiration; LUE, left under eye; RAE, right above eye.

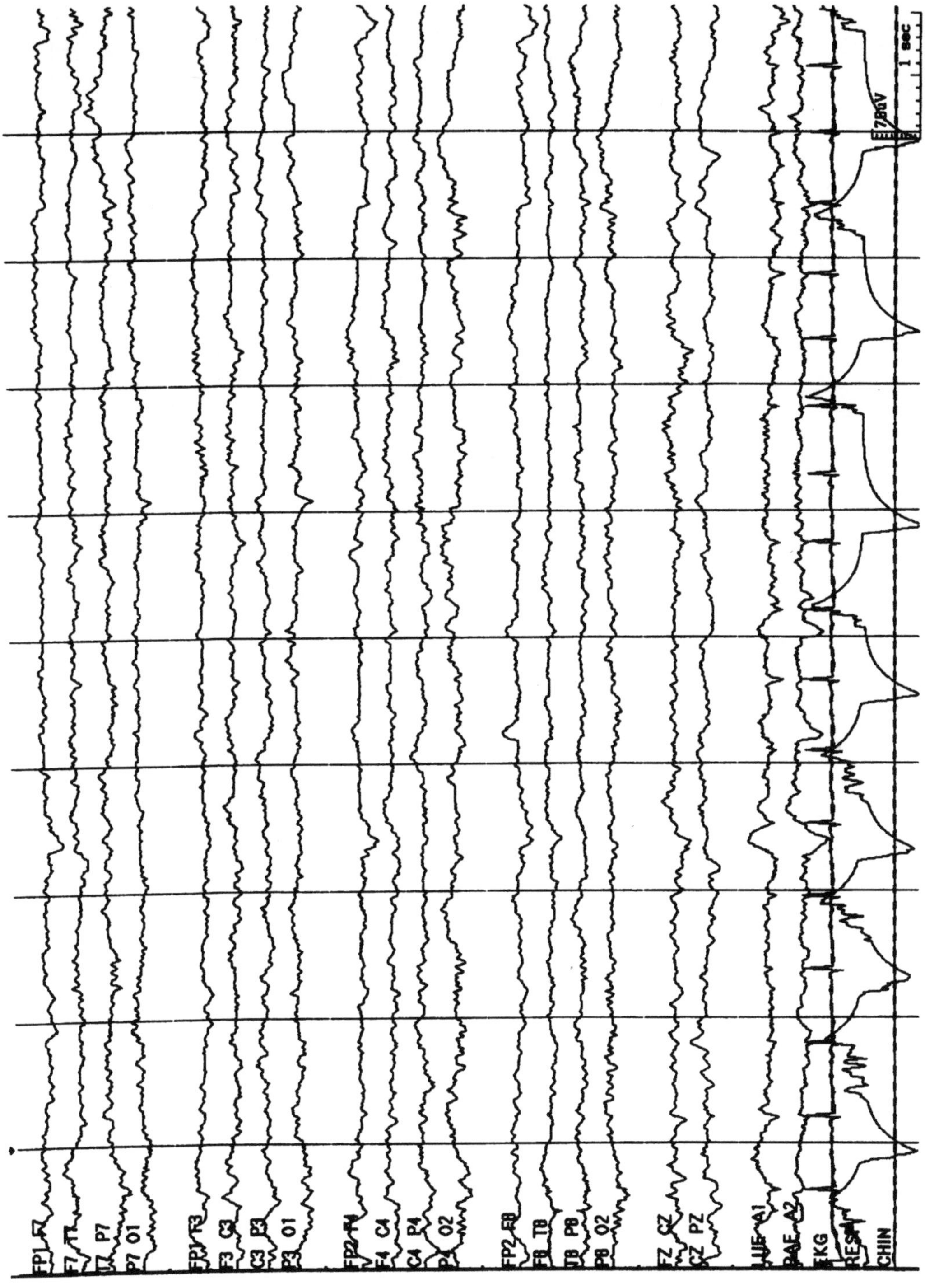

Fig. 9. Term. Active sleep, low-voltage irregular pattern. RESP, respiration; LUE, left under eye; RAE, right above eye.

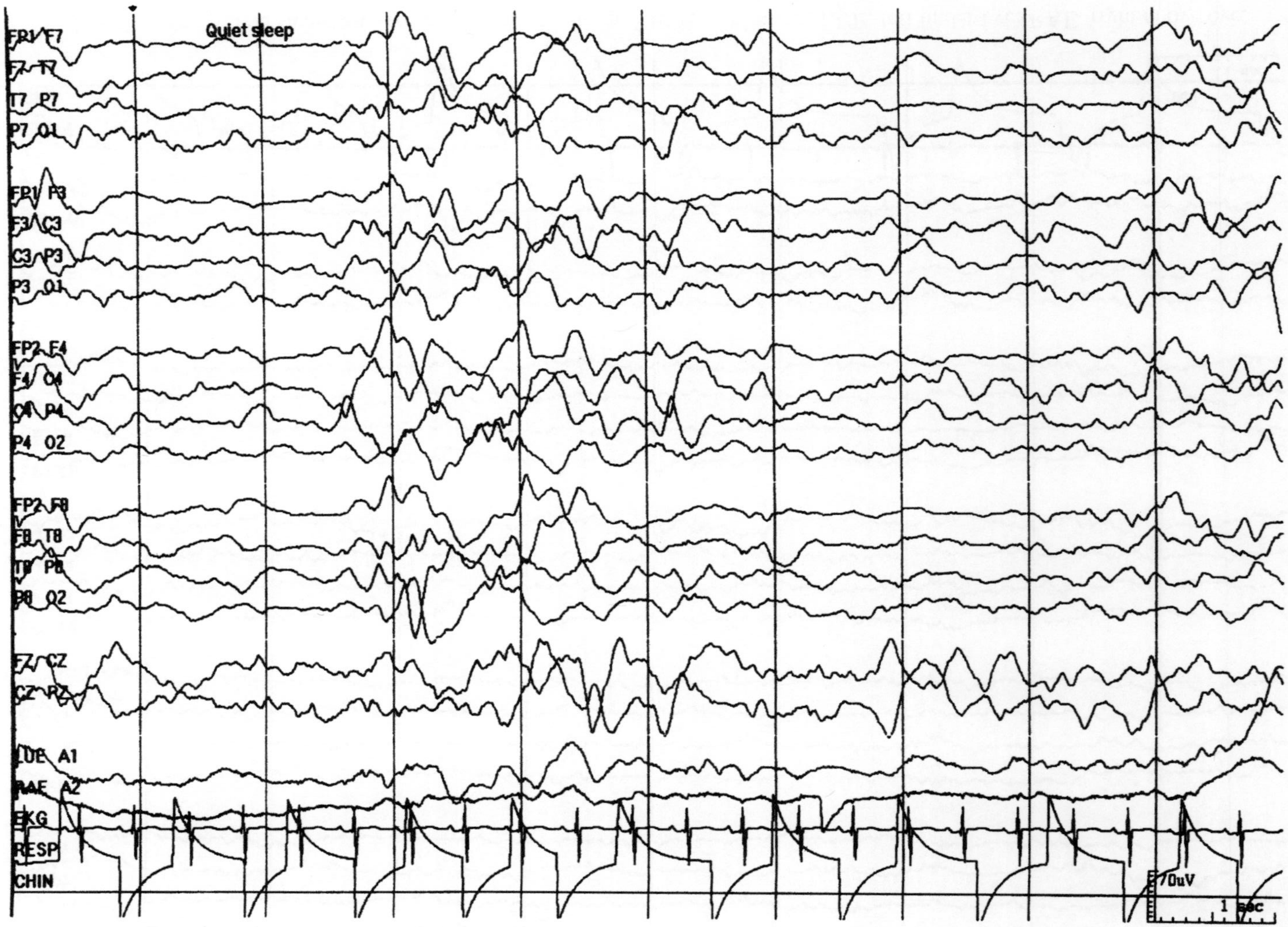

Fig. 10. Term. Quiet sleep, trace alternant pattern. RESP, respiration; LUE, left under eye; RAE, right above eye.

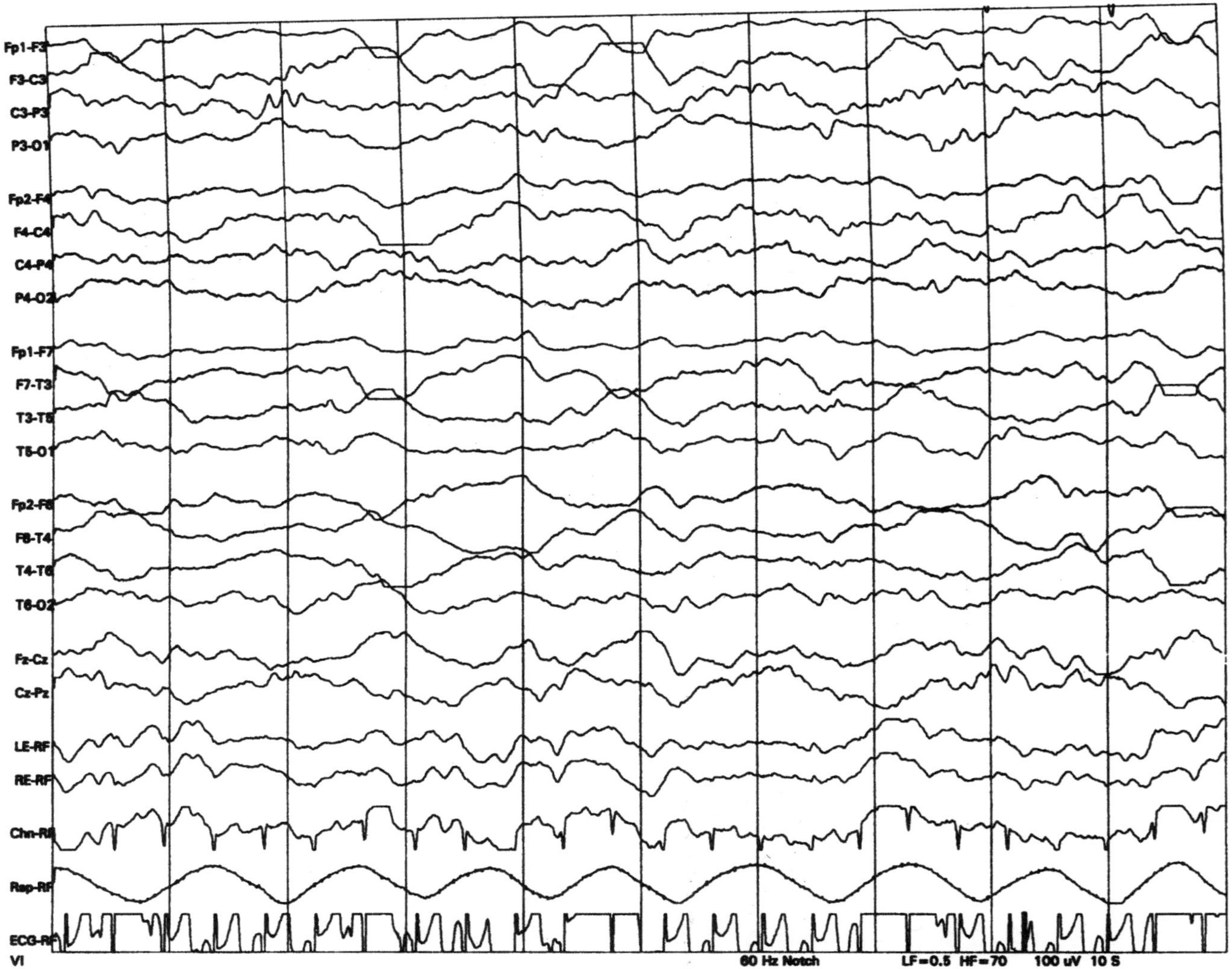

Fig. 11. Term. Quiet sleep, high-voltage slow pattern. Rsp, respiration; LF, low frequency filter; HF, high frequency filter; LE-RF, left eye to reference; RE-RF, right eye to reference; Chn-RF, chin to reference.

2.2.1. Infancy (<1 Yr)

With the end of the neonatal period (after 6-8 wk of age), trace alternant and frontal sharp waves are no longer observed in the healthy infant. In the awake state, an early, often poorly sustained, poorly reactive posterior dominant rhythm of three per second is first observed at 3 mo of age, and often increases to four per second at 4 mo (Fig. 12). Reactivity to eye closure emerges quickly. The posterior dominant rhythm may be up to 6 to 7 Hz by 12 mo of age (Fig. 13). The voltage of the dominant rhythm varies from 30 to 40 μV, up to 100 μV or higher in the first year. The waking background activity is dominated by delta frequencies throughout the first year, but there is a steady increase in the amount of theta activity present. Central theta activity of 4 to 5 Hz may be observed as early as 3 mo (Fig. 14), is usual by 6 mo, and faster (?) rhythms up to 8 Hz are observed in the central/Rolandic areas by 12 mo. Activation by photic stimulation is most likely to produce a driving response in the lower theta frequencies.

After the first to second month, infants move from wakefulness into QS (Fig. 15), instead of directly into AS. The drowsy pattern is characterized by nonspecific slowing and increase in amplitude in the first 6 mo. Thereafter, drowsiness is manifest by an increase in diffuse, highly synchronous and rhythmic, theta activity (hypnagogic hypersynchrony). Sleep spindles develop by 3 mo of age, although fragmentary forms may be observed shortly after term (Fig. 16). Spindles in infancy are typically comb-like, with rounded positive and a sharper negative component. They increase to a maximum duration (up to 10 s) at approx 6 mo of age, and are maximal in the central and parietal areas, rather than at the midline, at this age. They are commonly asynchronous. Vertex waves appear between 3 and 5 mo. They are broader, and less sharply contoured than later in childhood.

2.2.2. Early Childhood (>1 to 3 Yr)

The posterior dominant rhythm increases in frequency from 6 to 7 Hz in the second year to 7 to 8 Hz in the third year, and the blocking response to eye opening is now robust (Fig. 17). As in adults, the dominant rhythm may be of greater amplitude in the nondominant hemisphere. The difference should not be greater than 50%. In the waking background, delta activity remains prominent and may be observed diffusely or shifting in position throughout the record. There is a relative increase in the amount of theta activity, and this is visually the most striking frequency at this age. Throughout childhood, waking theta activity is prominent, often shifting in prominence from side to side. Children of this age are usually unable to cooperate with hyperventilation, but occasionally sobbing may induce diffuse slowing caused by a hyperventilation effect (Fig. 18). Occipital driving response to photic stimulation is still more likely at slower stimulation rates.

With drowsiness, diffuse, high-voltage, rhythmic theta (3–5 Hz) appears, mainly in the parasagittal areas. It is typically continuous, but may appear as discrete bursts in some children. It is often also present at arousal (hypnagogic and hypnopompic hypersynchrony) (Fig. 19). As the child progresses into sleep, diffuse irregular slow activity (1–3 Hz) develops, mixed with medium voltage theta activity. Slow activity has a maximal amplitude in the occipital leads (Fig. 20). Vertex sharp waves appear, which are now of higher voltage and more sharply contoured than previously. Runs of vertex sharp waves may occur (Fig. 21). Spindles, usually 12 to 14 Hz, may have a wider field, and are mostly synchronous by 2 yr of age.

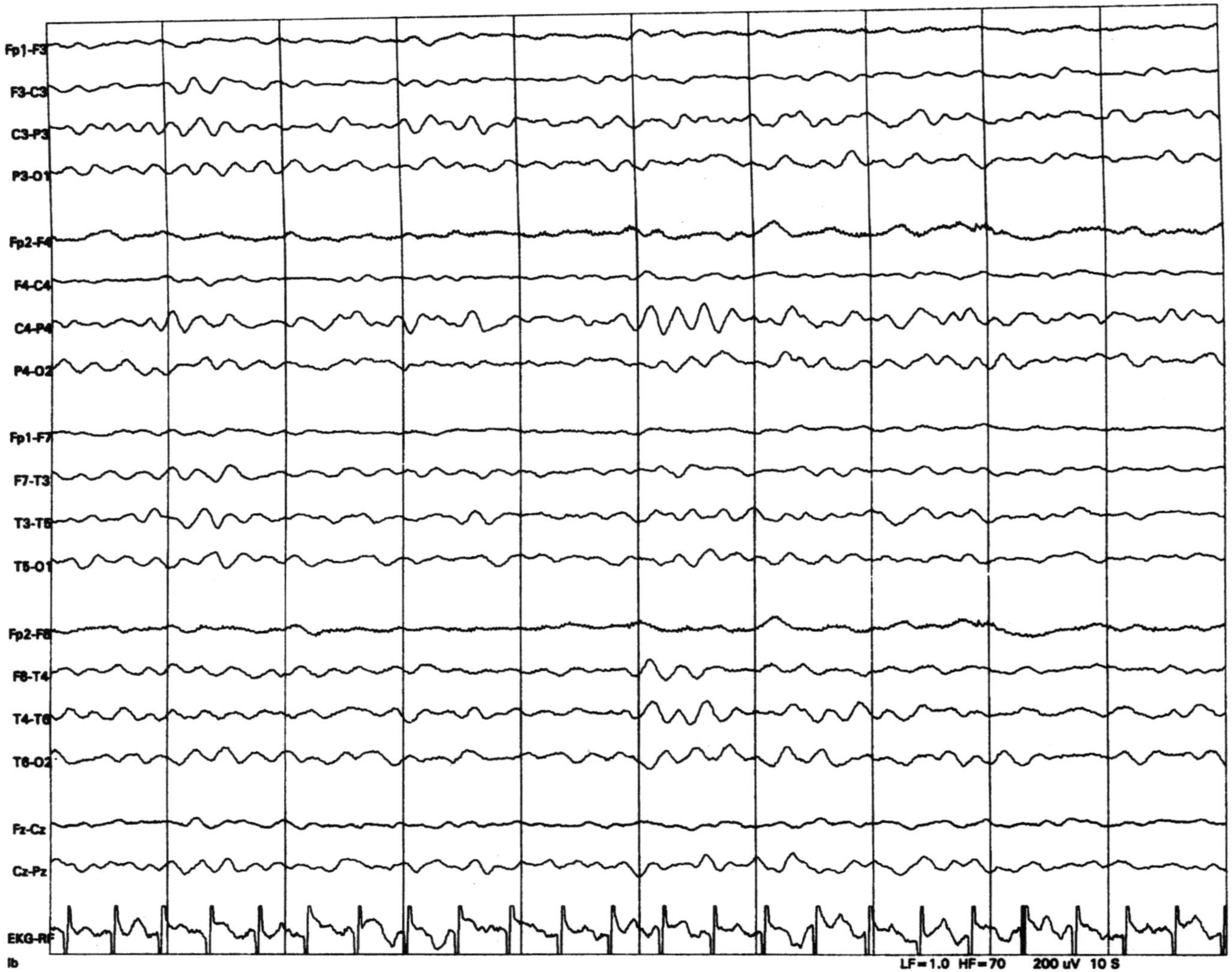

Fig. 12. Four months. Four per second posterior rhythm; LF, low frequency filter; HF, high frequency filter.

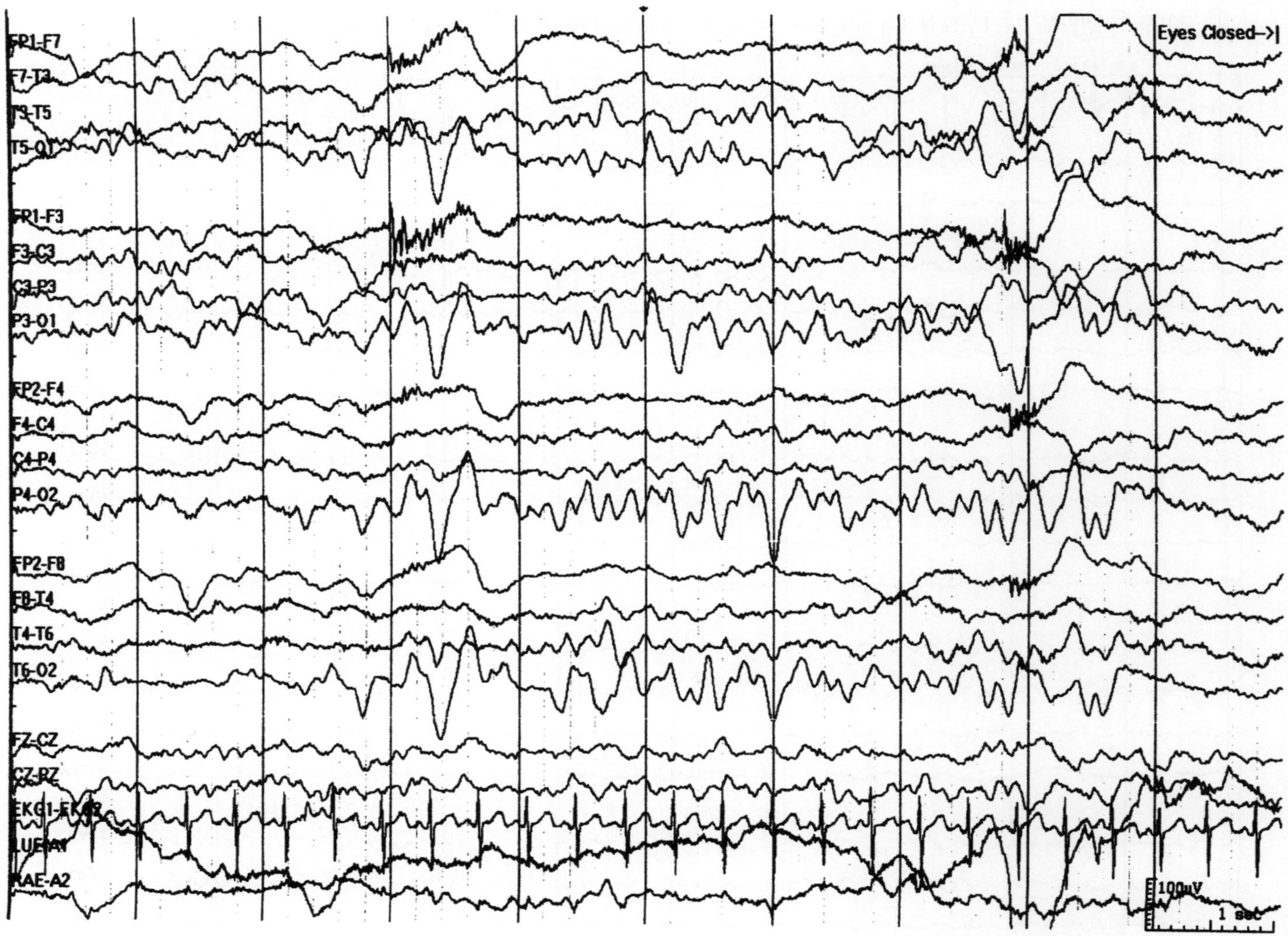

Fig. 13. Twelve months. Six per second reactive posterior dominant rhythm; LUE, left under eye; RAE, right above eye.

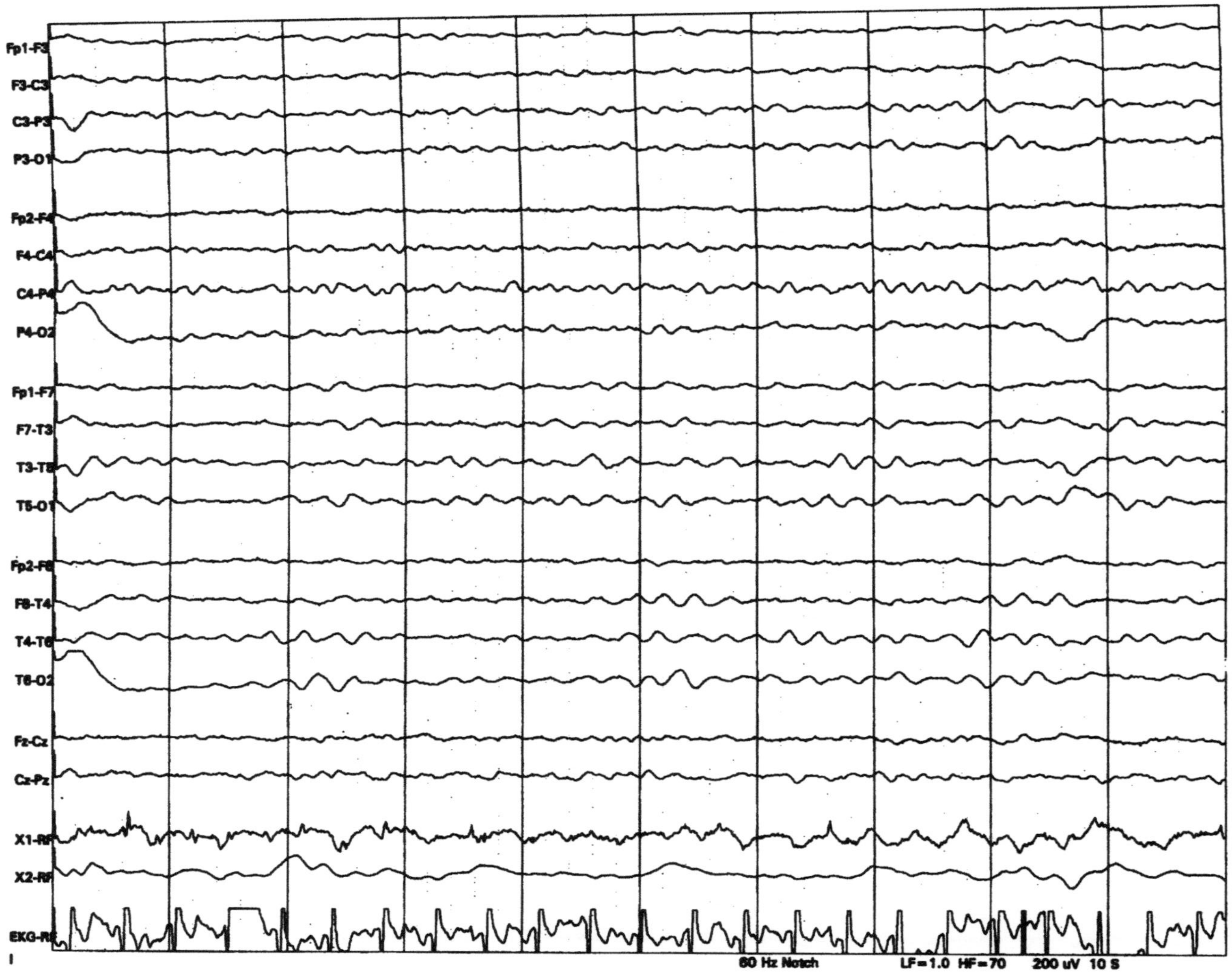

Fig. 14. Four and a half months. Five per second central activity, four per second posterior rhythm; LF, low frequency filter; HF, high frequency filter.

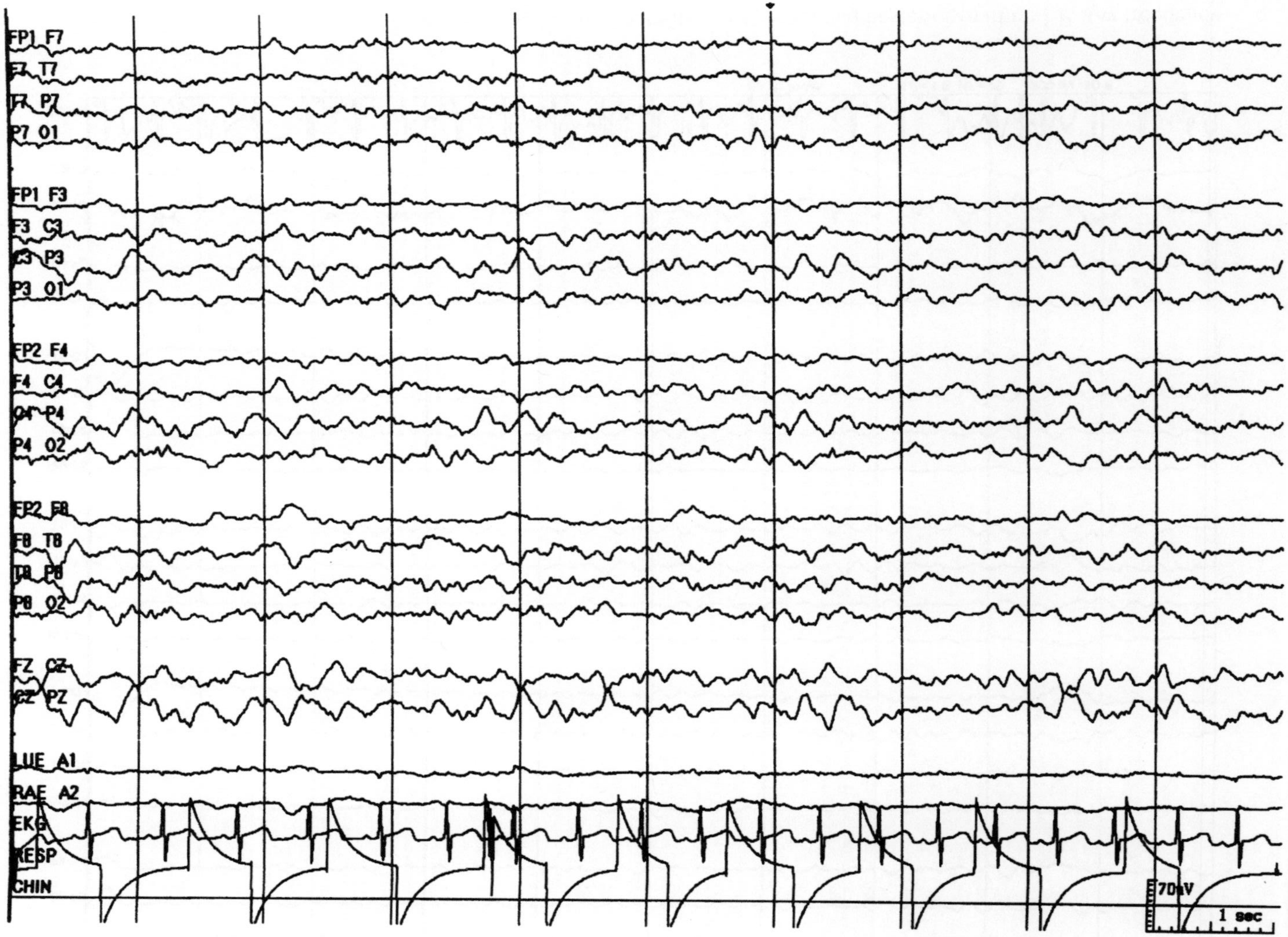

Fig. 15. One month. Quiet sleep. RESP, respiration; LUE, left under eye; RAE, right above eye.

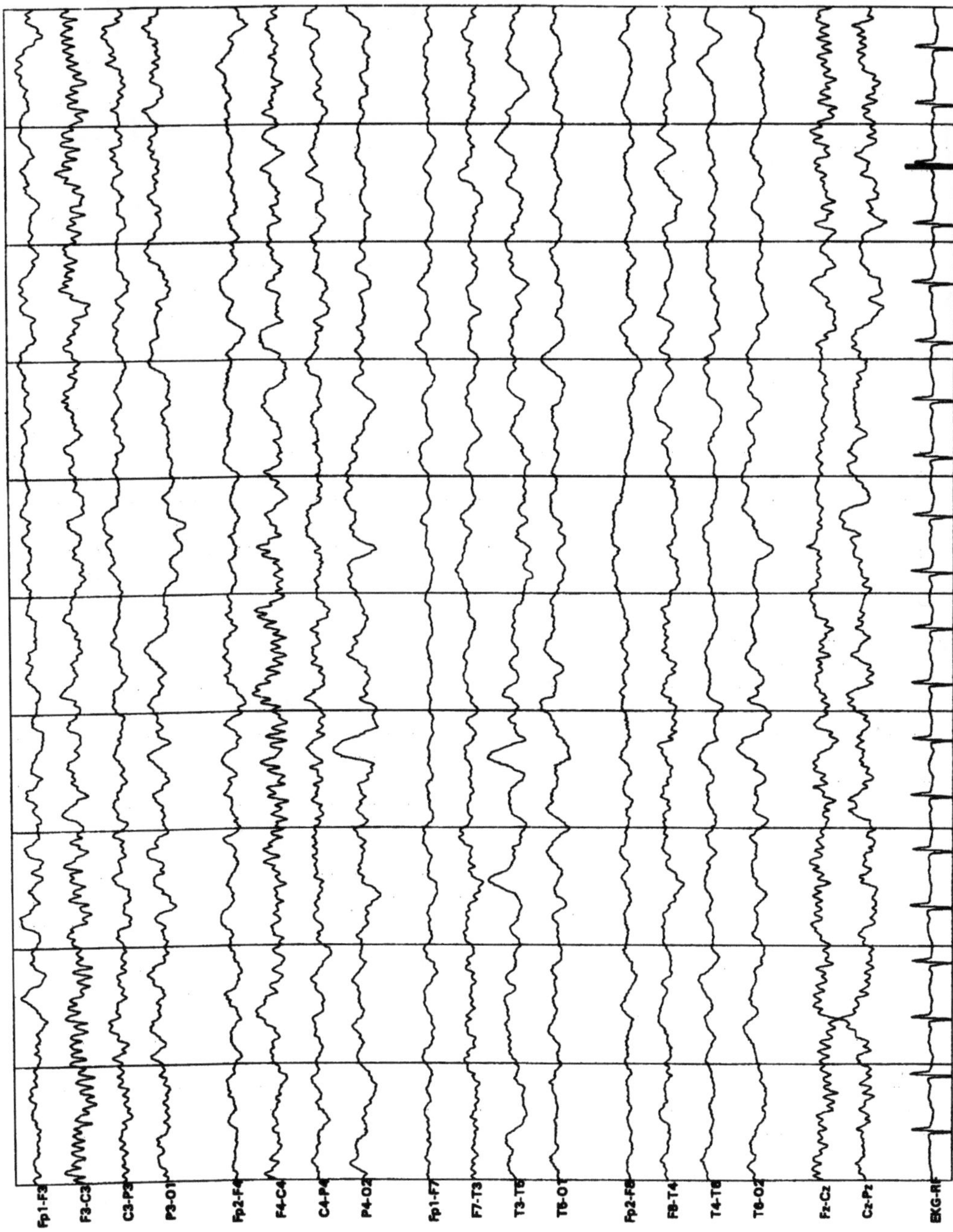

Fig. 16. Three months. Asynchronous sleep spindles; LF, low frequency filter; HF, high frequency filter.

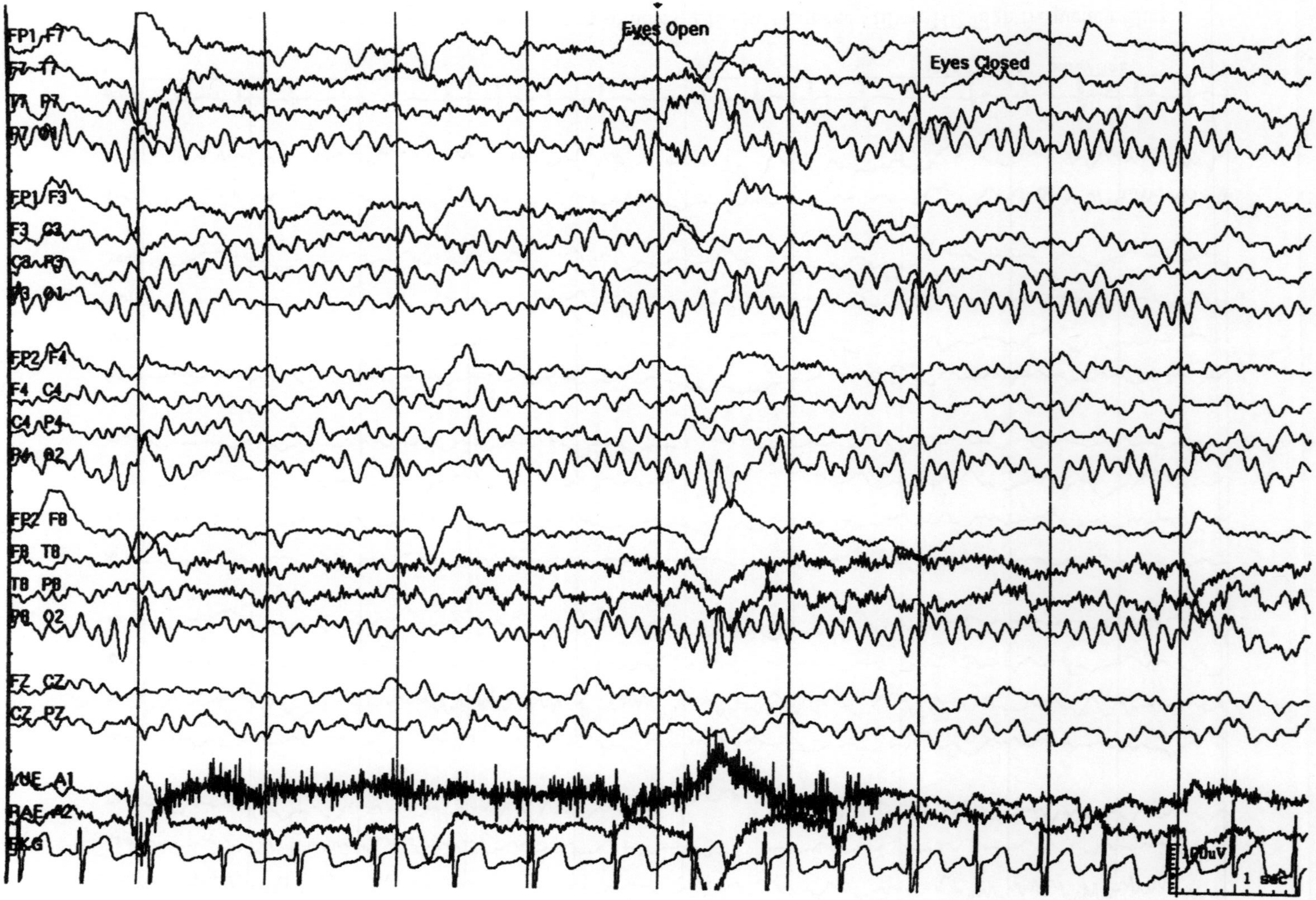

Fig. 17. Three years. Reactive posterior dominant rhythm in the alpha range (8/s); LUE, left under eye; RAE, right above eye.

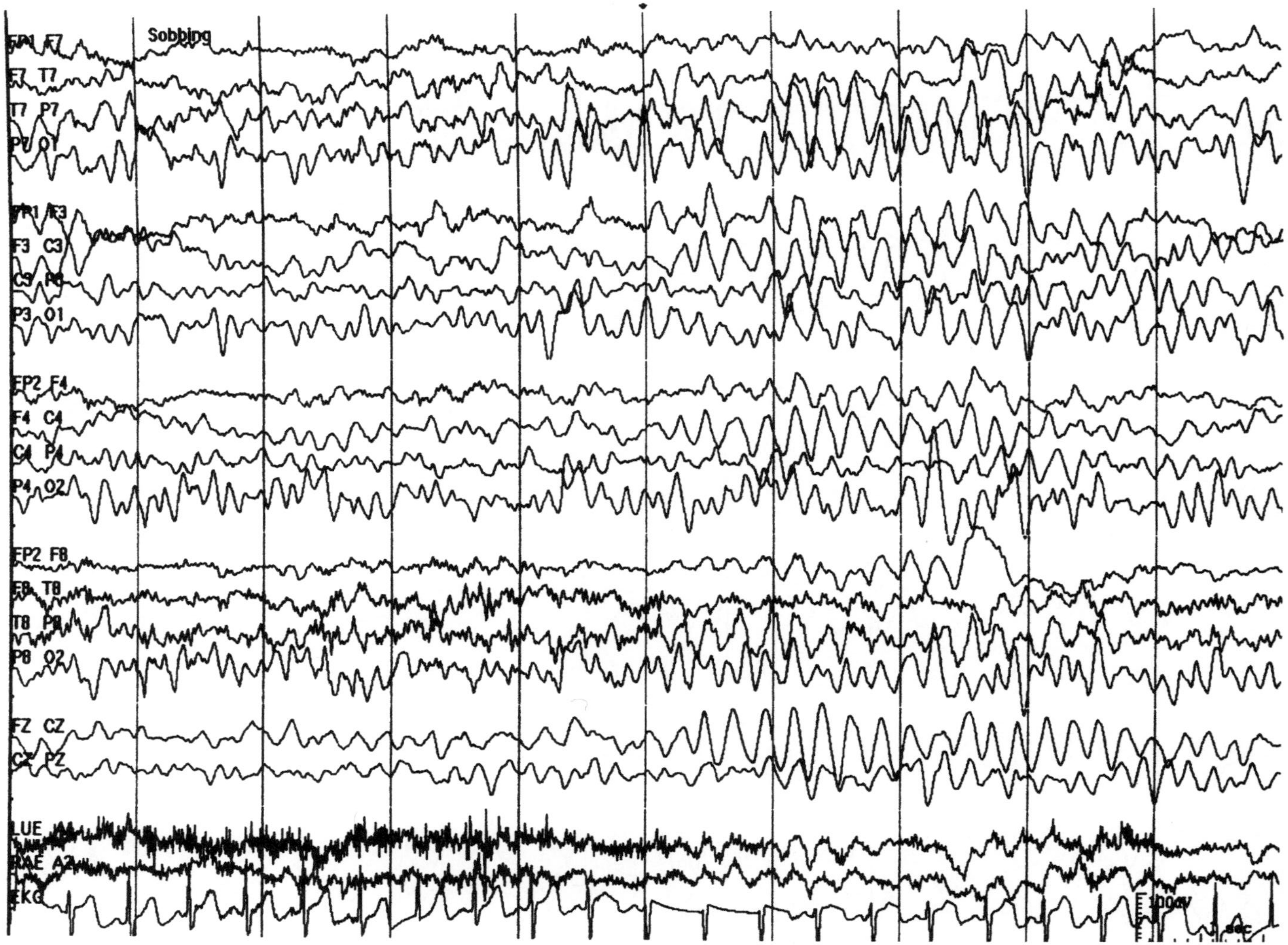

Fig. 18. Three years. Sobbing produces hyperventilation effect; LUE, left under eye; RAE, right above eye.

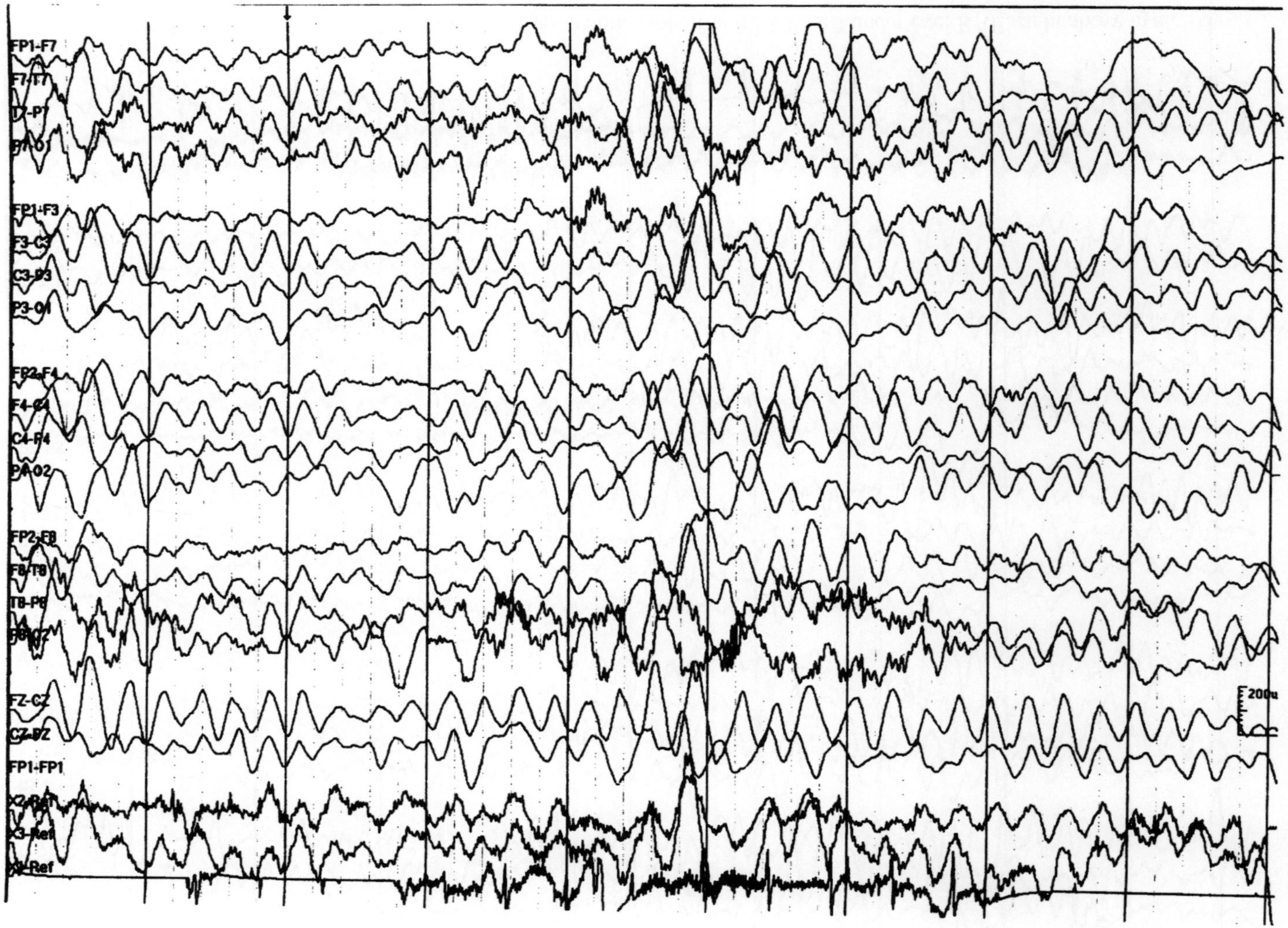

Fig. 19. Three years. Hypnagogic hypersynchrony with drowsiness; LUE, left under eye; RAE, right above eye.

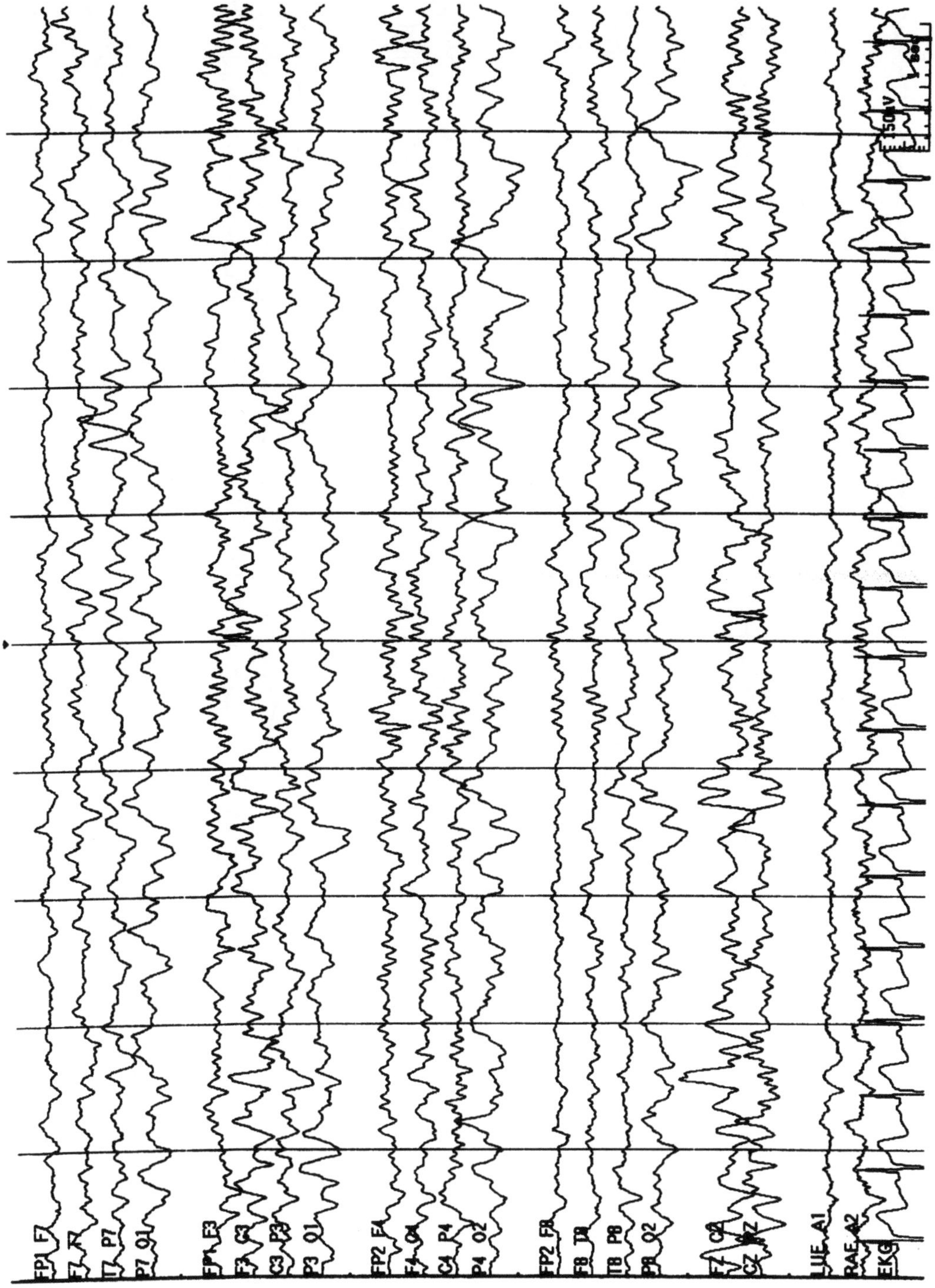

Fig. 20. Three years. Sleepspindles with frontal maximum, delta with posterior maximum; LUE, left under eye; RAE, right above eye.

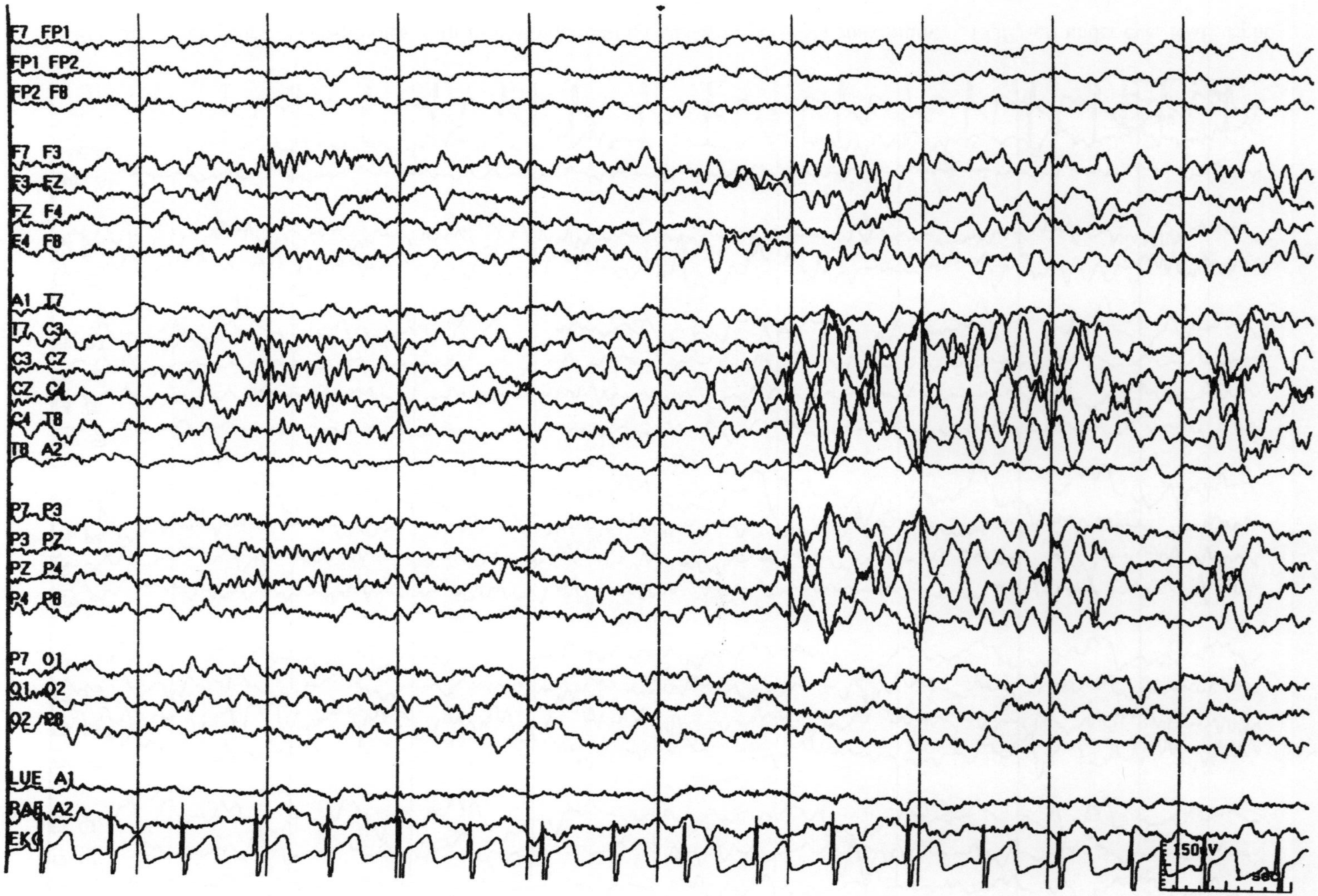

Fig. 21. Three years. Sleep ~~V~~ wave and run of V waves (transverse montage). Fourteen per second spindle with v ertex maximum; LUE, left under eye; RAE, right above eye.

2.2.3. Preschool Age (>3 to 6 Yr)

At this age, the posterior basic rhythm consistently reaches alpha frequency. It is still of high amplitude, often greater than 100 μV. Throughout early childhood, low voltage background (<30 μV) is abnormal. Posterior slow waves of youth emerge at this age (Fig. 22). These are 1.5- to 3-Hz waves, maximal in the occipital region. They are intermixed with posterior alpha, and, at times, fused slow waves can resemble occipital sharp waves, although lacking typical morphology and after-coming slow wave. Posterior slow waves, in common with the posterior dominant rhythm, block with eye opening. This pattern persists throughout childhood and adolescence, disappearing in young adulthood. Rolandic mu rhythm may be apparent at this stage, often shifting from side to side. Children can now cooperate with hyperventilation. Hyperventilation produces prominent diffuse slowing to 3 to 5 Hz, which may be more apparent on the left initially, although becoming symmetrical (Fig. 23). This response is enhanced by fasting. It may persist beyond apparent cessation of hyperventilation if the child continues to breathe deeply. Intermittent photic stimulation is still associated with a driving response at stimulation rates less than 8 Hz.

Drowsiness is often still associated with hypersynchrony, as described in Section 2.2.2. Vertex waves are increasingly sharply contoured. Spindles are now maximal in the midline, at 14 Hz, decreasing to 10 Hz with deeper sleep. Sleep-related slowing is still maximum in amplitude posteriorly. Positive occipital sharp transients of sleep (POSTS) are not yet expected, and, if present, are poorly formed. Frontal arousal rhythm in the theta range may be observed, but is more common later.

2.2.4. Late Childhood (>6 to 12 Yr)

Posterior dominant rhythm reaches 10 Hz by 10 yr of age, and reaches its maximum amplitude before that age. Posterior slow waves are prominent, and may be asymmetric, with higher amplitude on the right, as with the posterior dominant rhythm. Medium voltage semi-rhythmic frontal theta activity may be observed in healthy children at this age, and may persist into young adulthood (Fig. 24). There is increasing prominence of the mu rhythm (up to 15–16 yr) (Fig. 25). Lambda waves may be observed posteriorly with saccadic eye movements in response to patterned visual stimulus. Hyperventilation still produces high-voltage slowing (1.5–4 Hz). Intermittent photic stimulation now stimulates driving at 6 to 16 Hz.

Hypnagogic hypersynchrony is disappearing, and is rare after the age of 6 yr. The drowsy pattern at this age is gradual alpha dropout, with increasing amounts of theta and delta activity. Vertex sharp waves are still prominent, with large amplitude, and may have asymmetric field over the midline. Spindles are now disposed anteriorly over the frontal midline, and typically last less than 1 s. Fully developed POSTS are observed for the first time (Fig. 26), and other sleep patterns, such as 14- and/or 6-Hz positive sharp waves are more common.

2.2.5. Adolescence (>12 to 18 Yr)

At this age, the EEG begins to resemble the adult EEG more closely, as the amount of underlying delta activity wanes completely (Fig. 27). The amplitude of the posterior dominant rhythm also declines gradually, although it remains higher than in adults throughout this period in many children. The mu rhythm reaches its maximum prominence at 15 to 16 yr, waning thereafter. Hyperventilation-related slowing is less pronounced, and the response to intermittent photic stimulation is mature, with a driving response occurring over the range 6- to 20-Hz stimulation.

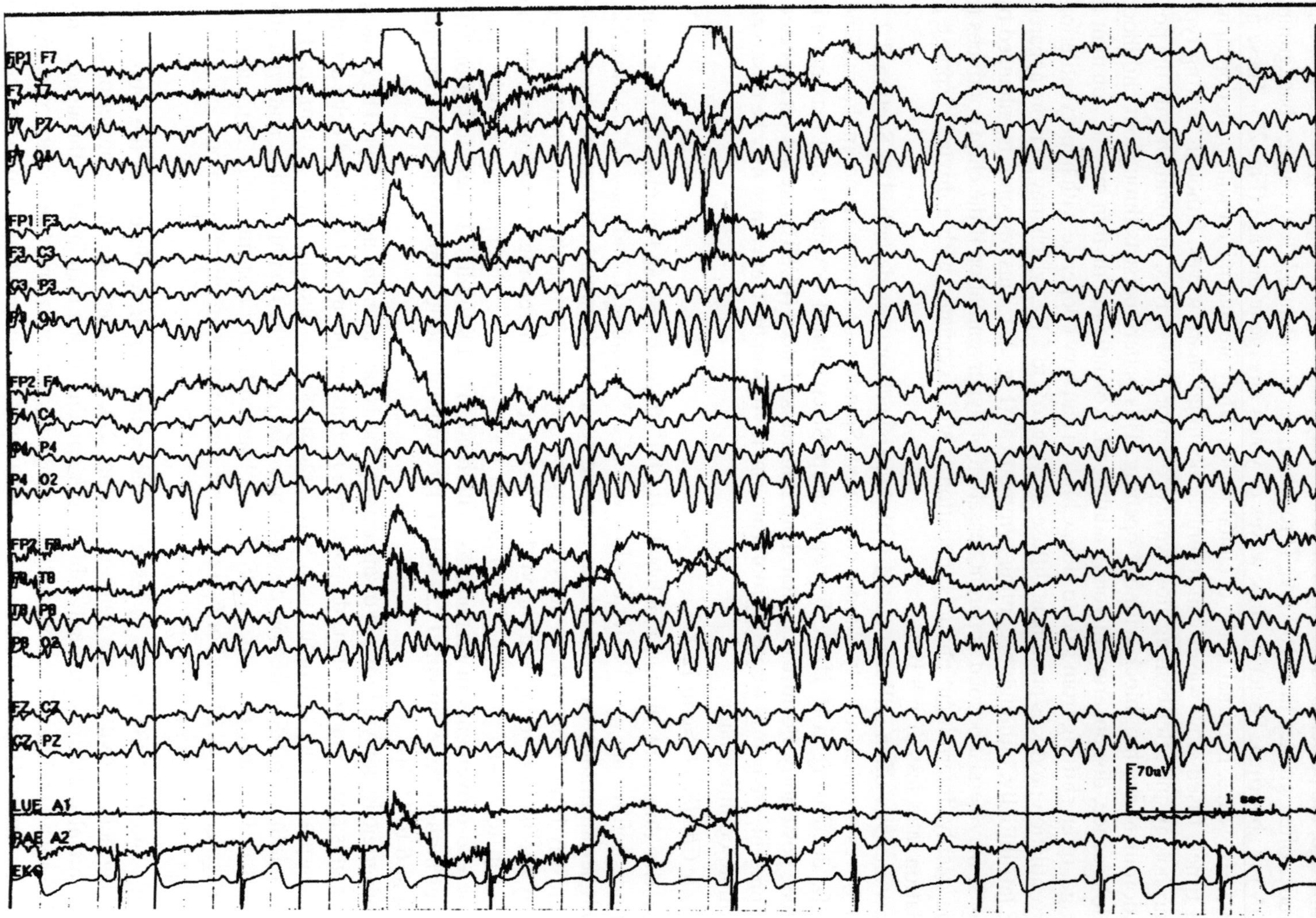

Fig. 22. Six years. Posterior slow waves of youth; LUE, left under eye; RAE, right above eye.

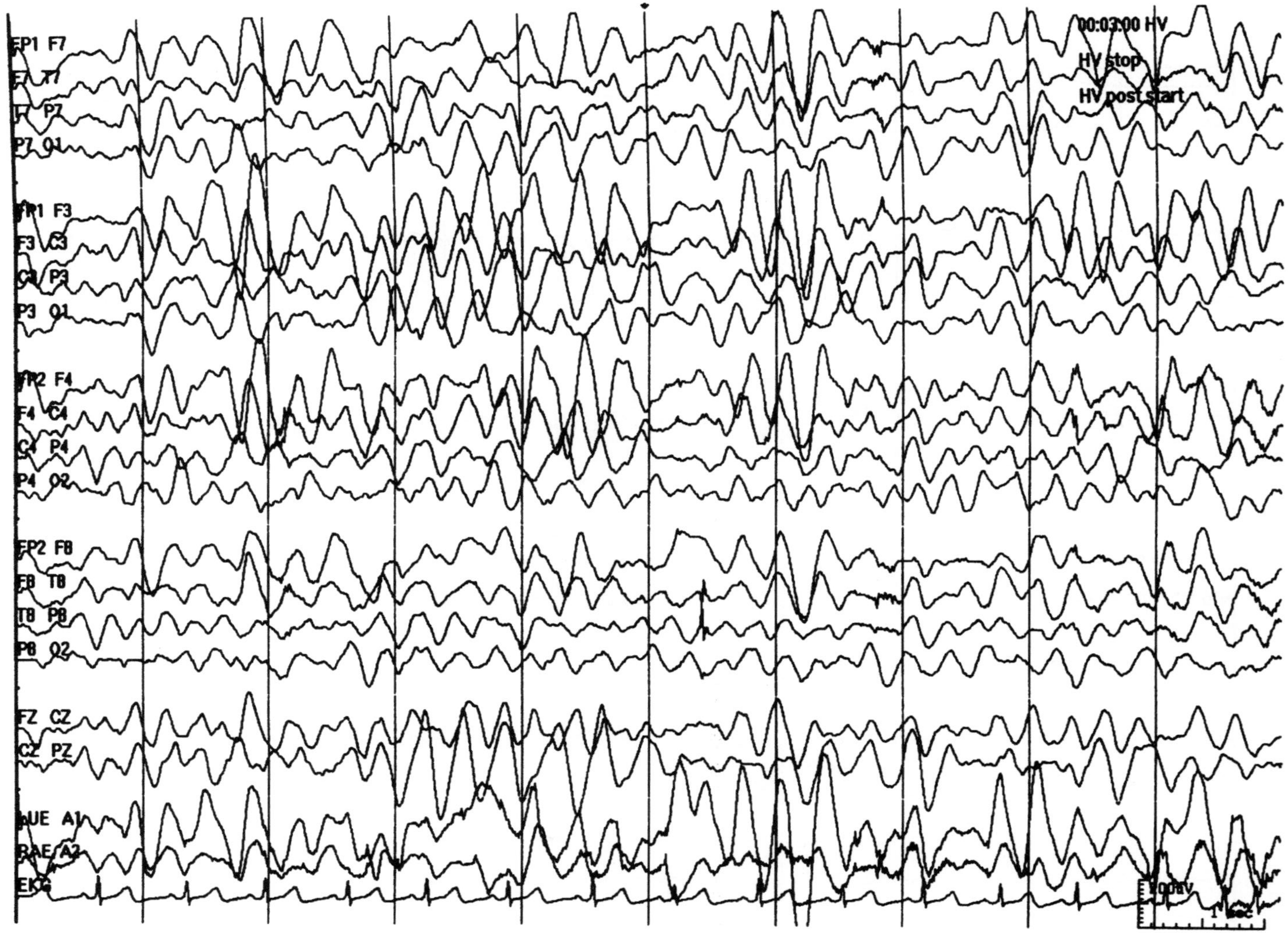

Fig. 23. Six years. Hyperventilationdiffuse slowing; LUE, left under eye; RAE, right above eye.

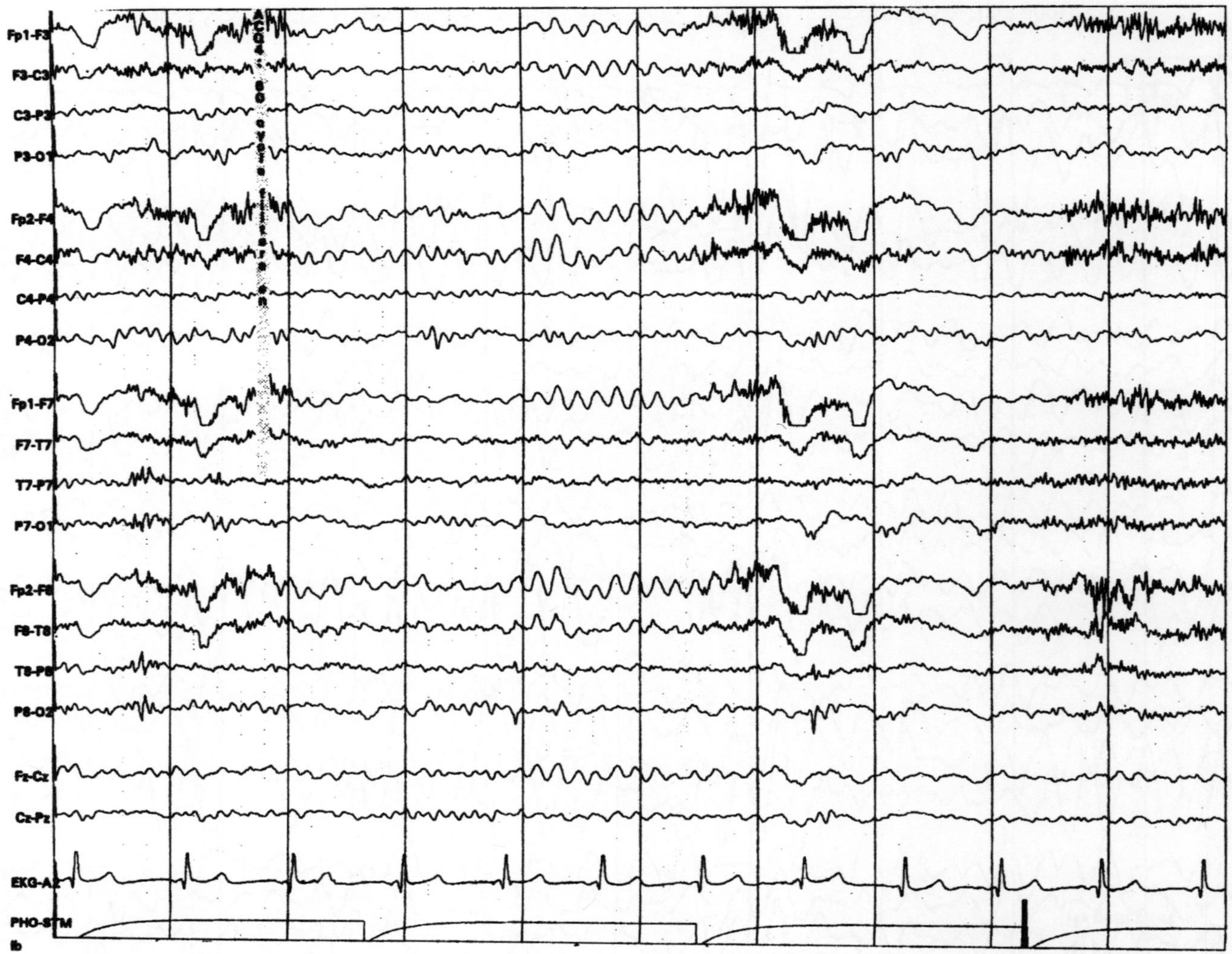

Fig. 24. Nine years. Frontal theta. PHO-STM, photic stimulation; LUE, left under eye; RAE, right above eye.

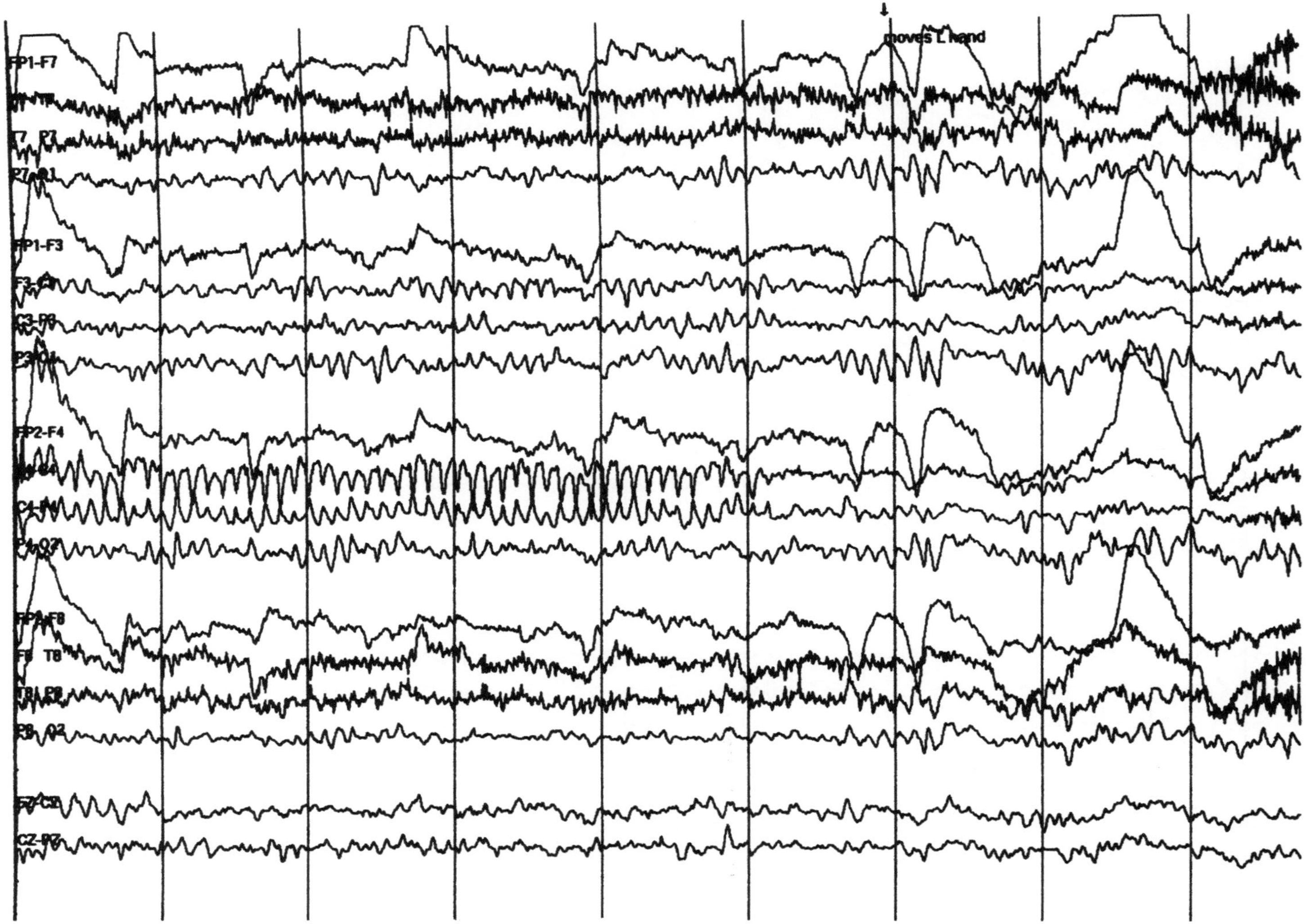

Fig. 25. Twelve years. Mu rhythm on right, responds to movement of left hand; LUE, left under eye; RAE, right above eye.

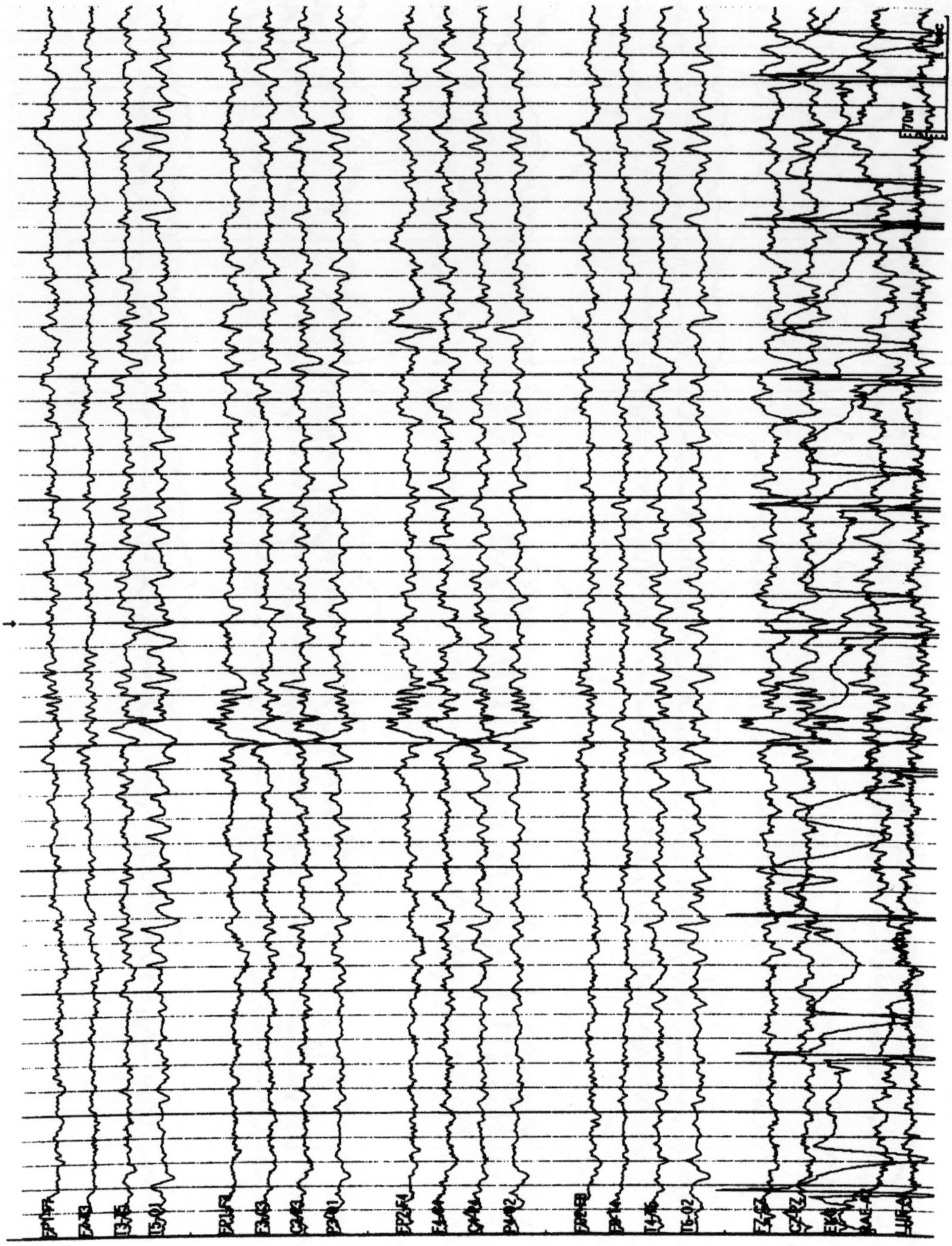

Fig. 26. Twelve years. Sleeppositive occipital sharp transients of sleep; LUE, left under eye; RAE, right above eye.

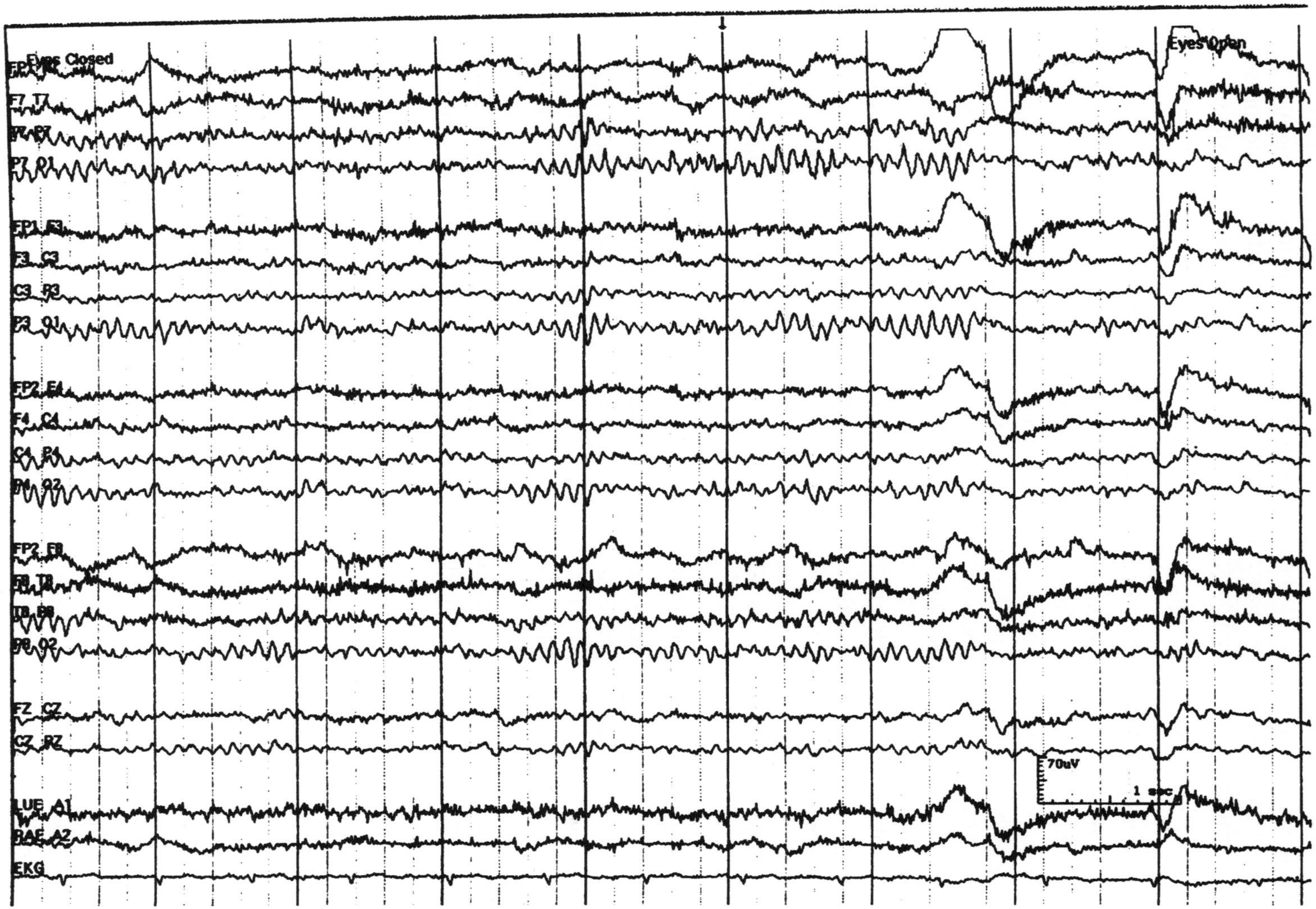

Fig. 27. Sixteen years. Awake EEG, little delta activity remains; LUE, left under eye; RAE, right above eye.

The drowsy pattern is now mature, with alpha dropout, and replacement with a low-voltage mixture of slow and fast activity. Prominent V waves remain, still sharper in contour than in adulthood, but less so than earlier in childhood. Spindles are mature. POSTS are abundant and mature. They may occur in semiregular runs and resemble an occipital rhythm. Fourteen- and/or 6-Hz positive sharp waves are not uncommon in sleep.

3. NORMAL VARIANT PATTERNS

3.1. Phantom Spike Wave (Six Per Second Spike Wave Complex)

This pattern looks like a miniature version of the typical three per second spike wave of typical absence. The spike component is often unimpressive (hence, the "phantom spike" designation). Although it may be diffuse, at times the distribution of this phenomenon and the state in which it occurs allow discrimination of two forms. It occurs in waking, at relatively higher amplitude and the anterior head regions in male subjects (Fig. 28), and, in female subjects, is typically occipital, of lower amplitude, and occurs in drowsiness. The discharge usually lasts 1 to 2 s, but may persist longer. It is usually symmetrical, but may have a lateral predominance. This pattern is present in adolescence and adulthood. Unlike more pathological spike wave patterns, it tends to disappear as sleep deepens. Although here described as a normal variant, the status of phantom spike wave as denoting increased risk of seizure has been controversial, particularly the anterior/frontal form, which some authors consider more likely to be associated with other epileptiform findings and with epileptic seizures.

3.2. Fourteen- and 6-Hz Positive Spikes (Fig. 29)

This benign pattern is first observed in early childhood, achieves maximum frequency in the adolescent age group, and wanes thereafter. It is observed in drowsiness and sleep, has a broad field, but is usually maximum in the posterior temporal area. At times, the 14-Hz or the 6-Hz component may predominate. The 14-Hz component may be observed more frequently in middle childhood or adolescence, the 6-Hz pattern in younger children and young adults. The discharge may last 1 to 2 s. It often occurs independently on both sides. Because the amplitude is low, it is best observed with referential montages, which, in general, use longer interelectrode distances.

3.3. Rhythmic Temporal Theta Bursts of Drowsiness (Psychomotor Variant) (Fig. 30)

This is a rare pattern. It is observed in adolescents and adults. It occurs in drowsiness, and, similar to other benign patterns, disappears as sleep supervenes. It may occur bilaterally or independently. It begins with rhythmic sharply contoured, often notched or flat-topped theta waves in the mid-temporal area. The amplitude and field increase, but there is no evolution in frequency or of waveform. These last features help distinguish this pattern from true seizure. This pattern is no longer thought to denote increased risk of seizure or epilepsy.

3.4. Alpha Variant Patterns

3.4.1. Slow Alpha Variant

This is most likely the result of the superimposition of two alpha rhythms producing the sudden appearance of a notched posterior rhythm at half the previous frequency. Similar to the individual's usual alpha frequency, it is blocked by eye opening. This change may occur in trains during normal alpha activity. It may be more likely to occur in drowsiness.

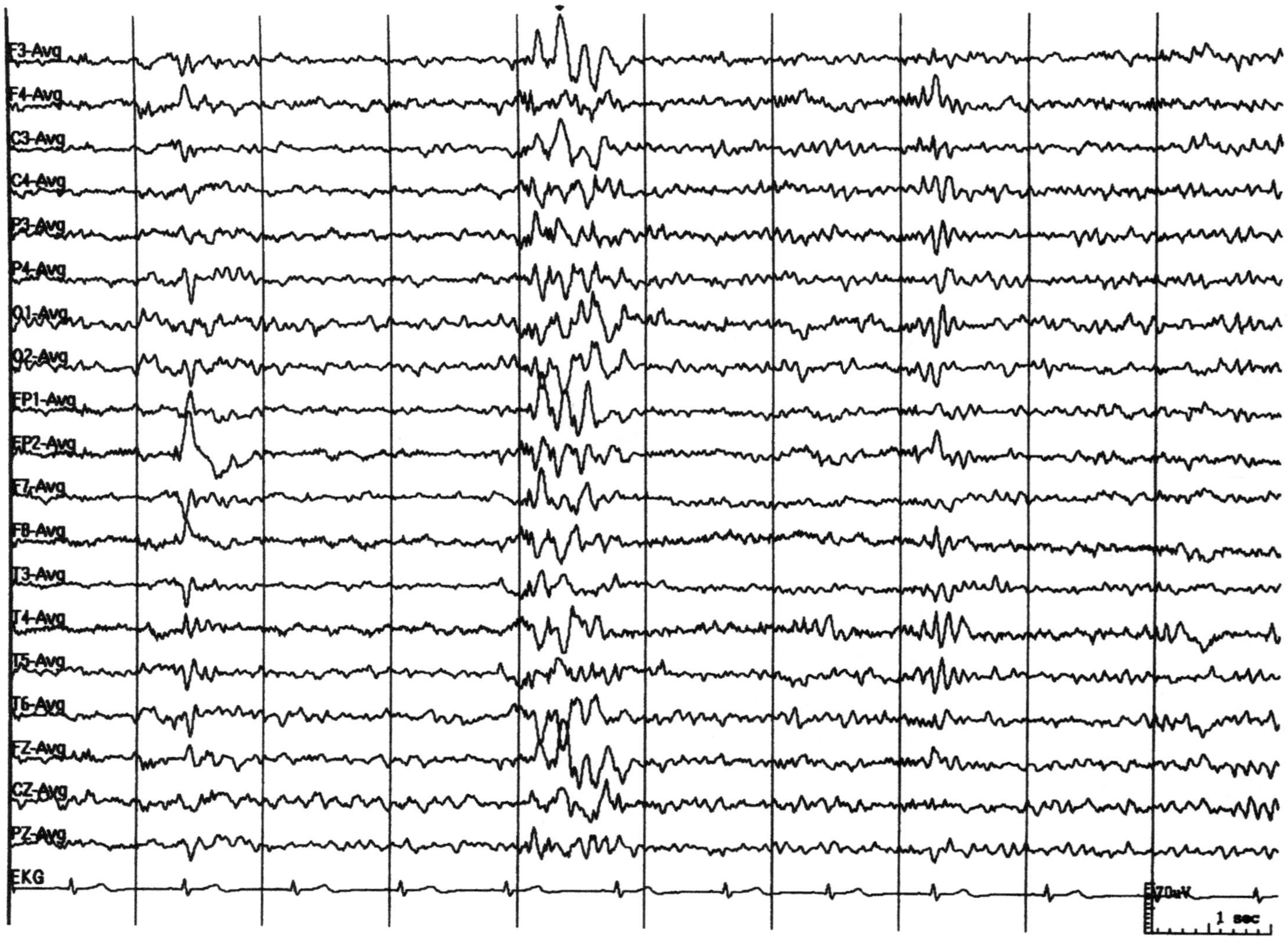

Fig. 28. Fourteen years. Phantom spike wave, "waking, at relatively higher amplitude and the anterior head regions in males" variant (*see* text). Note Fp1 and F3 electrodes (average referential montage).

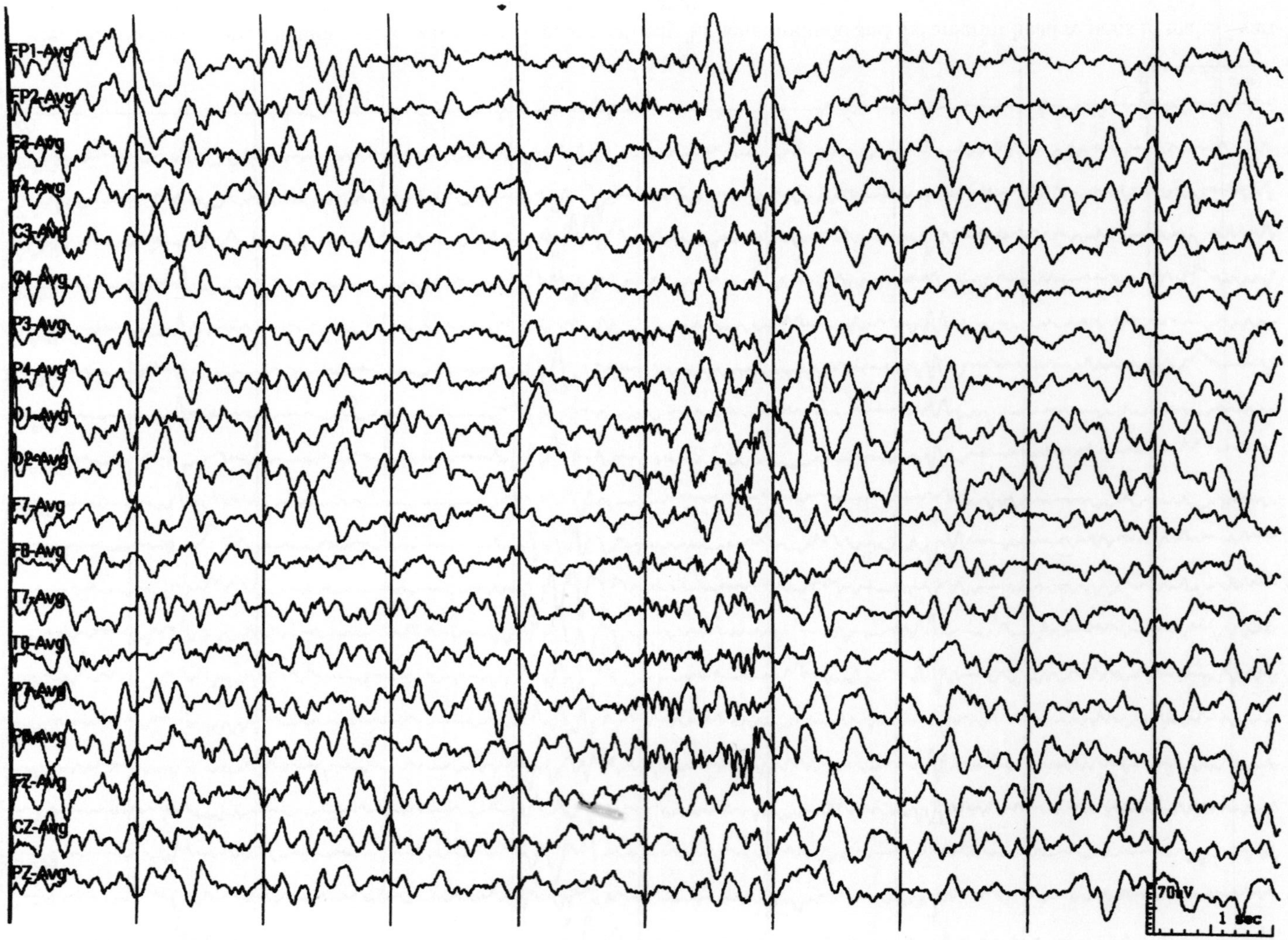

Fig. 29. Thirteen years. Fourteen- and 6-Hz positive spikes. Note P7 and P8 electrodes (average referential montage).

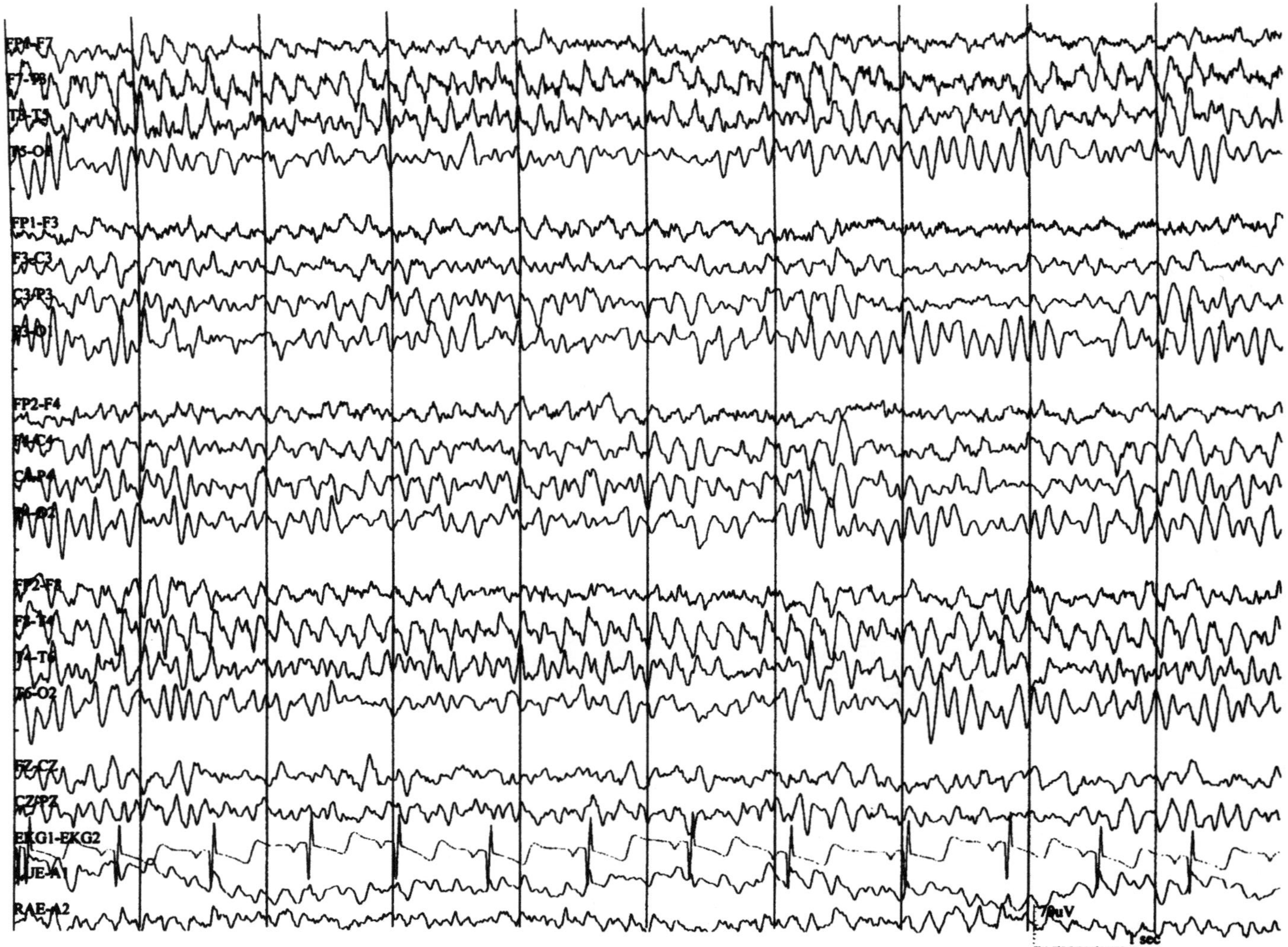

Fig. 30. Seventeen years. Rhythmic temporal theta bursts of drowsiness ("psychomotor variant"); LUE, left under eye; RAE, right above eye.

3.4.2. *Fast Alpha Variant*

This typically occurs on first eye closure, when a beta-range activity occurs in the posterior regions, which attenuates with eye opening. This is usually replaced within a second or two by the individual's usual alpha frequency. This fast occipital activity has been described as the "squeak" phenomenon. Both the slow and fast alpha variants are normal physiological phenomena.

SUGGESTED READING

Blume WT, Kaibara M. Atlas of Pediatric Electroencephalography, 2nd ed. Lippincott-Raven, Philadelphia, PA, 1999.

Daly DD, Pedley TA. Current Practice of Clinical Electroencephalography. 2nd ed. Lippincott-Raven, Philadelphia, PA, 1997.

Neidermeyer E. Maturation of the EEG: development of waking and sleep patterns. In: Electroencephalography. Basic Principles, Clinical Applications and Related Fields, 4th ed. (Neidermeyer E, Lopes da Silva F, eds). Williams & Wilkins, Baltimore, MD, 1999.

Scher MS. Electroencephalography of the newborn: normal and abnormal features. In: Electroencephalography. Basic Principles, Clinical Applications and Related Fields, 4th ed. (Neidermeyer E, Lopes da Silva F, eds). Williams & Wilkins, Baltimore, MD, 1999.

REVIEW QUESTIONS

1. To accurately assess the preterm EEG, all of the following facts must be known except:
   A. Chronological age.
   B. Conceptional age.
   C. Physiological data (EKG, respirations, etc.).
   D. State of infant (awake, asleep).
   E. Time of last feed.
2. Which of the following statements is true regarding the EEG in premature infants?
   A. EEG response to stimulation is typically observed as early as 26 wk.
   B. The EEG of the 30-wk premature infant is mostly synchronous.
   C. Cyclic state changes begin to correlate with changes in respiratory rate, heart rate, and eye movements at 34 to 35 wk.
   D. Frontal sharp waves are always abnormal at 36 wk.
   E. At 36 wk the infant responds to stimulation with high voltage synchronous activity.
3. At term:
   A. Reactive posterior dominant activity is observed with eye closure.
   B. Discontinuity of EEG is no longer present in any state.
   C. Asynchrony of EEG is no longer present in any state.
   D. Frontal sharp waves are normal in the first month.
   E. Infants fall directly from waking into QS.
4. The posterior dominant rhythm in childhood:
   A. Is poorly reactive to eye closure throughout early childhood.
   B. Rarely reaches alpha frequency before 8 yr.
   C. May exhibit voltage of greater than 100 μV in the first year.
   D. Should never by asymmetrical.
   E. Is commonly of low voltage (<30 μV).
5. As the EEG matures through childhood:
   A. Theta activity is unusual in the first year.
   B. Delta activity predominates in the first year.
   C. Frontal theta activity is never normal.
   D. Theta activity is synchronous throughout childhood.

6. The following are true of sleep spindles except:
   A. They are first observed at 3 mo of age.
   B. They may be prolonged up to 10 s in the first year of life.
   C. They are synchronous at 2 yr.
   D. They remain comb-shaped until adolescence.
   E. They are central and parietal in infancy and frontal midline by late childhood.
7. In childhood, during sleep:
   A. Vertex waves are first observed at 1 mo of age.
   B. Slow activity in sleep is maximal in the frontal area.
   C. Vertex sharp waves may be asymmetrical.
   D. POSTs are normally observed early childhood.
   E. Hypnagogic and hypnopompic hypersynchrony remain common in adolescence.
8. Activation procedures:
   A. There is no response to photic stimulation before 2 yr of age.
   B. Younger children demonstrate a driving response at higher flash frequencies than older age groups.
   C. Slowing associated with hyperventilation may be initially asymmetrical.
   D. High-voltage slowing with hyperventilation is exacerbated by hyperglycemia.
9. Normal variants:
   A. The slow alpha variant is more likely to be associated with pathology than the fast alpha variant.
   B. Alpha variant patterns are alpha patterns observed in posterior head regions in sleep.
   C. Phantom spike wave pattern occurring in the occipital areas is more likely to be abnormal than that occurring in the frontal areas.
   D. Fourteen- and/or 6-Hz positive sharp waves are high-amplitude phenomena, usually in the posterior temporal area.
   E. Rhythmic temporal theta bursts are distinguished from seizure by the absence of change in frequency and waveform.
10. In pediatric EEG:
    A. Lambda waves are occipital sleep phenomena related to REM movements.
    B. Mu activity is blocked by movement of the contralateral hand.
    C. Posterior slow waves of youth, unlike posterior dominant activity, do not block with eye opening.
    D. Delta brushes are occasionally observed after the first year.

REVIEW ANSWERS

1. The correct answer is E.
2. The correct answer is C.
3. The correct answer is D.
4. The correct answer is C.
5. The correct answer is B.
6. The correct answer is D.
7. The correct answer is C.
8. The correct answer is C.
9. The correct answer is E.
10. The correct answer is B.

11
Pediatric EEG Abnormalities

James J. Riviello, Jr.

Summary

EEG is an important tool in pediatric neurology and EEG abnormalities occur in many different disorders. EEG abnormalities are nonspecific and do not make a specific diagnosis. However, the EEG is especially useful in the diagnosis, differential diagnosis, classification, and management of seizures and epileptic syndromes. Once the diagnosis of epilepsy is established, specific EEG patterns help to define specific epileptic syndromes.

Key Words: EEG; EEG and epileptic syndromes; pediatric EEG; pediatric EEG abnormalities; pediatric EEG patterns.

1. INTRODUCTION

EEG is an important tool in pediatric neurology. EEG abnormalities occur in all categories of neurological disorders. These abnormalities are useful for lateralizing and localizing a neurological process, but are nonspecific and do not make a specific diagnosis. Epileptiform activity, described as an EEG waveform recorded in a proportion of those suffering from an epileptic disorder, may occur in conditions other than epilepsy. Therefore, the presence of overt epileptiform features suggests, but is not absolutely diagnostic of epilepsy. For example, in a sample of children referred for EEG who had spikes, Kellaway reported a location-dependent incidence of epilepsy, ranging from 38 to 91% (central the lowest, temporal the highest).

EEG interpretation starts with an assessment of the EEG background, which refers to the entire EEG. This starts with whether the background is continuous or symmetric. It is important to know the child's age, and, if a newborn, the conceptual age rather than the legal age, because background continuity decreases with a decreasing conceptual age. The most premature infants have a very discontinuous background, and may even have prolonged periods of electrocerebral inactivity, called trace discontinu. Even in infants and older children, specific waveforms, such as the posterior rhythm, have a developmental appearance. Next, evaluate the overall symmetry from side to side, and then look for the presence of generalized or focal abnormalities, such as slowing or epileptiform activity. If an actual seizure occurs during the EEG, it is called an ictal EEG. The majority of EEGs do not record actual seizures; these are called interictal EEGs.

The EEG is especially useful in the evaluation and management of seizures, epilepsy, and the epilepsy syndromes. A recent practice parameter on evaluating a first nonfebrile seizure in children recommended routine EEG. EEG findings may help to differentiate an epileptic

From: *The Clinical Neurophysiology Primer*
Edited by: A. S. Blum and S. B. Rutkove © Humana Press Inc., Totowa, NJ

from a nonepileptic event. Many paroxysmal disorders in childhood mimic epilepsy, but do not have an epileptic mechanism. These are called nonepileptic paroxysmal events. Although nonepileptic paroxysmal events typically have a normal interictal EEG, the EEG may change at the time of the event. With syncope, slowing and even electrocerebral inactivity may occur at the time of the event. It is especially important to include cardiac disorders in the differential diagnosis, because arrhythmias, such as the prolonged QT syndrome, may present with a "seizure." However, a normal EEG does not exclude the diagnosis of epilepsy, and an abnormal EEG does not, in itself, establish a diagnosis of epilepsy. In a longitudinal study of 3726 normal children, aged 6 to 13 yr, Cavazutti et al. reported that 3.5% had epileptiform patterns. The EEG is "diagnostic" of a seizure only when actual clinical manifestations occur at the time of the electrographic discharge; if not, this finding is only very suggestive of epilepsy.

With a definite diagnosis of epilepsy, EEG helps to classify seizure type or the associated epileptic syndrome. For example, with a generalized tonic–clonic seizure (GTCS), a focal interictal EEG suggests a focal onset. EEG monitoring is also useful in the presurgical evaluation of refractory epilepsy, because identifying the location of the seizure focus is a very important part of the presurgical evaluation.

There are epileptic disorders associated with EEG epileptiform activity, such as Landau–Kleffner syndrome (LKS) or continuous spike waves of sleep (CSWS), that may not always be associated with overt clinical seizures but have cognitive dysfunction with cognitive or language regression. These specific disorders typically have sleep-activated epileptiform activity, with regression related to the duration of this EEG activity. The entity transient cognitive impairment refers to altered cognitive function associated with epileptiform activity, although it may take specific computerized testing to detect. EEG is used in both the diagnosis and treatment of these disorders, because the treatment goal is both clinical and electrographic improvement.

EEG is also used to exclude epilepsy in disorders that mimic epilepsy, such as the various childhood migraine syndromes, attention deficit disorders, or other psychiatric disorders. The incidence of paroxysmal abnormalities in headaches and migraines varies from 9 to 30%. In a recent study by Richer et al. in children with ADHD, epileptiform activity occurred in 6.1% of 347 EEGs. However, only 3 of 21 children with epileptiform activity developed a seizure disorder. In the era before modern neuroimaging, EEG was used in the diagnostic evaluation for brain tumors. However, practice parameters for headache no longer recommend routine EEG. Certain EEG findings may also be associated with specific metabolic or genetic disorders. For example, a "comb-like" central rhythm has been observed in neonates with maple-syrup-urine disease, and bifrontal slow spike-and-wave discharges have been reported in the ring 20 chromosome syndrome.

The International League Against Epilepsy system for seizure classification, the International Classification of Epileptic Seizures (ICES), starts with seizure type, whether focal or generalized, and then classifies the epileptic syndrome. There is also an unclassified category that includes neonatal seizures. An epileptic syndrome refers to a complex of signs and symptoms that define a unique epilepsy (Table 1), and is useful for choosing antiepileptic drug (AED) treatment and for predicting prognosis. Etiology is also included in epilepsy classification, whether symptomatic, cryptogenic, or idiopathic, and either generalized or focal.

Epileptic syndromes are divided into benign or malignant, depending on the expected outcome. In general, the "idiopathic" epilepsies are analogous to the benign epileptic syndromes, and the "symptomatic" epilepsies are analogous to the malignant epileptic syndromes.

Table 1
Features Used to Classify Epileptic Syndromes

Seizure Type(s)
Partial onset
Simple, complex, secondarily generalized
Generalized onset
Absence, tonic, tonic-clonic, atonic, myoclonic
Specific
Spasms, gelastic, others
Cluster of Signs and Symptoms Customarily Occurring Together
Age of onset
Etiology
Anatomy
Precipitating factors
Severity: prognosis, benign or malignant
EEG, both ictal and interictal
Duration of epilepsy
Associated clinical features
Chronicity
Diurnal and circadian cycling

Benign syndromes are those in which seizures are successfully treated with AEDs, require no other specific treatment, or may even remit without sequelae. Malignant syndromes generally are resistant to treatment and have a poor prognosis. However, not every "benign" epileptic syndrome is associated with a normal outcome, and not every "malignant" epileptic syndrome has a poor outcome. For example, a small number of children with benign familial neonatal seizures will have developmental problems and persistent seizures, some children with absence epilepsy can have learning difficulties, and photosensitive epilepsy can occasionally be severe. The term catastrophic epilepsy is used for the malignant epileptic syndromes.

2. SPECIFIC EEG ABNORMALITIES

The ICES starts with determining whether the seizure onset is focal or generalized. A proposed new classification will instead focus on seizure semiology. Specific seizure semiology can be identified from temporal, frontal, parietal, and occipital regions. Within each, further subdivision is possible. Clinical manifestations by location are similar in both adult and childhood epilepsy.

3. SPECIFIC EPILEPTIC SYNDROMES

In contrast to seizure location and the resultant clinical manifestations, differences exist in the occurrence of epileptic syndromes in adult and pediatric epileptology. Most specific epileptic syndromes begin in childhood and may continue into adulthood. We shall start with the benign epileptic syndromes and then the malignant syndromes. Descriptions of the epileptic syndromes come from the ICES. Additional points within each syndrome are specifically referenced. The following specific epileptic syndromes are those associated with specific EEG findings.

3.1. Benign Epileptic Syndromes

3.1.1. Febrile Seizures

The conventional wisdom regarding the EEG in benign febrile seizures is that the EEG should be normal, whereas, in a child with epilepsy, the EEG will be abnormal, with epileptiform features. The ICES considers febrile seizures a situation-related seizure: the situation being the illness with fever. However, there may not be such a clear distinction. Alvarez and Lombroso reported hypnagogic paroxysmal spike wave activity (minimal epileptiform features, sharp waves embedded into hyperventilation, or hypnagogic hypersynchrony) occurring with a higher incidence in children with febrile seizures. This is similar to epileptiform activity admixed into vertex waves and sleep spindles, called dyshormia by Niedermeyer (Fig. 1). Dyshormia refers to an abnormal paroxysmal arousal features during sleep. These features may indicate a lower seizure threshold and explains why they occur in both the normal population and in those with epilepsy.

The recent practice parameter (1999) on benign febrile seizures does not require an EEG for evaluating a benign febrile seizure. This is defined as a short (<15 min) GTCS without significant postictal depression, in a child with a nonfocal neurological examination without a family history of epilepsy.

3.1.2. Benign Myoclonic Epilepsy in Infancy

This disorder is characterized by brief bursts of generalized myoclonus that begin in otherwise normal children during the first 2 yr of life. There is often a family history of epilepsy. EEG shows brief bursts of generalized spike waves during early sleep. No other seizures occur at onset, but GTCS may occur during adolescence and developmental delay may be present.

3.1.3. Childhood Absence Epilepsy

The peak age of onset of childhood absence epilepsy is from 6 to 7 yr. The seizures consist of an abrupt cessation of ongoing activity, with a change of facial expression and a blank gaze. The duration is short, rarely lasting longer than 30 s, there is no preceding aura or subsequent postictal depression, and there may be frequent automatisms. Seizures may be activated by hyperventilation and photic stimulation.

Absence seizures are divided into typical absence and atypical absence seizures. Typical absence seizures are classified as simple or complex; a simple absence has a sudden onset without motor activity; complex absences have associated motor activity or autonomic activity. The atypical absence seizure has a less clearly defined onset or cessation, so that it may be difficult to identify the beginning or the end of the seizure without EEG. Atypical absence seizures may have more pronounced tone changes, a longer duration, and are typically associated with other seizure types and mental retardation.

Penry et al. studied 374 absence seizures with video EEG in 48 patients; a simple absence occurred in only 9% of patients; automatisms occurred in at least one episode in 88% of patients, clonic components occurred in 41% of patients, and seizures lasted less than 20 s in duration in 85% of patients.

Automatisms may be perseverative or *de novo*, and autonomic abnormalities may consist of pupillary dilatation, pallor, flushing, salivation, or incontinence.

The EEG in a typical absence seizure shows a symmetric 3-Hz generalized spike-and-wave discharge, which has a faster frequency at the beginning and a slower frequency at the end (Fig. 2). The atypical absence has a frequency between 1.5 and 2.5 Hz (usually slower), may be asymmetric, and usually has a slow background.

A pattern reported in children is hyperventilation-induced high-amplitude rhythmic slowing (HIHARS), in which altered awareness may occur with HIHARS on EEG. Clinically, eye opening and eyelid flutter occur more frequently in absence seizures, whereas fidgeting, smiling, and yawning occurred more frequently with HIHARS. Arrest of activity, staring, and oral and manual automatisms were observed in both groups. A similar pattern has been described in adults. These two disorders emphasize the importance of clinical observation by the EEG technologist during the study.

3.1.4. Intermittent Rhythmic Delta Activity

Intermittent rhythmic delta activity (IRDA) consists of rhythmic, sinusoidal, delta activity, which may be notched and even have low amplitude admixed spikes. IRDA may be generalized or have an age-dependent specific location: an occipital location (OIRDA) is more common in children (Fig. 3), whereas a frontal location is more common in older children and adults. Although IRDA itself is nonspecific for etiology, it typically occurs in disorders associated with acute altered awareness. In children, OIRDA is seen most commonly in epilepsy, especially generalized absence epilepsy. In our study of IRDA in children, 80% had epilepsy and the majority had absence epilepsy. This was referred to as posterior paroxysmal activity by Holmes et al. OIRDA occurs usually as an interictal pattern, but may be an ictal pattern; we have seen OIRDA evolve into a 3-Hz spike-and-wave discharge.

3.1.5. Juvenile Absence Epilepsy

The onset is around puberty, but the seizures may be less frequent than in the childhood absence syndrome. GTCS and myoclonic seizures may occur and the EEG shows a fast spike-and-wave discharge, similar to that seen in juvenile myoclonic epilepsy (JME). There may also be electrographic discharges that are more prolonged, with less altered awareness.

3.1.6. Juvenile Myoclonic Epilepsy

JME of Janz is characterized by mild myoclonic, generalized tonic–clonic, sometimes clonic–tonic–clonic seizures, and absence seizures. The onset is typically around puberty. JME is also referred to as impulsive petit mal.

The myoclonic seizures are usually mild to moderate in intensity, may involve the entire extremity, rather than isolated muscles, and generally are bilateral. These occur after awakening and are aggravated by fatigue, sleep deprivation, or alcohol. The patient may drop objects or fall. Myoclonic status epilepticus has occurred. GTCS are frequent. Delgado-Escueta et al. reported that GTCS occurred in 41 of 43 patients, and Asconape and Penry reported GTCS in 83% of patients. Myoclonic jerks usually precede the GTCS and may evolve into a GTCS. GTCS usually occur shortly after awakening. Absence seizures occurred in 40% in the series of Delgado-Escueta et al. These usually occur in association with GTCS, and most commonly occur shortly after awakening.

The EEG in JME is characterized interictally by a fast spike-and-wave discharge, 3.5- to 6-Hz spike-and-wave and polyspike-and-wave complexes (Fig. 4). Ten- to 16-Hz rapid spikes with slow waves occur with myoclonic seizures, photosensitivity is common, and abnormalities may occur in sleep.

3.1.7. Epilepsy With GTCS on Awakening

GTCS occur exclusively or predominantly shortly after awakening. The onset is usually in the second decade. Absence or myoclonic seizures may occur. The EEG shows the fast

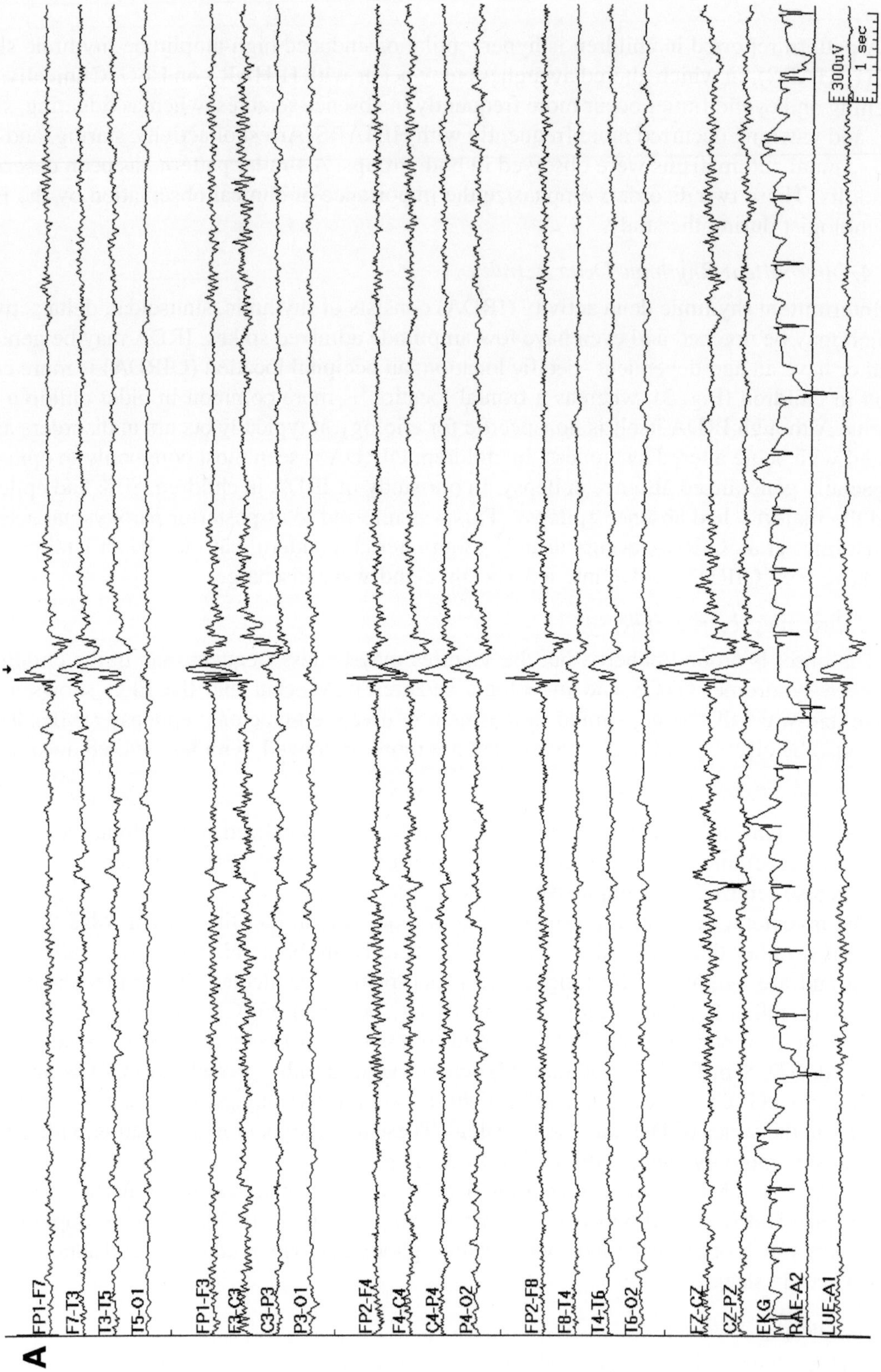
FP1-F7
F7-T3
T3-T5
T5-O1
FP1-F3
F3-C3
C3-P3
P3-O1
FP2-F4
F4-C4
C4-P4
P4-O2
FP2-F8
F8-T4
T4-T6
T6-O2
FZ-CZ
CZ-PZ
EKG
RAE-A2
LUE-A1
300uV
1 sec

A

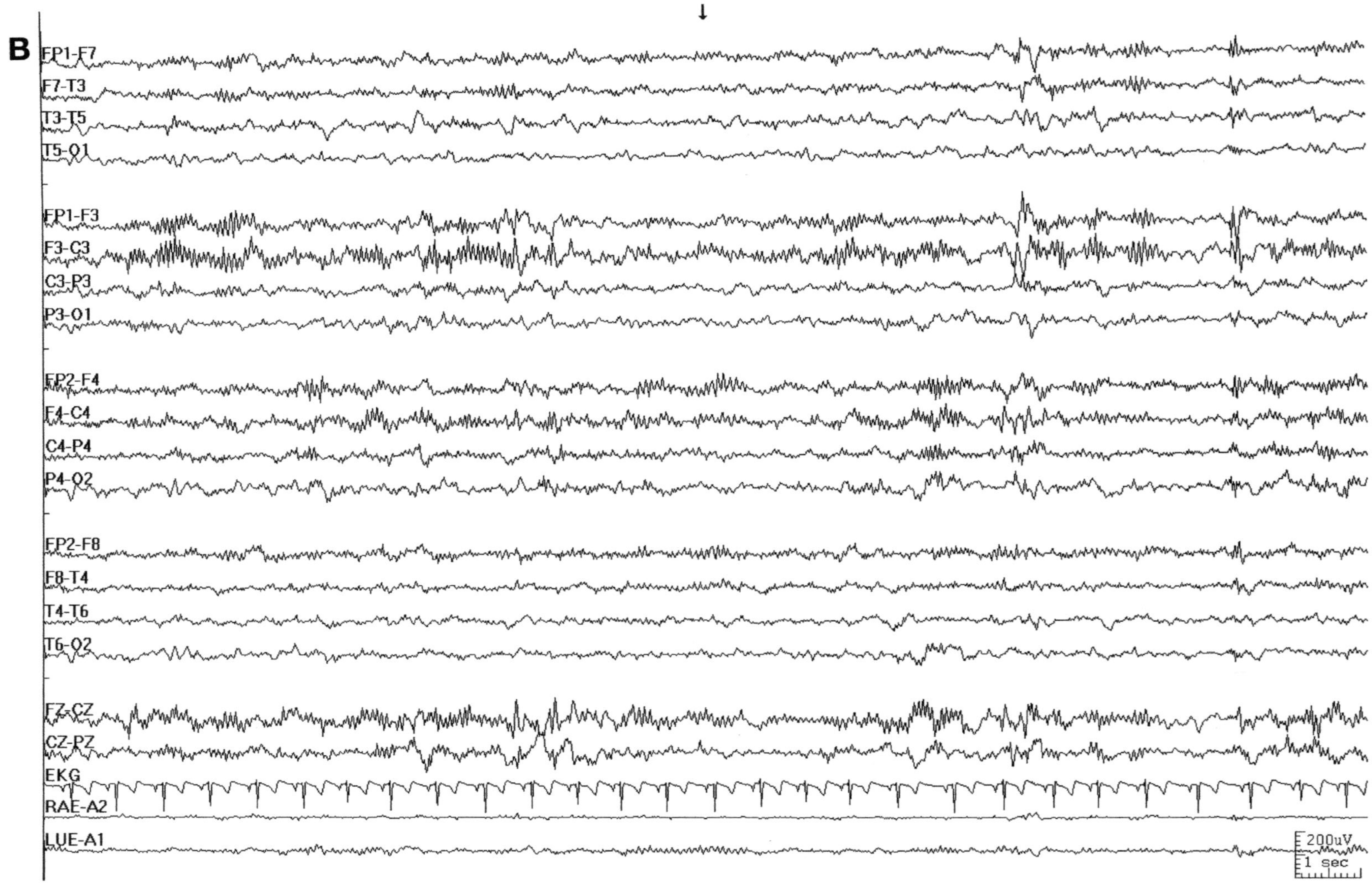

Fig. 1. Dyshormia. Six-year-old girl with childhood absence epilepsy. Note the spike component embedded with a vertex wave (**A**) and within spindles (**B**) during sleep.

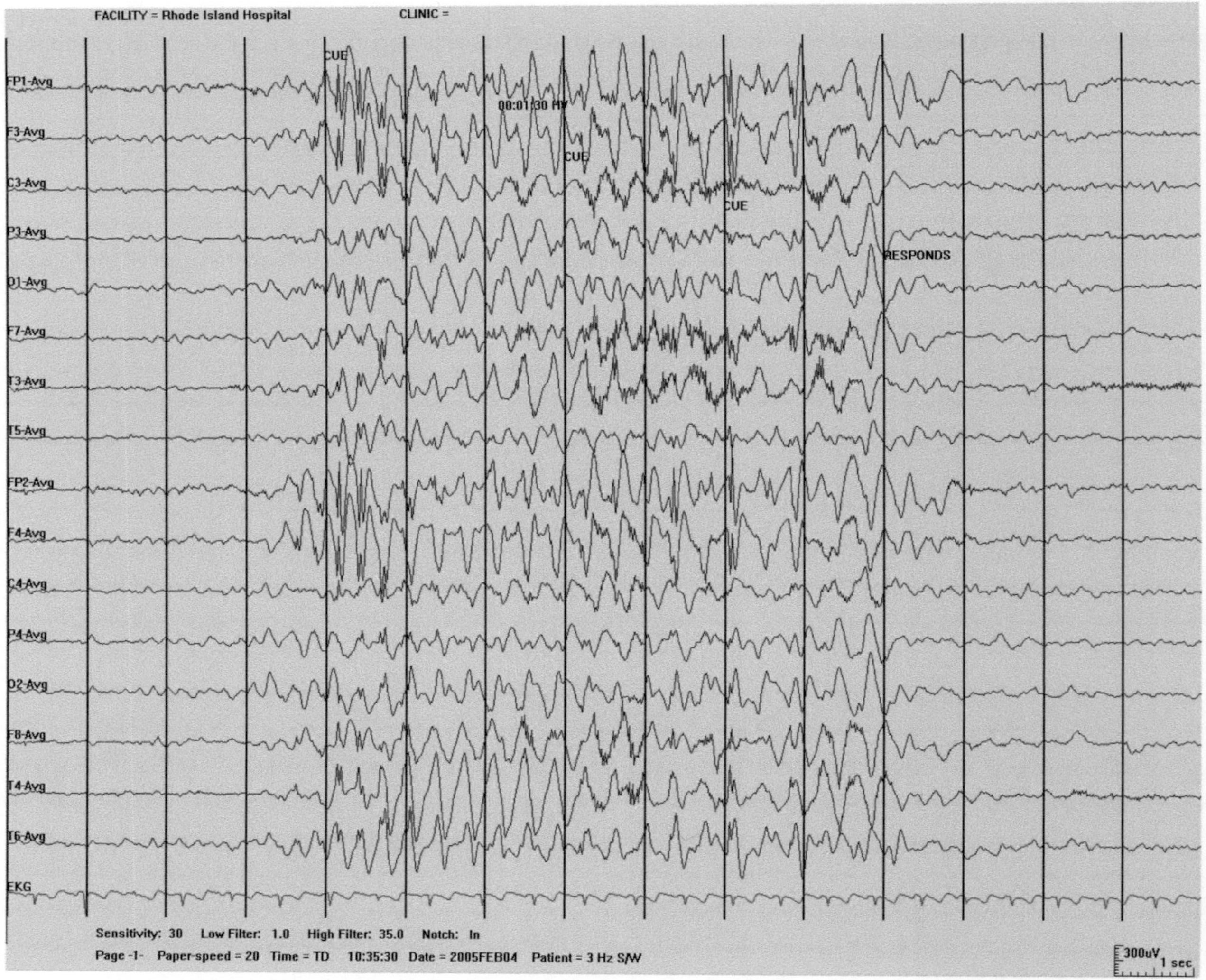

Fig. 2. Absence epilepsy. Twelve-year-old girl with seizures denoted by staring and a family history of seizure. Note the 3-Hz spike-and-wave pattern with responsiveness after its cessation.

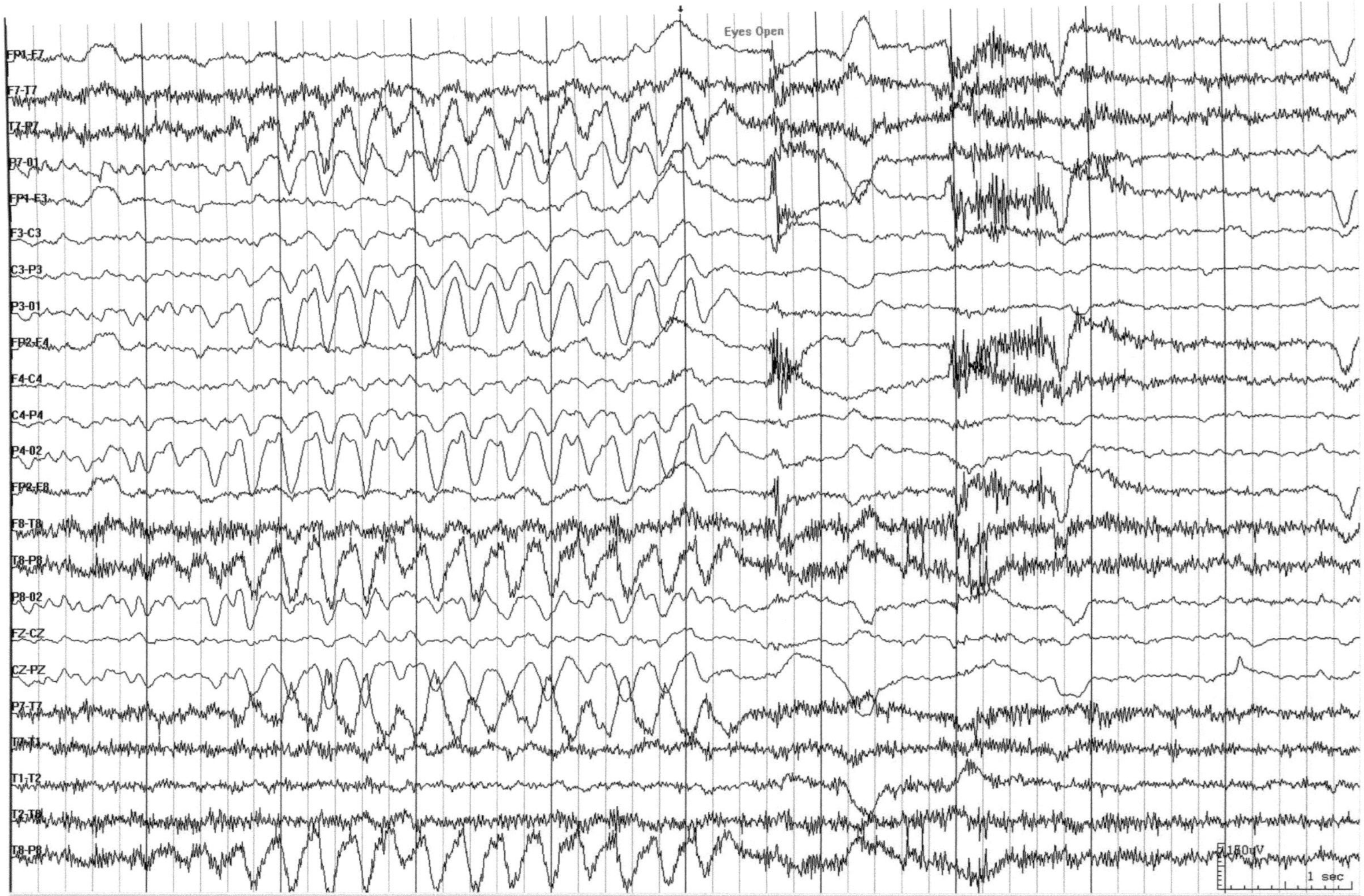

Fig. 3. OIRDA. Eight-year-old girl with childhood absence epilepsy. Note the high amplitude, notched, 3- to 3.5-Hz occipital activity that attenuates with eye opening.

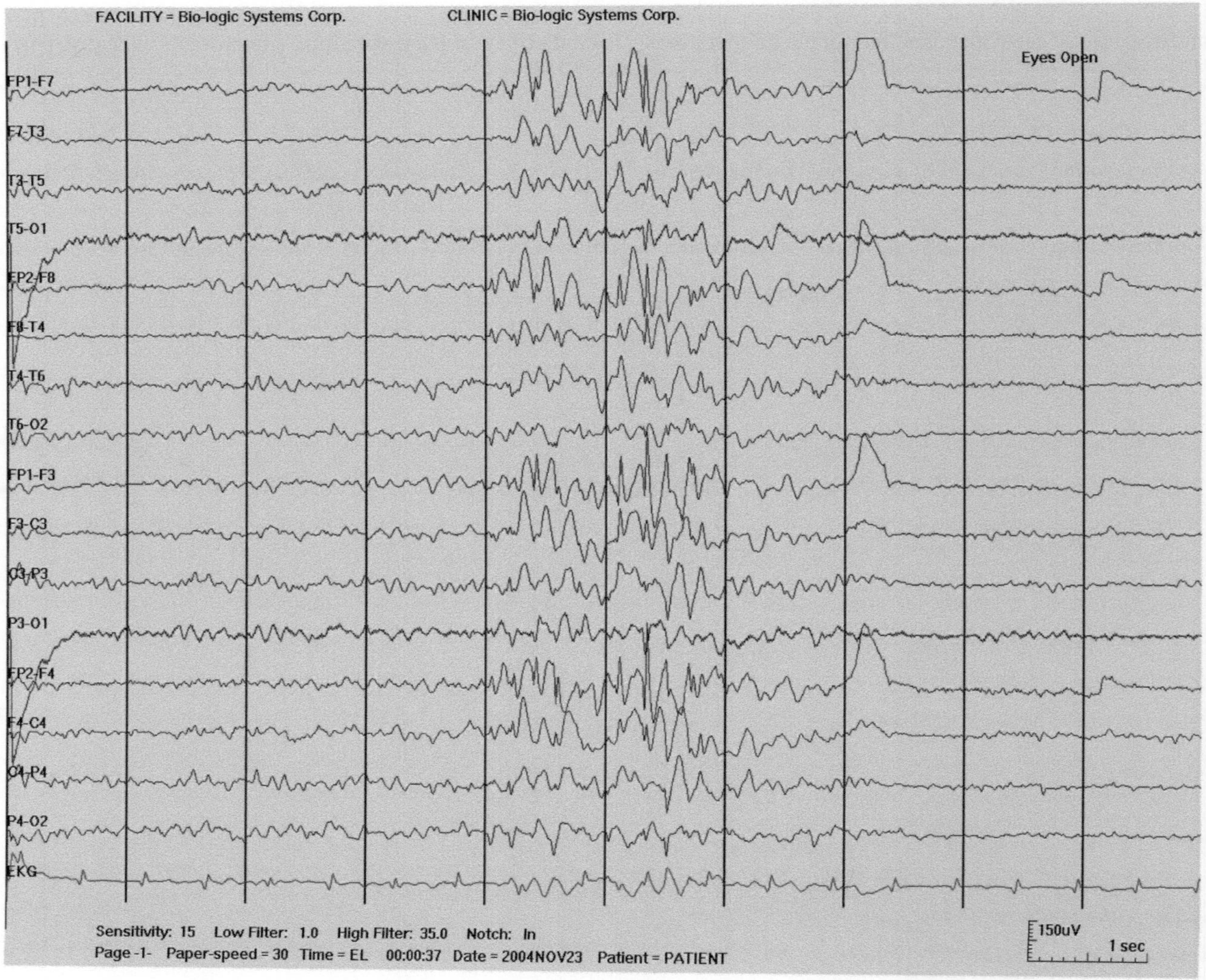

Fig. 4. Juvenile myoclonic epilepsy (JME). Seventeen-year-old girl with JME. Note the generalized fast spike-and-wave discharge at approx 4 Hz.

spike-and-wave discharge, similar to JME. Photosensitivity may also occur. This syndrome may represent JME without myoclonus.

3.1.8. Benign Focal Epilepsy of Childhood

This disorder is also called benign childhood partial seizures. There are different clinical syndromes, which depend on discharge location, centro–temporal (Rolandic, benign Rolandic epilepsy [BRE], or benign childhood epilepsy with centrotemporal spikes), occipital, or frontal. The centro–temporal location is the most common, followed by the occipital location. BRE is a common epileptic syndrome, occurring in 16 to 24% of childhood epilepsies. In Iceland, the incidence of BRE is 4.7 in 100,000 people. Although considered a benign syndrome, cognitive dysfunction may occur.

The centro–temporal location, or BRE, has an age of onset of 3 to 13 yr, with a peak at 9 to 10 yr, and recovery occurs before the ages of 15 to 16 yr. According to Lerman, the seizure frequency is low, 13% may have just one seizure, 66% have infrequent seizures, and 21% have frequent seizures. The typical seizure characteristics consist of oral–buccal–lingual paresthesias, speech arrest, preservation of consciousness, sialorrhea, and tonic or clonic facial movements. The seizures are nocturnal in 50% of patients, occur in both the waking and sleep states in 15% of patients, and the waking state in only 10 to 20% of patients. Status epilepticus is rare. The prognosis is excellent, seizures are usually well-controlled, and the seizure frequency, recurrence, and duration are similar in treated and untreated children. Lerman and Kivity reported EEG normalization by adulthood in all patients. Lombroso referred to these as "sylvian" seizures, given the location of the discharges near the Sylvian fissure.

The EEG in BRE has distinctive, high-amplitude, diphasic spikes or sharp waves with a prominent slow wave, in the midtemporal (T3, T4) and central (C3, C4) regions (Fig. 5). However, when additional scalp electrodes are placed, maximum negativity was in the central region, either in a "high" (C3, C4) or "low" (C5, C6) location. Marked sleep activation occurs, and a horizontal dipole is present. There is a common misconception that the presence of a horizontal dipole is pathognomonic of a benign focal epilepsy. Similar findings have been reported in different pathological processes, including tumors. Massa et al. identified the following EEG characteristics as predictive of a "complicated" course: intermittent slow wave focus, multiple, asynchronous spike wave foci, spike wave clusters, generalized 3-Hz spike wave discharges, discharges with positive or negative myoclonia, or abundance of interictal abnormalities during waking or sleep.

There are now two defined occipital epilepsy syndromes in children, also called childhood epilepsy with occipital spikes or childhood epilepsy with occipital paroxysms: the early onset form (Panayiotopoulos type) and the late-onset (Gastaut type). Panayiotopoulos estimates that the early onset form comprises approx 6% of childhood epilepsies. The typical age of onset is 2 to 12 yr, with a median onset of 5 yr. The typical seizure consists of autonomic and behavioral disturbances, with vomiting, eye deviation, and altered awareness. Headaches and vomiting may occur, with migraines in the differential. The late-onset type consists of visual seizures with elementary visual hallucinations, which may evolve to a feeling of ocular movements or pain, eye deviation, ictal blindness, or focal or secondarily generalized seizures. The typical age of onset is 3 to 16 yr, with a mean age of 8 yr. Migraine is in the differential diagnosis. Neuropsychological dysfunction is also seen in these two disorders. On EEG, the giant spikes typical for benign occipital epilepsy are similar morphologically to those seen in BRE (Fig. 6). These occur with closed eyes, or have "fixation-off" sensitivity; fixation-off refers to blocking central vision or closing the eyes.

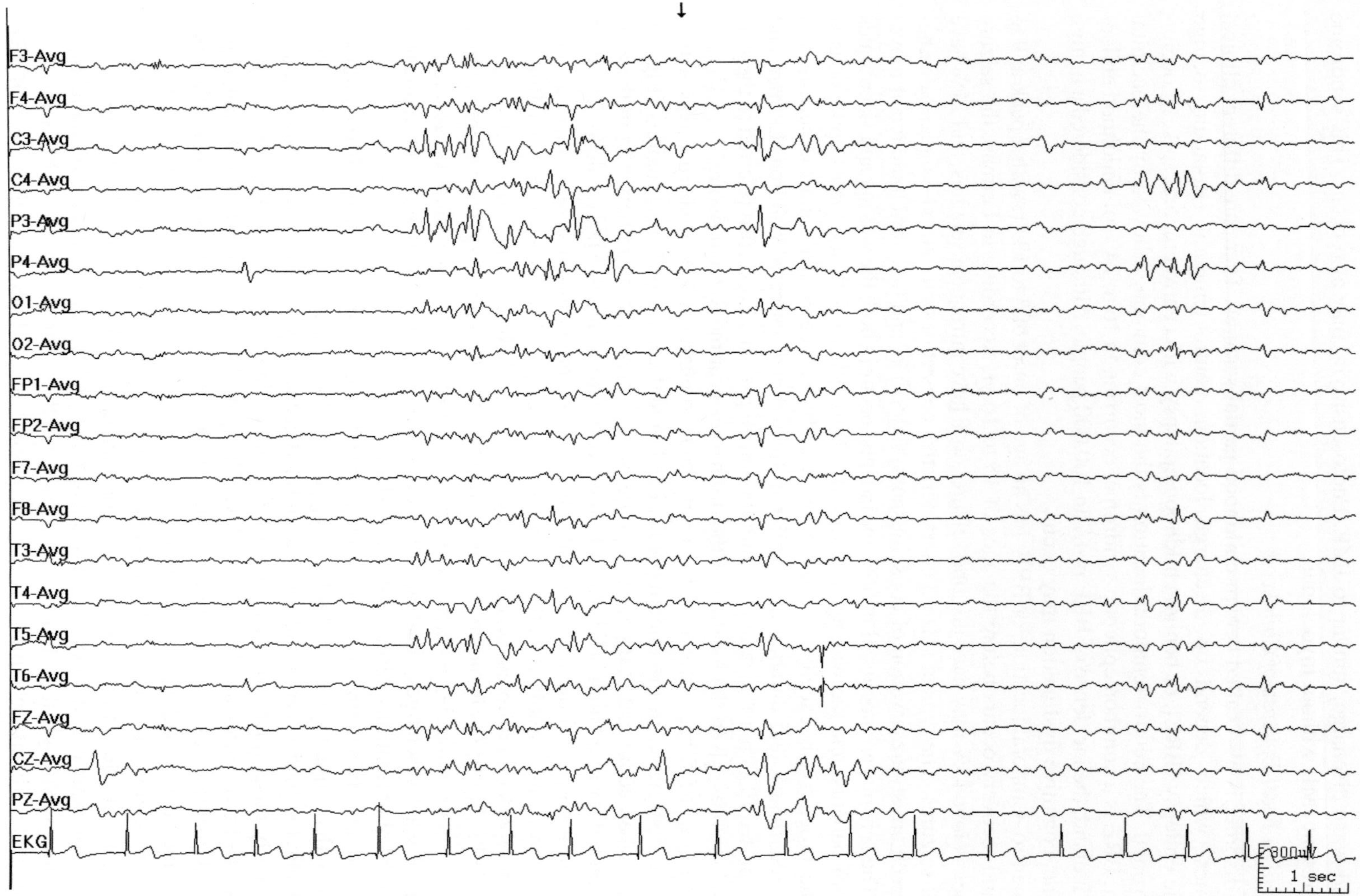

Fig. 5. Benign Rolandic epilepsy. Nine-year-old boy with nocturnal seizures, consisting of focal facial contractions and drooling. His EEG shows prominent triphasic sharp waves in the centro–parietal region, occurring independently bilaterally, and temporal sharp waves. In addition, a horizontal dipole is seen, with frontal positivity and posterior negativity.

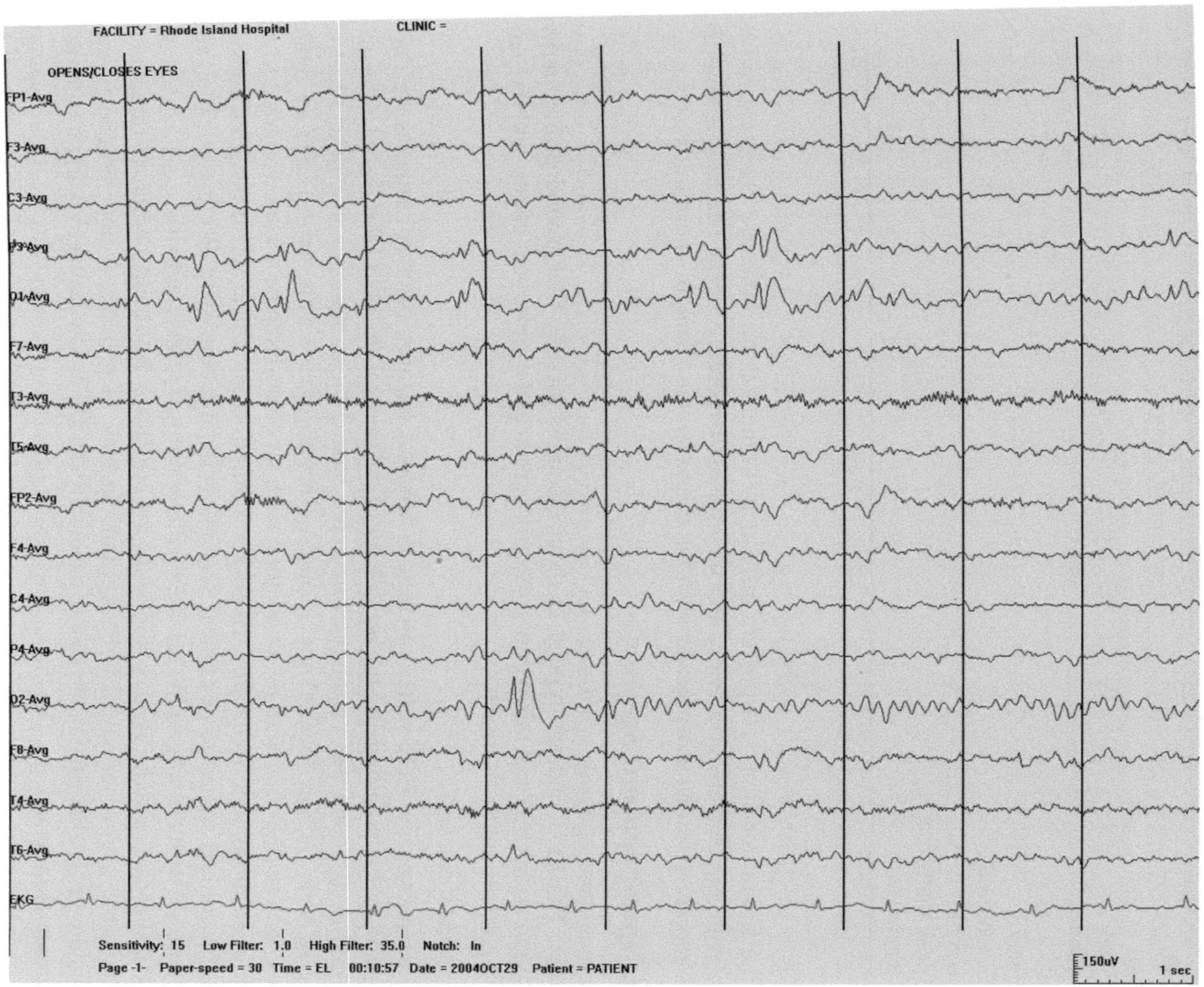

Fig. 6. Benign occipital epilepsy. Four-year-old girl with seizures. Note the independent, bi-occipital sharp and slow wave complexes with eyes closed.

Table 2
Neonatal Seizures: Classification System of Volpe

	Electrographic Seizure	
Clinical Seizure	*Common*	*Uncommon*
Subtle	+	
Clonic		
Focal	+	
Multifocal	+	
Tonic		
Focal	+	
Generalized		+
Myoclonic		
Focal, multifocal		+
Generalized	+	

With a frontal location, seizures are infrequent in the waking state, are characterized by staring, with automatisms or version, and may generalize during sleep. The EEG shows the typical, repetitive frontal sharp waves with the characteristic "benign" morphology.

3.2. Malignant Epileptic Syndromes

The malignant epilepsy syndromes include disorders most likely associated with a poor prognosis. The EEG findings typical in these disorders include background discontinuity, burst suppression (BS), hypsarrhythmia, electrodecremental events, beta bursts, and the generalized paroxysmal fast activity.

3.2.1. Neonatal Seizures

There is no specific EEG pattern present in neonatal seizures. Neonatal seizures are classified by Volpe essentially using the semiology of the clinical seizure (Table 2).

Mizrahi et al. found that certain seizure types have a close association with an electrographic seizure, such as focal clonic seizures, whereas other seizure types, such as subtle seizures, may not have as high an association with an electrographic seizure (referred to as having an electrographic signature). These are referred to as clinical seizures with a consistent electrocortical signature or clinical seizures without a consistent electrocortical signature. There are also electrographic seizures without clinical seizure activity.

The background activity is very important for prognosis. Rose and Lombroso reported a good outcome with a normal EEG background in 86% of patients. A poor prognosis was associated with an abnormal background, especially the BS or low-voltage invariant patterns, or electrocerebral inactivity. The burst-suppression pattern is the most extreme manifestation of background discontinuity and is measured by the interburst interval (IBI). However, neonatal EEG may have features suggestive of dysmaturity, such as an increased duration of the IBI. Hahn et al. studied the duration of the IBI in normal premature infants and found a correlation with the duration of the IBI with the gestational age. Neonatal seizures do not all fit under malignant epileptic syndromes, but these set the stage for the malignant neonatal epileptic syndromes.

3.2.2. Neonatal Myoclonic Encephalopathy

Neonatal myoclonic encephalopathy (NME) is also known as early myoclonic encephalopathy. The onset is typically within the neonatal period, but before 3 mo of age. The seizures are characterized by partial or fragmentary myoclonus, massive myoclonias, and tonic spasms developing later. EEG consists of complex bursts of spikes and sharp waves, with periods of discontinuity; and the burst-suppression pattern, which may later evolve into a hypsarrhythmia pattern. All patients with NME are severely retarded and there is a high mortality during the first year of life. The metabolic disorder nonketotic hyperglycinemia needs to be excluded in this group.

3.2.3. Early Infantile Epileptic Encephalopathy

Early infantile epileptic encephalopathy with suppression bursts is also known as Ohtahara's Syndrome. The onset is within the first few months of age, typically with tonic seizures (spasms), focal seizures, and a burst-suppression EEG. The prognosis is similar to that of NME, and these patients may evolve into infantile spasms (IS; West Syndrome).

3.2.4. Severe Myoclonic Seizures of Infancy

In Severe myoclonic seizures of infancy, there is typically a family history of epilepsy or febrile seizures, no antecedent neurological history, and seizures beginning in the first year of life with generalized or unilateral febrile clonic seizures, with myoclonic seizures typically occurring between 1 and 4 yr of age. Focal seizures may also occur. The febrile seizures tend to be long and recurrent. EEG shows generalized spike waves and polyspike waves, early photosensitivity, and focal abnormalities. Psychomotor retardation occurs, usually beginning in the second year of life, with ataxia and interictal myoclonus. Seizures are typically very resistant to treatment.

The EEG may be normal at the time of the initial febrile seizures, but when the myoclonus starts, generalized spike-and-wave or polyspike-and-wave discharges occur. It may be difficult early on to differentiate the benign from the malignant syndrome, although febrile seizures do not typically occur in the benign form, in which seizures respond well to valproic acid and psychomotor retardation is absent.

3.2.5. Infantile Spasms (West Syndrome)

IS is an epileptic syndrome characterized by the triad of myoclonic or tonic seizures, hypsarrhythmia, and mental retardation. This is also called West Syndrome. IS is considered an age-dependent malignant epileptic syndrome, with a peak onset between 4 and 7 mo and always before 12 mo of age.

IS is divided into idiopathic, symptomatic, and cryptogenic forms. Symptomatic refers to IS secondary to a known neurological insult; cryptogenic refers to a suspected, but not definitely identified neurological insult; idiopathic is used when no specific insult has been identified. The prognosis is worse if the spasms are either symptomatic or cryptogenic.

Clinical manifestations include brief head nods, with quick extension and flexion movements of the trunk, arms, and legs, occurring in clusters and during transitions from sleeping to awaking. These are also described as flexor, extensor, or mixed. The actual spasms may resolve, but other seizure types occur later. The prognosis depends on the etiology.

Evaluation includes neuroimaging to exclude anatomic lesions, but other disorders, especially metabolic and infectious, need to be excluded. Focal spasms may occur secondary to focal lesions, such as tumor, stroke, or focal cortical dysplasia. Treatment is with steroids or standard anticonvulsant medications. Adrenocorticotropic hormone has been considered the drug of choice by many, although benzodiazepines, especially nitrazepam, and valproic acid, and now vigabatrin have been used. Vigabatrin may be especially effective if spasms are caused by tuberous sclerosis, and recent studies show a similar efficacy between vigabatrin and ACTH, with a lower incidence of acute side effects from vigabatrin. However, vigabatrin has been associated with retinal dysfunction, and visual field defects have been reported. Vigabatrin has not yet been released in the United States because of the retinal dysfunction.

If a focal lesion is responsible, then resection may be curative. Even in the absence of a frank mass lesion, surgery may be performed on refractory cases if an epileptogenic focus has been identified. Most of these cases result from focal cortical dysplasias. Functional neuroimaging, especially positron emission tomography scan, have been used to localize the epileptogenic region.

The classic interictal EEG finding is the hypsarrhythmia pattern, which consists of a high-voltage, disorganized background with multifocal spike and sharp waves. This pattern may first occur during non-rapid eye movement sleep and may disappear during rapid eye movement (active sleep) sleep or the waking state. The typical ictal pattern is the electrodecremental event, in which a generalized spike or sharp wave is followed by a flattening of EEG activity (Fig. 7).

Hrachovy et al. described the variations of the pattern: increased interhemispheric synchronization, asymmetrical hypsarrhythmia, hypsarrhythmia with a consistent focus of abnormal discharge, hypsarrhythmia with episodes of attenuation, and hypsarrhythmia comprising primarily high-voltage slow activity with little sharp-wave or spike activity. From our department, Kramer et al. studied the hypsarrhythmia variant patterns: electrodecremental discharges, absence of normal sleep activity, relative normalization, hemihypsarrhythmia, BS, occipital hypsarrhythmia, interhemispheric asymmetry, and interhemispheric synchronization, and devised a scoring system. The variant patterns occurred in 69% of 53 consecutive EEGs. A better prognosis was not related to the presence of any one variant pattern, but was associated with faster background activity (<75% delta), a lower total hypsarrhythmia score (≤10), and with absence of electrodecremental discharges on the pre-ACTH EEG.

3.2.6. Lennox–Gastaut Syndrome

Lennox–Gastaut Syndrome (LGS) is a severe, mixed seizure disorder, with tonic, atonic, and myoclonic seizures, mental retardation, and a slow spike-and-wave pattern on the EEG. This is a typical example of what has been termed an "epileptic encephalopathy." The majority of children with LGS have frequent intractable seizures, starting between 1 to 8 yr.

Tonic seizures are a major component of LGS, and criteria that are more restrictive for LGS require tonic seizures for the diagnosis. Tonic seizures are typically sleep activated and repetitive, although long-term monitoring may be needed to discover these. Benzodiazepines may induce tonic status epilepticus in LGS.

The EEG hallmark of LGS is the slow spike-and-wave discharge (1.5–2.5 Hz) superimposed on a slow background (Fig. 8). Bursts of fast rhythms occur during sleep (Fig. 9). These discharges may be frequent, and may be very sleep activated.

LGS is typically resistant to therapy with standard AEDs. Alternative therapies, such as corticosteroid therapy, the ketogenic diet, or vagal nerve stimulation, may be helpful.

3.2.7. Atypical Partial Benign Epilepsy of Childhood

Aicardi and Chevrie described the atypical partial benign epilepsy of childhood syndrome, which has some features of BRE; nocturnal focal seizures and focal EEG abnormalities, with

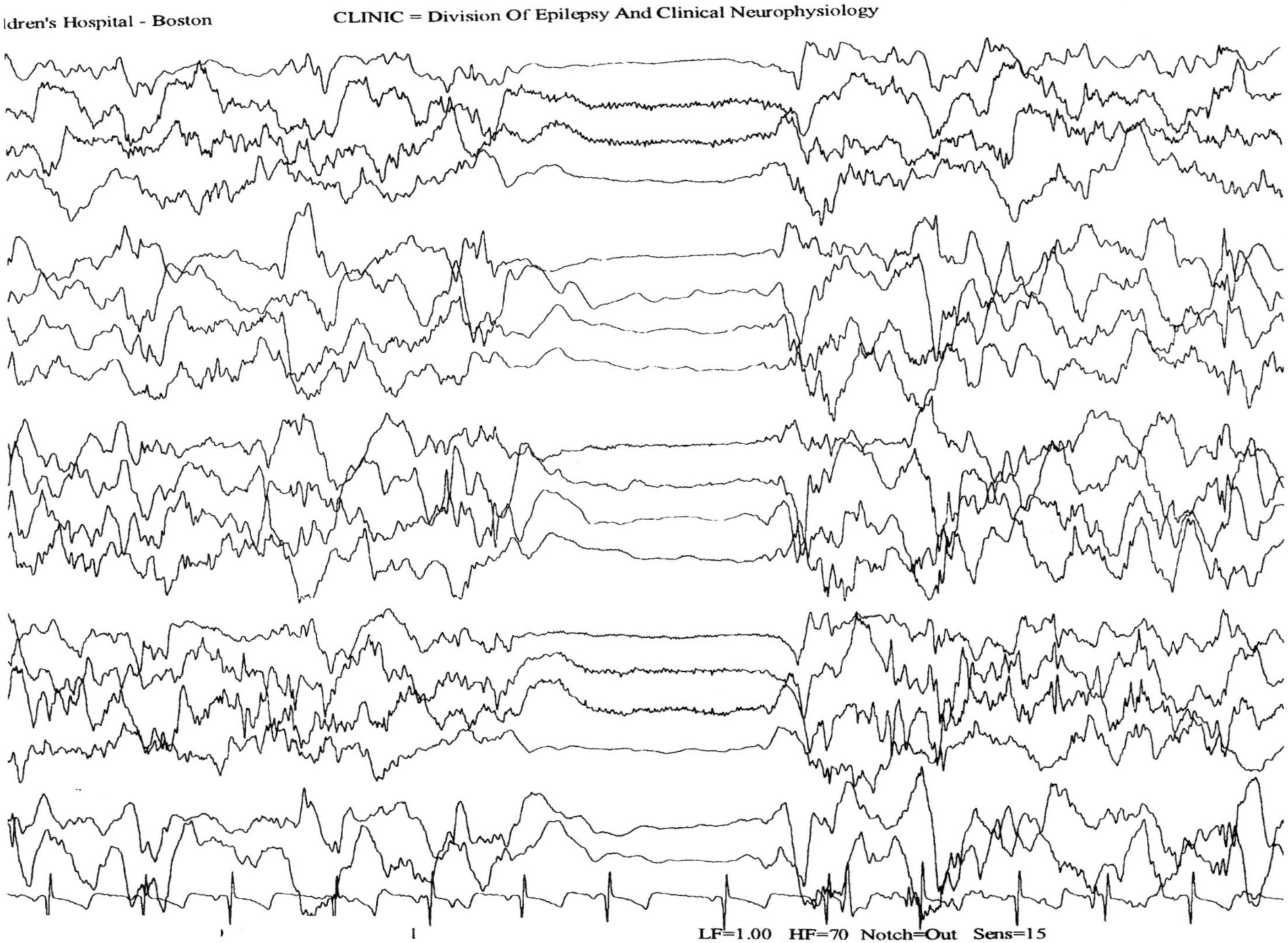

Fig. 7. Infantile spasms (IS). Seven-month-old child with IS. Note the high-amplitude, disorganized slow activity with multifocal spikes with an electrodecremental response. This sort of electrodecremental event may be associated with a clinical spasm.

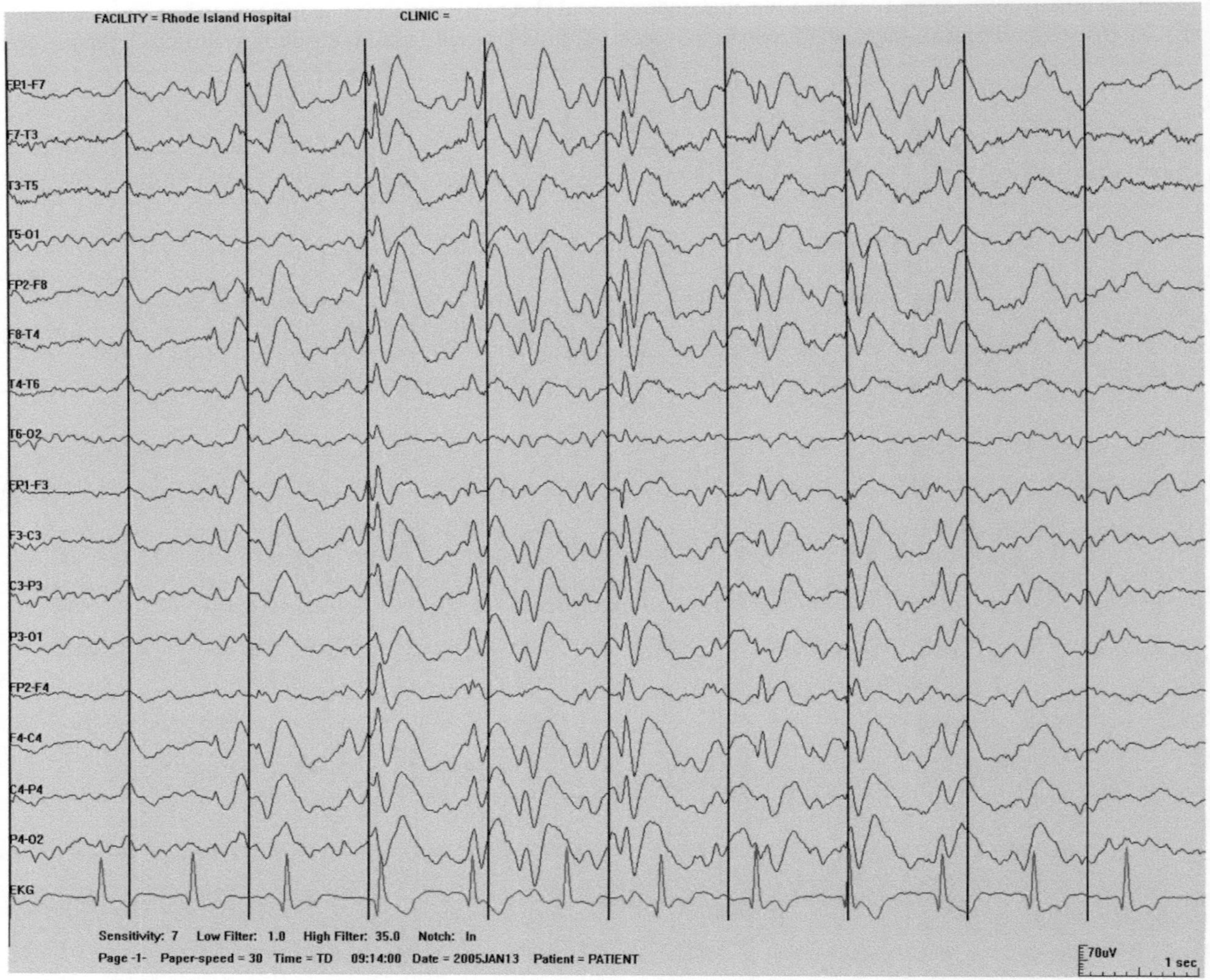

Fig. 8. Slow spike and wave of Lennox–Gastaut syndrome (LGS). Thirty-six-year-old man with LGS. Note the bilaterally synchronous 1.5- to 2-Hz spike-and-wave discharges.

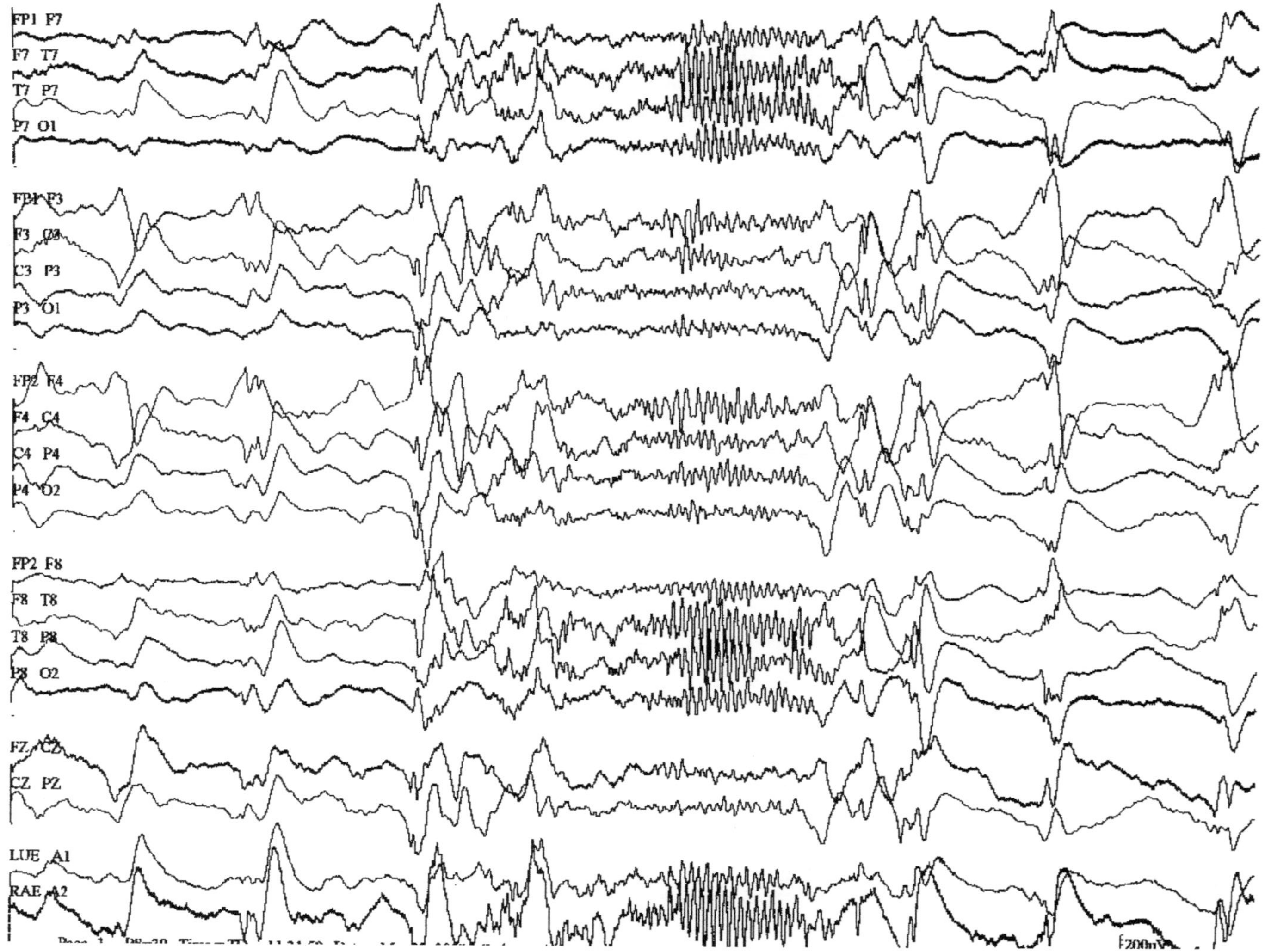

Fig. 9. Slow spike and wave with fast activity in sleep. Eleven-year-old girl with LGS. Her EEG shows slow spike-and-wave discharges, interrupted by a generalized burst of beta activity. Some of these beta bursts may be associated with tonic seizures.

initial seizures similar to BRE, but then evolving into periodic atonic seizures, with brief absences and GTCS. Atonic episodes correlate with the slow wave discharges on EEG. The EEG can be very sleep activated, similar to that seen in electrical status epilepticus of sleep (ESES), and time periods with marked sleep activation correlate with atonic seizures and intellectual regression. This fluctuates with the EEG findings. This disorder has been referred to as "pseudo-LGS," because EEG findings are similar to those of LGS.

3.2.8. Multiple Independent Spike Syndrome

Multiple independent spikes are defined as three or more independent spike foci, with at least one originating in both the right and left hemispheres. It is related to hypsarrhythmia and the slow spike-and-wave pattern seen in LGS. Background slowing and sleep-activated discharges are common, and similar to IS, increased synchrony may occur during sleep.

3.2.9. Epilepsy With Myoclonic–Astatic Seizures

Myoclonic astatic epilepsy is also called Doose Syndrome. Onset is typically between 7 mo and 6 yr of age. The seizures consist primarily of generalized myoclonic, astatic, or myoclonic–astatic seizures, short absences, and mostly GTCS, typically without tonic seizures or drop attacks. However, tonic seizures may occur in cases with a poor outcome. There are generalized spike-and-wave patterns on EEG with the background activity characterized by a 4- to 7-Hz rhythm with a parietal accentuation (Fig. 10). There is a high incidence of seizures or EEG abnormalities in relatives. Before the onset of the seizures, which may start as febrile seizures, development is normal. Valproic acid and ethosuximide are the drugs of choice. The prognosis is variable, some having a spontaneous remission.

3.2.10. Eyelid Myoclonia With Absences (Jeavons Syndrome)

Eyelid myoclonia with absences consists of eyelid myoclonia and absences, which may occur independently. The eyelid myoclonia typically occurs immediately after eye closure and is associated with brief bilateral spike-and-wave activity (Fig. 11). Eyelid myoclonia looks like rapid blinking, with upward eye deviation, and is more frequent in bright light and does not occur in the dark. However, eyelid myoclonia occurs in other idiopathic generalized epilepsies, especially with photosensitive generalized epilepsy, and has been reported in family members without seizures. These were called paroxysmal eyelid movements by Camfield et al. The frequency may help to differentiate these: the eyelid flutter with an absence seizure has a frequency of 3 Hz, eyelid myoclonia has a frequency of 4 to 6 Hz, and paroxysmal eyelid movements have a frequency of 10 Hz.

3.2.11. Photosensitive Epilepsy (A Reflex Epilepsy)

Photosensitive epilepsy is a specific reflex epilepsy, which refers to the induction of a seizure associated with a discharge of spikes or spike and slow waves with photic stimulation (referred to in the EEG laboratory as intermittent photic stimulation). The term photoparoxysmal is used if there is only an electrographic discharge, whereas photoconvulsive is used if there is an actual clinical event associated with the discharge. Photosensitive seizures are usually associated with idiopathic generalized epilepsy, but may also occur with focal or symptomatic epilepsies. Seizures may be self-induced, and this may be difficult to treat. Avoidance of precipitating stimuli may be important (e.g., polarized sunglasses).

3.2.12. LKS (Acquired Epileptic Aphasia)

LKS is a rare epileptic syndrome characterized by language regression and an abnormal EEG. Language regression usually begins in those older than 4 yr and may first manifest as

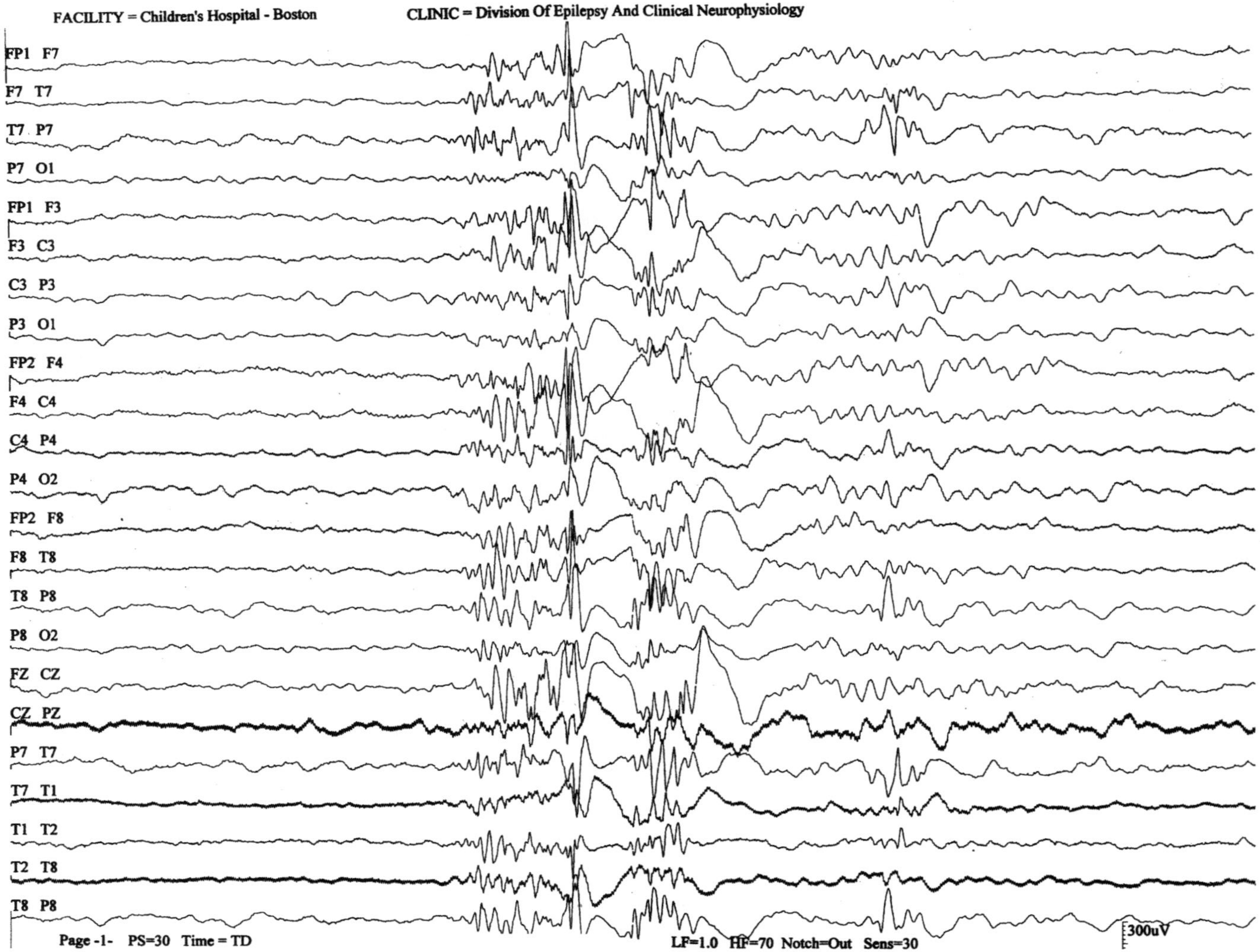

Fig. 10. Myoclonic astatic epilepsy (Doose syndrome). Five-year-old boy with generalized tonic–clonic seizures and myoclonic–astatic seizures. Note the generalized spike-and-wave discharge, with a frontal predominance, with a background of 6 Hz.

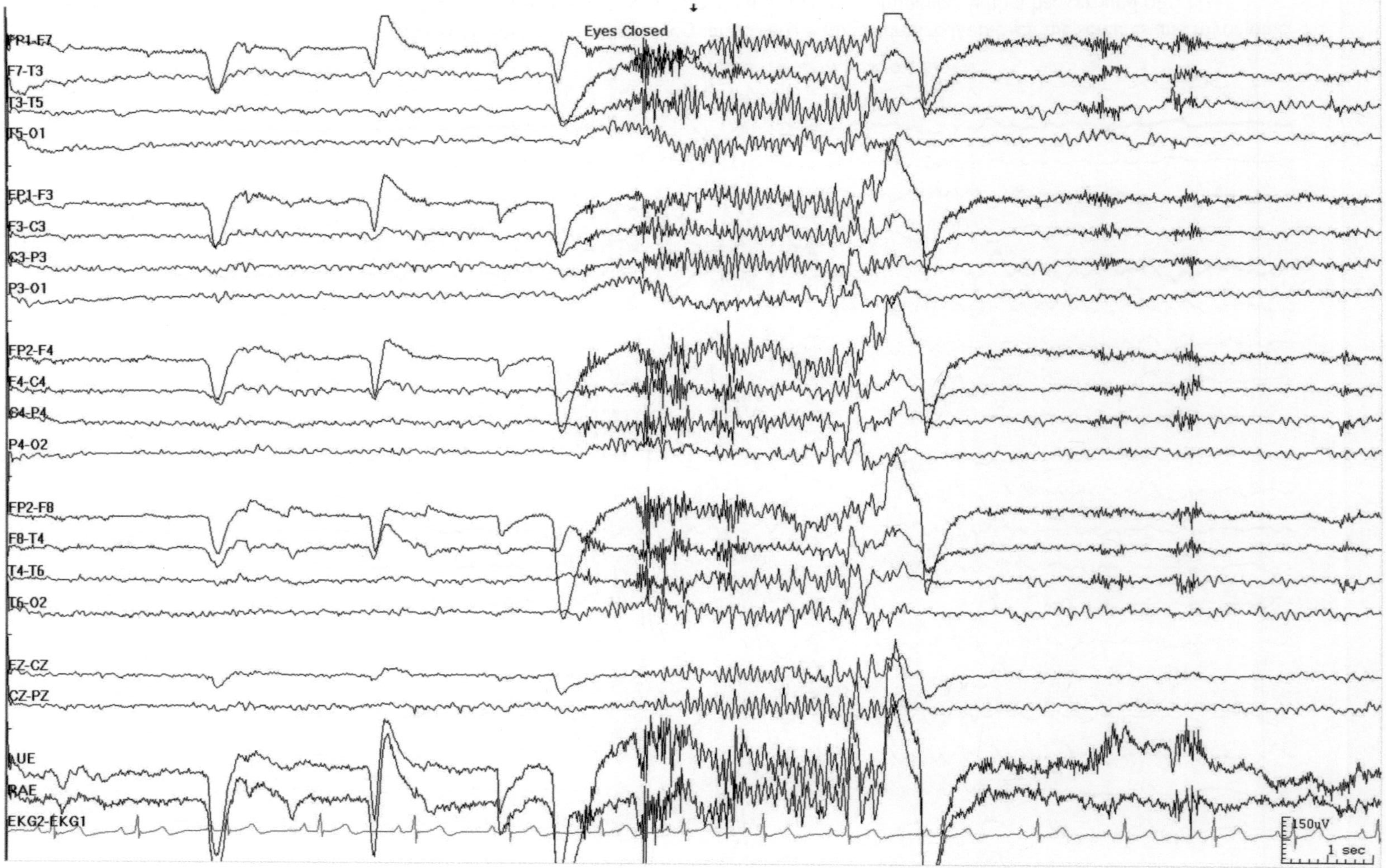

Fig. 11. Eyelid myoclonia with absences (Jeavons syndrome). Twelve-year-old girl with eyelid myoclonia associated with absence seizures and a rare, generalized tonic–clonic seizure, responsive to valproic acid. Note the generalized polyspikes induced by eye closure.

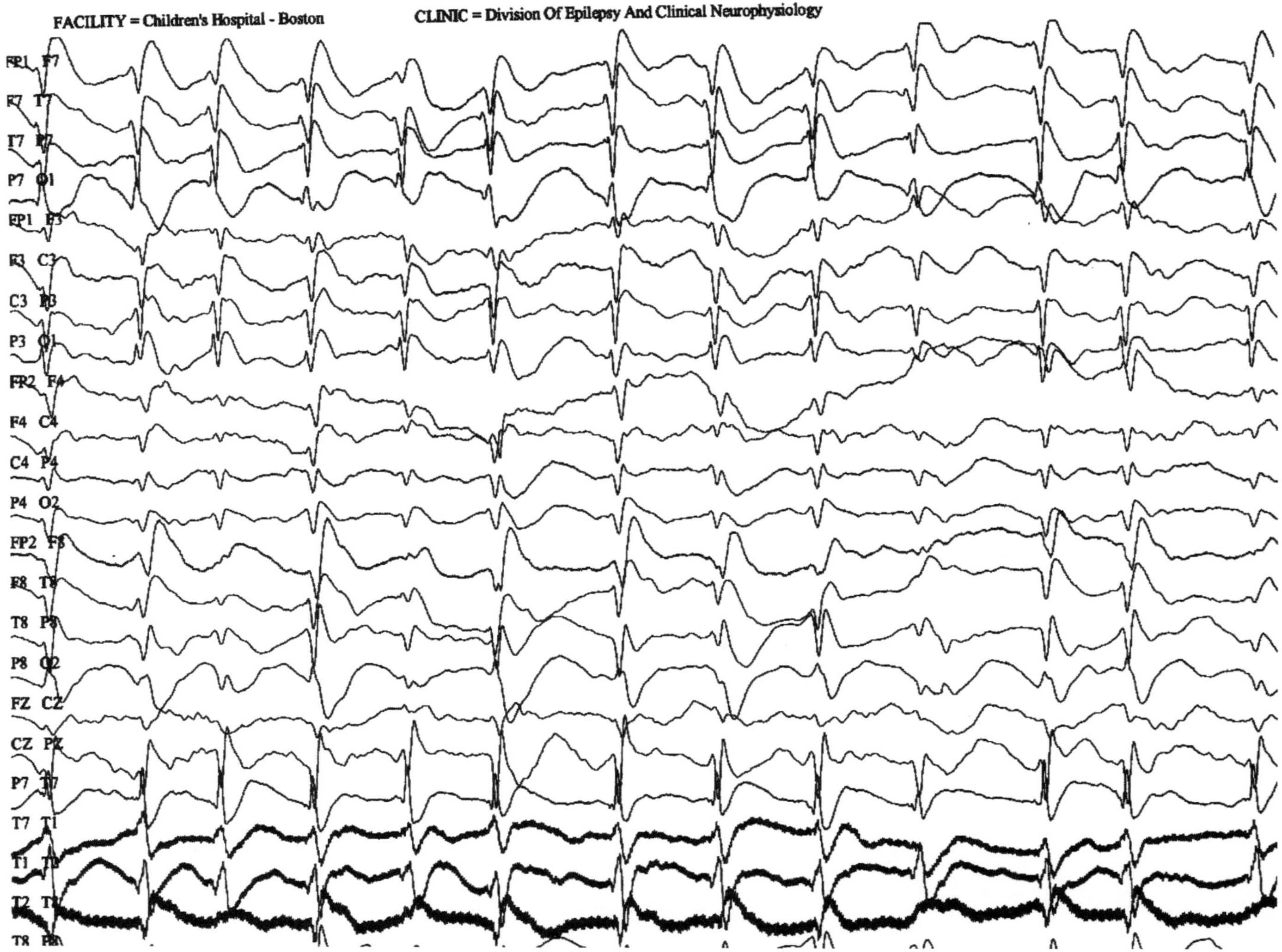

Fig. 12. Landau-Kleffner syndrome (LKS). Four-year-old girl with LKS. Language regression occurred at 3 yr of age. Note the generalized, high-amplitude, slow spike-and-wave discharge during sleep, representing electrical status epilepticus of sleep.

an apparent word deafness, called a "verbal auditory agnosia." Seizures and behavior disturbances, particularly hyperactivity, each occur in approximately two-thirds of children with LKS.

The EEG in LKS shows bilateral, multifocal spikes and spike-and-wave discharges, occurring usually in the posterior regions, especially the temporal region, with a marked activation during sleep (Fig. 12). However, discharges occur in many locations, and may even be generalized. Some authorities require the presence of ESES to make the diagnosis.

3.2.13. Epilepsy With CSWS

CSWS is an epileptic syndrome characterized by continuous spike and waves during slow wave sleep. This has also been called ESES. CSWS consists of a sleep-activated EEG pattern with paroxysmal features present during 85% of slow wave sleep, and is divided into symptomatic and cryptogenic groups, determined by whether normal neurological or psychomotor development was present before the onset of the CSWS. Seizures are common but may not be frequent. Regression occurs in a more global manner, rather than predominantly the language regression seen in LKS. The sleep activated EEG may also occur in LKS, and some require the presence of CSWS for making the diagnosis of LKS. Guilhoto and Morrell reported that a more focal EEG pattern is associated with language regression, whereas, if the EEG pattern is more generalized, the regression is more global, and not predominantly in language skills.

LKS and CSWS have similar treatments: anticonvulsants are used for seizures and corticosteroids are used for the language or neurobehavioral difficulties. Valproic acid, the benzodiazepines, and ethosuximide have been the most successful anticonvulsants, and there is data suggesting that carbamazepine may worsen the EEG. Steroids may have a better therapeutic efficacy than standard AEDs for the aphasia in LKS.

It is important to distinguish a sleep-activated EEG pattern from the epileptic syndromes of LKS and CSWS. The EEG could show the pattern of marked sleep activation but the patient need not have the specific epileptic syndromes with either language or global regression.

SUGGESTED READING

@Refs:Alvarez N, Lombroso CT, Medina C, Cantlon B. Paroxysmal spike and wave activity in drowsiness in young children: its relationship to febrile convulsions. Electroencephalogr Clin Neurophysiol 1983;56:406–413.

Aicardi J, Chevrie JJ. Atypical benign partial epilepsy of childhood. Dev Med Child Neurol 1982;24:281–292.

Asconape J, Penry JK. Some clinical and EEG aspects of benign juvenile myoclonic epilepsy: Epilepsia 1984;25:108–114.

Astradsson A, Olafsson E, Ludvigsson P, Bjorgvinsson H, Hauser WA. Rolandic epilepsy: an incidence study in Iceland. Epilepsia 1998;39:884–886.

Camfield CS, Camfield PR, Sadler M, et al. Paroxysmal eyelid movements: a confusing feature of generalized photosensitive epilepsy. Neurology 2004;63:40–42.

Cavazzuti GB, Cappella L, Nalin A. Longitudinal study of epileptiform EEG patterns in normal children. Epilepsia 1980;21:43–55.

Delgado-Escueta AV, Enrile-Bacsal F. Juvenile myoclonic epilepsy of Janz. Neurology 1984;34:285–294.

Guilhoto LMFF, Morrell F. Electrophysiological differences between Landau–Kleffner Syndrome and other conditions showing the CSWS electrical pattern. Epilepsia 1994;35(Suppl 8):126.

Hahn JS, Monyer H, Tharp BR. Interburst interval measurements in the EEGs of premature infants with normal neurological outcome. Electroencephalogr Clin Neurophysiol 1989;73:410–418.

Hirtz D, Ashwal S, Berg A etal. Practice parameter: evaluating a first nonfebrile seizure in children: report of the quality standards subcommittee of the American Academy of Neurology, The Child Neurology Society, and The American Epilepsy Society. Neurology 2000;55:616–623.

Holmes GL, McKeever M, Adamson M. Absence seizures in children: clinical and electroencephalographic features. Ann Neurol 1987;21:268–273.

Hrachovy RA, Frost JD Jr, Kellaway P. Hypsarrhythmia: variations on the theme. Epilepsia 1984;25:317–325.

Kellaway P. The incidence, significance, and natural history of spike foci in children. In: Current Clinical Neurophysiology. Update on EEG and Evoked Potentials (Henry CE, ed). Elsevier/North Holland, New York, NY, 1980, pp. 151–175.

Kotagal P. Multifocal independent Spike syndrome: relationship to hypsarrhythmia and the slow spike-wave (Lennox–Gastaut) syndrome. Clin Electroencephalogr 1995;26:23–29.

Kramer U, Sue WC, Mikati MA. Hypsarrhythmia: frequency of variant patterns and correlation with etiology and outcome. Neurology 1997;48:197–203.

Lerman P. Benign partial epilepsy with centro-temporal spikes. Epileptic syndromes in infancy, childhood and adolescence, 2nd ed. (Roger J, Bureau M, Dravet C, et al., eds). John Libbey and Company, London, UK 1992, pp. 189–200.

Legarda S, Jayakar P, Duchowny M, Alvarez L, Resnick T. Benign rolandic epilepsy: high central and low central subgroups. Epilepsia 1994;35:1125–1129.

Lerman P, Kivity S. Benign focal epilepsy of childhood. A follow-up study of 100 recovered patients. Arch Neurol 1975;23:261–264.

Lewis DW, Ashwal S, Dahl G, et al. The Quality Standards Subcommittee of the American Academy of Neurology; The Practice Committee of the Child Neurology Society. Practice parameter: evaluation of children and adolescents with recurrent headaches: report of the Quality Standards Subcommittee of the American Academy of Neurology and the Practice Committee of the Child Neurology Society. Neurology 2002;59:490–498.

Lombroso CT. Sylvian seizures and mid temporal spike foci in children. Arch Neurol 1967;17:52 59.

Lum LM, Connolly MB, Farrell K Wong PK. Hyperventilation-induced high-amplitude rhythmic slowing with altered awareness: a video-EEG comparison with absence seizures. Epilepsia 2002;43:1372–1378.

Massa R, de Saint Martin A, Carcangiu R, et al. EEG criteria predictive of a complicated evolution in idiopathic rolandic epilepsy. Neurology 2001;57:1071–1079.

Mizrahi EM, Kellaway P. Characterization and classification of neonatal seizures. Neurology 1987;37:1837–1844.

Niedermeyer E. Primary (idiopathic) generalized epilepsy and underlying mechanisms. Clin Electroencephalogr 1996;27:1–21

Panayiotopoulos CP. Benign childhood epileptic syndromes with occipital spikes: new classification proposed by the International League Against Epilepsy. J Child Neurol 2000;15:548–552.

Penry JK, Porter RJ, Dreifuss RE. Simultaneous recording of absence seizures with video tape and electroencephalography. A study of 374 seizures in 48 patients. Brain 1975;98:427–440.

Proposal for Revised Classification of Epilepsies and Epileptic Syndromes, Epilepsia 1989;30:389–399.

Practice parameter: long-term treatment of the child with simple febrile seizures. American Academy pf Pediatrics. Committee on Quality Improvement, Subcommittee on Febrile Seizures. Pediatrics 1999;103:1307–1309.

Richer LP, Shevell MI, Rosenblatt BR. Epileptiform abnormalities in children with attention-deficit-hyperactivity disorder. Pediatr Neurol 2002;26:125–129.

Rose AL, Lombroso CT. A study of clinical, pathological, and electroencephalographic features in 137 full-term babies with a long-term follow-up. Pediatrics 1970;45:404 425.

Volpe JJ. Neurology of the Newborn, 3rd ed. WB Saunders Co., Philadelphia, PA, 1995, pp. 172–207.

Wirrell EC, Camfield PR, Gordon KE, Dooley JM, Camfield CS. Benign rolandic epilepsy: atypical features are very common. J Child Neurol 1995;10:455–458

REVIEW QUESTIONS

1. Do epileptiform abnormalities oblige the diagnosis of epilepsy?
2. What is dyshormia?
3. What is the distinction between simple and complex typical absence seizures?
4. What is the distinction between typical and atypical absence seizures?
5. What is the significance of OIRDA?
6. What is the characteristic EEG finding in JME?
7. What are features of the two benign occipital epilepsies of childhood?
8. What defines the syndrome of IS (West Syndrome)? What are its classical interictal and ictal EEG findings?
9. What defines LGS? What is the EEG of LGS?
10. What features are common to both LKS and CSWS?

REVIEW ANSWERS

1. No, epileptiform abnormalities may be seen in as many as 3.5% of healthy children. Therefore, epilepsy is a clinical diagnosis, supported by appropriate EEG findings.
2. Dyshormia is an abnormal paroxysmal feature intermixed with normal sleep elements in the EEG. Epileptiform features embedded within vertex waves, for example, represents dyshormia. It has been suggested to indicate the possibility of a lower than normal seizure threshold.
3. Simple absence seizures lack associated motor phenomena. Complex absence seizures have associated motor or autonomic elements.
4. Typical absence seizures have a sudden onset and offset and tend to be briefer. Their EEG illustrates a 3-Hz generalized spike-and-wave pattern. Atypical absence seizures have less clearly defined beginning and end, more change in tone, tend to last longer, and often are in the company of mental retardation and other seizure types. Their EEG typically has a slower frequency of 1.5 to 2.5 Hz, is more apt to be asymmetric, and usually has a slow background.
5. OIRDA is a childhood pattern of paroxysmal rhythmic delta activity, sometimes with admixed notched or spike elements. It is a nonspecific pattern, but seems to occur with great frequency in children with epilepsy, mostly absence epilepsy. However, it can be seen in other scenarios and need not be an epileptic finding.
6. JME is associated with a fast spike-and-wave pattern, 3.5- to 6-Hz spike-and-wave or polyspike-and-wave complexes. Photosensitivity occurs quite frequently.
7. The early form of benign occipital epilepsy involves children ages 2 to 12 yr having seizures involving vomiting, eye deviation, and altered awareness. The later type (age 3–16 yr) has visual seizures with visual hallucinations, feeling of ocular movement or pain, eye deviation, ictal blindness, and focal or generalized seizures. Both show giant spikes with morphologies resembling those of BRE.
8. IS is defined by a triad of myoclonic or tonic seizures, hypsarrhythmia, and mental retardation. The classic interictal finding is the hypsarrhythmia pattern involving high voltage and a disorganized background with multifocal spike and sharp waves. It is usually sleep activated. The classic ictal finding is the electrodecremental pattern with an initial generalized spike or sharp wave, then a diffuse attenuation of the EEG of variable duration.
9. LGS is a disorder characterized by mixed seizure types, mental retardation, and a slow (1.5–2.5 Hz) spike-and-wave pattern on EEG. Background slowing is typical. It is commonly a highly refractory epilepsy. Tonic seizures and interictal epileptiform discharges are both sleep-activated features in LGS.
10. Both LKS and CSWS are uncommon epilepsies with profound sleep activation. In both, the sleep EEG may become dominated by epileptiform activity, even to the level of ESES (>85% of slow wave sleep with paroxysmal features). In LKS, the discharges are thought to be more bitemporal and/or biparietal, whereas in CSWS discharges are more diffusely seen. Seizures accompany both syndromes, but the more devastating clinical problems are language regression (in LKS) and global behavioral regression (in CSWS). Treatments involve anticonvulsants and often corticosteroids for both.

III

Nerve Conduction Studies and Electromyography

12

Neurophysiology of Nerve Conduction Studies

James B. Caress, Gregory J. Esper, and Seward B. Rutkove

Summary

The methodology for performing standard nerve conduction studies has been established by identifying the most helpful and consistent physiological data obtainable while being constrained by a variety of technical and practical limitations. Nerve stimulation occurs underneath the negatively charged anode of the applied stimulator and simultaneous hyperpolarization of the nerve occurs beneath the positively charged cathode. Referential or bipolar recording techniques are used for all types of measurements. Sensory conduction studies can be performed either antidromically or orthodromically, although, for technical reasons, the former are usually preferred; the recorded sensory nerve action potential is made up of the simultaneous depolarization of all of the cutaneous sensory axons. In motor studies, the compound motor action potential is recorded from the motor point of the muscle of interest and represents the depolarization of the underlying muscle fibers rather than the nerve itself and is, thus, of considerably greater amplitude and duration. F-waves and H-reflexes represent the two most commonly evaluated forms of late responses and assist with assessing the entire length of the neurons, from spinal cord to distal muscle.

Key Words: Compound motor action potential; depolarization; late responses; nerve conduction study; sensory nerve action potential; stimulation.

1. PHYSIOLOGY OF STIMULATION

Stimulators used in routine nerve conduction studies (NCS) have a cathode and an anode and are, therefore, bipolar. The cathode is negatively charged, whereas the anode is positively charged. The depolarization of axons occurs under the cathode because the negativity in the region of the cathode leads to a reduction in the potential difference between the inside and the outside of the cell (the inside of the cell is relatively negative at baseline). On the other hand, the extracellular environment under the anode is positively charged, leading to hyperpolarization of the underlying axons.

Much of the current supplied by the stimulator travels in the very low resistance extracellular space because current follows the path of least resistance. The cross-sectional resistance of an axon will determine whether some of that current will enter and depolarize the nerve. Cross-sectional resistance is reduced as diameter of the axon increases, resulting in a greater area in which current can flow. The result relevant to NCS is that large axons will depolarize with relatively less stimulus current than smaller axons. Hence, with low stimulus intensities, large axons will be preferentially stimulated.

From: *The Clinical Neurophysiology Primer*
Edited by: A. S. Blum and S. B. Rutkove © Humana Press Inc., Totowa, NJ

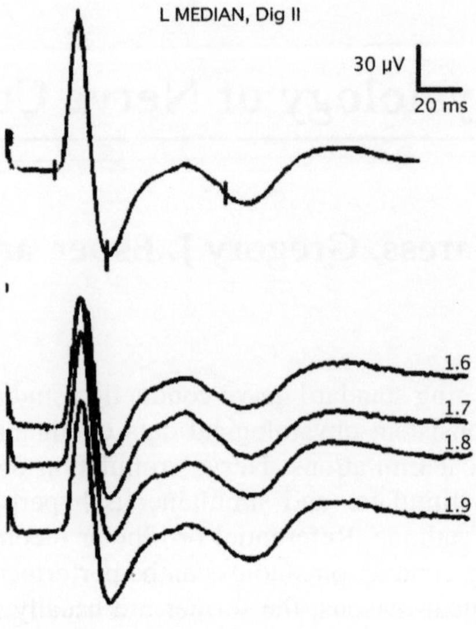

Fig. 1. A median sensory nerve action potential recorded from digit 2. The digitally averaged waveform is on top.

2. MEASUREMENT OF SENSORY POTENTIALS IN NCS

Waveforms that are displayed during sensory NCS reflect the passage of current beneath surface electrodes that are at least a few millimeters distant from the current generators (that is, the depolarizing axons). Routine studies usually use a bipolar recording technique in which both the active and reference electrodes lie above the nerve. The potential measured by the active electrode is compared with that measured by the reference electrode, and both are compared with a ground electrode lying elsewhere on the patient. The reference electrode makes an important contribution to the observed compound sensory nerve action potential (SNAP) (Fig. 1). The nerve current is not directly measured in NCS but overlying loops of current in the soft tissue reflect the actual action potentials. This situation is referred to as volume conduction and is further described in Chapter 4.

Temporal dispersion refers to the phenomenon that, as a sensory conduction study is performed over longer and longer segments of nerve, the recorded SNAP loses its sharpness (high amplitude and short duration) and becomes a broader, lower-amplitude potential. This occurs normally in any sensory NCS because individual sensory axons conduct at slightly different rates and because these slight differences are magnified when a NCS is performed over a large segment of nerve. In diseased states, such as polyneuropathies, temporal dispersion can be enhanced (abnormal temporal dispersion). Because sensory NCS are especially vulnerable to the effects of temporal dispersion in both health and disease, sensory studies are usually performed over shorter segments of nerve than motor studies, which tend to be less susceptible to these effects (described in Section 3).

Sensory responses can be recorded with an orthodromic or an antidromic technique. In orthodromic studies, the recording electrodes are proximal to the stimulation site, and, in antidromic studies, stimulation is proximal to the recording position. In the hands, ring electrodes are used around the fingers to stimulate (orthodromic) or record (antidromic). Although orthodromic

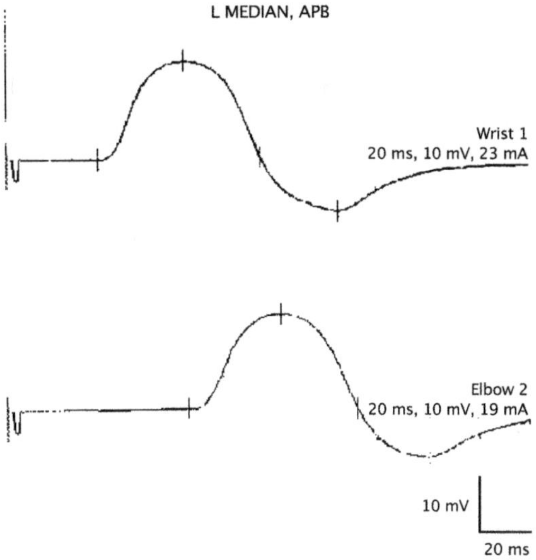

Fig. 2. A median compound muscle action potential. Recordings made with wrist and elbow stimulating APB, abductor pollicis brevis.

responses reflect more accurately the physiology of sensory responses via afferent conduction, upper extremity orthodromic responses are usually considerably smaller than antidromic responses.

3. MEASUREMENT OF MOTOR POTENTIALS IN NCS

The active electrode is placed over the *motor point* of the muscle, where the majority of the motor axons synapse with the end plates of the muscle fibers. Rather than measuring a compound nerve action potential, the summation of the muscle fiber action potentials is measured. This is called the compound motor action potential (CMAP) (Fig. 2). In this case, there is no leading edge of a dipole to cause an initial downward deflection, because the depolarization initiates directly beneath the recording electrode. The influx of Na^+ at the muscle end plates generates extracellular negativity that is displayed as an abrupt upward deflection from the baseline. The waveform returns to the baseline as the resting membrane potential of the muscle fibers is reestablished. The distal latency is the time at which the depolarization of the fastest nerve fiber is recorded; this response is routinely recorded but does not directly measure the conduction velocity, as is the case with sensory recordings. This is because of the additional time it takes for acetylcholine to traverse the synapse, to attach to receptors, and to generate muscle fiber action potentials. Conduction velocity is calculated only in the proximal segments of nerve, by subtracting the distal latency and distance from the values obtained with proximal stimulation. However, by establishing normal values for distal latencies for specific distances, meaning can be given to the measurement that can assist in the evaluation of a variety nerve disorders, including distal compression neuropathies, such as carpal tunnel syndrome (median neuropathy at the wrist).

One point that is frequently overlooked is that the recorded CMAP does not reflect muscle contraction. The CMAP is a purely electrical signal from the summation of muscle fiber action potentials, which occurs well before the actual contraction of the muscle takes place.

The visible muscle contraction requires Ca^{2+} release cross-linking of actin and myosin fibers (so-called excitation–contraction coupling), which occurs much more slowly and is not measured in routine NCS. Also, as in sensory NCS, the reference electrode is not electrically inactive, and it contributes to the waveform, even though it is often placed over a bone, a region generally considered electrically silent. This is particularly clear during ulnar or tibial motor NCS, where the waveform characteristically exhibits a double-peaked negative phase. The second peak results from activity recorded at the reference electrode.

Similar to sensory responses, normal temporal dispersion of the responses does occur because smaller motor neurons will conduct at slower velocities than larger, causing the recorded CMAP to spread out (increase in duration and decrease in amplitude) the longer the segment of nerve studied. However, the effects are not nearly as dramatic as those observed in sensory studies, because the duration of the CMAP is much larger than of the SNAP (~6 times longer). Hence, small increases in duration are not be readily apparent, and the amplitude of the response generally declines only modestly with increasing distance.

4. NEUROPHYSIOLOGY OF THE LATE RESPONSES

4.1. F-Waves

The "F" stands for foot, because these responses were first recorded from intrinsic foot muscles. On stimulation of a single motor axon, the wave of depolarization will travel distally to be recorded as part of the CMAP, or "M-wave," but also will travel proximally to its anterior horn cell (AHC). Retrograde depolarization of AHCs will result in regeneration of an action potential at the axon hillock in a small subset of neurons (~5–10%), which then travels back down the motor axon to the innervated muscle, recorded at the electrodes as the F-wave (Fig. 3). Although, for theoretical reasons, it is desirable to reverse the polarity of the stimulator, placing the anode distally and cathode proximally to avoid "anodal block," this remains more of a theoretical rather than a practical consideration. F-waves are easily generated with the cathode in the distal position. At the level of the spinal cord, no synapse is involved, and an F-wave from a single axon has the same morphology and almost identical latency each time. However, a single axon will not generate an F-wave with each successive depolarization. The probability that an F-wave will be generated from any neuron is dependent on the variable excitability of the AHC membrane that can be increased with reinforcement maneuvers (clenching teeth or making a tight fist) and decreased by sleep and anesthesia. F-waves may be absent in sleeping, sedated, or comatose persons, and should not be interpreted as a sign of peripheral nervous system disease in these patients. During NCS, supramaximal stimulation of the peripheral nerve ensures that all motor axons capable of generating an F-wave are depolarized each time. Under normal conditions, several axons generate an F-wave, and the summation of these responses is recorded. The varying probability that a single axon will produce an F-wave results in variation of the morphology and latency of the summated F-waves. In severe neuropathic conditions, when only one axon capable of generating an F-wave is surviving, it is recorded as an "all or none" response without variability. Also under neuropathic conditions, unusually high amplitude F-waves may be recorded because of reinnervation of the motor units.

At least 10 F-waves are usually collected for analysis. The minimum F-wave latency is the parameter measured in most labs, but reflects only the fastest motor axon contributing to the F-wave; this value may be spared in pathological situations and, importantly, in acquired demyelinating neuropathies. Mean or median F-wave latencies provide better information concerning the proximal sections of the nerves in diseased states. However, most EMG

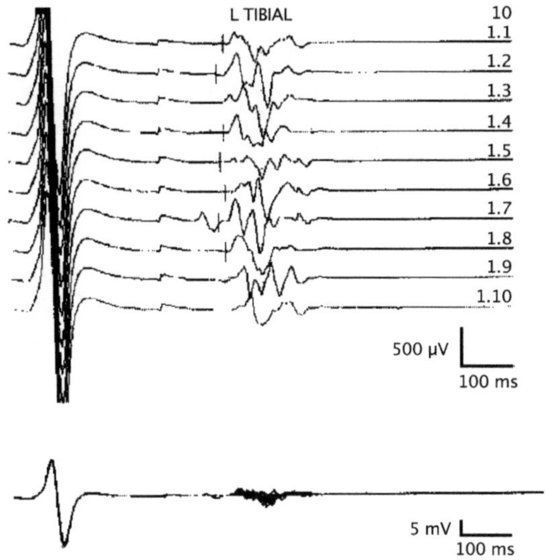

Fig. 3. F-responses recorded from the tibial nerve; the bottom tracing shows the superimposed data.

laboratories do not have normal values from which to interpret abnormalities in these parameters. Other F-wave parameters that may also be useful include *chronodispersion* and *persistence*. Chronodispersion is a measure of the range of minimum latencies and is normally less than 5 ms. This range can be exceeded in demyelinating neuropathies or radiculopathies. Persistence is a measure of the frequency of obtaining an F-wave after supramaximal stimulus, and varies for different motor nerves. Persistence is 80 to 100% for most nerves, but peroneal F-waves may be difficult to elicit, even in healthy persons.

F-wave minimum latency is dependent on limb length, and nomograms are used to judge abnormality. If a height-adjusted table is not available, the appropriate F-wave latency corrected for height (±2.5 ms) can be estimated using the formula:

$$\textit{F-estimate}\ (\text{ms}) = [2 \times \textit{F distance}\ (\text{mm})/\textit{CV}\ (\text{m/s})] + \textit{distal latency}\ (\text{ms}) + 1\ \text{ms}$$

where *F distance* is from the stimulus site to the C7 spinal process (median, ulnar) or the xiphoid process (peroneal, tibial), and *CV* is the conduction velocity. The addition of 1 ms allows for the central conduction delay. F-estimates can also be useful for identifying proximal pathology in the proximal segments of a nerve. Because the calculation of the estimate relies on the conduction velocity, which is obtained distally, proximal pathology will create a longer measured F-wave latency than would be anticipated by the *F-estimate* calculation.

Because F-waves probe the proximal portions of nerves, it seems reasonable that they should be useful in studying the plexus and nerve roots, which otherwise can only be directly stimulated using special techniques. Unfortunately, F-waves do have limitations for common pathology, such as structural radiculopathies, for several reasons. For example, the standard median and ulnar motor NCS evaluate only the C8 and T1 roots, which are not commonly damaged in structural radiculopathies, C5, C6, and C7 radiculopathies being much more common. Further, the shared root innervation (C8, T1) of abductor digiti quinti and abductor pollicis brevis muscles means that, in the setting of a completely transected C8 root, the F-wave latency may still be normal because of sparing of the T1 root. F-waves are likely to be more useful in detecting radiculopathies affecting the lower extremities, where standard peroneal and tibial motor conduction studies evaluate L5 and S1 derived neurons. Finally, because the F-response is a pure motor phenomenon, radiculopathies affecting only the sensory root cannot be detected (although this also remains a limitation of needle EMG).

Recall that the F-wave evaluates the proximal and distal segments of the nerve because it travels to the root level and then back to the distal muscle to be recorded. This implies that F-waves may be prolonged in distal entrapment neuropathies and polyneuropathies; thus, abnormal F-waves are not specific for proximal nerve pathology.

F-waves have their greatest usefulness in the setting of demyelinating radiculoneuropathies, especially Guillain–Barré Syndrome (GBS). Early GBS is frequently characterized by prominent involvement of the spinal roots. In these cases, the distal motor nerve segments may be electrophysiologically normal despite clinical weakness and areflexia caused by demyelination at the root level. F-wave minimum latency is commonly prolonged, and persistence can be severely reduced or absent in this setting; this suggests demyelination with conduction block in the proximal segments of the nerves. The combination of a normal distal CMAP and absent F-waves is highly specific for proximal demyelination. F-wave latencies are most prolonged 3 to 5 wk after the onset of GBS, secondary to further demyelination throughout the length of the nerve.

4.2. H-Reflex

"H" stands for Hoffman, after the investigator who first recorded the late response in 1918. The H-reflex is often said to be the NCS parallel of the clinical ankle tendon reflex, but the two tests are probably not evaluating the exact same group of nerve fibers. The tibial nerve is stimulated in the popliteal fossa, with the cathode proximal to the anode, and the response is recorded with the active electrode overlying the belly of the soleus muscle and the reference electrode placed at the calcaneal tendon. As with F-waves, the "reversed" cathode–anode orientation avoids the possibility of anodal block. The nerve is stimulated with a long-duration (1 ms) pulse rather than the short-duration pulse commonly used in routine NCS (0.05–0.2 ms) because the longer pulse is more effective at selectively stimulating the type 1A sensory fibers from muscle spindle organs that initiate the response.

In clinical practice, the H-reflex in adults is generally mostly recorded from the soleus; however, it is also readily obtainable form flexor carpi radialis with medial nerve stimulation, but is generally not performed because it provides little additional information over conventional EMG. However, in children and in diseased states, H-reflexes may also be elicitable from many other nerves and muscles.

The H-reflex is best studied by generating a recruitment curve, accomplished by gradually increasing stimulus intensities such that the variability of the H-reflex can be measured (Fig. 4). At very low stimulus intensities, no response can be elicited. At slightly higher intensity, a small H-reflex can be seen without the M-wave. This occurs because the type 1A fibers are being stimulated without direct depolarization of the adjacent alpha motor neurons. The 1A fibers conduct the stimulus to the spinal cord, where a monosynaptic reflex occurs, as it does with the clinical ankle jerk, where the afferent volley is initiated by the Golgi tendon organs. From the spinal cord, the response travels orthodromically down the motor neuron to the soleus, where the triphasic H-wave is recorded. With further increasing stimulation, the H-reflex continues to grow in amplitude as more and more 1A afferents are stimulated. Then, as motor fibers are directly depolarized, an M-wave begins to appear. The H-reflex then begins to decline in amplitude with increasing stimulus intensity caused by collision with the

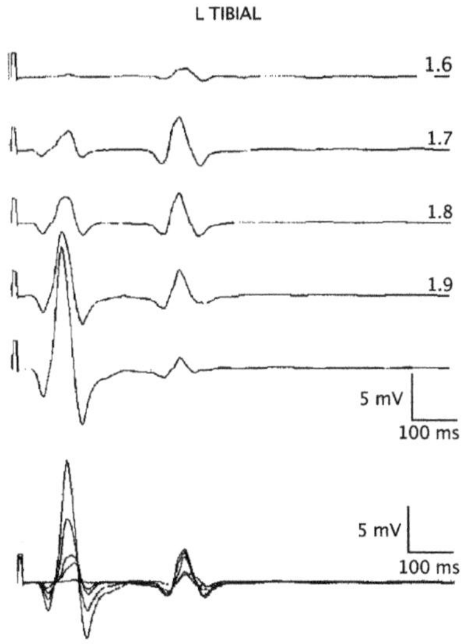

Fig. 4. H-reflex recruitment. Note how the potential comes in initially before the M-wave, maximizes just as the M-wave appears, and then gradually decreases in size as the M-wave becomes supramaximal. Once the M-wave is supramaximal, the only recorded late responses will be the F-waves. The bottom tracing shows the superimposed data.

antidromic impulse being conducted along the motor neurons. Eventually, the H-reflex is completely abolished, as a supramaximal motor response is elicited and antidromic motor responses collide with all the descending motor input arriving via the sensory pathways. At this point, any late responses recorded will be F-waves only.

The minimum onset latency (usually 25–34 ms) is the only routinely measured aspect of the H-reflex and, similar to the F-wave, is dependent on the height of the subject and the integrity of the nerves. Comparing the affected and unaffected leg reflexes is more useful than analysis of a unilateral response, and most laboratories consider up to a 1.5 ms difference as normal. Also, comparing bilateral H-reflex amplitudes can be informative, as can comparing the ratio of the H-reflex amplitude with the M-wave amplitude. The ratio may reflect the degree of AHC excitability. The ratio usually increases in upper motor neuron lesions and can be used to evaluate spasticity. Amplitude can be measured as long as care is taken to move the recording electrodes to ensure optimum positioning.

The H-reflex suffers from a lack of specificity for similar reasons as outlined for F-waves. The absolute latency is increased in polyneuropathies (axonal and demyelinating), or the reflex may be absent. Comparison of the healthy and diseased legs may be useful for diagnosing S1 radiculopathies.

4.3. *A-Waves*

The A-wave (Fig. 5) is another late response; however, it is only prominent in pathological states. It can be distinguished from F-waves by its invariable latency and morphology with each stimulus. The A-wave latency is typically shorter than the F-wave but can be longer. A-waves can be seen in the setting of any neuropathic process, but are quite common in polyneuropathies and radiculopathies.

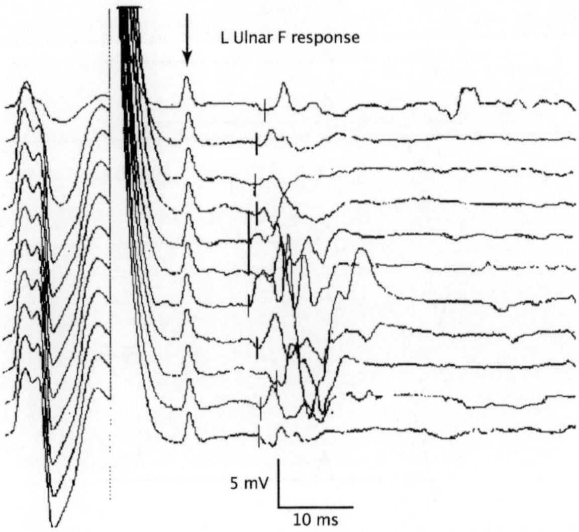

Fig. 5. An example of A-wave (arrow), obtained while evaluating ulnar F-waves.

A number of potential explanations have been suggested for the occurrence of A-waves. One possibility is that an abnormal axon may give off a branch proximally. Stimulation of the nerve may result in antidromic excitation of the branch, leading to a descending orthodromic depolarization. This would essentially represent a so-called "axon reflex." Another possibility is that ephaptic transmission is occurring between the axons proximally because of abnormal (crosstalk) demyelination. An impulse can then travel antidromically up one neuron and orthodromically down another, exciting the muscle at a constant latency.

SUGGESTED READING

Albers JW, Donofrio PD, McGonagle TK. Sequential electrodiagnostic abnormalities in acute inflammatory demyelinating polyradiculoneuropathy. Muscle Nerve 1985;8:528–539.

Dumitru D. Physiologic basis of potentials recorded in electromyography. Muscle Nerve 2000; 23:1667–1685.

Fraser JL, Olney RK. The relative diagnostic sensitivity of different F-wave parameters in various polyneuropathies. Muscle Nerve 1992;15:912–918.

Kincaid JC, Brasher A, Markand ON. The influence of the reference electrode on CMAP configuration. Muscle Nerve 1993;16:392–396.

Preston DC, Shapiro BE. Electromyography and Neuromuscular Disorders. Butterworth-Heinemann, Boston, MA, 2000.

Roth G. Clinical Motor Electroneurography: Evoked Responses Beyond the M-Wave Eectopic Activity. Elsevier, Amsterdam, Holland, 2000.

REVIEW QUESTIONS

1. Which of the following statements concerning temporal dispersion is true?
   A. Abnormal temporal dispersion is typically observed only in sensory nerves.
   B. Normal temporal dispersion is independent of the length of nerve being studied.
   C. Abnormal temporal dispersion is the hallmark of axonal polyneuropathies.
   D. Abnormal temporal dispersion is usually accompanied by a drop in response amplitude.
   E. Marked abnormal temporal dispersion is typically seen in the median nerve of patients with carpal tunnel syndrome.

2. Which of the following comments is true regarding nerve stimulation?
   A. Nerve depolarization typically occurs beneath the anode.
   B. Little of the applied electrical current typically flows through the extracellular space.
   C. Large and small nerve fibers have an equal propensity toward initiating an action potential with stimulation.
   D. Anodal black can theoretically interfere with the acquisition of F wave data.
   E. Standard nerve conduction study techniques rely on the use of submaximal stimulation.
3. Which of the following statements concerning the F-estimate is true?
   A. The F-estimate can assist in identifying the presence of a distal neuropathic process.
   B. The F-estimate is useful only in lower extremity studies where the F-waves are longer.
   C. In a patient with an isolated L5 radiculopathy, but otherwise a normal nerve conduction study, the F-estimate would likely provide a value of shorter latency than the measured F-waves.
   D. The distal latency is not used to help calculate the F-estimate.
   E. The F-estimate is likely to be elevated falsely in patients with axonal polyneuropathies.
4. The following statements concerning sensory nerve conduction studies are true, EXCEPT:
   A. They can be performed orthodromically or antidromically.
   B. They are more susceptible to normal temporal dispersion than motor studies.
   C. They rely on depolarization of the nerve beneath the anode.
   D. They rely on measurements from both the E1 and E2 electrodes in reference to the ground.
   E. They preferentially stimulate larger diameter nerve fibers at low stimulus intensities.
5. Motor nerve conduction studies:
   A. Require that E2 be replaced over the belly of the muscle and E1 be placed at a relatively inactive site.
   B. Result in waveforms that are the summation of muscle fiber action potentials.
   C. Allow nerve conduction velocity to be measured with distal stimulation.
   D. Are representative of muscle contraction.
   E. Are typically performed in the antidromic fashion.
6. F waves are:
   A. Late responses measured during sensory nerve conduction studies.
   B. Indicative of pathology at the root level only.
   C. Typically less than 31–32 ms in latency in the upper extremities and less than 56 ms in latency in the lower extremities.
   D. Independent of height.
   E. Of identical morphology with each stimulus.
7. A patient with Guillain Barré Syndrome, or acute inflammatory demyelinating polyneuropathy, might be expected to have:
   A. Prolonged distal latencies.
   B. Conduction velocity slowing.
   C. Abnormal temporal dispersion.
   D. Prolonged F wave latencies.
   E. All of the above.
8. The tibial H-reflex:
   A. Is electrophysiologically similar to the ankle reflex.
   B. Represents the contraction of gastrocnemius.
   C. Represents the contraction of the soleus.
   D. May be abnormal in an L5 radiculopathy.
   E. Is independent of height.
9. Sensory conduction studies are typically abnormal in:
   A. Myopathy.
   B. Compressive radiculopathies.
   C. Polyneuropathy.
   D. Myasthenia gravis.
   E. All of the above.
10. A patient with a decreased median nerve sensory response amplitude, prolonged median nerve F wave latency, and prolonged median nerve motor distal latency may have:

A. Median neuropathy at the wrist.
B. Generalized polyneuropathy.
C. Generalized myopathy.
D. C8-T1 radiculopathy from herniated disc.
E. A and B.

REVIEW ANSWERS

1. Answer: D. Abnormal temporal dispersion is usually accompanied by a drop in response amplitude. Normal temporal dispersion is very dependent on the length of the neuron. Abnormal temporal dispersion is the hall mark of demyelinating polyneuropathies and can be observed easily in motor conduction studies. It is not typically seen in patients with carpal tunnel syndrome.
2. Answer: D. Anodal block can theoretically interfere with the acquisition of F wave data, although this is usually not a practical concern. All the other statements are false. Nerve depolarization occurs beneath the cathode, most of the electrical current travels through the extracellular space, large diameter nerve fibers have the greater propensity toward action potential generation, and standard nerve conduction studies are supramaximal stimuli.
3. Answer C: In an isolated L5 radiculopathy, a peroneal F-estimatate would likely provide a shorter value than the actual F-waves, since the F-estimate would be calculated from the relatively unaffected distal latency and conduction velocity. F-estimates are most helpful in identifying the presence of a superimposed proximal process on a distal, can be performed in both the arms and legs, includes the distal latency and will not be adversely affected by the presence of an axonal neuropathy.
4. Answer: C. The nerve is depolarized beneath the cathode, not the anode. All the other statements are true.
5. Answer: B. The CMAP is the summation of all the muscle fiber action potentials. E1 is placed over the muscle belly, not E2. Nerve conduction velocity cannot be measured with distal stimulation alone. The muscle contraction is not being measured in standard motor conduction studies. Motor conduction studies cannot be performed antidromically since the muscle depolarization is being measured.
6. Answer: C. They are typically less than 31–32 ms in latency in the upper extremities and less than 56 ms in latency in the lower extremities. F waves are the late responses of motor, not sensory, nerve conduction studies. While prolonged F wave latencies are possibly indicative of pathology at specific root levels only, they can also be prolonged in focal compressive neuropathies (e.g. carpal tunnel syndrome) or generalized polyneuropathies. They are dependent on height, as taller people will have longer F wave latencies (increased distance for the action potential to travel).
7. Answer: E. All the statements are true.
8. Answer: A. The tibial H-reflex results from stimulation of Ia afferent tibial nerve fibers, which synapse in the spinal cord, travel back down the tibial nerve, and cause depolarization of the soleus muscle. It is therefore electrically similar to the ankle reflex (although not identical). It does not represent the actual contraction of a muscle, but rather the summed muscle fiber potentials of the recorded muscle. As the neurophysiologic equivalent of the ankle reflex, it may be abnormal in an S1 radiculopathy but not L5 radiculopathy. The H-reflex latency varies with different height: the latency is longer in people of taller stature, just like F-waves.
9. Answer: C. Sensory studies are typically abnormal in most, but not all, polyneuropathies. In myopathies, the motor responses can be normal or reduced. In compressive radiculopathies, the dorsal root ganglion, which houses the pseudounipolar sensory cell bodies, lies outside of the zone of compression. Because sensory nerves do not have a neuromuscular junction, sensory responses will be normal in myasthenia gravis.
10. Answer: E. These findings can be seen in both focal compressive lesions of the median nerve at the wrist and generalized polyneuropathy. Sensory responses in a generalized should not be abnormal in myopathy. In addition, the sensory responses should not be abnormal in a C8-T1 radiculopathy for two reasons: 1) the sensory component of the median nerve is derived from the C6-7 dermatomes and 2) sensory responses are typically normal in compressive radiculopathies.

13

Technical, Physiological and Anatomic Considerations in Nerve Conduction Studies

James B. Caress

Summary

Nerve conduction studies and their interpretation are subject to a variety of factors. First, technical factors including submaximal stimulation, environmental electrical noise, inaccurate placement of the recording electrodes, and stimulus artifact can substantially interfere with accurate recording of nerve and muscle responses. Second, physiological factors, such as the effects of body height and age, can cause profound variation in all nerve conduction parameters, and studies require interpretation keeping these individual variations in mind. Another physiological factor is temperature, in which cooling can produce a variety of changes, including slowing of conduction velocity and increase of response amplitude. Anatomic factors are also important, the most common being the Martin–Gruber anastomosis, usually presenting with a reduction in response amplitude with proximal stimulation of the ulnar nerve, and the second most common being the presence of an accessory peroneal nerve. Paying constant attention to all of these details is a critical element to the accurate performance and interpretation of nerve conduction studies.

Key Words: Conduction velocity; distal latency; Martin–Gruber anastomosis; nerve conduction study; stimulation; temperature.

1. INTRODUCTION

Nerve conduction studies (NCS) are considerably more technically demanding than EMG. Correct interpretation of NCS data is dependent on re-creating the conditions that prevailed when the normal data tables were generated. This means that close attention to temperature and electrode placement are crucial. Physiological potentials are tiny in comparison with ambient electrical noise and must be amplified many times for evaluation. Artifacts may be similarly amplified and need to be recognized and minimized.

2. SKIN PREPARATION

Good skin preparation is vital to performing NCS. The skin acts as a barrier to the measurement of electrical signals of interest and the effect of this barrier can be minimized by application of conductive gel that provides a low-resistance pathway to the electrode. Thick or edematous skin adds additional distance between the recording electrodes and the signal generator (i.e., the nerve or muscle), resulting in lower amplitudes. When the skin is callused, there may be different amounts of resistance to measuring the charge over the skin between the active and reference recording electrodes. *Impedance* is the term used to describe resistance

From: *The Clinical Neurophysiology Primer*
Edited by: A. S. Blum and S. B. Rutkove © Humana Press Inc., Totowa, NJ

to current flow in NCS. The impedance of recording electrodes must be similar, or ambient electrical noise will appear different at the electrodes. When this occurs, 60-Hz artifact or a broad stimulus artifact may obscure the waveforms. Skin lotions or perspiration can provide a conductive medium from the stimulator to the recording electrodes and can also result in a broad stimulus artifact. Soap and water cleansing is usually effective, but alcohol and acetone may be used to obtain a clean surface.

3. STIMULUS ARTIFACT

Spread of excess current along the skin and deeper tissues results in a stimulus artifact. Modern instruments use a variety of technical mechanisms to minimize this, but, even so, stimulus artifact can obscure potentials with very short latencies, especially palm-to-fingers or palm-to-wrist recordings. Reducing stimulus duration and intensity can reduce stimulus artifact, but it is imperative to obtain supramaximal responses. Placing the ground electrode between the stimulating and recording electrodes can reduce stimulus artifact by providing a low-impedance pathway for excess current to flow through. Rotating the anode of the stimulator while leaving the cathode in place can minimize the stimulus artifact. Increasing the distance between stimulating and recording electrodes or decreasing the distance between the active and reference recording electrodes will reduce stimulus artifact at the cost of lengthening the latency and, possibly, reducing the amplitude of the waveforms.

4. RECORDING ELECTRODES

Recording electrodes used in routine NCS include disc electrodes, ring electrodes, and bar electrodes that connect two discs at a fixed distance. Disc electrodes are commonly used to record motor potentials and nondigital sensory potentials. The discs are connected to separate wires that allow the distance between the active and reference electrode to be altered. Standard interelectrode distance is 3 to 4 cm, but there are occasions when adjustments must be made because of anatomic or physiological conditions. Any variation from standard positioning can change the characteristics of the recorded potentials, therefore, understanding the underlying electrophysiology is helpful in interpreting the data when alterations are necessary.

NCS and EMG electrodes record all potentials using a differential amplifier system that magnifies differences between the active and reference electrodes at each point in time. Identical signals appearing simultaneously at both electrodes undergo *common mode rejection* and are not displayed in the waveform. This is an ideal situation to minimize noise, but can also reduce physiological potentials.

In sensory NCS, the active and reference recording electrodes are placed linearly along the course of the nerve, approx 3 cm apart. It is optimal for the traveling depolarized zone to pass completely under the active electrode before beginning to pass under the reference electrode, because this maximizes the displayed amplitude. Interelectrode differences of less than 3 cm may result in reduction of the amplitude. Increasing interelectrode difference beyond 3 to 4 cm does not increase the amplitude further and improves the chance of introducing electrical noise that may appear different at the two points. When the interelectrode distance is necessarily altered (as occurs with an amputation of a distal phalanx or to avoid bandages or similar obstacles), these concepts need to be kept in mind because normal values depend on specific interelectrode distances. Amplitudes vary inversely with the recording electrode's distance from the nerve, thus, it is advisable to take steps to minimize edema, particularly when recording from the sural, superficial peroneal, and radial nerves with disc electrodes. Placement of subdermal monopolar needle electrodes can circumvent this problem.

In motor NCS, recording electrodes are placed in the "belly-tendon" arrangement. The active electrode is placed directly over the spot where most of the axons terminate in the neuromuscular junction, called the *motor point*. When the nerve is stimulated, proper placement over the motor point will result in an initial negative (upwards) deflection from the baseline. If the electrode is not over the motor point, depolarization adjacent to the electrode results in a "positive dip" with an initially positive deflection. This downward deflection is a volume-conducted response akin to the positive deflection at the onset of a triphasic sensory waveform. If a positive dip is noted, the active electrode should be moved until an initial negative deflection is observed. On occasion, repositioning the electrodes cannot correct a positive dip because of unusual anatomy (particularly tibial motor potentials) or because of severe neuromuscular pathology.

The reference electrode is placed distally over a tendinous region. Older texts have stated that the reference electrode is electrically inactive or "indifferent" and, thus, some practitioners may be unaware that the reference electrode makes important contributions to the measured motor potential. This is most apparent with ulnar motor studies recording from abductor digiti quinti (ADQ). The ulnar motor waveform characteristically exhibits a double-peaked morphology in its negative phase, and the second peak is usually of higher amplitude. In an elegant study, Kincaid et al. demonstrated that the reference electrode generates the second peak. When the reference electrode was moved to the contralateral index finger, the second peak was not recorded. In median nerve studies, the contribution of the reference electrode has a more subtle impact. The effect of the reference electrode should be kept in mind when double-peaked waveforms appear in unusual situations or when inclusion of the second peak would produce seemingly inaccurate results.

5. VIRTUAL CATHODE

A tenet of NCS is that depolarization occurs directly under the cathode of the stimulator but, with overzealous attempts to maximize the amplitude of potentials, excess current traveling in the overlying superficial tissues may result in depolarization distal to the cathode. When this occurs, the cathode is "virtually" in a different place than its actual anatomic location. This can be readily observed by keeping the stimulator in place and increasing the intensity beyond maximal values (*see* Fig. 1). The amplitude stays relatively stable but the distal latency is shortened. The erroneously short latency may mask pathological slowing. Virtual cathode can easily occur when the nerve is quite deep to the stimulator. For example, during ulnar motor NCS at the below-elbow site, the deep location of the nerve results in high-intensity stimulation to elicit the supramaximal response. Depending on the anatomy of the individual person, some fibers distal to the stimulator may be stimulated to achieve a maximal response. The result is an inaccurately fast forearm velocity. If this is combined with correctly performed above-elbow stimulation, an artifactual relative slowing across the elbow is created.

6. ANODAL BLOCK

When depolarization under the cathode occurs, there is also hyperpolarization of the nerve beneath the anode. The hyperpolarization is without significance in NCS unless the cathode and anode are inadvertently reversed. When this happens, depolarization occurs 2 to 3 cm proximal to the point that has been marked for depolarization, resulting in apparent prolonged

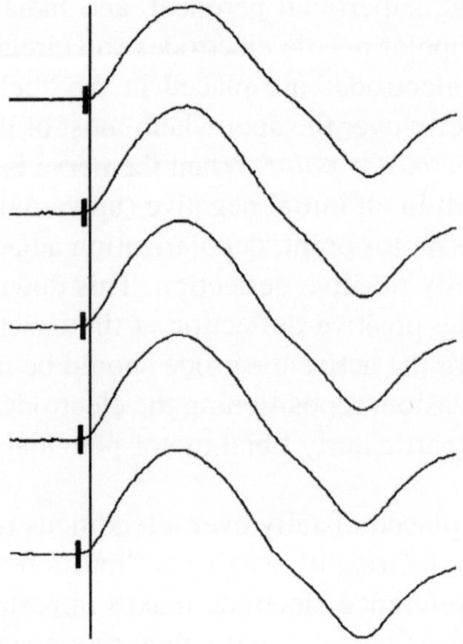

Fig. 1. Virtual cathode. From top to bottom, the first waveform shows a slightly submaximal motor response. The second stimulus produces a maximal response. The third stimulus is of 25% higher intensity and generates a supramaximal response that is unchanged in amplitude or latency. A further increase in stimulus intensity produces a response that is of slightly faster latency without a change in amplitude (virtual cathode). The virtual cathode effect is demonstrated again in the bottom tracing after another increase in intensity.

distal latencies, and the area of hyperpolarization may block some of the impulses coming from the proximal depolarized zone (anodal block).

Reversing the orientation of the stimulator (cathode proximal) is routinely used to avoid anodal block during F-wave measurements. However, a distal compound motor action potential (CMAP) is still generated despite the hyperpolarized zone, although there may be a subtle reduction in amplitude that goes unnoticed. The corollary is that F-waves are readily obtained even if the operator fails to reverse the position of the stimulator.

7. AVERAGING

Computerized software has made the "average" key on NCS machines a routine fixture, but it continues to be used relatively sparingly in clinical practice. All physiological waveforms recorded during NCS are contaminated by noise (ambient electrical fields, EKG artifact, etc.). The differential amplifier and proper use of filters attenuate most of this noise so that the averaging is only needed for recording very small amplitude potentials (sensory and mixed nerve potentials) in the microvolt range. By averaging repeated stimuli of identical intensity, noise is reduced, resulting in a flatter baseline and a more clearly defined physiological waveform. Mathematically, averaging potentials increases the signal-to-noise ratio by a factor equal to square root of the number of potentials averaged. During routine NCS on healthy individuals, the signal is evident above the noise, tempting the operator to omit averaging the potential. However, averaging will improve identification of the onset latency and provide a more accurate and reproducible waveform. The value of averaging becomes more apparent when it used to define potentials not evident on initial inspection because of surrounding noise. For example, consider a sural sensory potential that is 1 μV in amplitude but cannot be identified because the baseline "rumble" is 2 μV (signal-to-noise ratio of 1:2). Without averaging, the sural potential would be erroneously recorded as absent. If the potential is averaged four times (square root of 4 = 2), the signal-to-noise ratio is increased to 2:2 (or 1:1). If the potential is averaged 16 times, this increases the signal-to-noise ratio to 4:2 (2:1), and the potential will become visible above the noise.

8. PHYSIOLOGICAL FACTORS

Physiological factors, including age and height but particularly temperature, can significantly influence NCS responses. The effects of temperature on NCS can be explained by sodium channel physiology.

8.1. Temperature

Diminishing temperature results in slower opening and closing of the Na^+ channel pore but, because closing time is relatively more affected, the net result is that cooler temperatures produce a net increase in the total time the Na^+ channel remains open. The slower opening time causes delay in regeneration of the action potential at each node of Ranvier and results in prolonged distal latencies and slower conduction velocities. The net increase in sodium channel open time results in a longer duration action potential and, as more Na^+ ions enter the intracellular space, a larger amplitude potential is produced. Warming beyond the physiological range reverses this pattern and can result in faster conduction velocities and reduced amplitude. Studies attempting to document these changes in during routine NCS conditions have produced variable results because of differences in how cooling was applied. However, in clinical practice, the important effects of cold temperature in routine NCS are longer distal latencies, slower conduction velocities, and higher amplitudes (*see* Fig. 2). This pattern of findings can mimic demyelination, resulting in erroneous diagnoses of median or ulnar neuropathy at the wrist, and demyelinating polyneuropathies. Cooling also results in reduced decrement on repetitive stimulation and can mask the diagnosis of myasthenia gravis.

Skin temperatures between 32°C and 36°C are standard for NCS, but typical ambient skin temperature is between 28°C and 32°C. Therefore, the limbs of most patients need to be warmed before testing if standard tables of normal values are to be used. Conduction velocity increases by 1.5 to 2 m/s per 1°C as skin temperature increases throughout the physiological range. If adequate warming is not possible, correction factors using this data can be used provided there is understanding that these equations do not correct for amplitude changes, and, importantly, that these correction factors may be inaccurate in diseased nerves.

8.2. Age

Peripheral nerve myelination begins at approximately the 15th week of gestation and continues through 3 to 5 yr of age. Therefore, for children younger than the age of 5 yr, special tables are required to evaluate NCS data. In term infants, the latency and conduction velocity values are approximately half those recorded in adults, and premature infants have even slower conduction velocities; at the beginning of the third trimester, velocities are one-third of those measured in term infants.

NCS parameters are fairly stable from age 5 yr through 40 yr, and then there is a tendency towards slower latencies and velocities with lower amplitudes. In the healthy older

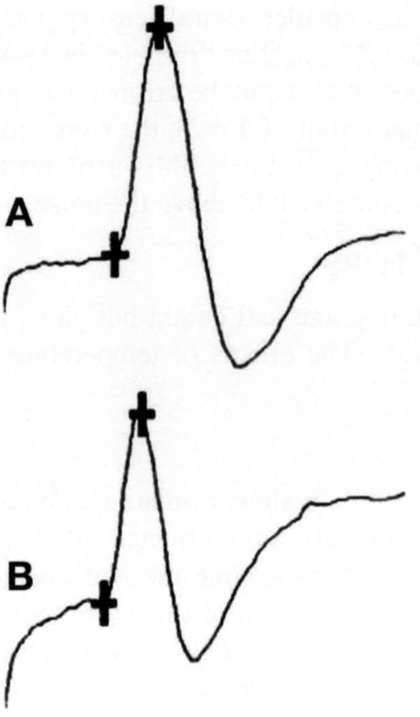

Fig. 2. The effect of warming. **(A)** At ambient temperature, the median sensory response is of longer latency, higher amplitude and longer duration than the same response **(B)** after the hand has been warmed.

population older than age 60 yr, upper extremity NCS latencies are relatively unchanged when compared with younger adult populations. Between the ages of 60 to 90 yr, median, radial, and ulnar sensory amplitudes fall slightly. Lower extremity latencies tend to increase and velocities tend to decrease compared with younger adults. Lower extremity motor and sensory amplitudes decrease throughout middle age, and the trend continues in the elderly. The thought that sural sensory responses may be absent in the elderly is widely circulated, but one study found recordable sural potentials in 98% of individuals older than age 60 yr, and superficial peroneal sensory responses in 90% of the same group.

Studies on the effect of age on NCS parameters consistently show that amplitude is more affected by this process than is velocity. Possible explanations for the changes in NCS with aging include "normal" axon loss caused by neuronal senescence, higher prevalence of asymptomatic, generalized neuropathy, increased lower extremity edema, and a higher prevalence of subclinical entrapment neuropathies.

8.3. Height

As opposed to the effects of aging, increasing height influences latencies and velocities more than amplitude. The effect is particularly evident in the lower extremities, and, in general, increasing height results in longer distal latencies, slower conduction velocities, and slightly reduced amplitude. The effect of height may not cause collected data to deviate from commonly used normal ranges except for very tall or very short individuals. Unusually fast velocities and latencies are not usually given consideration because the values are considered supernormal rather than pathological, however, in very short persons, latencies at the upper end of commonly used "normal" ranges may be pathological if not adjusted for height.

Similarly, very tall people may seem to have mild generalized slowing of NCS parameters that does not reflect demyelination or other pathology. It is easy to see that, at the extremes of height, mild neuropathic processes could be overlooked or mistakenly identified.

One theory purported to explain the effect of height on NCS is that, in taller people, the axons are more tapered and smaller in diameter in the distal segments than in people of common height. Another proposed explanation is that the distal segments are relatively cooler despite attempts at adequate warming.

9. ANATOMIC VARIATIONS

9.1. Martin–Gruber Anastomosis

The Martin–Gruber anastomosis (MGA) is a common anatomic variant of peripheral nerve pathways in the forearm. It involves primarily motor fibers that cross over from the median nerve in the proximal forearm to the ulnar nerve at the wrist. Although the MGA is identifiable in up to 30% of people, many instances of MGA do not affect the interpretation of the NCS data and, thus, are not identified. The MGA can be inherited in an autosomal dominant pattern and can be unilateral or bilateral. In the MGA, the course of the ulnar fibers from the plexus is altered, such that a portion of the ulnar fibers descends with the median nerve to the elbow. These fibers branch off the median nerve in the forearm (often via fibers from the anterior interosseous nerve) to join the ulnar nerve proximal to the wrist. The intrinsic hand muscles are innervated by the "usual" nerve (unlike the "all-ulnar hand"); it is only the pathway that particular ulnar fibers take to get to the wrist that is anomalous. There are three types of MGA to consider, but combinations of the types may occur and generally make the anomaly easier to recognize.

Type I: Crossover fibers innervate the ADQ. When the crossover involves the ADQ, the median studies appear normal, but ulnar stimulation at the wrist (recording from ADQ) produces a much higher-amplitude waveform than on stimulation below the elbow. The drop in amplitude may exceed the 10 to 20% reduction that is expected over this distance because of normal phase cancellation and temporal dispersion. The type I MGA can be confirmed by recording from ADQ while stimulating the median nerve at the elbow and demonstrating an initially negative (upward) waveform that is similar in amplitude to the "drop" seen between the distal and proximal ulnar stimulation sites. If the MGA goes unrecognized in this situation, conduction block of the ulnar nerve in the forearm may be erroneously diagnosed. The clinical situation and EMG results may alert the diagnostician to investigate the possibility of MGA rather than conduction block, but, in cases in which demyelinating neuropathy is clinically suspected, the possibility of MGA may be overlooked.

Type II: Crossover fibers innervate the first dorsal interosseous (FDI). This is the most common type of MGA. Ulnar NCS recording from the FDI can be used to evaluate lesions of the deep palmar branch or to investigate ulnar neuropathy at the elbow. The pattern of amplitude change is similar to that seen in type I and can mimic a conduction block of the ulnar nerve in the forearm. As with type I, type II MGA is proven by stimulating the median nerve at the wrist and elbow, and noting a higher amplitude response from the proximal site while still recording from FDI. During routine median motor studies, recording from APB, a type II MGA is often suggested by a slightly enlarged or altered median CMAP appearance with elbow stimulation when compared with the CMAP obtained at the wrist. Because NCS recording from FDI are not routinely performed in most labs, type II MGA is often detected in the setting of carpal tunnel syndrome (CTS) (*see* next paragraph).

In cases of CTS with prolonged distal motor Type III: Crossover fibers innervate thenar eminence muscles that have ulnar innervation (adductor pollicis and/or flexor pollicis brevis–deep head). Type III MGA is recognized during median motor NCS by an increase in amplitude at the proximal site. Submaximal wrist stimulation or co-stimulation of the ulnar nerve from the antecubital fossa produces a similar pattern and should be initially investigated. The MGA is confirmed by stimulating the ulnar nerve at the wrist (still recording from the thenar eminence over the abductor pollicis brevis) and then at the elbow. In persons with and without MGA, wrist stimulation produces a waveform with an initial positive deflection caused by volume conduction from ulnar innervated muscles thenar eminence. In individuals without MGA, stimulation of the ulnar nerve at the elbow, produces a waveform of similar appearance without much reduction of amplitude, but, in those with MGA, stimulation at the elbow results in a much lower amplitude than at the wrist, often with altered morphology.

In cases of CTS with prolonged distal motor latencies, type II and type III MGA are easier to detect, because median stimulation at the elbow produces a positive deflection before the CMAP, which was not observed with wrist stimulation. The delay through the carpal tunnel prevents the abductor pollicis brevis muscle from depolarizing before the muscles innervated by the crossover fibers. The initial volume-conducted response from the crossover muscles results in the "positive dip." If the onset latency is marked at the beginning of the "positive dip," the median forearm conduction velocity will be erroneously fast and usually nonphysiological (>80 m/s). An accurate velocity cannot be measured because the takeoff for the "true" median CMAP is lost in the positive deflection. This pattern is so distinctive that the diagnosis of CTS plus MGA can be made by inspection of the median waveforms without determining the type of MGA. Distal and proximal ulnar nerve stimulation recording from the thenar eminence and FDI, as described in this section, will differentiate between type II and type III MGA (Fig. 3). Ulnar studies recording from ADQ are unaffected in this situation.

In the setting of ulnar neuropathy at the elbow or proximal median neuropathy, serious errors in EMG interpretation are easily made if the MGA is unrecognized during NCS. Sparing of the certain ulnar muscles innervated by crossover fibers in ulnar neuropathy at the elbow produces patterns that may mimic a lesion of the deep palmar branch. Similarly, a lesion of the proximal median nerve affecting crossover fibers could produce a pattern suggesting C8 radiculopathy on EMG, because abnormalities may be identified in both median- and ulnar-innervated hand muscles.

9.2. Accessory Peroneal Branch to Extensor Digitorum Brevis

The extensor digitorum brevis (EDB), the recording location for peroneal motor NCS, is usually exclusively innervated by the deep branch of the peroneal nerve. In 13 to 22% of people, a branch of the superficial peroneal nerve traveling behind the lateral malleolus innervates the lateral portion of the EDB. The anomalous innervation is recognized when proximal peroneal stimulation produces a higher amplitude CMAP than that obtained by stimulation at the ankle. Submaximal stimulation at the ankle or co-stimulation of the tibial nerve in the popliteal fossa may cause a similar pattern, and should be considered initially. The presence of an accessory peroneal branch is confirmed by low-intensity stimulation behind the lateral malleolus and demonstrating a CMAP with an initial negative deflection. Rarely, the EDB is completely supplied by the accessory branch.

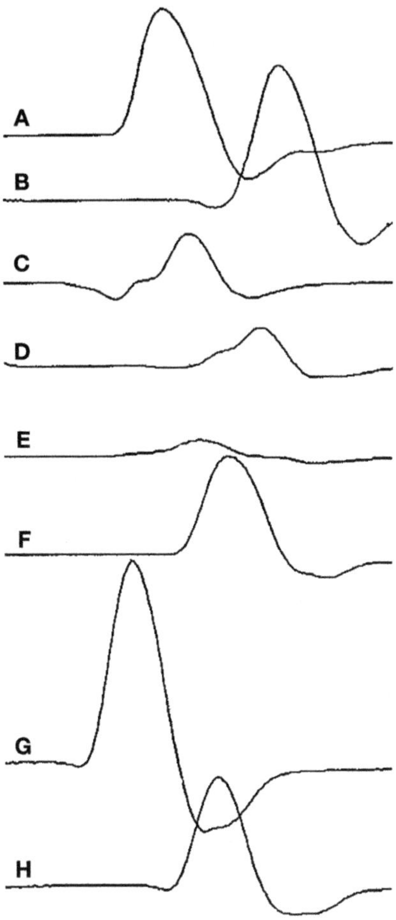

Fig. 3. Type II Martin-Gruber anastomosis (MGA) to first dorsal interosseous (FDI) with carpal tunnel syndrome. In **(A)**, the median nerve is stimulated at the wrist and a response with a prolonged latency is recorded. Median nerve stimulation at the elbow generates a slightly higher amplitude waveform preceded by a positive deflection **(B)**. Stimulation of the ulnar nerve at the wrist **(C)** and elbow **(D)** recording over the abductor pollicis brevis muscle does not show a significant difference in amplitude, excluding a type III MGA. In **(E–H)**, the recording electrodes are moved over the FDI. The median nerve is stimulated at the wrist **(E)** and elbow **(F)**, demonstrating the type II MGA. Then, the ulnar nerve is stimulated at the wrist **(G)** and elbow **(H)**, demonstrating the drop in amplitude expected from this type of crossover.

REVIEW QUESTIONS

1. All of the following interventions may reduce stimulus artifact EXCEPT:
   A. Increasing the distance between stimulating and recording electrodes.
   B. Cleaning the skin with acetone.
   C. Reducing the sensitivity on the display.
   D. Rotating the anode of the stimulator while leaving the cathode stationary.
   E. Placing the ground electrode between the stimulating and recording electrodes.
2. In sensory NCS, signal-to-noise ratio can be improved by all of the following EXCEPT:
   A. Placement of subdermal monopolar needle electrodes for recording.
   B. Increasing the interelectrode difference beyond 3 to 4 cm.
   C. Averaging.

D. Cleaning the skin with alcohol.
E. Decreasing the distance between the stimulating and recording electrodes.

3. All of the following can cause a "positive dip" on routine NCS EXCEPT:
   A. Co-stimulation.
   B. Improper recording electrode placement.
   C. MGA.
   D. MGA with CTS.
   E. Submaximal stimulation.
4. Virtual cathode commonly occurs during:
   A. Stimulation at intensity above supramaximal.
   B. Submaximal stimulation.
   C. MGA.
   D. Axonal neuropathy.
   E. Improper recording electrode placement.
5. Virtual cathode can result in all of the following EXCEPT:
   A. An erroneously short latency.
   B. An erroneously fast conduction velocity.
   C. Significantly higher amplitudes.
   D. Loss of sensitivity to entrapment neuropathy distal to the stimulation point.
   E. More patient discomfort.
6. If the signal-to-noise ratio is 1:4, how many times would the potential need to be averaged to produce a signal-to-noise ratio of 2:1?
   A. 64.
   B. 4.
   C. 128.
   D. 16.
   E. 2.
7. During NCS, inadequate warming of a limb will have the most marked effect on:
   A. Amplitude.
   B. Distal latency.
   C. Proximal conduction velocity.
   D. Decrement.
   E. Duration.
8. In the MGA:
   A. Some muscles in the thenar eminence that are typically innervated by the median nerve are innervated by the ulnar nerve.
   B. The proximal median amplitude is always higher than the distal median amplitude.
   C. A pseudoconduction block of the ulnar nerve in the forearm may occur.
   D. The sensory potential recording from digit two has contributions from the median and ulnar nerves.
   E. Individuals who have this variant may have relative protection from median neuropathy at the wrist.
9. In accessory peroneal nerve branch to the EDB:
   A. Responses may mimic a conduction block in the calf.
   B. Responses may mimic a conduction block across the fibular head.
   C. Anamolous innervation is by stimulating behind the lateral malleolus.
   D. Anamolous innervation is when the bulk of the EDB is reduced.
   E. The accessory branch involves only sensory fibers.
10. After recording of the motor CMAP, if the stimulator is not "turned around" with the cathode where the anode had been, the F-waves will:
    A. Be absent because of anodal block.
    B. Have a longer latency.
    C. Will show more chronodispersion.

D. Have a shorter latency.
E. Be reduced in amplitude.

REVIEW ANSWERS

1. The correct answer is C. Reducing the sensitivity on the display does not change the degree of stimulus artifact relative to the recorded potential of interest.
2. The correct answer is B. Increasing the interelectrode difference beyond 3 to 4 cm does not result in higher amplitude signals. Reducing the distance between the stimulator and recording electrodes will result in higher amplitudes.
3. The correct answer is C. When recording over the thenar eminence and stimulating the median nerve at the elbow, a positive dip is not usually seen unless there is concomitant CTS slowing down the action potentials as they enter the hand.
4. The correct answer is A. Virtual cathode occurs when stimulation intensity is sufficient to depolarize the nerve distal to the location of the stimulating cathode.
5. The correct answer is C. Virtual cathode is unlikely to have an appreciable effect on the amplitude of the response. The most noticeable effect is a shorter latency, resulting in faster conduction velocities and erroneously fast velocities across entrapment point immediately distal to the stimulation point.
6. The correct answer is A. The square root of the number of stimuli is the factor that improves the signal-to-noise ratio. In this case, the square root of 64 is eight times the given signal-to-noise ratio of 1:4 = 8:4 = 2:1.
7. The correct answer is B. Inadequate warming of a limb usually produces a marked effect on distal latency because the distal parts of the limb tend to be the coldest. Expected lesser effects include increased amplitude, decrease proximal conduction velocity, decreased decrement (if present), and increased duration of the response.
8. The correct answer is C. An MGA involving the ADQ would be expected to produce a drop in amplitude when comparing the distal site with the proximal site. This will have an appearance of conduction block in the forearm and not across the elbow.
9. The correct answer is C. The accessory branch of the peroneal nerve to the EDB is confirmed by stimulating behind the lateral malleolus. This anatomic variant is suspected when the amplitude at the below fibular head site is higher than the amplitude recorded at the ankle. Therefore, it does not mimic a conduction block.
10. The correct answer is B. When F-waves are recorded with the stimulator in position for routine motor studies, there is a risk that some potential F-waves will be blocked under the anode. However, F-waves are typically recordable with the cathode distal to the anode, although they have a slightly longer latency than if the cathode and anode positions had been reversed.

REFERENCES

1. Kincaid JC, Brasher A, Markand ON. The influence of the reference electrode on CMAP configuration. Muscle Nerve 1993;16:392–396.
2. Hodgkin AL, Katz B. The effect of temperature on the electrical activity of the giant axon of the squid. J Physiol (Lond) 1949;109:240–249.
3. Rutkove SB. The effects of temperature in neuromuscular electrophysiology. Muscle Nerve 2001;24:867–882.
4. Rivner MH, Swift TR, Malik K. Influence of age and height on nerve conduction. Muscle Nerve 2001;24:1134–1141.
5. Falco FJE, Hennessey WJ, Braddom RL, Goldberg G. Standardized nerve conduction studies in the upper limb of the healthy elderly. Am J Phys Med Rehabil 1992;71:263–271.
6. Falco FJE, Hennessey WJ, Goldberg G, Braddom RL. Standardized nerve conduction studies in the lower limb of the healthy elderly. Am J Phys Med Rehabil 1994;73:168–174.
7. Preston DC, Shapiro BE. Electromyography and Neuromuscular Disorders. Butterworth–Heinemann, Boston, MA, 1998.

14
Introduction to the Needle Electrode Examination

Gregory J. Esper and Seward B. Rutkove

Summary

Needle EMG remains an important part of the evaluation of the peripheral nervous system and can assist substantially in characterizing a variety of disease states. The needle electrode examination generally consists of three parts: evaluation of spontaneous activity, evaluation of motor unit potential morphology, and evaluation of motor unit potential recruitment. Abnormal spontaneous activity includes the commonly observed fibrillation potential and positive sharp wave and the less frequent myotonic, myokymic, neuromyotonic, and complex repetitive discharges. Neurogenic disease produces prolonged duration and increased amplitude of motor unit potentials with reduced recruitment, whereas myopathic disease generally produces low-amplitude, short-duration motor unit potentials, with normal or early recruitment. An understanding of the basic mechanisms of these disease changes can greatly aid in the interpretation of EMG data. Finally, the distribution of abnormalities across muscles can also aid in diagnosis, especially in the assessment of radiculopathy and plexopathies.

Key Words: Fibrillation potential; motor unit potential; motor unit recruitment; myotonic discharge; needle electromyography; positive sharp wave.

1. INTRODUCTION

Needle electromyography (EMG) complements nerve conduction studies and serves a critical role in the evaluation of both muscle and nerve diseases. With experience, needle EMG can be rapidly performed and interpreted. However, to the uninitiated, the procedure can seem very complicated and the interpretation of the waveforms arbitrary. In this chapter, we provide a background for a general understanding of how needle EMG is performed and interpreted, but leave its detailed application to specific disorders for discussion in other chapters of this text.

2. BASIC CONCEPTS

2.1. The Needle Electrode

There are several different types of needle electrodes. The monopolar needle electrode provides referential recordings. The needle is Teflon covered, except for the tip, which is exposed and where electrical activity is recorded; these signals are compared with those recorded from the surface via a separated electrode placed on the skin at a distance from the muscle. Although monopolar needle electrodes continue to be widely used, concentric needle electrodes have become the favored choice among most electromyographers. With this type of electrode, a barrel that surrounds the active needle electrode serves as the reference, allowing the recorded signal from the active needle electrode to be compared with that of the

From: *The Clinical Neurophysiology Primer*
Edited by: A. S. Blum and S. B. Rutkove © Humana Press Inc., Totowa, NJ

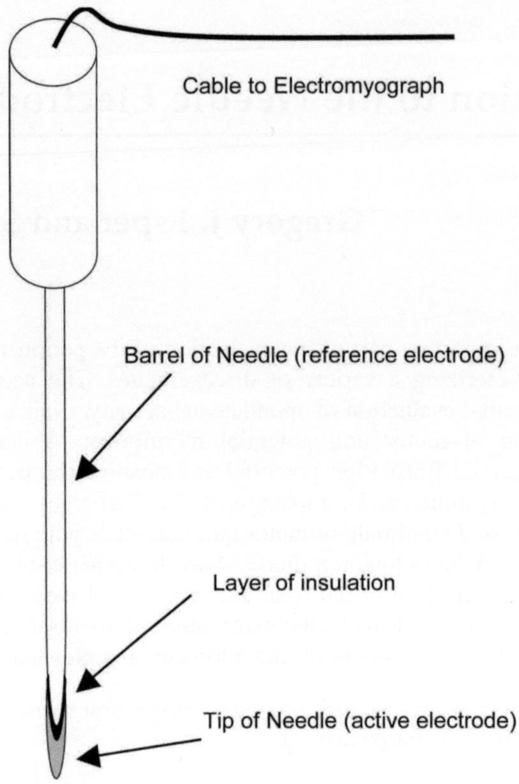

Fig. 1. A simple drawing of a concentric needle electrode.

barrel, without requiring a separate surface electrode (*see* Fig. 1). An insulating material separates these two surfaces. Because the placement of the reference electrode is practically adjacent to the active site, motor unit potentials (MUPs) using a concentric needle tend to be smaller in duration and amplitude than those same MUPs measured with a monopolar electrode.

2.2. The Oscilloscope

Waveforms representing MUPs are visually displayed on the oscilloscope in real time, and both qualitative and quantitative assessments can be made of MUPs. Qualitative visual analysis of the MUPs on the oscilloscope allows the real-time assessment of the duration, amplitude, and number of phases of each waveform, along with the recruitment ratio (discussed in more detail in Section 7 below). MUP morphology determination can be also be confirmed by a more accurate, but usually more time consuming, quantitative assessment of waveforms, which are analyzed by available software. Standard EMG is generally performed in a qualitative fashion, because an experienced electromyographer can often assess abnormalities rapidly and accurately; quantitation of MUP morphology is usually reserved for subtle processes (e.g., a mild myopathy) and research purposes.

2.3. The Loudspeaker/Amplifier

Listening to the electrical activity can tell the electromyographer a great deal regarding the condition of the muscle. Musical concepts apply in this regard: the *volume* (loud vs soft) of a given element tells you about its amplitude, the *pitch* (high vs low) tells you about the duration of the potential, and the *timbre* (quality of the sound) tells you about its phases. Other electrical discharges, such as myotonia, neuromyotonia, and complex repetitive discharges (CRDs) have characteristic sounds that are helpful in determining the type of disease present. Similar to the visual interpretation of waveforms, the auditory interpretation gradually develops with time and experience. These issues are discussed in further detail below, in the section on the evaluation of voluntary activity.

3. THE PROCESS

Needle EMG evaluates the electrical activity of muscle, through which neuropathic or myopathic processes can be identified and characterized. The needle electrode examination (NEE) generally consists of three distinct parts: evaluation of spontaneous activity, evaluation of MUP morphology, and evaluation of activation and recruitment.

3.1. Evaluation of Spontaneous Activity

The first part of the NEE is the evaluation of spontaneous activity—the activity that is present when the muscle is completely relaxed. When a needle electrode is inserted into an electrically active tissue, an initial burst of electrical activity from the depolarizing muscle fibers is to be expected. These bursts generally last no longer than approx 0.5 s (500 ms). With every movement of the needle electrode, a similar burst of electrical activity will be generated. However, when the movement ceases, the electrical activity should end and only electrical silence should normally be identified. This is what is described as "normal spontaneous activity."

A relaxed, normal muscle should be electrically silent; the exception to this rule occurs when the needle comes into close proximity to the end-plate region of the muscle. Intermittent bursts of sharp spike activity, called "end-plate spikes," might be observed. They are generally easy to recognize because they have a sputtering quality, are biphasic, and usually have an initial negativity (*see* Fig. 2). Another helpful hint is the sound of "end-plate noise"—a very high-frequency hissing sound attributed to the depolarization of the end-plate via the release of individual quanta of acetylcholine. Often end-plate spikes will be seen superimposed on a very irregular baseline (also visible in Fig. 2), the irregular baseline being caused by these individual quanta of acetylcholine. When the needle is near the end-plate, patients generally note an abrupt increase in discomfort. Movement of the needle a short distance away generally relieves the pain, the spontaneous discharges abruptly stop, and the muscle returns to its usual quiet state.

In most pathological states, spontaneous activity increases. In both neuropathic and myopathic diseases, the resting potential of the muscle fibers is elevated (is less negative) because of changes in the ion conductances, and the muscle fibers, thus, are more excitable. The most minor increase in spontaneous activity might be an increase in the length of a needle-induced discharge (i.e., it lasts longer than 500 ms). A number of abnormal spontaneous potentials can be observed with the muscle at rest. These potentials can be divided into two groups: those generated by the depolarization of a single muscle fiber and those generated by the depolarization of a motor neuron. It should also be noted that in some cases (mainly chronic myopathies), spontaneous activity can actually decrease because of the loss of myocytes.

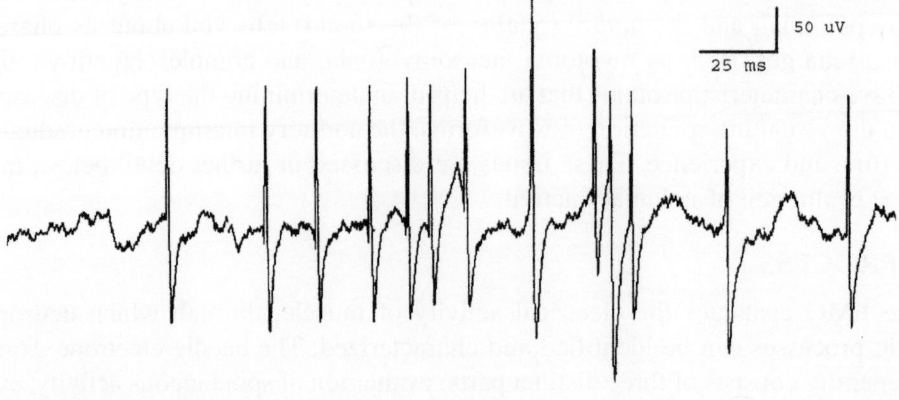

Fig. 2. End-plate spikes. Notice that the end-plate spike waveforms have an initial negativity. The irregular baseline is observed as part of "end-plate noise" and represents the electrical equivalent of individual quanta of acetylcholine exciting receptors in the end plate.

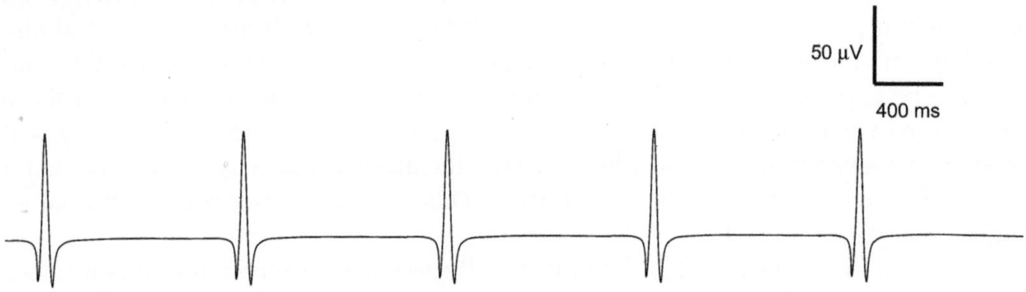

Fig. 3. Fibrillation potential. Note the triphasic character (an initial positive phase followed by a negative phase followed again by a positive phase).

4. SPONTANEOUS POTENTIALS GENERATED BY SINGLE MUSCLE FIBERS

4.1. Fibrillation Potentials

A fibrillation potential represents the electrical activity generated by the depolarization of a single muscle fiber and is depicted in Fig. 3. These potentials are typically triphasic, with an initial small positivity, followed by a large negative spike, and ending with another small positive spike. Generally, these potentials are thought of as being recorded by the needle electrode a short distance from the spontaneously depolarizing muscle fiber. The triphasic morphology of this potential is explained by the principles of volume conduction (*see* Chapter 4). The first phase represents the distant approach of the depolarization toward the needle; the large negative spike is caused by the transit of the depolarization in the immediate vicinity of the needle, and the final positive phase is caused by the potential receding into the distance. Fibrillation potentials may not always have this classic morphology, especially because the shape depends on the exact position of the needle in relationship to the muscle. They generally have an amplitude of 20 to 200 µV, are 1 to 5 ms in duration, and fire at a rate of 0.5 to 15 per second. The classic view of fibrillation potentials is that they fire in a very regular, clock-like pattern, with the depolarization occurring consistently. However, many fibrillation potentials, if observed for a long enough time, will slow down until they eventually stop.

The change in frequency can be quite prolonged, moving from a rate of two per second to one every 2 s during a period of 30 to 40 s. Fibrillation potentials can fire substantially slower than motor units (discussed below), often reaching firing frequencies of 0.5 Hz. Alternatively, MUPs generally cannot fire more slowly than approx 4 to 5 Hz and do so slightly irregularly. Hence, the slow, regular, steady "tick" of a fibrillation potential can usually help in its identification. This becomes more difficult if the screen is filled with multiple fibrillation potentials with different morphologies and firing frequencies, such that the individually firing potentials can no longer be identified. Some have described this sound as "frying bacon," which, although helpful, may be a little misleading. In situations in which there are many fibrillation potentials occurring at once, the presence of positive sharp waves (PSWs) (described below in Section 4.2.) can also help confirm the identity of these waves. Alternatively, simply waiting for the electrical activity to decrease until only a couple of potentials are present can help in accurately assessing these waveforms.

Although fibrillation potentials are undoubtedly caused by the partial depolarization of the muscle membrane, their significance is rather nonspecific. They are typically found in any nerve disease in which there is axonal injury causing muscle fibers to lose their innervation (i.e., states of denervation). As reinnervation proceeds after injury, a given muscle fiber will have its resting membrane potential restored, and the spontaneous depolarization will cease. For this reason, fibrillation potentials are generally considered evidence of subacute injury. However, in any condition in which reinnervation is incomplete and muscle fibers remain denervated, the resting membrane potential will be abnormally elevated, and spontaneous depolarization of the muscle fiber is to be expected. The prototypic illness in which this is observed is remote poliomyelitis, in which reinnervation of all the muscle fibers does not occur. Although the disease may have occurred decades earlier, fibrillation potentials will persist in the fibers that have not been reinnervated. Fibrillation potentials are also commonly observed in most muscle diseases. The reasons for this are debated and are explored more in Chapter 20. However, the most likely explanations are:

1. The muscle disease is accompanied by some distal neuronal injury.
2. The muscle disease itself is causing splitting of muscle fibers (leaving a partial muscle fiber no longer attached to the part receiving innervation).
3. That the disease itself is affecting the resting potential of the muscle membrane.

The morphology of fibrillation potentials is one aspect of disease that is not commonly evaluated. Nonetheless, noting the size of fibrillation potentials can be helpful. Small fibrillation potentials are commonly observed in muscle diseases and chronic neurogenic states (e.g., old polio). The reasons for this likely relate to morphological and physiological changes in both of these types of disease states. In muscle disease, split or atrophic muscle fibers simply will be smaller, and there will be fewer Na^+ channels activated to produce a given fibrillation potential. Hence, the amplitude of the spike will be smaller. A similar argument holds for chronic neurogenic disease, in which a reduction in the amplitude of the response is likely caused by a reduction in the number of active Na^+ channels or decreased function of the Na^+–K^+ ATPase. One possible mechanism through which this may occur is via downregulation of protein synthesis induced by the longstanding absence of innervation.

4.2. Positive Sharp Waves

Similar to fibrillation potentials, PSWs (Fig. 4) represent the depolarization of single muscle fibers; however, their morphology is quite distinct. Unlike fibrillation potentials, they are

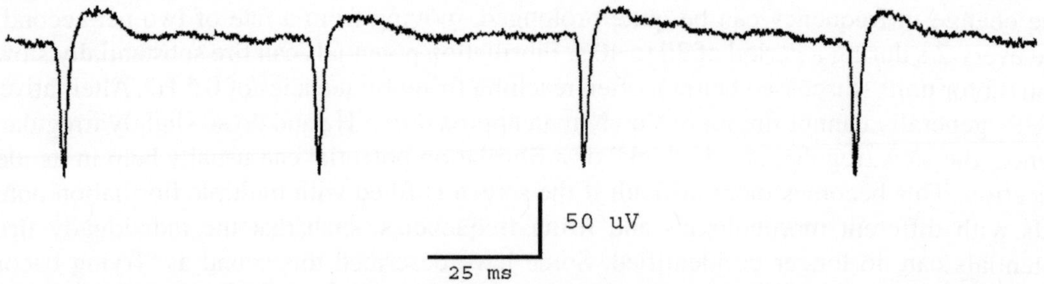

Fig. 4. Positive sharp wave. The initial sharp positivity is followed by a broad negative wave that slowly fades away.

biphasic with an initial sharp positivity, followed by a more prolonged negativity. The firing pattern is also more variable than that of a fibrillation potential. Although they can be clock-like and similar to a fibrillation potential, one can also observe a rapid run of PSWs that gradually slows down until it stops.

The mechanism underlying the generation of PSWs has been debated. The most widely accepted explanation for PSWs is that they represent a discharge of a single muscle fiber in which the discharge is abruptly aborted. The proposed pathophysiology is that the needle electrode is in contact with the spontaneously depolarizing muscle fiber. Like a fibrillation potential, the distantly approaching depolarization produces an initial positivity. The depolarization approaches the needle and the negativity commences. However, because the needle is in contact with the membrane itself, the depolarization abruptly ends near the electrode because the mechanical presence of the needle against the membrane causes the membrane to become resistant to the depolarization, resulting in a prolonged dying out of the negative phase.

PSWs, similar to fibrillation potentials, can occur in neurogenic and myopathic disease. However, PSWs are perhaps slightly more sensitive but less specific for the presence of muscle or nerve disease. This is possibly because PSWs can be triggered by the mechanical action of the needle touching the muscle fiber. There are several examples of this. First, in patients with recent onset neurogenic injury, short runs of PSWs may be observed in the muscle before fibrillation potentials actually develop. Second, PSWs can often be found in the intrinsic foot muscles and in the lumbosacral paraspinal muscles of otherwise healthy individuals. In these groups, fibrillation potentials tend to be much less common. Third, PSWs can be diffusely present in muscles of people who are otherwise healthy. This syndrome, which has been termed "EMG disease," may represent a channelopathy in which the muscle membrane is mildly depolarized at baseline and, hence, more easily irritated by needle movement.

4.3. Myotonic Discharges

Similar to fibrillation potentials and PSWs, myotonic discharges represent the repetitive spontaneous depolarization of single muscle fibers and are the electrical equivalent of clinical myotonia. However, the character of the myotonic discharge is different from that of either a fibrillation potential or PSW, because the frequency and amplitude of the discharge waxes and wanes over time (Fig. 5). Although the specific mechanism underlying the variation in these parameters is controversial, the individual waveforms that make up the discharges will have the appearance of either a fibrillation potential or a PSW that is changing in size over time. A "classic" myotonic discharge will generally start at a low frequency, gradually speed up, and then

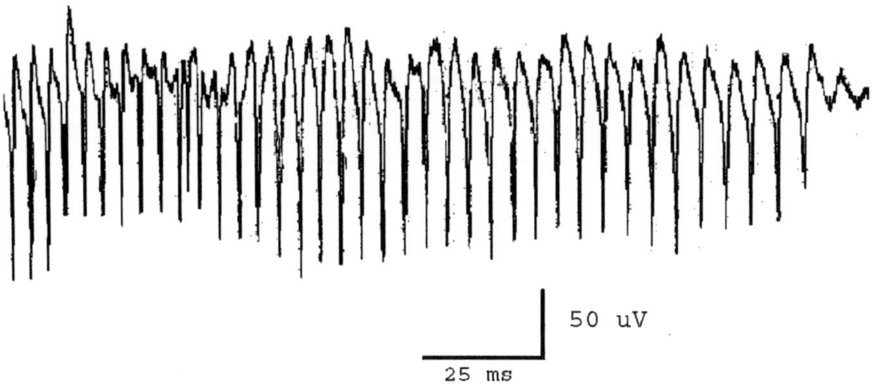

Fig. 5. A myotonic discharge; note the slow decrease in frequency and amplitude of the potential. In this example, a clear "revving" up of the potential is not obvious.

gradually slow down before it stops. The amplitude may similarly increase and decrease during this cycle. The sound over the loudspeaker is most accurately described as similar to that of a revving motorcycle engine. However, many myotonic discharges may be observed that do not have the initial speeding up and only demonstrate a gradual reduction in frequency and amplitude as the discharge slows down, giving a "dive-bomber" sound to the discharge. Strictly speaking, these may be difficult to separate from a short run of PSWs. Nonetheless, usually after assessing spontaneous activity for some time in a patient with these partial myotonic discharges, a few with a clear initial waxing of frequency and amplitude will also be found.

Myotonic discharges are helpful because they reduce the differential diagnosis of a patient presenting with symptoms of muscle disease. The list of disorders that are characterized by diffusely prominent myotonic discharges is relatively limited and includes myotonia congenita, the myotonic dystrophies (types 1 and 2), paramyotonia congenital/hyperkalemic paralysis, drug-induced myotonia (observed with colchicine, chloroquine, and with some cholesterol-lowering drugs). However, occasional myotonic discharges can also be present in patients with many other muscle diseases, including inflammatory myopathies, acid maltase deficiency (discharges are observed most prominently in the paraspinal muscles), myotubular myopathy, and occasionally even in patients with neurogenic disease. It should be noted that, in some situations, muscle cooling can enhance the number of myotonic discharges, and applying ice to the skin overlying a muscle can sometimes be helpful.

4.4. Complex Repetitive Discharges

CRDs are unlike the other three types of muscle fiber discharge, because they represent the depolarization of a group of muscle fibers rather than a single fiber firing repetitively. Understanding the EMG appearance and sound of a CRD is best achieved by first learning the pathological mechanism that underlies these unusual discharges. The initial denervation of muscle fiber is subsequently followed by reinnervation of that muscle fiber by the axon of a neighboring muscle fiber. If a denervating process continues, that group of muscle fibers may subsequently lose axonal contact, leaving a group of denervated muscle fibers of the same type adjacent to one another. The initial depolarization of a denervated, individual muscle fiber may then initiate the depolarization of its neighboring fiber, which, in turn, can initiate firing in adjacent fiber. This type of linked electrical depolarization is termed *ephaptic*

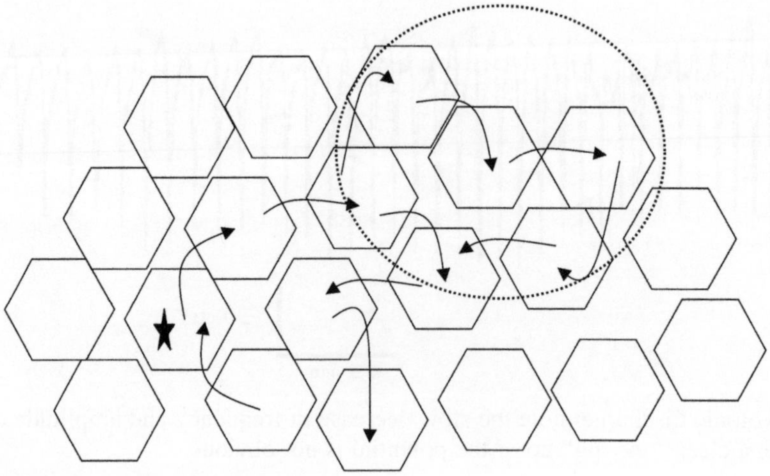

Fig. 6. Complex repetitive discharge pathophysiology. The starred fiber is the pacemaker. Note the additional reentrant loop, shown in a circle.

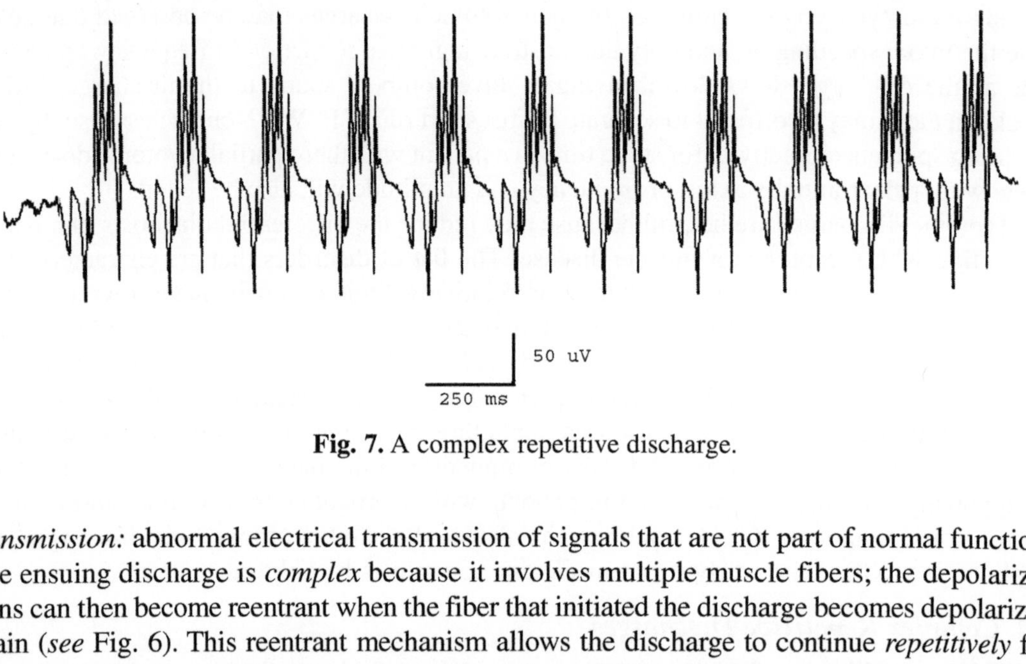

Fig. 7. A complex repetitive discharge.

transmission: abnormal electrical transmission of signals that are not part of normal function. The ensuing discharge is *complex* because it involves multiple muscle fibers; the depolarizations can then become reentrant when the fiber that initiated the discharge becomes depolarized again (*see* Fig. 6). This reentrant mechanism allows the discharge to continue *repetitively* for some time. CRDs can become more complex when additional reentrant circuits are added or removed (*see* Fig. 6). A needle movement causing the first fiber to depolarize may initiate a CRD. On the other hand, CRDs may simply be present without requiring direct contact to a muscle fiber by a needle. Occasionally a "complex-fibrillation potential" may be identified: a single repetitively depolarizing muscle fiber that has several other fibers electrically linked to it. Such a potential will not be a classic CRD because it is not caused by a reentrant mechanism.

Thus, based on an understanding of this pathophysiology, the EMG appearance of a CRD can be easily understood. CRDs appear as multiphasic waves (Fig. 7) that can abruptly change morphology when additional loops of muscle fibers join in or drop out. The multiphasic appearance is caused by the depolarizing fibers associated with the reentrant

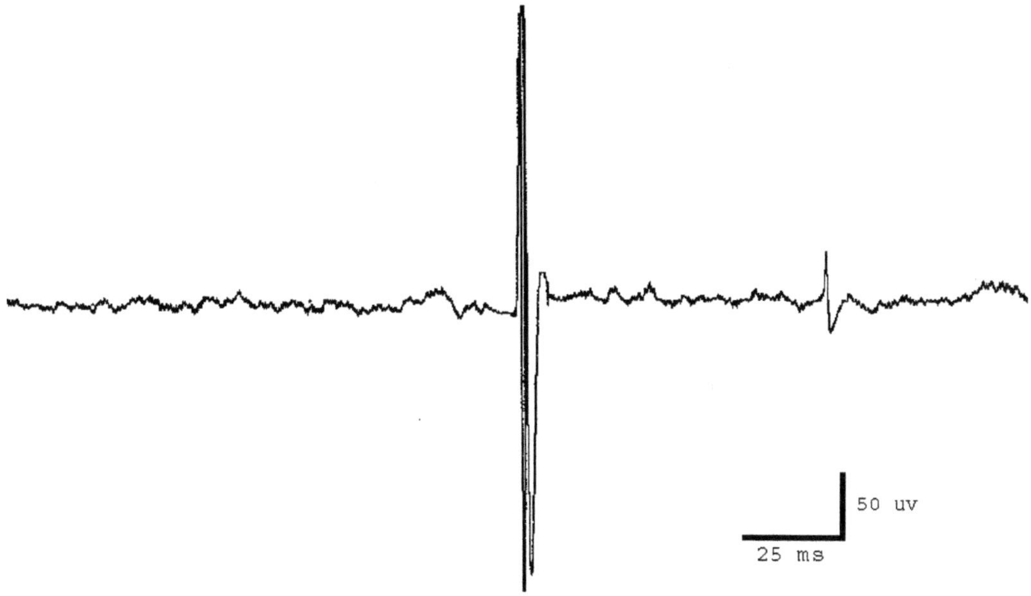

Fig. 8. A fasciculation potential.

circuit. CRDs have a machine-like sound and can also abruptly stop if a fiber in this circuit fails to depolarize.

5. SPONTANEOUS POTENTIALS GENERATED BY MOTOR NEURONS

5.1. Fasciculation Potential

Fasciculation potentials (Fig. 8) result from the spontaneous depolarization of a motor neuron and its associated muscle fibers. The morphology of a fasciculation potential is that of a MUP. Fasciculation potentials are substantially larger than fibrillation potentials, are polyphasic, and fire erratically. Fasciculation potentials will appear only intermittently. In fact, one of the best ways for identifying a fasciculation potential is to simply place the needle into the muscle of interest and for the examiner to remove his/her hand from the needle and simply wait 30 s or longer to see whether a fasciculation potential occurs.

The mechanism underlying a fasciculation potential is that of a reduced threshold for depolarization of the motor neuron. The reduced threshold can be caused by neurogenic disease or can be related to reduced concentration of a serum ion (most notably calcium) leading to intermittent depolarizations of the motor neuron. In addition, potassium channels are important membrane stabilizers; if these channels become dysfunctional, spontaneous depolarization of the motor neuron will occur, potentially leading to fasciculation potentials.

Not surprisingly, the clinical significance of fasciculation potentials is not specific. Although fasciculation potentials commonly occur in severe neurogenic diseases, including motor neuron disease, occasional fasciculation potentials can also occur normally in many individuals. In fact, patients are frequently referred to physicians for evaluation of clinical fasciculations that are often benign—that is, there is no associated neurogenic disease. Fasciculation potentials can be part of the "cramp-fasciculation" syndrome, a disorder of mild nerve hyperexcitability in which patients experience intermittent cramping (usually

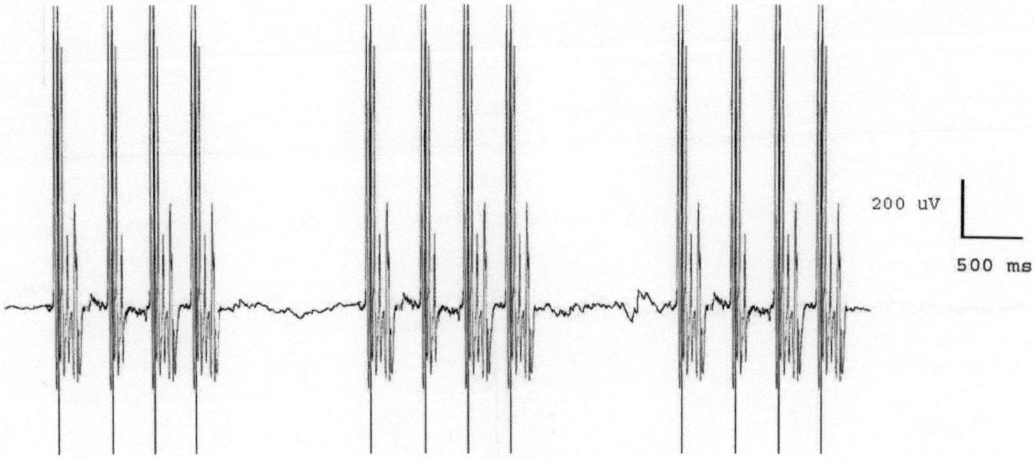

Fig. 9. A myokymic discharge.

more pronounced in the legs than arms) in association with symptomatic muscle twitching (fasciculations). Clinically, such benign fasciculations can be discriminated from the "malignant" fasciculations of amyotrophic lateral sclerosis by their character. Benign fasciculations tend to affect one muscle at a time and cause the muscle to fire frequently; malignant fasciculations tend to affect multiple muscles simultaneously throughout the body. Malignant fasciculation potentials tend to be larger, in part because the motor units that are activated with each depolarization have already undergone substantial reinnervation.

5.2. Doublets/Triplets/Multiplets

These discharges are grouped fasciculations that fire irregularly and may be observed in patients with peripheral nerve hyperexcitability. They most commonly occur in situations of reduced ion concentrations (again, most notably calcium). At one moment a doublet (two fasciculation potentials in a row from the same motor unit), then a triplet, and yet later a single fasciculation potential might be identified. These grouped fasciculations have a firing frequency of 0.1 to 10 Hz.

5.3. Myokymic Discharge

A myokymic discharge (Fig. 9) represents another form of repetitive fasciculation potential, but this time the repetitive depolarization is more "hard wired" than in a doublet or triplet. A disease state within the axon causes a grouped repetitive discharge. There are essentially two firing frequencies of interest in myokymic discharges. The intergroup firing rate is often very slow, with each group appearing once every 2 to 3 s; the intragroup firing rate is substantially higher, with a rate of 50 to 150 Hz. The morphology of these individual waveforms is that of a grouped, repetitive MUP. The MUP itself is usually polyphasic and large suggesting neurogenic injury (*see* discussion of MUPs in Section 7). The sound of myokymic discharges has been described as "marching soldiers," although the gaps in between the appearance of each group would make a description of "marching soldiers starting and stopping frequently" more accurate. Multiple myokymic discharges can also occur simultaneously, complicating the electrophysiological interpretation.

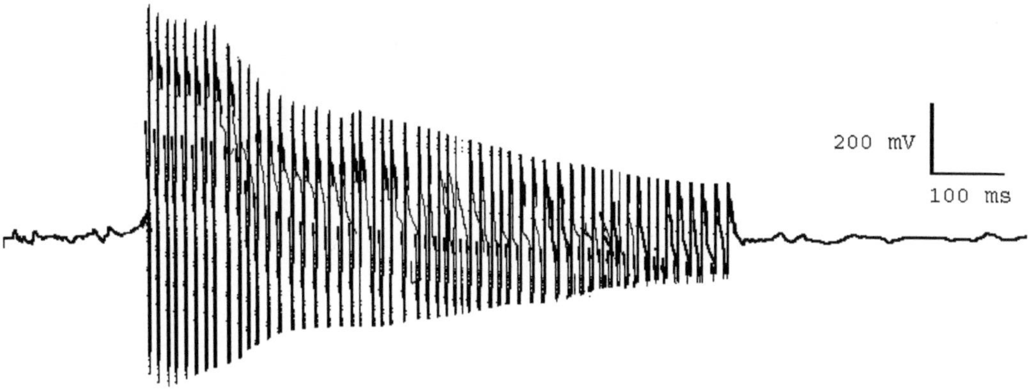

Fig. 10. A neuromyotonic discharge.

Myokymic discharges represent the electrical equivalent of clinical myokymia (the bag of worms appearance of muscle). These discharges are most commonly associated with demyelinating lesions that affect the exiting motor axons from the spinal cord or brainstem, and are also observed after radiation injury to nerves. However, these discharges can be observed uncommonly in any neurogenic disease.

5.4. Neuromyotonic Discharge

Neuromyotonic discharges (Fig. 10) are the most rarely observed of any discharge. They have a characteristic "ping" sound to them, reflecting their very rapid rate of firing (up to 300 Hz), with a rapid decrescendo, lasting less than a second in duration. Visually, a neuromyotonic discharge appears on the oscilloscope as a high-frequency waveform of rapidly decreasing amplitude before it stops, similar to a tornado lying on its side. Unlike a myotonic discharge, a "revving up" is not observed. Although these discharges may seem to have some similarities to a myotonic discharge, there are two obvious differences: first, these discharges are considerably larger than myotonic discharges because they represent the entire motor unit and not just a single muscle fiber. Second, the firing rate is much faster and the duration of the entire discharge is very short. These discharges can be observed very rarely in any neurogenic disorder, but are most commonly associated with axonal potassium channelopathies (e.g., Isaac's syndrome). Dysfunction in the potassium channel leads to an elevated resting potential and subsequent instability of the cell membrane, such that recurrent spontaneous depolarizations of the membrane occur.

5.5. Cramp Discharge

Cramp discharges are most commonly observed as an involuntary discharge that is produced during a muscle contraction. For example, while asking a patient to contract a muscle with an inserted needle electrode, the patient may suddenly complain that their muscle is very painful and they are getting a cramp. What may be observed on the oscilloscope is a repetitive, rapidly firing (40–60 Hz) single motor unit or group of motor units that stops as the muscle is stretched out. Cramp discharges can be very difficult to distinguish from activity produced by a normal contraction, except that the patient will complain of pain and that, in some situations, a single motor unit will fire very frequently for a short time, before a more normal recruitment pattern is re-established.

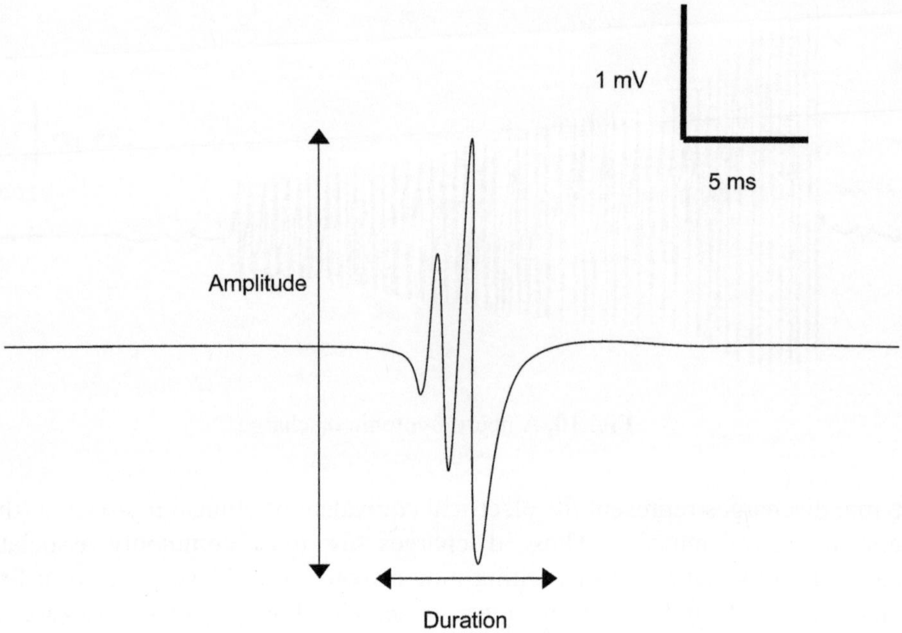

Fig. 11. A motor unit potential with five phases.

6. EVALUATION OF VOLUNTARY MOTOR ACTIVITY

After evaluating spontaneous activity, the electromyographer studies the morphology and firing characteristics of the MUPs while asking the patient to voluntarily activate the muscle. The morphology of the motor unit can most clearly separate neurogenic from myopathic disease while also adding substantial information concerning the severity and chronicity of the disorder. The firing pattern of the MUPs is also critical to identify because it enhances the interpretation of the MUP morphology. Although evaluation of the motor units can be quantified by using one of several techniques, most EMG laboratories rely on qualitative assessment and grading by the electromyographer performing the study. Although seemingly arbitrary to the casual observer, an experienced electromyographer uses a number of developed skills, both visual and auditory, to interpret the condition of the motor units.

7. BASIC MUP PARAMETERS

7.1. Duration

The duration of the motor unit is perhaps its most important characteristic (Fig. 11). The duration reflects how dispersed the motor unit is in time and space and is the least affected by proximity of the needle electrode to the motor unit being recorded. Generally, short-duration motor units are commonly observed in myopathic conditions, whereas long-duration motor units are observed in neurogenic disorders. In any muscle, motor unit size will vary within a distribution of durations; however, some muscles (e.g., those of the quadriceps or triceps) have longer-duration units than those of others (e.g., iliopsoas and biceps). Duration of the MUP is also age-dependent, because larger motor units are observed more prominently in older people, due to the normal drop out of motor neurons in the spinal cord with advancing age. Although normal values for this parameter have been painstakingly obtained for people of different ages, with experience, one can learn to be comfortable identifying what is probably normal for someone and what is not. To give a sense of these changes, a normal MUP from biceps femoris in a 20-yr-old patient is approx 11 ms in duration, whereas a normal MUP from the vastus lateralis of an 80-yr-old patient is approx 16 ms. From an auditory standpoint, longer-duration MUPs have a lower pitched, "thuddy" sound, consistent with the fact that the prolonged duration equates to a longer wavelength and, thus, a lower frequency. Short-duration MUPs are higher pitched and have a "tinny" quality to them.

7.2. *Amplitude*

The amplitude of a MUP (Fig. 11) is much more subject to variation than duration, and it is strongly dependent on the distance of the needle from the motor unit. One way of determining how close the needle is to a given motor unit is to evaluate the *rise time* of the motor unit—that is, the time it takes for the first phase of the potential to reach its maximum after the initial departure from baseline. By freezing the oscilloscope screen, the rise time can be easily evaluated visually and should be generally less than approx 500 μs for the needle to be considered in close proximity to the motor unit. Nonetheless, with small movements of a needle in close proximity to the muscle, the amplitude will vary considerably. Even so, in patients with neurogenic disorders, the absolute highest amplitude motor units will be distinctly higher than those from a healthy patient. For example, a patient with a chronic neurogenic disorder could have amplitudes for a single motor unit reaching 10 mV or more, whereas in a healthy individual, MUPs may not reach above 3 mV.

Amplitude equates to volume when listening to the EMG activity. Hence, for a given volume setting on the EMG, a higher-amplitude motor unit will sound louder than a lower-amplitude motor unit will. Because chronic neurogenic disease produces both a prolongation in duration and increase in amplitude, loud, low-pitched motor units are commonly observed in neurogenic disease. Quiet, high-pitched, "crackling" motor units are commonly observed in myopathic conditions.

7.3. *Phases*

A phase is an individual "peak" or "trough" of the motor unit and can be calculated as the number of baseline crossings minus 1 (Fig. 11). Phases are differentiated from *turns or serrations*, in that a turn is a peak or trough, whether or not it involves a baseline crossing. Most motor units usually consist of four phases or less, although some units with more phases will be observed approx 5% of the time. Electrically, a phase can be thought of as the electrical signature of one or more muscle fibers that make up the overall motor unit, which are in relatively close proximity to one another.

Polyphasia is defined as an excess of motor units with five or more phases, and it is a common occurrence in both myopathic and neuropathic neuromuscular conditions. Phases will increase if the conduction velocity in the terminal nerve twigs to individual muscle fibers is slowed (leading also to a longer-duration MUP) and if the MUP is spread out over a larger area of the muscle because of reinnervation. In a muscle undergoing reinnervation, polyphasia is usually most apparent relatively early to midway through the recovery process, and gradually decreases as time passes (discussed in more detail below in Section 8). Injury to the distal axon is also common in myopathies, and may also contribute to the polyphasia in these disorders; however, the fact that each muscle fiber's depolarization time shortens can lead to a decrease in the normal "bundling" of individual muscle fiber potentials into a single phase and cause the potential to appear polyphasic.

From an auditory standpoint, the number of phases affects the timbre of the sound—the quality that allows one to distinguish, for instance, a violin from a flute. The timbre of a sound is dependent on the overtones in a sound. MUPs, however, have a very limited range of overtones; polyphasic MUP will sound like a "ratchet" produced by the higher frequencies embedded within it.

7.4. Recruitment

Recruitment refers to the orderly participation of MUPs that occurs with a muscle contraction. When an individual is asked to just barely contract a muscle, a single MUP will begin to fire at a rate of approx 4 to 5 Hz. As the patient is asked to push harder, the motor unit will reach approx 10 Hz in firing frequency before the next motor unit begins to fire. People refer to this as the rule of 5s: When two units are firing, the fastest is firing at 10 Hz; when three units are firing, the fastest is firing at a normal frequency of 15 Hz. However, this rule does not hold perfectly well for all muscles. In the cranial nerve musculature, a more rapid firing rate is often observed with a single unit normally firing at up to 15 Hz.

As more motor units join into the contraction, the screen gradually fills up with motor units, such that individual units can no longer be observed. This is what is called a "full interference pattern." In neurogenic disease, this pattern is disturbed, such that a single motor unit will fire more quickly to achieve sufficient force, essentially trying to make up for its absent brethren. Hence, in cases of severe neurogenic disease in which only a single motor unit is present, that MUP may be found to fire at rates of 30 to 40 Hz without other accompanying MUPs. In cases of less-profound axon loss, several individual motor units will fire at 15 to 20 Hz each, more rapidly than observed normally, producing a "picket fence" appearance to the EMG activity rather than the full interference pattern normally generated.

In myopathic disease, the "rule of 5s" still holds; however, MUPs fill the screen so rapidly that they make such an orderly appearance of MUPs impossible to observe. This has been described as "early recruitment," but perhaps should more accurately be called "compressed recruitment" or "condensed recruitment." Because the muscle fiber itself is disrupted and unable to contribute adequately to the force required to contract a muscle, more muscle fibers must be recruited earlier than normal. Thus, in severe myopathies, a barely contracting muscle may produce the firing of many MUPs at once, leading to a full interference pattern, something that is usually only observed with high force levels.

7.5. Activation

Activation describes the CNS drive that produces a contraction. Activation can be reduced if there is a true CNS lesion, such as an infarction or area of demyelination, or if the patient is simply not trying hard to contract the muscle, perhaps because of pain in the limb. The oscilloscope screen in patients with reduced activation demonstrates motor units that preserve the 5:1 ratio; however, a full interference pattern is never achieved. Rather, a couple of normal-appearing motor units will fire slowly and inconsistently. Perhaps, with encouragement, this firing pattern will improve momentarily before lapsing.

The presence of reduced activation, although implying the presence of a CNS lesion or simply poor effort, does *not* exclude the presence of superimposed peripheral neurogenic or myopathic disease. For example, patients with severe sensory ataxia, such as one due to a sensory neuronopathy, may have difficulty moving their limbs simply secondary to impaired sensory feedback, leading to reduced activation. Similarly, patients with amyotrophic lateral sclerosis often have a combination of central and peripheral motor neuron loss, leading to a picture of reduced activation superimposed on reduced recruitment and MUP enlargement.

8. UNDERSTANDING THE TIME COURSE OF DEVELOPMENT OF MUP ABNORMALITIES

8.1. Neurogenic Disorders

Let us take a situation in which axonal injury occurs to a nerve such that some, but not all of it, will regrow and eventually reinnervate a muscle it controls. For simplicity, this muscle is normally only innervated by three neurons. Figure 12A shows the initial state with a healthy muscle and nerve, with individual muscle fibers having a checkerboard pattern of innervation from three different motor neurons. Figure 12B shows the injured nerve and the associated denervated muscle shortly after the injury. The motor neuron has its basement membrane intact and can regrow, whereas the other two do not, and cannot regrow. The nerve is now not connected, and if a needle were placed in the muscle, marked spontaneous activity would be observed (as shown), but no MUPs would be identified. Figure 12C shows early reinnervation. Only one of the three axons in the nerve has managed to grow back, and it has attached to only three muscle fibers. If an EMG were performed at this stage, one motor unit would appear, but it would be made up of only three muscles and would be recruited very rapidly because there are no other motor neurons to assist. This unusual situation of a low-amplitude, short-duration motor unit with reduced recruitment is called a "nascent" MUP (a MUP in the process of being born). Gradually more muscle fibers will be added to the motor unit, including muscle fibers that are more distant that were not normally part of the initial unit. These developing nerve twigs are poorly myelinated (dotted lines) and are long and, hence, the resulting potential is demonstrated in Fig. 12D—polyphasic with low amplitude but long duration. In this example, there is a single satellite potential attached to the end of MUP; this potential represents the firing of a distant recently reinnervated muscle fiber. The recruitment remains reduced, because there are no other associated motor units able to contribute to the muscle contraction. Finally, in Fig. 12E, reinnervation is complete. The terminal nerve twigs are well-myelinated and a new, single, large motor unit has formed, now doing the work of nearly three motor units (not all of the fibers were successfully reinnervated in this example). This MUP has a larger amplitude, longer-duration, and, perhaps, some mild polyphasia; recruitment remains reduced. Should all of the fibers be successfully reinnervated, spontaneous activity will also normalize and no fibrillation potentials or PSWs will be identified.

In summary, not all neurogenic disorders are the same, and the appearance of the MUP will depend on the severity of the lesion, when in the course of recovery the neurophysiological study is being completed, and the relative success of reinnervation. Of course, if no reinnervation occurs (as does happen in some severe nerve injuries), than no MUPs will be identified and fibrillation potentials will persist in the muscle indefinitely.

8.2. Myopathic Disorders

Myopathic disorders also produce changing characteristics on needle EMG over time. Acute and subacute myopathies generally are identified by short-duration, low-amplitude MUPs, usually accompanied by fibrillation potentials and PSWs. The spontaneous activity itself is often of low amplitude (small fibrillation potentials). The reason for the MUP

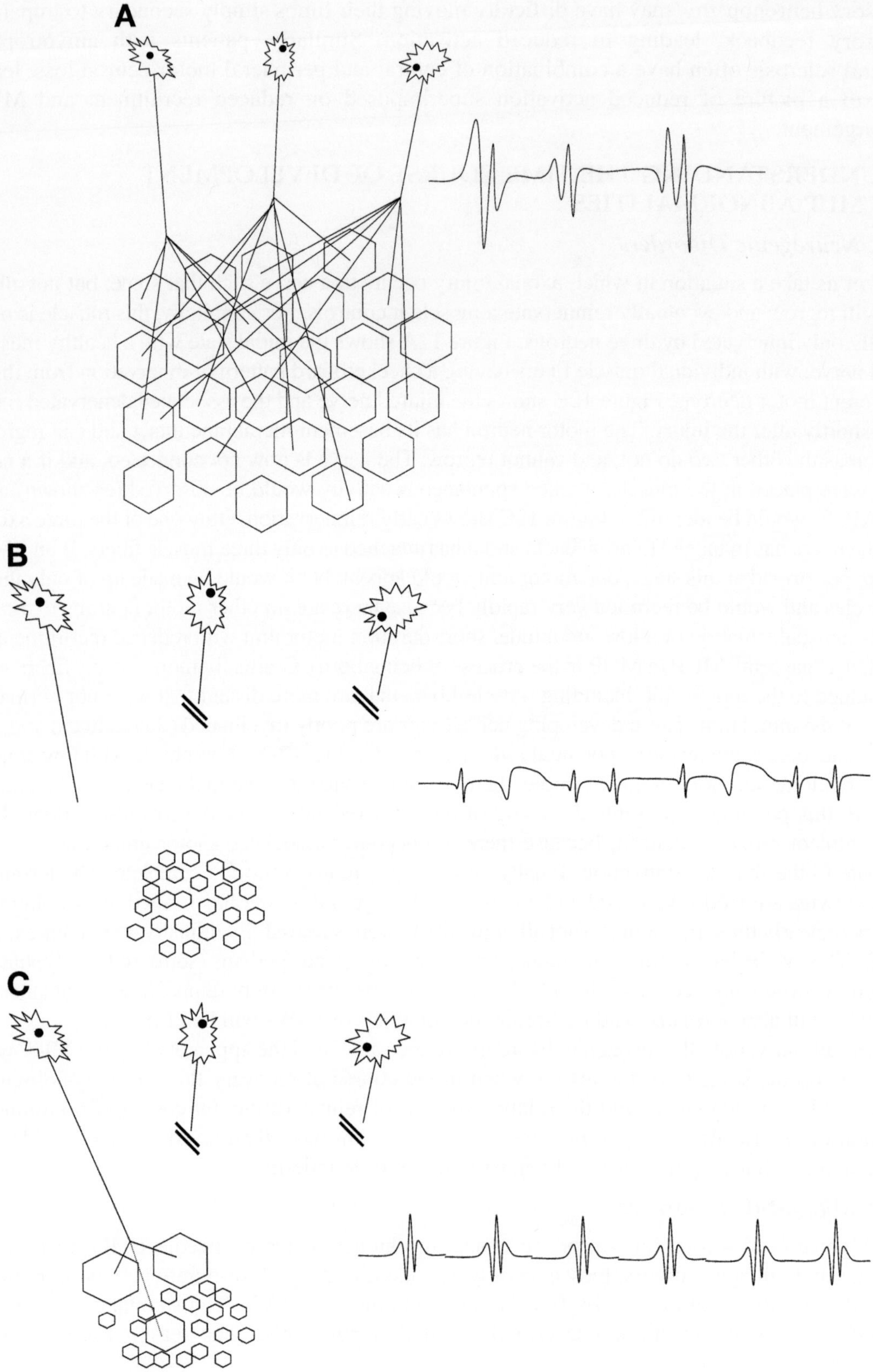
A
B
C

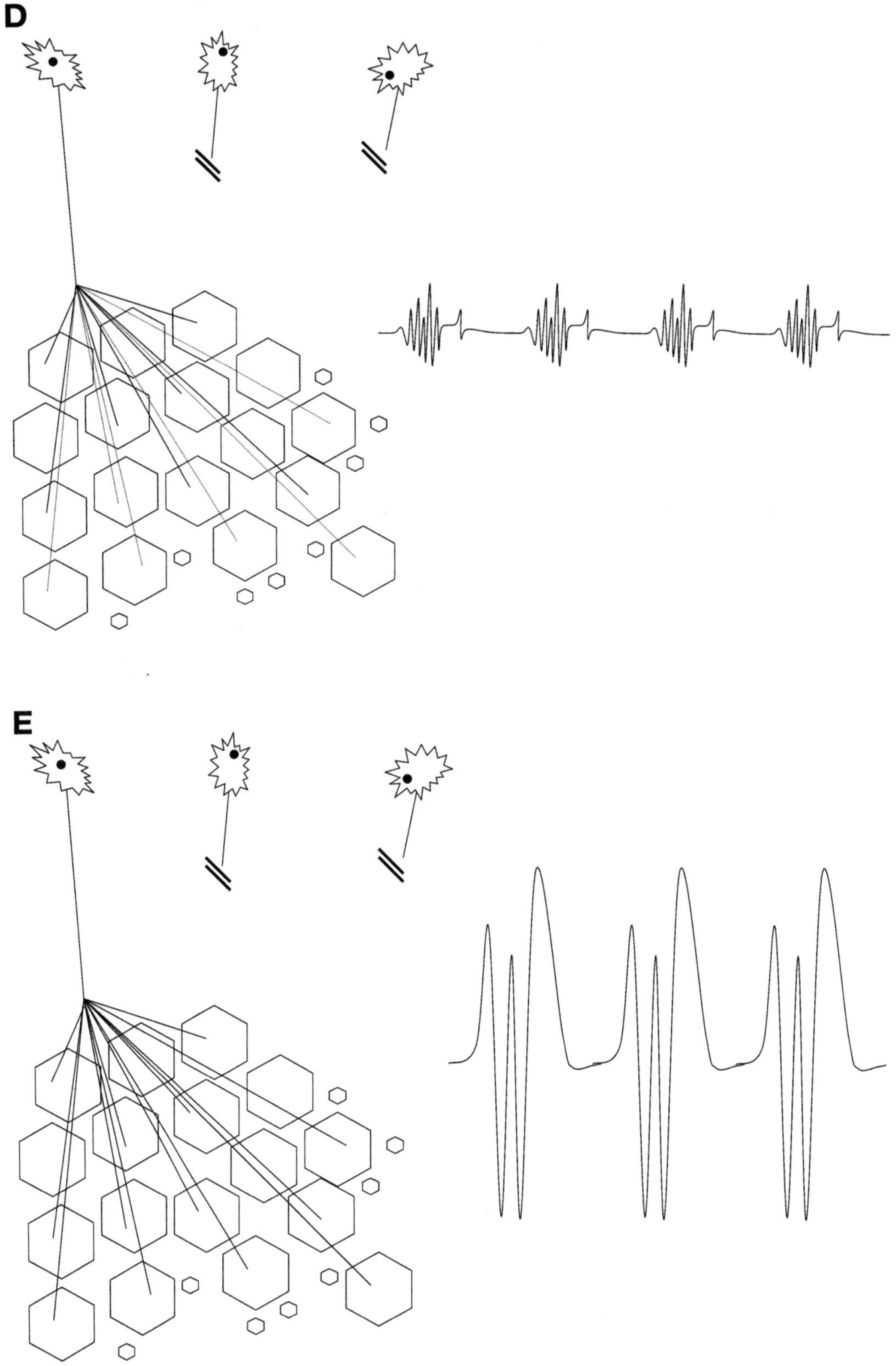

Fig. 12. Motor unit remodeling with nerve injury. *See* text for details.

changes is demonstrated in Fig. 13A and B, which shows a MUP before and after a myopathic process has begun. Several processes likely contribute to the development of small motor units, including destruction of muscle fibers associated with a given motor unit, truncation of muscle fibers leading to smaller individual muscle fiber action potentials, and destruction of intramuscular nerve twigs by inflammation. Ischemia and other pathological processes may also play a role. The result, however, is that the number of individual motor units is preserved, but their amplitude and duration are markedly reduced. In addition to these changes, the force-generating capacity of a single motor unit will also decrease, making it necessary to recruit additional motor units earlier than normal to produce even small movements of the extremity. Hence, short-duration, low-amplitude motor units with "early" recruitment—many motor units participating to produce small degrees of force—are seen in subacute or ongoing myopathy, as detailed above in Section 7.4.

As a myopathic condition persists and there is ongoing destruction of muscle fibers and some motor nerve terminals are lost, a gradual degree of reinnervation occurs, such that motor neurons sprout twigs to still-functioning individual muscle fibers. In time, the previous clear myopathic picture may become more and more clouded by the presence of some occasional large motor units. In some situations, the myopathic motor units become almost impossible to find as they are replaced by a few, usually very polyphasic, high-amplitude, long-duration units. If there are muscle fibers or segments of muscle fiber that remain without reinnervation, then ongoing denervation will persist.

Muscle diseases are either hereditary or acquired. Hereditary muscle diseases, including the muscular dystrophies, can show a more chronic pattern of large MUPs mixed with short-duration, low-amplitude, polyphasic MUPs. This reflects the chronicity of the disease process, which, in adult cases, is longstanding. Acquired muscles diseases, especially inflammatory myopathies such as dermatomyositis or polymyositis, typically have short-duration, low-amplitude MUPs with prominent denervating features. In contrast to there inflammatory myopathies, inclusion body myositis, an acquired long-standing disease, characteristically has a mixture of large and small MUPs and prominent fibrillation potentials and positive sharp waves.

9. LIMITATIONS AND FINER ASPECTS OF NEEDLE EMG

Needle EMG, although providing valuable insights into a patient's peripheral pathology, has a number of limitations that are important to identify. Similarly, these limitations go hand-in-hand with several often misunderstood aspects of the procedure.

9.1. Needle Placement and MUP Morphology

Whereas we would like to think of an individual MUP as the electrical signal of a single motor unit, in fact, a single MUP is actually the electrical signature of a single motor unit as recorded by a needle electrode in a specific position relative to that motor unit. In other words, the orientation and distance of the needle to an individual motor unit determines the exact configuration of the MUP. Small movements of the needle may turn a polyphasic motor unit into one with normal phases; amplitude will also fluctuate. At least one study has demonstrated that MUPs will appear smaller when evaluated with the needle inserted toward the periphery of a muscle then when inserted more deeply into it. Duration of a given MUP usually is the most stable characteristic, is less dependent on needle position, and should be considered the parameter of choice to evaluate.

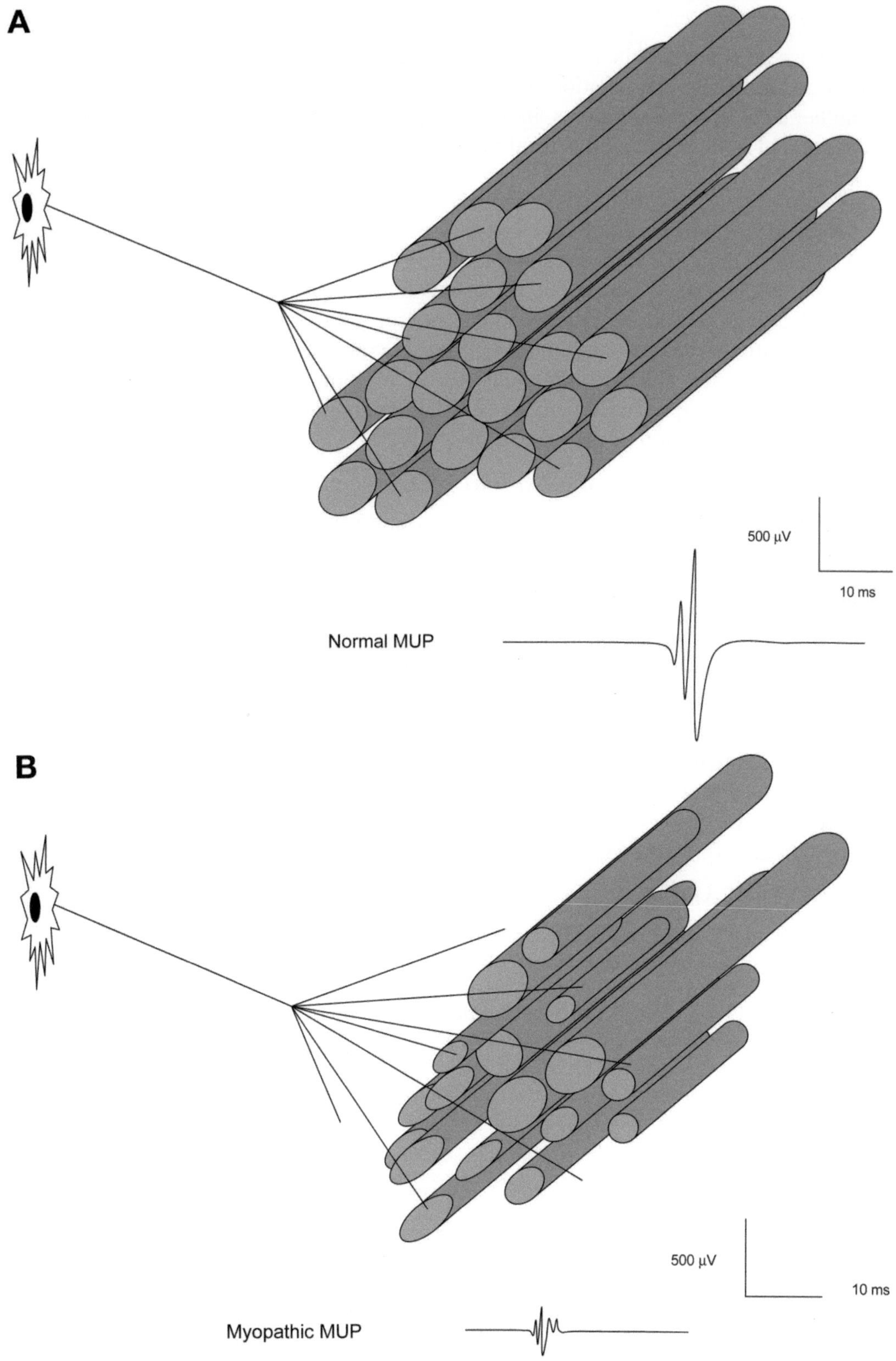

Fig. 13. Motor unit remodeling in myopathy. *See* text for details. MUP, motor unit potential.

9.2. *Sampling Error*

When a needle is inserted in a muscle, only a small area of the muscle is evaluated. Hence, for mild lesions, needle examination in one area of the muscle might show results different from another area. In some situations, in which the findings are very subtle, examining several different parts of a muscle can be helpful in establishing the overall condition of that muscle.

9.3. *Age-Dependent Changes*

As noted, MUPs will gradually enlarge as a person ages, and most people who undergo EMG who are in their 80s will demonstrate MUPs that are relatively large compared with those of younger people. Whether this represents true disease or whether it is merely a consequence of aging, caused by normal motor neuron loss, is never clear in an individual patient. This situation is analogous to finding degenerative changes on imaging of the spine of asymptomatic elderly individuals. Such "abnormalities" may not be meaningful. Ultimately, how to interpret mild degrees of MUP enlargement in a given elderly individual is really up to the discretion of the electromyographer.

9.4. *Weakness in the Setting of a Normal EMG*

It must be recalled that although EMG is very useful test, it does not evaluate all aspects of the muscle. It only evaluates the electrical activity of the muscle fiber—that produced by the depolarization of the muscle fiber membrane. Although strength depends on this being normal, it also depends on normal release of Ca^{2+} into the sarcoplasm and normal function of actin and myosin, as well as other associated proteins, to produce a muscle contraction. Strength also depends on having an intact connection between the muscle and bone (i.e., the presence of an intact tendon). Another cause of milder degrees of weakness in the setting of relatively normal EMG is mildly reduced activation or recruitment that is simply difficult to observe on the oscilloscope screen. Patients with subtle weakness, in the 4+ to 5–range, may have relatively normal-appearing EMG studies. This is especially true in mild myopathies, in which subtle changes in MUP morphology may not be readily observed and in which computer-assisted techniques at quantifying the results may be helpful.

SUGGESTED READING

Brown WF, Bolton CF, eds. Clinical Electromyography. Butterworths, Boston, MA, 1987.

Daube JR. AAEM minimonograph 11: needle examination in clinical electromyography. Muscle Nerve 1991;14:685–670

Daube JR. Assessing the Motor Unit With Needle Electromyography. In: Clinical Neurophysiology (Daube JR, ed.). FA Davis, Philadelphia, PA, 1996, pp. 257–265.

Kimura J, ed. Electrodiagnosis in Diseases of Nerve and Muscle. FA Davis, Philadelphia, PA, 1983.

Bonner FJ Jr, DevlescHoward AB. AAEM minimonograph #45: the early development of electromyography. Muscle Nerve 1995;18:825–833.

REVIEW QUESTIONS

1. Which of the following statements is true?
   A. PSWs are unrelated to fibrillation potentials.
   B. PSWs are produced by groups of muscle fibers, whereas fibrillation potentials are produced by individual muscle fibers.
   C. PSWs are a more specific abnormality than fibrillation potentials.
   D. PSWs are uncommon in most muscle disorders.
   E. None of the above.

2. Recruitment will be reduced in which of the following conditions?
   A. Subacute myopathy.
   B. Chronic myopathy.
   C. Acute Guillain–Barré syndrome.
   D. Myasthenia gravis.
   E. B and C.
3. Which of the following is *not* generated by nerve?
   A. CRD.
   B. Cramp potential.
   C. Doublet.
   D. Fasciculation.
   E. Myokymic discharge.
4. In distinguishing a neuromyotonic discharge from myotonic discharge, which of the following is true?
   A. A neuromyotonic discharge has a faster firing frequency than a myotonic discharge.
   B. A myotonic discharge has a larger amplitude than a neuromyotonic discharge.
   C. A true neuromyotonic discharge has a "revving" up phase (increasing frequency), whereas a myotonic discharge does not.
   D. A neuromyotonic discharge may have the form of a PSW.
   E. Myotonic discharges are a characteristic finding in Isaac's syndrome.
5. Which of the following characteristics is most important in evaluating a MUP?
   A. Amplitude.
   B. Duration.
   C. Phases.
   D. Turns.
   E. Activation.
6. Abnormalities in activation on needle EMG can be observed in which of the following?
   A. Stroke.
   B. Spinal cord injury.
   C. Sensory neuronopathy.
   D. Depression.
   E. All of the above.
7. Which of the following statements concerning motor unit recruitment is FALSE?
   A. Recruitment of motor units usually proceeds from smaller to larger motor units.
   B. Recruitment can be difficult to judge in someone with poor activation.
   C. In neurogenic disorders, recruitment of the second motor unit may be delayed and not occur until the first motor unit is firing at rates of 20 to 30 Hz.
   D. Early recruitment of MUPs, as occurs in myopathies, indicates that MUPs fire at unusually low frequencies.
   E. In patients with chronic neurogenic injury, enlargement of MUP size is accompanied by a reduction in recruitment.
8. Which statement is FALSE concerning the sounds of needle EMG?
   A. A myotonic discharge sounds like a motorcycle revving its engine.
   B. Polyphasia will give a MUP a ratchety timbre.
   C. A sputtering sound is often associated with end-plate spikes.
   D. Fasciculation potentials "tick" in a clock-like fashion.
   E. "Marching soldiers" suggests a myokymic discharge.
9. Which of the following statements is true?
   A. MUP duration and amplitude generally increase with increasing age.
   B. A single MUP can have a great many morphologies, depending on the orientation of the needle relative to the motor unit.
   C. Chronic myopathies may have substantial numbers of larger MUPs.

D. Fibrillation potentials are commonly present in myopathic conditions.
E. All of the above.

10. Which of the following discharges would cease in a patient exposed to curare (a blocker of the muscle end-plate acetylcholine receptor)?
A. CRD.
B. Myotonic discharge.
C. Positive wave.
D. Fasciculation.
E. None of the above.

REVIEW ANSWERS

1. The correct answer is E. None of the above. PSWs are depolarizing single muscle fibers, just like fibrillation potentials, but are more nonspecific and can be found variably in many muscle disorders.
2. The correct answer is E. In both Guillain–Barré syndrome and chronic myopathies, recruitment may be reduced, whereas in myasthenia gravis and subacute myopathy, recruitment will be normal or early.
3. The correct answer is A. A CRD is caused by a reentrant path in a group of muscle fibers, without any nerve involvement. All of the other potentials described are induced by neuronal discharges.
4. The correct answer is A. Neuromyotonic discharges have the fastest firing frequency of any spontaneous waveform, reaching rates of up to 300 per second. Because myotonic discharges are caused by the depolarization of a single muscle fiber, they cannot be polyphasic and they are smaller than neuromyotonic discharges, which are motor unit discharges. Only myotonic discharges "rev up." Because they are motor units, *neuromyotonic discharges* can never have the morphology of a PSW.
5. The correct answer is B. The single most important MUP characteristic is the duration, because it is the least likely to be affected by electrode positioning.
6. The correct answer is E. Any condition that may reduce the overall descending drive to recruit motor units will lead to reduced activation. These conditions can be organic (stroke and spinal cord injury), behavioral (depression, malingering), or caused by reduced sensory feedback (sensory loss caused by sensory neuronopathy).
7. The correct answer is D. In myopathies, motor unit recruitment is not truly early, but rather is "compressed," in that many motor units are activated quickly, because each motor unit generates substantially less force than normal. MUPs never fire more slowly than approx 4 to 5 Hz. All of the other answers are true.
8. The correct answer is D. Fasciculation potentials occur irregularly; fibrillation potentials, on other hand, can tick like a clock. All of the other answers are correct.
9. The correct answer is E. All of the answers are correct.
10. The correct answer is D. A fasciculation potential is the only discharge that is neuronally generated and, hence, requires the activation of the muscle end plate.

15

Mononeuropathies of the Upper and Lower Extremity

Kevin R. Scott and Milind J. Kothari

Summary

Nerves of both the upper and lower extremities are frequently injured for a variety of reasons. In the arms, median neuropathy at the wrist is by far the most common disorder; ulnar neuropathy also occurs with a relatively high frequency. Other mononeuropathies affecting the upper extremities, including anterior and posterior interosseous neuropathies, and musculocutaneous neuropathies, are very infrequent. In the legs, peroneal neuropathy at the fibular neck and lateral femoral cutaneous neuropathy are the most common disorders, and other focal neuropathies, such as tibial neuropathy at the ankle (tarsal tunnel syndrome) are often sought but rarely identified. Although all mononeuropathies can be preliminarily diagnosed by history and clinical examination, neurophysiological testing provides a key component to confirming the diagnosis and assisting with treatment planning.

Key Words: Carpal tunnel syndrome; median neuropath; meralgia paresthetic; peroneal neuropathy; radial neuropathy; tarsal tunnel syndrome; tibial neuropathy; ulnar neuropathy.

1. INTRODUCTION

Focal compression mononeuropathies of the upper and lower extremities are commonly encountered in clinical practice. These neuropathies serve as a major source of referrals to a neurophysiology laboratory. In this chapter, common entrapments of the upper and lower limbs will be discussed and correlated with the electrophysiological findings.

2. MEDIAN NEUROPATHY

The median nerve can be compressed at various levels; the most common site being at the wrist. This neuropathy is the most common entrapment neuropathy affecting the upper extremity.

2.1. *Anatomy*

The median nerve arises from the medial and lateral cords of the brachial plexus. The lateral cord is made up of C6–C7 fibers, and supplies median sensory fibers to the thenar eminence, thumb, index, and middle fingers, as well as, the majority of motor fibers to the proximal median-innervated forearm muscles. The medial cord is made up of C8–T1 fibers, and supplies the majority of motor fibers to the distal median-innervated muscles of the forearm and hand, as well as sensory fibers to the lateral half of the ring finger.

After the medial and lateral cords join, the median nerve courses distally, along the medial side of the humerus. No muscular branches are given off during its course through the proximal arm.

From: *The Clinical Neurophysiology Primer*
Edited by: A. S. Blum and S. B. Rutkove © Humana Press Inc., Totowa, NJ

The nerve then enters the region of the antecubital fossa, where it lies medial to the biceps tendon and brachial artery. At this point, the median nerve pierces the pronator teres muscle, providing branches to innervate this muscle, as well as the flexor carpi radialis, palmaris longus, and flexor digitorum superficialis muscles of the forearm. Posteriorly, the median nerve gives off the *anterior interosseous nerve* (AIN), which innervates the flexor pollicis longus, flexor digitorum profundus (to digits 2 and 3), and the pronator quadratus muscles. The AIN provides no cutaneous sensory innervation, but does supply sensation to the interosseous membrane and wrist joint.

Proximal to the wrist, in the distal portion of the forearm, the median nerve gives off the *palmar sensory cutaneous nerve*, which provides sensation over the thenar eminence. The median nerve, itself, along with the nine flexor tendons of the hand, continues through the carpal tunnel formed by the transverse carpal ligament and carpal bones. After passing into the palm, the nerve divides into sensory and motor trunks. The sensory trunk divides further, providing *digital sensory nerves* to innervate the first three fingers and lateral portion of the fourth. The motor trunk gives off the *recurrent thenar branch*, which supplies the muscles of the thenar eminence, including the opponens pollicis, the abductor pollicis brevis (APB), and the superficial potion of the flexor pollicis brevis muscles, before continuing distally and supplying the first and second lumbricals.

2.2. Clinical Features of Median Nerve Dysfunction

2.2.1. Median Neuropathy at the Wrist (Carpal Tunnel Syndrome)

This is the most common entrapment neuropathy affecting the upper extremity. Patients may present with a variety of signs and symptoms, including weakness and sensory disturbances. The most common complaint is that of pain and paresthesias. Pain is usually localized to the wrist and fingers, but may radiate to the forearm, arm, or even the shoulder. Certain patients may report a diffuse aching sensation involving the entire arm. Sensory disturbances usually involve the lateral three fingers. However, patients may report paresthesias involving the entire hand or even in regions supplied by the ulnar nerve. It is very important to note that sensation over the thenar eminence will be spared in carpal tunnel syndrome (CTS). This region is supplied by the palmar cutaneous branch, which arises proximal to the carpal tunnel. Patients usually report an increase in symptoms during activities that cause the hand to be placed in either a flexed or extended posture (typing, driving, holding a telephone, reading a newspaper). A hallmark of CTS is nocturnal paresthesias. Patients frequently awaken from sleep to shake their hands. This is likely secondary to nerve ischemia resulting from a persistent flexed or extended posture of the wrist. Late in the course, there may be frank weakness of thumb abduction and opposition with atrophy over the thenar eminence. Activities requiring fine motor dexterity may be impaired (buttoning shirts, writing, dropping utensils, and opening jars). Provocative testing (Tinel's and Phalen's) may be helpful. The clinician must be aware of the sensitivity and specificity issues. In CTS, a Tinel's sign may be present in approximately half of the cases. Of importance is that normal people may also exhibit a false-positive Tinel's sign.

2.2.1.1. Differential Diagnosis of CTS

Many peripheral or central nervous system lesions may produce symptoms similar to CTS. In evaluating the patient with CTS, the clinician must consider a proximal median nerve lesion, a lesion affecting the brachial plexus, a cervical radiculopathy, and even stroke. The examination can be helpful in excluding these conditions. With lesions that are more proximal,

Table 1
Potential Causes of Median Mononeuropathy at the Wrist (Carpal Tunnel Syndrome)[a]

1. Inflammatory/infectious
   A. Connective tissue disease (e.g., rheumatoid arthritis or lupus)
   B. Sarcoidosis
   C. Lyme
   D. Tuberculosis
   E. Complications of septic arthritis
2. Traumatic/overuse
   A. Repetitive use injuries
   B. Wrist or hand trauma
3. Metabolic/endocrine
   A. Diabetes mellitus
   B. Hypothyroidism
   C. Pregnancy
   D. Acromegaly
4. Congenital
   A. Congenitally small carpal tunnel
   B. Anomalous muscles
   C. Persistent median artery
5. Idiopathic

[a]From ref. *1*.

patients should have weakness involving the proximal muscles and may also have reflex abnormalities.

There is a long list of potential conditions that may lead to the development of CTS (Table 1). CTS may be the result of repetitive use injuries or wrist trauma. Inflammatory or infectious causes may include connective tissue disease, sarcoid, lyme, tuberculosis, or compression as the result of a septic joint. Metabolic causes, such as diabetes, hypothyroidism, or acromegaly, lead to gains in weight or tissue edema that may lead to compression within the tunnel. Less common causes include a variety of different tumors, such as lipoma, schwannoma, neurofibroma, or ganglion cysts; or congenital disorders, such as anomalous muscles, congenitally small tunnel, or persistent median artery.

2.2.1.2. ELECTROPHYSIOLOGY IN CTS

When performing electrodiagnostic studies, limb temperature must be controlled and adequately maintained. The pathophysiology in CTS is usually that of demyelination, which may be associated with secondary axonal degeneration. In patients with typical CTS, the median distal motor and sensory latencies are prolonged, whereas the ulnar studies are normal. However, there is a small group (10–25%) of patients, in whom these routine nerve conduction studies are normal. In this group, comparison studies should be performed and may demonstrate abnormalities of the median nerve.

Routine nerve conduction studies evaluating for CTS should include:

1. Median motor study—stimulating at the wrist and elbow; recording from the APB muscle.
2. Ulnar motor study—stimulating at the wrist, below elbow, and above elbow sites; recording from the abductor digiti minimi muscle.
3. Ulnar and median F-responses.

4. Median sensory study—stimulating at the wrist; recording from the index finger.
5. Ulnar sensory study—stimulating at the wrist; recording from the fifth digit.

If routine nerve conduction studies are normal, comparison studies are used to evaluate for compression by comparing median nerve to ulnar nerve function. Three frequently used comparison studies are:

1. Median vs ulnar palm-to-wrist mixed nerve latencies.
2. Median vs ulnar wrist-to-digit 4 sensory latencies.
3. Median (second lumbrical) vs ulnar (interosseous) distal motor latencies.

In each case, identical distances are used between the stimulating and recording electrodes for the median and ulnar nerves. These sensitive comparison studies are considered abnormal if differences between the median and ulnar latencies exceed 0.4 to 0.5 ms. Other studies have also been described, including inching across the wrist, median vs radial comparison study to the thumb, median vs ulnar F-wave minimal latency difference, and wrist-to-palm vs wrist-to-digit 2 sensory latency. These latter studies are less frequently used in most laboratories.

Needle examination is usually performed in the patient being evaluated for CTS. The APB muscle and at least two proximal C6–C7 muscles, that is, extensor digitorum communis, pronator teres, and flexor carpi radialis, should be sampled to exclude a C6–C7 radiculopathy. If the APB muscle is abnormal:

1. At least two other lower trunk/C8–T1 muscles, for example, first dorsal interosseous or extensor indicis proprius, should be examined to exclude a C8–T1 radiculopathy, lower trunk brachial plexopathy, or polyneuropathy.
2. At least two proximal, median nerve-innervated muscles, for example, flexor carpi radialis, pronator teres, or flexor pollicis longus, should be sampled to exclude a proximal median neuropathy in the forearm.

2.2.1.3. Treatment

The conservative management of CTS involves withdrawal of provoking factors, use of a neutral wrist splint, and local corticosteroid injections. If this approach fails, surgical decompression may be necessary. Surgery usually provides benefit in 85 to 90% of patients.

2.2.2. Proximal Median Neuropathy

There are several syndromes involving proximal median nerve dysfunction. These include entrapments at different locations, such as:

1. The ligament of Struthers.
2. The lacertus fibrosus.
3. The heads of the pronator teres muscle.
4. The sublimis bridge of the flexor digitorum sublimis muscle.

Patients with proximal median neuropathy will usually present with complaints of pain in the arm and forearm, paresthesias in the distribution of the median nerve, and weakness. The distinguishing feature on sensory examination is that the thenar eminence is involved in these cases, whereas it is spared in CTS. Patients will also demonstrate weakness of proximal median-innervated muscles (flexor digitorum sublimis, flexor carpi radialis, and/or pronator teres). However, the pronator teres may be spared in pronator syndrome. Pronation/supination type movements may aggravate the symptoms. The sole finding of increased pain on provocative testing is unreliable. A Tinel's sign may be present in approx 50% of cases.

The electrodiagnostic study is quite helpful in these patients. The median sensory study is usually abnormal. The median motor study may show slowing of the forearm conduction velocity. There may also be a drop in the amplitude proximally as compared with the distal site. The needle examination will demonstrate abnormalities in median-innervated muscles involving the wrist and forearm. Thus, if testing for a proximal median neuropathy, at least two proximal muscles should be examined. It is also important to remember that, in a lesion at the level of the pronator teres, the pronator teres muscle itself may be spared. MRI or ultrasound may be useful to visualize the nerve.

2.2.3. *Anterior Interosseous Neuropathy (AIN)*

The AIN is the largest branch of the median nerve and originates just distal to the pronator teres in the forearm. The nerve has no cutaneous sensory fibers, thus, patients will not experience cutaneous sensory symptoms. The nerve supplies three muscles: the flexor pollicis longus, the flexor digitorum profundus subserving digits 2 and 3, and the pronator quadratus. To test the pronator quadratus muscle, the elbow should be flexed to avoid the effect of the pronator teres. Involvement of the AIN is easily tested by having the patient perform the "OK" sign. Patients with AIN injury will be unable to flex the distal phalanges of the thumb and forefinger. Instead of forming a normal circular "OK" sign, these patients maintain an extended position at the distal interphalangeal joints and form a pincer-type grasp. The electrodiagnostic study will usually demonstrate normal nerve conduction studies. The needle examination should demonstrate abnormalities in muscles supplied by the AIN, and is critical in making this diagnosis. Patients may present with AIN as a manifestation of an autoimmune brachial plexopathy. Treatment usually involves conservative measures, although, occasionally, surgical intervention is warranted.

3. ULNAR NEUROPATHY

3.1. Anatomy

The ulnar nerve is derived from the C8–T1 nerve roots. Nearly all ulnar nerve fibers travel through the lower trunk and medial cord of the brachial plexus. As the ulnar nerve descends through the medial arm, it provides no branches until it reaches the elbow. Here, muscular branches to the flexor carpi ulnaris and flexor digitorum profundus (subserving digits 4 and 5) arise. The nerve then descends to the wrist. Proximal to the wrist, the *dorsal ulnar cutaneous sensory branch* exits, and just before entering Guyon's canal, the nerve gives off the *palmar cutaneous sensory branch*, providing sensation over the hypothenar area. The nerve then enters Guyon's canal and provides sensation to the volar fifth and medial fourth digits, as well as motor innervation to the hypothenar muscles, palmar and dorsal interossei, third and fourth lumbricals, adductor pollicis, and deep head of the flexor pollicis brevis muscles.

3.2. Clinical Features of Ulnar Nerve Dysfunction

3.2.1. *Ulnar Neuropathy at the Elbow*

This neuropathy is the second most common entrapment neuropathy. Typical symptoms include numbness and tingling in the distribution of the ulnar nerve. Some patients may report elbow pain that radiates into the ulnar aspect of the hand. In some cases, only sensory symptoms are present. Impaired cutaneous sensation in the volar fingertips is the most

Table 2
Common Causes of Ulnar Neuropathy at the Elbow[a]

1. Old fracture with joint deformity
2. Recent elbow trauma without fracture
3. Habitual leaning on elbow
4. Occupational repetitive flexion/extension
5. Congenital variations of HUA architecture
   A. Absent HUA with nerve prolapse
   B. Hypertrophy of retinaculum
   C. Anconeus epitrochlearis muscle
6. Diabetes mellitus
7. Hereditary neuropathy with liability to pressure palsies
8. Rheumatoid arthritis
9. Iatrogenic
   A. Malpositioning during surgery
   B. Nerve infarction during transposition
10. Amyotrophic lateral sclerosis

[a]HUA, humeroulnar aponeurotic arcade.

common sensory deficit. Sensory loss in the ulnar palm is less frequent. An early motor sign may be an inability to adduct the fifth digit (Wartenberg's sign). In more involved cases, there will be weakness of handgrip and atrophy of the intrinsic hand muscles. Weakness of the first dorsal interosseous muscle (84%) is more frequent than weakness of the abductor digiti quinti muscle (76%). Weakness of the flexor digitorum profundus and flexor carpi ulnaris muscle occurs in 56% and 20% of patients, respectively. In severe cases, clawing of digits 4 and 5 can develop. Deep tendon reflexes are usually preserved in this focal neuropathy. Various provocative maneuvers have been described to increase the diagnostic yield. These include a Tinel's sign at the elbow, sustained manual pressure over the cubital tunnel, sustained elbow flexion, and flexion combined with manual pressure. Combined flexion with manual pressure over the cubital tunnel has been reported to have the highest sensitivity (91%). The differential diagnosis in a patient suspected of an ulnar neuropathy at the elbow includes a lower trunk or medial cord brachial plexopathy, a C8–T1 radiculopathy, or an ulnar neuropathy at some location other than the elbow. Some common causes of ulnar neuropathy at the elbow are listed in Table 2.

3.2.1.1. Electrophysiology

As with other mononeuropathies, the electrodiagnostic study should localize the abnormalities to a specific nerve, in this case, the ulnar nerve. The ulnar motor study should demonstrate evidence of focal demyelination across the elbow, characterized by either focal slowing or conduction block. When performing the ulnar motor study, elbow position is crucial. The flexed elbow position should be used, and has been shown to be more sensitive than testing in the extended position. In certain cases, inching across the elbow can be performed to demonstrate a focal area of demyelination. Recording over the first dorsal interosseous muscle may be slightly more sensitive than recording over the abductor digiti quinti muscle.

Evaluation of the ulnar nerve with nerve conduction studies should include the routine studies described in Table 3. Should routine nerve conduction studies not localize the lesion, additional techniques may be helpful to consider. These may include:

Table 3
Electrodiagnostic Evaluation of Ulnar Neuropathy at the Elbow[a]

Nerve conduction studies

1. Ulnar nerve studies
   A. Ulnar motor study, stimulating at the wrist, below elbow, and above elbow sites; recording from the abductor digiti minimi muscle
   B. Ulnar F-responses
   C. Ulnar sensory study stimulating at the wrist; recording from the fifth digit
2. Median nerve studies
   A. Median motor study stimulating at the wrist and below elbow sites; recording from the abductor pollicis brevis muscle
   B. Median F-responses
   C. Median sensory study stimulating at the wrist; recording from the first digit

EMG

1. Routine
   A. At least one ulnar-innervated muscle distal to the wrist (e.g., first dorsal interosseous and abductor digiti minimi)
   B. Two ulnar-innervated muscles of the forearm (e.g., FDP and FCU)
2. If testing of any of the routine muscles is abnormal, then additional needle examination should include:
   A. At least two non-ulnar, lower trunk, C8–T1 muscles (e.g., APB, FPL, and EIP)
   B. C8 and T1 paraspinal muscles

[a]FDP, flexor digitorum profounder; FCU, flexor carpi ulnaris; APB, abductor pollicis brevis; FPL, flexor pollicis longus; EIP, extensor indicis proprius; ECU, extensor carpi ulnaris; EDC, extensor digitorum communis; ECR extensor carpi radialis; FCR, flexor carpi radialis; PT, pronator teres

From ref. *1*.

1. Repeating the ulnar motor study; recording from the first dorsal interosseous muscle.
2. Ulnar motor study using inching across the elbow segment.
3. Sensory or mixed-nerve studies across the elbow.
4. Comparing the dorsal ulnar cutaneous sensory responses between the affected and asymptomatic contralateral limb.
5. Comparing the medial antebrachial cutaneous sensory response between affected and asymptomatic sides if there is reason to suspect a brachial plexopathy.

In most cases, the lesion is at the elbow; however, lesions at the wrist or more proximal locations (brachial plexus or root) should be excluded by the electrodiagnostic study.

3.2.1.2. Treatment

The treatment of patients with ulnar neuropathy at the elbow may consist of conservative or surgical measures. Nonoperative management should include avoidance of pressure on the elbow and/or prolonged elbow flexion, and use of an elbow splint. In certain cases, steroid injections into the cubital tunnel may be helpful. In patients who have significant or progressive neurological deficits, surgical decompression is recommended. There are a number of procedures used; however, which procedure is most effective remains controversial.

3.2.2. Ulnar Neuropathy at the Wrist

Entrapment of the ulnar nerve at the wrist is rare compared with compression at the elbow. The common site of entrapment occurs within the Canal of Guyon. Five different syndromes

Table 4
Clinical Syndromes Produced by Ulnar Nerve Compression Within the Canal of Guyon[a]

1. Combined motor and sensory syndrome (Type 1)
   A lesion at the proximal portion of the canal may involve both motor and sensory divisions. Weakness of all ulnar innervated hand muscles and loss of sensation over the palmar fifth and medial fourth fingers occurs. Cutaneous sensation over the hypothenar and dorsomedial surfaces of the hand should be spared.
2. Pure sensory syndrome (Type 2)
   Clinically, there is loss of sensation over the palmar surface of the fifth and medial fourth fingers. Sensation is spared over the hypothenar eminence. Motor fibers are not affected. There is no weakness associated with this lesion.
3. Pure motor syndromes
   A. Lesion affecting the deep palmar and hypothenar motor branches (Type 3)
   This lesion affects the motor trunk proximal to the takeoff of the hypothenar branches. As a result, all ulnar innervated muscles of the hand are involved. Because the sensory branch is not affected, sensation is spared.
   B. Lesion affecting the deep palmar motor branch only (Type 4)
   Clinically, there is weakness of lumbricals 1 and 2, as well as ulnar-innervated muscles of the thenar eminence. This type of lesion spares the muscles of the hypothenar eminence.
   C. Lesion affecting only the distal deep palmar motor branch (Type 5)
   This type of lesion occurs just proximal to the branches innervating the adductor pollicis and first dorsal interosseous muscles, resulting in weakness of these muscles.

[a]From refs. *1* and *16*.

have been described with entrapment in this region (Table 4). Patients may present with sensory and/or motor involvement confined to the distal ulnar nerve distribution. They may have sensory loss, paresthesias, or pain in the region supplied by the distal ulnar sensory branch. The region supplied by the dorsal ulnar cutaneous sensory branch may be spared. Motor deficits are limited to the muscles of the hand with sparing of the proximal ulnar-innervated muscles. Examination may demonstrate weakness with atrophy or fasciculations in the intrinsic hand muscles. A Tinel's sign may be present over Guyon's canal.

The electrophysiology in this entrapment is often complex. Table 5 outlines a protocol when testing for an ulnar nerve lesion at the wrist. Table 6 summarizes typical electrodiagnostic findings in each of the various syndromes. MRI may be useful in detecting structural abnormalities that may compress the ulnar nerve in Guyon's canal. A variety of different causes have been described. A ganglion cyst or traumatic wrist injury account for the majority of cases. In cases in which a structural lesion is identified, surgical removal is recommended. In certain cases, surgical exploration may be considered even if MRI fails to identify a lesion.

4. RADIAL NEUROPATHY

4.1. Anatomy

The radial nerve receives fibers from all three trunks of the brachial plexus (C5–T1 roots). The posterior divisions of the three trunks unite to form the posterior cord, which gives off the radial nerve. The radial nerve exits the lateral wall of the axilla, and travels distally through the proximal arm, just medial to the humerus. Proximally, three sensory nerves: the *posterior cutaneous nerve of the arm*, the *lower lateral cutaneous nerve of the arm*, and the *posterior cutaneous nerve of the forearm*, join the radial nerve, providing cutaneous sensation over the

Table 5
Electrodiagnostic Evaluation of Ulnar Neuropathy at the Wrist[a]

Nerve conduction studies

1. Ulnar nerve studies
   A. Ulnar motor study stimulating at the wrist, below elbow, and above elbow sites; recording from the abductor digiti minimi muscle
   B. Ulnar motor study (bilateral) stimulating at the wrist; recording from the first dorsal interosseous muscle
   C. Ulnar F-responses
   D. Ulnar sensory study stimulating at the wrist; recording from digit 5
   E. Dorsal ulnar cutaneous sensory study stimulating forearm; recording from the dorsolateral hand
2. Median nerve studies
   A. Median motor study stimulating at the wrist and elbow sites; recording from the abductor pollicis brevis muscle
   B. Median F-responses
3. Ulnar–median comparison studies
   Lumbrical (2nd)–interosseous (first palmar) comparison study

EMG

1. Routine
   A. One deep palmar motor muscle (e.g., first dorsal interosseous)
   B. One hypothenar branch muscle (e.g., abductor digiti minimi)
   C. Two forearm muscles (e.g., FCU and FDP)
2. If testing of any of the routine muscles is abnormal, then additional needle examination should include:
   A. At least two nonulnar, lower trunk, C8–T1 muscles (e.g., APB, FPL, and EIP)
   B. C8 and T1 paraspinal muscles

[a]FDP, flexor digitorum profounder; FCU, flexor carpi ulnaris; APB, abductor pollicis brevis; FPL, flexor pollicis longus; EIP, extensor indicis proprius; ECU, extensor carpi ulnaris; EDC, extensor digitorum communis; ECR extensor carpi radialis; FCR, flexor carpi radialis; PT, pronator teres

From ref. *1*.

Table 6
Nerve Conduction Study Findings in Ulnar Neuropathy at the Wrist[a]

1. Combined motor and sensory syndrome (Type 1)
   Decreased ulnar sensory amplitude. Decreased ulnar motor amplitude with prolonged distal latency. EMG shows denervation of all intrinsic hand muscles.
2. Pure sensory syndrome (Type 2)
   Decreased ulnar sensory amplitude. Ulnar motor study will be normal. EMG is normal.
3. Pure motor syndromes
   A. Lesion affecting the deep palmar and hypothenar motor branches (Type 3)
      Ulnar sensory response is normal. Ulnar motor amplitude is decreased with prolonged distal latency. EMG shows denervation of all intrinsic hand muscles.
   B. Lesion affecting the deep palmar motor branch only (Type 4)
      Ulnar sensory response is normal. Ulnar motor amplitude is decreased with prolonged distal latency when recording from the first dorsal interosseous muscle. EMG shows denervation of the first dorsal interosseous muscle with sparing of the hypothenar muscles.
   C. Lesion affecting only the distal deep palmar motor branch (Type 5)
      Ulnar sensory response is normal. Ulnar motor amplitude is decreased with prolonged distal latency when recording from the first dorsal interosseous muscle. EMG shows denervation of the first dorsal interosseous muscle with sparing of the hypothenar muscles.

[a]From ref. *17*.

posterolateral portions of the arm and a small strip along the middle posterior aspect of the forearm. Muscular branches are provided next to the long, lateral, and medial triceps muscles, as well as the anconeus muscle. Moving distally, the radial nerve wraps around the humerus, traveling in the spiral groove, before giving off additional branches to the supinator, long head of the extensor carpi radialis, and brachioradialis muscles. A few centimeters distal to the lateral epicondyle, the radial nerve divides into the *superficial radial sensory nerve* and the *posterior interosseous nerve*. The superficial radial sensory nerve travels distally along the radius providing cutaneous sensation over the dorsolateral hand and proximal portions of the dorsal aspect of the thumb, index, middle, and ring fingers. The posterior interosseous nerve travels through the supinator muscle passing under the arcade of Frohse. The posterior interosseous nerve supplies muscular branches to the short head extensor carpi radialis, extensor digitorum communis, extensor carpi ulnaris, abductor pollicis longus, extensor indicis proprius, extensor pollicis longus, and extensor pollicis brevis muscles.

4.1.1. Radial Neuropathy at the Axilla

This entrapment results from prolonged compression of the nerve as it courses through the axilla. A common presentation is the patient on crutches, who uses them incorrectly, thereby, applying prolonged pressure to the axilla. Because the lesion occurs proximal to muscular branches supplying the triceps muscle group, the clinical presentation is similar to radial neuropathy at the spiral groove, with the addition of triceps muscle weakness. Additionally, sensory disturbance extending into the posterior arm and forearm caused by compression of the posterior cutaneous sensory nerves of the forearm and arm is commonly seen.

4.1.2. Radial Neuropathy at the Spiral Groove

This is the most common site of compression of the radial nerve. This commonly occurs when a person has draped an arm over a chair or bench during deep sleep or intoxication ("Saturday Night Palsy"). Other cases may occur after strenuous muscular effort or fracture of the humerus. Patients with this particular entrapment typically present with wrist and finger drop in combination with decreased sensation over the posterolateral hand in the distribution of the superficial radial sensory nerve. Patients typically have weakness of supination and elbow flexion. However, elbow extension (triceps muscle) will be spared.

4.1.3. Posterior Interosseous Neuropathy

In this condition, patients also present with wrist drop. However, there are several distinct features of this particular entrapment that distinguish it from lesions at the spiral groove. In a posterior interosseous neuropathy (PIN), there is sparing of radial-innervated muscles proximal to the takeoff of the posterior interosseous nerve (triceps, anconeus, brachioradialis, and long head of the extensor carpi radialis muscles). Entrapment usually occurs at the Arcade of Frohse. When the patient extends the wrist, they may do so weakly, and with radial deviation; this occurs because the extensor carpi ulnaris is weak but the extensor carpi radialis is preserved. These patients typically do not experience cutaneous sensory deficits. Patients, however, may complain of forearm pain, which results from dysfunction of the deep sensory fibers of the posterior interosseous nerve that supplies the interosseous membrane and joint capsule.

4.1.4. Superficial Radial Sensory Neuropathy

In the forearm, the superficial radial sensory nerve travels subcutaneously next to the radius. Its superficial location makes it quite susceptible to compression. Sensory disturbances occur over the dorsolateral surface of the hand, and dorsal, proximal, portions of the

Table 7
Electrodiagnostic Evaluation for Evaluating Radial Neuropathy[a]

Nerve conduction studies

1. Radial nerve studies
   A. Radial motor study stimulating at the forearm, elbow, below spiral groove, and above spiral groove sites; recording from extensor indicis proprius muscle
   B. Superficial radial sensory study. Stimulating at the forearm; recording over the extensor tendons of the thumb. Bilateral studies are recommended
2. Median nerve studies
   A. Median motor study stimulating at the wrist and below elbow sites; recording from the abductor pollicis brevis muscle
   B. Median sensory study stimulating at the wrist; recording from the first digit
   C. Median F-responses
3. Ulnar nerve studies
   A. Ulnar motor study, stimulating at the wrist, below elbow, and above elbow sites; recording from the abductor digiti minimi muscle
   B. Ulnar F-responses
   C. Ulnar sensory study stimulating at the wrist; recording from the fifth digit

EMG

1. At least two PIN-innervated muscles (e.g., EIP, ECU, and EDC)
2. At least two radial-innervated muscles proximal to the PIN, but distal to the spiral groove (e.g., brachioradialis and long-head ECR)
3. At least one radial-innervated muscle proximal to the spiral groove (e.g., triceps)
4. At least one nonradial, posterior cord-innervated muscle (e.g., deltoid)
5. At least two nonradial C7-innervated muscles (e.g., FCR, PT, FPL, and cervical paraspinal muscles)

[a]PIN, posterior interosseous neuropathy; EIP, extensor indicis proprius; ECU, extensor carpi ulnaris; EDC, extensor digitorum communis; ECR, extensor carpi radialis; FCR, flexor carpi radialis; PT, pronator teres; FPL, flexor pollicis longus.

From ref. *1*.

fingers. Various objects, such as tight fitting bands, watches, bracelets, or handcuffs may lead to a superficial radial neuropathy. Because this is a pure sensory neuropathy, these patients do not develop weakness.

4.1.4.1. Differential Diagnosis of Wrist Drop

The differential diagnosis of a wrist drop should include the various radial nerve lesions that have been discussed. In addition, lesions that are more proximal, such as a posterior cord brachial plexopathy, a C7–C8 radiculopathy, or even a central lesion, should be considered. A careful clinical examination can be invaluable in localizing the lesion causing wrist drop.

4.1.4.2. Electrophysiology

The electrodiagnostic study should identify the presence of a radial neuropathy and properly localize the level of dysfunction. The radial motor study should be performed and compared with the contralateral side. A protocol outlining electrodiagnostic recommendations for evaluating radial neuropathy is outlined in Table 7.

Needle examination should focus on localizing the lesion level. Typically, one should examine at least: two PIN-innervated muscles (e.g., extensor indicis proprius, extensor carpi ulnaris, and extensor digitorum communis muscles); two radial-innervated muscles that are proximal to the PIN but distal to the spiral groove (e.g., long head of extensor carpi radialis

Table 8
Electrodiagnostic Findings in Radial Neuropathy[a]

1. Posterior interosseous neuropathy
   A. Nerve conduction studies—superficial radial sensory response is normal. Radial motor study may show low-amplitude response (if axonal) or conduction block at the elbow (if demyelinating)
   B. EMG—denervation in the extensor indicis proprius, extensor digitorum communis, and extensor carpi ulnaris muscles
2. Radial neuropathy at the spiral groove
   A. Nerve conduction studies—superficial radial sensory response is low (if axonal). Radial motor study may show low-amplitude response (if axonal) or conduction block at the spiral groove (if demyelinating)
   B. EMG—denervation as in PIN plus long head extensor carpi radialis, brachioradialis, and supinator muscles
3. Radial neuropathy at the axilla
   A. Nerve conduction studies—superficial radial sensory response is low (if axonal). Radial motor study may show low-amplitude response (if axonal)
   B. EMG—denervation as in spiral groove, plus triceps muscle
4. Posterior cord brachial plexopathy
   A. Nerve conduction studies—superficial radial sensory response is low (if axonal). Radial motor study may show low-amplitude response (if axonal)
   B. EMG—denervation as in axilla, plus deltoid, and latissimus dorsi muscles
5. C7 Radiculopathy
   A. Nerve conduction studies—radial motor study may show low-amplitude response (if axonal)
   B. EMG—denervation as in axilla, plus flexor carpi radialis and cervical paraspinal muscles, but sparing the brachioradialis and supinator muscles

[a]PIN, posterior interosseous neuropathy; From ref. *1*.

and brachioradialis); two nonradial nerve, C7-innervated muscles (e.g., pronator teres, flexor pollicis longus, flexor carpi radialis, and cervical paraspinal muscles); one radial-innervated muscle proximal to the spiral groove (e.g., triceps muscle); and one nonradial, posterior cord-innervated muscle (e.g., deltoid). Table 8 summarizes the different electrophysiological abnormalities that are encountered in the various radial nerve lesions discussed.

4.2. Other Focal Mononeuropathies of the Upper Extremity

4.2.1. Suprascapular Neuropathy

This entrapment most commonly occurs at the suprascapular notch, under the transverse scapular ligament. Less frequently, the nerve can be entrapped distally at the spinoglenoid notch. The most common symptom is shoulder pain. Shoulder pain is typically deep and boring, and occurs along the superior aspect of the scapula, with radiation into the shoulder. The pain may be exacerbated by adduction of the extended arm.

Patients may demonstrate weakness of shoulder abduction (supraspinatus) and external rotation (infraspinatus). Atrophy may be present in cases that are more severe. If the nerve is entrapped distally at the spinoglenoid notch, the deficit is limited to the infraspinatus muscle only. In this case, pain is usually absent because the deep sensory fibers to the shoulder joint arise proximal to the lesion.

Various conditions that should be excluded include C5–C6 radiculopathy, upper trunk brachial plexopathy, rotator cuff injury, and other orthopedic conditions of the shoulder.

In rare cases, a focal suprascapular neuropathy may be the primary manifestation of a more diffuse autoimmune-mediated brachial plexopathy (neuralgic amyotrophy).

Electrodiagnostic studies should identify the involvement of suprascapular-innervated muscles, and exclude other causes. Routine nerve conduction studies of the upper extremity should be normal. Motor studies of the suprascapular nerve can be performed, but require careful technique, and, therefore, are not commonly performed. Compound muscle action potentials may be recorded using a monopolar needle electrode placed in the spinatii muscles while stimulating over Erb's point. Needle examination is very useful, and will help identify abnormalities limited to the spinatii muscles, thus, supporting the diagnosis of suprascapular neuropathy. An MRI of the shoulder is usually recommended to exclude a ganglion cyst causing compression of the nerve.

4.2.2. *Axillary Neuropathy*

This neuropathy typically occurs in the setting of trauma (dislocation of the shoulder or fracture of the humerus). Clinically, patients have a well-demarcated "patch" of numbness along the lateral aspect of the shoulder, and will have weakness of shoulder abduction (deltoid) and external rotation (teres minor). Differential diagnosis is similar to that of a suprascapular neuropathy. Routine nerve conduction studies are, again, normal. Motor study of the axillary nerve can be performed but, again, requires careful technique and experience. The needle examination usually demonstrates abnormalities in the deltoid and teres minor muscles only.

4.2.3. *Long Thoracic Neuropathy*

The long thoracic nerve arises directly from the C5, C6, and C7 nerve roots proximal to the brachial plexus. The nerve supplies the serratus anterior muscle exclusively. Most commonly, this neuropathy is seen in patients with neuralgic amyotrophy. Dysfunction of the long thoracic nerve can result from traumatic injuries that cause widespread damage to multiple cervical nerve roots. Isolated cases of long thoracic neuropathy, although rare, are observed. Patients will exhibit scapular winging with their arms outstretched. Electrodiagnostic studies should be performed to determine if a more widespread process exists.

4.2.4. *Musculocutaneous Neuropathy*

An isolated musculocutaneous neuropathy is rare. Most commonly, it occurs as part of a widespread traumatic lesion involving the shoulder and arm. Patients will have weakness of elbow flexion, sensory loss over the lateral forearm (lateral antebrachial cutaneous nerve), and an absent biceps reflex. The electrodiagnostic study should localize the lesion to this nerve, and exclude a brachial plexopathy or cervical radiculopathy. The lateral antebrachial cutaneous sensory study should be abnormal as compared with the contralateral side. Motor study of the musculocutaneous nerve can be performed, but careful technical skill is required to ensure accuracy of the result. Needle examination should demonstrate abnormalities in the biceps, brachialis, and coracobrachialis muscles.

4.2.5. *Spinal Accessory Neuropathy*

Isolated lesions of this nerve occur in the region of the posterior cervical triangle and result in isolated weakness of the trapezius muscle. Stretch injuries may occur, but usually this neuropathy results after local surgical procedures. Patients may have shoulder drop caused by weakness of the trapezius muscle. If the lesion is more proximal, there may be weakness of the sternocleidomastoid muscle as well. The needle examination provides the greatest usefulness with this type of entrapment syndrome. Routine nerve conduction studies of the upper extremity should be normal. A careful needle examination should exclude brachial plexus dysfunction.

5. PERONEAL NEUROPATHY

Peroneal neuropathy is the most common focal neuropathy affecting the lower extremity. It is a common cause of foot drop, and, most often, is compressed at the level of the fibular head.

5.1. Anatomy

The peroneal nerve fibers are derived from the L4–S1 nerve roots. These fibers travel through the lumbosacral plexus, and continue on through the sciatic nerve. Within the sciatic nerve, fibers of the *common peroneal nerve* run separately from *tibial nerve* fibers. The *common peroneal nerve* first gives off the *lateral cutaneous nerve of the calf*. Near the fibular neck, the common peroneal nerve divides into its terminal branches, the *superficial* and *deep peroneal nerves*.

The *superficial peroneal nerve* courses distally along the fibula. It provides muscular branches to the peroneus longus and peroneus brevis muscles, controlling ankle eversion. The nerve then divides into the *medial* and *lateral cutaneous branches*, which provide sensation over the distal anterolateral calf, dorsum of the foot, and dorsum of the first three to four toes, sparing a small wedge-shaped area over the webspace between the medial two toes.

The *deep peroneal nerve* travels distally along the anteromedial aspect of the tibia, and mediates dorsiflexion of the ankle and toes. Proximal to the ankle, it provides muscular branches to the tibialis anterior, extensor digitorum longus, extensor hallucis longus, and peroneus tertius muscles. After crossing the ankle, muscular branches are supplied to the extensor digitorum brevis and first dorsal interosseous pedis. A small cutaneous sensory branch, the *dorsal digital cutaneous nerve*, arises distally and provides sensation over the webspace between the first two toes.

5.2. Clinical Features of Peroneal Nerve Dysfunction

5.2.1. Peroneal Neuropathy at the Fibular Head

Patients with a peroneal neuropathy at the fibular head have a characteristic clinical presentation. Many cases occur postoperatively and are thought to occur because of patient positioning that resulted in prolonged compression of the nerve. Other factors that may predispose patients to peroneal neuropathies at the fibular head include trauma resulting in fracture of the fibula or knee dislocation, weight loss, frequent leg crossing, prolonged hospitalization in an ICU setting, diabetes mellitus, preexisting polyneuropathy, and compression caused by braces, casts, or splints.

Involvement of the superficial peroneal nerve results in weakness of foot eversion, as well as sensory disturbance in the mid- and lower-lateral calf and over the dorsum of the foot. Involvement of the deep peroneal fibers results in weakness of toe and ankle dorsiflexion (foot drop). Ankle inversion and toe flexion are spared, because the tibial nerve mediates these functions. Foot drop may be partial or complete, and, in some instances, may develop during a period of days or weeks. In most cases, the lesion is unilateral, but, in 10% of patients, bilateral lesions may be present. In exclusively deep peroneal neuropathies, which are much less frequent than common peroneal neuropathies, sensory disturbances are lacking (except for sensory loss in the first web space), and ankle eversion is normal. The clinical examination should help localize the lesion and identify whether a more proximal lesion is present.

Table 9
Recommended Electrodiagnostic Evaluation of Peroneal Neuropathy[a]

Nerve conduction studies

1. Peroneal nerve
   A. Peroneal motor study stimulating at the ankle, below fibular neck, and popliteal fossa; recording from the extensor digitorum brevis muscle. If there is evidence of focal slowing/conduction block, this study should be repeated stimulating at the fibular neck and popliteal fossa; recording from the tibialis anterior muscle
   B. Peroneal F-responses
   C. Superficial peroneal sensory study stimulating over the lateral calf; recording from the ankle
2. Tibial nerve
   A. Tibial motor study stimulating at the ankle and popliteal fossa; recording from the abductor hallucis muscle
   B. Tibial F-responses
3. Sural nerve sensory study

EMG

1. Two deep peroneal-innervated muscles (e.g., tibialis anterior and extensor hallucis longus)
2. At least one superficial peroneal-innervated muscle (e.g., peroneus longus and peroneus brevis)
3. Two tibial-innervated muscles (e.g., tibialis posterior, medial gastrocnemius, and flexor digitorum longus)
4. Short head of the biceps femoris

[a]From ref. *1*.

5.2.1.1. ELECTROPHYSIOLOGY

Electrodiagnostic studies are extremely useful in the evaluation of patients who present with foot drop. The differential diagnosis of foot drop includes peroneal neuropathy at the fibular head, deep peroneal neuropathy, sciatic neuropathy primarily or exclusively involving peroneal fibers, lumbar plexopathy, L5 radiculopathy, motor neuron disease, or a cortical lesion. Electrodiagnostic study can establish the diagnosis of peroneal mononeuropathy, or suggest one of these more proximal causes. In addition, the study may not only establish the site (fibular head, thigh, or deep peroneal nerve) of the lesion, but assist in making prognostic judgements.

Peroneal motor studies recording over the extensor digitorum brevis and tibialis anterior muscles should be performed. A protocol for this neuropathy is outlined in Table 9. Needle examination of the short head of the biceps is crucial in identifying lesions at the fibular neck (should be normal) or more proximal causes (if abnormal).

6. FEMORAL NEUROPATHY

6.1. Anatomy

The femoral nerve is derived from the lumbar plexus, originating from the posterior divisions of the L2, L3, and L4 nerve roots. The nerve provides motor branches to the psoas and iliacus muscles before traveling underneath the inguinal ligament. Subsequently, the nerve divides into motor and sensory branches. Motor branches supply the sartorius, pectineus, and the four heads of the quadriceps muscles. Sensory branches supply sensation to the medial thigh (*medial cutaneous nerve of the thigh*), anterior thigh (*intermediate cutaneous nerve of the thigh*), and medial calf (*saphenous nerve*).

Table 10
Electrodiagnostic Evaluation of a Femoral Neuropathy

Nerve conduction studies

1. Femoral nerve
   A. Femoral motor study stimulating at the below inguinal ligament site; recording from the rectus femoris. Comparison with the asymptomatic side is necessary.
   B. Saphenous sensory study stimulating the medial calf; recording from the medial ankle. Comparison with the asymptomatic side is necessary.
2. Peroneal nerve
   A. Peroneal motor study stimulating at the ankle, below fibular neck, and popliteal fossa; recording from the extensor digitorum brevis muscle
   B. Peroneal F-responses
3. Tibial nerve
   A. Tibial motor study stimulating at the ankle and popliteal fossa; recording from the abductor hallucis muscle
   B. Tibial F-responses
4. Sural nerve sensory study

EMG

1. Iliopsoas and at least two quadriceps muscles (e.g., vastus lateralis and vastus medialis)
2. At least one obturator-innervated muscle (e.g., adductor longus)
3. L2–L4 paraspinal muscles
4. At least two nonfemoral muscles to exclude a polyneuropathy (e.g., tibialis anterior, medial gastrocnemius, and biceps femoris)

6.2. Clinical Features of Femoral Nerve Dysfunction

Patients with a femoral neuropathy usually complain of leg weakness when walking (because of buckling of the knee) and frequent falls. Numbness and paresthesias may be reported in the anterior thigh or medial calf. Examination may demonstrate weakness of the quadriceps muscles and an absent knee jerk reflex. Atrophy may or may not be present. Hip adduction should be normal, because this is mediated by the obturator nerve. If there is weakness of hip flexion, the lesion localizes proximal to the inguinal ligament. Sensory examination should demonstrate abnormalities in the anteromedial thigh and medial calf.

6.3. Electrophysiology

Electrodiagnostic studies are very helpful in identifying a femoral neuropathy and also in determining prognosis. Nerve conduction studies of the femoral nerve can be performed and should be compared with the contralateral side. Table 10 outlines a protocol for evaluating a suspected femoral neuropathy. The needle examination is often more useful than the femoral motor study.

6.4. Other Diagnostic Studies

Imaging of the pelvis with either CT or MRI should be performed to exclude a retroperitoneal hematoma. Often, lumbar spine imaging studies are also performed to exclude a possible root lesion. Laboratory studies should exclude diabetes and vasculitic disorders.

7. SCIATIC NEUROPATHY

7.1. Anatomy

The sciatic nerve derives its supply from the L4–S2 nerve roots. The nerve arises from the lumbosacral plexus, and exits the pelvis through the greater sciatic foramen before traveling under the piriformis muscle. The nerve itself consists of two distinct trunks, the lateral trunk (peroneal) and medial trunk (tibial). Branches originating in the proximal thigh arise predominantly from the tibial division. The tibial division supplies the hamstring muscles, with the exception of the short head of the biceps femoris, which receives its supply from the peroneal division. All muscles below the knee receive innervation from the sciatic nerve through one of its two divisions (peroneal or tibial). Afferent sensory input from the leg also travels in the sciatic nerve, except for that region supplied by the saphenous nerve.

7.2. Clinical Features of Sciatic Nerve Dysfunction

Weakness, numbness, and paresthesias in the lower extremity are common symptoms of sciatic neuropathy. A flail foot is commonly seen with this condition. Weakness of knee flexion and impairment of all ankle and toe movements will be present. Sensory loss may be appreciated in various areas over the posterior calf, ankle, and sole of the foot. Sciatic neuropathy may arise from a myriad of causes, including compression during surgery or coma, tumors (compression or direct nerve invasion), vascular abnormalities, fibromuscular bands, mononeuritis multiplex, piriformis compression, and traumatic injuries arising from gunshot or knife wounds or injections.

7.3. Electrophysiology

The electrodiagnostic evaluation should confirm a sciatic neuropathy and exclude other conditions that mimic this lesion. The nerve conduction studies/needle examination that should be performed are summarized in Table 11.

7.4. Other Diagnostic Studies

Imaging studies, especially MRI, are often very helpful in identifying a structural lesion, if present. Angiography may be indicated if an arterial aneurysm or iliac artery thrombosis is a possible cause.

8. TIBIAL NEUROPATHY

8.1. Anatomy

The tibial nerve descends to the level of the medial malleolus before coursing under the flexor retinaculum on the medial side of the ankle. The distal tibial nerve then divides into three or four terminal branches. The *medial and lateral calcaneal sensory nerves* are purely sensory, and supply sensation to the heel of the foot. The *medial and lateral plantar nerves* contain both motor and sensory fibers that supply the medial and lateral sole, respectively.

8.2. Clinical Features of Tibial Nerve Dysfunction at the Ankle (Tarsal Tunnel Syndrome)

The most frequent symptom in patients with tarsal tunnel syndrome (TTS) is perimalleolar pain. Ankle and sole pain is often described as burning, and is often worse with

Table 11
Electrodiagnostic Evaluation for Sciatic Neuropathy[a]

Nerve conduction studies

1. Peroneal nerve
   A. Peroneal motor study stimulating at the ankle, below fibular neck, and popliteal fossa; recording from the extensor digitorum brevis muscle. If there is evidence of focal slowing/conduction block, this study should be repeated stimulating at the fibular neck and popliteal fossa; recording from the tibialis anterior muscle
   B. Peroneal F-responses
   C. Superficial peroneal sensory study stimulating the lateral calf; recording from the ankle
2. Tibial nerve
   A. Tibial motor study stimulating at the ankle and popliteal fossa; recording from the abductor hallucis muscle
   B. Tibial F-responses
3. Sural nerve sensory study
4. H-reflex (bilateral studies are necessary)

EMG

1. Two peroneal-innervated muscles (e.g., tibialis anterior and extensor hallucis longus)
2. At least one superficial peroneal-innervated muscle (e.g., peroneus longus and peroneus brevis)
3. Two tibial-innervated muscles (e.g., tibialis posterior and medial gastrocnemius)
4. Short and long heads of the biceps femoris
5. At least one superior and one inferior gluteal-innervated muscle
6. L5–S1 paraspinal muscles
7. At least two nonsciatic, non-L5–S1 innervated muscles to exclude more widespread processes (e.g., vastus lateralis, iliopsoas, and adductor longus)

[a]From ref. *1*.

weight bearing or at night. Patients usually do not describe weakness. Some patients may exhibit a positive Tinel's sign over the ankle. Differential diagnosis of TTS should include local trauma, proximal tibial neuropathy, mild polyneuropathy, or S1 radiculopathy.

8.3. Electrophysiology

When performing electrodiagnostic studies in these cases, it is helpful to perform a side-to-side comparison, particularly in the patient with unilateral complaints. Those patients with bilateral complaints may be more difficult to assess. A recommended electrodiagnostic protocol is provided in Table 12. The needle examination can be problematic because the examination may be uncomfortable to the patient. Needle examination is further complicate by the presence of nonspecific abnormalities that are often present in the intrinsic foot muscles of normal individuals.

9. CONCLUSION

We reviewed the most common entrapment neuropathies that practicing neurologists are likely to see in clinical practice. A good history, physical examination, and neurological examination are essential to guiding one's diagnostic evaluation, and ultimately, in assessing each patient's prognosis and treatment options.

Table 12
Electrodiagnostic Protocol to Evaluate for Tarsal Tunnel Syndrome

Nerve conduction studies
1. Peroneal nerve A. Peroneal motor study stimulating at the ankle, below fibular neck, and popliteal fossa; recording from the extensor digitorum brevis muscle B. Peroneal F-responses 2. Tibial nerve A. Tibial motor study stimulating at the ankle and popliteal fossa; recording from the abductor hallucis muscle. B. Tibial F-responses C. Medial and lateral plantar motor and sensory studies 3. Sural nerve sensory study 4. H-reflex (bilateral studies are necessary)
EMG
1. One peroneal-innervated muscle (e.g., tibialis anterior and extensor hallucis longus) 2. At least two tibial-innervated muscles distal to the tarsal tunnel (e.g., abductor hallucis and abductor digiti minimi) 3. At least two tibial-innervated muscles proximal to the tarsal tunnel (e.g., tibialis posterior and medial gastrocnemius) 4. If any muscle proximal to the tarsal tunnel is abnormal, additional muscles may need to be sampled to exclude more proximal causes (e.g., sciatic neuropathy, polyneuropathy, or lumbosacral plexopathy)

REVIEW QUESTIONS

1. Which of the following would not be an expected clinical finding in a patient with median mononeuropathy at the wrist?
   A. Paresthesias involving the first three fingers (and possible the lateral fourth) of the hand.
   B. Numbness involving the first three fingers of the hand and thenar eminence.
   C. Weakness of thumb abduction.
   D. Worsening of symptoms at night.
   E. Reproduction of pain and paresthesias with forced flexion of the wrist.
2. Which of the following is not a potential site of entrapment of the median nerve?
   A. Wrist.
   B. Ligament of Struthers.
   C. Guyon's canal.
   D. Sublimis bridge.
   E. Lacertus fibrosus.
3. Common clinical findings suggesting ulnar nerve entrapment at the elbow include all of the following EXCEPT:
   A. Numbness involving the fourth and fifth fingers of the hand.
   B. Weakness of the hand.
   C. Pain at the elbow that radiates along the medial aspect of the forearm.
   D. Loss of deep tendon reflexes in the arm.
   E. Pain with percussion of the area over the cubital tunnel.
4. One easy clinical finding that can help to exclude ulnar nerve compression at the wrist as compared to a more proximal lesion (elbow, brachial plexus, or nerve roots) would be:

A. Sensory deficits involving the palmar aspect of the fourth and fifth fingers.
B. Atrophy of the hypothenar and intrinsic hand muscles.
C. Sensory deficit involving the dorsomedial aspect of the hand.
D. Weakness of wrist flexion.
E. C and D are correct.

5. Which of the following would not be an expected finding with a radial neuropathy at the spiral groove?
A. Wrist drop.
B. Weakness of finger extension.
C. Weakness of elbow flexion.
D. Numbness over the dorsolateral hand.
E. Weakness of elbow extension.
6. A lesion of the posterior interosseous nerve would cause all of the following EXCEPT:
A. Weakness of wrist extension (possibly with radial deviation).
B. Numbness of the dorsolateral hand.
C. Forearm pain.
D. Weakness of finger extension.
E. Sparing of elbow flexion and extension.
7. Compression of the peroneal nerve at the fibular head is the most common entrapment neuropathy of the lower extremity. Clinical findings that would suggest this diagnosis include all of the following EXCEPT:
A. Weakness of foot dorsiflexion.
B. Weakness of foot inversion.
C. Weakness of foot eversion.
D. Numbness over the lateral calf and dorsum of the foot.
E. Tinel's sign at the fibular head.
8. In localizing a peroneal lesion to the fibular head, EMG study of which muscle is essential?
A. Peroneus longus.
B. Tibialis anterior.
C. Short head of biceps femoris.
D. Extensor hallucis longus.
9. Compression of the femoral nerve at the inguinal ligament can result in which of the following:
A. Weakness of knee flexion.
B. Weakness of hip adduction.
C. Weakness of hip flexion.
D. Sensory disturbance over the anteromedial thigh and medial calf.
10. Which of the following is consistent with tarsal tunnel syndrome (TTS)?
A. Sensory loss over the dorsum of the foot.
B. A Tinel's sign in the region of the medial malleolus.
C. Weakness of foot plantar flexion.
D. Ankle jerk loss on the affected side.
E. All of the above.

REVIEW ANSWERS

1. The correct answer is B. Patients presenting with symptoms suggestive of a median mononeuropathy at the wrist often describe paresthesias involving the first three fingers and lateral fourth finger, with numbness that is often worse at night. In fact, patients may report being awakened from sleep by wrist/hand pain, requiring them to "shake-out" their hands or run them under water. Because the median nerve innervates the APB muscle, weakness of thumb opposition may be present, as well as atrophy of the thenar eminence, in more advanced cases. Asking the patient to hold the wrists in a flexed position for approx 1 min is known as Phalen's

maneuver, and although not 100% sensitive and specific, may be a useful test to confirm one's clinical suspicion of CTS. One would not expect numbness to extend to the thenar eminence in CTS, because the palmar cutaneous sensory branch, which arises proximal to the carpal tunnel and would be spared in an entrapment neuropathy at the wrist, innervates this area. Involvement of the thenar eminence or extending proximal to the wrist would suggest a more proximal median nerve injury.

2. The correct answer is C. The median nerve can potentially be entrapped at six different sites along its course through the arm. The ligament of Struthers refers to a fibrous band that may arise from the medial humerus and inserts on the medial epicondyle. The lacertus fibrosus arises from the biceps tendon and inserts on the forearm flexor muscles. The median nerve can also be compressed as it pierces the pronator teres muscle, and as it passes under the fibrous sublimis bridge of the flexor digitorum sublimis muscle. In addition, a branch of the median nerve, the AIN, can be compromised as it travels through the forearm. Isolated involvement of the AIN may occur in cases of brachial plexitis (Parsonage–Turner syndrome). Finally, the median nerve can be compressed at the wrist for a variety of reasons. Guyon's canal is a small anatomic tunnel through which the *ulnar* nerve passes through the wrist to innervate the hand and is not a site of median nerve entrapment.
3. The correct answer is D. Compromise of the ulnar nerve as it passes through the cubital tunnel is the second most common type of entrapment neuropathy after CTS. Patients may report pain that radiates into the ulnar forearm and hand. In some cases, only sensory symptoms may be present, manifesting as sensory loss primarily involving the fourth and fifth fingers. An early motor sign may be an inability to adduct the fifth finger (Wartenberg's sign) causing patients to report frequently "catching" of the finger when placing their hands in their pockets. Patients may develop atrophy and weakness of intrinsic hand muscle strength. Various provocative maneuvers, such as percussion (Tinel's sign) or sustained manual pressure over the cubital tunnel, can be helpful in confirming one's clinical suspicion. Because the ulnar nerve arises from the C8–T1 nerve roots and travels through the lower trunk and medial cord of the brachial plexus, none of the deep tendon reflexes commonly tested in the arm (biceps, triceps, or brachioradialis) should be affected.
4. The correct answer is E. Ulnar nerve entrapment at the wrist can present in a variety of ways. Commonly, some combination of sensory deficit involving the fourth and fifth fingers with weakness of the hypothenar and intrinsic hand muscles is present. Sensory loss over the dorsal aspect of the hand would not be a finding consistent with ulnar neuropathy at the wrist. The dorsomedial aspect of the hand is primarily innervated by the dorsal ulnar sensory cutaneous nerve, which is a sensory component of the ulnar nerve, but arises 6 to 8 cm *proximal* to the wrist and would not be involved in an entrapment neuropathy at the wrist. Similarly, weakness of the wrist flexors would indicate a more proximal lesion.
5. The correct answer is E. Radial neuropathy at the spiral groove commonly results from trauma to the proximal arm or from prolonged compression of the nerve that can occur when a patient falls asleep with the arm draped over the edge of a chair or bathtub. Weakness of wrist and finger extension, as well as elbow flexion, are all findings consistent with a lesion of the radial nerve at the spiral groove because these muscles are all innervated distal to the groove. A lesion at the spiral groove would also be expected to affect the function of the radial sensory nerve of the forearm, leading to numbness over the dorsal aspect of the lateral hand. Because the triceps are innervated proximal to the groove, elbow extension would be spared.
6. The correct answer is B. The posterior interosseous nerve is primarily a motor branch of the radial nerve that innervates most of the wrist and finger extensors. Because the extensor carpi radialis (long head) arises proximal to the PIN, the wrist may deviate radially with attempted extension. Elbow flexion and extension (brachioradialis and triceps) are both innervated proximally to the takeoff of the PIN and are, thus, spared. Forearm pain is common because the PIN provides some sensation to the deep interosseous membrane of the forearm; however, no cutaneous sensory branches arise from it. One should reconsider the diagnosis of PIN entrapment if cutaneous sensation is impaired.

7. The correct answer is B. Weakness of foot inversion would not be an expected finding because this action is primarily mediated by the tibialis posterior muscle, which is innervated by the tibial nerve. This muscle is critical in determining whether foot drop is solely a peroneal nerve problem or a more proximal sciatic neuropathy or L5 root lesion. The other signs are all consistent with a peroneal neuropathy localizing to the fibular head involving both the superficial and deep branches.
8. The correct answer is C. The short head of the biceps femoris is the only muscle that is innervated by the peroneal nerve proximal to the fibular head and is, thus, critical to localizing a peroneal lesion to the fibular neck. The short head of biceps femoris should be normal in a compressive lesion of the peroneal nerve at the fibular head. The peroneus longus, tibialis anterior, and extensor hallucis longus are all peroneal innervated muscles distal to the fibular head and would be expected to be abnormal on EMG examination.
9. The correct answer is D. Weakness of knee flexion is a sciatic nerve mediated movement. Likewise, weakness of hip adduction, although also suggesting involvement of the L2–L4 nerve roots, is mediated by the obturator nerve. Hip flexion is mediated by the iliopsoas muscle group; however, this muscle group arises proximal to the inguinal ligament and would be expected to be spared. The medial and intermediate cutaneous nerve of the thigh and the saphenous nerves, all of which arise from the femoral nerve and would be affected in compression at the level of the inguinal ligament, mediate sensory disturbance over the anterior and medial aspect of the distal thigh as well as the medial calf. Weight gain, preexisting obesity, pregnancy, and the wearing of tight work belts may predispose to this type of injury.
10. The correct answer is B. A Tinel's sign over the tibial nerve in the region of the medial malleolus is relatively common. In TTS, sensory loss is relatively confined to the plantar aspect of the foot, not the dorsum. There is no associated weakness of foot plantar flexion, and the ankle jerk should be preserved.

SUGGESTED READING

Bradshaw DY, Shefner JM. Ulnar neuropathy at the elbow. Neurol Clin 1999;17(3):447–461.

Busis NA. Femoral and obturator neuropathies. Neurol Clin 1999;17(3):633–653.

Campbell WW, Pridgeon RM, Sahni SK. Short segment incremental studies in the evaluation of ulnar neuropathy at the elbow. Muscle Nerve 1992;15:1050–1054.

Carlson N, Logigian EL. Radial neuropathy. Neurol Clin 1999;17(3):499–523.

Clarke AM, Stanley D. Prediction of the outcome 24 hours after carpal tunnel decompression. J Hand Surg (Br) 1993;18:180–181.

Fricker R, Fuhr P, Pippert H, et al. Acute median nerve compression at the distal forearm caused by a thrombosed aneurysm of an epineural vessel: case report. Neurosurgery 1996; 38(1):194–196.

Goslin KL, Krivickas LS. Proximal neuropathies of the upper extremity. Neurol Clin 1999; 17(3):525–548.

Gross PT, Tolomeo EA. Proximal median neuropathies. Neurol Clin 1999;17(3):425–445.

Katirji B. Peroneal Neuropathy. Neurol Clin 1999;17(3):567–591.

Katirji MB, Wilbourn AJ. Common peroneal mononeuropathy: a clinical and electrophysiologic study of 116 lesions. Neurology 1988;38:1723–1728.

Kothari MJ, Preston DC. Comparison of the flexed and extended elbow positions in localizing ulnar neuropathy at the elbow. Muscle Nerve 1995;18:336–340.

Kothari MJ, Heistand M, Rutkove SB. Three ulnar nerve conduction studies in patients with ulnar neuropathy at the elbow. Arch Phys Med Rehabil 1998;79:87–89.

Kothari MJ. Ulnar neuropathy at the wrist. Neurol Clin 1999;17(3):463–476.

Kuntzer T, van Melle G, Regli F. Clinical and prognostic features in unilateral femoral neuropathies. Muscle Nerve 1997;20:205.

Novak CB, Lee GW, MacKinnon SE, et al. Provocative testing for cubital tunnel syndrome. J Hand Surg 1994;19A:817–820.

Oh SJ, Meyer RD. Entrapment neuropathies of the tibial (posterior tibial) nerve. Neurol Clin 1999;17(3):593–615.

Preston DC. Distal median neuropathies. Neurol Clin 1999;17(3):407–424.

Preston DC, Logigian EL. Lumbrical and interossei recording in carpal tunnel syndrome. Muscle Nerve 1992;15:1253–1257.

Rennels GD, Ochoa J. Neuralgic amyotrophy manifesting as anterior interosseous nerve palsy. Muscle Nerve 1980;3:160–164.

Preston DC, Ross MH, Kothari MJ, et al. The median-ulnar latency difference studies are comparable in mild carpal tunnel syndrome. Muscle Nerve 1994;17:1469–1471.

Preston DC, Shapiro BE. Electromyography and Neuromuscular Disorders: Clinical-Electrophysiologic Correlations. Butterworth-Heinemann, Boston, MA, 1998.

Shea JD, McClain EJ. Ulnar-nerve compression syndrome at and below the wrist. J Bone Joint Surg 1969;51A:1095–1103.

Stewart JD. The variable clinical manifestations of ulnar neuropathies at the elbow. J Neurol Neurosurg Psychiatry 1987;50:252–258.

Wu JS, Morris JD, Hogan GR. Ulnar neuropathy at the wrist. Case report and review of the literature. Arch Phys Med Rehabil 1985;66:785–788.

Yuen EC, Olney RK, So YT. Sciatic neuropathy: clinical and prognostic features in 73 patients. Neurology 1994;44:1669–1674.

Yuen EC, So YT. Sciatic neuropathy. Neurol Clin 1999;17(3):617–631.

16
Electrophysiology of Polyneuropathy

Annabel K. Wang and Seward B. Rutkove

Summary

As a class, polyneuropathies are some of the most commonly presented disorders to a neuromuscular clinician. Polyneuropathies generally are of one of two forms, axonal or demyelinating. This classification is often helpful in identification of potential causes. Axonal polyneuropathies are most commonly associated with diabetes, alcohol, or toxins, including side effects of medication. Demyelinating polyneuropathies, as a whole, are much less common and can be acute or chronic, and idiopathic or associated with a monoclonal gammopathy. Finally, hereditary neuropathies, including the various forms of Charcot–Marie–Tooth disease, have similar but somewhat distinct characteristics compared with acquired forms of polyneuropathy. Nerve conduction studies, and, to some extent, needle eletromyography (EMG), can assist with the evaluation of polyneuropathies, and can help to shorten the differential diagnosis regarding potential causes.

Key Words: Axonal polyneuropathy; Charcot–Marie–Tooth; demyelinating polyneuropathy; diabetic polyneuropathy; Guillain–Barré Syndrome; polyneuropathy.

1. INTRODUCTION

Electrophysiological testing serves a critical role in the evaluation of polyneuropathies (PNs). In conjunction with the information obtained from the neurological history and examination, electrophysiology can be used to assist in isolating a specific diagnosis. Testing can (1) help identify whether the PN involves sensory or motor fibers; (2) determine whether the underlying pathophysiology is predominantly demyelinating, axonal, or mixed (demyelinating and axonal); (3) establish duration and severity of the PN. Importantly, the combination of nerve conduction studies and needle eletromyography (EMG) can also be used to evaluate superimposed problems, such as polyradiculopathy or focal mononeuropathies, which may potentially contribute to the patient's neuropathic symptoms.

2. NERVE CONDUCTION STUDIES

In the evaluation of PNs, routine motor and sensory nerve conduction studies are performed, usually starting with the lower extremities. If asymmetry is suggested by history or examination, the more-affected limb is evaluated first, followed by the contralateral lower extremity. Nerve conduction studies of the upper extremities can provide additional information regarding the pathophysiology of PN (e.g., demyelinating features), especially if the lower extremity motor and sensory responses are markedly reduced in amplitude or unobtainable. Evaluation of both the lower and upper extremities can also provide information regarding whether the

From: *The Clinical Neurophysiology Primer*
Edited by: A. S. Blum and S. B. Rutkove © Humana Press Inc., Totowa, NJ

disorder could be caused by multiple individual mononeuropathies (confluent mononeuropathy multiplex) rather than true PN.

When studying the lower extremity, tibial and peroneal motor studies with F-waves are performed. The tibial nerve is recorded from the abductor hallucis brevis, with electrical stimulation at the ankle and popliteal fossa. The peroneal nerve is recorded from the extensor digitorum brevis, with electrical stimulation at the ankle and below and above the fibular head sites. If the response from extensor digitorum brevis is unobtainable, recording from tibialis anterior may help to differentiate between demyelinating and axonal lesions in the peroneal nerve. Sural sensory studies are usually performed antidromically, recording at the ankle, with electrical stimulation at the calf. Although other distal nerves of the lower extremities can also be studied (e.g., the plantar nerves), such methods are more difficult to perform and interpret, in part because of substantial variation in normal values among individuals of different ages. In fact, in older individuals, absent plantar sensory responses are often considered a normal finding.

In the upper extremity, median and ulnar motor studies with F-waves are performed, as well as sensory studies. Radial sensory study, with electrical stimulation in the forearm and recording from the anatomical snuffbox, can also be informative, because the superficial radial nerve is generally not affected by entrapment or compression. Comparing the amplitude of the superficial radial sensory response with the amplitude of the sural nerve can also be helpful in identifying early axonal PN, especially in individuals with "borderline normal" sural amplitudes.

Additional studies can also be performed, including H-reflex testing. H-reflex latencies can be abnormal early, in patients with both axonal and demyelinating PNs. Blink reflex studies are also occasionally helpful, especially in patients with early acute demyelinating PN (Guillain–Barré syndrome [GBS]).

While performing nerve conduction studies, careful temperature monitoring is of particular importance. Decreased temperature of the distal limbs may lead to the erroneous conclusion that a demyelinating PN is present because of reductions in conduction velocity and prolongation of distal latencies. These effects are discussed in more detail in Chapter 13.

3. NEEDLE EMG

Needle EMG in the evaluation of PN is helpful for several reasons. First, needle examination of lower extremity muscles can assist in establishing the presence of motor axonal injury by identifying the presence of a gradient of abnormality. In this situation, the distal muscles (e.g., extensor hallucis longs) are most severely affected, with proximal muscles (e.g., tibialis anterior and gastrocnemius) showing more minor changes, and muscles that are even more proximal (e.g., vastus lateralis) appearing entirely normal. Comparing the severity of distal reinnervation between the two legs can also be helpful in some situations. For example, symmetric reinnervation restricted to extensor hallucis longus may be present. Second, needle examination is helpful to evaluate for superimposed processes, most notably lumbosacral radiculopathy that may be contributing to a patient's clinical picture. In fact, in a proportion of elderly individuals referred for symptoms of PN (distal sensory loss), the major pathology identified is polyradiculopathy, often related to lumbar stenosis. Third, in patients with acute demyelinating PNs, reduced recruitment of motor unit potentials is one of the first abnormalities observed, and is often present before nerve conduction abnormalities develop.

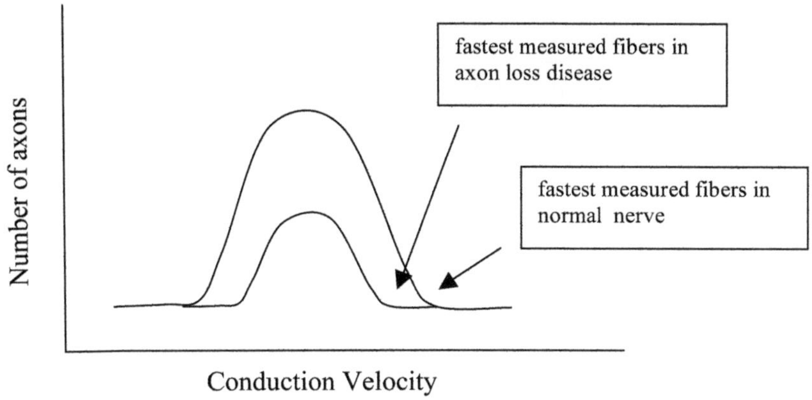

Fig. 1. Slowing of conduction velocity in axon loss lesions. In normal nerve, conduction velocity is measured from the potential onset. When the largest fibers are lost, the fastest conducting fibers can no longer be detected, leading to a slowing of the measured conduction velocity.

4. DIFFERENTIAL DIAGNOSIS OF PN

4.1. Usefulness of Electrophysiological Testing

4.1.1. Axonal vs Demyelinating Injury

One of the most common reasons for performing electrophysiological testing in patients with suspected PN is to differentiate axonal from demyelinating injury. The general rules for separating these two types of disorders are relatively straightforward: axonal PN reduces amplitudes of the recorded motor and sensory responses, whereas demyelinating PN increases the distal latencies, slows conduction velocities, and prolongs F-wave latencies. In addition, conduction block and abnormal temporal dispersion are also present in demyelinating PN.

Unfortunately, however, there is more overlap than this simple description suggests. In fact, some slowing of conduction velocity does occur in axonal lesions, and reductions in amplitude often occur in demyelinating neuropathies. The reason for the reduction in recorded conduction velocity is demonstrated in Fig. 1. A given nerve contains myelinated fibers with a range of conduction velocities, the largest diameter fibers having the fastest velocities and the fibers with the smallest diameters having the slowest velocities. In performing nerve conduction studies, the velocity is typically measured from the onset of the potential, which represents those fibers with the fastest conduction velocities. Now, if an axonal lesion occurs, equally reducing all populations of fibers, the few fastest fibers contributing to the onset of the potential may no longer be present and detectable, and, hence, the measured velocity will be slower.

Likewise, predominantly demyelinating processes can also produce reductions in amplitude in addition to reductions in conduction velocity. These can occur for a number of reasons. First, conduction block may occur. In this situation, the response amplitude is reduced proximally as compared with distally; a reduction in area of greater than 50% is generally considered a conservative estimate. In truth, reductions in area or amplitude of 20% or greater, or increases in duration of 15% or greater, if over short distances, are suggestive of conduction block. If conduction block occurs distal to the most distal stimulation site (e.g., in the hand, if undergoing standard median or ulnar nerve conduction studies), the recorded response amplitude will be low, giving the appearance of axon loss. Second, abnormal

Table 1
Criteria for Acute Inflammatory Demyelinating Polyradiculoneuropathy

At least three of the following in motor nerves:		
	>50% NL CMAP	*<50% NL CMAP*
CV[b]	<90% LLN	<80% LLN
DL[a]	>115% ULN	>125% ULN
F-wave[b]	>125% ULN	—
Ratio[b]	<0.70	—

Abbreviations: CMAP, compound motor amplitude potential; CV, conduction velocity; DL, distal latency; F-wave, F-wave latency; LLN, lower limit of normal; NL, normal; ratio, proximal/distal CMAP area ratio; ULN, upper limit of normal.

[a]More than two nerves.

[b]More than one nerve.

Modified from ref. 2.

Table 2
Criteria for CIDP

At least three of four in motor nerves:

1. Reduction in nerve conduction velocity in at least two nerves (*see* chart)
2. Partial conduction block in at least one nerve
3. Prolonged distal latencies in at least two nerves (*see* chart)
4. Absent F-waves or prolonged F-waves in at least two nerves (*see* chart)

	>80% NL CMAP	*<80% NL CMAP*
CV[a]	<80% LLN	<70% LLN
DL[a]	>125% ULN	>150% ULN
F-waves[a]	>120% ULN	>150% ULN

Abbreviations: CMAP, compound motor amplitude potential; CV, conduction velocity; DL, distal latency; F-wave, F-wave latency; LLN, lower limit of normal; NL, normal; ULN, upper limit of normal.

Modified from ref. 1.

[a]More than two nerves.

temporal dispersion of recorded compound motor action potentials can also reduce the amplitude of the recorded response. The duration of the response is prolonged, and the amplitude drops, so that the area remains relatively constant. Third, and perhaps most commonly, reductions in amplitude occur because there is secondary axonal loss with many demyelinating processes. Hence, many demyelinating lesions are not purely demyelinating, but, rather, are "mixed" (demyelinating and axonal).

Several attempts have been made to develop criteria to separate demyelinating from axonal processes (Tables 1 and 2). Unfortunately, these criteria have substantial limitations because they were developed mainly for research purposes and for specific disorders, such as acute inflammatory demyelinating polyradiculoneuropathy (AIDP; also known as GBS) or chronic inflammatory demyelinating polyradiculoneuropathy (CIDP). CIDP has stricter criteria than AIDP because the longstanding nature of the disorder leads to more substantial axonal injury.

Needle EMG does not serve a major role in differentiating acute axonal from demyelinating disorders. In acute demyelinating disorders, reduced recruitment will be present often diffusely, whereas, in acute axonal neuropathies, abnormalities may be restricted to nerve

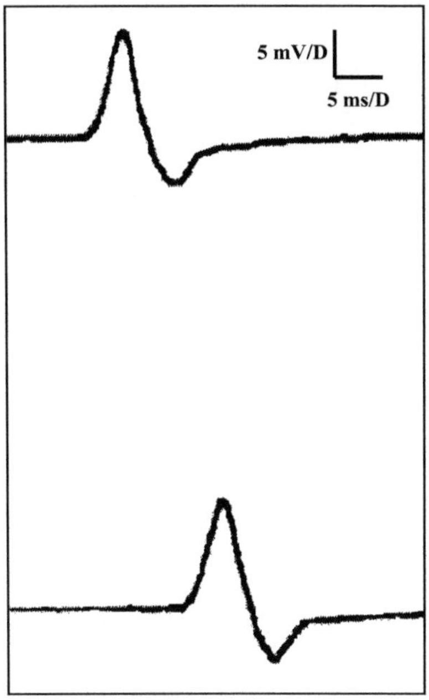

Right Median Motor-Nerve Conduction Study

Segment	Latency (ms)	Amplitude (mV)	Area (ms*mV)	Distance (cm)	CV (m/s)
APB-wrist	9.8	9.2	32.4	7.8	22.2
Wrist-elbow	20.8	9.2	33.6	24.5	

APB, abductor pollicis brevis; CV, conduction velocity.

Fig. 2. Typical motor conduction studies in a patient with Charcot–Marie–Tooth IA. Although the waveform morphology appears normal, there is a marked prolongation of distal latency and severe slowing of conduction velocity.

segments that are more distal or individual nerves, as occurs in monoeuropathy multiplex. Fibrillation potentials often develop in both types of PN and, hence, are not clearly helpful in establishing the diagnosis. Myokymic discharges often occur early in AIDP, and these potentials, if present, can also be helpful in the diagnosis.

4.1.2. Acquired vs Inherited Demyelinating PNs

Nerve conduction studies are also valuable in differentiating whether a demyelinating process is acquired or inherited. Although the clinical history may make such a distinction inconsequential (e.g., the abrupt onset of numbness and tingling and generalized weakness would make Charcot–Marie–Tooth disease an unlikely consideration), certain features in the electrophysiology may help guide the clinician. In hereditary illnesses, demyelination tends to be distinctly uniform, with the same abnormally slow conduction velocity in all upper extremity nerves. Conduction block is not present and although some abnormal temporal dispersion may occur, the waveforms themselves have a relatively normal morphology. This is demonstrated in Fig. 2, which represents typical motor conduction studies for a patient with Charcot–Marie–Tooth disease IA. Although the motor responses seem normal, the latency is

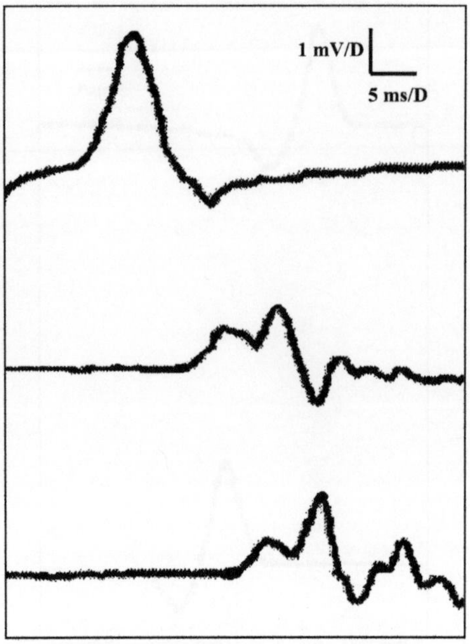

Right Peroneal Motor-Nerve Conduction Study

Segment	Latency (ms)	Amplitude (mV)	Area (ms*mV)	Distance (cm)	CV (m/s)
EDB-ankle	7.5	2.67	18.9	8.5	
Ankle-fib head	19.6	1.26	6.7	29.8	23.9
Fib head-pop fossa	24.5	1.55	6.4	18.5	21.4

CV, conduction velocity; EDB, extensor digitorum brevis; fib, fibular; pop, popliteal.

Fig. 3. Typical motor conduction studies in a patient with acquired demyelinating neuropathy. Conduction block and temporal dispersion are demonstrated in this study.

actually very prolonged (normal <4.4 ms) and the conduction velocity very slow (normal >50 m/s). In patients with acquired demyelinating PNs, one nerve can demonstrate only modest changes, whereas another may be profoundly abnormal, with conduction block and temporal dispersion (Fig. 3). Recently, there have been reports of individuals with hereditary diseases, such as X-linked Charcot–Marie–Tooth disease, with nerve conduction study abnormalities that are more consistent with an acquired picture and, hence, even this simple dichotomy does not seem to hold universally. Caution regarding the interpretation of a given patient's electrophysiological results is warranted.

4.1.3. Evaluation of Small Fiber Neuropathies

Standard electrophysiological testing generally does not involve a detailed analysis of small myelinated and unmyelinated nerve function. Special testing is required to evaluate small fiber PNs. Tests that are useful in the assessment of PNs that specifically or predominantly affect the small sensory fibers include quantitative sensory testing, quantitative sudomotor axon reflex test, thermoregulatory sweat test, sympathetic skin response, and epidermal skin biopsy. An approach to the evaluation of these disorders is discussed in Chapter 24 (Autonomic Nervous System Testing).

5. DIFFERENTIAL DIAGNOSIS OF PNs

5.1. Motor PNs

Pure motor PNs are uncommon (in fact, many have some sensory component), and can be axonal or demyelinating in nature. One example of an acute axonal motor PN is the axonal form of GBS, or acute motor axonal neuropathy. Lead neuropathy can occur acutely or chronically, and is predominantly motor in nature. Multifocal motor neuropathy with conduction block is a chronic and demyelinating neuropathy that is often associated with high ganglioside (GM1) antibody titers, but generally presents as multiple mononeuropathies, rather than a confluent motor neuropathic process.

5.2. Sensory PNs

Pure sensory PNs can be either acute or chronic and are usually axonal in nature. Many PNs that seem to be predominantly sensory at the time of presentation are also found to have a motor component on electrophysiological testing. However, true sensory neuropathies occur, including early in diabetes mellitus, and are often related to chemotherapeutic medications, such as thalidomide and *cis*-platinum.

Isolated small fiber sensory PNs represent an important subset of pure sensory neuropathies. Key features include hyperesthesia, allodynia with loss of pinprick sensation, as well as symptoms of autonomic dysfunction, including orthostatic hypotension, decreased sweating, and bladder/gastrointestinal symptoms. Large fiber modalities, including vibration and joint position sense, along with deep tendon reflexes are preserved, although some limited involvement of these may also be identified. As noted, routine sensory nerve conduction studies are normal. The small fiber PNs have a number of causes and can be associated with systemic disorders (diabetes and amyloidosis), with inherited disorders (familial amyloidosis, hereditary sensory and autonomic neuropathy, Fabry's Disease, and Tangier's Disease), with infectious etiologies (e.g., HIV), with toxins (alcohol, chemotherapy, or antiretroviral), or can be idiopathic in nature.

Sensory neuronopathies are disorders that present with pure sensory loss and pain and often seem similar to PNs clinically, except that the sensory loss may be patchier and involve proximal as well as distal areas at the time of presentation. Symptoms within the trigeminal nerve territory occur often as well. In these disorders, sensory neuron cell bodies within the dorsal root ganglia degenerate, so that "neuronopathies" rather than "axonopathies" are the problem. However, from an electrophysiological perspective, the isolated reduction in sensory responses can make it very difficult to differentiate a sensory neuronopathy from a classic PN. However, the time course, clinical involvement of the trigeminal territory, lack of a clear distal-to-proximal gradient of abnormalities (that is, sensory potentials in the arms are as abnormal as those in the legs), and some asymmetry to the responses can suggest a non-PN entity. Testing of the blink reflex can also be helpful in identifying trigeminal nerve dysfunction. The most common causes of sensory neuronopathy include paraneoplastic disorders (associated with the antibody to Hu), vitamin B6 (pyridoxinel) as less as 200 toxicity—usually in doses mg/d, and Sjogren's syndrome.

5.3. Acute Sensorimotor PNs

5.3.1. Acute Demyelinating Sensorimotor PNs

The prototypic acute demyelinating sensorimotor PN is GBS (or AIDP). Onset is usually 2 to 4 wk after a viral infection or other predisposing factor, such as childbirth or surgery. Although the initial symptoms usually include prominent weakness and sensory loss, the examination can be remarkably variable. Weakness and sensory loss are usually distally predominant, but, in some cases, the weakness may be more proximal and the asymmetries of weakness can be substantial. CSF generally reveals a normal cell count with an elevated protein level (so-called albuminocytological dissociation). Cranial and autonomic nerve function can also be involved, often to a dramatic extent. Electrophysiological testing can usually be very helpful in identifing the disorder, although the typical features of demyelination and conduction block may take a week or longer to become apparent. Early abnormalities may include only absent or delayed F-responses, prolonged distal latencies, and, most importantly, reduced motor unit potential recruitment on needle EMG.

Other causes of acute demyelinating PNs are uncommon, but include diphtheria and exposure to n-hexane.

5.3.2. *Acute Axonal Sensorimotor PNs*

Acute axonal PNs are generally rare, but can represent a form of GBS, known as acute motor sensory axonal neuropathy. Acute motor axonal neuropathy mainly affects motor fibers. Nerve biopsies reveal virtually no demyelinating features, only severe axonal pathology. Other causes of acute axonal sensorimotor PNs include toxin exposure (e.g., heavy metals and organic solvents) and porphyria. However, most axonal sensorimotor PNs tend to be subacute to chronic in nature.

5.4. Chronic Sensorimotor PNs

5.4.1. *Chronic Demyelinating PNs*

CIDP is the most common acquired demyelinating disorder. CIDP can occur in isolation or in association with a monoclonal gammopathy or other disorders, such as HIV. Hereditary demyelinating PNs, including the various forms of Charcot–Marie–Tooth disease, are also included in this category. Patients with diabetes mellitus can also develop a form of diabetic PN that may have a substantial "demyelinating" component, although the conduction velocity slowing is more likely caused by ion channel abnormalities and Na–K ATPase dysfunction than demyelination.

5.4.2. *Chronic Axonal PNs*

Most chronic sensorimotor PNs are predominantly axonal in nature, and the list of causes is lengthy, including diabetes mellitus, vitamin B_{12} deficiency, as well as other metabolic, infectious, toxic, and inherited disorders. A brief list is included in Table 3.

5.5. Multiple Mononeuropathies

In patients presenting with an acute or subacute course of asymmetric sensory loss and weakness, the diagnosis of mononeuropathy multiplex should be considered. This group of disorders usually has a course distinct from that of standard PNs, in that lesions seem to involve individual nerve territories rather than a length-dependent process. However, as the disease progresses, confluent sensory loss and weakness can develop, making it difficult to discern individual nerve lesions. There is a similarity to the clinical picture of sensory neuronopathy (e.g., related to Sjogren's or paraneoplastic disorder), because both disorders present with a patchy pattern of painful sensory complaints. There is no weakness in sensory neuronopathy, a feature essentially required to make a diagnosis of mononeuropathy multiplex. The most common nerves involved include the sciatic nerve, usually in the mid-thigh region, and the median and ulnar nerves, usually just proximal to the elbow. The nerve

Table 3
Differential Diagnosis of PNs

1. Motor (acute and chronic)
   - A. Axonal Guillain–Barré syndrome (acute motor axonal neuropathy)
   - B. Lead
   - C. Multifocal motor neuropathy with conduction block
2. Sensory
   - A. Acute axonal
     - Paraneoplastic sensory neuronopathy
     - Postinfectious inflammatory sensory polyganglionopathy
     - Pyridoxine (B6) toxicity
     - Sjogren's syndrome
   - B. Chronic axonal
     - *Cis*-platinum toxicity
     - Hereditary and sensory autonomic neuropathies
     - Monoclonal protein-associated neuropathy
     - Paraneoplastic sensory neuronopathy
     - Spinocerebellar ataxias (including Friedreich's ataxia)
     - Syphilis
     - Thalidomide
     - Vitamin E deficiency
   - C. Small fiber (acute or chronic)
     - Amyloidosis (inherited and acquired)
     - Diabetes
     - Fabry's Disease
     - Hereditary and sensory autonomic neuropathies
     - HIV (AIDS)
     - Idiopathic
     - Porphyria
     - Tangier's disease
     - Toxins
       - Alcohol
       - Chemotherapy
       - Antiretroviral
3. Multiple mononeuropathies (acute or chronic)
   - Diabetes
   - Hereditary tendency to pressure palsy
   - HIV
   - Hypothyroid
   - Leprosy
   - Lyme
   - Lymphomatoid granulomatosis
   - Multifocal mononeuropathy with conduction block
   - Neurofibromatosis
   - Paraneoplastic
   - Perineuroma
   - Sarcoidosis
   - Vasculitis
4. Sensorimotor
   - A. Acute demyelinating
     - Guillain–Barré syndrome

(Continued)

Table 3 *(Continued)*

Diphtheria
HIV variant of Guillain–Barré syndrome
Tick paralysis
Toxic
Amiodarone
Perhexiline
Glue-sniffing (n-hexane)
B. Acute axonal
Critical illness PN
Porphyria
Toxic
Arsenic
Organophosphates
Thallium
Vasculitis
C. Chronic demyelinating (uniform)
Hereditary motor sensory neuropathy type I (Charcot–Marie–Tooth disease IA)
Hereditary motor sensory neuropathy type III (Dejerine Sottas)
Hereditary motor sensory neuropathy type IV (Refsum's disease)
Cockayne's disease
Krabbe's globoid leukodystrophy
Metachromatic leukodystrophy
D. Chronic demyelinating (nonuniform)
Anti-myelin associated glycoprotein
Chronic inflammatory demyelinating polyradiculoneuropathy
Hereditary tendency to pressure palsy
Monoclonal protein-associated neuropathy
n-hexane
Osteosclerotic myeloma
Perhexiline toxicity
X-linked Charcot–Marie–Tooth disease
Waldenstom's macroglobulinemia
E. Chronic axonal
Amyloidosis
Connective tissue disease
Diabetes
Hereditary motor sensory neuropathy type II (Charcot–Marie–Tooth disease 2)
HIV
Idiopathic
Lyme
Paraneoplastic
Sarcoidosis
Syphilis
Toxic
Arsenic
Alcohol
Paclitaxel
Vincristine
Thiamine deficiency
Thyroid
Vasculitis
Vitamin B_{12} deficiency

involvement is generally not at a site of compression but, rather, at a site of more tenuous blood supply or watershed region. Mononeuropathy multiplex is usually related to one of the vasculitic disorders, including polyarteritis nodosa, Churg–Strauss, Wegener's granulomatosis, hypersensitivity vasculitis, hepatitis C with cryoglobulinemia, HIV, or one of the connective tissue disorders. However, multiple mononeuropathies can also occur at sites of compression in patients with underlying PNs, diabetes, and hypothyroidism, as well as in individuals with a hereditary disposition to pressure palsies. It is worth making a distinction between multiple mononeuropathies and mononeuritis multiplex (of the vasculitic kind), because of the distinctly different clinical implications.

SUGGESTED READING

Ad Hoc Subcommittee of the American Academy of Neurology AIDS Task Force. Research criteria for diagnosis of chronic inflammatory demyelinating polyneuropathy (CIDP). Neurology 1991;41:617–618.

Albers JW, Kelly JJ Jr. Acquired inflammatory demyelinating polyneuropathies: clinical and electrodiagnostic features. Muscle Nerve 1989;12:435–451.

Auger RG, Windebank AJ, Lucchinetti CF, Chalk CH. Role of the blink reflex in the evaluation of sensory neuronopathy. Neurology 1999;53:407–408.

Chalk CH. Acquired peripheral neuropathy. In: Acquired Neuromuscular Diseases, Vol. 15 (Pourmand R, ed.). WB Saunders, Philadelphia, PA, 1997, pp. 501–528.

Donofrio PD, Albers JW. AAEM minimonograph #34: polyneuropathies: classification by nerve conduction studies and electromyography. Muscle Nerve 1990;13:889–903.

England JD, Gronseth GS, Franklin G, et al. Distal symmetric polyneuropathy:a definition for clinical research. Neurology 2005;64(2):199–207.

Kimura J. Electrodiagnosis in Diseases of Nerve and Muscle: Principles and Practice, Oxford University Press, Oxford, 2001.

Lewis RA, Sumner AJ, Shy ME. Electrophysiological features of inherited demyelinating neuropathies: a reappraisal in the era of molecular diagnosis. Muscle Nerve 2000;23:1472–1487.

Mendell JR, Kissel JT, Cornblath JR. Diagnosis and Management of Peripheral Nerve Disorders. University Press, Oxford, England, 2001.

Stewart JD, Low PA, Fealey RD. Distal small fiber neuropathy: results of tests of sweating and autonomic cardiovascular reflexes. Muscle Nerve 1992;15(6):661–665.

REVIEW QUESTIONS

1. What is the role of electrophysiology in the evaluation of PNs?
   A. Identify nerve fiber type.
   B. Determine pathophysiology.
   C. Establish chronicity.
   D. Ascertain severity.
   E. All of the above.
2. Reduced temperature of the limbs:
   A. Falsely elevates sensory nerve potential amplitudes.
   B. Falsely reduces sensory nerve potential amplitudes.
   C. Falsely increases of conduction velocity.
   D. Falsely reduces sensory nerve potential durations.
   E. Has no effect on nerve conduction studies.
3. A practical definition for partial conduction block over a *short* distance would include which of the following:
   A. Drop in amplitude of greater than 20% and a greater than 15% increase in duration.
   B. Drop in amplitude of greater than 50% and a less than 15% increase in duration.
   C. Drop in amplitude of greater than 50% and a greater than 15% increase in duration without temporal dispersion.

D. Drop in amplitude of greater than 20% and a less than 15% increase in duration.
E. Drop in amplitude of greater than 20% and a greater than 50% increase in duration.

4. Needle EMG at the onset of symptoms in a demyelinating PN typically reveals:
   A. Fibrillation potentials.
   B. Neuromyotonic discharges.
   C. Reduced recruitment.
   D. Fasciculations.
   E. Positive sharp waves.
5. Tests that evaluate small nerve fiber function include all of the following EXCEPT:
   A. Quantitative sudomotor axon reflex test.
   B. Sympathetic skin response.
   C. Quantitative sensory testing.
   D. Epidermal skin biopsy.
   E. Standard nerve conduction studies.
6. Which feature on needle EMG best characterizes chronicity?
   A. Fasciculations.
   B. Reduced recruitment.
   C. Polyphasia.
   D. Enlarged motor unit potential amplitudes.
   E. Positive sharp waves.
7. Which of the following is characteristic of nonuniform demyelination?
   A. Charcot–Marie–Tooth disease I.
   B. Charcot–Marie–Tooth disease II.
   C. Charcot–Marie–Tooth disease III.
   D. X-linked Charcot–Marie–Tooth disease.
   E. Pyridoxine deficiency.
8. All of the following have predominantly sensory symptoms and signs EXCEPT:
   A. Paraneoplastic syndrome.
   B. Vitamin B_6 deficiency.
   C. Sjogren's syndrome.
   D. Lead exposure.
   E. *Cis*-platinum.
9. What feature is not observed in pure small fiber PNs?
   A. Allodynia.
   B. Loss of pinprick sensation.
   C. Absent reflexes.
   D. Normal nerve conduction studies.
   E. Hyperesthesia.
10. Which of the following is not a criterion for CIDP (CMAP > 80%)?
    A. Decreased conduction velocity at less than 70% of the lower limit of normal.
    B. Distal latency greater than 125% of the upper limit of normal.
    C. Prolonged F-waves greater than 120% of the upper limit of normal.
    D. Absent F-waves.
    E. Partial conduction block.

REVIEW ANSWERS

1. The correct answer is E. Nerve conduction studies are helpful in determining whether the PN is axonal, demyelinating, or mixed. Needle EMG provided information regarding reinnervation of motor unit potentials or the chronicity of the disorder. Both nerve conduction studies and needle EMG ascertains the severity of the PN.

2. The correct answer is A. For every 1°C decrease in temperature, the distal latency increases by 0.2 ms and the conduction velocity decreases between 1.5 and 2.5 m/s. The amplitude also increases, due to prolonged opening of the sodium channels.
3. The correct answer is D. In normal patients, the typical drop in amplitude is less than 20% and the increase in duration is less than 15% across a short segment of nerve (e.g., less than 10 cm). Thus, any greater drop of amplitude and duration is suggestive of conduction block.
4. The correct answer is C. Reduced recruitment is one of the earliest abnormalities observed in demyelinating PN. Demyelinative changes on nerve conduction studies can lag behind by more than 1 wk. PSWs and fibrillation potentials typically develop 1–2 weeks after onset. Neuromyotonic discharges are not typically seen.
5. The correct answer is E. All of the other tests are used to evaluate small nerve fiber function. Nerve conduction studies evaluate the fastest and largest diameter nerve fibers.
6. The correct answer is D. Reduced recruitment and positive sharp waves can occur in acute PN. Fasciculations represent damage to the motor neuron or axon. Polyphasic motor unit potentials can be observed with subacute lesions. Enlarged motor unit potentials are a result of reinnervation, which typically requires substantial time to develop.
7. The correct answer is D. Charcot–Marie–Tooth disease I and Charcot–Marie–Tooth disease III (Dejerine-Sottas) are examples of uniform demyelination. Patients with Charcot–Marie–Tooth II have an axonal PN, whereas pyridoxine (vitamin B_6) deficiency typically presents as a sensory neuronopathy.
8. The correct answer is D. Lead typically causes a predominantly motor neuropathy and can often present with a wrist drop. All of the others cause mainly sensory dysfunction.
9. The correct answer is C. Signs of large fiber dysfunction including absent vibration and depressed reflexes, are typically not observed in small fiber PN.
10. The correct answer is E. See criteria in Table 2.

17
Radiculopathy and Motor Neuron Disorders

Juan A. Acosta and Elizabeth M. Raynor

Summary

Although radiculopathies and motor neuron disorders have vastly different underlying mechanisms and clinical presentations, the electrodiagnostic examination for these disorders demonstrate remarkable similarities. In all of these conditions, the electrophysiological examination reflects motor neuron injury with virtually no involvement of sensory neurons, even in the face of substantial sensory symptoms, as may occur in radiculopathy. The needle examination is a critical part of electrophysiological testing, helping to define the extent and severity of the abnormality, often more effectively than is possible with nerve conduction studies alone. Late responses, including F-waves and the H-reflexes, can also assist in the diagnosis of radiculopathy to some extent. The electrodiagnostic evaluation of motor neuron disease requires special consideration and care.

Key Words: Amyotrophic lateral sclerosis; F-wave; motor neuron disease; myotome; needle electromyography; radiculopathy.

1. INTRODUCTION

The electrodiagnostic (EDX) examination has an established role in the evaluation of radiculopathy and motor neuron disorders (MNDs). An extension of the neurological examination, the EDX examination must always be interpreted in the proper clinical context. Radiculopathy results from injury to nerve roots producing sensory and/or motor symptoms and signs in the distribution of the corresponding dermatomes and/or myotomes. On the other hand, MNDs are characterized by injury and loss of the anterior horn cell, affecting only motor fibers. Although radiculopathies and MNDs are clinically dissimilar, their electrophysiological features are alike, because, in both, the nerve injury occurs proximal to the dorsal root ganglion (DRG). For this reason, they will be discussed together. The fact that polyradiculopathy and MNDs are often indistinguishable from an electrophysiological perspective, yet have dissimilar symptoms and clinical course, underscores the importance of clinical context in assessing EDX findings.

In evaluating radiculopathy, EDX studies are useful in determining the functional consequence of a given structural lesion. They specifically are used to delineate the distribution of the affected muscles, localize the level, and establish the chronicity and extent of the problem. Although EDX studies are generally sensitive and specific, a normal EMG result in the face of signs and symptoms consistent with radiculopathy does not exclude this diagnosis.

In evaluating MND, EDX studies are used to establish objective evidence of lower motor neuron degeneration in multiple body segments and are an essential diagnostic procedure in the workup of these disorders.

From: *The Clinical Neurophysiology Primer*
Edited by: A. S. Blum and S. B. Rutkove © Humana Press Inc., Totowa, NJ

2. ANATOMIC CONSIDERATIONS

Thirty-one pairs of spinal nerves are formed from dorsal and ventral roots (8 cervical, 12 thoracic, 5 lumbar, 5 sacral, and 1 coccygeal). Ventral roots arise from motor neurons in the anterior and lateral gray columns of the spinal cord. Dorsal roots extend proximally from sensory neurons in the DRG, which lie within the neural foramen (outside the spinal canal). Ventral and dorsal roots join together just distal to the DRG, forming a spinal nerve; on exiting the foramen, the nerve divides into dorsal and ventral rami. The dorsal rami supply the paraspinal muscles and skin overlying the paraspinous region; ventral rami form the plexus, which branches into the individual peripheral nerves that supply the upper and lower limb and the sacral region.

The muscles supplied by a single spinal segment constitute a myotome; the skin region supplied by a single spinal segment is a dermatome. There is significant variability in individual myotomal and dermatomal representation for a particular muscle or skin region. Each muscle receives innervation from multiple contiguous nerve roots, and each dermatome overlaps extensively with neighboring dermatomes.

3. PHYSIOLOGICAL CONSIDERATIONS

3.1. *Radiculopathies*

The majority of radiculopathies result from nerve root compression, either from disc herniation or as a consequence of spondylotic arthropathy; inflammatory and immunological lesions are less common. The most commonly involved roots are L5 and S1, and C6 and C7 (*see* Table 1 for a list of radiculopathies and associated symptoms and signs). These lesions may affect nerve roots by causing axonal degeneration, focal demyelination, or both. The electrophysiological findings in a patient are determined by the degree to which each of these pathophysiological processes occurs. Regardless of the underlying cause, injury to the roots occurs at a location that is proximal to the DRG. This is a critical point in terms of establishing the electrophysiological pattern of abnormalities, and allows differentiation of radicular lesions from those distal to the DRG. Notably, only a portion of the nerve roots fibers are injured in the majority of cases of radiculopathy, an important factor with regard to the pattern of abnormalities observed on EDX testing. The most commonly used EDX test for radiculopathies, the needle examination of muscle, is used to identify the extent and severity of motor axonal injury in a patient, which is a reflection of the nerve root fiber injury.

3.2. *Motor Neuron Disorders*

Amyotrophic lateral sclerosis (ALS) is the most common of the MND evaluated in the EMG laboratory. ALS is characterized by upper and lower motor neuron degeneration involving multiple body regions (i.e., cranial, cervical, thoracic, and lumbosacral). The loss of motor neurons usually begins in one area, is asymmetrical, and later becomes evident in other areas. The clinical and electrophysiological findings are dependent on the body segment involved and the severity of involvement. Whereas recognition of upper motor neuron involvement depends on clinical signs, routine EDX studies are of major help in identifying lower motor neuron abnormalities, even before they are clinically recognizable.

A single motor neuron innervates a group of muscles fibers; the motor unit is comprised of the motor neuron, its axon, and all of the muscle fibers innervated by it. The innervation ratio is the number of muscle fibers innervated by a single motor neuron. In ALS, degeneration

Table 1
Neurological Examination Findings in Monoradiculopathies

Root level	*Pain*	*Sensory loss (paresthesias)*	*Motor abnormalities/ weakness*	*Deep tendon reflex abnormalities*
C5	Pain lateral shoulder	Shoulder	Deltoid, supraspinatus, and infraspinatus	Biceps reflex
C6	Neck radiating to the arm	Radial side of the arm to thumb	Brachioradialis, flexor carpi radialis, biceps	Biceps reflex
C7	Neck to the fingers	Between second and fourth finger	Triceps, wrist extensors	Triceps reflex
C8–T1	Neck to the fingers	Ulnar aspect of hand and/or forearm	All hand intrinsics, flexor digitorum profundus	None
L1	Inguinal region	Inguinal region	None	None
L2	Groin, anterior thigh	Anterolateral thigh	Iliopsoas	None
L3	Anterior thigh to knee, anterior leg	Medial thigh and knee	Quadriceps, iliopsoas, hip adductors	Knee jerk
L4	Medial foreleg	Medial lower leg	Tibialis anterior, quadriceps, hip adductors	Knee jerk
L5	Lateral thigh and lower leg, dorsum foot	Lateral lower leg, dorsum foot, great toe	Toe extensors, ankle dorsiflexor, everter and inverter, hip abductors	None, unless S1 involved
S1	Posterior thigh, calf, heel	Sole, lateral foot and ankle, lateral two toes	Toe flexors, gastrocnemii, hamstrings, gluteus maximus	Ankle jerk
S2–S4	Medial buttocks	Medial buttocks, perineal, perianal region	None, unless S1–S2 involved	Bulbocavernosus, anal wink Ankle jerk if S1 involved

of anterior horn cells leads to loss of motor neurons and denervation of muscle fibers that are part of these motor units; nearby, unaffected axons attempt to reinnervate denervated muscle fibers, increasing the innervation ratio in surviving motor neurons. These pathophysiological events are associated with active denervation/chronic reinnervation changes observed on EMG examination, as well as reduced recruitment related to loss of motor units. EDX abnormalities are frequently observed in muscles that are not clinically involved in this disease. Similarly, in polio, another MND, chronic reinnervation and reduced motor unit recruitment is generally more widespread than the clinical examination alone would suggest.

4. ELECTRODIAGNOSIS

4.1. Sensory Nerve Conduction Studies

Sensory nerve conduction study (SNCS) results are nearly always normal in radiculopathy as well as MND. The sensory nerve action potential (SNAP) amplitude is normal even when patients have clinical sensory loss because the lesion occurs proximal to the DRG (i.e., is preganglionic) and the peripheral sensory axons are intact. This is the most useful piece of information for differentiating radiculopathies from lesions involving the plexus or individual peripheral nerves. The most common, albeit rare, exception to this rule is involvement of the L5 DRG with L5 radiculopathy caused by a far lateral disc herniation. In MND, the sensory study results are normal because the disorder affects only motor neurons.

There are two essential points that are mostly relevant to radiculopathy with associated sensory symptoms. First, it is most helpful to compare the SNAP from the symptomatic limb with the corresponding SNAP in the contralateral uninvolved limb (with reduction in amplitude of ≥50% considered abnormal) rather than relying purely on reference values for the amplitudes; this is even more essential if there is coexistent polyneuropathy. Second, in evaluating radiculopathy, it is important to attempt recordings from nerves most closely representing the clinically affected dermatome (e.g., superficial peroneal recording for suspected L5 lesions or sural recording for S1 lesions).

4.2. Motor Nerve Conduction Studies

Motor nerve conduction study (MNCS) results are frequently normal in radiculopathy, for several reasons. Myotomal overlap of root innervation to individual muscles makes it likely that, from each muscle, a significant number of uninvolved nerve roots will contribute to maintenance of a normal compound muscle action potential (CMAP); additionally, only a fraction of nerve root fibers are injured in most cases of compressive radiculopathy. If a nerve root lesion causes focal segmental demyelination, the CMAP, which is recorded distal to the lesion, will also be normal. In an injury that is characterized by significant axonal loss, there will be distal axonal degeneration that may result in reduced CMAP amplitude. If degeneration of a large number of motor axons occurs, conduction velocity may be decreased and distal latency prolonged, because of loss of the fast-conducting axons, although, generally, this is mild. All CMAP abnormalities reflect the degree of axon loss and, therefore, are more likely to occur with severe lesions. As with sensory studies, a CMAP should be recorded from a muscle in the clinically relevant myotomes, if possible. This is readily achievable in the legs, where the most common radiculopathies affect L5 and S1 nerve fibers, but less relevant in the arms, where the most common nerve roots involved, C6 and C7, are not generally evaluated with motor nerve conduction studies.

In MND, CMAP may be normal or amplitudes may be reduced in proportion to the loss of motor units. As with radiculopathy, changes in conduction velocity and/or distal latency are generally mild. However, thorough nerve conduction studies are extremely important for identifying peripheral neuropathies that may mimic ALS clinically, such as multifocal motor neuropathy with conduction block. In this disorder, evidence of conduction block is identified, often in unusual segments of nerve (e.g., not at standard compression sites). Despite the prominent conduction block and often some associated axon loss of motor studies, sensory study results are normal or nearly normal. Sometimes the block can be difficult to identify, prompting the electromyographer to pursue proximal testing of nerves, including root stimulation using needle electrodes.

4.3. Late Responses

4.3.1. F-Responses

These late responses are produced by antidromic activation of a subpopulation of motor neurons and are useful because they provide an assessment of proximal motor pathways.

However, F-responses also reflect conduction along the distal motor pathway and may be prolonged if axonal loss or demyelination along any portion of the motor nerve has occurred. Unfortunately, the sensitivity of F-responses in radiculopathy seems low. This may be because F-responses are mediated by more than one nerve root (i.e., L5/S1). It may also be because nerve root injury in most radiculopathies is partial, leaving a reasonable number of motor axons intact, resulting in a normal minimal F-latency. For this reason, a prolonged minimum–maximum latency range (chronodispersion) or evaluating mean F-latency are likely to be more sensitive indicators of radiculopathy. Nevertheless, F-responses also have a low specificity, because axon loss or demyelination anywhere along the entire length of the motor fiber being studied may prolong the F-latency.

In MND, F-responses are generally of normal latency until marked axon loss has occurred, in which case, they are usually only mildly prolonged. Their main usefulness in evaluation of these disorders is in excluding polyneuropathies that may produce a clinical syndrome indistinguishable from ALS, particularly those characterized by proximal conduction abnormalities, such as the multifocal motor neuropathy with conduction block, discussed above.

4.3.2. H-Reflexes

H-reflexes are useful in the evaluation of radiculopathy because they assess both motor and sensory pathways involving the nerve root; however, standard studies evaluate only the S1 root level, a major limitation. H-reflex latency is the most useful parameter to measure. Ipsilateral prolongation (>2 ms longer than the contralateral, uninvolved side) or absence of the H-reflex is observed in unilateral S1 radiculopathy. In bilateral lesions, the reflexes may be prolonged or absent bilaterally. One advantage of the H-reflex over needle EMG is that it becomes abnormal immediately after compression injury of the nerve root, although it remains abnormal thereafter. There are important disadvantages of the H-reflex, including the lack of specificity. They may be abnormal with any lesion that causes depression of the ankle jerks and are generally absent in patients with polyneuropathy. Although uncommonly attempted, an H-reflex can usually be obtained from the flexor carpi radialis in healthy individuals; abnormalities in this response may, thus, be helpful in identifying a C6–C7 radiculopathy.

4.3.3. Needle EMG

Needle EMG is overall the most useful procedure in the electrodiagnosis of radiculopathy. Needle EMG is a sensitive tool for demonstrating axonal loss in motor fibers secondary to nerve root injury; however, EMG is insensitive for demonstrating purely demyelinating lesions unless severe enough to produce substantial reductions in motor unit potential recruitment. The needle EMG in suspected radiculopathies must be extensive and, at the very least, several muscles must be examined. Timing of the EMG in relation to the injury must be considered when analyzing EDX data. Denervation (i.e., fibrillation potentials and positive sharp waves) is expected to develop in paraspinal muscles within 7 to 10 d of a root injury; in the proximal limbs in 2 to 3 wk, and in the distal limbs in 3 to 6 wk, although such abnormalities are often observed sooner. Reinnervation changes generally develop in involved muscles in a similar fashion as the denervation changes (i.e., proceeding distally from the point of injury); usually, these changes take 3 to 6 mo to develop after the acute injury. The evaluation of at least five to seven muscles of the involved extremity, including the paraspinals, is generally required to adequately screen for radiculopathy (Table 2). It is important to verify that muscles in adjacent myotomes are normal, to localize a radiculopathy to a particular level; it is usually not possible to be more precise than to indicate two adjacent root levels of involvement (e.g., C6–C7).

Table 2
Nerves and Spinal Root Supply of Representative Muscles Commonly Tested During Needle Evaluation of Radiculopathy[a]

Nerve	*Muscle*	*Root*
Brachial Plexus		
Spinal accessory nerve	Trapezius	C3–C4
Dorsal scapular nerve	Rhomboid major/minor	C4–C5
Suprascapular nerve	Supraspinatus	**C5–C6**
	Infraspinatus	**C5–C6**
Axillary nerve	Deltoid	**C5–C6**
Musculocutaneous nerve	Biceps brachii	**C5–C6**
Median nerve	Pronator teres	**C6–C7**
	Flexor carpi radialis	**C6–C7**
	Flexor pollicis longus	**C6–C7**
	Abductor pollicis brevis	C8–**T1**
Ulnar nerve	Flexor carpi ulnaris	C7–**C8**–T1
	FDP IV–V	C7–**C8**
	First dorsal interosseous	C8–**T1**
	Abductor digiti quinti	C8–**T1**
Radial nerve	Triceps	C6–**C7**–C8
	Brachioradialis	C5–**C6**
	Extensor carpi radialis	C5–**C6**
Posterior interosseous nerve	Extensor digitorum communis	**C7**–C8
	Extensor indicis proprius	**C8**
Lumbosacral plexus		
Superior gluteal nerve	Gluteus maximus	L5–**S1**
Inferior gluteal nerve	Gluteus medius	**L5**–S1
	Tensor fasciae latae	**L5**–S1
Femoral nerve	Iliopsoas	**L2–L3**
	Quadriceps	L2–**L3–L4**
Obturator nerve	Adductor longus	**L3–L4**
Sciatic nerve	Hamstrings	L5–**S1**–S2
Tibial nerve	Gastrocnemius and soleus	**S1**–S2
	Tibialis posterior	**L5**–S1
Superficial peroneal nerve	Peroneus longus	**L5**–S1
Deep peroneal nerve	Tibialis anterior	L4–**L5**
	Extensor hallucis longus	**L4–L5**

[a]FDP IV–V, flexor digitorum profundus of the fourth and fifth digits.

Characteristic needle EMG findings in radiculopathy include ongoing denervation and/or chronic reinnervation in limb muscles sharing the same root (myotome) distribution but innervated by different peripheral nerves, as well as ongoing denervation in paraspinal muscles. Abnormalities should be sought in distal and proximal limb muscles, although sometimes only distal muscles will be affected. Compromised muscles initially show reduced recruitment followed by fibrillation potentials and/or positive sharp waves, and, in chronic cases, high-amplitude, long-duration motor unit potentials. Understanding of the time frame for development of these EMG abnormalities in relationship to the clinical nerve root injury is important in interpretation of the findings (*see* preceding paragraphs).

The presence of denervation in paraspinal muscles is an important localizing finding indicating axonal loss in the dorsal rami (i.e., at nerve root level). However, there are several points regarding the paraspinal muscle examination that require further discussion:

1. Paraspinal muscles may be normal in cases of clear-cut radiculopathy.
2. Ongoing denervation, particularly isolated positive sharp waves, may be observed in paraspinal muscles in healthy subjects, particularly those older than the age of 50 yr. Generally, however, abundant denervation changes are not observed in healthy subjects.
3. Paraspinal denervation changes are not specific for compressive radiculopathy and may be observed in inflammatory myopathy and MND, and may be persistent after posterior laminectomy procedures.
4. There is considerable overlap of paraspinal muscle innervation from the dorsal rami; thus, the finding of paraspinal abnormalities at a particular level does not localize the radiculopathy to that level, only limb muscle abnormalities should be used for this purpose. Conversely, paraspinal muscles may require sampling at several adjacent levels for abnormalities to be detected (e.g., examine C6 and C8 paraspinals if C7 paraspinals are normal in a C7 radiculopathy).

In generalized MND, the needle EMG is of critical importance for demonstrating evidence of widespread lower motor neuron degeneration. The findings include ongoing denervation in the form of fibrillation potential, as well as fasciculation potentials, in addition to chronic reinnervation changes. These abnormalities are frequently present in a myotomal distribution (therefore, are identical to polyradiculopathy), but the characteristic that differentiates MNDs from radiculopathy is the finding of abnormalities in multiple body segments: cranial, cervical (either arm and associated paraspinal musculature), thoracic, and lumbar (either leg and associated paraspinal musculature). The evaluation in MND should be focused at demonstrating abnormalities involving more than one nerve as well as more than one root level of abnormality in the various segments with the least discomfort to the patient. This may be possible by examination of only two or three carefully chosen muscles in each of these regions. The finding of fasciculation potentials, although characteristic of MND, is not specific and may also occur with radiculopathy or disorders of peripheral nerve hyperexcitability (e.g., cramp-fasciculation syndrome).

5. SUMMARY OF EDX FINDINGS IN RADICULOPATHY

5.1. Nerve Conduction Studies

1. Sensory: usually normal even if clinical sensory loss is present, an important differentiating feature from nerve and plexus lesions. Care should be taken to record from clinically relevant sensory nerves and to compare with the contralateral side if appropriate.
2. Motor: usually normal even in the presence of weakness. CMAP may be decreased with lesions causing severe axon loss, particularly if multiple, adjacent nerve roots are affected.

5.2. Needle EMG

Overall, needle EMG is the most useful EDX test for evaluating radiculopathy. Timing of the study in relation to symptom onset is very important for interpreting EMG findings. Ongoing denervation takes an average of 2 to 3 wk to develop in affected limb muscles; chronic reinnervation changes take 3 to 6 mo to develop. Paraspinal denervation is useful, although nonspecific, and may be absent in nearly half of all lesions.

6. SUMMARY OF EDX FINDINGS IN MND

6.1. Nerve Conduction Studies

SNCSs: normal.

MNCS: reduced CMAP amplitudes may be observed in muscles experiencing severe axonal loss. In less-involved muscles, CMAP amplitudes may be normal.

6.2. Needle EMG

Ongoing denervation and chronic reinnervation are present in varying degrees in different areas, reflecting abnormalities in multiple nerve and root distributions. The presence of fasciculation potentials should be sought; although they are characteristic of the disorder, fasciculation potentials may be absent in MND and are not specific to MND. Abnormalities should be sought in the most clinically affected body regions, but at least three body regions must be examined, and abnormalities demonstrated in muscles innervated by different peripheral nerves and nerve roots.

SUGGESTED READING

Aminoff MJ. Electrophysiological evaluation of root and spinal cord disease. Semin Neurol 2002; 22:197–204.

Raynor EM, Kleiner-Fisman G, Nardin R. Lumbosacral and thoracic radiculopathies. In: Neuromuscular Disorders in Clinical Practice (Katirji B, Kaminski H, Preston D, Ruff R, Shapiro B, eds.). Butterworth-Heinemann, Boston, MA 2002.

Stewart JD. The lower cervical spinal nerves. In: Focal Peripheral Neuropathies (Stewart JD, ed.). Raven Press, New York, NY, 1993; pp. 93–109.

REVIEW QUESTIONS

1. The best single sensory study to differentiate L5 radiculopathy vs lumbosacral plexopathy is:
   A. Sural nerve.
   B. Saphenous nerve.
   C. Superficial peroneal nerve.
   D. Lateral femoral cutaneous nerve.
   E. None of the above.
2. The sensitivity of F-responses in radiculopathy is generally considered low because:
   A. F-responses are rarely performed.
   B. F-responses are mediated by more than one root.
   C. Nerve injury in radiculopathy is partial.
   D. B and C.
   E. None of the above.
3. The most useful procedure in evaluation of radiculopathy is:
   A. F-responses.
   B. H-reflexes.
   C. Motor nerve conduction studies.
   D. Needle examination.
   E. A and B.
4. The finding of fibrillation potentials in paraspinal muscles:
   A. Clearly localizes the affected root.
   B. Is typical of plexopathy.
   C. Suggests radiculopathy but does not localize the root involved.
   D. Indicate chronic disease.
   E. None of the above.

5. Sensory studies in MND show:
   A. Reduced conduction velocity.
   B. Reduced amplitude.
   C. Prolonged distal latency.
   D. Abnormal morphology.
   E. Normal morphology.
6. Chronic reinnervation on needle examination is characterized by:
   A. Fibrillation potentials.
   B. Positive sharp waves.
   C. Short duration motor unit potentials.
   D. Long duration motor unit potentials.
   E. Nascent motor unit potentials.
7. A 42-yr-old man presents with pain in his lower back radiating down to his toes. You suspect an S1 radiculopathy. Which study will be more likely to be abnormal?
   A. Needle exam of the tibialis anterior.
   B. H-reflex latency.
   C. Peroneal motor response.
   D. Sural sensory amplitude.
   E. None of the above.
8. The following pair of muscles is innervated primarily by the S1 root:
   A. Tibialis posterior and gastrocnemius.
   B. Gastrocnemius and gluteus maximus.
   C. Tibialis anterior and peroneus longus.
   D. Quadriceps and hamstring.
   E. Long head biceps femoris and extensor digitorum brevis.
9. A pure, subacute L5 radiculopathy is more likely to show fibrillation potentials in the following muscles:
   A. Obturator and tibialis anterior.
   B. Vastus lateralis and peroneus longus.
   C. Gastrocnemius and tibialis anterior.
   D. Tibialis anterior and tibialis posterior.
   E. Gluteus maximus and foot intrinsics.
10. The best way to differentiate ALS from a severe ongoing and chronic polyradiculopathy is:
   A. The presence of fasciculation potentials in paraspinal muscles.
   B. The absence of motor unit potentials in abductor pollicis brevis.
   C. The clinical history and neurological exam.
   D. Needle examination of upper and lower extremities.
   E. MRI of the entire spine.

REVIEW ANSWERS

1. The correct answer is C. Sensory nerves tend to be affected in plexopathy and spared in radiculopathy. The saphenous nerve will be abnormal if the L4 plexus is involved and the sural nerve will be abnormal if the S1 branches of the plexus are compromised.
2. The correct answer is D. The sensitivity of F-responses in radiculopathy seems low. This may be because F-responses are mediated by more than one nerve root (i.e., L5/S1). It may also be because nerve root injury in most radiculopathies is partial, leaving a reasonable number of motor axons intact, resulting in a normal minimal F-latency.
3. The correct answer is D. Needle exam is the most useful EDX study to evaluate radiculopathies.
4. The correct answer is C. The finding of paraspinal abnormalities at a particular level dose not localize the radiculopathy to that level, only the limb abnormalities should be used for this purpose.
5. The correct answer is E. Sensory nerve conduction studies are normal in MND.

6. The correct answer is D. Long duration, large amplitude, and polyphasic motor units are sign of chronic reinnervation.
7. The correct answer is B. H-reflexes are useful in the evaluation of radiculopathy because they assess both motor and sensory pathways involving the nerve root; standard studies evaluate only the S1 root level, which is the root involved in this patient.
8. The correct answer is B.
9. The correct answer is D. Tibialis anterior and tibialis posterior are primarily L5 muscles (although tibialis anterior has a substantial contribution from L4 as well).
10. The correct answer is C. ALS and a severe ongoing and chronic polyradiculopathy look almost the same on EDX studies and sometimes can be challenging for diagnosis. The history will be fundamental. MND is painless and does not have associated sensory symptoms. Therefore, clinical history and examination will be the first step in defining the underlying pathology.

18

Electrophysiology of Brachial and Lumbosacral Plexopathies

Juan A. Acosta and Elizabeth M. Raynor

Summary

Brachial and lumbosacral plexopathies represent a heterogeneous group of disorders including traumatic injury as well as infiltrating and inflammatory lesions. The anatomy of both regions is complex, creating challenges in their evaluation; electrophysiological testing is a key tool in the assessment of plexopathies. Accurate electrodiagnosis requires a comfortable knowledge of the relevant plexus anatomy and applicable nerve conduction techniques, as well as recognition of the underlying pathophysiological processes affecting these regions. The sensory nerve conduction studies are used to differentiate lesions of the plexus from radiculopathies, which they may closely resemble clinically and electrophysiologically. Motor nerve conduction studies, late responses, and needle EMG, in conjunction with the sensory studies, allow precise localization and often provide clues to the underlying pathophysiology of a given lesion.

Key Words: Brachial plexopathy; lumbosacral plexitis; lumbosacral plexopathy; Partonage–Turner syndrome; radiation plexopathy.

1. INTRODUCTION

Accurate electrodiagnosis of plexopathies requires a fundamental understanding of the complex regional anatomy. As in other disorders, the electrodiagnostic study represents an extension of the neurological examination. Plexus lesions may cause motor, sensory, and/or sympathetic disturbances and may be indistinguishable from lesions involving multiple nerves or nerve roots based on physical examination alone. In addition to providing information regarding the localization, extent, pathophysiology, and severity of a given lesion, the electrodiagnostic study may be useful in determining the etiology and/or prognosis in some cases, and is, therefore, a critical tool in the evaluation of suspected lesions of the lumbosacral and brachial plexus.

Lesions of the brachial plexus are more common than those involving the lumbosacral plexus for several reasons: the brachial plexus is relatively superficially located, and is closely related to the bony structures of the shoulder and neck that are freely mobile (making it susceptible to trauma); it is also closely related to major blood vessels, the apex of the lung and lymph nodes draining the lung and breast (making it susceptible to metastatic lesions).

2. BRACHIAL PLEXOPATHY

2.1. *Brachial Plexus Anatomy*

The brachial plexus is formed from the ventral roots of C5 through T1 (Fig. 1). It has three trunks (upper, middle, and lower), three cords (medial, lateral, and posterior), and a number of

From: *The Clinical Neuropysiology Primer*
Edited by: A. S. Blum and S. B. Rutkove © Humana Press Inc., Totowa, NJ

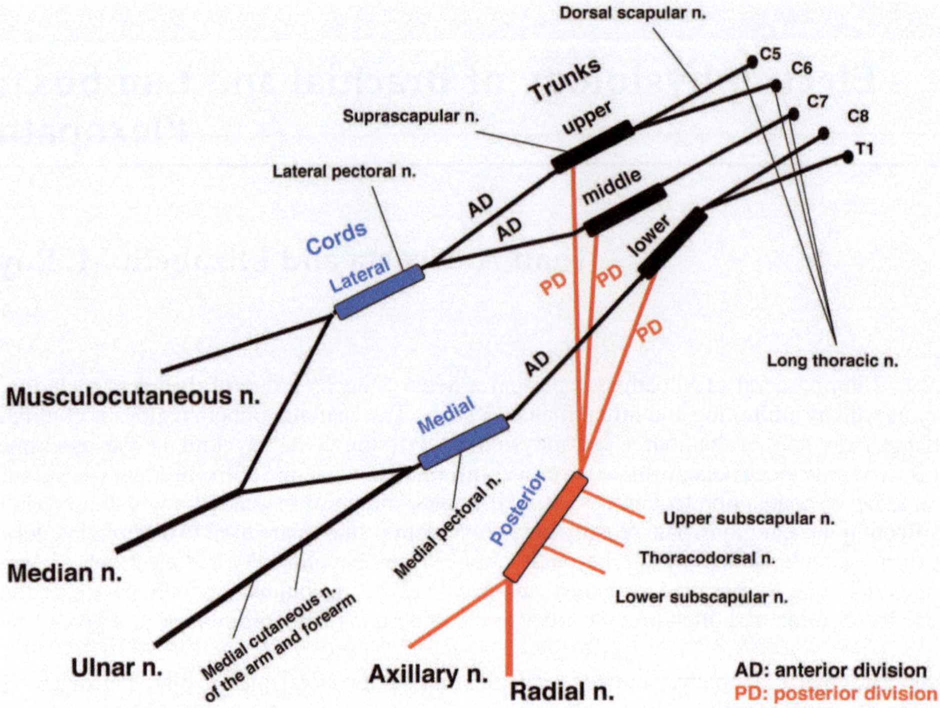

Fig. 1. Anatomy of the brachial plexus, showing the eventual destinations of all root components. Ext, external; extn, extension.

terminal nerves, the most substantial being the median, ulnar, and radial nerves. The ventral rami of C5 and C6 merge to form the upper trunk, C7 ventral ramus forms the middle trunk and C8-T1ventral rami join to form the lower trunk. The anterior divisions of the upper and middle trunks form the lateral cord; the anterior division of the lower trunk forms the medial cord; and posterior divisions from all three trunks form the posterior cord. Terminal branches from the cords are responsible for the motor and sensory innervation of the upper extremity and shoulder girdle. These terminal branches include the musculocutaneous nerve (lateral cord), median nerve (lateral and medial cord), ulnar nerve (medial cord), and radial and axillary nerves (posterior cord). Several smaller nerves also arise directly from the plexus.

Useful points to remember are:

- The long thoracic nerve (C5,6,7) and dorsal scapular nerve (C5,6) arise directly from the nerve roots and therefore their involvement implies a very proximal lesion, usually of the roots themselves.
- The most likely differential diagnostic consideration for an upper trunk or lateral cord brachial plexus lesion is a C5-6 radiculopathy; for a lower trunk or medial cord lesion is a C8-T1 radiculopathy and for a middle trunk lesion is a C7 radiculopathy.
- The lateral cord contribution to the median nerve is mainly *sensory* and derived from C5 and C6 ventral rami through the upper trunk.
- The medial cord contribution to the median nerve is mainly *motor* and originates from C8 and T1 ventral rami through the lower trunk.

2.2. *Pathophysiology*

Brachial plexopathies present with a variety of clinical syndromes depending on the anatomical part of the plexus involved. In terms of frequency, the upper trunk is the most often involved, likely due to its superficial location, with trauma being the most frequent cause. Less common but potentially more serious (due to the incidence of infiltrating tumors in this region) are lesions of the lower trunk; isolated middle trunk lesions are rare. Regardless of etiology or the portion of the plexus involved, axonal loss is the underlying pathophysiology in the majority of brachial plexus lesions.

3. ELECTRODIAGNOSIS: BRACHIAL PLEXOPATHIES

3.1. *Motor Nerve Conduction Studies*

Motor nerve conduction studies (MNCS) are useful mostly for demonstrating significant axonal loss in motor fibers. However, routine MNCS will be abnormal in plexopathy *only* if the nerve under study is derived from the involved trunk or cord of the brachial plexus. Thus, routine MNCS of the median and ulnar nerves provide information relevant only to the lower trunk/medial cord and C8-T1 nerve roots. Unusual conduction studies may be performed for the evaluation of other plexus lesions; for example, information regarding the upper trunk/lateral cord and C5-6 roots may be obtained from MNCS of the musculocutaneous nerve, recording from the biceps brachii. These studies are technically more difficult than routine MNCS and the range of response amplitudes is large; a good rule of thumb when performing these or other technically difficult or unusual studies is to obtain results from the uninvolved contralateral limb and compare the responses for significant asymmetries in amplitude (>50% reduction considered abnormal).

MNCS are often normal in plexopathies but a reduced CMAP amplitude, often associated with mild conduction slowing (due to loss of fast conducting axons) may be observed when axonal loss is significant. Side-to-side comparisons of amplitude are sometimes useful in demonstrating more subtle abnormalities, with a 50% reduction compared to the uninvolved side considered abnormal. MNCS are important for identifying the presence of mononeuropathies that could complicate the diagnosis or clinically mimic brachial plexopathy, such as ulnar neuropathy at the elbow.

3.2. *Sensory Nerve Conduction Studies*

Sensory nerve conduction studies are essential in the assessment of plexopathies. In contrast to radiculopathies, in which sensory studies are invariably normal, sensory nerve action potentials (SNAP) amplitudes are abnormally reduced in plexopathies as the sensory fibers are injured distal to the dorsal root ganglia.

As with MNCS, amplitude is the most important parameter for demonstrating axonal loss and side to side comparison is important for demonstrating subtle abnormalities. Reduction in SNAP amplitude of 50% or more compared to the contralateral, uninvolved nerve is considered abnormal, even if the absolute value of the SNAP amplitude falls within the normal range. Most important is that the sensory nerve from which the recording is made should reflect the clinically involved dermatome. The choice of which sensory studies to perform in evaluating suspected brachial plexopathy depends on a directed physical examination to determine suspected anatomical level of involvement based on careful mapping of sensory loss, in conjunction with reflex changes and muscle weakness.

The upper extremity nerves most commonly evaluated are the median, ulnar, superficial radial, and lateral antebrachial cutaneous (LAC) nerves; medial antebrachial cutaneous (MAC) nerves are examined less frequently. Relevant sensory nerves for evaluating suspected upper trunk lesions include median (recorded from digits 2 and 1), superficial radial (recorded from snuffbox or digit 1), and LAC. For suspected upper trunk brachial plexopathies, LAC and median SNAPs recorded from digit1 are more sensitive for detecting abnormalities than routine median recordings from digit two or superficial radial recordings from the snuffbox. This is due to the contribution from middle trunk fibers to median nerves innervating digit 2 and superficial radial nerves innervating the snuffbox. Radial SNAPs can help differentiate upper trunk from lateral cord lesions, as in the latter the radial SNAP should be preserved.

For evaluation of suspected lower trunk lesions, the MAC and ulnar nerves are the relevant nerves to study. Both the MAC and the ulnar SNAP are derived from the lower trunk and medial cord. In general, the ulnar SNAP is at least as sensitive as the MAC recording in the identification of lower trunk brachial plexus lesions; however, MAC recordings may be helpful in cases where an ulnar neuropathy is present and a superimposed lower trunk or medial cord lesion must be ruled out. The MAC as well as the LAC are best performed on a contralateral asymptomatic limb to allow side to side comparison with the involved limb.

3.3. *F-Responses*

As in evaluation of radiculopathies, late responses are complementary to the routine conduction studies but do not increase diagnostic sensitivity in most cases as these lesions are predominantly axonal. In addition, the F-responses are non-localizing as F-response latency prolongation may occur from a lesion anywhere from cervical spine to the hand. Like MNCS, F responses are generally only recorded from the motor fibers of the median and ulnar nerves, allowing evaluation of only the lower trunk and C8-T1 nerve roots.

3.4. *Needle Electromyography*

Needle Electromyography (EMG) is an essential tool for electrodiagnosis of suspected brachial plexus lesions as there is no restriction to the anatomic level and extent of the plexus which can be examined by needle EMG, as opposed to nerve conduction studies. In addition to demonstrating the extent of a given lesion, it is useful in documenting the severity of the lesion, including the presence or absence of axonal continuity to a particular muscle (fundamental in determining prognosis). As in radiculopathy, EMG may demonstrate axonal loss in muscles that appear clinically normal. Conversely, EMG may be used to later demonstrate evidence of reinnervation before it can be identified by physical examination.

Generally, extensive EMG study is required to evaluate suspected brachial plexus lesions as muscles innervated by the relevant trunks, cords, and nerves should be examined. Additionally, the presence of paraspinal muscles abnormalities should be sought in order to rule out associated nerve root injury, particularly in cases of trauma where nerve root avulsion may be a consideration. The aim of the EMG is to demonstrate involvement of multiple muscles innervated by a common brachial plexus structure(s) but different nerves. Muscles outside of the area of suspected brachial plexus involvement that are innervated by these nerves should be examined, in addition to paraspinal muscles, with the expectation that these are uninvolved, thus ruling out radiculopathy or multiple mononeuropathies as the cause for a given clinical syndrome. Table 1 outlines an electrodiagnostic approach to suspected brachial plexus lesions.

Table 1
Electrophysiological Patterns of Brachial and Lumbosacral Plexopathies

Plexopathy	*SNAP*	*CMAP*	*F-response*	*Needle exam*	*Others*
Upper trunk	LAC abnormal Median, radial D1 +/– abnormal	Median and ulnar normal	Normal	Abnormal: (C5 innervated) deltoid, biceps supraspinatus, infraspinatus Partially involved: triceps, brachioradialis pronator teres flexor carpi radialis	Rhomboids, serratus anterior and cervical paraspinal muscles normal
Lower trunk (neurogenic thoracic outlet syndrome)	Ulnar abnormal MAC abnormal	Ulnar and median +/- reduced (related to axon loss)	May be abnormal	Abnormal: (C8-T1 innervated median, ulnar, and radial) abductor pollicis brevis, first dorsal interosseous, extensor indicis propius	Pronator teres, triceps, biceps, and C8-T1 paraspinals normal
Middle trunk	Median abnormal Radial +/- abnormal	Normal	Normal	Abnormal: (C7 innervated) triceps pronator teres flexor carpi radialis	
Lateral cord	LAC abnormal Median D1, D2 abnormal	Normal	Normal	Abnormal: biceps pronator teres flexor carpi radialis	
Medial cord	Ulnar abnormal MAC abnormal	Median and ulnar +/- reduced	+/- abnormal (related to axon loss)	Abnormal C8-T1 muscles (median, and ulnar) Abductor pollicis brevis, first dorsal interosseous	Radial innervated C8 muscles (extensor indicis proprius, digitorum communis) normal

(Continued)

Table 1 (Continued)

Plexopathy	*SNAP*	*CMAP*	*F-response*	*Needle exam*	*Others*
Posterior cord	Radial abnormal	Radial abnormal Median and ulnar normal	Normal	Abnormal: Proximal and distal radial-innervated muscles triceps, brachioradialis extensor carpi radialis extensor indicis proprius	
Lumbosacral Plexopathy					
Lumbar plexus (L2-L4)	Saphenous (L4) abnormal Lateral femoral cutaneous (L2-3) abnormal	Tibial normal Peroneal normal	Normal	Abnormal: iliopsoas quadriceps hip adductors	
Sacral plexus (L5-S3)	Superficial peroneal (L5) abnormal Sural (S1-2) abnormal	Tibial, peroneal +/– reduced (related to axon loss)	+/– abnormal	Abnormal: L4-5: tensor fasciae lata, gluteus medius, tibialis anterior, peroneus longus, and tibialis posterior; S1: gluteus maximus, biceps femoris, and gastrocnemius.	

SNAP, sensory nerve action potential; CMAP, compound muscle action potential; LAC, lateral antebrachial cutaneous; MAC medial antebrachial cutaneous.

As in all axonal injury involving motor fibers, the earliest finding on needle exam is decreased motor unit recruitment (immediate) followed by denervation (i.e., fibrillation potentials, positive sharp waves) within the first 3 wk or so. Reinnervation changes (long-duration, high-amplitude, and polyphasic motor unit potentials) are expected in involved muscles within 3–6 mo after the acute injury. In cases where axonal continuity cannot be demonstrated initially (i.e., lesions characterized by complete denervation in some muscles), serial electrodiagnostic studies may be used to search for any evidence of recovery so that, in cases where no reinnervation is occurring, decisions may be made regarding surgical exploration or nerve grafting.

4. SPECIFIC DISORDERS OF THE BRACHIAL PLEXUS

Brachial plexopathies are most commonly caused by trauma, including closed traction injuries as well as penetrating trauma and dislocation of the humerus. The position of the arm and head is important in determining susceptibility of the brachial plexus structures during a closed injury: with the arm down at the side, a force causing excessive shoulder depression will be transmitted to the upper trunk (C5-6 roots), whereas with the arm overhead, force applied to the axilla is transmitted to the lower trunk (C8-T1 roots). Iatrogenic brachial plexopathies are not uncommon; for example, lower trunk plexopathies can be encountered as a complication of median sternotomy procedures. Obstetric paralysis may occur in newborns following difficult vaginal deliveries; upper trunk lesions (Erb's palsy) are more common than isolated lower trunk lesions (Klumpke's palsy). Infiltrating tumors and other masses, including those in the apex of the lung, can also cause severe, usually very painful, plexopathies.

Neuralgic amyotrophy goes by many names, including idiopathic brachial neuritis and Parsonage-Turner syndrome. This immune-mediated disorder has a characteristic clinical and electrodiagnostic pattern which is distinctly different from structural lesions involving the brachial plexus. Rather than an upper trunk brachial plexopathy, the electrodiagnostic pattern in neuralgic amyotrophy is one of multifocal peripheral nerve involvement, with a predilection for motor nerves innervated by the upper trunk, as well as the anterior interosseous nerve, long thoracic nerve, and proximal median nerve. Often there are minimal sensory abnormalities on both the physical examination and the sensory nerve conduction studies may be normal, in contrast to structural lesions of the brachial plexus.

Radiation plexopathy is a rare complication of radiation therapy. It is usually seen as a late sequelae, years after treatment, most often in women receiving axillary irradiation for breast cancer. Either the upper or lower trunk may be involved; the absence of pain differentiates this from recurrent tumor. If present, myokymic discharges may be identified on EMG and are a useful diagnostic feature distinguishing radiation plexopathy from metastatic tumor.

True neurogenic thoracic outlet syndrome (TOS) is a rare disorder, distinct from the more commonly diagnosed symptomatic TOS, in which pain is a prominent feature, without focal neurological symptoms. In neurogenic TOS a fibrous band compresses the proximal aspect of the lower trunk of the brachial plexus, producing a clinical picture of intermittent paresthesias progressing to intrinsic hand weakness and atrophy with sensory loss in the medial forearm and hand. Electrodiagnostic studies in neurogenic TOS demonstrate a lower trunk brachial plexopathy.

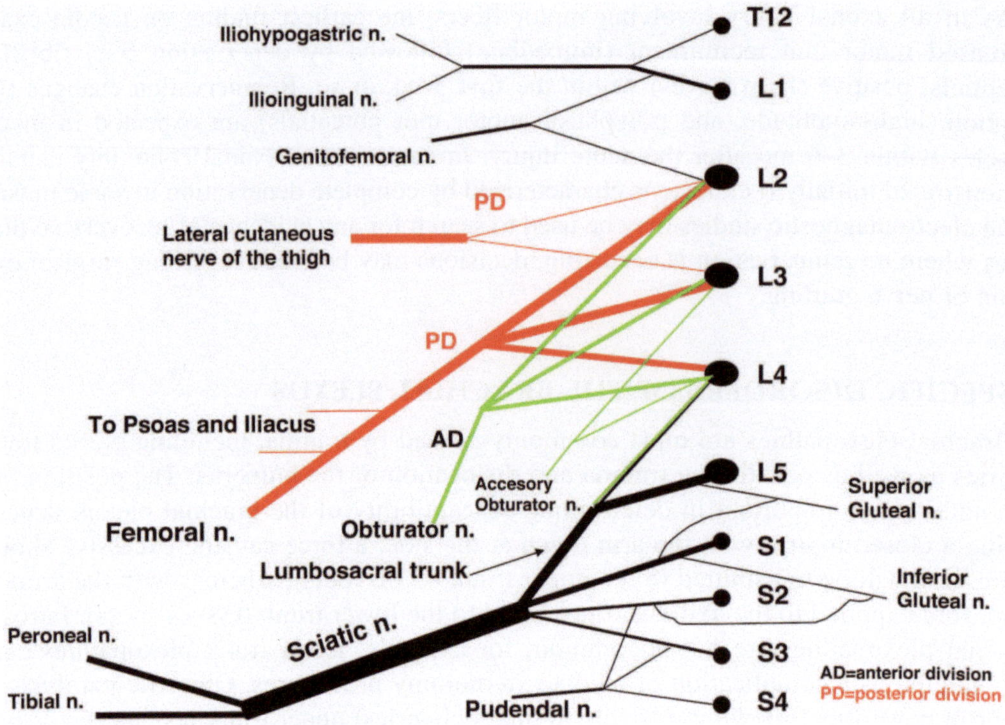

Fig. 2. Schematic of anatomy of the lumbosacral plexus.

5. LUMBOSACRAL PLEXOPATHY

5.1. Lumbosacral Plexus Anatomy

The lumbosacral plexus may best be considered as two distinct plexuses, the lumbar and the sacral (Fig. 2). The lumbar plexus is derived from L1-L4 nerve roots and gives rise to two major nerves: the *femoral* nerve (L2-L4) that innervates the quadriceps muscle and the skin of the anterior thigh and medial lower leg, and the *obturator* nerve (L2-L4) innervating the thigh adductor muscles and the skin of the medial thigh. Branches to the iliopsoas muscle arise directly from the plexus, traveling alongside the femoral nerve. The lumbar plexus also gives rise to several sensory nerves: the iliohypogastric (L1) supplying the anterior and lateral lower abdominal wall, the ilioinguinal (L1) to the upper medial thigh and root of the penis or labium majus, the genitofemoral nerve (L1-L2) to the upper anterior thigh and scrotum, and the lateral cutaneous nerve of the thigh (L2-L3) to the lateral thigh.

The sacral plexus derives from L4 through S4 nerve roots and gives rise to the largest nerve of the plexus, the *sciatic* nerve. The sciatic nerve supplies the hamstring muscles and all the muscles below the knee through its two branches: the *peroneal* and *tibial* nerves. The superior and inferior *gluteal* nerves arise from the sacral plexus proximal to the sciatic nerve and supply the gluteus muscles.

5.2. Pathophysiology

The lumbosacral plexus is well protected by the pelvis, making it less susceptible to traumatic injury than the brachial plexus. It is, however, a favored site for involvement in diabetes, in contrast to the brachial plexus, which is rarely affected by diabetic amyotrophy. The LSP can be also affected by malignant invasion, radiation, retroperitoneal hematoma, and less frequently by trauma. As in the brachial plexus, the lesions that can be identified on electrophysiological testing are largely axonal in nature and occur distal to the dorsal root ganglion, as opposed to radiculopathies. Hence, as with brachial plexopathy, sensory NCS abnormalities are typical and help differentiate lumbosacral plexopathy from radiculopathy.

6. ELECTRODIAGNOSIS: LUMBOSACRAL PLEXOPATHIES

Several principles of electrodiagnosis in lumbosacral plexopathy are similar to those described in the evaluation of brachial plexus lesions. The anatomy of the lumbosacral plexus is more straightforward; again, considering lumbar and sacral plexus separately is helpful. In general, lesions tend to affect either one but rarely both the lumbar and sacral plexus. In cases of lumbar plexopathy, the patient generally presents with quadriceps weakness and the differential diagnostic considerations are femoral neuropathy versus high lumbar radiculopathy (L2-3). The involvement of obturator-innervated muscles localizes a lesion outside of the femoral nerve territory; to distinguish lumbosacral plexopathy from radiculopathy, the saphenous SNCS should be performed and paraspinal EMG is important.

A patient with a lesion of the sacral plexus will often present with foot drop (inability to dorsiflex the foot) or a flail foot (inability to dorsiflex or plantar flex the foot), and the differential diagnostic considerations include peroneal neuropathy, sciatic neuropathy, and lumbosacral radiculopathy. Weakness outside of the ankle dorsiflexors and foot evertors (involving tibial-innervated muscles such as tibialis posterior and gastrocnemius) indicates a lesion proximal to the peroneal nerve while weakness of gluteal muscles indicates a lesion proximal to the sciatic nerve, involving either plexus or nerve roots.

6.1. Motor Nerve Conduction Studies

Routine MNCS of the peroneal and tibial nerves evaluate only the sacral plexus. As lumbosacral plexopathies may superficially resemble peroneal neuropathy, it is important to exclude focal lesions of this nerve, most commonly occuring at the fibular neck. This includes peroneal MNCS recording tibialis anterior if no CMAP is obtainable from extensor digitorum brevis.

Because of their proximal location, the muscles innervated by the lumbar plexus are technically difficult to evaluate with MNCS. A CMAP may be recorded from quadriceps, stimulating femoral nerve at the level of the inguinal ligament. Side-to-side comparison is the most reliable means of obtaining a relative normal value for amplitude. While this may be helpful in demonstrating axonal loss, one can obtain evidence for this from the needle EMG of the quadriceps; thus, femoral MNCS is not commonly performed in the evaluation of lumbar plexus lesions.

6.2. Sensory Nerve Conduction Studies

As in brachial plexopathy, SNCS are a key tool in the assessment of lumbosacral plexus lesions; however, there are important issues which complicate both the performance and the interpretation of SNCS in the lower extremities. First, it is technically difficult and almost impossible to elicit responses from some of the nerves arising from the lumbar plexus, such as the iliohypogastric, genitofemoral, and ilioinguinal nerves. While sensory responses from the lateral femoral cutaneous and saphenous nerves can be obtained, it is often with significant difficulty, even in normal subjects; as previously emphasized, side-to-side comparison is necessary. Sensory responses representing sacral plexus dermatomes (i.e., superficial peroneal and sural) are more easily obtainable and extremely helpful in the evaluation of sacral plexus lesions.

Another complicating issue which frequently influences the electrodiagnostic evaluation of this region is the presence of underlying polyneuropathy, a common finding in the population of patients seen in the EMG laboratory. Abnormalities due to polyneuropathy will affect SNCS most significantly; in some cases, SNAPs are absent, significantly hindering lesion localization. However, when milder polyneuropathy is present, the finding of an asymmetrically reduced SNAP in the involved lower extremity is a useful, localizing finding.

6.3. F-Responses

As in brachial plexopathy, late responses may provide information about proximal conduction in the tibial and peroneal nerves; however, abnormalities are not specific for localization to the plexus. Still, F-response abnormalities may be an early finding in lumbosacral plexopathy.

6.4. Needle EMG

EMG is one of the most important electrodiagnostic tests in the evaluation of lumbosacral plexopathy. As in the upper extremity, a plexus lesion will produce abnormalities in multiple nerve and root territories, largely sparing related paraspinal muscles. Therefore, muscles from different myotomes supplied by different peripheral nerves need to be evaluated, and the sampling should include those innervated by both the lumbar and sacral plexus. When evaluating the lumbar plexus it is useful to examine muscles innervated by the femoral nerve and obturator nerves, as well as the iliopsoas and the high lumbar paraspinal muscles (Table 1). Lumbar plexus lesions are differentiated from femoral neuropathy by the presence of EMG abnormalities in the obturator-innervated adductor longus muscle, in addition to the quadriceps. The finding of abnormalities restricted to the iliopsoas and quadriceps is considered evidence of a femoral neuropathy, because the nerves innervating these muscles run in very close proximity one another, even though, strictly speaking, the branch to iliopsoas comes directly from the lumbosacral plexus rather than the femoral nerve.

In the evaluation of the sacral plexus, the electromyographer examines muscles innervated by peroneal and tibial divisions of the sciatic nerve in multiple myotomes both below and above the knee, followed by gluteal-innervated muscles, and followed finally by lower paraspinal muscles. For example, to distinguish between peroneal neuropathy and more proximal lesions causing foot drop, i.e., sciatic neuropathy or lumbosacral plexopathy, EMG abnormalities are sought in the short head of the biceps femoris (peroneal division of the sciatic nerve, above the knee) in addition to the tibial-innervated tibialis posterior. Additional abnormalities in gluteal-innervated muscles then help localize the lesion to the sacral plexus, proximal to the sciatic nerve.

7. SPECIFIC DISORDERS OF THE LUMBOSACRAL PLEXUS

Diabetic amyotrophy is the most common lesion affecting the lumbosacral plexus. The exact localization of this disorder remains subject to considerable controversy; electrodiagnostically, it is most consistent with an asymmetric, radiculoneuropathy affecting high lumbar regions. Clinically, however, its appearance most closely resembles that of a lumbosacral plexopathy.

Patients typically present with severe thigh pain, followed by weakness and wasting involving quadriceps and hip adductors/abductors. Sensory abnormalities are clinically less prominent. Careful electrophysiological examination will nearly always demonstrate involvement outside of the femoral and lumbar plexus region, with abnormalities in sensory and motor branches of peroneal and tibial nerves, frequently accompanied by abnormalities in the paraspinal muscles. The quadriceps, iliopsoas, adductors, and glutei are often most prominently affected. The presence of fibrillation potentials in the paraspinal muscles certainly brings into the question the exact localization of the disorder; despite this, for simplicity, most electromyographers continue to think of diabetic amyotrophy as a form of plexopathy. The electrodiagnostic evaluation is more helpful than physical examination for demonstrating abnormalities outside of a single myotome or nerve territory, and the widespread nature of this lesion often differentiates it from structural lesions involving the nerve roots or plexus. In cases where uncertainty remains, appropriate imaging studies are helpful in establishing diagnosis.

Idiopathic lumbosacral plexitis is another uncommon disorder that often affects elderly men and is associated with considerable pain, weakness, and sensory loss of the involved leg. Like idiopathic brachial neuritis, its cause is unknown, and the physiology is mainly axonal in nature. Unlike brachial neuritis, however, this disorder has a tendency toward recurrence and persistence with prominent pain. Electrodiagnostic studies reveal abnormalities in tibial and peroneal nerves and in the saphenous, sural, and superficial peroneal sensory nerves. Abnormalities on EMG correspond with areas of weakness. Unlike diabetic amyotrophy, distal involvement of the leg and sensory disturbances are common.

Other specific disorders include radiation plexopathy, compression (for example, due to an expanding mass or hematoma in the retroperitoneum), and trauma, (including pelvic fractures and difficult obstetrical delivery).

FURTHER READING

Eisen AA. The electrodiagnosis of plexopathies. In: Clinical Electromyography. Eds. Brown WF, Bolton CF. Butterworth-Heinemann, Boston 1993: pp. 211–225.

Ferrante MA, Wilbourn AJ. Electrodiagnostic approach to the patient with suspected brachial plexopathy. Neurol Clin 2000;20:423–450.

Rutkove SB, Sax TW. Lumbosacral plexopathies. In: Neuromuscular Disorders in Clinical Practice. Eds. Katirji B, Kaminski H, Preston D, Ruff R, Shapiro B. Butterworth-Heinemann, Boston 2002.

QUESTIONS

1. The presence of fibrillation potentials in the paraspinal muscles is most consistent with
   A. Median neuropathy at the wrist
   B. Upper trunk brachial plexopathy
   C. Crush injury to the posterior cord
   D. Radiculopathy
   E. None of the above
2. Electromyography of the rhomboid major muscle is important to evaluate for:
   A. Long thoracic neuropathy
   B. Lower trunk brachial plexopathy
   C. Medial cord brachial plexopathy
   D. C5 radiculopathy
   E. C8-T1 radiculopathy
3. A 38-year-old carpenter comes to the EMG lab for evaluation of right hand weakness and paresthesias in the 4th and 5th digits. On examination you find normal reflexes, mild sensory loss over

digit 5 and hand intrinsic muscle weakness. In addition to ulnar neuropathy, what else is in your differential diagnosis?
   A. Upper trunk brachial plexopathy
   B. Posterior cord brachial plexopathy
   C. Lower trunk brachial plexopathy
   D. C8-T1 radiculopathy
   E. C and D
4. Which of the following tests would be most helpful to further localize the lesion in the previous patient?
   A. Median sensory response recorded from digit 3
   B. Musculocutaneous motor study
   C. Medial antebrachial cutaneous sensory study
   D. Lateral antebrachial cutaneous sensory study
   E. Needle electromyography of the deltoid muscle
5. Which of the following is commonly seen in brachial neuritis?
   A. Winging of the scapula due to trapezius weakness
   B. Weakness in muscles innervated by the anterior interosseous nerve
   C. Deltoid weakness
   D. A radial neuropathy like picture
   E. Diffuse plexopathy with weakness of muscles whose fibers traverse the lower, middle, and upper trunks of the plexus
6. Which of the following is *not* commonly seen in diabetic amyotrophy?
   A. Prominent weakness of foot plantar flexion
   B. Weakness of hip flexion
   C. Fibrillation potentials in iliopsoas, adductor longus, and quadriceps muscles
   D. Reduced deep tendon reflex at the knee
   E. Fibrillation potentials in the paraspinal muscles
7. Which of the following statements is not true regarding brachial plexopathies?
   A. Medial cord plexopathies may clinically look similar to C8-T1 radiculopathies
   B. Involvement of triceps would be anticipated in a lateral cord plexopathy
   C. Sensory loss involving the lateral aspect of the forearm would be expected in right upper trunk plexopathy.
   D. Weakness of serratus anterior would be atypical in a lateral cord plexopathy
   E. In a lower trunk plexopathy, the ulnar sensory response amplitude may be reduced
8. Which of the following is correct?
   A. Traumatic plexopathies are much more common in the upper extremities than the lower
   B. The lower extremity proximal sensory nerves are difficult to test electrophysiologically
   C. In both brachial and lumbosacral plexopathies, F-wave latencies may be prolonged
   D. Both idiopathic brachial neuritis and lumbosacral plexitis are accompanied by prominent pain
   E. All of the above
9. Which of the following pairs is incorrect?
   A. C5 nerve root/upper trunk
   B. C7 nerve root/middle trunk
   C. L3 nerve root/sciatic nerve
   D. Posterior cord of brachial plexus—radial nerve
   E. All are correct
10. Which of the following statements is incorrect?
   A. Median F responses can be helpful in the diagnosis of an upper trunk plexopathy
   B. Peroneal F responses may assist in the diagnosis of a lumbosacral plexopathy
   C. Median motor response amplitude may be reduced in true neurogenic thoracic outlet syndrome
   D. Ulnar sensory response from digit 5 is likely to be normal in vascular thoracic outlet syndrome

ANSWERS

1. Answer: D. The presence of fibrillation potentials in the paraspinal muscles is more consistent with radiculopathy. This is an important finding used to differentiate a plexopathy from a radiculopathy.
2. Answer: D. The dorsal scapular nerve which takes off directly from the C5 motor nerve root before joining the upper trunk of the brachial plexus innervates the rhomboid muscle. Abnormalities seen in the rhomboid muscle generally support a C5 radiculopathy over a plexopathy although a very proximal upper trunk plexopathy could potentially produce abnormalities in this muscle as well.
3. Answer: E. The distribution of weakness and sensory findings are consistent with either a lower trunk/medial cord brachial plexopathy or a C8-T1 radiculopathy.
4. Answer: C. Abnormalities in this sensory nerve will indicate that the problem is at the level of the lower trunk or medial cord and will rule out C8 and T1 roots. In addition, the medial antebrachial cutaneous sensory study will be normal in isolated ulnar neuropathy.
5. Answer: B. Weakness of anterior interosseous muscles is commonly seen in brachial neuritis. Winging of the scapula is usually seen in brachial neuritis due to weakness of serratus anterior. The other clinical pictures would be distinctly atypical for brachial neuritis.
6. Answer: A. Prominent foot flexion weakness would be distinctly uncommon in diabetic amyotrophy. All of the other items are commonly seen in this disorder.
7. Answer: B. Nerve fibers to triceps leave the upper, middle, and lower trunks to form the posterior cord, from which the radial nerve arises; fibers destined for the radial nerve do not travel through the lateral cord. All of the other statements are correct.
8. Answer: E. All the statements are correct.
9. Answer: C. The L3 nerve root provides fibers that are enter the obturator and femoral nerves; the sciatic nerve fibers are derived from the L4-S2 roots.
10. Answer: A. Median F responses evaluate only the C8-T1 roots, the lower trunk, and the medial cord. While the sensory fibers from the median nerve traverse other areas of the plexus, the motor fibers from which the F waves are obtained are restricted to these regions.

19
Evaluation of the Cranial Nerves

George Sachs

Summary

Electrophysiological testing of the cranial nerves remains an important and perhaps under utilized area of neurophysiological evaluation. Motor responses from individual branches of the facial nerve can be obtained to assist in the diagnosis and prognosis of facial neuropathies. In blink reflex testing, the first division of the trigeminal nerve is stimulated and responses from the orbicularis oculi are obtained. This form of testing can be used for a variety of purposes including helping to localize disorders of the Vth and VIIth cranial nerves, assisting in the evaluation of Guillain–Barré syndrome, and even in the assessment of brainstem disorders. It is also useful in the assessment of hemifacial spasm. The masseter reflex and master silent period can also be assessed when clinically indicated. Although less commonly pursued, other cranial nerves can also be evaluated electrophysiologically, including the XIth and XIIth cranial nerves. Assessment of these cranial nerves can be important in assessing other diseases, including amyotrophic lateral sclerosis.

Key Words: Bell's palsy; blink reflex; facial neuropathy; hemifacial spasm; trigeminal neuropathy.

1. INTRODUCTION

Although clinical neurophysiological techniques can be applied to nearly all cranial nerves, this chapter will focus on those readily studied in standard EMG laboratories. Evoked potential studies of the optic and auditory nerves are discussed in Chapters 25 and 26, and only brief mention will be made of specialized techniques applied to extraocular and laryngeal muscles. Standard nerve conduction studies and EMG can assess the facial, trigeminal, spinal accessory, and hypoglossal nerves and the muscles they innervate. However, a few general caveats apply to the study of these cranial nerves. First, normal values for conduction studies are not as well-established as those for limb nerves and, therefore, side-to-side comparisons become particularly important. Second, the short or inaccessible extracranial course of some nerves prohibits direct assessment of conduction velocity. In such cases, reflex studies and latency measurements provide the next best option. Finally, EMG of cranial muscles requires a delicate approach and familiarity with specific motor unit potential (MUP) characteristics.

2. CONDUCTION STUDIES

2.1. *Facial Nerve*

2.1.1. *Relevant Anatomy*

Of all cranial nerves, the facial nerve (Fig. 1) is the one most frequently studied in EMG labs. It emerges from the pons, coursing across the cerebello-pontine angle to the internal

From: *The Clinical Neurophysiology Primer*
Edited by: A. S. Blum and S. B. Rutkove © Humana Press Inc., Totowa, NJ

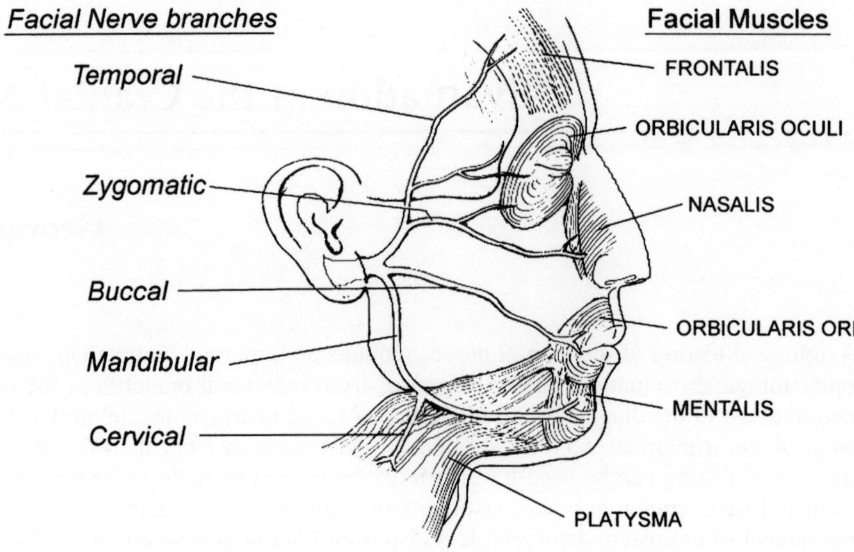

Fig. 1. Major branches of the facial nerve and muscles that they innervate.

acoustic meatus, where it enters the facial canal in the petrous bone. Branches given off within the petrous bone innervate the stapedius muscle, salivary, and lacrimal glands as well as taste buds on the anterior tongue. Soon after it exits the skull through the stylomastoid foramen, the nerve sends small branches to the digastric, stylohyoid, and occipital scalp muscles. The main nerve trunk courses through the parotid gland and then divides into five branches innervating facial muscles:

1. The temporal branch supplying the frontalis.
2. The zygomatic branch supplying the orbicularis oculi and nasalis.
3. The buccal branch supplying the orbicularis oris.
4. The mandibular branch supplying the mentalis.
5. The cervical branch supplying the platysma.

2.1.2. Conduction Studies

Motor conduction studies can be performed for any of the facial muscles mentioned in Subheading 2.1.1. Percutaneous stimulation underneath the ear at the tip of the mastoid process excites all five facial branches. A more distal site, just anterior to the tragus, stimulates the zygomatic and temporal branches. The nasalis muscle is a particularly attractive recording site because it tends to produce a compound muscle action potential (CMAP) with a sharp initial negative deflection. Motor conduction studies to the orbicularis oculi are convenient because the same recording site is shared with the blink reflex. However, flat circular muscles, such as the orbicularis oculi and oris may produce CMAPs that are more dispersed, with initial positive deflections. Furthermore, artifact from direct co-stimulation of the masseter is more likely to complicate recording from these muscles. Facial muscles normally produce CMAPs with amplitudes ranging between 2 and 4 mA, although a value less than 50% of the normal contralateral is the most widely accepted criterion for abnormality.

2.2. Trigeminal Nerve

2.2.1. Relevant Anatomy

Unlike the facial nerve, the trigeminal nerve contains a large number of somatosensory afferent fibers in addition to motor axons. Primary neurons relaying pain/temperature and touch modalities reside within the gasserian ganglion. Their distal axons leave the ganglion in the three sensory divisions: ophthalmic (V1), maxillary (V2), or mandibular (V3). Proximal axons of touch fibers project to the principal trigeminal nucleus of the pons, whereas pain/temperature fibers send their proximal axons to the spinal trigeminal nuclei (extending from the pons to the cervical spinal cord). The cell bodies of trigeminal muscle spindle afferents are thought to occupy the mesencephalic trigeminal nucleus. This nucleus extends rostrally to the level of the colliculi, serving as a relay center for proprioceptive reflexes.

Motor fibers of the trigeminal nerve originate from their nucleus in the mid pons. They pass through Meckel's cave at the tip of the petrous bone and then under the Gasserian ganglion to emerge from the skull via the foramen ovale. The motor fibers join the mandibular division of the nerve to supply the masticatory muscles (masseter, temporalis, and pterygoids) as well as palatal, mylohyoid, and the anterior belly of the digastric muscle.

2.2.2. Conduction Studies

The trigeminal nerve's extracranial course is not suitable for the performance of motor conduction studies. Branches to masticatory muscles originate from the mandibular division soon after its exit from the foramen ovale and take short, deep trajectories to their targets. Many sensory branches are also inaccessible, but orthodromic sensory conduction studies of the supraorbital nerve have been described. Stimulation at the upper forehead elicits a sensory nerve action potential recordable at the supraorbital notch. The latency of this response is usually shorter than 1 ms, which makes it prone to distortion from stimulus artifact.

2.2.3. Blink Reflex

Reflex studies provide an alternative method of assessing sensory conduction in the trigeminal nerve. Of these, the blink reflex is the most commonly used. It represents the electrophysiological correlate of the glabellar tap reflex. The technique involves recording responses bilaterally from orbicularis oculi after electrical stimulation of the supraorbital trigeminal branch. This produces two separate responses. The earlier response (termed R1) appears with a latency of 8 to 13 ms and is normally limited to the ipsilateral orbicularis oculi. Appearing later, and bilaterally, is the R2 response. It is temporally more dispersed, with a normal onset latency ranging from 24 to 41 ms. R2 responses recorded contralateral to stimulation have a slightly longer average latency, but the side-to-side latency difference for any single stimulation should not exceed 5 ms.

Paired stimuli at 3 to 5 ms intervals will, at times, evoke an R1 response when single stimuli fail. The blink reflex can also be elicited by a mechanical glabellar tap delivered using a reflex hammer with a microswitch to trigger the trace. This technique stimulates both sides, leading to bilateral R1 responses with slightly longer latencies than those evoked electrically. Electrical stimulation of the infraorbital or mental nerves often elicits blink responses, which can assess sensory transmission through these branches.

Blink reflexes can delineate pathology in the trigeminal nerve, the facial nerve, and brainstem pathways connecting trigeminal and facial nuclei (Fig. 2). Interpretation requires comparison

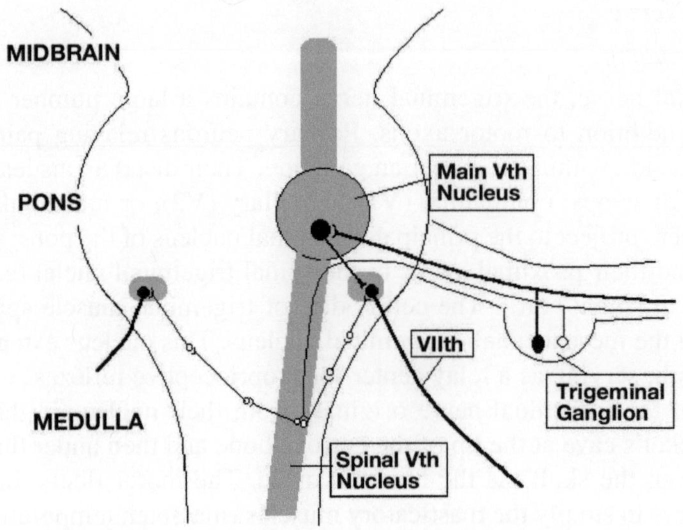

Fig. 2. Schematic diagram illustrating the neural pathways of the blink reflex. Large-caliber fibers form a reflex arc through the main trigeminal nucleus to the ipsilateral facial nucleus, producing the R1 response. Thinner fibers form multisynaptic pathways through the spinal trigeminal nucleus to produce bilateral R2 responses.

of responses from both right- and left-sided stimulation. Further localization involves an understanding of the divergent pathways underlying the R1 and R2 components. The R1 response reflects conduction through the main trigeminal nucleus and is a good indictor of pontine function. The R2 responses are mediated by the spinal trigeminal nuclei and can identify pathology at pontine or medullary levels. Table 1 summarizes the patterns of abnormality used to localize various lesions.

2.2.4. *Masseter Reflex*

The masseter reflex is an electrophysiological analog of the jaw jerk. It is used less commonly than the blink reflex for evaluation of trigeminal sensory function. However, it can provide complementary information, in that it assesses muscle spindle afferents and their connections through the mesencephalic trigeminal nucleus. Additionally, it is one of the few tests assessing motor conduction within the trigeminal nerve.

The reflex is elicited by a reflex hammer with a microswitch that triggers recording on percussion. Response to a midline tap of the chin is recorded bilaterally from percutaneous electrodes over the masseter muscles. Because the response can be difficult to obtain in normal subjects, the only meaningful abnormality is a consistent unilateral absence or delay.

2.2.5. *Masseter Silent Period*

This reflex, also known as the masseter inhibitory reflex, is mediated through both sensory and motor divisions of the trigeminal nerve. Stimulation of the mental or infraorbital branch is performed during tonic, voluntary closure of the jaw. Surface EMG activity recorded over the masseter muscle is typically interrupted by two silent periods. The first, termed SP1, has an onset latency of 10 to 15 ms. The later SP2 silent period normally occurs at 40 to 50 ms. Both SP1 and SP2 occur bilaterally after unilateral stimulation. Successful recording of this reflex requires forceful dental occlusion for 2 to 3 s while relaxing other facial musculature. Background activity from

Table 1
Interpretation of Blink Reflex Studies

		R supraorbital stim		*L supraorbital stim*	
	Record	*R1*	*R2*	*R1*	*R2*
Normal	Right	nl	nl		nl
	Left		nl	nl	nl
R CN VII lesion	Right	Ab	Ab		Ab
	Left		nl	nl	nl
R CN V lesion	Right	Ab	Ab		nl
	Left		Ab	nl	nl
R pons lesion	Right	Ab	±		±
	Left		±	nl	nl
R medulla lesion	Right	nl	Ab		nl
	Left		Ab	nl	nl

Expected blink reflex results for trigeminal and facial neuropathies as well as brainstem lesions. R, right; stim, stimulus; L, left; nl, normal; CN, cranial nerve; ab, abnormal or absent; ±, possibly abnormal depending on precise location of lesion within the pons.

surrounding muscles may mask the silent periods and, if this is the case, recording the EMG activity with a concentric needle in the masseter may prove more successful.

The SP1 silent period likely reflects conduction through the sensory trigeminal branch to inhibitory interneurons projecting bilaterally to the motor trigeminal nucleus. This arc lies within the midpontine level. The SP2 is mediated through the spinal trigeminal nucleus and involves upper medullary as well as pontine levels. Similar to the masseter (jaw jerk) reflex, the masseter inhibitory reflex provides an assessment of both trigeminal sensory and motor conduction, but it supplies complementary information regarding smaller caliber sensory fibers and pontomedullary connections that are more caudal.

2.3. *Spinal Accessory Nerve*

2.3.1. *Relevant Anatomy*

This largely motor nerve emerges from the brainstem and cervical spine as a long, linear array of rootlets. The spinal portion of the nerve ascends through the foramen magnum to merge with the cranial portion and then exits the skull through the jugular foramen. Coursing down the neck between the carotid artery and jugular vein to innervate the sternocleidomastoid, the nerve penetrates that muscle before traversing the posterior triangle of the neck. It then terminates by innervating the trapezius. Communications from cervical roots C3 and C4 join the spinal accessory nerve along its distal course. These innervate the trapezius to a variable degree. The lower portion of the trapezius more commonly receives direct innervation from the C3 and C4 roots, but cervical innervation of the entire trapezius has been reported.

2.3.2. *Conduction Studies*

Motor conduction studies of the spinal accessory nerve most often evaluate the distal segment innervating the trapezius. The nerve is stimulated percutaneously at the posterior border of the sternocleidomastoid muscle, with CMAP recorded from the upper portion of the trapezius. Conduction studies assessing portions of the nerve that are more proximal can be performed. These involve stimulation anterior to the mastoid process with recording via needle electrodes in the sternocleidomastoid and trapezius.

2.4. Hypoglossal Nerve

2.4.1. Relevant Anatomy

The hypoglossal nerve derives from 10 to 15 rootlets exiting the ventral medulla. These form two bundles that pass through the hypoglossal canal and merge into a single nerve on leaving the skull. The nerve descends toward the angle of the mandible, receiving communications from C1 and C2 ventral rami and giving off a branch to the ansa cervicalis that supplies the infrahyoid muscles. The hypoglossal nerve proper turns medially past the hyoid bone and divides into branches supplying the extrinsic (styloglossus, hypoglossus, genioglossus, and geniohyoid) muscles as well as the intrinsic muscles of the tongue.

2.4.2. Conduction Studies

Hypoglossal nerve stimulation is achieved with a bipolar surface electrode, placing the cathode 1 cm medial to the inner aspect of the mandible, at a point one-third the distance from the angle to the apex. CMAPs are recorded from the surface of the tongue using disc electrodes affixed to a bite bar or tongue depressor. The electrodes are placed along the midline, with the active electrode 1 cm behind the incisors and the reference electrode 2 cm more posterior. The amplitude of the CMAP may be variable because of tongue movement, but latencies are reportedly quite stable. As with other cranial nerve conduction studies, clear asymmetries on bilateral stimulation provide the most convincing abnormalities.

2.5. Recurrent Laryngeal Nerve

2.5.1. Relevant Anatomy

This branch of the vagus nerve originates at the base of the neck and takes a different proximal course on the two sides. On the left, it loops under the aortic arch. On the right, it loops under the subclavian artery before ascending to the larynx in the sulcus between the trachea and esophagus. All laryngeal muscles except the cricothyroid receive innervation from the recurrent laryngeal nerve.

2.5.2. Conduction Studies

Conduction studies of this nerve require specialized techniques. Recording from laryngeal muscles is most often accomplished with a concentric needle placed under laryngoscopic guidance. Needle electrodes are used to stimulate two sites: the descending vagus trunk at the posterior border of the sternocleidomastoid and the recurrent laryngeal branch lateral to the trachea. Conduction velocity can be estimated using measurements of nerve length between these two sites, derived from cadavers.

3. EMG

The needle exam of cranial musculature involves largely the same principles and technique used in studying limb muscles. However, a few points deserve special attention. Because facial muscles are small and located in sensitive areas, many examiners prefer to use very thin recording needles (0.3-mm diameter or 30-gauge concentric needles, often referred to as "facial needles"). Their impedance is higher than standard concentric needles, raising the amplitude of both signal and noise. Other signal characteristics, most importantly the duration and complexity of MUPs, are not altered by the use of facial needles.

In general, motor units in cranial muscles tend to be brief and of low amplitude, making assessment of myopathies difficult. Relaxation of cranial muscles may be hard to achieve but is most important for detection of fibrillation potentials, which can be confused with small brief motor units. Furthermore, activity from surrounding muscles may contaminate the recording unless selective activation of a given muscle is attempted on a background of complete relaxation. This is particularly relevant to examination of the orbicularis oris and orbicularis oculi, thin muscles in close proximity to the masseter. Every effort should be made to relax the jaw when recording from these muscles.

The cranial muscles most commonly studied by EMG include the temporalis and masseter for the trigeminal nerve; the frontalis, orbicularis oculi, and orbicularis oris for the facial nerve; the trapezius and sternocleidomastoid for the spinal accessory nerve; and the tongue for the hypoglossal nerve. One approach to hypoglossal muscles involves needle insertion through the surface of the tongue while holding it protruded with a gauze pad. An alternative is to assess the genioglossus through the under surface of the chin, 2 cm back from the mental apex. With either approach, it is important to allow the tongue to settle back within the floor of the mouth to achieve relaxation, which can often be very difficult to accomplish.

EMG of extraocular muscles is performed rarely and is usually carried out by ophthalmologists familiar with the details of the anatomy. The approach is through the eyelid, sometimes with the use of a topical anesthetic. Assessment for spontaneous activity is difficult because extraocular muscles maintain some baseline motor unit activation even in primary position.

4. CLINICAL CORRELATIONS

4.1. Bell's Palsy—Initial Evaluation

The diagnosis of Bell's Palsy is usually obvious from its clinical features. Electrodiagnostic studies serve mainly to quantify the severity and determine prognosis. Assessing the relative contribution of demyelinating and axonal components is crucial to prognosis. Because Wallerian degeneration does not fully develop until 5 to 8 d after axonal injury, motor conduction studies provide accurate prognostic data only after that time. CMAP amplitude is the most useful prognostic parameter. If the CMAP amplitude is less than 10% of that on the healthy side, maximum recovery will be delayed 6 to 12 mo, usually leaving moderate-to-severe weakness. If the amplitude is 10 to 30% of the healthy side, recovery may take 2 to 8 mo, with mild-to-moderate residua. If the CMAP amplitude is greater than 30% of normal, there is usually full recovery within 2 mo. CMAP amplitude is most accurate in predicting a good prognosis, and less accurate in those patients with greater than 90% loss of amplitude, because a substantial number (up to 47%) will still have a good recovery. The CMAP amplitude method is of limited use in bilateral facial neuropathies.

The blink reflex does not add much to prognosis in Bell's palsy. Although it evaluates conduction along the entire nerve, its assessment of axonal degeneration offers little beyond direct facial nerve studies and is limited by the same time constraints. Absence of the R1 component within 2 wk of onset indicates a 45% chance of satisfactory recovery, compared with 94% in patients with normal R1. A prolonged R1 latency only reduces the chances of satisfactory recovery to 87%.

Needle EMG will show abnormalities in motor unit recruitment from the outset of Bell's palsy, but this finding does not help differentiate demyelinating from axonal lesions and, consequently, has little prognostic usefulness. However, the presence of even a few voluntary MUPs in a patient with complete clinical paralysis indicates the nerve remains in continuity and is consistent with a better prognosis than those patients with no MUPs. This finding needs to be interpreted cautiously in the orbicularis oris because there may be some degree of crossed innervation in this midline muscle. Fibrillation potentials and positive sharp waves indicate the presence of axonal degeneration but are difficult to quantify and do not necessarily imply poor recovery. They usually do not appear earlier than 1 to 2 wk after onset and may not be observed for up to 3 wk.

4.2. Bell's Palsy—Later Studies

Regeneration after Bell's palsy can lead to a number of electrophysiological changes. Findings on needle exam include low-amplitude, polyphasic MUPs typical of newly regenerated motor units (nascent motor unit potentials). In addition to abnormal volitional recruitment, these motor units may also generate spontaneous discharges, either single or grouped (myokymic). This presumably reflects a hyperexcitable state of facial motor neurons or their axons.

Synkinetic movements often develop after regeneration of the facial nerve, manifesting as jaw winking and other clinical phenomena. An electrophysiological counterpart can be observed in the blink reflex, in which supraorbital nerve stimulation elicits R1 and R2 responses not only from the orbicularis oculi but also from the orbicularis oris, platysma, or other lower facial muscles. A combination of aberrant axonal branching, ephaptic transmission between axons, and neuronal hyperexcitability underlies these synkinetic discharges.

4.3. Hemifacial Spasm

This unilateral paroxysmal contraction of facial muscles may follow Bell's palsy or occur with a mass compressing the facial nerve, but, in most instances, hemifacial spasm arises without a clear cause. Surgical exploration often reveals an arterial loop impinging the nerve in such "idiopathic" cases. EMG activity during spasm shows bursts of MUP firing at rates of 80 to 150 Hz. Brief bursts often repeat at rates of 5 to 20 per second but prolonged tonic discharges also occur. During voluntary blinking, synkinetic discharges appear in orbicularis oris or other facial muscles normally not involved in blinking. The synkinesis of idiopathic hemifacial spasm may vary from moment to moment, unlike the constant synkinesis observed after Bell's palsy.

Nerve conduction and reflex studies reveal abnormal responses in cases of hemifacial spasm. Blink reflexes demonstrate synkinetic R1 and R2 responses in muscles such as the orbicularis oris (that is, those not normally activated by supraorbital nerve stimulation), that are more variable than those observed after Bell's palsy. Motor conduction studies with selective stimulation of individual facial nerve branches will evoke CMAPs from muscles supplied by other divisions. This may reflect either ephaptic transmission or hyperexcitability of facial neurons. Hyperexcitability may also manifest as after-discharges following CMAP.

4.4. Trigeminal Neuralgia/Neuropathy

Neuralgia is the most common affliction of the trigeminal nerve. It manifests as paroxysmal pain within one or more branches, affecting (in order of decreasing frequency) the maxillary, mandibular, and ophthalmic divisions. Although a sensation of mild residual numbness can persist in the affected area, the neurological exam shows no deficit. Trigeminal neuralgia complicates many cases of multiple sclerosis; however, it most often occurs in isolation. Surgical exploration of such idiopathic cases frequently reveals a vascular loop contacting proximal portions of the nerve.

Electrophysiological evaluation is normal in cases of trigeminal neuralgia, except for rare prolongation of the R1 component of the blink reflex. Other abnormalities in reflex studies or EMG of trigeminal muscles argue for the alternative diagnosis of trigeminal neuropathy. Trigeminal neuropathy with evidence of denervation on EMG often reflects a neoplastic or traumatic etiology.

Pure sensory trigeminal neuropathy occurs in patients with connective tissue disorders (sensory neuronopthy). These cases show abnormalities in R1 and R2 components of the blink reflex as well as abnormal masseter silent periods. The sensory trigeminal neuropathy complicating Sjogren's syndrome produces blink reflex and silent period abnormalities, but spares proprioceptive neurons within the mesencephalic nucleus and consequently the masseter (jaw jerk) reflex remains normal, because these neurons are located within the central nervous system.

4.5. *Spinal Accessory Neuropathy*

The most common causes of spinal accessory neuropathies are surgical trauma and tumors, but occasionally they occur without clear etiology. Many of these spontaneous cases appear akin to brachial neuritis, likely involving disimmune mechanisms. Motor nerve conduction studies and EMG confirm the diagnosis and localization, with assessment of the sternocleidomastoid important in determining the proximal extent of pathology. On the whole, electrophysiology is less helpful in prognosis than it is for cases of facial neuropathy. The amplitude of trapezius CMAPs on early conduction studies does not predict outcome well. Cases with very low initial CMAP amplitudes and marked denervation on EMG can recover surprisingly well. In part, this may reflect compensatory reinnervation directly from cervical roots.

4.6. *ALS*

EMG of cranial muscles plays an important role in the diagnosis of ALS. Detection of denervation outside the cervical and lumbar regions becomes particularly significant in patients with extensive spondylosis or other spinal pathology that could confound denervation identified in appendicular muscles. EMG abnormalities can be observed well before clinical evidence of weakness in facial or bulbar muscles. Fibrillations and positive waves provide the most definitive evidence of denervation but may be difficult to recognize. Reported studies of cranial muscles in ALS have found fibrillations (in order of decreasing frequency) in the tongue, facial nerve muscles, and trigeminal nerve muscles. Evidence of reinnervation (increased MUP duration and amplitude) is less definitive but more common. Quantitative analysis of MUP parameters has revealed abnormalities in up to 60% of cranial nerve muscles in ALS patients without bulbar signs or symptoms.

SUGGESTED READING

Ongerboer de Visser BW, Cruccu G. Chapter 3: Neurophysiologic examination of the trigeminal, facial, hypoglossal, and spinal accessory nerves in cranial neuropathies and brain stem disorders. In: Clinical Electromyography, 2nd ed. (Brown WF, Bolton CF, eds.). Butterworth Heinemann, Boston, MA, 1993, pp. 63–64.

Kimura J. Chapter 16: The blink reflex. In: Electrodiagnosis in Diseases of Nerve and Muscle: Principles and Practice, 2nd ed. FA Davis, Philadelphia, PA, 1989, pp. 307–331.

Preston DC, Shapiro BE. Chapter 5: The blink reflex. In: Electromyography and Neuromuscular Disorders: Clinical–Electrophysiologic Correlations. Butterworth-Heinemann, Boston, MA, 1998, pp. 57–62.

REVIEW QUESTIONS

1. Poor prognosis in Bell's palsy is indicated by:
   A. Ipsilateral R1 on the blink reflex prolonged by 10%.
   B. Reduced motor unit recruitment in frontalis.
   C. Nasalis CMAP amplitude less than 10% of contralateral.
   D. A and C.
2. Which is least likely to be observed with hemifacial spasm:
   A. R1 response from orbicularis oris on blink reflex.
   B. Denervation in the masseter.
   C. Response in frontalis from stimulation of the zygomatic branch.
   D. After-discharges on conduction studies.
3. On the blink reflex testing, stimulation of the left supraorbital nerve produces abnormal R2 responses in orbicularis oculi bilaterally. All other responses are normal. The lesion is most likely in:
   A. The left trigeminal nerve.
   B. The left pons.
   C. The left medulla.
   D. The right medulla.
4. With the blink reflex, R1 response from the ipsilateral mentalis is observed with:
   A. Aberrant regeneration after Bell's palsy.
   B. Medullary lesions.
   C. Hemifacial spasm.
   D. A and C.
5. Which is least likely to be observed with the trigeminal neuropathy of Sjogren's syndrome:
   A. Denervation in the masseter.
   B. Abnormal blink reflex.
   C. Abnormal masseter silent period.
   D. Normal masseter (jaw jerk) reflex.
6. Compared with larger concentric EMG needles, facial (30 gauge) concentric needles:
   A. Increase MUP duration.
   B. Increase MUP amplitude.
   C. Increase MUP polyphasia.
   D. Increase MUP firing rate.
7. Synkinesis of facial muscles after Bell's palsy reflects:
   A. Aberrant branching of axons.
   B. Hyperexcitability within the Gasserian ganglion.
   C. Ephaptic transmission.
   D. A and C.
8. Which is least likely to occur with a compressing mass in the jugular foramen:
   A. Denervation in the sternocleidomastoid.
   B. Denervation in the geniohyoid.
   C. Denervation in the trapezius.
   D. Scapular winging.
9. Common electrodiagnostic findings in idiopathic trigeminal neuralgia include:
   A. Denervation in the temporalis.
   B. Denervation in the masseter.
   C. Absent R1 and R2 responses of the blink reflex.
   D. None of the above.
10. In patients with ALS:
   A. Denervation is observed more often in temporalis than in the tongue.
   B. Denervation is observed more often in the mentalis than the tongue.
   C. Cranial EMG abnormalities are observed in patients without bulbar symptoms.
   D. There is often early recruitment of motor units in cranial muscles.

REVIEW ANSWERS

1. The correct answer is C. If the CMAP amplitude is less than 10% of the contralateral facial muscles, there is almost always severe axonal degeneration, affording a poor prognosis in Bell's palsy. Delay of R1 response has little prognostic value because slowing of conduction does not cause significant weakness. Reduced motor unit recruitment does not necessarily imply axonal damage. It may reflect conduction block, which is reversible.
2. The correct answer is B. The masseter is supplied by the trigeminal nerve and should not be affected by this disorder of the facial nerve. Answers A and C describe lateral spread of response among facial muscles. This is common in hemifacial spasm and may reflect ephaptic transmission or hyperexcitability within the facial nucleus. After discharges, another manifestation of hyperexcitability, can be observed after CMAPs in motor conduction studies of the facial nerve in cases of hemifacial spasm.
3. The correct answer is C. R2 responses are mediated through the spinal trigeminal tract and nuclei in the medulla. The pathway leading to the contralateral R2 response crosses at the level of the medulla. Although trigeminal nerve and pontine lesions can affect the R2 response, they would most likely also affect the R1 response. A lesion in the right medulla would not disrupt the pathway mediating the ipsilateral R2 from left-sided stimulation.
4. The correct answer is D. R1 responses are normally restricted to the orbicularis oculi. An R1 response from the mentalis represents a synkinetic response, which can be observed after regeneration in Bell's palsy or in hemifacial spasm. A medullary lesion is too caudal to affect the pathway underlying the R1 response.
5. The correct answer is A. The trigeminal neuropathy observed in Sjogren's syndrome (and other collagen vascular disorders) is purely sensory. Sparing of the masseter reflex is characteristic of Sjogren's syndrome. The cell bodies of 1A afferents involved in this reflex reside in the mesencephalic nucleus, which may be less susceptible to immune attack than other trigeminal sensory neurons within the Gasserian ganglion.
6. The correct answer is B. Facial EMG needles are thinner than standard EMG needles and, therefore, have a higher impedance. This increases the amplitude of recorded signals but does not affect other motor unit action potential features.
7. The correct answer is D. Both aberrant branching of regenerated axons and ephaptic transmission between axons cause synkinesis in Bell's palsy. Hyperexcitability in the Gasserian ganglion should not affect facial nerve muscles, nor should it occur in Bell's palsy.
8. The correct answer is B. Compression at the jugular foramen should injure spinal accessory nerve fibers destined for both the sternocleidomastoid and trapezius. Trapezius weakness causes one form of scapular winging. The geniohyoid is supplied by the hypoglossal nerve, which leaves the skull through the hypoglossal canal. Although skull base tumors will frequently compress both cranial nerves XI and XII, a "jugular foramen syndrome" spares the latter.
9. The correct answer is D. Idiopathic trigeminal neuralgia very rarely prolongs R1 latency in the blink reflex, but is not associated with any other electrodiagnostic abnormalities.
10. The correct answer is C. Evidence of ongoing denervation and reinnervation in cranial muscles is frequently observed in ALS patients without bulbar signs or symptoms. EMG abnormalities are found in the tongue, facial muscles, and trigeminal muscles, in order of decreasing frequency.

20

Electrophysiology of Myopathy

Approach to the Patient With Myopathy in the EMG Laboratory

Nithi S. Anand and David Chad

Summary

The myopathic disorders represent a heterogeneous group of diseases with a variety of causes. Although electrodiagnostic testing rarely allows an entirely specific diagnosis to be made, such testing can be extremely helpful in first confirming the presence of myopathy and therefore helping to appropriately categorize it. Standard motor nerve conduction studies generally do not demonstrate substantial abnormalities, except occasional reductions in compound motor potential amplitude in severe cases or where predominantly distal disease is present. Needle EMG remains the most important part of neurophysiological examination. In most myopathic conditions, spontaneous activity is increased, although it is most prominently increased in inflammatory or necrotizing myopathic processes. The presence of myotonic discharges can be very helpful in limiting the differential diagnosis. Complex repetitive discharges, which are present in myopathic conditions, are nonspecific. Motor unit potentials in myopathy are generally of low-amplitude and short duration, except in very chronic conditions, in which the motor unit potentials can actually become of long duration and high amplitude. Finally, it is critical to remember that an entirely normal EMG does not exclude the presence of myopathy.

Key Words: Dermatomyositis; fibrillation potential; myopathy; myositis; myotonic discharge; myotonic dystrophy; polymyositis.

1. INTRODUCTION

The term myopathy is used to denote diseases of the striated muscle. Diseases of muscle are common in neurological practice, and EMG has an important role in their diagnosis and management. We suggest that most patient encounters in the EMG laboratory start with a brief history. This is performed to ascertain the mode of onset, progression, pattern of muscle involvement, and to inquire about relevant medical history, including medications (e.g., statin drugs, coumadin). The history is followed by a focused neurological examination to identify the weak muscle groups and confirm the pattern of muscle involvement. In general, the pattern of muscle involvement in myopathy is proximal more than distal, although there are exceptions to this general rule. In this context, it is useful to have in mind a classification of muscle diseases (Table 1). It is worth remembering that lesions in other parts of the nervous system, such as watershed cerebral infarctions, motor neuron diseases, neuromuscular junction (NMJ) disorders, and motor neuropathies might mimic the clinical picture of myopathy. It is also useful to remember key clinical findings of common

From: *The Clinical Neurophysiology Primer*
Edited by: A. S. Blum and S. B. Rutkove © Humana Press Inc., Totowa, NJ

Table 1
Classification of Myopathies

Muscular dystrophies
Duchenne muscular dystrophy
Becker muscular dystrophy
Emery–Dreifuss muscular dystrophy
Limb girdle muscular dystrophy
Facioscapulohumeral muscular dystrophy
Oculopharyngeal muscular dystrophy
Myotonic muscular dystrophy
Proximal myotonic myopathy
Congenital
Centronuclear myopathy
Congenital fiber type disproportion
Nemaline rod myopathy
Central core myopathy
Metabolic
Glycolytic pathway myopathy (acid maltase deficiency, McArdle's, and Taruis)
Lipid storage myopathy (carnitine palmitoyltransferase deficiency and carnitine deficiency)
Mitochondrial myopathy (chronic progressive external ophthalmoplegia, Kearn–Sayre syndrome, mitochondrial encephalopathy with lactic acidosis, and myoclonus epilepsy with ragged red fibers)
Ion channelopathies
Periodic paralyses (hypokalemic and hyperkalemic)
Myotonia congenita
Malignant hyperthermia
Inflammatory
Polymyositis
Dermatomyositis
Inclusion body myositis
Connective-tissue disorders
Infections
Toxic
Chloroquine
Colchicine
Statins
ICU related
Endocrine
Corticosteroid excess
Hypothyroid
Hyperthyroid

myopathies because these distinguishing features provide important clues to the final diagnosis (Table 2). Although EMG plays an important role in the diagnosis of myopathic disorders, it should always be considered as an extension of the clinical examination. The final diagnosis of myopathy rests not only on the EMG findings but also on patient's history, the clinical examination, laboratory tests, such as creatine kinase, and muscle biopsy. Before delving into details of EMG testing, it is helpful to review the anatomy and physiology of normal and myopathic muscle.

Table 2
Clinical Patterns in Myopathy

Myopathies with ocular palsies (ptosis and ophthalmoparesis)
Myotonic dystrophy (ptosis only)
Facioscapulohumeral muscular dystrophy (FSHD) (ptosis only)
Mitochondrial myopathies
Oculopharyngeal muscular dystrophy
Grave's disease[a]
Myasthenia gravis[a]
Myopathies with facial weakness
Myotonic dystrophy
FSHD
Congenital myopathies
Congenital facial diplegia (Mobius syndrome)[a]
Myopathies with dysarthria/dysphagia/dysphonia
Myotonic dystrophy
Inflammatory myopathies
Oculopharyngeal muscular dystrophy
Thyroid myopathy[a]
Myopathies with hypertrophied muscles
Duchenne's and Becker's dystrophy
Limb–girdle dystrophy
Hypothyroid myopathy
Myotonia congenita
Cysticercosis[a]
Malignant hyperpyrexia[a]
Myopathies with muscle pain and tenderness
Polymyositis and dermatomyositis
Myopathies with connective-tissue disease
Polymyalgia rheumatica[a]
Acute rhabdomyolysis[a]
Infection (influenza, trichinosis)
Myopathies with notable distal weakness
Myotonic dystrophy (mainly type 1 but also type 2)
Inclusion body myositis
Distal myopathies
Scapuloperoneal syndromes
Miscellaneous patterns
"Popeye arms" and inverted axillary folds in FSHD
Skin rash and Gottron's papules in dermatomyositis
Lymphadenopathy in malignancy, infection, and sarcoidosis[a]
Episodic hypoglycemia/encephalopathy in carnitine deficiency[a]

[a]Conditions not primarily diseases of the muscle.

2. NORMAL MUSCLE ANATOMY, HISTOLOGY, PHYSIOLOGY, AND THE ORIGIN OF THE NORMAL ELECTRICAL ACTIVITY OF MUSCLE

2.1. *Anatomy/Histology*

Muscle fibers in adults are approx 50 μm in size. They are polygonal in shape and are bundled into fascicles (Fig. 1). Each fascicle contains approx 20 to 60 muscle fibers, and each

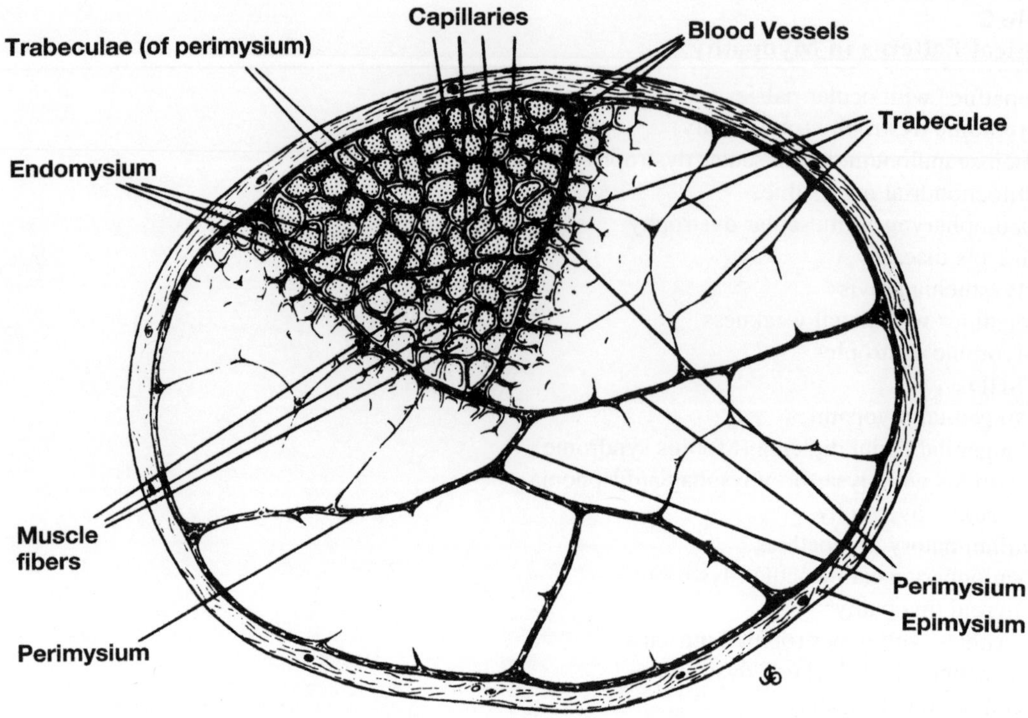

Fig. 1. Schematic diagram of the connective tissue sheaths of muscle. Muscle fibers are bundled into fascicles bordered by perimysial connective tissue. From DeGirolami and Smith, 1982 with permission.

muscle fiber consists of 50 to 100 myofibrils (Fig. 2). The myofibrils are divided into sarcomeres, which are the smallest contractile units. The sarcomere is that portion of the myofibril that extends from Z band to Z band (Fig. 3). The sarcomere consists of two filaments, the thick (myosin) filaments alternating with the thin (actin) filaments. The four major contractile proteins that are present in a myofilament are actin, myosin, troponin, and tropomyosin. The T tubules (Fig. 2) are located near the junction of the A and I bands and transmit the initial depolarization at the motor end plate. They are opposed on each side by the terminal cisterns of the sarcoplasmic reticulum. The terminal cisterns act as the storage space for calcium ions. This sarcoplasmic reticulum-T tubule-sarcoplasmic reticulum complex is called the "triad" and is responsible for converting electrical signals (membrane action potentials) to chemical signals (calcium release).

In a normal muscle, the muscle fibers are organized into functional groups called motor units. The term "motor unit," first used by Liddell and Sherrington in 1929, refers to a single lower motor neuron and the muscle fibers it innervates. One motor neuron innervates multiple muscle fibers, but each muscle fiber is innervated by only one motor neuron. The average number of muscle fibers in a motor unit determines the innervation ratio of the muscle. This ratio varies greatly from one muscle to another and is related to the degree of dexterity required of that muscle. For instance, the innervation ratio of the external ocular muscles is approx 10 muscle fibers to 1 motor neuron, that of intrinsic hand muscles is approx 100 muscle fibers to 1 motor neuron, and that of gastrocnemius is approx 2000 muscle fibers to 1 motor neuron. The muscle fibers belonging to a motor unit are typically distributed widely, reaching over as many as 100 fascicles. They are randomly distributed over a circular or oval

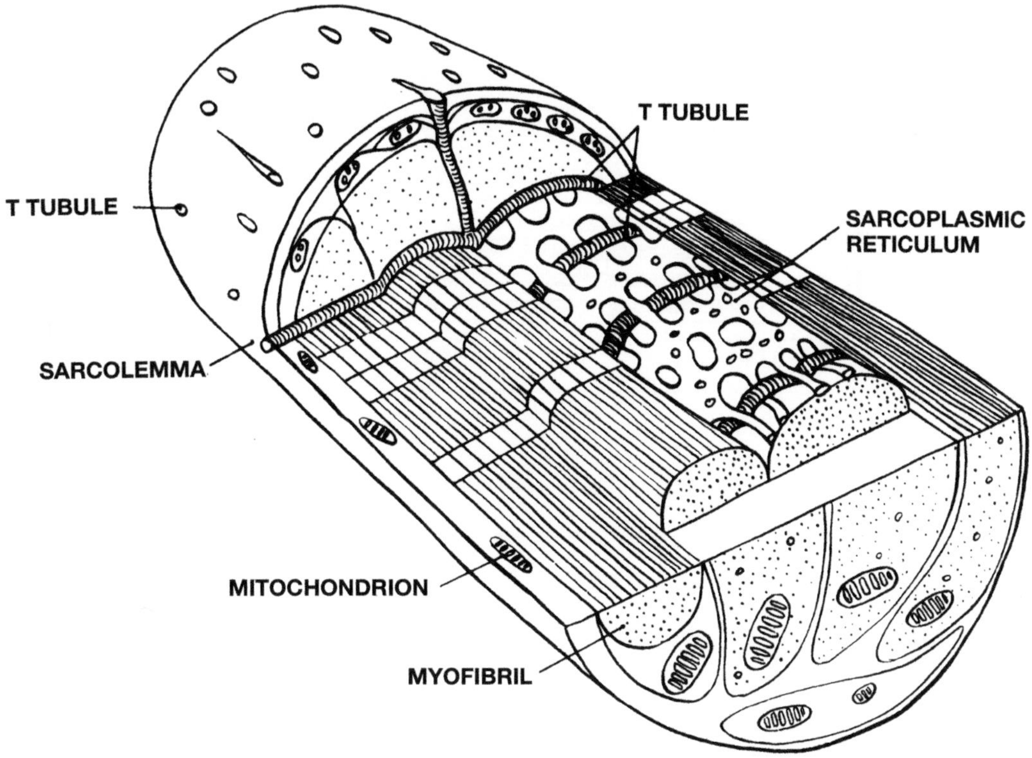

Fig. 2. Structure of a single muscle fiber cut both horizontally and longitudinally. Individual myofibrils are surrounded and separated by sarcoplasmic reticulum. T tubules are continuous with extracellular fluid and interdigitate with sarcoplasmic reticulum. Note the regular association of T tubule with sarcoplasmic reticulum to form membranous triads. Note the location of myonuclei at the periphery of the muscle fiber and the presence of many mitochondria. From Westmoreland et al., 1994 with permission.

region that can reach approx 20 to 30% of the muscle's cross sectional area, with an average diameter of 5 to 10 mm. In general, there is a mosaic pattern of as many as 20 to 50 overlapping motor units in the same muscle, although a slightly higher density of muscle fibers belonging to the same motor unit is observed at the center of the motor unit.

2.2. *Physiology*

Muscle contraction occurs as a result of a series of steps. The process starts with an action potential traveling along the motor nerve axon and reaching the axon terminal (Fig. 4). This (presynaptic) depolarization opens voltage-gated calcium channels, which increases calcium conductance and leads to release of acetylcholine (ACh) into the synaptic cleft. The ACh receptors (ligand-gated channels) at the postsynaptic membrane bind to this ACh, resulting in an end-plate potential, which, when reaching threshold, triggers opening of the adjacent muscle fiber voltage-gated sodium channels (Fig. 4). This ionic flow induces a voltage change, producing an action potential that initiates the excitation-contraction coupling.

With this outline in mind, let us now consider each step in more detail. In the resting state, small fluctuations in membrane potentials occur at the end-plate region called miniature end-plate potentials. The miniature end-plate potentials are produced by the spontaneous release of quanta of ACh from the motor nerve ending that are not sufficient to produce an action

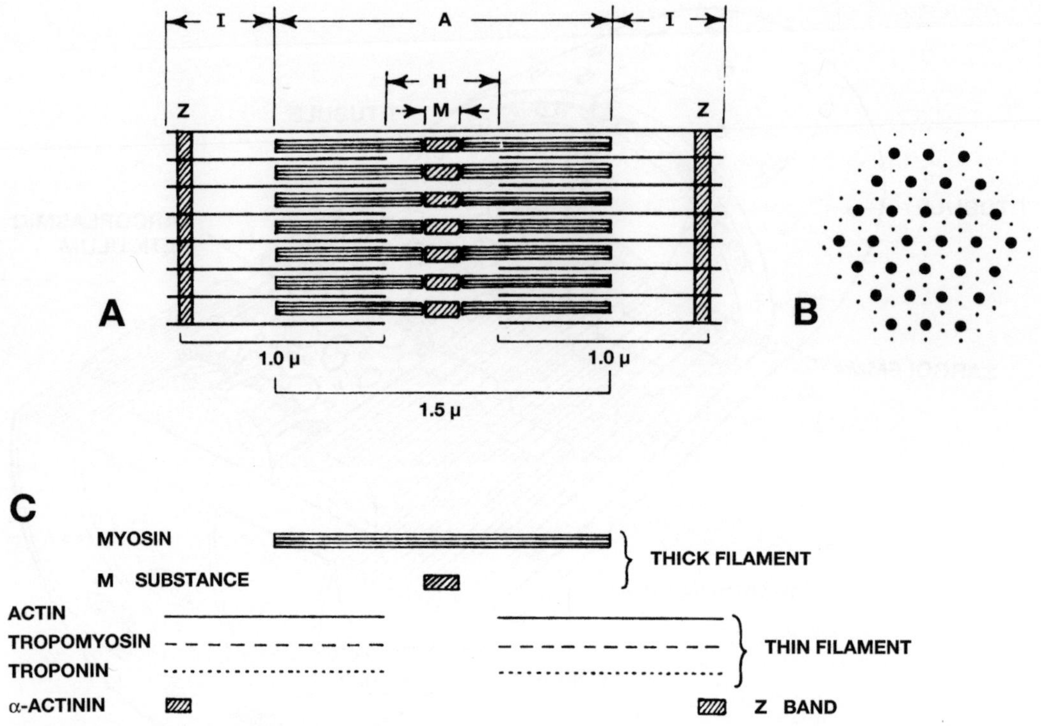

Fig. 3. Organization of protein filaments in a myofibril. (**A**) Longitudinal section through one sarcomere (Z disc to Z disc) showing the overlap of actin and myosin. (**B**) Cross section through the A band, where the thin actin filaments interdigitate with the thick myosin filaments in a hexagonal formation. (**C**) Location of specific proteins in the sarcomere. From Westmoreland et al., 1994 with permission.

potential. An action potential is generated if this end-plate potential reaches sufficient amplitude. Postsynaptic depolarization causes the action potential to travel along the muscle fiber in both directions at a velocity of approx 4 m/s and to come in contact with the terminal cisternae of the sarcoplasmic reticulum, thus, gaining entry into the contractile apparatus.

The T tubules transmit this action potential inward, penetrating the triad complex (sarcoplasmic reticulum-T tubule-sarcoplasmic reticulum complex). This results in a signal transduction through ryanodine receptors releasing calcium ions. The calcium ions then bind to the troponin subunits, causing a conformational change in the tropomyosin and the actin helix configuration. Sliding of the thin actin filaments over the thick myosin filaments produces muscle contraction. The shortening of the sarcomeres and the I band during contraction is not caused by any change in the absolute length of the filaments but rather by the sliding of the filaments themselves. Contraction ceases when calcium is removed from the sarcoplasmic reticulum by active transport.

2.3. EMG Correlates of Normal Muscle Physiology

In muscle at rest, the advancing needle records discharges described as insertional activity and the stationary needle electrode records discharges described as spontaneous activity. When the muscle is voluntarily activated, the needle electrode records groups of individual motor unit action potentials (MUAPs).

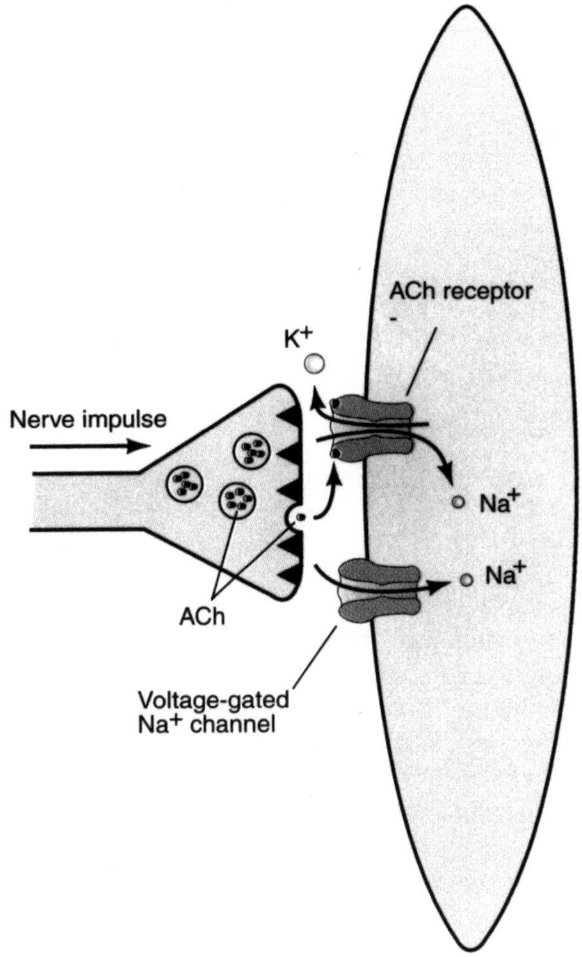

Fig. 4. The binding of acetylcholine (ACh) at transmitter-gated channels opens channels permeable to both Na^+ and K^+. The flow of these ions into and out of the cell depolarizes the cell membrane, producing the end-plate potential. This depolarization opens neighboring voltage-gated Na^+ channels. To elicit an action potential, the depolarization produced by the end-plate potential must open sufficient Na^+ channels to reach the threshold for initiating the action potential. From Kandel et al., 1995 with permission.

2.3.1. *Insertional Activity*

In normal muscle, the act of inserting the needle electrode generally evokes only a brief discharge that lasts a little longer than the actual movement of the needle.

2.3.2. *Spontaneous Activity*

Physiological spontaneous activity is usually restricted to the end-plate regions, observed as end-plate noise and spikes. A more prolonged discharge can occur when the needle electrode is in the end-plate zone (Fig. 5).

2.3.3. *Motor Unit Action Potential*

The standard concentric needle electrode used in EMG practice has an active recording surface of approx 150×600 μm. This recording surface captures activity of muscle fibers that

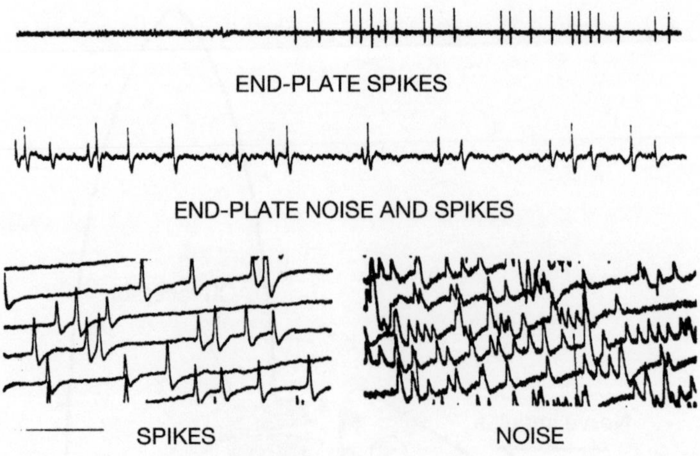

Fig. 5. Normal spontaneous activity in the end-plate region. Electrical activity recorded from the muscle at rest after insertional activity has subsided, and there is no voluntary muscle contraction. From Daube, 1991 with permission.

are located within a 10-mm diameter. The electrical activity of a motor unit recorded by a needle electrode represents the summation of action potentials of muscle fibers that are firing near the electrode. Approximately 8 to 20 muscle fibers belonging to the same motor unit contribute to the recorded MUAP. It is important to appreciate that the MUAP represents the summated activity of *some* muscle fibers in a motor unit but not activity of *all* fibers of a motor unit. Type I muscle fibers are primarily responsible for the generation of the MUAP because low-threshold small motor neurons are preferentially activated on initial minimal voluntary contraction.

Normally, there is some variation of size and form of the MUAPs in a single muscle and of the average size and form of action potential in different muscles. The morphology of the MUAPs is influenced by a number of technical and physiological factors, such as type of needle electrode used, patient's age, muscle temperature, and so on. In general, MUAP duration is shorter in proximal than in distal muscles and their size is larger in adults than in children. MUAPs recorded in normal muscles of the extremities are commonly diphasic or triphasic waves, and they produce a thumping or knocking sound over the loudspeaker.

A variety of parameters define the MUAP (Fig. 6). The *amplitude* is primarily derived from the muscle fiber action potentials of the motor unit residing within 500 μm of the active recording surface and usually measures between 100 μV and 2 mV. It is influenced by the number and diameter of fibers involved, proximity of the needle electrode to motor units, and the synchronicity of their action potentials. The *rise time* is measured from the initial positive peak to the subsequent negative peak, and is an indicator of the distance between the recording tip of the electrode and the depolarizing muscle fibers. The rise time should be less than 500 μs before a MUAP is accepted as a genuine near-field MUAP. A short rise time is characterized by a sharp, crisp sound. If the MUAP is associated with a dull sound, the electrode should be adjusted until MUAPs with sharp and crisp sounds are detected. The amplitude and rise time are inversely proportional to the distance between the electrode and the muscle fibers.

The *duration* of the MUAP depends on the depolarization of many muscle fibers that are both away from and close to the tip of the needle. It best reflects the number of muscle fibers

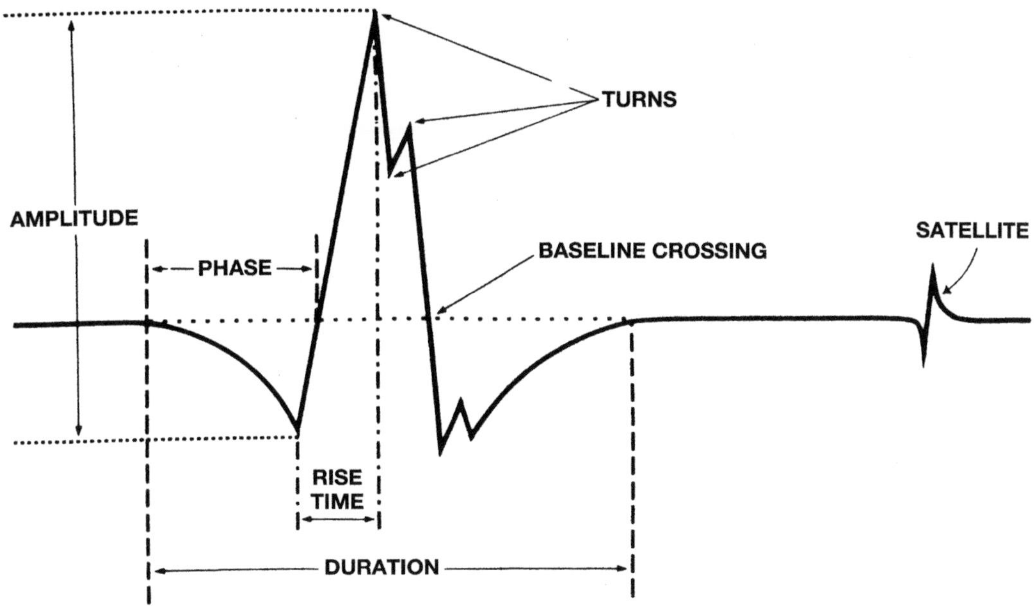

Fig. 6. A schematic motor unit potential with characteristics that can be measured. The motor unit potential (also known as the motor unit action potential) is the compound action potential of a single motor unit whose muscle fibers lie within the recording range of an electrode. From Daube, 1991 with permission.

in a motor unit, and typically measures approx 5 to 15 ms. It is the time from the initial deflection away from the baseline to its final return to the baseline. It varies with the muscle tested, patient's age, and the muscle temperature. The initial deflection reflects the arrival of the fastest muscle fiber action potential and, thus, the arrival time of various action potentials determine the final duration of that particular MUAP. Factors that influence the arrival time of various action potentials include length of individual terminal nerve branches, the distance of individual NMJs from the recording electrode, the diameter of individual muscle fibers, and the conduction velocity along individual muscle fibers (Fig. 7). Thus, unlike the amplitude, the duration of the MUAP is significantly affected by the distant muscle fibers. The number of *phases* in the MUAP depends on the synchrony of depolarization of the muscle fibers, that is, the extent to which the muscle fibers within a motor unit fire at the same time. MUAPs with more than four phases are called polyphasic. Polyphasia may be observed normally, and up to 12% polyphasic MUAPs has been described in normal muscles.

2.3.4. *Recruitment in Normal Muscle*

Motor unit recruitment refers to the successive activation of the same and additional motor units with increasing strength of voluntary muscle contraction. The Henneman size principle refers to the orderly successive activation of motor units during an increasing voluntary muscle activation, with the activation of small, weak (type 1) motor units first in an early contraction, and the sequential addition of larger, stronger (type 2) motor units to provide a smooth increase in muscle power. Recruitment frequency refers to the firing rate of a MUAP when a different MUAP appears during gradually increasing voluntary muscle contraction. Recruitment interval refers to the time elapsed between consecutive discharges of a MUAP when a different MUAP first appears during gradually increasing muscle contraction.

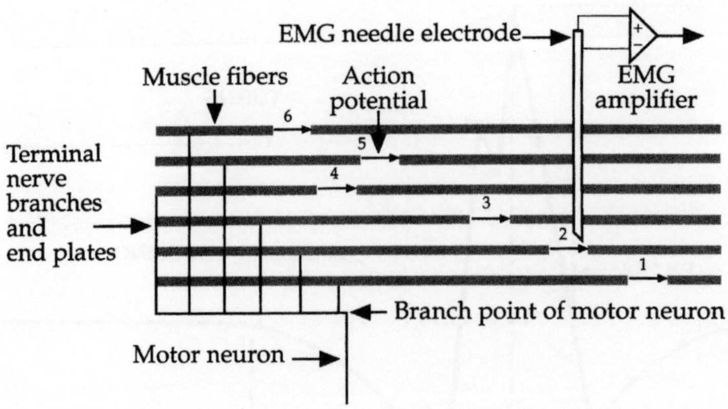

Fig. 7. Schematic representation of motor unit action potential generation and recording by a concentric needle electrode. Individual muscle fiber action potentials numbered one through six arrive at the recording electrode at different times, depending on factors such as terminal nerve branch length, distance between the muscle fiber's neuromuscular junction and the recording electrode, diameter of the individual muscle fiber, and the conduction velocity of muscle fiber action potentials along individual muscle fibers. From Ball 1985 with permission.

In the EMG laboratory, motor unit recruitment is commonly assessed with the rule of fives. Let us consider a situation in which the first motor unit begins to fire at 5 Hz. A new MUAP is recruited when the first motor unit reaches a firing frequency of 10 Hz. Subsequent new motor units are added as the previous motor units reach their maximal firing frequency. For instance, if the MUAPs are firing at 25 Hz, there should be at least five different MUAPs on the EMG monitor. If the ratio exceeds six, this indicates that few motor units are present and those are firing rapidly, at more than the usual frequency of 5 Hz. Each muscle also has a characteristic maximal MUAP recruitment frequency. For instance, the facial muscles have high maximal recruitment frequencies (20–30 Hz) in contrast to the extremity muscles, which have maximal recruitment frequencies in the 10- to 12-Hz range.

The other commonly used term for recruitment in the EMG laboratory is interference pattern. A full interference pattern implies that no individual MUAP can be clearly identified. In healthy individuals, one would expect a full interference pattern on voluntary muscle contraction. A reduced interference pattern is one in which some of the individual MUAPs may be identified, whereas other individual MUAPs cannot be identified because of superimposition of waveforms. The importance of early or increased recruitment observed in myopathy will be discussed in Section 3.2.4. If documenting recruitment pattern or the interference pattern, it is important to specify the force of muscle contraction associated with that pattern.

3. MYOPATHIC MUSCLE: HISTOPATHOLOGY AND THE ORIGIN OF ABNORMAL ELECTRICAL ACTIVITY DETECTED BY EMG

3.1. Histopathology

Myopathic states result in a variety of structural alterations in the muscle. Muscle fiber atrophy is typically produced by denervation, although it can be observed in the late stages of severe myopathies (Fig. 8). Muscle fiber hypertrophy (Fig. 8) is observed in chronic muscle diseases, notably in the muscular dystrophies and in other longstanding disorders (e.g., hypothyroid myopathy). Muscle fiber necrosis results in disruption and fragmentation

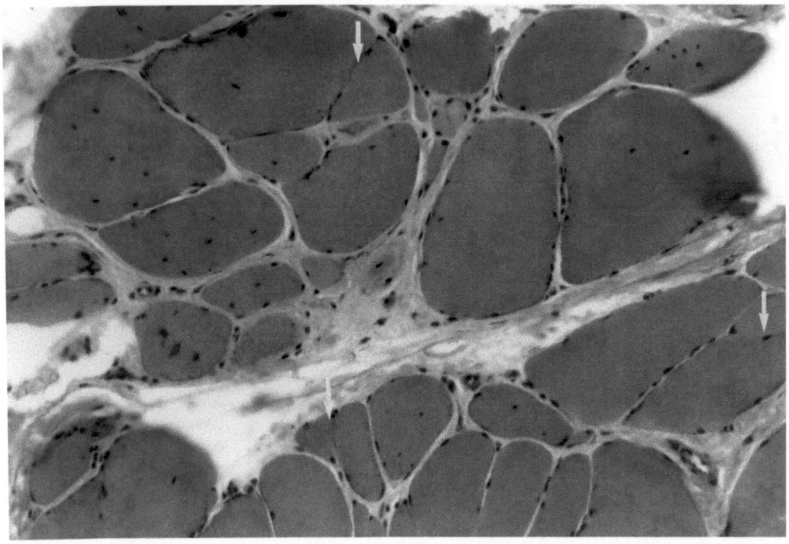

Fig. 8. Chronic myopathic changes (Emery–Dreifuss muscular dystrophy). Cross-section demonstrating pronounced variation in muscle fiber size with enlarged (hypertrophic) and small (atrophic) fibers. Additionally, there are prominent internalized myonuclei, fiber splitting (arrows), and an excess of connective tissue. This increase in fiber size variation compared with healthy muscle contributes to the temporal dispersion of muscle fiber action potentials arriving at the recording electrode and, hence, to the polyphasic MUAP of myopathic disorders [*see* Fig. 7]. By creating two fibers from one, fiber splitting has the effect of increasing MUAP amplitude. H&E stain.

of the sarcoplasm and loss of normal cross striations (Fig. 9). Regenerating fibers stain dark blue with hematoxylin stain because of their increased RNA content. Segmental necrosis may result in separation of a portion of a muscle fiber from the NMJ, resulting in denervation of the detached portion of the muscle fiber. Interstitial mononuclear cell infiltrates (Fig. 10) are important markers for inflammatory muscle disease, such as polymyositis. Vacuolar degeneration may be observed in a number of myopathies, including glycogen and lipid storage disorders, IBM (Fig. 11), and hypokalemic periodic paralysis. By electron microscopy, the vacuoles may contain normal or abnormal material and fluid. Various structural changes may be observed with congenital myopathies, such as nemaline rod myopathy, centronuclear myopathy (Fig. 12), and central core disease. Additionally, mitochondrial abnormalities (Fig. 13) are observed in a variety of myopathies, including the Kearns-Sayre syndrome.

3.2. EMG Correlates of Myopathic Muscle

3.2.1. Insertional Activity in Myopathic Muscle

A marked reduction of insertional activity occurs if muscle fibers are severely atrophied or replaced by fibrous tissue or fat, or if they become unexcitable, for example, during severe paralysis in familial periodic paralysis. An increase in insertional activity is the first EMG clue to the presence of abnormal spontaneous potentials in the muscle (Fig. 14).

3.2.2. Spontaneous Activity in Myopathic Muscle

The general rule of thumb is that any spontaneous potential with an initial positive deflection should be regarded as abnormal. It is also important to pause and allow sufficient time

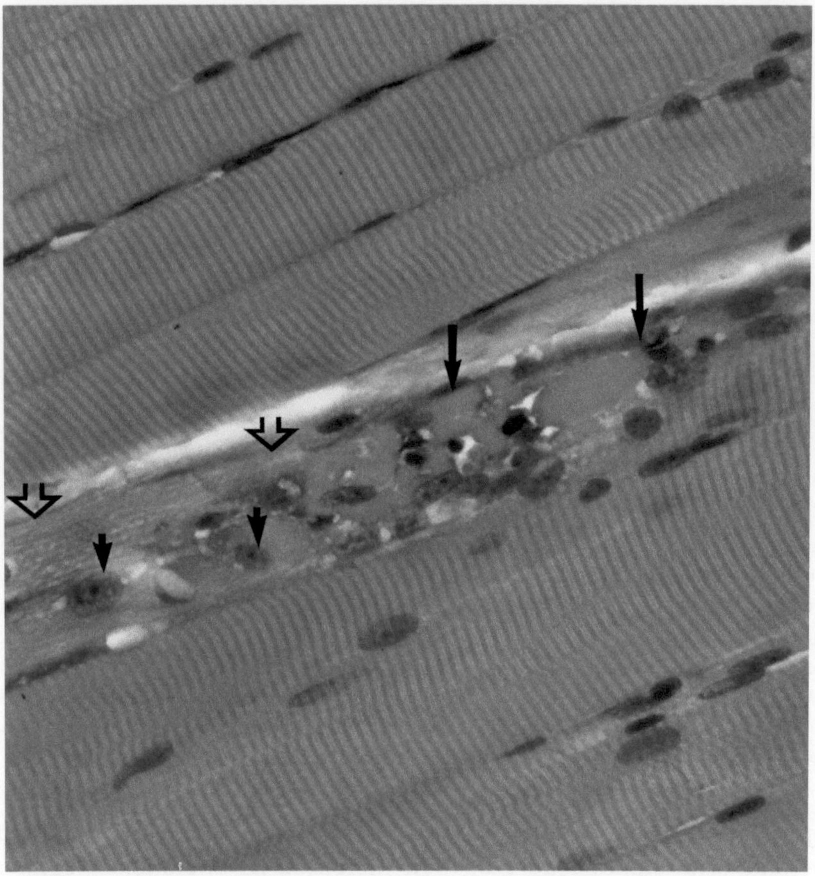

Fig. 9. Muscle fiber necrosis. Longitudinal section showing a necrotic muscle fiber segment (arrows) with disruption/fragmentation of the sarcoplasm and loss of normal cross striations. Myonuclei (arrowheads), with prominent nucleoli of a regenerating muscle fiber (open arrowheads) are also noted. Such a pathological state may be associated with fibrillation potential activity because still healthy muscle fiber segments may have become separated from the muscle end-plate region by segmental fiber necrosis. H&E stain.

(≥1–2 s) between each needle electrode advancement to look for abnormal spontaneous potentials. The four abnormal spontaneous potentials to look for in a patient with myopathy comprise: fibrillation potentials, positive sharp waves (PSWs), myotonic potentials, and complex repetitive discharges (CRDs).

Fibrillation potentials are the most common abnormal insertional/spontaneous activity observed in myopathies. They seem to arise spontaneously from either a single muscle fiber or a few muscle fibers and are not associated with visible contractions. They are biphasic or triphasic waves with an initial positive deflection (Fig. 15). They are usually 1 to 2 ms in duration and less than 100 μV in amplitude. They are identified by their regular frequency, with a typical sound on the loudspeaker that is likened to "raindrops on tin roof." PSW, as the name implies, are biphasic waves with an initial PSW followed by a long negative wave. They are usually 30 ms in duration, with amplitudes of 50 μV to 1 mV. They are regular in frequency and are identified by their "dull thud" or "ticking of clock" sounds on the loudspeaker. Fibrillation potentials in myopathic disorders were described as early as 1949 by

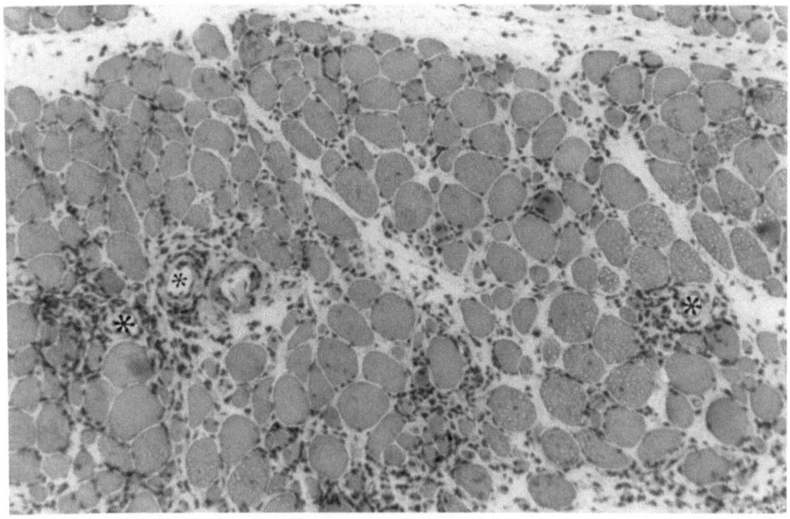

Fig. 10. Mononuclear cell infiltration and excessive muscle fiber size variation in inflammatory myopathy. The inflammation surrounds small vessels (*) and spills out into the endomysium to surround some muscle fibers. H&E stain.

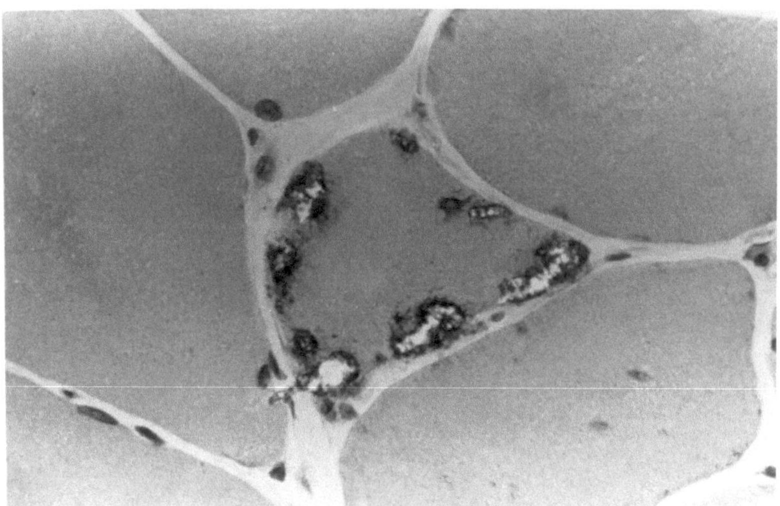

Fig. 11. Vacuolated muscle fiber in the setting of inclusion body myositis. The vacuoles are lined or "rimmed" with granular material. H&E stain.

Kugelberg. They are thought to result from segmental necrosis of muscle fibers, in which the necrosis leaves a distal muscle fiber segment separated from the part carrying the motor plate. Both fibrillation potentials and PSWs have the same clinical significance and can be observed with both neurogenic and myopathic disorders. However, when observed in excess in a myopathic illness, the muscle biopsy almost always shows some necrotizing features.

Myotonic discharges are the second most common abnormal insertional/spontaneous activity observed in myopathies. They are the EMG hallmark of myotonia and are often associated with clinical myotonia. When observed diffusely on needle examination, the patient is considered to have a myotonic disorder unless proven otherwise. They are generated by single

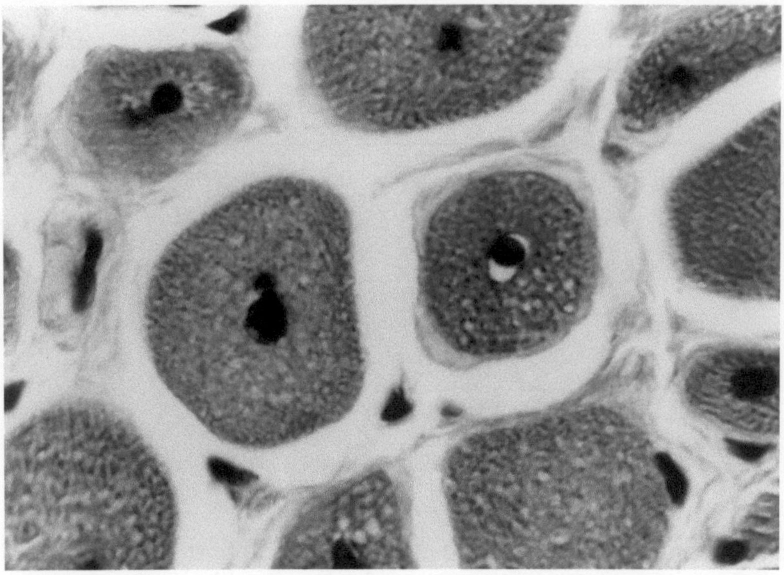

Fig. 12. Centronuclear myopathy. Instead of their normal peripheral location, muscle nuclei are located in the central regions of the muscle fibers. H&E stain.

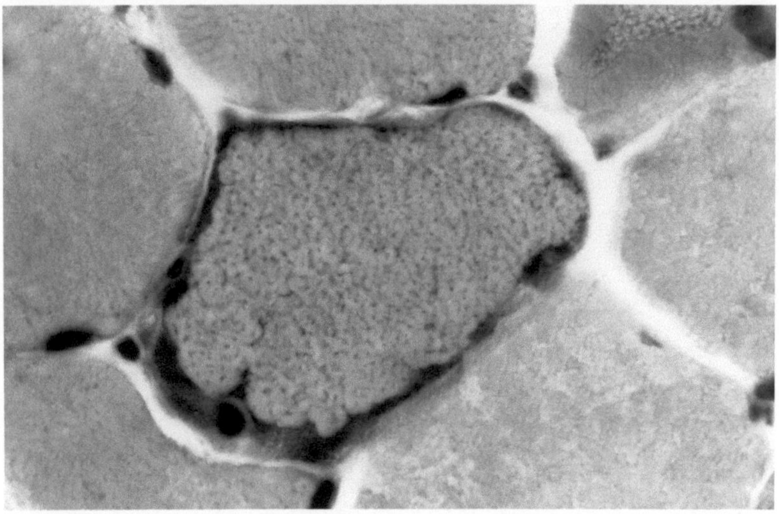

Fig. 13. Mitochondrial myopathy in the context of Kearns-Sayre syndrome. Note the accumulation of darkly staining material in the subsarcolemmal regions that prove by electron microscopy to be abnormal mitochondria. H&E stain.

muscle fibers and can occur either spontaneously, by stimulation of the nerves supplying the recorded muscle, by voluntary MUAP activation, or by mechanical stimulation. Cold is thought to enhance myotonic discharges in most conditions; the one exception to this rule is paramyotonia congenita, in which prolonged cooling will lead to complete electrical silence of the muscle. The frequency of their discharge ranges between 15 and 150 Hz, and their amplitude ranges between 10 µV and 1 mV. Their waxing and waning amplitudes and the "dive-bomber sound" on the loudspeaker are characteristic (Fig. 16). Although the precise

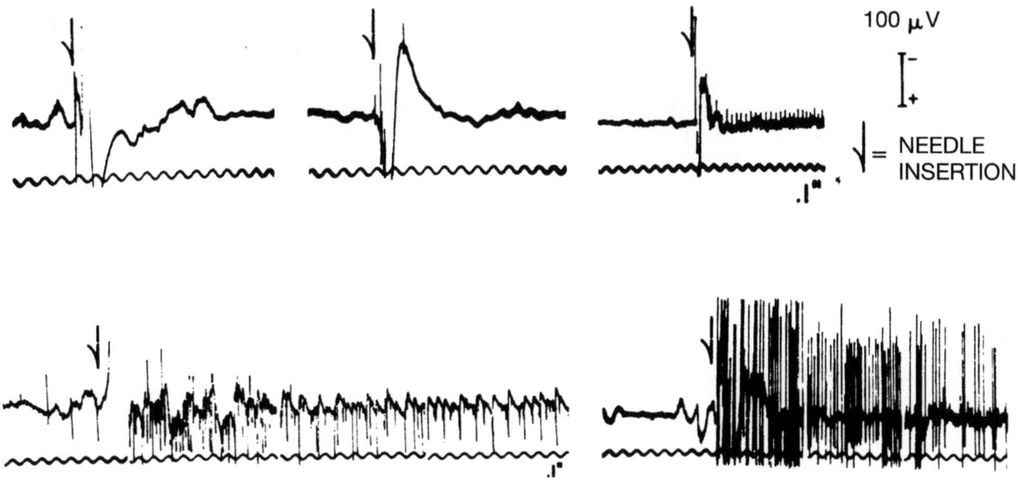

Fig. 14. Insertion activity. Top, normal insertion activity. Bottom, increased insertion activity. From Daube, 1991 with permission.

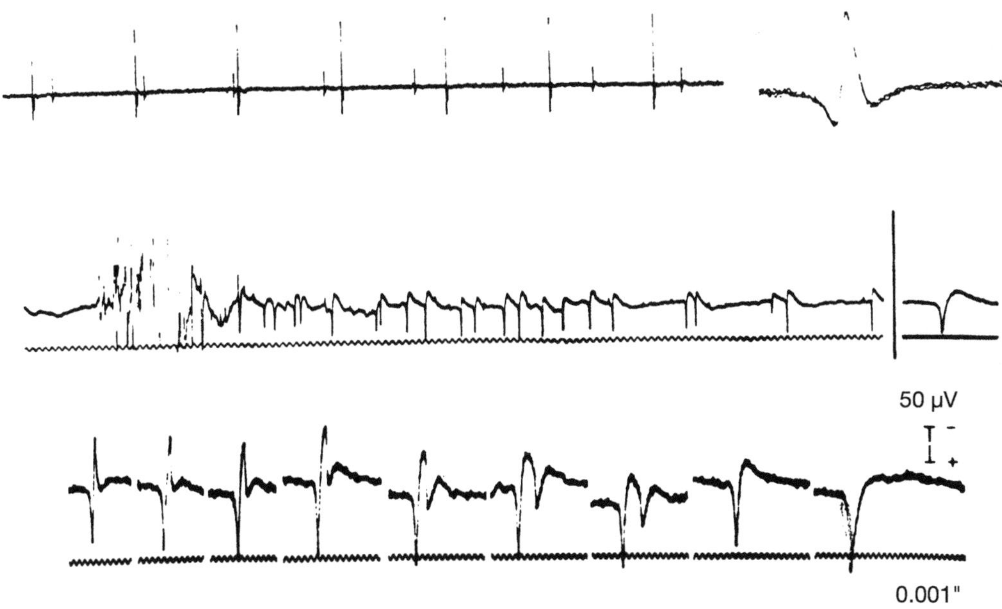

Fig. 15. Fibrillation potentials. Top, spike form. Center, positive waveform. Bottom, development of a positive wave form (serial photographs taken after insertion of needle electrode). From Daube, 1991 with permission.

mechanism of myotonia is unclear, it is thought to be secondary to a defect in ion channels on the muscle membrane (the chloride channel in myotonia congenita, and the sodium channel in paramyotonia and hyperkalemic periodic paralysis).

CRDs are repetitive discharges of polyphasic or serrated action potentials characterized by their uniform frequency, shape, and amplitude, with abrupt onset, cessation, or change in configuration (Fig. 17). In contrast to myotonic discharges, CRDs do not wax and wane in amplitude or frequency. Formerly referred to as "bizarre high-frequency waves" or the confusing

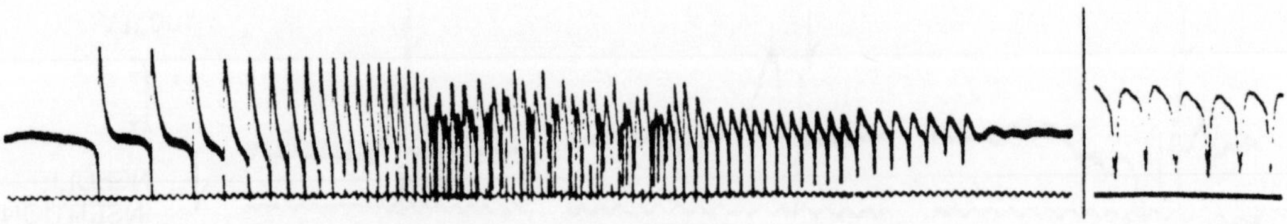

Fig. 16. Myotonic discharge. Note that the potentials wax and wane in amplitude and frequency. From Daube, 1991 with permission.

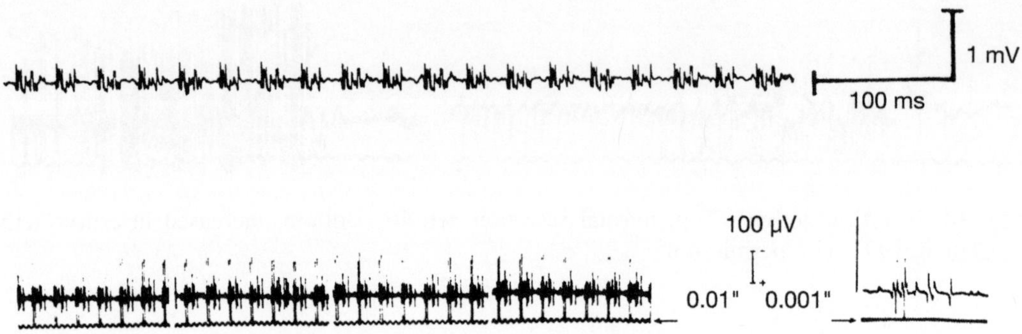

Fig. 17. Complex repetitive discharges. These are the action potentials of groups of muscle fibers discharging in near synchrony at high rates. Note that they are characterized by abrupt onset and cessation. From Daube, 1991 with permission.

term "pseudomyotonic discharges," they are indicative of a hyperirritable muscle membrane with a group of muscle fibers firing in near synchrony. CRDs are observed in both neurogenic and myopathic disorders. Their presence is suggestive of a more chronic or longstanding process. They may be found in patients with muscular dystrophy, hyperkalemic periodic paralysis, glycogen storage disorders, and hypothyroid myopathy, but also in chronic denervating diseases.

Clinically, fasciculations are visible muscle twitches and they represent the spontaneous contractions of a group of muscle fibers belonging to a single motor unit. The electrical activity associated with the twitch is called a fasciculation potential, and it has the configuration of a MUAP. Although fasciculation potentials generally arise in the wake of an underlying neurogenic process, rarely, they are observed in myopathies, such as thyrotoxic myopathy and inclusion body myositis (IBM). Even in such rare instances, an underlying neurogenic component may be the cause of fasciculations. Likewise, myokymic potentials are not usually observed in diseases of the muscle in the absence of a coexisting or underlying neurogenic process. In fact, the presence of fasciculations or myokymia should prompt the clinician to rethink the diagnosis of a primary muscle disease.

3.2.3. MUAP Changes in Myopathic Muscle

In myopathy, there is a reduction in the average number of functional muscle fibers per motor unit. Some of the fibers within a motor unit may be nonfunctional because of muscle fiber necrosis and, hence, the electrode may be less likely to record activity from distant fibers. The initial and terminal portions of the MUP are not recorded, thereby producing short-duration MUAPs. The shortening does not depend on the type of myopathy but is greater in patients with weakness that is more advanced, and is more likely to be found in

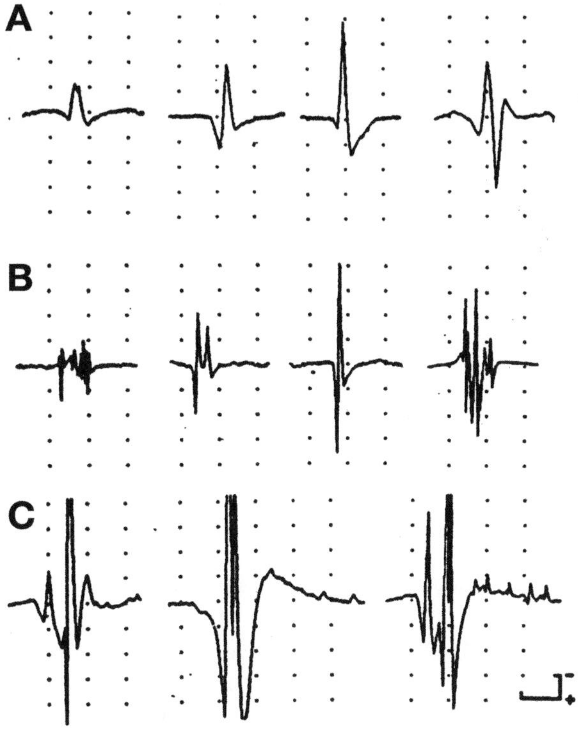

Fig. 18. Normal and abnormal motor unit action potential (MUAP) configurations. **(A)** Normal MUAPs. **(B)** Myopathic MUAPs. Short-duration, low-to-moderate amplitude, "spikey" and polyphasic MUAPs. The reduced number of muscle fibers per motor unit leads to a MUAP that is shorter in duration and lower in amplitude than normal. Because of the increased variation in size of remaining muscle fibers, there is less synchronous firing of individual muscle fibers, leading to increased polyphasia. **(C)** Large-amplitude, long-duration (including satellite potentials), polyphasic MUAPs, as observed in neurogenic disorders. Time marker, 10 ms; voltage marker, 200 µV. From Bromberg and Albers, 1988 with permission.

proximal than distal muscles. If the number of active fibers lying close to the electrode is also reduced, the mean amplitude of the potentials may also be diminished because of loss of contributions of fibers that have disappeared. In adult limb muscles, MUAPs of shorter than 6 ms are considered to be short in duration. Short-duration MUAPs are usually also low in amplitude (<500 µV) and recruited early (*see* below) (an inappropriately large number of MUAPs in the face of a weak voluntary muscle contraction). Thus, small-amplitude, short-duration (SASD) MUAPs are the EMG hallmarks of myopathy (Fig. 18). There are likened to "newspaper rubbing between two fingers" on the loudspeaker. These potentials (SASDs) are often referred to as "myopathic." As discussed below, it is important to remember that the so-called "myopathic" potentials are not diagnostic of myopathy and can be observed with a number of other conditions.

In chronic or end-stage myopathies, some MUAPs may be increased rather than decreased in amplitude and duration. These alterations in MUAP appearance, observed mostly in necrotizing myopathies (Fig. 9), probably result from reinnervation of previously denervated muscle fiber segments by secondary collateral sprouts, fiber splitting, and from the increase in numbers of hypertrophic fibers (Fig. 8). An additional point to keep in mind is that in severe,

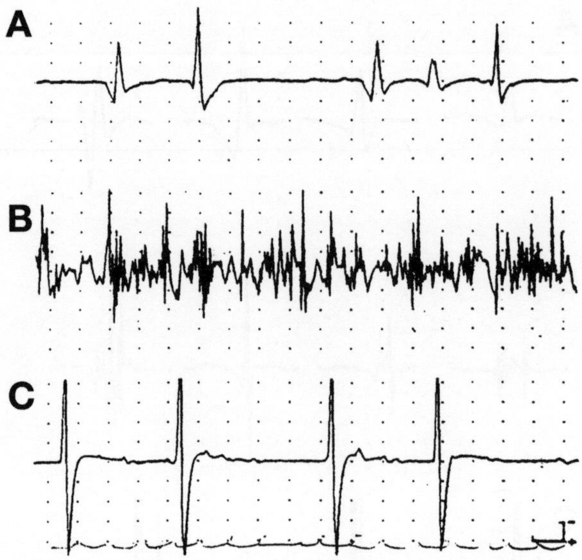

Fig. 19. Pattern of motor unit action potential (MUAP) recruitment at low levels of muscle force. **(A)** Normal recruitment with 3 MUAPs recruited. **(B)** Myopathy (myositis). Early recruitment with many short duration polyphasic MUAPs despite low level of force. Because each motor unit contains fewer muscle fibers, it generates less force than normal. To develop a given level of force, an increased number of MUAPs is required compared with normal. **(C)** Decreased recruitment with only one high-amplitude MUAP firing rapidly, as observed in neurogenic disorders. Time marker, 10 ms; voltage marker, 200 μV. From Bromberg and Albers, 1988 with permission.

end-stage disease, the process of recruitment (*see* next section) may actually be reduced (Fig. 19) rather than early or increased because the extensive loss of muscle fibers leads to the functional equivalent of a loss of motor units.

3.2.4. Recruitment Pattern in Myopathic Muscle

In myopathic disorders, individual muscle fibers drop out of the motor unit without affecting the integrity of the motor unit's connection with the central nervous system. Accordingly, the number of functional motor units remains unaltered until an advanced stage of the disease. Because each motor unit contains fewer functioning muscle fibers, it generates less force than normal. As a result, more motor units than normal are required to generate a certain level of force; this phenomenon is called early recruitment (Fig. 19). Hence, in myopathic diseases, an inappropriately large number of MUAPs is firing in the face of muscle contractions that generate only weak force. In this situation, the interference pattern remains full (but compared with normal, the amplitude is reduced because the pattern is generated by SASD MUAPs). Because activation, a central nervous system process—the ability to fire motor units faster—is unaffected in myopathy, the ratio of firing frequency to number of MUAPs will typically be less than normal (<5).

3.2.5. Clinical Significance and Differential Diagnosis of "Myopathic" MUAPs

Overall, just as fibrillation potentials are indicative of the disease *activity*, MUAP changes are indicative of disease *severity*. Of the various MUAP alterations, an increased proportion of polyphasic MUAP is often reported to be the earliest recognizable change with myopathies. Decreased MUAP duration and early recruitment are considered the most reliable

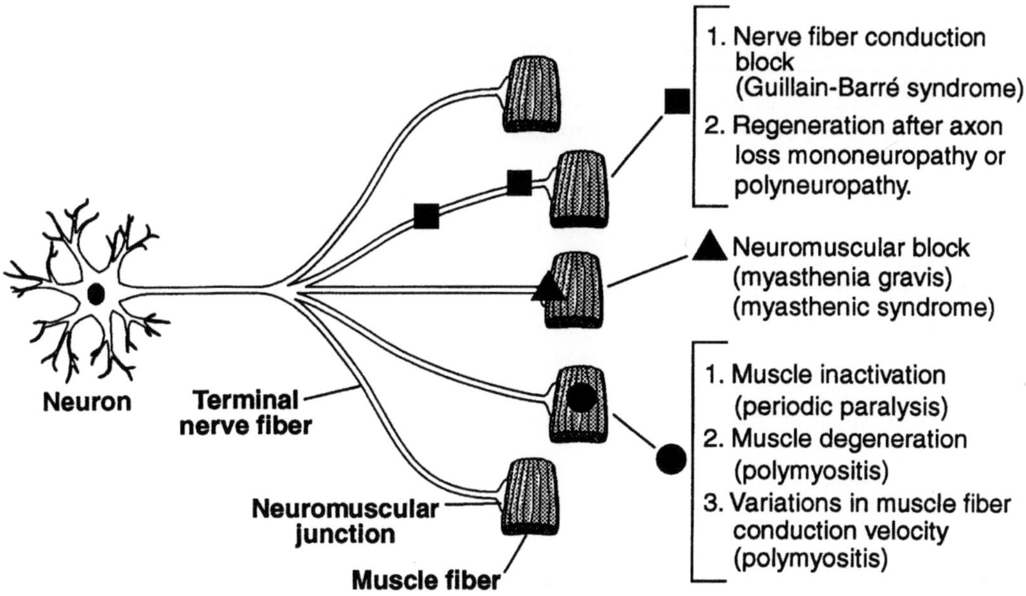

Fig. 20. Differential diagnosis of the short duration MUAP. Schematic representation of the motor unit and specific targets of disease processes that lead to failure to generate muscle fiber action potentials. From Ferrante and Wilbourn, 2000 with permission.

changes. Decreased MUAP amplitude is considered least reliable, because it is dependent on various needle electrode factors.

In addition to myopathies, short-duration MUAPs can also be observed in NMJ disorders such as myasthenia gravis, Lambert-Eaton myasthenic syndrome (LEMS), and botulism (Fig. 20). In these cases, the "myopathic" MUAP occur because of the failure of action potential generation secondary to a defect in the NMJ. SASD MUAPs may also be observed with nascent motor units after denervation and reinnervation. However, nascent MUAPs reveal markedly reduced recruitment in contrast to the early recruitment observed in myopathies.

4. THE WORKUP FOR MYOPATHY

4.1. Advantages of Electrodiagnosis

EMG provides a much wider source of sampling than other diagnostic procedures, such as muscle biopsy. EMG also helps enormously in the differential diagnosis of weakness, distinguishing myopathic disorders from diseases of peripheral nerve/anterior horn cell or NMJ. For example, IBM and amyotrophic lateral sclerosis may share clinical features of asymmetric muscle atrophy and weakness, but the former would be more likely to demonstrate SASD MUAPs and early recruitment. EMG is also useful in identifying certain electrophysiological abnormalities, such as myotonic discharges, that may not have a clinical counterpart, thereby providing an important clue for the diagnostic process. In addition, EMG findings may provide important information that helps in the selection of a muscle for biopsy. For example, a muscle involved by an inflammatory myopathy might be expected to reveal needle electrode findings of fibrillation potentials and SASD MUAPs. Finally, EMG findings may be the only objective evidence of motor unit dysfunction and is especially useful in the early stages of a disease or if the disease is mild. For example, early in the course of an inflammatory myopathy,

Table 3
Recommended Electrodiagnostic Studies for the Patient With a Suspected Myopathy[a]

Recommended nerve conduction studies for myopathy:
- At least one motor and one sensory conduction, with corresponding F-wave from the arm (e.g., median sensory and motor, median F-wave)
- At least one motor and one sensory conduction, with corresponding F-wave from the leg (e.g., sural sensory and tibial motor, tibial F-wave) (if generalized decrease in CMAP, proceed with workup for neuromuscular junction disorders)

Recommended needle electrode examination for myopathy (perform examination on muscles on one side of the body, leaving the other side for muscle biopsy, if needed):
- At least two proximal and two distal muscles in the lower extremity (e.g., tibialis anterior, gastrocnemius, vastus lateralis, and iliopsoas)
- At least two proximal and two distal muscles in the upper extremity (e.g., first dorsal interosseus, flexor carpi radialis, biceps, and deltoid)
- At least one paraspinal muscle

[a]Adapted from Preston and Shapiro.

when weakness is minimal and limited, fibrillation potential activity might be mild but widespread, including paraspinal muscles.

4.2. *Limitations of Electrodiagnosis*

The findings of EMG are not diagnostic of any one illness. No waveforms are pathognomonic of a specific disease. The EMG findings are diverse; some disorders, such as congenital or endocrine myopathies, do not produce extensive EMG changes (probably because they mainly affect the contractile properties of the muscle fibers without modifying their electrical activity), whereas others, such as inflammatory myopathy, may produce striking findings. The EMG appearance may differ depending on the stages of the disease. A coexisting neurogenic disorder also limits the usefulness of the EMG. Therefore, before arriving at a specific diagnosis, EMG results should be integrated with the clinical findings.

4.3. *Planning the EMG Study*

Just before the EMG examination we explain the purpose of the EMG and the procedure itself to the patient. We tailor testing conditions to patient's comfort, ensuring that the room temperature is warm and the patient appropriately clothed. Once the clinical impression of a myopathy is confirmed, it is reasonable to proceed with the first of the two-part EMG, the nerve conduction study (NCS) (Table 3). It is a common practice to perform the NCS before the needle examination.

4.3.1. *Nerve Conduction Studies in Myopathy*

4.3.1.1. Nerve Conduction Studies

Because most myopathies tend to be characterized by proximal more than distal muscle involvement, the results of routine motor studies recording from distal small muscles are usually normal, unless the myopathy is severe. Proximal and intermediate muscles (deltoid, biceps, and tibialis anterior) certainly may be evaluated, and some reductions in compound muscle action potentials (CMAPs) may be found more readily in those muscles. The reduction in CMAP amplitude is explained by the fact that the CMAP is a summation of individual muscle action potentials. More importantly, a generalized reduction in the CMAP amplitudes should alert the electromyographer to the possibility of disorders involving either the presynaptic portion of the NMJ, motor polyradiculopathies, or motor neuron diseases. It is then essential to perform repetitive nerve stimulation (RNS) testing and postactivation facilitation testing to look for LEMS. Another reason to perform RNS in the workup of a myopathy is that myasthenia gravis may simulate myopathy in some patients if it presents with proximal muscle weakness without symptomatic cranial nerve signs.

Late responses, such as H-reflexes and F-waves, are usually normal in myopathies. They may be rarely reduced if the recorded muscles are severely affected. Sensory NCS results are generally normal in myopathy and when attenuated or absent suggest a coexisting neuropathy that may be part of a systemic disease.

4.3.1.2. Repetitive Nerve Stimulations

RNS are generally expected to be normal in disease of the muscle alone. The usefulness of RNS in myopathies is mainly to differentiate myopathies from diseases such as myasthenia gravis, especially when it presents with proximal muscle weakness alone without any ocular or bulbar symptoms, or LEMS. A few points regarding RNS in myopathies are worth noting:

- A decremental response may be observed with the myotonic disorders. However, it is more prominent, with repetition rates of 5 Hz, rather than the usual 2-Hz repetition rate.
- An incremental response may be observed during attacks of periodic paralysis (where a low-amplitude CMAP increases with repetitive rates of 10 Hz).
- Both decremental and incremental responses have been reported with polymyositis.
- Decremental responses have also been noted with myophosphorylase deficiency (McArdle's disease).

4.3.2. Needle Electrode Examination in Myopathy

The NCS is followed by the needle electrode examination (NEE). Here, a monopolar or concentric bipolar needle is inserted into selected muscles to assess the insertional activity, spontaneous activity, and the motor unit morphology on voluntary activity. Certain key points are worth remembering during the needle examination, taking inflammatory muscle disease as an example. In this regard, we find the approach of Dr. Wilbourn very useful. The most proximal muscles, such as iliacus, glutei, and paraspinal muscles, are the ones most likely to show abnormalities, rather than biceps, deltoid, and the vasti. Certain mid-limb muscles, particularly brachioradialis and tibialis anterior, more often show abnormalities than muscles that are more proximal, such as the vasti. In addition, whenever possible, the needle examination should be confined to one side of the body, leaving the other for a muscle biopsy, if needed. It is important to avoid any needle-induced muscle fiber trauma that could potentially influence the muscle biopsy results.

4.3.3. Single-Fiber EMG in Myopathy

Single-fiber EMG is not commonly pursued in the evaluation of myopathy. However, abnormal jitter and blocking will be identified because these findings are the result of nonspecific damage to the end terminal.

5. PUTTING IT ALL TOGETHER

Now that we have an understanding of the muscle physiology in myopathy and its electrical correlates, it is time to make the final diagnosis. Again, as we mentioned in Section 1., EMG is

Table 4
Electrodiagnostic Clues to the Diagnosis of Specific Myopathies

Myopathies that may present with normal EMG
Metabolic myopathies
Endocrine myopathies
Hypokalemic periodic paralysis (in between attacks)
Myopathy with fiber type disproportion
Myopathies that may present with fibrillation potentials only
Inflammatory myopathies
Chloroquine myopathy
Acute rhabdomyolysis
Myopathies that may present with myotonia
Myotonic dystrophy
Myotonia congenita (MUAPs normal)
Proximal myotonic myopathy
Paramyotonia congenita
Colchicine myopathy
Acid maltase Deficiency
Hyperkalemic periodic paralysis (in between attacks)
Centronuclear myopathy
Myopathies with denervating features (fibrillation potentials and positive sharp waves)
Inflammatory myopathies
Muscular dystrophies
Congenital myopathies
Metabolic myopathies
Toxic myopathies
Infectious myopathy

only one part of the puzzle. We have to blend the clinical history and physical examination findings with the EMG results to arrive at a final diagnosis.

Although the examiner is generally expected to wear the "electromyographer hat" during the study, taking care not to be biased with the available clinical information, the final report should be formulated after considering all of the technical factors and the inherent subjective nature of analyzing the MUAPs on the EMG monitor. Some caveats are worth remembering when interpreting abnormal EMG findings in a patient with suspected myopathy.

- Small MUAPs in the early stages of reinnervation after severe denervation may be misinterpreted as caused by myopathy.
- Individual MUAPs may have a normal or even high amplitude in patients with mild myopathies or in clinically unaffected muscles from patients with myopathy; for instance, if the needle electrode is close to normal or hypertrophic muscle fibers (as a compensatory reaction in early stages).
- EMG mainly reflects activity of type I fibers (diseases of the adrenal or pituitary glands as well as steroid-related myopathy show type 2 atrophy without degenerative change).
- The spike duration may sometimes seem extremely prolonged because of late satellite potentials. (In Duchenne's dystrophy, polyphasic MUAPs with prolonged spike duration are rather typical (Fig. 6).

It is also important to remember that although EMG is useful in the diagnosis of myopathies, there are a number of diseases of muscle that may present as challenges to the electromyographer (Table 4). Metabolic and endocrine myopathies may not have abnormal EMG findings.

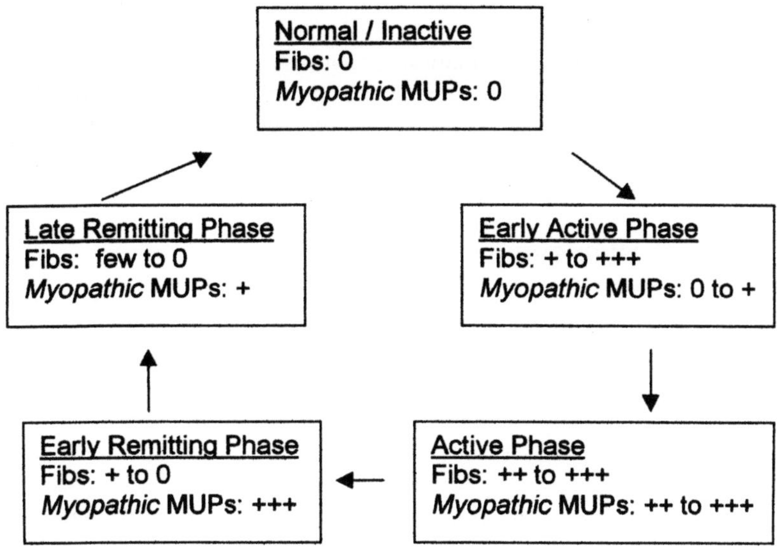

Fig. 21. The cycle of needle electrode examination changes that can be observed during the various stages of a reversible necrotizing myopathy, such as polymyositis. Fibs, fibrillation potentials; MUPs, motor unit potential. From Wilbourn, 1993 with permission.

Fibrillation potentials may be the only abnormal findings in the early stages of necrotizing myopathies, such as inflammatory myopathies. Likewise, myotonic discharges may be the only abnormal EMG finding in a number of myotonic myopathies.

6. EMG IN SELECTED MYOPATHIES

Before concluding, we highlight a few important features (and EMG caveats) of some important groups of muscle diseases. Inflammatory myopathies are important because some are eminently treatable. EMG plays a critical role not only in their diagnosis but also for monitoring the activity of the disease and its response to treatment. Although IBM is not responsive to immunosuppressive therapies, unlike polymyositis and dermatomyositis, it is considered here because it may simulate motor neuron disease clinically, and EMG plays an essential roll in establishing the correct diagnosis. Certain myopathies with myotonia are often missed clinically because of the absence of clinical myotonia. Endocrine and congenital myopathies may have a normal EMG and it is important for the electromyographer to be aware of these possibilities.

6.1. Polymyositis/Dermatomyositis

- Characterized by abundant fibrillation potentials and PSW, CRD, and MUAPs of short duration and low amplitude.
- Serial EMG studies during treatment showing a reduction in fibrillation potential activity is considered a marker of disease improvement (Fig. 21).

6.2. Inclusion Body Myositis

- A mixed pattern of high-amplitude, long-duration MUAPs and SASD MUAPs accompanied by abnormal spontaneous activity is highly suggestive of this condition.
- Approximately 15 to 30% of patients with IBM may have an associated axonal neuropathy.

6.3. *Myotonic Myopathies*

- Myotonic discharges on needle examination is characteristic.
- Myotonic discharges may be restricted to paraspinal muscles in patients with adult-onset acid maltase deficiency and to intrinsic hand muscles in asymptomatic patients with myotonic dystrophy.
- MUAP changes may be more prominent in distal muscles than in proximal muscles. Hence, several distal limb muscles must be assessed in addition to proximal muscles.

6.4. *Acute Illness/ICU Myopathy*

- Generalized low-amplitude or absent CMAPs and normal sensory nerve action potentials on NCS.
- Widespread fibrillation potentials on NEE (this is found variably).
- "Myopathic MUAPs" if voluntary activity possible.
- RNS necessary to exclude NMJ disorders.
- Post-tetanic facilitation (for LEMS) unreliable because muscles too weak.

6.5. *Endocrine Myopathies*

- Fasciculation potentials and myokymic discharges may be the only spontaneous activity in hyperthyroidism.
- Hyperkalemic periodic paralysis may be associated with increased fibrillation potentials (myotonic discharges and CRDs are found between attacks).
- EMG in familial hypokalemic periodic paralysis is often normal between attacks.
- Steroid myopathy is a misnomer, because, in the absence of denervation, steroids do not cause histological signs of myopathy but, rather, a selective atrophy of type 2 muscle fibers.

6.6. *Congenital Myopathies*

- A normal EMG does not exclude a congenital myopathy.
- Scant fibrillation potentials and short-duration MUAPs are observed.
- Among all congenital myopathies, centronuclear myopathy has the most prominent changes in spontaneous activity (fibrillations, PSW, CRD, and myotonia), sometimes mimicking myotonic dystrophy.

SUGGESTED READING

Aminoff MJ. Electromyography in Clinical Practice, 3rd ed. Churchill Livingstone, New York, NY, 1998.

Ball RD. Basics of needle electromyography. An AAEE Workshop. AAEM, 1985, Rochester, MN.

Bromberg MD, Albers JW. Electromyography in idiopathic myositis. Mt Sinai J Med 1988;55(6): 459–464.

Daube JR. AAEM minimonograph #11: needle examination in electromyography. Muscle Nerve 1991;14:685–700.

Daube JR. Electrodiagnosis of muscle disorders. In: Myology (Engel AG, Franzini-Armstrong C, eds.). McGraw-Hill, New York, NY, 1994.

DeGirolami U, Smith TW. Pathology of skeletal muscle. Am J Pathol 1982;107:235–276.

Dumitru D. Myopathies. In: Electrodiagnostic Medicine (Dumitru D, ed). Hanley and Belfus, Philadelphia, PA 1994.

Ferrante MA, Wilbourn AJ. In: Comprehensive Clinical Neurophysiology (Levin K, Luders H, eds.). WB Saunders, Philadelphia, PA, 2000, pp. 268–280.

Kandel ER, Schwartz JH, Jessell TM. Essentials of Neural Science and Behavior. Appleton and Lange, Stamford, CT, 1995, pp. 212.

Katirji B. Electromyography in Clinical Practice. St. Louis, Mosby, 1998.

Kimura J. Electrodiagnosis in Diseases of Nerve and Muscle: Principles and Practice. Oxford University Press, New York, NY, 2001. WB Saunders, Philadelphia, 19.

Oh SJ. Principles of Electromyography. Williams and Wilkins, Baltimore, MD, 1998.

Phillips LH, Litchy WJ, Auger RG, et al. AAEM Glossary of terms in electrodiagnostic medicine. Muscle Nerve 2001;Suppl 10.

Preston DC, Shapiro BE. Electromyography and Neuromuscular Disorders. Clinical–Electrophysiological Correlations. Butterworth-Heinemann, Boston, MA, 1998.

Westmoreland BE, Benarroch EE, Daube JR, Reagan TJ, Sandok BA. Medical Neurosciences. An Approach to Anatomy, Pathology and Physiology by Systems and Levels. Little Brown and Co., Boston, MA, 1994, pp. 353.

Wilbourn AJ. The electrodiagnostic examination with myopathies. J Clin Neurophysiol 1993; 10:132–148.

REVIEW QUESTIONS

1. Which one of the following structures comprises the essential contractile unit of a muscle fiber?
   A. Sarcomere.
   B. Myosin.
   C. Actin.
   D. Myofibril.
2. All of the following statements regarding the motor unit are true except:
   A. One motor unit innervates multiple muscle fibers.
   B. Each muscle fiber is innervated by only one motor unit.
   C. Muscle fibers belonging to a motor unit are contiguous.
   D. The innervation ratio varies with each muscle.
3. Essential steps involved in the process of muscle contraction include all of the following except:
   A. Depolarization induced opening of voltage gated calcium channels.
   B. Signal transduction of ryanodine receptors.
   C. Sliding of actin and myosin filaments.
   D. Decrease in absolute length of thick and thin filaments.
4. Which one of the following statements regarding the MUAP is false:
   A. MUAP amplitude depends on number and size of the muscle fibers closest to the recording electrode.
   B. The presence of any polyphasic MUAPs is generally indicative of an underlying pathological process.
   C. Duration depends both on fibers close to and away from the recording electrode.
   D. Rise time should ideally be shorter than 500 μs.
5. Match the following:

A. Fibrillation potentials	i. Waxing and waning amplitudes
B. Myotonic discharges	ii. "Raindrops on roof"
C. Complex repetitive discharges	iii. Thyrotoxic myopathy
D. Fasciculation potentials	iv. Normally found in Iliacus

6. EMG findings most consistent with the diagnosis of myopathy are:
   A. Absent motor responses on NCS, fibrillation potentials, and SASD MUAPs.
   B. Fibrillation potentials, fasciculation potentials, and large-duration MUAPs.
   C. Fibrillation potentials, complex repetitive potentials, and SASD MUAPs.
   D. Normal EMG.
7. All statements regarding NEE findings in myopathy are true EXCEPT:
   A. Decreased MUAP amplitude is the most reliable finding.
   B. MUAP findings indicate disease severity.
   C. Increased or early recruitment is characteristic.
   D. Fibrillation potentials indicate disease activity.
8. A generalized reduction in CMAP amplitudes in NCS may indicate all EXCEPT:
   A. Presynaptic defect of the NMJ.
   B. Motor polyradiculopathies.
   C. Motor neuron disease.
   D. Myopathy in early stages.

9. Myopathies that may present with "denervating features" on EMG include all EXCEPT:
   A. Inflammatory myopathies.
   B. Muscular dystrophies.
   C. Corticosteroid myopathy.
   D. Toxic myopathies.
10. All of the following myopathies may present with myotonic discharges EXCEPT:
   A. Myotonia congenita.
   B. The myopathy of intensive care.
   C. Colchicine myopathy.
   D. Paramyotonia congenita.

REVIEW ANSWERS

1. The correct answer is A. The sarcomere is the smallest contractile unit in a myofibril. It extends from Z band to Z band. The sarcomere consists of two filaments, the thick (myosin) filaments alternating with the thin (actin) filaments. The four major contractile proteins that are present in a myofilament are actin, myosin, troponin, and tropomyosin.
2. The correct answer is C. The term "motor unit" refers to a single lower motor neuron and the muscle fibers it innervates. One motor neuron innervates multiple muscle fibers but each muscle fiber is innervated by only one motor neuron. The muscle fibers belonging to a motor unit are typically distributed widely, extending over as many as 100 fascicles. Muscle fibers are randomly distributed over a circular or oval region that can reach approx 20 to 30% of the muscle's cross-sectional area, with an average diameter of 5 to 10 mm.
3. The correct answer is D. Sliding of the thin actin filaments over the thick myosin filaments produces muscle contraction. The shortening of the sarcomeres and the I band during contraction is not caused by any change in the absolute length of the filaments but rather by the sliding of the filaments themselves.
4. The correct answer is B. MUAPs with more than four phases are referred to as polyphasic. Up to 12% of recorded MUAPs in normal muscles are polyphasic.
5. The audio-amplified correlate of fibrillation potentials have been likened to the sound of "raindrops on a tin roof." Complex repetitive potentials may be found in the iliacus muscle of healthy individuals and, hence, should not be considered abnormal if restricted to that muscle. Fasciculations may be observed in thyrotoxic myopathy. Electrical myotonia (waxing and waning of discharges) is characteristically observed in myotonic myopathies.
6. The correct answer is C. Fibrillation potentials are usually observed in necrotizing myopathies. SASD MUAPs are not specific for myopathies (they are also observed in neurogenic disorders and diseases of the NMJ). Motor responses on NCS are expected to be normal in diseases of muscle, unless it is end-stage disease. Although CRDs are nonspecific, they may occur in myopathies. The EMG may be normal in some myopathies—those that are endocrine or metabolic in nature.
7. The correct answer is A. Just as fibrillation potentials are indicative of disease *activity*, MUAP changes are indicative of disease *severity*. Of the various MUAP alterations, an increased proportion of polyphasic MUAP is often reported to be the earliest recognizable change with myopathies. Decreased MUAP duration and early or increased recruitment are considered the most reliable changes. Decreased MUAP amplitude is considered the least reliable, because it is dependent on various needle electrode factors.
8. The correct answer is D. A generalized reduction in the CMAP amplitudes should alert the electromyographer to the possibility of disorders involving either the presynaptic portion of the NMJ, motor polyradiculopathies, or motor neuron diseases. Early stage myopathies generally do not present such an EMG profile.
9. The correct answer is C. "Denervating features" on EMG generally refer to a mixture of high-amplitude, long-duration MUAPs accompanied by abnormal spontaneous activity. Inflammatory

myopathies, IBM, muscular dystrophies, and toxic myopathies may all produce such an EMG picture. Corticosteroids usually produce a rather selective atrophy of type 2 muscle fibers and do not cause denervation.

10. The correct answer is B. Myotonic discharges are characterized by the waxing and waning of amplitude and frequency on EMG. These discharges are found in myopathies with defects in the sarcolemma. They are not observed in the myopathy of intensive care, a disorder characterized by loss of myosin filaments with sparing of the sarcolemmal membrane.

21
Neurophysiology of Neuromuscular Transmission and Its Disorders

James M. Gilchrist

Summary

Despite several antibody tests being available for the assessment of disorders of neuromuscular transmission, electrophysiological testing of the neuromuscular junction remains a very important part of clinical practice. The neuromuscular junction is a complex structure and an understanding of its anatomy and physiology can assist in better understanding the value of electrodiagnostic testing. The most common disorders include myasthenia gravis, Lambert–Eaton myasthenic syndrome, and botulism, and are usually readily identified using several electrophysiological techniques including slow (2–3 Hz) and fast (20- to 50-Hz stimulation). Single-fiber needle EMG remains an additional powerful and sensitive test for patients with disorders that are more mild, in whom repetitive stimulation testing is negative or indeterminate.

Key Words: Botulism; Lambert–Eaton myasthenic syndrome; myasthenia gravis; neuromuscular junction; repetitive nerve stimulation.

1. NEUROPHYSIOLOGY OF NEUROMUSCULAR TRANSMISSION

1.1. Anatomy of the Terminal End-Plate Region

The physiology of the neuromuscular junction (NMJ) derives from the anatomy of the terminal axon and motor end plate, also referred to as the presynaptic and postsynaptic regions (Fig. 1). Motor nerve fibers end in an arborization of fine intramuscular twigs ending at the terminal bouton. The motor twigs are myelinated until the very terminus, with a Schwann cell covering all but the synaptic interface. The terminal axon is separated from the motor end-plate region of the muscle fiber by an extracellular space approx 70-nm wide, called the primary synaptic cleft. On the far side of this cleft, in the motor end-plate region, are a series of invaginations of the muscle membrane, which are called the secondary synaptic clefts. These junctional folds are unique to NMJs.

The terminal axon membrane contains voltage-gated calcium channels (VGCC) of the P/Q type, which are arranged in active zones in a semigeometric array. Within the terminal axon, synaptic vesicles containing acetylcholine collect at the active zones (Fig. 1). Acetylcholine is synthesized from choline and acetate in the cytoplasm by the enzyme choline acetyltransferase. Acetylcholine enters the synaptic vesicles using the vesicular acetylcholine transporter, which also exports the acetylcholine during exocytosis. Choline is actively transported by an energy-dependent reuptake mechanism.

From: *The Clinical Neurophysiology Primer*
Edited by: A. S. Blum and S. B. Rutkove © Humana Press Inc., Totowa, NJ

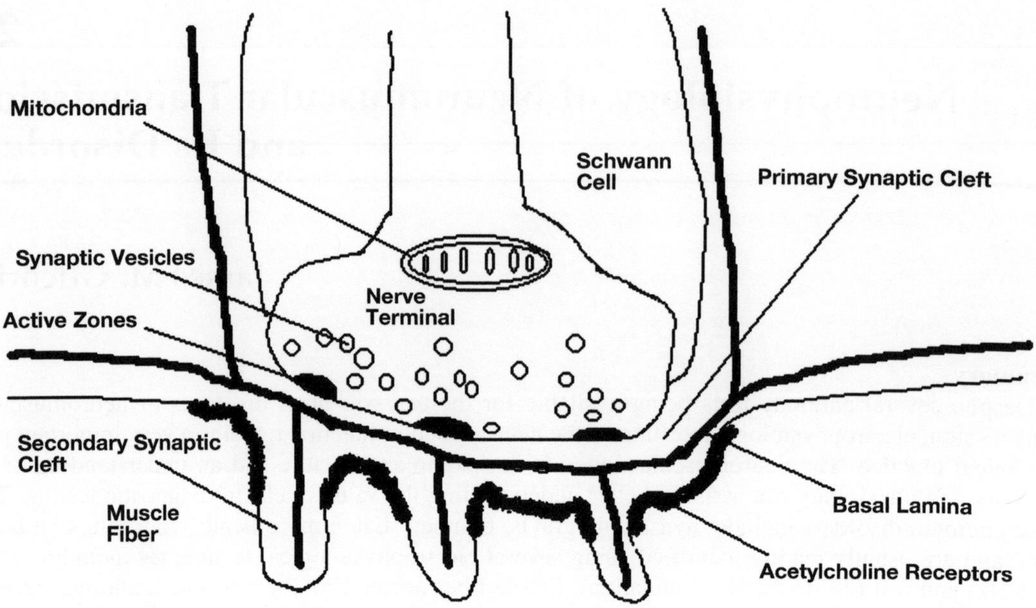

Fig. 1. A cartoon of a neuromuscular junction (NMJ). The synaptic vesicles contain acetylcholine. The active zones contain voltage-gated calcium channels, and part of the apparatus necessary for exocytosis. The basal lamina in the primary synaptic cleft contains acetylcholinesterase. Voltage-gated sodium channels necessary for propagation of the action potential generated at the neuromuscular junction are in the muscle fiber membrane, including the depths of the secondary synaptic clefts.

The primary synaptic cleft is divided by a basal lamina, a loosely organized, porous boundary containing acetylcholinesterase, which catabolizes acetylcholine as it diffuses across the primary synaptic cleft. Approximately 50% of the acetylcholine released from the presynaptic membrane is catabolized before it reaches the postsynaptic membrane.

The most important components of the postsynaptic membrane are the acetylcholine receptors. These receptors are of the nicotinic type, and contain a ligand-activated cation channel. At the base of the secondary clefts are voltage-gated sodium channels, essential for transmitting any action potential along the muscle membrane. Acetylcholine receptors are manufactured by membrane-bound ribosomes in the cytoplasm and then inserted in the postsynaptic membrane. There are approx 10,000 receptors per square micrometer at the terminal and upper areas of the secondary clefts. Each adult acetylcholine receptor (Fig. 2) is a tetramer containing 2 α-subunits, and one each of the β-, δ-, and ε-subunits. Fetal acetylcholine receptors substitute a γ-subunit for the ε-subunit. Of note, ocular muscles, which differ from other skeletal muscles in a number of ways, have an enriched population of fetal-type receptors. The half-life for adult acetylcholine receptors is 8 to 11 d. Ligand sites for acetylcholine are located on each of the α-subunits, and both most be engaged to activate the receptor channel. The main immunogenic region (MIR) is also located on each α-subunit, but separate from the ligand-binding site. After acetylcholine dissociates from the receptor, it is catabolized by acetylcholinesterase. Application of agents that inhibit the activity of acetylcholinesterase will prolong the activity of acetylcholine at the postsynaptic receptors.

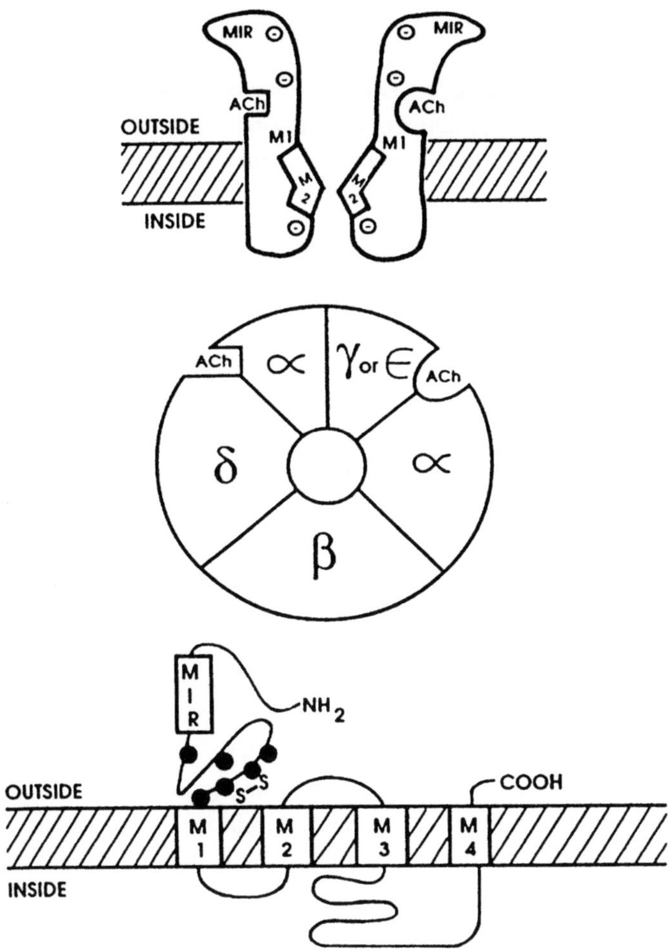

Fig. 2. A cartoon of the acetylcholine receptor, showing a transverse section, a top view, and a deconstructed view of the α-subunit. MIR, the main immunogenic region of the receptor; ACh, the ligand-binding site for acetylcholine. From Engel AG, 1999 with permission.

1.2. Presynaptic Physiology of Neuromuscular Transmission

The motor nerve action potential generated in the cell body is transmitted to the terminal axon membrane. As it traverses the presynaptic membrane, it activates the VGCCs, which open, allowing the movement of calcium into the terminal axon. The influx of calcium activates calmodulin-dependent protein kinase II, which binds with synapsin I, resulting in the docking of the synaptic vesicle with the terminal membrane. Release of the acetylcholine by exocytosis involves a complicated interaction of many proteins, including synaptobrevin, syntaxin, and SNAP-25. The synaptic membrane remains part of the axonal membrane and is recycled.

Acetylcholine-containing vesicles are organized into at least two pools. The immediately available pool is the smallest and consists of those vesicles actually lined up at the active zones for release. During a series of axonal depolarizations, such as when attempting to contract a muscle, the immediately available store will become depleted, and fewer synaptic vesicles will be released. The second pool is considerably larger than the immediately available store and is called the mobilization or reserve pool. During the course of a few seconds, it will replenish the immediately available pool.

The influx of calcium via the VGCCs is by passive diffusion. In contrast, calcium egress is by active transport and takes longer. During a sustained series of action potentials, the calcium concentration in the terminal axon will continue to increase and facilitate release of synaptic vesicles, partially countering the effects of depletion of the immediately available pool.

1.3. Postsynaptic Electrophysiology of Neuromuscular Transmission

The acetylcholine molecules passively diffuse across the primary synaptic cleft. Those that escape catabolism bind with acetylcholine receptors on the postsynaptic membrane. The activated receptors undergo a conformational change, allowing sodium to enter the cell and potassium to leave. This causes a small depolarization of the immediately adjacent muscle membrane. Release of single synaptic vesicles from the presynaptic membrane causes a reproducible level of depolarization, approx 1 mV, called a miniature end-plate potential (MEPP). This is the basis for the quantal theory of neuromuscular transmission (NMT). Because many synaptic vesicles are released with each depolarization of the terminal axon membrane, many MEPPs are produced. The MEPPs summate temporally and spatially to form an end-plate potential (EPP). If this EPP is sufficient to depolarize the membrane to threshold, an action potential is generated, which is then propagated along the muscle membrane by voltage-gated sodium channels in the sarcolemmal membrane, eventually resulting in muscle fiber contraction, the defining purpose of NMT.

An important concept is that of the safety factor. The EPP normally produced after each action potential activation of the presynaptic membrane is approximately four times that necessary to reach threshold. It is calculated by the formula $m = n \times p$, where m is the quantal of the EPP, and n is the percentage of vesicles released from the immediately available pool, which is determined by p, the probability of release. This excessive EPP derives from the abundance of acetylcholine released presynaptically and of acetylcholine receptors postsynaptically. Thus, even with the normally encountered depletion of the immediately available pool of acetylcholine-containing synaptic vesicles during a rapid train of motor nerve action potentials, the safety factor always provides successful NMT in the healthy NMJ. When healthy muscle fatigues, it is not because of failure of NMT but, rather, because of muscle metabolic issues, such as lactic acid buildup and failure of energy pathways. However, in the abnormal NMJ, if either the amount of acetylcholine released or the number of acetylcholine receptors declines, the safety factor will begin to fall as the size of the EPP falls. If the EPP safety factor falls below 1, that is, that needed to just reach threshold for action potential generation, then NMT will fail. This is the basis for fatigable weakness in disorders of postsynaptic NMT, such as myasthenia gravis (MG). As the number of receptors falls, the EPP also decreases. In many NMJs, the safety factor will fall below 1, and that muscle fiber will be, in essence, denervated. In many other fibers, the safety factor will hover just at or above 1, and initially, there will be successful NMT. However, during a train of motor nerve action potentials, the normal depletion of the immediately available store of acetylcholine-containing synaptic vesicles will result in a decrease in acetylcholine reaching the decreased number of receptors on the postsynaptic membrane. This will cause the safety factor to fall below 1, and NMT will fail, the muscle fiber will not contract, and the muscle will weaken.

The safety factor is also important in understanding incremental strength during sustained effort in disorders of presynaptic NMJ. In these disorders, there are abundant postsynaptic acetylcholine receptors, but release of the acetylcholine-containing synaptic vesicles is impaired. In a weak muscle, the fibers do not contract because the lack of acetylcholine release causes the safety factor to fall below 1, with resultant failure of NMT. In the case of Lambert–Eaton syndrome (LEMS), this is caused by loss of VGCCs and decreased influx of calcium to start the process of synaptic vesicle release. In botulism, it is impairment of the release of vesicles themselves. In both cases, with rapid repetitive activation of the presynaptic membrane, such as during strong effort, calcium will build up in the terminal axon, potentially increasing to levels that may approximate that observed during normal presynaptic function. At this point, acetylcholine release will improve toward normal levels, as will the EPP. As the safety factor increases and surpasses 1, NMT will successfully resume, though it will quickly fail as soon as the rapid train of motor action potentials ceases. This effect is much more evident in LEMS than botulism, but, if observed, is diagnostic of a presynaptic disorder.

2. NEUROPHYSIOLOGICAL TECHNIQUES TO STUDY NMT

2.1. EMG

The first described clinical neurophysiological abnormality in MG was by Harvey and Masland in 1941, when they reported variability in motor unit potential (MUP) amplitudes. The defect in MG is widespread but not universal; individual motor end plates are affected to varying degrees, even within the same motor unit. Some NMJs may be nonfunctional, some may have marginal safety factors, and others may be healthy. Because of this variable involvement, during muscle contraction, NMT may fail at a variable number of NMJs with each MUP firing, resulting in variation in amplitude and area of the MUP. This is not specific to MG, because it can be observed in any disorder of NMT, including presynaptic and postsynaptic diseases. It is a common finding in ongoing reinnervation after nerve injury, but can also be observed in acute denervation, old polio and postpolio syndrome, amyotrophic lateral sclerosis, and myopathic injury and recovery.

2.2. Repetitive Stimulation

Repetitive stimulation of nerve (RNS) while recording from muscle has been a valuable tool in the assessment of NMT since 1941, and remains the technique in widest use. It is easy to learn, easy to perform, not invasive, and requires no special equipment or training. On the negative side, it is often painful and poorly tolerated, is prone to artifact if not performed properly, and has a limited sensitivity, especially in localized diseases affecting NMT, such as ocular MG. The usual technique is to give a train of supramaximal electrical stimulations to a motor or mixed nerve while recording from an appropriate muscle. The train is usually four to nine stimuli long. For most indications, a rate of 2 to 3 Hz is most appropriate. This should be performed at rest, with the amplitude of the first compound motor action potential (CMAP) compared with the fourth or fifth. Significant decrement is usually defined as exceeding 10% (Fig. 3). After this, the patient should maximally contract the muscle for 1 min, if possible, which is followed by trains of stimuli immediately and at 30-s intervals out to at least 3 min. Decrement may repair immediately after exercise but should reach a maximum amount between 2 and 3 min after exercise, a phenomenon known as postactivation exhaustion. The improvement in decrement is called facilitation and is caused by increased calcium concentration in the terminal axon leading to enhanced release of acetylcholine. Small increments in CMAP amplitude can also be observed in healthy people.

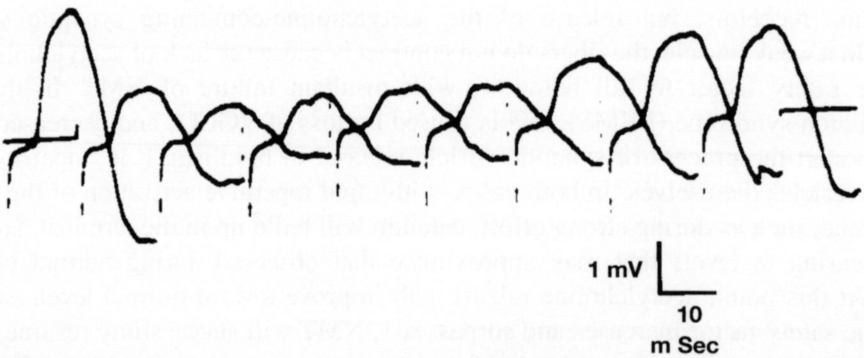

Fig. 3. Repetitive stimulation in a patient with myasthenia gravis demonstrating typical U-shaped decrement: maximal by the third or fourth stimulation in the train of nine, with return toward normal by the ninth stimulation. Decrement, calculated by comparing the fourth evoked response amplitude to that of the first, exceeds 65%.

There are several caveats worth noting regarding RNS:

1. Two- to 5-Hz stimulation is preferred if looking for decrement. One-hertz stimulation is usually too slow to produce decrement, and rates faster than 5 Hz may produce facilitation of the response, masking any decrement.
2. Maximal voluntary contraction for 10 to 60 s (depending on the strength of the patient) followed by a supramaximal nerve stimulus is the preferred method for looking for increment. In patients who cannot cooperate, or are too weak to voluntarily contract the muscle, rapid stimulation up to 50 Hz can be performed but this is exquisitely painful and should not be performed for more than 10 s unless the patient is deeply comatose.
3. Proximal muscles are more likely to show decrement in MG than are distal muscles, but proximal muscles are more prone to technical artifact and are more painful.
4. Decrement disappears as muscle temperature drops, therefore, a cool limb can result in failure to elicit decrement even if there is a defect in NMT. This is a greater problem with distal muscles, which should be warmed to at least 32°C. On the other hand, warming a limb above standard temperature may enhance a mild decrement.
5. In healthy individuals, CMAPs can increase up to approx 40% in amplitude simply because of the phenomenon of pseudofacilitation. Hence, increments of greater than 40% should be considered abnormal, although most patients with LEMS have considerably greater increments, in the range of 100 to 400% (*see* below).
6. Stimulation site and intensity must remain constant throughout the test because decreases in intensity may mimic decrement.
7. A healthy NMJ should have no decrement, but, because of the technical limitations of RNS, a decrement of up to 10% is within normal limits.

There have been several studies of the sensitivity of RNS in MG and LEMS, including comparisons to other techniques. In summary, RNS in MG is more likely to be abnormal in generalized disease than in ocular MG, is marginally more sensitive than measurement of acetylcholine receptor antibodies, is less sensitive than single-fiber EMG (SFEMG) at all levels of disease, and the diagnostic yield increases as more muscles are studied. The yield for a distal muscle RNS in generalized disease is 40% and approaches 70% for a proximal muscle.

NMT is an energy-dependent activity and ischemia will affect it adversely. This is the basis for an uncommonly used procedure, called double-step repetitive stimulation. This method

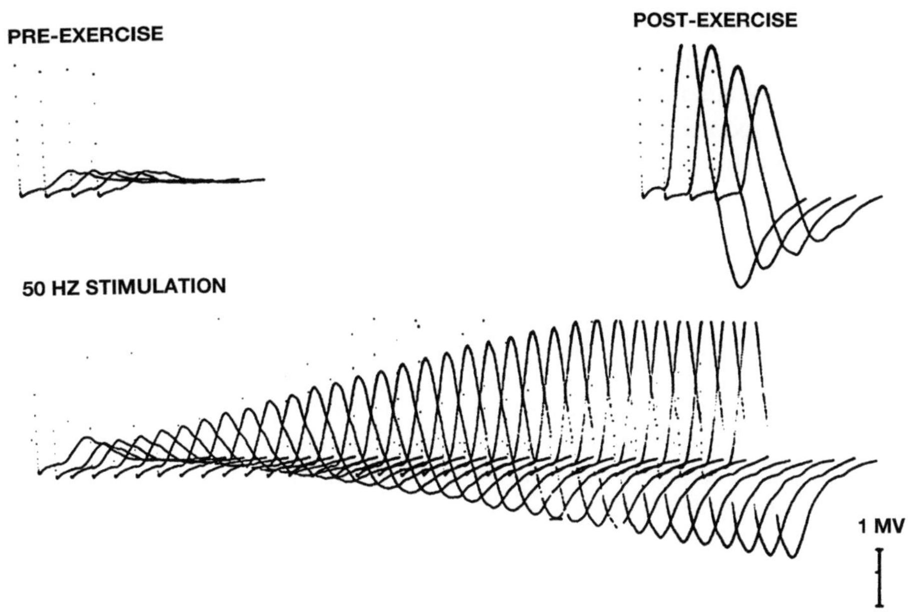

Fig. 4. Repetitive stimulation in a patient with Lambert–Eaton myasthenia syndrome. Pre-exercise shows a very small CMAP with decrement at low rates of stimulation. When repeated after exercise, the CMAP amplitude has increased by several hundred fold but decrement persists. The bottom trace is at 50 Hz stimulation and shows initial decrement followed by dramatic increment in CMAP amplitude. From Maselli R, 1998.

requires near-nerve needle stimulation of the ulnar nerve at the wrist at 3 Hz for 4 min while recording from the abductor digiti quinti muscle, measuring decrement, then repeating the 4 min of 3-Hz stimulation with a sphygmomanometer inflated above systolic blood pressure, measuring decrement after the cuff is deflated. This technique has been shown to increase sensitivity in MG comparable to proximal muscle RNS, but still lags considerably behind SFEMG.

Eaton and Lambert first described RNS in LEMS in 1956. At a low rate of stimulation, these patients will have decrement indistinguishable from MG. At high rates of stimulation, or after maximal voluntary activation of the muscle, there will be an increment in the response reaching and exceeding 100% in most patients (Fig. 4). This facilitation of the CMAP is not unique to LEMS, and up to 90% CMAP facilitation has been reported in MG. However, increase in the CMAP by 100% should be considered diagnostic of a presynaptic defect. A recent large study of LEMS found 98% of patients had decrement with 3-Hz stimulation, 88% of patients had CMAP potentiation greater than 100% in at least one muscle, but only 39% had potentiation greater than 100% in all three muscles studied.

2.3. *Single-Fiber EMG*

SFEMG is the selective recording of a limited number of single muscle fiber action potentials from one motor unit in vivo. This requires a needle electrode with different specifications from a concentric or monopolar needle electrode, and SFEMG needle electrodes have a dramatically smaller recording area than either. A SFEMG recording surface is 25 μm in diameter, with an effective recording area of 300 μm^3, as compared with a concentric needle electrode, which records from approx 1 cm^3. A smaller electrode emphasizes the amplitude difference between close and distant fiber potentials. A smaller recording surface will also restrict the number of recordable muscle fiber potentials. In addition, muscle fiber potentials adjacent to the recording electrode will have high amplitudes and short duration, and relatively more high-frequency components compared with more distant potentials. By using a high-pass filter of 500 Hz, much of the amplitude of distant muscle fiber potentials will be attenuated while preserving that of the nearby potentials. This allows single muscle fiber potentials to be selectively studied while the rest of the MUP is effectively dampened to nil.

By counting the number of single muscle fiber potentials observed with each MUP firing, the number of muscle fibers from that MUP within the small recording territory of the SFEMG needle electrode can be determined. For the most part, this should be one or two. Sampling 20 different sites in a muscle allows calculation an average number of single muscle fiber potentials per recording site. This is called the fiber density. In conditions with loss of random distribution of MUP muscle fibers, such as reinnervation, fiber density will increase. Specific disorders that can increase fiber density include anterior horn cell diseases, such as spinal muscular atrophy, polio, postpolio muscular atrophy, and amyotrophic lateral sclerosis; and any peripheral or cranial neuropathy with axonal loss, specifically, those caused by diabetes, alcohol, uremia, toxins, amyloidosis, Guillain–Barré syndrome, chronic demyelinating inflammatory polyneuropathy, and multiorgan failure. The only study to compare etiologies found that alcoholic polyneuropathy produced higher fiber densities than did uremic or diabetic neuropathy, despite (or because of) better nerve conduction velocities. A variety of muscle disorders will also have increased fiber densities, especially as the disease progresses and chronic disability ensues, including muscular dystrophies, inflammatory myopathies, mitochondrial myopathies, and congenital myopathies.

In those instances in which two or more fiber potentials from a single MUP are recorded, an interpotential interval (IPI) can be calculated. By recording multiple consecutive firings of the muscle fiber potentials, the difference between consecutive IPIs can be calculated. The variation amongst these consecutive IPIs is called jitter. Jitter is most accurately determined by calculating a mean consecutive difference using the formula:

$$\text{Mean consecutive difference} = [(\text{IPI}_1 - \text{IPI}_2) + \ldots + (\text{IPI}_{n-1} - \text{IPI}_n)]/(n-1)$$

Jitter is thought to derive from variation in the time in takes the NMJ EPP to reach threshold for action potential generation at the postsynaptic membrane. In disorders with disturbed NMT, there will be an increased variation in the time taken to attain an EPP capable of reaching threshold. This will lead to increased jitter (Fig. 5). Therefore, abnormal jitter is an indicator of abnormal NMT.

In those instances in which an EPP fails to reach threshold for action potential generation, one of the muscle fiber potentials in the pair will be absent. This is called impulse blockade (Fig. 5). It is another indicator of abnormal NMT, usually indicating a more severe disturbance than increased jitter alone. Impulse blocking, often referred to simply as blocking, is usually intermittent, with the affected fiber potential appearing and disappearing in an unpredictable pattern. Blocking is uncommonly observed with jitter less than 100 μs.

Increase jitter, and even blocking, are not specific to MG or LEMS, and can be observed in any disorder of NMT. It is an early finding in any neuropathy with axonal loss, including

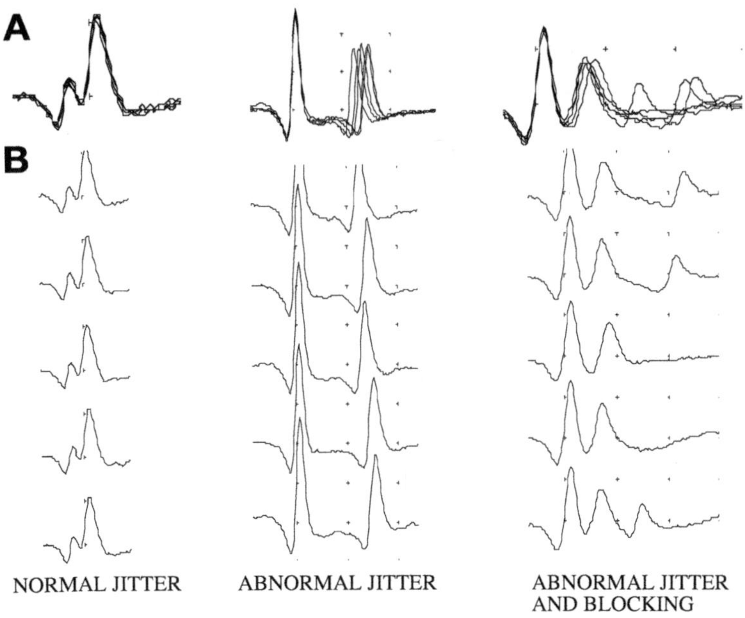

Fig. 5. Single-fiber EMG examination for jitter in normal and abnormal motor end plates. (**A**) Superimposed and (**B**) serial traces indicating (from left to right) normal jitter, increased jitter without impulse blocking, and increased jitter as well as blocking.

acute transection. Wallerian degeneration after nerve fiber transection transpires over 11 to 14 d, and sensory nerve action potentials become unobtainable below the transection by 11 d. However, CMAPs are lost within 7 d of nerve transection, because NMT fails before the nerve fiber becomes inexcitable. Increased jitter will be observed in anterior horn cell disorders, acute and chronic peripheral neuropathies, and myopathies. It is a reflection of acute denervation and subsequent failure of NMT, as the nerve terminus degenerates, and reinnervation, as the nerve terminus regenerates and the NMJ matures. In muscle disease, jitter indicates degeneration of motor end plates caused by myofiber degeneration as well as myofiber regeneration with immature motor end plates. However, it is in the primary disorders of NMT that SFEMG jitter studies are most useful.

SFEMG has been used in the diagnosis of MG since at least 1971. Since then, there have been numerous studies of the sensitivity of SFEMG in diagnosing MG, and comparisons to other diagnostic techniques. The largest series of SFEMG studies in MG reported the results of 788 patients. Results of SFEMG of the extensor digitorum communis muscle was abnormal in 85% of all patients with MG at the time of initial examination. If the extensor digitorum communis muscle was healthy, and a second muscle was studied, 85% of those patients had abnormal jitter studies. Thus, if two muscles were studied when the first was normal, results of SFEMG for jitter analysis were abnormal in 98% of all patients with MG. This far exceeds the sensitivity of all other diagnostic tests for MG, as further illustrated by several comparative studies, which, in one study, found SFEMG to be the most sensitive, at 92% (testing a single muscle); with RNS at 77% (testing multiple muscles) and acetylcholine receptor antibody testing at 73%. SFEMG was sensitive regardless of whether disease was generalized or ocular; the yield was higher than 90%, even in ocular disease, if more than one muscle was studied.

The enhanced sensitivity of SFEMG makes physiological sense: RNS results will not be abnormal until at least 10% of muscle fiber end plates undergo impulse blockade and, therefore, fail to generate or propagate a muscle fiber action potential. Muscle fibers with slowed and unstable NMT, but not so affected that they are blocked, will count as normal. SFEMG not only can determine the fibers with impulse blockade, but, by assessing jitter, will allow those fibers with disturbed but still functional NMT to be measured.

SFEMG studies of jitter and impulse block also correlate well with the clinical severity of the disease. Mean jitter, percentage of fiber pairs with increased jitter, and percentage of fiber pairs with blocking all increase with worsening disease. Mean jitter worsens by at least 10% in two-thirds of patients when their disease worsens, and mean jitter improves by at least 10% in 80% of patients who clinically improve. Despite the sensitivity of SFEMG to changes in NMT, SFEMG does not predict progression of ocular MG to generalized disease.

Results of SFEMG jitter studies are also abnormal in LEMS, often more so than would be expected from the clinical picture. Large case series with SFEMG studies are not available, but virtually all patients reported have had markedly abnormal jitter and large percentages of blocking fibers. Because of the presynaptic nature of LEMS, a relationship between jitter and firing rate would be predicted, but, in fact, this is variable. It is safe to say that a dramatic improvement in jitter and blocking with increasing firing rate is suggestive of a presynaptic defect in NMT.

Botulism arises from defective presynaptic release of acetylcholine caused by the toxin of *Clostridium botulinum.* Results of SFEMG are abnormal in 95% of patients with botulism, and in 100% of botulism patients with clinical weakness. Results of SFEMG studies improve as patients improve. Initially, fiber density is normal but, because botulism causes an irreversible block of acetylcholine release, patients improve by reinnervation and fiber density increases.

3. DISEASES AFFECTING NMT

3.1. Presynaptic Disorders

3.1.1. Lambert–Eaton Myasthenic Syndrome

LEMS was first described in 1951 and received its eponym from the investigators who first described its clinical neurophysiological characteristics. LEMS is caused by a polyclonal antibody attack directed against the P/Q VGCCs located on the presynaptic membrane of acetylcholine terminal nerve axons. VGCCs contain $\alpha 1$-, β- and $\alpha 2/\delta$-subunits, with the $\alpha 1$-subunit containing the calcium conductance channel, as well as being the ligand-binding site. The autoimmune attack results in loss of calcium channels and disorganization of the active zones, leading to inhibition of release of acetylcholine-containing synaptic vesicles. This inability to fully release acetylcholine creates MEPPs that are normal in amplitude but decreased in number, resulting in a decreased EPP. If the EPP is decreased below the level necessary to reach threshold for action potential generation, then NMT is unsuccessful and the muscle remains noncontractile.

The major clinical feature of LEMS is weakness, often generalized and symmetric, and usually affecting proximal muscles more than distal. Affected patients will be areflexic or markedly hyporeflexic, with normal sensation. Signs and symptoms of autonomic dysfunction are also present, such as dry mouth, impotence, constipation, bladder retention, and abnormal papillary reactions, indicating that the autoantibody attack is not limited to VGCCs at NMJs.

Unlike MG, oculomotor function is uncommon and the onset is usually insidious, although acute presentations, including respiratory crisis, are reported. The most striking clinical feature is the improvement in muscle contraction and strength with continued effort. Muscles that were too weak to resist gravity will become nearly normal with sustained effort and reflexes that at first are absent, will steadily improve to normal with repetitive striking of the tendon. All such improvements prove transient and disappear rapidly with cessation of effort. This facilitation with sustained effort is pathognomonic of a presynaptic disorder of NMT.

Unlike MG, the typical patient with LEMS is male and older, often in the fifth or sixth decade of life. This is likely because 50% of LEMS patients have a paraneoplastic syndrome, often associated with a small cell carcinoma of the lung. The 50% of patients without an associated carcinoma are younger, but again with a strong male preponderance. The clinical characteristics do not differ between the cancer and noncancer groups. Survival is determined by the underlying cancer in those patients with one, but is not otherwise shortened.

The diagnosis is never made without a high index of suspicion. Antibodies against the P-type VGCC are found in up to 85% of patients, but the diagnosis is most frequently made in the EMG laboratory. Any patient with small CMAP amplitudes and normal conduction velocities but normal sensory nerve action potential studies should be considered a possible case of LEMS. To screen for LEMS, the muscle being recorded should then be exercised for 10 to 30 s, depending on how weak it is, and an immediate supramaximal stimulus administered. If there is no increment in the CMAP amplitude, a presynaptic defect is not present. If there is more than a 40% increment, a presynaptic defect becomes a strong possibility, and, if the increment exceeds 100%, a presynaptic defect is definite. Any such finding should be confirmed in other muscles by a similar method. Repetitive stimulation (Fig. 4) will show decrement at low rates of stimulation before exercise, with immediate increase in CMAP amplitude and repair of decrement after exercise, and return to decrement after 2 to 3 min. At high rates of stimulation, which are not recommended in the conscious patient because of great discomfort, there will be a dramatic facilitation of the CMAP amplitude. SFEMG will show increased jitter, impulse blockade, and, in some fibers, a decrease in jitter when there is an increase in firing rate.

In patients with cancer, treatment is obviously directed at the cancer, which will often provide improvement and even remission. Symptomatic treatments may also provide clinical benefit. Guanidine enhances release of synaptic vesicles and was shown to be effective, but had severe side effects and is no longer available. Pyridostigmine enhances the amplitude of MEPPs, much as it does in MG, and can provide some benefit. The most effective symptomatic treatment comes from the aminopyridines, which inhibit voltage-gated potassium conductance, lengthening the action potential and prolonging calcium conductance into the terminal axon; 4-aminopyridine is effective but crosses the blood–brain barrier and causes seizures, tremors, and anxiety; 3,4-diaminopyridine has been found to be much safer because it is less capable of crossing the blood–brain barrier and more potent in enhancing release of acetylcholine. The industrial solvent, 3,4-diaminopyridine, is available by the barrel for that purpose, but not approved by the Food and Drug Administration except for individual compassionate use.

3.1.2. Botulism

Botulism is caused by the neurotoxin secreted by the anaerobic bacteria Clostridium botulinum and takes three forms, food-borne, wound, and infantile. Food-borne is the most common worldwide, but in the United States, infantile botulism is most common. The toxin consists of a heavy and a light chain, and enters the terminal motor axon via receptor-mediated endocytosis. Inside the axon, the light chain interferes with proteins involved in fusion of the synaptic vesicle with the terminal membrane, including synaptobrevin (part of the synaptic vesicle membrane), SNAP-25, and syntaxin (both part of the presynaptic membrane). The effect on the nerve terminal is irreversible and, if severe enough, results in denervation of the myofiber. Recovery in those cases is prolonged, because it requires nerve regeneration and reinnervation.

Infantile botulism occurs between 2 wk and 6 mo of age, and presents as hypotonia, constipation, poor feeding, and dyspnea during the course of hours or days. Food-borne botulism affects older children and adults and begins abruptly, with diplopia, ptosis, dysphagia, limb weakness, and respiratory compromise. External ophthalmoplegia, papillary paralysis, areflexia or hyporeflexia, and limb weakness are found on examination. Respiratory failure can lead to death, but supportive care and use of botulism antitoxin provide for a good prognosis.

The diagnosis should be suspected clinically and can be made by documenting the presence of botulinum toxin. The diagnosis is usually confirmed using electrodiagnosis. CMAPs are always small at rest. Facilitation greater than 100% is observed in 90% of patients but requires a more sustained effort or a longer period of high-rate stimulation. Repetitive stimulation will show a decrement if the CMAP is not too low for decrement to be accurately measured. Routine EMG will often reveal fibrillations and positive waves, with a myopathic recruitment pattern consisting of small, brief, polyphasic MUPs that are early recruited to a full interference pattern. SFEMG will show increased jitter and blocking, which may inversely correlate with MUP firing rate. Fiber density will increase as reinnervation proceeds.

3.2. Postsynaptic Disorders of NMT

3.2.1. Myasthenia Gravis

MG is an autoimmune disorder in which polyclonal antibodies are directed against the nicotinic acetylcholine receptor of skeletal muscle. This results in degradation of the NMJ, with simplification of the secondary synaptic clefts, loss of acetylcholine receptors, and failure of NMT. The loss of receptors results in a MEPP that has decreased amplitude, leading to a decreased EPP. The clinical hallmark of the disease is fatigable weakness, usually after repetitive action, causing intermittent symptoms, such as ptosis, diplopia, dysphagia, dysarthria, and facial and limb muscle weakness. Respiratory compromise can occur in severe cases. The disease has a bimodal peak incidence, affecting older men and young women of childbearing age.

Although the clinical history and examination are often typical and highly suggestive, confirmation of the diagnosis rests on pharmacological, immunological, and electrodiagnostic grounds. Edrophonium (Tensilon) administered intravenously will quickly but briefly reverse the signs of MG and serves as a good bedside test. Assay for the presence of serum acetylcholine receptor antibodies is very specific for MG, and is abnormal in 70 to 90% of cases. Sensitivity is lower in patients with only ocular signs.

Electrodiagnostic methods most useful in the diagnosis of MG are RNS and SFEMG. The results of RNS are most likely to be abnormal in patients with generalized disease, and when testing proximal muscles, but even then sensitivity is only in the 70% range if multiple muscles are tested. Decrement is most likely with low rates of stimulation, 3 to 5 Hz, after a period of maximal muscle contraction. In patients with decrement at rest, exercise will increase the amount of decrement. At high rates of stimulation, facilitation of the CMAP may mask decrement. The decrement tends to be maximal by the fourth stimulus in a train and often will repair by the ninth. This U-shaped decrement is common in MG but not specific.

SFEMG remains the most sensitive test for dysfunction of NMT and the most sensitive diagnostic test for MG, reaching a 98% yield if two muscles are tested when the first is normal. This is true for ocular as well as generalized disease. SFEMG is also useful in the management of MG because it is a faithful and sensitive indicator of the status of NMT, unlike receptor antibody assay and RNS.

Treatment may address symptoms only or may be curative. Anticholinesterases can be used to briefly abate or improve symptoms attributable to MG, but will not affect the underlying immunological dysfunction. These drugs work by inhibiting the breakdown of acetylcholine, the neurotransmitter released by terminal motor nerve fibers. Edrophonium, neostigmine, and pyridostigmine (Mestinon) are all anticholinesterases, the latter most commonly used because of its longer duration of action (2–4 h) and lesser muscarinic side effects. Mestinon is commonly used by itself in mild cases and in conjunction with immune suppression in more severe cases. Side effects include diaphoresis, hypersalivation, diarrhea, nausea, abdominal cramping, bradycardia, and fasciculations. Intravenous dosages of pyridostigmine and neostigmine are 1/30 of the oral dose for both drugs.

Suppression of the immune system attack on the acetylcholine receptor is indicated when the disease is generalized, involves vital functions, such as ventilation or swallowing, or is not amenable to symptomatic treatment alone. Various treatments can be used, including corticosteroids, immune suppressants, such as azathioprine, mycophenylate, mofetil and cyclosporine, plasmapheresis, intravenous human immunoglobulin, and thymectomy. Corticosteroids can cause worsening of symptoms at initiation of therapy and patients must be carefully watched early on, preferably as inpatients. This worsening can be limited by starting patients on very low doses, with a slow titration upward in dose, although this delays clinical benefit. Plasmapheresis is indicated in the severely compromised patient, in the patient refractory to other treatment modalities, and in the patient in whom an immediate response is required.

Thymomas are present in 10 to 15% of patients with MG, and MG occurs in 30% of patients with thymomas. Thymic hyperplasia is present in another 70% of patients with MG. The thymus gland is the likely site for initial sensitization to the acetylcholine receptor. Removal of thymic tissue increases remission rate from 15 to 30%, and results in significant clinical improvement in two-thirds of patients, although the improvement may take up to 5 yr.

3.2.2. Congenital Myasthenic Syndromes

Congenital myasthenic syndromes are genetic disorders affecting components of the NMJ important in the normal function of NMT. The first clinical description was in 1937, but the realization that they were not related to MG occurred only after the autoimmune nature of MG was defined in the 1970s. Many such syndromes have now been defined either by their physiology, their clinical manifestations, or their ultrastructural characteristics, affecting presynaptic, synaptic, and postsynaptic function. Postsynaptic syndromes account for 76% of cases, with end-plate acetylcholinesterase deficiency accounting for 13% of cases and presynaptic syndromes, 8%. There is neither the space nor the inclination to detail each of the syndromes, but a notation of the actual defect usually suffices to illustrate why there is a problem (Table 1).

Onset of symptoms may be in the neonatal period, childhood, or adulthood. The hallmark of all is fatigable weakness, that is, increasing weakness with continued exertion. This can involve ocular, bulbar, limb, and respiratory muscles. Antibodies to acetylcholine receptors are always absent, by definition. The Tensilon test is variably useful depending on the defect, and obviously not useful in end-plate acetylcholinesterase deficiency. A decremental

Table 1
Congenital Myasthenic Syndromes With Location and Description

Presynaptic	Paucity of synaptic vesicles and reduced quantal release
	Defect in ACh resynthesis or packaging
	CMS resembling Lambert–Eaton syndrome
Synaptic	End-plate acetylcholinesterase deficiency
Postsynaptic	Increased response to ACh: slow-channel syndromes
	Decreased response to ACh: low-affinity fast-channel syndromes
	Decreased response to ACh: fast-channel syndrome secondary gating abnormality
	Decreased response to ACh: mode-switching kinetics
	Receptor deficiency secondary recessive mutations in receptor subunits
Unknown	Familial limb–girdle myasthenia
	Benign CMS with facial malformations

ACh, acetylcholine; CMS, congenital myasthenic syndromes.

response to RNS at low rates of stimulation is also important in diagnosis, as are abnormal jitter and blocking during SFEMG. At times, patients with impaired resynthesis of acetylcholine or vesicular packaging will have normal electrodiagnostic studies when asymptomatic. Patients with end-plate acetylcholinesterase deficiency or slow-channel syndrome will have repetitive CMAPs to single stimuli.

SUGGESTED READING

AAEM Professional Practice Committee, Chiou-Tan FY, Gilchrist JM, Tim RW. Practice parameter for repetitive nerve stimulation and single fiber electromyographic evaluation of adult patients with suspected myasthenia gravis or Lambert–Eaton myasthenic syndrome: summary statement. Muscle Nerve, 2001.

Engel AG, ed. Myasthenia Gravis and Myasthenic Disorders. Oxford University Press, New York, NY, 1999.

Fon EA, Edwards RH. Molecular mechanisms of neurotransmitter release. Muscle Nerve 2001;24:581–601.

Gilchrist JM. Myasthenia gravis. In: Current Diagnosis in Neurology (Feldmann E, ed.). Mosby, St. Louis, MO, 1994, pp. 350–352.

Lindstrom JM. Acetylcholine receptors and myasthenia. Muscle Nerve 2000;23:453–477.

Maselli RA. Electrodiagnosis of disorders of neuromuscular transmission. Ann NY Acad Sci 1998; 841:696–711.

Maselli RA, Bakshi N. Botulism. AAEM case report #16. Muscle Nerve 2000;23:1137–1144.

Oh SJ, Doo EM, Kuruoglu R, Bradley RJ, Dwyer D. Diagnostic sensitivity of the laboratory tests in myasthenia gravis. Muscle Nerve 1992;15:720–724.

Sanders DB, Stalberg EV. AAEM minimonograph #25: Single fiber electromyography. Muscle Nerve 1996;19:1069–1083.

Stalberg E, Trontelj JV. Single Fiber Electromyography. Studies in Healthy and Disease Muscle. 2nd Ed. Raven Press, New York, NY, 1994.

REVIEW QUESTIONS

1. Normal presynaptic release of acetylcholine is dependent on:
   A. A VGCC of the P/Q type.
   B. Synaptobrevin.
   C. SNAP-25.

D. Vesicular acetylcholine transporter.
E. All of the above.

2. The immediately available pool of acetylcholine vesicles is:
A. Rapidly depleted by a train of presynaptic action potentials.
B. Mobilized to replenish the active zones.
C. Expanded in proportion to the postsynaptic depolarization.
D. Held constant by the vesicular acetylcholine transporter.
E. Irrelevant to the normal function of NMT.

3. NMT safety factor:
A. Is equal to the number of receptor activations necessary to reach threshold for action potential activation.
B. Is not a factor in normal NMT.
C. Refers to the EPP in excess of that necessary to reach threshold for action potential activation.
D. Fails frequently during exercise in healthy people.
E. Is determined solely by the number of available acetylcholine receptors.

4. Lambert–Eaton syndrome:
A. Is a postsynaptic disorder of VGCCs.
B. Is caused by a defect in acetylcholine vesicle release.
C. Is not readily diagnosed by repetitive nerve stimulation.
D. Improves with continued exercise.
E. Is less likely than botulism to show changes with continued exercise.

5. EMG:
A. Is irrelevant to the examination of disorders of NMT.
B. Will show MUP variation only with strong contractions.
C. Will show abnormalities of motor unit variation that are specific to MG.
D. Will show denervation potentials in most cases of LEMS.
E. Can reveal MUP amplitude variation suggestive of defective NMT.

6. Repetitive nerve stimulation:
A. Is the most sensitive test for disorders of NMT.
B. Is best performed at high rates of stimulation.
C. Should be performed both before and after a period of maximum voluntary muscle contraction.
D. Most commonly shows facilitation 2 to 3 min after exercise.
E. Does not require impulse blockade to be abnormal.

7. SFEMG:
A. Is usually performed with a low filter of 20 Hz, using a concentric needle electrode.
B. Performed by axonal stimulation allows the examiner to determine fiber density.
C. Is a sensitive measure of NMT safety factor.
D. Can differentiate presynaptic from postsynaptic disorders.
E. Is abnormal only in primary disorders of NMT.

8. The clinical characteristics of LEMS:
A. Are indistinguishable at the bedside from MG.
B. Often involve oculomotor function.
C. Always indicate an underlying lung cancer of the small cell type.
D. Include areflexia and autonomic dysfunction.
E. Are obvious and rarely missed.

9. Botulism:
A. Is caused by a toxin directed against acetylcholinesterase in the basal laminar matrix of the primary synaptic cleft.
B. Is a presynaptic disorder of the VGCCs.
C. Does not show abnormalities on repetitive nerve stimulation.
D. Will reveal facilitation after exercise.
E. Does not involve oculomotor function.

10. MG:
    A. Is characterized by fatigable weakness.
    B. Shows maximal decrement by the fourth stimulus in a train.
    C. Does not require SFEMG for diagnosis.
    D. Is characterized by a MEPP of decreased amplitude.
    E. All of the above.

REVIEW ANSWERS

1. The correct answer is E. VGCCs, synaptobrevin, SNAP-25, and vesicular acetylcholine transporter all participate in the release of acetylcholine from the presynaptic terminal.
2. The correct answer is A. The immediate pool is rapidly depleted by a train of presynaptic action potentials and is not expanded in proportion to the postsynaptic depolarization. The mobilization or reserve pool is mobilized to replenish the immediate pool. The immediate pool is not held constant by the acetylcholine transporter and is, in fact, very relevant to normal NMT.
3. The correct answer is C. The NMT safety factor refers to the EPP in excess of that necessary to reach threshold. It is not the number of receptor activations needed to reach threshold and it is a major factor in normal NMT. However, it does not fail during exercise in healthy people (there continues to be some safety factor normally present even after intense exercise and rapid rates of stimulation). It is determined, in part, by the number of acetylcholine receptors as well as the amount acetylcholine in the NMJ.
4. The correct answer is D. LEMS improves with repeated exercise. It is a presynaptic disorder of VGCC, it is not caused by a defect in acetylcholine release, it is readily diagnosed by repetitive nerve stimulation (with a decrement at slow rates of stimulation and an increment with high rates), and is *more* likely than botulism to show improvements with continued exercise.
5. The correct answer is E. EMG will show unstable MUPs with varying amplitude in patients with defective NMT of any cause and are not necessarily dependent on the strength of contraction (hence, EMG is not irrelevant to examination of disorders of NMT, but there are no abnormalities specific to any one disorder). Denervation potentials would be distinctly uncommon in LEMS (although they are relatively common in botulism).
6. The correct answer is C. To elicit abnormalities on repetitive stimulation, it should be performed both before and after a period of maximum voluntary muscle contraction. It is not the most sensitive test (SFEMG is). It should be performed at slow rates for postsynaptic disorders and rapid rates for presynaptic disorders. Facilitation should occur immediately after exercise (after exercise exhaustion usually is maximal at 2–3 min after exercise). It does require impulse blockade to be abnormal.
7. The correct answer is C. SFEMG provides an excellent measure of the safety factor of NMT. It is usually performed with a single-fiber electrode (not a concentric needle electrode) with a low-frequency filter set to approx 2 kHz. Axonal stimulation is not necessary to determine fiber density. It cannot differentiate presynaptic from postsynaptic disorders. It is a nonspecific finding and can be abnormally elevated in almost any form of neuromuscular disease.
8. The correct answer is D. Areflexia and autonomic dysfunction are usually associated with LEMS. LEMS and MG usually are quite distinct, although, in some rare cases, they can look similar. Oculomotor function is rarely if ever involved in LEMS. It is not always associated with small cell carcinoma of the lung (it can occur as an independent autoimmune disorder). LEMS can be difficult to diagnosis and the disorder can be overlooked.
9. The correct answer is D. Botulism usually demonstrates facilitation after exercise, although in severe cases facilitation may not occur. It is not caused by antibodies directed against acetylcholinesterase or the VGCCs, but rather against the proteins involved in fusion of the synaptic vesicle with the terminal membrane. Abnormalities on repetitive stimulation are to be expected and involvement of oculomotor function is quite common.
10. The correct answer is E. All of the choices are typically observed in MG.

22

Pediatric Nerve Conduction Studies and EMG

Peter B. Kang

Summary

A variety of neuromuscular conditions affect children, ranging from severe, usually fatal disorders, such as spinal muscular atrophy type I (Werdnig–Hoffman syndrome) to relatively mild problems, such as benign congenital hypotonia. The evaluation of children in the EMG laboratory requires special care because of the discomfort of the tests. Moreover, other considerations, such as slower baseline nerve conduction velocities and conditions that generally do not present in adulthood, such as congenital myasthenic syndromes, can make the pediatric neurophysiological examination especially challenging. This chapter reviews both the common pediatric neuromuscular conditions and their assessment in the EMG laboratory.

Key Words: Electromyography; Guillain–Barré syndrome; hereditary neuropathies; pediatrics; root avulsion; spinal muscular atrophy.

1. INTRODUCTION

EMG is a useful diagnostic tool in children suspected of having acquired or inherited neuromuscular disease and complements the advances in molecular genetic testing during the past decade. The technical limitations of performing EMG in children require that the electromyographer be selective in deciding how to approach the study. In evaluating for generalized processes, examination of one or two extremities is often adequate to narrow the differential diagnosis or, in some cases, suggest a specific disorder. EMG may be used to evaluate hypotonia in infants, assess the severity and localization of perinatal brachial plexus injuries, and distinguish between different possible causes of gait difficulties in older children. In children, EMG may contribute to the diagnosis of many disorders, including spinal muscular atrophy (SMA), brachial plexus injury, hereditary polyneuropathy, acquired polyneuropathy, disorders of neuromuscular transmission, myopathy, muscular dystrophy, and myotonic disorders.

2. TECHNICAL ISSUES

2.1. *General Approach*

Electrodiagnostic studies are highly dependent on technique in adults, but are even more dependent on technique in children. Infants and toddlers do not understand the purpose of the study, have a limited tolerance for discomfort, and often squirm and withdraw during testing. In selected cases, sedation or general anesthesia is required to obtain adequate data. Topical anesthetic creams are used in some pediatric EMG laboratories, which can help reduce pain,

From: *The Clinical Neurophysiology Primer*
Edited by: A. S. Blum and S. B. Rutkove © Humana Press Inc., Totowa, NJ

but they require that the electromyographer either guess which sites are most likely candidates for needle EMG before the nerve conduction studies, or impose a stressful half-hour wait between the nerve conduction studies and needle EMG on the patient. Despite these limitations, it is usually possible to perform a successful and informative study.

As in adults, a brief history and focused physical examination is critical in directing the tests performed, especially because they may be terminated prematurely. If the child is old enough to understand the procedure, it is important to explain it to the child as well as the parent. It is sometimes possible to convince younger children that the EMG machine is a "tickling" machine for nerve conduction studies—the power of suggestion being quite effective. Infants may be more comfortable sitting in a parent's lap on the examination table or in a chair. Pacifiers and toys are often effective in calming infants and toddlers. Even school-age children and adolescents may feel more comfortable if a parent sits or stands near them during the study.

2.2. *Nerve Conduction Studies*

The temperature of the extremity during nerve conduction studies is as important in children as in adults. An excessively cold extremity will yield falsely slow conduction velocities and increased amplitudes. Upper extremity skin temperatures, measured at first dorsal interosseous, should be at least 32°C. Lower extremity skin temperatures, measured at lower gastrocnemius, should be at least 30°C. Warming may be achieved with the use of towels dampened with hot water and wrapped around the distal extremity for approx 5 min. Care must be taken not to overheat the towels; a child's skin is more delicate than that of an adult, and is more susceptible to scalding injury. It is also important to wring out the towel before applying to the skin and to dry the skin after warming; a wet extremity will rapidly cool. Some electromyographers use disposable hot packs that produce heat via a chemical reaction. If used in children, these packs should be wrapped in towels to prevent burns, for they may become very hot. The temperature should be measured again after warming to confirm that it is in the acceptable range.

Surface active and reference recording electrodes may need to be trimmed or even cut in half for infants. Full-sized ground electrodes should be used whenever possible, however, because their placement is more flexible. Pediatric stimulators with small cathodes and anodes are useful for this age group. Standard adult distal distances cannot be used in infants and young children because of the small size of the extremities involved; thus, evaluation of motor distal latencies must take the patient's age into account (*see* Table 1). The initial stimulation should always be less than 10 mA. It can be very reassuring to a child when the first stimulation is barely perceptible.

Questions of generalized processes, such as polyneuropathies and myopathies, are common in pediatric EMG laboratories. Limited patient tolerance often makes it practical to perform only a motor and sensory study in an upper and lower extremity. In such cases, the electromyographer may perform median motor and sensory studies in an upper extremity, and peroneal motor and a sural or medial plantar (depending on the age) sensory study in a lower extremity. In a child who is uncooperative or anxious, it may be best to perform the sensory studies first, because those require less stimulation intensity and are, thus, better tolerated.

The median motor study is performed as in an adult, with the E1 recording electrode placed over the abductor pollicis brevis and stimulation occurring at the wrist and cubital

Table 1
Motor Nerve Conduction Studies, Suggested Values

Age	*A (mV)*	*CV (m/s)*	*DL (ms)*
Median nerve			
Preterm (33–39 wk)	—	≥18	—
0–1 mo	≥2.5	≥20	≤3.5
1–6 mo	≥3.5	≥25	≤3.0
7–12 mo	≥2.5	≥30	≤3.0
1–2 yr	≥3.5	≥35	≤2.5
2–3 yr	—	≥40	≤2.5
3–4 yr	—	≥45	≤2.5
4+ yr	—	≥50	≤3.0
Adult	≥4.0	≥50	≤4.0
Ulnar nerve			
Preterm (33–39 wk)	—	≥18	≤3.3
0–1 mo	≥1.5	≥20	≤3.0
1–6 mo	≥2.5	≥25	≤3.3
7–12 mo	≥3.0	≥35	≤2.5
1–2 yr	≥2.5	≥40	≤2.5
2–3 yr	—	≥40	—
3–4 yr	—	≥45	—
4+ yr	—	≥50	—
Adult	≥6.0	≥50	≤3.3
Peroneal nerve			
0–1 mo	≥1.5	≥20	≤3.0
1–6 mo	≥1.5	≥25	≤2.5
7–12 mo	≥2.0	≥30	≤3.5
1–2 yr	≥1.5	≥35	≤3.5
2–3 yr	—	≥40	—
3–4 yr	—	≥40	—
4+ yr	—	≥40	—
Adult	≥2.0	≥40	≤6.5
Tibial nerve			
Preterm (33–36 wk)	—	≥14	—
Preterm (37–39 wk)	—	≥18	—
0–1 mo	—	≥20	≤4.5
1–6 mo	—	≥20	≤4.0
7–12 mo	—	≥25	≤3.5
1–2 yr	—	≥30	≤3.0
2–3 yr	—	≥35	≤4.0
3–4 yr	—	≥40	≤4.0
4–6 yr	—	≥40	≤4.5
6+ yr	—	≥40	≤5.0
Adult	≥4.0	≥40	≤5.8

A, suggested amplitude; CV, suggested conduction velocity; DL, suggested distal latency; —, data not available.

[a]Adapted from refs. *1, 2, 4–6, 23–25.*

fossa. The peroneal motor study is performed as in an adult, with the E1 recording electrode placed over the extensor digitorum brevis and stimulation occurring at the ankle, fibular head, and popliteal fossa. Recording of tibialis anterior may be used when no response or a very small response is obtained from extensor digitorum brevis. One of the proximal stimulation sites may be omitted in children, unless, of course, there is a question of a peroneal neuropathy at the fibular head, such neuropathy being quite rare in children.

In infants, the most difficult aspect of the median antidromic sensory study involves the placement of the active and reference electrodes. The active electrode may be placed on the second digit and the reference on the third, or the electrodes can be cut in half and both placed on the second or third digit. If this is unsuccessful, ring electrodes may be used instead. The stimulation site is, as usual, at the wrist. In neonates and some infants, surface recordings of sural nerve sensory action potentials are often obscured by artifact caused by high skin impedance and short interelectrode distances. In this age group, the most technically reliable sensory study to obtain in the lower extremity is the orthodromic medial plantar study, recording the tibial nerve at the ankle and stimulating the medial plantar region of the sole.

Full-term newborns typically have nerve conduction velocities that are approximately half those expected in an adult. Nerve conduction velocities increase steadily during the first 3 to 5 yr of life, because of growth in axon diameter and thickening of myelin. Thus, adult nerve conduction values cannot be reliably expected until 3 to 5 yr of age, although some children's responses reach those values earlier. Normal amplitudes are also diminished in children compared with adults, especially in the first year of life. In a child younger than 3 to 5 yr, it is important to consult age-matched reference values in interpreting the results of nerve conduction studies. Tables 1 and 2 list suggested normal values adapted and summarized from published data on surface recordings, and also include adult normal values for comparison. Most studies seeking to establish normal values in children focus on motor nerve conduction studies, thus, Table 1 is more comprehensive than Table 2.

Elicitation of late responses may produce significant discomfort in a child, and should not be performed unless they can help answer a specific question, for example, when Guillain–Barré syndrome is a possible diagnosis. If a number of late response studies are necessary, sedation or anesthesia should be considered. Despite incomplete myelination, short stature results in F-response latencies that are shorter in children than in adults. Age-matched normal values, such as the ones listed in Table 3 should be used. H-reflexes may be obtained in the upper as well as the lower extremities in the first year of life, but are rarely necessary in the pediatric population.

2.3. *Needle EMG*

For the needle examination, a 30-gauge, 25-mm disposable concentric needle (the smallest commercially available) is almost always the best choice. School-age children often ask if "it will hurt." One can reply that it will "pinch" or say that it will be like a blood draw, while avoiding the term "needle." The needle itself can be described as a "microphone." Older children and adolescents may often be examined as thoroughly as adults, but it is still advisable to examine the fewest muscles necessary to answer the question at hand. These patients sometimes reach a limit of tolerance unexpectedly during the needle examination and become upset with little warning.

Except in cases of neonatal brachial plexus injury or trauma, extensive root screens are rarely required in children. In many cases, the question revolves around a generalized process,

Table 2
Sensory Nerve Action Potentials, Suggested Values

Age	*A (mV)*	*CV (m/s)*
Median nerve (antidromic)		
0–1 mo	≥5	≥25
1–6 mo	≥10	≥35
7–12 mo	≥15	≥30
1–2 yr	≥15	≥40
2–3 yr	—	—
3–4 yr	—	—
4+ yr	—	—
Adult	≥20	≥50
Sural nerve (antidromic)		
0–1 mo	≥6	≥20
1–6 mo	≥6	≥20
7–12 mo	≥6	≥25
1–2 yr	≥6	≥30
2–3 yr	≥6	≥35
3–4 yr	≥6	≥40
4+ yr	≥6	≥40
Adult	≥6	≥40
Medial plantar nerve orthodromic in children		
0–1 mo	≥10	—
1–6 mo	≥15	≥35
7–12 mo	≥15	≥35
1–2 yr	≥15	≥35
2–4 yr	—	—
4+ yr	—	—
Adult (antidromic)	≥2	≥35

A, suggested amplitude; CV, suggested conduction velocity; —, data not available. Sural sensory responses may be obscured by subcutaneous tissue in neonates and infants; medial plantar studies may be more accurate in this age group.

[a]Adapted from refs. *1–3, 6, 25.*

therefore, a limited needle examination often yields the necessary data. In infants and toddlers, poor cooperation often makes it impossible to evaluate both insertional activity and voluntary activity in the same muscle. In the upper extremities, insertional activity is most easily observed in triceps and first dorsal interosseous, whereas motor unit activity is better assessed in biceps and flexor carpi radialis. In the lower extremities, medial gastrocnemius and vastus lateralis are better for insertional activity, whereas tibialis anterior and iliopsoas are preferred for motor unit potential (MUP) analysis.

As in adults, if there is a question of a polyneuropathy, it is important to examine distal muscles, although it is rarely necessary to study intrinsic foot muscles in young children. Children with possible myopathy should have several proximal muscles evaluated, but one extremity (preferably lower) should be spared so that histological examination of a muscle biopsy performed at a later time will be free of potential artifacts from needle injury.

Table 3
F-Wave Responses, Suggested Values

Age	*Latency (ms)*
Median nerve	
0–1 mo	≤20
1–6 mo	≤20
7–12 mo	≤21
1–2 yr	≤21
2–4 yr	≤21
4–6 yr	≤23
6–14 yr	≤29
Adult	≤31
Ulnar nerve	
0–1 mo	—
1–6 mo	≤17
7–12 mo	≤17
1–2 yr	≤17
2–4 yr	—
4–6 yr	—
6–14 yr	—
Adult	≤32
Peroneal nerve	
0–1 mo	≤26
1–6 mo	≤27
7–12 mo	≤29
1–2 yr	≤30
2–4 yr	≤34
4–6 yr	≤36
6–14 yr	≤43
Adult	≤56
Tibial nerve	
0–1 mo	—
1–6 mo	—
7–12 mo	≤24
1–2 yr	≤26
2–4 yr	—
4–6 yr	—
6–14 yr	—
Adult	≤56

—, data not available.
[a]Adapted from refs. *1, 3, 25, 26.*

3. COMMON REFERRALS TO THE PEDIATRIC EMG LABORATORY

Deciding whether to request or perform an EMG in a particular case is sometimes difficult, because the number of available diagnostic modalities has increased dramatically in recent years. Genetic testing is available for many neuropathies and muscular dystrophies, and may make EMG unnecessary in certain patients. However, there are many children with acquired conditions and unusual presentations of inherited disorders in whom EMG is an essential test, and the volume of EMG referrals, in our experience, has remained steady.

The differential diagnosis of hypotonia in infancy and early childhood is vast, and includes a number of peripheral nervous system processes. However, it is important to remember that the majority of cases of infant hypotonia arise from central causes. An extensive peripheral nervous system evaluation should be pursued only if supported by the history and examination. Peripheral nervous system lesions may localize to the anterior horn cell, peripheral nerve, neuromuscular junction, or muscle. If a congenital myopathy or muscular dystrophy lies in the differential diagnosis, it is often useful to obtain a set of muscle enzymes (creatine kinase, aldolase, alanine aminotransferase, aspartate aminotransferase, and lactate dehydrogenase) before requesting or performing an EMG or muscle biopsy. It is important to remember that muscle enzymes may be artifactually elevated immediately after needle EMG is performed. A significant elevation of one or more enzymes (the elevated enzymes should include either creatine kinase or aldolase, because these are the ones that are most specific to muscle) will justify a muscle biopsy as the next step. Mild abnormalities or normal levels of these enzymes do not always exclude a congenital myopathy; in those cases, an EMG may be helpful in directing further studies. The precise sensitivity of EMG in the evaluation of the hypotonic infant is unclear. One study reported an 80% overall sensitivity for the correct diagnosis, whereas another recorded a 65% sensitivity for SMA and 10% for myopathy.

A neonate with a birth injury to the brachial plexus may present with a weak or flaccid upper extremity. It is rare for alternative diagnoses to be plausible, therefore, the question for the electromyographer usually revolves around issues of severity and prognosis. Except in the most severe cases, an EMG is most helpful after the first few months of life, when the presence or absence of chronic reinnervation and axonal continuity may best be assessed.

Toddlers and older children may present with delayed motor milestones or difficulty walking. As in the evaluation of hypotonia, the differential diagnosis is broad, and may include almost any segment of the peripheral nervous system. Proximal weakness, often detected by the presence of Gowers sign, Trendelenburg gait, or both, typically suggests the possibility of Duchenne, Becker, or limb–girdle muscular dystrophy. If such a patient has significantly elevated muscle enzyme levels, either molecular genetic testing or muscle biopsy will yield a diagnosis in most cases. When such an evaluation is unrevealing, an EMG may be useful in assessing the child for a neuropathic process, such as SMA type III (also known as Kugelberg–Welander disease). SMA type III may resemble a muscular dystrophy with a proximal distribution of weakness and a mild elevation of the creatine kinase level.

Abrupt onset gait difficulty may be caused by acute inflammatory demyelinating polyneuropathy (AIDP; Guillain–Barré syndrome), an inflammatory myopathy, tick paralysis, spinal cord conditions (mass lesion or transverse myelitis), or central causes. Although poliomyelitis, fortunately, has almost been eradicated in much of the world, West Nile virus has also been found to produce a similar syndrome of acute motor paralysis. A detailed skin and scalp examination is important in such acutely weak children, because tick paralysis is readily cured by removing the tick. Because ataxia is often observed in AIDP, it may be difficult in some cases to distinguish between that condition and other causes of ataxia, such as acute cerebellitis. If a particular diagnosis cannot be supported by other findings, such as areflexia, cytoalbuminological dissociation, rash, elevated creatine kinase level, or nerve root enhancement on MRI imaging of the spine, EMG may be helpful in distinguishing between the possibilities.

Chronic gait difficulty associated with areflexia, sensory loss, pes cavus, distal weakness, or a combination of these suggests the presence of an isolated inherited polyneuropathy, such as Charcot–Marie–Tooth (CMT) disease, also called hereditary motor sensory neuropathy. In some instances, a demyelinating polyneuropathy may be associated with a more generalized neurodegenerative disorder, such as metachromatic leukodystrophy, especially if there is an associated cognitive decline.

4. IMPORTANT FINDINGS IN THE PEDIATRIC EMG LABORATORY

4.1. *Motor Neuron Disease*

Because of the development of the polio vaccine, SMA has become the dominant anterior horn cell disease affecting children in developed countries. The onset of the most common variant, type I (Werdnig–Hoffman disease), is typically in the first 6 mo of life. The infants present with hypotonia, weakness, and delayed motor milestones. Mothers may report decreased fetal movements. Examination is notable for areflexia or hyporeflexia, hypotonia, weakness, and preserved extraocular movements. Tongue fasciculations may sometimes be observed, but are not a reliable finding, especially in young infants, and the lack of tongue fasciculations should never be used to exclude SMA. These children never sit independently and never walk. The majority do not survive the second year of life.

The onset of SMA type II is typically from 6 to 18 mo, with some cases beginning as early as the neonatal period. These children often present with motor delays. They eventually sit independently but never walk, and survive only to adolescence or early adulthood. The gait difficulties that are often the first signs of SMA type III usually begin after 18 mo. These patients walk independently for most of their lives. Life expectancy is normal in most cases of SMA type III, although some patients require wheelchair assistance as adults.

SMA is caused by a deletion in the survival motor neuron (*SMN*) gene on chromosome 5q13. Until this discovery, EMG and muscle biopsy were the principal means of confirming the diagnosis during life. Genetic testing for deletions in *SMN* is now readily available, but EMG is still helpful in many cases, especially if the presentation is mild or otherwise unusual, or if an infant develops respiratory failure before the results of genetic testing become available.

As in adult motor neuron disease, it is critical to document normal sensory responses on nerve conduction studies; if these are abnormal, the diagnosis should remain in serious doubt. Compound motor action potential (CMAP) amplitudes are often diminished in SMA type I, but may be minimally reduced or normal in SMA types II and III. During the needle examination, at least three limbs should be studied. Cranial muscles may be substituted for one limb, especially if tongue fasciculations are observed. Examination of paraspinal muscles is not generally helpful and should be avoided. Both ongoing denervation and chronic reinnervation may be observed. In adults, chronic reinnervation should be demonstrated in most muscles of at least three limbs to meet criteria for motor neuron disease, but in children it is not always possible to do this extensive a needle examination, and any findings should be confirmed by genetic testing. Fasciculation potentials are observed more rarely in SMA as compared with amyotrophic lateral sclerosis, and the MUP abnormalities are generally symmetrical in SMA.

4.2. *Sensory Neuropathy and Neuronopathy*

It is rare to find an isolated sensory neuropathy or sensory neuronopathy in childhood. In infants, the sural and superficial peroneal sensory responses are often difficult to record. Thus, in children of that age, the medial plantar sensory responses and upper extremity sensory responses must be checked before concluding that the patient has a true loss of sensory responses. Age-specific normal values must be used in children younger than 3 to 5 yr.

If a true sensory neuropathy or sensory neuronopathy is present, the most likely causes are Friedreich's ataxia and the hereditary sensory and autonomic neuropathies. Because genetic testing for the triplet-repeat expansion in Friedreich's ataxia is available, patients with the typical clinical presentation are often diagnosed without the assistance of an EMG. Associated findings include ataxia, absent lower extremity reflexes, extensor plantar responses, thinning of the spinal cord on spine MRI, normal brain MRI early in the course, onset before 20 yr, pes cavus, dysarthria, distal sensory loss, optic atrophy, diabetes, and cardiomyopathy.

4.3. Brachial Plexus

Brachial plexus injuries in children most commonly occur at birth, as a result of shoulder dystocia (difficulty in extracting the shoulder from the birth canal). Risk factors include difficult delivery and large birthweight. As might be expected from the mechanism of injury, upper plexus injuries (Erb's palsy) predominate. Lower brachial plexus lesions (Klumpke's palsy) and total plexus injuries (Erb-Klumpke paralysis) are less common. The neonate will have flaccid weakness of one upper extremity.

The history and examination are, in most cases, diagnostic of perinatal brachial plexus injury. A careful examination will suggest the root levels involved. In upper plexus lesions, the posture known as the "waiter's tip" is usually found, consisting of arm adduction, elbow extension, forearm pronation, and wrist flexion. In lower plexus lesions, wrist and finger flexors and intrinsic hand muscles are weak. A Horner's syndrome is typically present. A combination of these findings or complete upper extremity paralysis suggests total plexus involvement.

Initial management includes gentle immobilization of the affected extremity against the abdomen and range of motion exercises, preferably under the direction of a trained physical therapist. In almost all cases, observation and conservative management are recommended until 3 to 6 mo of age. If there is no improvement at that point, evaluation for possible microsurgical repair may be indicated, including EMG and MRI.

In this setting, the purpose of performing an EMG study is to determine the localization and severity of the injury. The severity is largely dependent on the localization, because discontinuity of the axon, most commonly caused by avulsion of the nerve root, is associated with a much poorer prognosis than "stretching" of the brachial plexus structures.

If brachial plexus injury and root avulsion were mutually exclusive conditions, the study would be relatively simple, because, in the former condition, sensory responses would be absent, whereas, in the latter, they would be preserved. However, in many cases, the two lesions coexist, making abnormal sensory nerve action potentials (SNAPs) less reassuring than they would be under other circumstances.

The needle study is the key to determining prognosis. Because examination of the cervical paraspinal muscle is impractical in most infants, the presence of MUPs on needle EMG at various root levels confirms continuity of the axon and, thus, the likelihood of at least partial recovery of function. This assessment can only be made several months after birth, after reinnervation has begun to occur, because a premature examination may yield an inaccurately grim prognosis.

Precise localization among the root, trunk, division, cord, and nerve structures requires the evaluation of a large number of muscles, but unless the infant is anesthetized, this is not usually possible. Thus, because of the high likelihood that the study will be terminated prematurely, the choice of muscles is critical. If an infant has a typical upper plexus presentation, the deltoid, biceps, and triceps muscles should be studied first. Comparing deltoid and biceps can help distinguish between root and plexus involvement; both muscles will be abnormal when a C5–C6 root lesion is present, but only one may be abnormal in lesions of the brachial plexus. Unfortunately, examination of the rhomboids and serratus anterior is rarely practical in infants, limiting the precision of localization within the brachial plexus.

When a Horner's syndrome or other signs of lower plexus involvement are present, a distal muscle, such as the first dorsal interosseous, should be among the first muscles examined. Depending on the clinical picture and the tolerance of the patient, other muscles that may be studied include the supraspinatus, extensor digitorum communis, flexor digitorum superficialis, and abductor digiti quinti. If the triceps has been successfully examined, needle EMG of extensor digitorum communis may not be necessary. As for median-innervated nerve muscles, examination of the flexor digitorum superficialis is usually more informative than that of abductor pollicis brevis, because the infant is more likely to activate the former.

Muscle activation in infants is often fleeting. For flexor muscles, tickling or gentle pinching of the fingers and hand may stimulate withdrawal when the needle itself fails to induce this response. The infant is less likely to activate extensor muscles under these circumstances. It is sometimes difficult to decide how long to examine a muscle, especially if the child is crying and it seems that the time remaining in the study is rapidly diminishing. When no MUPs are present, tickling or pinching to activate muscle contractions should be performed at least once for each major level (C5–C6, C7, and C8–T1), and preferably in multiple muscles. Because a decision whether or not to perform surgery may hinge partly on the study results, any examination that is too truncated to be reliable should be repeated under anesthesia.

4.4. Polyneuropathies

AIDP (Guillain–Barré syndrome) may occur at almost any age in childhood, from 1 mo to adulthood. The classic clinical presentation is the same as in adults, with an acute onset of ascending paralysis, areflexia, and cytoalbuminological dissociation in the cerebrospinal fluid, with protein values often ranging from 80 to 200. An EMG may be very helpful in confirming or casting doubt on the diagnosis. The use of spine MRI with gadolinium enhancement to detect nerve root enhancement is also becoming more widespread. The initial symptoms may be atypical in some patients, consisting primarily of numbness, paresthesias, or pain.

Nerve conduction studies are most useful in the evaluation of possible AIDP. In the classic demyelinating form of AIDP, the diagnostic findings are temporal dispersion and conduction block in the setting of slow conduction velocities and prolonged distal latencies. The criteria for these findings are generally the same in children as in adults: at least 15% prolongation of CMAP duration on proximal vs distal stimulation for temporal dispersion, at least 50% diminution of CMAP amplitude on proximal vs distal stimulation for partial conduction block, and an absent CMAP on proximal stimulation for complete conduction block. The presence of temporal dispersion excludes partial or complete conduction block. Another criterion for nonuniform slowing suggesting an acquired demyelinating process is a significant disparity in nerve conduction velocities between different nerves, usually at least 10 m/s difference within the upper or lower extremities, and at least 15 m/s difference between an upper and a lower extremity. The usual adult standards stipulating the number of nerves involved may not always be practical to apply in children, because the study may be terminated early. Evidence for axonal loss may be present, but unless severe, this is usually a secondary phenomenon.

In some early or mild cases, temporal dispersion, conduction block, and conduction velocity slowing may not be present. Prolonged or absent F-responses and H-reflexes may be early signs of AIDP, as may be reduced recruitment on an otherwise normal needle EMG study. If tolerated, F-waves should be performed on several nerves in multiple extremities. H-reflexes are extremely uncomfortable, especially for a child, and, thus, should be avoided. These early electrophysiological signs of AIDP may be subtle and nonspecific; if the diagnosis cannot clearly be supported by other data, a repeat study later in the course may provide more definite evidence.

The less common, primarily axonal form of Guillain–Barré, known as acute motor axonal neuropathy, was first recognized in China, and most commonly occurs there, but has also been described in the United States, Europe, and other parts of Asia. In adults, the prognosis is clearly worse in axonal variants. Children with acute motor axonal neuropathy have a worse course acutely (many require assisted ventilation) and recover more slowly than those with AIDP, but the long-term prognosis is generally good if they survive the acute phase. Conduction velocities are normal or mildly slow. There is evidence for axonal loss both on nerve conduction studies and needle EMG.

A generalized sensorimotor polyneuropathy in the setting of long-standing symptoms may be caused by chronic inflammatory demyelinating polyradiculoneuropathy (CIDP) or an inherited polyneuropathy, such as CMT (also known as hereditary sensory and motor neuropathy), metachromatic leukodystrophy, Cockayne syndrome, Pelizaeus-Merzbacher disease, or adrenomyeloneuropathy. An EMG can determine whether a polyneuropathy is present, and may, in some cases, help distinguish between acquired and genetic causes. Traditionally, signs of nonuniform slowing of nerve conduction, such as temporal dispersion, conduction block, and significant disparities in nerve conduction velocities between different nerves have indicated the presence of an acquired polyneuropathy, whereas uniform slowing has been associated with an inherited polyneuropathy. However, numerous reports have demonstrated nonuniform slowing in some instances of inherited polyneuropathies, including metachromatic leukodystrophy, hereditary neuropathy with liability to pressure palsies, the X-linked variant of CMT (CMTX), and adrenomyeloneuropathy.

Children with CIDP generally have manifestations of motor delay or impairment. The weakness may show a distal predominance or may be diffusely distributed. Other findings include sensory loss in some patients, normal cranial nerve function, and hyporeflexia or areflexia. CIDP may develop at any age, and has even been reported to be present at birth.

The electrophysiological findings in CIDP are similar to those in fully developed AIDP. Nerve conduction velocities may be as slow as 2.3 m/s. Temporal dispersion and conduction block are classic findings. Needle EMG reveals evidence for secondary axonal loss. The chronicity or relapsing nature of the course distinguishes CIDP from AIDP. It is sometimes difficult to confirm CIDP based on electrophysiology alone, especially in light of the exceptions to the uniform vs nonuniform slowing rule that have recently been described. Palpable enlarged nerves and pes cavus may be present in CIDP as well as in CMT. Elevated protein in cerebrospinal fluid is found in CIDP, but also in inherited disorders such as metachromatic leukodystrophy and CMT. The absence of cognitive involvement, MRI abnormalities, and long-tract signs is suggestive of CIDP rather than an inherited leukodystrophy, but nerve biopsy is still required to dispel remaining doubts, especially if there is a question of CMT.

CMT type I is the most common inherited polyneuropathy. It is characterized by a predominantly demyelinating pattern on EMG, with nerve conduction velocities typically in the teens. The pattern of inheritance is autosomal dominant. Initial symptoms may include distal lower extremity weakness and wasting, with slow progression over several decades. Hyporeflexia, palpable enlarged nerves, pes cavus, distal weakness and wasting, and distal sensory loss are classic physical findings.

Primary axonal loss is found in CMT type II. Nerve conduction velocities are normal or mildly slow. Onset is typically in early adulthood, and the course is mild. Inheritance is usually autosomal dominant, although autosomal recessive kindreds with moderately slow nerve conduction velocities have been described.

CMTX is characterized by primary axonal loss combined with secondary demyelination. The most prominent EMG findings are reduced CMAP and SNAP amplitudes, but nerve conduction velocities are moderately slow, which helps distinguish CMTX from CMT type II electrophysiologically, and there may be heterogeneity of conduction parameters, including conduction block, giving it an "acquired" appearance. Heterozygote females have mild symptoms, whereas affected males have the full spectrum of manifestations, including pes cavus, distal muscle weakness and atrophy, and distal sensory loss.

Type III (Dejerine–Sottas disease) is associated with severe demyelination. EMG demonstrates profound slowing of nerve conduction velocities, typically in the single digits, with temporal dispersion of CMAPs. Onset may occur anytime between birth and 2 yr. Delayed motor milestones, weakness, areflexia, and hypotonia typically occur. Elevated cerebrospinal fluid protein may be present. Palpably enlarged nerves are a classic but inconsistent finding.

4.5. Mononeuropathies

Mononeuropathies are rare in children compared with adults. Median neuropathies at the wrist may be associated with idiopathic carpal tunnel syndrome, but may also occur in the setting of acute trauma, chronic trauma via sports, mucopolysaccharidosis (Hurler/Scheie, Hunter, and Maroteux–Lamy), mucolipidosis (types II or III), or scleroderma. In some cases, the median neuropathy may be the first specific finding that indicates a systemic condition. Thus, the possibility of a mucopolysaccharidosis or mucolipidosis should always be considered if a child is diagnosed with a distal median neuropathy. Proximal median neuropathies seem to occur at least as often as median neuropathies at the wrist, and are typically caused by trauma, such as a fracture of the humerus or a laceration. The classic clinical picture of a median neuropathy occurs in some of these patients but is not universal. A careful needle EMG is important in cases of median neuropathy to investigate the possibility of a proximal lesion.

An isolated ulnar neuropathy may occur at the level of the elbow, forearm, wrist, or hand. It is most commonly caused by trauma, including fracture in the supracondylar area or forearm, laceration anywhere along the course of the nerve, acute compression during surgery, and chronic compression during activity (e.g., pressure from bicycle handlebars and wheelchair armrests). Cubital tunnel syndrome occurs occasionally. Because direct trauma is more common in children than compression, axonal pathophysiology is correspondingly observed more often in these patients.

In contrast to median and ulnar mononeuropathies, radial neuropathy may be observed at any age, from newborns to 17-yr-old children, and generally has a good prognosis. Because of the location of the radial nerve, it is susceptible to prenatal or perinatal intrauterine compression. Other causes of compression and entrapment, as well as direct trauma from fractures and lacerations have also been identified. Weakness of extensor muscles and wristdrop are the most common signs. Radial sensory nerve conduction studies should be performed in such cases. If normal values are not available for children younger than 3 to 5 yr and the symptoms are unilateral, a study in the contralateral arm may be used for comparison. Radial motor studies are difficult because of the technical challenges in young children and their limited tolerance for the strong stimuli that may be required. The needle EMG study is critical to localize the lesion. With adequate sedation or anesthesia, detailed needle examination is possible even in neonates. Unless symptoms progress or there is evidence for entrapment, surgical exploration is generally unnecessary, and most patients may be managed conservatively.

Peroneal neuropathy is the most common lower extremity mononeuropathy. It may be caused by direct trauma during sports activities, compression from casts and other devices attached to the leg, or entrapment from fibrous bands. Patients whose lesions were predominantly demyelinating had a better outcome than those with significant axonal loss, as indicated by a low-amplitude or absent peroneal CMAP.

4.6. Neuromuscular Junction

Disorders of neuromuscular transmission are caused by impaired transmission of acetylcholine across the neuromuscular junction. In children, the most common causes of these disorders are myasthenia gravis (of which there are three types: neonatal, congenital, and juvenile) and botulism. Lambert–Easton myasthenic syndrome is rare in this age group.

In neonatal myasthenia gravis, maternal antibodies to the acetylcholine receptor cross the placenta and enter the fetal circulation. This occurs in 12% of infants born to mothers with symptomatic or quiescent autoimmune myasthenia gravis. Soon after birth, an affected neonate develops respiratory distress, feeding difficulties, and weakness. In one-third of cases, ventilatory support and nasogastric tube feedings are required until the baby clears the antibodies from the circulation, which usually takes several weeks. Oral pyridostigmine may be helpful in severe cases.

Congenital myasthenic syndrome is caused by genetic mutations affecting the neuromuscular junction. The most common mutations are in genes coding for the acetylcholine receptor, but presynaptic lesions may also occur. The inheritance is usually autosomal recessive. These patients do not have antibodies to the acetylcholine receptor. Onset is typically in the first year. Compared with juvenile myasthenia gravis, there are fewer fluctuations in weakness and more prominent ocular features. Episodes of apnea may be present in some subtypes such as choline acetyltransferase deficiency and rapsyn deficiency. Cholinesterase inhibitors may alleviate symptoms, but immune modulating therapies are ineffective, as would be expected from the pathophysiology.

Juvenile myasthenia gravis has the same autoimmune origins as adult myasthenia gravis, but the onset is during childhood or adolescence. Symptoms such as ophthalmoplegia, ptosis (bilateral or unilateral), orbicularis oculi or facial weakness, dysphagia, dysarthria, and dyspnea in a fluctuating pattern may suggest the diagnosis. Isolated generalized weakness may present a challenge, because the differential diagnosis is broad. The evaluation is essentially the same as in adult myasthenia gravis. Diagnosis may be made through acetylcholine receptor antibody testing, EMG, the edrophonium test, or a combination of these. Once the diagnosis is confirmed, a chest CT study should be performed, although the incidence of thymoma is lower in children than in adults. Symptomatic therapy with acetylcholinesterase inhibitors and immune modulating medications may be used, as in adults. Medications known to exacerbate myasthenia gravis should be avoided.

The most common form of botulism in childhood is infant botulism, caused by ingestion of *Clostridium botulinum* spores from the soil. Endemic states include Pennsylvania, Utah, and California. Botulinum toxin inhibits the release of acetylcholine from the presynaptic terminal. Infants typically present with the acute onset of constipation, extremity weakness, bulbar weakness, sluggish pupillary responses, and oculomotor palsies. Reflexes may be either preserved or diminished. Because the infants ingest spores rather than preformed toxin as in adult botulism, the illness is caused by low levels of subacute toxin production rather than an overwhelming single dose. Thus, stool samples rather than blood samples should be sent for botulinum toxin testing in patients suspected of having infant botulism.

Routine nerve conduction studies are an important component of an evaluation for disorders of neuromuscular transmission. Not only must a generalized polyneuropathy be excluded, but certain findings may be suggestive of particular neuromuscular transmission defects, such as low-amplitude CMAPs in presynaptic disorders and repetitive CMAPs on single supramaximal stimulation in slow channel syndrome (a form of congenital myasthenic syndrome). During repetitive stimulation, the accessory nerve should be the first one studied, unless there is focal or maximal weakness in the distribution of another nerve, such as the ulnar or median nerve. Accessory nerve studies are more sensitive than ulnar studies, and in children, the study may need to be terminated at any time because of patient discomfort. In infants, repetitive nerve stimulation should only be performed with the patient under sedation or anesthesia.

In postsynaptic disorders, such as neonatal, juvenile, or postsynaptic congenital myasthenia, the characteristic finding on low-frequency (2–3 Hz) repetitive stimulation is a greater than 10% decrement in the CMAP amplitude. The nadir typically occurs at the fourth or fifth stimulus. If the nadir occurs earlier or later, the decrement is most likely artifactual. In presynaptic disorders, such as botulism and the presynaptic forms of congenital myasthenia, low-frequency repetitive stimulation may also produce a decremental response, but it is not observed as consistently as in postsynaptic disorders. The more sensitive test in suspected disorders of presynaptic transmission is high-frequency repetitive stimulation at 20 to 50 Hz, which yields an incremental response in nearly all such patients. This pattern may also be observed occasionally in patients with severe postsynaptic defects. The threshold for determining abnormal facilitation of the CMAP amplitude on high frequency repetitive stimulation is typically considered a 100% increment. On the rare occasion when an older child is suspected of having a presynaptic disorder, such as adult botulism or Lambert–Eaton myasthenic syndrome, a single supramaximal stimulus pre- and then again post- 10 seconds of maximal contraction of the muscle being tested may substitute for high-frequency repetitive stimulation. Because high-frequency stimulation is even more uncomfortable than low-frequency stimulation, this procedure is generally performed under anesthesia in any age group.

Abnormalities in at least two nerves should be identified before diagnosing a child with a disorder of neuromuscular transmission. This may be difficult in a child who is uncomfortable or uncooperative. Single-fiber EMG, a technique used often in adults, is impractical in most children. If the study is technically inadequate, other diagnostic modalities (acetylcholine receptor antibody levels and edrophonium testing) may be considered, or the study may be repeated under anesthesia.

Low-amplitude, short-duration MUPs suggestive of a myopathy are often observed on needle EMG in patients with disorders of neuromuscular transmission. Fibrillation potentials may be observed in cases of botulism and severe cases of autoimmune myasthenia gravis. Some investigators have found neurogenic findings in patients with botulism, which may be explained by the functional denervation that can occur in this disorder. In a child with myopathic EMG findings of unclear etiology, it is important to consider the possibility of a disorder of neuromuscular transmission and, if indicated, perform repetitive stimulation studies.

4.7. Muscle

Before the discovery of dystrophin and the subsequent development of genetic testing for Duchenne and Becker muscular dystrophy, EMGs were commonly used in the diagnostic evaluation of boys who had gait difficulties. The question of possible myopathy or muscular dystrophy does still arise in the pediatric EMG laboratory, but the patients who are referred often have atypical presentations or rare conditions. Congenital myopathies include centronuclear myopathy, nemaline myopathy, and central core disease. Muscular dystrophies include congenital muscular dystrophy, Emery–Dreifuss dystrophy, facioscapulohumeral muscular dystrophy, limb–girdle muscular dystrophy, Duchenne muscular dystrophy, and Becker muscular dystrophy. Children may also present with symptoms suggestive of a metabolic myopathy, categorized as glycogenosis or fatty acid oxidation disorder. Mitochondrial myopathies are also considered a form of metabolic myopathy. Glycogenoses include acid maltase deficiency (glycogenosis type II, Pompe's disease), myophosphorylase deficiency (glycogenosis type V, McArdle's disease), and phosphofructokinase deficiency (glycogenosis type VII). Inflammatory myopathies include dermatomyositis and polymyositis.

The evaluation of a possible primary muscle disorder in the EMG laboratory can be difficult. Because normal MUPs are smaller in infants and children than in adults, the threshold for defining a unit as myopathic is different. In infants, MUPs are typically biphasic, with durations of 1 to 4 ms and amplitudes less than 100 μV. In addition, inconsistent muscle activation in some children may make it difficult to detect short-duration MUPs and early recruitment with confidence. When in doubt, it is always more prudent to conclude that the results of a study are normal rather than overemphasize potentially false positive findings. The EMG report should indicate that normal study results do not exclude a myopathic process.

In cases of possible myopathy or muscular dystrophy, it is important to examine proximal muscles and any other muscles that are weak. Iliopsoas should be studied whenever possible—in infants and younger children, it is crucial to locate the femoral pulse and ensure that the needle is not placed in the femoral artery. If the patient is so agitated that this cannot be guaranteed, it is advisable to defer study of that muscle. Other proximal muscles in the upper and lower extremities should also be studied.

In many cases of noninflammatory myopathy and muscular dystrophy, needle EMG will demonstrate low-amplitude, short-duration MUPs, sometimes accompanied by fibrillation potentials and positive sharp waves. These findings may also be observed in disorders of neuromuscular transmission. Abnormal spontaneous activity will typically be more prominent in cases of inflammatory myopathy. Early recruitment, in which multiple MUPs are activated despite minimal force generation, is another sign of myopathy, but is very difficult to assess in infants and children. Myopathic findings do not generally help to distinguish between the various myopathic disorders, unless there is very prominent spontaneous activity, myotonia, or both.

A myotonic discharge is a distinct finding of abnormal spontaneous activity that may best be described as a series of rapidly firing fibrillation potentials or positive sharp waves, with fluctuating amplitude and frequency and a duration of 2 to 20 s. The fluctuations produce a sound that has been described as resembling that of a "dive bomber," but the sound of a motorcycle engine may be a better analogy. Myotonic discharges are often elicited with small insertions of the EMG needle. These discharges may be missed, however, especially in children, if the needle EMG study is technically limited or only performed on a few muscles. The absence of myotonic discharges on a needle EMG study does not entirely exclude that entity.

Classically, myotonic discharges are associated with myotonic dystrophy, myotonia congenita, paramyotonia congenita, and hyperkalemic periodic paralysis. However, because it is, in a sense, a severe manifestation of positive sharp waves and fibrillation potentials, it may also be observed in some myopathic disorders not caused by channelopathies, such as acid maltase deficiency, centronuclear myopathy, polymyositis, and dermatomyositis.

4.8. Myotonic Disorders

There are four classic primary myotonic disorders: myotonic dystrophy, myotonia congenita, paramyotonia congenita, and hyperkalemic periodic paralysis. There are two types of myotonic dystrophy (type 1 and type 2). Type 1 is the most common form in childhood; type 2 is mainly an adult disorder. Onset of symptoms of type 1 myotonic dystrophy is in late childhood or adolescence; there is, however, a variant that is symptomatic at birth. A fully developed case of myotonic dystrophy has many distinctive features, including frontal balding, temporal wasting, distal weakness, and myotonia on examination. The neonatal variant may present with arthrogryposis multiplex congenita. These patients are often diagnosed directly by DNA testing. However, the findings may be more subtle early in the course, therefore, an EMG may be helpful in those cases. Routine motor and sensory nerve conduction studies are usually normal, although some reports demonstrate mild slowing of conduction velocities. Repetitive stimulation produces CMAP decrements. Needle EMG reveals, in addition to myotonic discharges, other abnormal spontaneous activity, and short-duration, low-amplitude MUPs, demonstrative of an underlying myopathic process.

There are autosomal dominant (Thomsen's disease) and recessive (Becker's disease) variants of myotonia congenita, which is caused by mutations in the chloride channel. Muscle stiffness may be present at rest, but is relieved on activity. Muscle hypertrophy occurs in both disorders. Motor strength is normal in Thomsen's disease, but weakness typically occurs in Becker's disease. Routine motor and sensory nerve conduction studies are normal, but repetitive stimulation and short exercise will lead to CMAP decrements. Needle EMG demonstrates myotonic discharges, but there are usually no other abnormal findings.

Paramyotonia congenita and hyperkalemic periodic paralysis are both caused by mutations in the sodium channel. Signs of paramyotonia congenita are often present in infancy. There may be delayed eye opening after sneezing. In later years, the child may experience stiffening of the extremities, weakness, and falling with activity. This pattern of "paradoxical" myotonia with activity is the reverse of what is typically found in other myotonic disorders, leading to the name paramyotonia. Symptoms may also be triggered by cold weather. Eating hard-to-chew or cold foods, such as bagels or ice cream, may trigger dysphagia. The results of standard motor and sensory nerve conduction studies are normal. The response to repetitive stimulation or exercise is normal in most cases, but some patients may develop CMAP decrements, especially with cooling of the extremity. Myotonic discharges and other abnormal spontaneous activity may be accentuated by briefly cooling the extremity before needle EMG. There may be electrical silence on needle EMG during an episode.

ACKNOWLEDGMENT

The author thanks Basil T. Darras, M.D., for helpful comments on the manuscript.

REVIEW QUESTIONS

1. You are attempting to perform a sural sensory nerve conduction study in a 6-mo-old infant, but despite several attempts cannot record a reproducible SNAP. The best course at this point is to:
   A. Move on to the next patient.
   B. Record a superficial peroneal sensory response.
   C. Record a saphenous sensory response.
   D. Record a medial plantar sensory response.
   E. Proceed to the motor studies.
2. You are performing nerve conduction studies and EMG on a full term neonate with hypotonia and hyporeflexia. The upper extremity nerve conduction study results are as follows:

	Distal latency	*Amplitude*	*Velocity*
Median motor	2.3 ms	3.1 mV	27 m/s
Median sensory		14 μV	30 m/s

   You conclude that:
   A. The study is normal so far.
   B. The patient has a demyelinating neuropathy.
   C. The patient has an axonal neuropathy.
   D. The patient could have either a motor neuron disorder or a myopathy.
   E. Electrode placement is deficient.
3. Nerve conduction velocities are different in infants and children compared with adults mainly because of:
   A. The shorter distances between electrodes.
   B. Persistence of fetal isoforms of ion channels in the axons.
   C. Incomplete myelination of axons.
   D. Smaller surface area of electrodes used in infants and children.
   E. Differences in axon diameters.
4. While performing needle EMG in infants:
   A. It is almost always possible to record information on both spontaneous activity and voluntary activity from each muscle.
   B. Insertional activity is best recorded in extensor muscles, whereas voluntary units are best observed in flexor muscles.
   C. Fibrillation potentials and positive sharp waves may often be regarded as normal findings.
   D. It is important to examine all relevant muscles despite the patient's level of discomfort.
   E. Needle placement is often incorrect because the infant cannot follow directions.

5. A 4-mo-old full term infant whose birthweight was 9 lb, 11 oz has a flaccid, pronated left upper extremity with flexed fingers. He becomes agitated during nerve conduction studies and you anticipate that he will only tolerate needle EMG on one muscle. Which muscle would you choose?
   A. Biceps.
   B. Triceps.
   C. Flexor digitorum superficialis.
   D. First dorsal interosseous.
   E. Abductor pollicis brevis.
6. A 10-yr-old girl has had difficulty walking with frequent falls for 4 yr, and is slowly worsening. On examination, she has proximal weakness with a Gowers sign. Her serum creatine kinase level is normal. Nerve conduction studies reveal low-amplitude CMAPs and normal SNAPs, and needle EMG demonstrates large motor units with significantly decreased recruitment. The next step in the evaluation should be:
   A. Muscle biopsy.
   B. Genetic testing for PMP22 duplication.
   C. Genetic testing for Duchenne muscular dystrophy deletion.
   D. Genetic testing for *SMN* deletion.
   E. Lumbar puncture for protein level.
7. A 15-yr-old boy is teased in school for lifting his legs up high "like a robot" when he walks. His mother reports that he has always walked this way, but that his symptoms have been worsening. He has calf atrophy, high arches, and foot drop on examination, with loss of vibration sensation and proprioception in his toes. His mother also has high arches and uses a cane, and says that "stumbling" runs on her side of the family. Based on the subtype of his suspected disease that is most common in the general population, the most likely findings on nerve conduction studies will be:
   A. Mildly slow conduction velocities, in the 35 to 40 m/s range.
   B. Low amplitude CMAPs.
   C. Low amplitude SNAPs.
   D. Conduction block.
   E. Markedly slow conduction velocities, in the 20 to 30 m/s range.
8. An 18-yr-old woman with a long history of diabetes has developed progressive difficulty walking and numbness in her toes during the past 5 yr. She was initially thought to have diabetic neuropathy and was referred with that presumptive diagnosis, but when you examine her before nerve conduction studies and EMG, you find that she has marked dysarthria, absent lower extremity reflexes, and extensor plantar responses. What electrophysiological findings would you expect?
   A. Low-amplitude motor and sensory responses.
   B. Low-amplitude or absent sensory responses.
   C. Markedly slow conduction velocities.
   D. Myopathic motor units on needle EMG.
   E. Temporal dispersion.
9. A 12-yr-old girl is referred for numbness in her fingers. She has markedly prolonged median motor distal latencies, abnormal median SNAPs, and evidence for active denervation and chronic reinnervation at the abductor pollicis brevis. The remainder of her electrodiagnostic examination is normal, but you notice during the study that she has a large-appearing head with coarse facial features, short hands and fingers, and a protruding abdomen. She wears hearing aids and sometimes has difficulty understanding your directions during the study. Her parents tell you that previous doctors have just said that she is "different." If you were to include a clinical correlation in your report, you would say:
   A. That median neuropathies are quite common in childhood.
   B. She might have CMT disease.

C. It may be helpful to send urine for mucopolysaccharides.
D. A chromosome analysis may be indicated.
E. She should have serum amino acids and urine organic acids checked.

10. You are in practice in Pennsylvania when you are asked to perform nerve conduction studies and EMG on an inpatient at your local hospital, a 4-mo-old infant boy with an acute onset of weakness and hypotonia. On examination, he is intubated and quite flaccid, with ophthalmoparesis, sluggish pupillary responses, and hyporeflexia. His mother tells you that the first sign of difficulty was constipation 4 d ago. His father is not present because he is a foreman on a crew breaking ground for the new hospital outpatient building across the street. The highest yield studies on nerve conduction studies and EMG will be:
A. Median and peroneal motor studies.
B. Sensory studies.
C. 2- to 3-Hz repetitive stimulation.
D. 20- to 50-Hz repetitive stimulation.
E. Needle EMG.

REVIEW ANSWERS

1. The correct answer is D. The medial plantar sensory response is often easier to obtain than the sural sensory response in infants, in part, because of the substantial amount of subcutaneous tissue over the sural nerve at that age.
2. The correct answer is A. Nerve conduction studies and EMG normal values in infants and children differ markedly from those in adults, but follow a consistent age-dependent pattern. A general rule of thumb is that nerve conduction velocities in newborns are half those of adults.
3. The correct answer is C. Myelination of axons continues after birth, often not reaching maturity until the third to fifth year of life. Smaller axon diameters in infants and children also slow conduction velocities to some extent, but have a more modest effect than incomplete myelination.
4. The correct answer is B. An infant's typical reflexive response to a painful stimulus is withdrawal of the extremity, thus, needle insertion in an extensor muscle will usually result in relaxation of that muscle, whereas insertion in a flexor muscle will lead to activation. Fibrillation potentials and positive sharp waves are abnormal findings in infants, just as in adults. The needle study in an infant is often abbreviated compared with an equivalent study in an adult. It is unnecessary to subject an infant to excessive discomfort, and the electromyographer must carefully triage the muscles studied. In the hands of a skilled electromyographer, needle placement is almost always correct. Anatomical landmarks are critical in infant studies, and larger muscles are studied whenever possible.
5. The correct answer is A. The infant most likely has an Erb's palsy affecting the upper trunk of the brachial plexus caused by traction on the neck and shoulders during a difficult delivery, which can be deduced from the large birthweight and clinical presentation. Thus, a muscle innervated by the upper trunk of the brachial plexus would be ideal if only one muscle can be studied. The biceps, innervated by the musculocutaneous nerve, lateral cord, upper trunk, and C5–C6, is ideal. The other options all indicate muscles innervated by lower portions of the brachial plexus. The deltoid is another upper trunk muscle frequently studied in this setting, but is less likely than the biceps to be activated by an uncooperative infant. At the age of 4 mo, the most likely abnormal finding is chronic reinnervation rather than active denervation, making the biceps a better choice. If the infant were 1-mo old, the deltoid might be preferable. In an infant who tolerates the study well, as full a root/plexus screen as possible should be performed, because the extent and severity of injury can vary significantly.
6. The correct answer is D. This patient most likely has SMA type III (Kugelberg–Welander disease), with the proximal weakness, normal creatine kinase levels, and evidence for motor neuron disease on electrophysiological studies. In almost all cases, the patient will have a homozygous deletion in exon 7, exon 8, or both, of the *SMN* gene. A muscle biopsy should demonstrate fiber type grouping suggesting a neurogenic lesion, but is more invasive than

genetic testing on blood lymphocytes and does not reveal a molecular diagnosis. The patient does not fit the picture of hereditary motor and sensory neuropathy (CMT), which is associated with distal weakness, sensory loss, and a duplication in the *PMP22* gene. This presentation is also not consistent with Duchenne or Becker muscular dystrophy, nor with an inflammatory demyelinating polyneuropathy, which is associated with elevations in cerebrospinal fluid protein. If this patient were a severely hypotonic infant with the same creatine kinase levels and electrodiagnostic results, she would most likely have SMA type I (Werdnig–Hoffman disease).

7. The correct answer is E. This patient most likely has CMT disease. Type 1, a demyelinating form, is most common and statistically would be most likely in this patient without further demographic information, especially with the history of autosomal dominant inheritance. Conduction velocities are markedly slow in most cases. Amplitudes are typically normal in this form, although they can be markedly abnormal in type 2, the axonal form, of CMT. Evidence for multifocal slowing, such as conduction block, is classically found with acquired polyneuropathies, although multifocal slowing in inherited polyneuropathies, including some forms of CMT, does occur.
8. The correct answer is B. This patient seems to have Friedreich's ataxia, which is often associated with diabetes and cardiomyopathy. Onset occurs before age 20 yr. A sensory neuronopathy, with severe reduction in sensory response amplitudes is common, and preserved CMAPs and conduction velocities help distinguish it from CMT.
9. The correct answer is C. This girl most likely has Hurler disease, a mucopolysaccharidosis. A urine screen for mucopolysaccharides is indicated, followed by or concurrent with blood lysosomal enzyme testing. Median neuropathies, including carpal tunnel syndrome, are exceedingly rare in children in the absence of a systemic underlying condition. Assuming that she has no lower extremity symptoms, this presentation is inconsistent with CMT. Chromosome analysis, an amino acid panel, and an organic acid panel would not diagnose Hurler disease.
10. The correct answer is D. Pennsylvania, along with Utah and California, are states with high incidences of infant botulism, because of the presence of botulinum spores in the soil. A parent of an affected infant typically works in construction or is otherwise in close and frequent contact with newly disrupted soil. Constipation is a frequent early sign. The most important studies to perform are high-frequency repetitive stimulation studies at 20 to 50 Hz, which should yield an incremental response. It is important to perform such studies with the patient under anesthesia, because the studies are quite painful. Because this infant is already intubated, this should not be difficult. There may also be a decrement on low-frequency repetitive stimulation, but this finding may not always be present. Other electrodiagnostic studies, including routine nerve conduction studies and needle EMG, abnormalities such as diminished CMAP amplitudes will yield normal results or nonspecific abnormalities.

SUGGESTED READING

Baer RD, Johnson EW. Motor nerve conduction velocities in normal children. Arch Phys Med Rehabil. 1965;46:698-704.

Cornblath DR, Sladky JT, Sumner AJ. Clinical electrophysiology of infantile botulism. Muscle Nerve. 1983;6:448-452.

Cruz Martinez A, Perez Conde MC, Ferrer MT. Motor conduction velocity and H-reflex in infancy and childhood: 1.-study in newborns, twins and small-for-dates. Electromyogr Clin Neurophysiol. 1977;17:493-505.

Darras BT, Jones HR. Diagnosis of pediatric neuromuscular disorders in the era of DNA analysis. Pediatr Neurol. 2000;23:289-300.

Deymeer F, Jones HR, Jr. Pediatric median mononeuropathies: a clinical and electromyographic study. Muscle Nerve. 1994;17:755-762.

Escolar DM, Jones HR, Jr. Pediatric radial mononeuropathies: a clinical and electromyographic study of sixteen children with review of the literature. Muscle Nerve. 1996;19:876-883.

Felice KJ, Royden Jones H, Jr. Pediatric ulnar mononeuropathy: report of 21 electromyography-documented cases and review of the literature. J Child Neurol. 1996;11:116–120.

Gamstorp I. Normal Conduction Velocity of Ulnar, Median and Peroneal Nerves in Infancy, Childhood and Adolescence. Acta Paediatr. 1963;14:SUPPL146:168–176.

Jones HR, Jr., Bolton CF, Harper CM, Jr. Pediatric Clinical Electromyography. Philadelphia-New York: Lippincott-Raven, 1996:1–36.

Jones HR, Jr., Felice KJ, Gross PT. Pediatric peroneal mononeuropathy: a clinical and electromyographic study. Muscle Nerve. 1993;16:1167–1173.

Jones HR, Jr. Guillain-Barre syndrome in children. Curr Opin Pediatr. 1995;7:663–668.

Kang PB, Finkel RS. Myasthenia gravis. In: Burg FD, Ingelfinger JR, Polin RA, Gershon AA, eds. Current Pediatric Therapy. Philadelphia: Saunders Elsevier, 2006;1028–1033.

Martinez AC, Ferrer MT, Conde MC, Bernacer M. Motor conduction velocity and H-reflex in infancy and childhood. II. -Intra and extrauterine maturation of the nerve fibres. Development of the peripheral nerve from 1 month to 11 years of age. Electromyogr Clin Neurophysiol. 1978; 18: 11–27.

Miller RG, Gutmann L, Lewis RA, Sumner AJ. Acquired versus familial demyelinative neuropathies in children. Muscle Nerve. 1985;8:205–210.

Miller RG, Kuntz NL. Nerve conduction studies in infants and children. J Child Neurol. 1986;1: 19–26.

Namba T, Brown SB, Grob D. Neonatal myasthenia gravis: report of two cases and review of the literature. Pediatrics. 1970;45:488–504.

Nicolas G, Maisonobe T, Le Forestier N et al. Proposed revised electrophysiological criteria for chronic inflammatory demyelinating polyradiculoneuropathy. Muscle Nerve. 2002;25:26–30.

Packer RJ, Brown MJ, Berman PH. The diagnostic value of electromyography in infantile hypotonia. Am J Dis Child. 1982;136:1057–1059.

Parano E, Uncini A, De Vivo DC, Lovelace RE. Electrophysiologic correlates of peripheral nervous system maturation in infancy and childhood. J Child Neurol. 1993;8:336–338.

Preston DC, Shapiro BE. Electromyography and Neuromuscular Disorders. 2nd ed. Boston: Butterworth-Heinemann, 2005;704.

Russell JW, Afifi AK, Ross MA. Predictive value of electromyography in diagnosis and prognosis of the hypotonic infant. J Child Neurol. 1992;7:387–391.

Sladky JT. Neuropathy in childhood. Semin Neurol. 1987;7:67–75.

Sladky JT, Brown MJ, Berman PH. Chronic inflammatory demyelinating polyneuropathy of infancy: a corticosteroid-responsive disorder. Ann Neurol. 1986;20:76–81.

Streib EW. AAEE minimonograph #27: differential diagnosis of myotonic syndromes. Muscle Nerve. 1987;10:603–615.

Swoboda KJ, Edelbol-Eeg-Olofsson K, Harmon RL, et al. Pediatric electromyography. In: Jones HR, Jr., De Vivo DC, Darras BT, eds. Neuromuscular Disorders of Infancy, Childhood, and Adolescence: a clinician's approach. Boston: Butterworth-Heinemann, 2003;35–74.

Zerres K, Rudnik-Schoneborn S. Natural history in proximal spinal muscular atrophy. Clinical analysis of 445 patients and suggestions for a modification of existing classifications. Arch Neurol. 1995;52:518–523.

IV
Autonomic Testing, Evoked Potentials, and Sleep

23
Polysomnography and Sleep Disorders

Jean K. Matheson, Randip Singh, and Andreja Packard

Summary

The classification of sleep disorders is based both on clinical and neurophysiological criteria and is undergoing constant refinement. Sleep disorders can be caused by either a primary disorder of a mechanism controlling sleep or inadequate function of an end organ, such as the upper airways and lungs. Understanding the physiology and pattern of normal sleep is an important foundation for interpreting the clinical symptoms, signs, and neurophysiological abnormalities observed in patients with sleep disorders. The term polysomnography refers to the simultaneous recording of multiple sleep parameters, including a limited electroencephalogram, respiratory parameters, chest excursion, limb movements, and the electrocardiogram. Polysomnography is important for assessing a variety of sleep disturbances, including disorders such as sleep-related breathing disorders (including obstructive sleep apnea), rapid eye movement behavior disorder, and periodic movements of sleep. The multiple sleep latency test and maintenance of wakefulness test are studies that are especially useful in the evaluation of narcolepsy and other hypersomnias.

Key Words: Multiple sleep latency test; narcolepsy; obstructive sleep apnea; periodic movements of sleep; polysomnography; sleep.

1. INTRODUCTION

The understanding of sleep disorders and the development of sleep disorders medicine as a discipline has depended on the ability to simultaneously measure and record multiple physiological parameters during sleep. The term "polysomnography" was introduced in the early 1970s to describe both the recording and the interpretation of these variables. The technique of polysomnographic recording evolved from EEG with critical modifications that allowed researchers to correlate changes in EEG with changes in other physiological systems. In the late 1930s, Alfred L. Loomis, E. Newton Harvey, and Garret Hobart *(1)* introduced the concept of recording electroencephalographic activity continuously through the night and, in doing so, identified five sleep stages. In 1953, Eugene Aserinsky, working in Nathaniel Kleitman's laboratory at the University of Chicago, added electrodes to record the eye movements he had observed through the lids of sleeping subjects. By waking subjects during intense periods of rapid eye movement (REM), these researchers discovered that the REMs were associated with dreaming and, thus, discovered REM sleep *(2)*. Soon thereafter, William Dement, working in the same laboratory, established that REM alternated with non-REM (NREM) sleep in cycles of approx 90 min *(3)*. Later, after research showed that REM was associated with loss of muscle tone (REM sleep atonia), surface EMG electrodes on the chin

From: *The Clinical Neurophysiology Primer*
Edited by: A. S. Blum and S. B. Rutkove © Humana Press Inc., Totowa, NJ

were added to the EEG and the electro-oculogram (EOG) recordings to better define the REM sleep state. A committee of early sleep researchers refined the rules for recording and scoring sleep stages in 1968. The product of that effort, known as the "Rechtschaffen and Kales criteria" (R and K criteria) remains the international "gold-standard" in the field *(4)*. In the progression of the development of the polysomnogram (PSG), the pivotal discovery of sleep apnea in the 1960s resulted in the addition of simultaneous respiratory and cardiac monitoring. Both REMs and apneas are evident with direct observation but were not appreciated as important until studied with neurophysiological techniques. It is now difficult to understand how these obvious human behaviors could have been ignored for so long; there are perhaps no better examples of clinical neurophysiology amplifying human observation.

This chapter will focus on the standardized techniques used to measure sleep-related physiological parameters and their practical application to the diagnosis of specific sleep disorders.

2. CLINICAL ASSESSMENT OF SLEEP DISORDERS

A sleep disorder generally occurs for one of two reasons. It may represent a primary disorder of a mechanism controlling sleep or failure of a specific end organ, such as the upper airways and lungs. As in all of clinical medicine, testing must be ordered and interpreted within the context of the patient's clinical presentation, with a clear understanding of the questions to be answered and the inherent limitations of the study proposed. Most patients present with complaints of excessive daytime sleepiness, difficulty initiating or maintaining sleep, or some sort of unpleasant event that occurs during sleep. Detailed medical and sleep histories with careful attention to underlying medical and psychiatric illness, daytime schedules, lifestyle issues, medications, and drug use are prerequisites for the intelligent analysis of the problem and planning of the appropriate testing. A complete physical examination should always be obtained before referral for study. Sleep logs, which document daily sleep–wake behaviors, may be valuable tools that complement both the office history and the interpretation of the objective data acquired in the sleep laboratory.

2.1. *Classification of Sleep Disorders*

The American Academy of Sleep Medicine (AASM) has established a classification of sleep disorders, the "International Classification of Sleep Disorders," 2nd ed. (ICSD-2) *(5)*. A summary of this classification is given in Table 1.

A detailed discussion of these disorders is outside the scope of this chapter. Later subheadings will discuss the sleep disorders that are best suited to analysis by neurophysiological testing.

3. OVERVIEW OF SLEEP

REM sleep, sometimes called paradoxical sleep or dreaming sleep, and NREM sleep are the two sleep states. NREM and REM sleep alternate in recurring cycles of approx 90 min.

NREM sleep is divided into four stages (stages 1–4), which represent progressive deepening of sleep. Stage 1 sleep is characterized by the gradual disappearance of the alpha rhythm of quiet wakefulness, which is replaced by theta activity and some fast activity. The emergence of sleep spindles and K-complexes establishes the onset of stage 2 sleep. Stage 3 and stage 4 sleep are defined by the presence of high-voltage slow activity. Stage 3 and stage 4 sleep are often described together as slow-wave sleep or delta sleep.

The normal young adult descends in an orderly progression through the four NREM stages. Slow-wave sleep appears approx 30 to 40 min after sleep onset. The first REM period

Table 1
International Classification of Sleep Disorders-2: Diagnostic Categories

I	Insomnias
II	Sleep-related breathing disorders
III	Hypersomnias of central origin not due to a circadian rhythm sleep disorder, sleep-related breathing disorder, or other cause of disturbed nocturnal sleep
IV	Circadian rhythm sleep disorders
V	Parasomnias
VI	Sleep-related movement disorders
VII	Isolated symptoms, apparently normal variants, and unresolved issues
VIII	Other sleep disorders

follows this slow-wave sleep, approx 70 to 90 min after sleep onset. The PSG during REM sleep shows dramatic changes. A sudden loss of EMG activity occurs in the chin muscles, which is indicative of generalized skeletal muscle atonia. REMs occur in phasic bursts and the EEG shows mixed frequencies similar to waking and stage 1 sleep, sometimes with a characteristic sawtooth pattern.

The first REM period is short, lasting approx 10 min. The end of the first REM period completes the first sleep cycle. Thereafter, NREM sleep continues to alternate with REM sleep; the healthy adult goes through 4 to 6 cycles. Slow-wave sleep is concentrated in the first third of the night, whereas REM sleep episodes become progressively longer later in the night. Slow-wave sleep is prominent in adolescence and decreases significantly with age, whereas REM sleep duration tends to remain stable throughout adulthood. Newborns, however, demonstrate up to 50% REM sleep.

REM and NREM sleep differ physiologically. REM sleep is characterized by both phasic and tonic changes in physiology. The drop in baseline EMG correlates with a tonic change. REMs correlate with phasic changes. Tonic physiological changes also include impaired thermoregulation, reduction in ventilatory chemosensitivity, hypotension, bradycardia, increased cerebral blood flow, and intracranial pressure, increased respiratory rate, and penile erection. Phasic changes include vasoconstriction, increased blood pressure, tachycardia, and further increases in cerebral blood flow and respiratory rate. During NREM sleep, the physiological state is more stable, with an overall reduction in blood pressure, heart rate, cardiac output, and respiratory rate. One characteristic feature of NREM, slow-wave sleep is the secretion of growth hormone.

The control of sleep onset, duration, and stage changes is poorly understood. It is thought that there are two sleep drives, one homeostatic and the other circadian. The homeostatic drive increases with the duration of wakefulness, whereas the circadian signals are controlled by the suprachiasmatic nucleus of the hypothalamus. Recent evidence has shown that neurons of the ventrolateral preoptic nucleus of the hypothalamus are sleep active and sleep promoting, confirming old observations that lesions in this area induce insomnia. These neurons express the inhibitory transmitters, galanin and GABA, and innervate wake-promoting areas, including the hypocretin (also known as orexin) containing neurons of the posterolateral hypothalamus, histaminergic neurons of the tuberomammillary nucleus, the serotonergic dorsal raphe, norepinephrine containing neurons of the locus ceruleus, and the cholinergic neurons of the dorsal midbrain and pons. In turn, monoaminergic wake-promoting areas inhibit the ventrolateral preoptic nucleus, thereby resulting in reciprocal inhibition that self-reinforces stable periods of sleep and wake. In this model, homeostatic and circadian drives are hypothesized to shift the balance between states by still unknown mechanisms. Adenosine that accumulates during wakefulness, the effect of which is antagonized by caffeine, may be one of the factors that signals the homeostatic drive to sleep. Neurons critical for generating REM sleep are found in the lateral pons and adjacent midbrain, including the cholinergic cells of the pedunculopontine and lateral dorsal tegmental nuclei. These neurons are inhibited by norepinephrine, serotonin, and histamine. Hypocretin-1 cells in the posterolateral hypothalamus stimulate aminergic cells and, thus, also contribute to the inhibition of REM sleep while simultaneously promoting wakefulness. Loss of hypocretin-1-containing cells is now known to underlie the pathophysiology of narcolepsy (*see* Section 6.2.1.) (*see* Espana and Scammell for a recent review of sleep neurobiology; ref. *6*).

Some disorders are exacerbated by or occur only during certain sleep stages. Sleepwalking, for example, occurs with arousal from slow-wave sleep. Epileptic seizures tend to be facilitated by NREM sleep but inhibited by REM sleep. Obstructive sleep apnea (OSA) is typically worse in REM sleep because of REM atonia and alteration of respiratory chemosensitivity.

Because of this changing physiological template, a careful neurophysiological study of sleep is useful in understanding a wide variety of disease processes.

4. POLYSOMNOGRAPHY

Polysomnography is the term applied to the simultaneous and continuous measurement of multiple physiological parameters during sleep. In practice, the term PSG has come to mean a specific type of polysomnographic study in which measurements allow for:

1. The identification of sleep stage.
2. Monitoring of cardiopulmonary function.
3. Monitoring of body movements during sleep.

This study is typically obtained at night in a sleep laboratory for the purpose of identifying, as best as possible given the novel environment, the patient's typical sleep and its associated pathologies. The AASM has developed guidelines for the indication for polysomnography *(7)*. These indications include:

1. Suspicion of sleep-related breathing disorders.
2. Treatment and followup of sleep-related breathing disorders.
3. In combination with the MSLT for suspected narcolepsy.
4. Evaluation of sleep-related behaviors that are violent, potentially injurious, or do not respond to conventional therapy.
5. To assist in the diagnosis of paroxysmal arousals that are suggestive of seizure disorder (with additional video and EEG).
6. Evaluation of sleep-related movement disorders.

Many experienced clinicians also obtain PSGs in the evaluation of insomnia, circadian rhythm disorders, nocturnal angina, arrhythmia, hypertension, gastroesophageal reflux, and headache, because these entities are commonly exacerbated by coexisting sleep disorders that may not be evident on clinical grounds.

Standard PSG measurements in current clinical practice include (*see* Table 2):

1. EEG (C4–A1 and/or C3–A2).
2. Eye movement recording (EOG).
3. EMG of chin (surface).

Table 2
Parameters Recorded in Standard PSG, MSLT, and MWT

	Parameters recorded
PSG	EEG (C4–A1, C3–A2)
	Additional EEG if indicated
	EOG
	EMG (chin)
	Airflow
	Respiratory effort
	O_2 saturation est. (S_pO_2)
	EKG
	EMG limb, anterior tibialis muscles; extensor digitorum muscles when indicated
	Body position
	Esophageal pH (rarely)
MSLT and MWT	EEG (C4–A1, C3–A2)
	EEG (O1–A1, O2–A2)
	EOG
	EMG (chin)
	EKG
	Optional: respiratory monitoring

EOG, electro-oculogram; MWT, maintenance of wakefulness test; MSLT, multiple sleep latency test; PSG, polysomnogram; S_pO_2, pulse oximetry.

4. Respiratory effort (chest and abdomen).
5. Airflow (nasal/oral).
6. Oxygen saturation (pulse oximetry).
7. EKG (one lead).
8. EMG (surface) of anterior tibialis muscles.
9. Body position.

Other measurements may be performed as dictated by the clinical question. Video monitoring and additional EEG electrodes are commonly added for the question of nocturnal seizure disorder (Video EEG–PSG), whereas continuous blood pressure monitoring, penile tumescence recordings, and gastroesophageal pH measurements are only rarely performed in clinical situations. PSGs performed in the sleep laboratory are termed "attended" PSGs, which indicates that a technician is available throughout the study who may intervene to assure quality or to initiate therapies, such as the application of positive pressure for the treatment of sleep apnea. Full polysomnography is available at the bedside or for at-home testing, and although of high quality, is not generally accepted as equal in reliability to attended studies.

Four-channel studies geared for screening of respiratory cardiorespiratory abnormalities but lacking reliable methods of identifying whether the patient is awake or asleep, are poorly validated against formal polysomnography. However, these may have some usefulness when full polysomnographic study is not available.

Polysomnographic techniques with fewer recording channels are also applied to two other sleep laboratory tests: the multiple sleep latency test (MSLT) and the maintenance of wakefulness test (MWT). The MSLT measures the tendency to fall asleep during the day and screens for the occurrence of inappropriate daytime episodes of REM sleep. The MWT measures the ability to stay awake in multiple daytime naps.

4.1. The PSG

4.1.1. Sleep Stage Scoring/Sleep Architecture

Sleep stage "scoring" or "staging" refers to the lengthy process of identifying sleep stages recorded on the PSG. Identification of sleep stages requires the simultaneous assessment of three channels of the PSG: EEG, EOG, and chin EMG. The continuously recorded EEG is, by convention, analyzed in 30-s intervals, known as epochs. Sleep stages are identified or "scored" using the R and K criteria outlined in Section 4.1.1.1. (*see* Table 3). Because one epoch can show features of more than one sleep stage, the sleep stage that occupies more than 50% of the 30-s interval is assigned to that epoch. Although stage scoring can be performed in a limited fashion by computer, the accuracy is poor and most laboratories rely on visual scoring by trained sleep technicians.

Sleep architecture refers to the distribution and temporal sequence of sleep stages (Table 4), which contribute to a quantitative understanding of the night's sleep. Most analyses also include a hypnogram, a graphic display of sleep stages on a time line throughout the night that assists in the qualitative assessment of sleep. Hypnograms and sleep architecture parameters are produced by computer analysis of the visually derived data. Figure 1 shows representative hypnograms across ages. Table 5 shows ranges of normative values for sleep stages in adults. Architecture and stage distribution are influenced by commonly used medications. A summary of some of these medication effects is given in Table 6.

4.1.1.1. Sleep Stages: Technique and Scoring Rules

"A Manual of Standardized Terminology, Techniques and Scoring System for Sleep Stages of Human Subjects," edited by Rechtschaffen and Kales, is the internationally accepted standard for sleep staging (R and K criteria) *(4)*. Although those who intend to master sleep scoring need to study the manual directly, the following text and Table 3 provide useful summaries abstracted directly from the R and K manual.

4.1.1.1.1. Technical Requirements for Scoring

4.1.1.1.1.1. EEG A single EEG channel is used for the purposes of scoring sleep stages. The acceptable derivations are C4 or C3 (International 10–20 System) referred to the opposite ear or mastoid; two channels are recorded with one used as a backup for technical problems during the night. This sleep derivation tends to confuse electroencephalographers who do not typically record using an opposite side reference. The opposite side derivation was chosen to maximize inter-electrode distance. Because the amplitude of the EEG waveform is dependent on inter-electrode distance, and because amplitude is included in the criteria for stages 3 and 4 sleep (*see* below), the convention has remained. Filtering also influences the amplitude of the waveform. Traditional EEG recording uses a low filter frequency of 1 Hz, but the EEG of sleep studies is filtered at a low-frequency setting of 0.3 Hz to maximize amplitude of lower-frequency waves. A 1-Hz low-frequency filter reduces the amplitude of a 1-Hz signal by 20%. Using an incorrect low-frequency filter will, therefore, result in underscoring of stages 3 and 4 sleep. The high-frequency settings are typically the same for traditional EEG and

Table 3
Outline of Scoring Criteria According to "R and K" Manual

Stage/state	*Electroencephalogram (EEG)*	*Electrooculogram (EOG)*	*Electromyogram (EMG)*
Relaxed wakefulness	Eyes closed: rhythmic alpha (8–13 cps); prominent in occipital; attenuates with attention Eyes open: relatively low voltage mixed frequency	Voluntary control; REMs or none; blinks; slow eye movements when drowsy (SEMs) when drowsy	Tonic activity, relatively high; voluntary movement
Non-rapid eye movement sleep (NREM)			
Stage 1	Relatively low voltage, mixed frequency May be theta (3–7cps) activity with greater amplitude Vertex sharp waves Synchronous high voltage theta bursts in children	SEMs	Tonic activity, may be a slight decrease from waking
Stage 2	Background: relatively low voltage, mixed frequency Sleep spindles: waxing, waning, 12–14 cps (≥0.5 s) K-complex: negative sharp wave followed by slower positive component (0.5 s); spindles may ride on Ks; Ks maximal in vertex; spontaneous or in response to sound	Occasionally SEMs near sleep onset	Tonic activity, low level

(Continued)

Table 3 *(Continued)*

Stage/state	*Electroencephalogram (EEG)*	*Electrooculogram (EOG)*	*Electromyogram (EMG)*
Stage 3	≥20%, ≤50% high amplitude (>75 μV), slow frequency (≤2 cps); maximal in frontal	None, picks up EEG	Tonic activity, low level
Stage 4	>50% high amplitude, slow frequency	None, picks up EEG	Tonic activity, low level
Rapid eye movement sleep (REM)	Relatively low voltage mixed frequency Saw tooth waves Theta activity Slow alpha	Phasic REMs	Tonic suppression; phasic twitches
Movement time	Obscured	Obscured	Very high activity

Reprinted from Carskadon MA, Monitoring and staging human sleep. In: Kryger, MH, Roth, T, Dement WC, Rechtschaffen A, eds., Principles and Practice of Sleep Medicine, 4th ed., p.1364, 2005, with permission from Elsevier.

Table 4
Sleep Variables

Variable	*Abbreviation*	*Definition*
Time in bed	TIB	The time from lights out until the subject chooses to end the study
Sleep period time	SPT	TIB minus time awake after lights out before sleep onset, and minus time in bed after awakening in the morning
Total sleep time	TST	Total time the subject actually slept. This is SPT minus any time awake during the night
Sleep efficiency index	SE	TST/TIB. Often reported as a percentage. This is an important measure of sleep quality
Percentage of each stage		This may be reported as either a percentage of the SPT or the TST. The usual convention is to report percentage of SPT with stage W included as a sleep stage
Sleep-onset latency	SL	Time from lights out until the onset of sleep. Sleep onset is usually defined as the onset to the first epoch of any sleep stage. Sometimes reported as latency to stage 2
Number of awakenings		Records the number of times the subject returns to stage W after sleep onset
Number of arousals		Records the number of EEG arousals

polysomnography and are set at 70 Hz. Central lead placements are used because sleep-specific EEG waveforms (K-complexes, spindles, vertex waves, high-voltage slow activity, and sawtooth waves, *see* Section 4.1.1.1.2.) tend to be maximal in these areas. The posterior predominant alpha rhythm of the relaxed wakeful state is adequately, but not ideally, observed in the central leads. To better visualize the alpha rhythm, many laboratories add occipital channels, but R and K scoring rules are nonetheless based on the central derivations.

Most laboratories now use digital recording techniques that allow for manipulation of montages, time base, filter settings, and channel sensitivity. The EEG sampling rate should be at least 200 Hz and the monitor display should also allow appropriate resolution of the waveforms.

4.1.1.1.1.2. Eye Movement Recording Eye movement recording is necessary for the identification of REM sleep. Slow eye movements are characteristic of the onset of stage 1 sleep and aid in the identification of that sleep stage. Two channels are devoted to the EOG. Standard EEG electrodes are applied 1 cm above and lateral to the outer canthus of one eye and 1 cm below and lateral to the outer canthus of the other eye (E1 and E2, also called ROC and LOC).

The R and K manual recommends using the same ear or mastoid as a reference electrode for both channels, but, in usual practice, the eye electrodes are referred to the opposite ear or mastoid. As a result of the eye's natural dipole (cornea positive), vertical or lateral eye movements are identified as out-of-phase deflections on the two channels. This arrangement makes eye movements easily distinguishable from frontal EEG activity inadvertently picked up by the EOG electrodes. Frontal EEG activity will appear as in- phase activity. Typical filter settings are 0.3 Hz and 15 Hz, which allow for good resolution of both slow and fast eye movements.

4.1.1.1.1.3. Chin EMG EMG activity can be easily recorded from surface EEG-type electrodes applied to muscles on and under the chin. The EMG activity so measured is an indication of muscle activity, which is called "muscle tone" in the sleep literature. Analysis of chin muscle

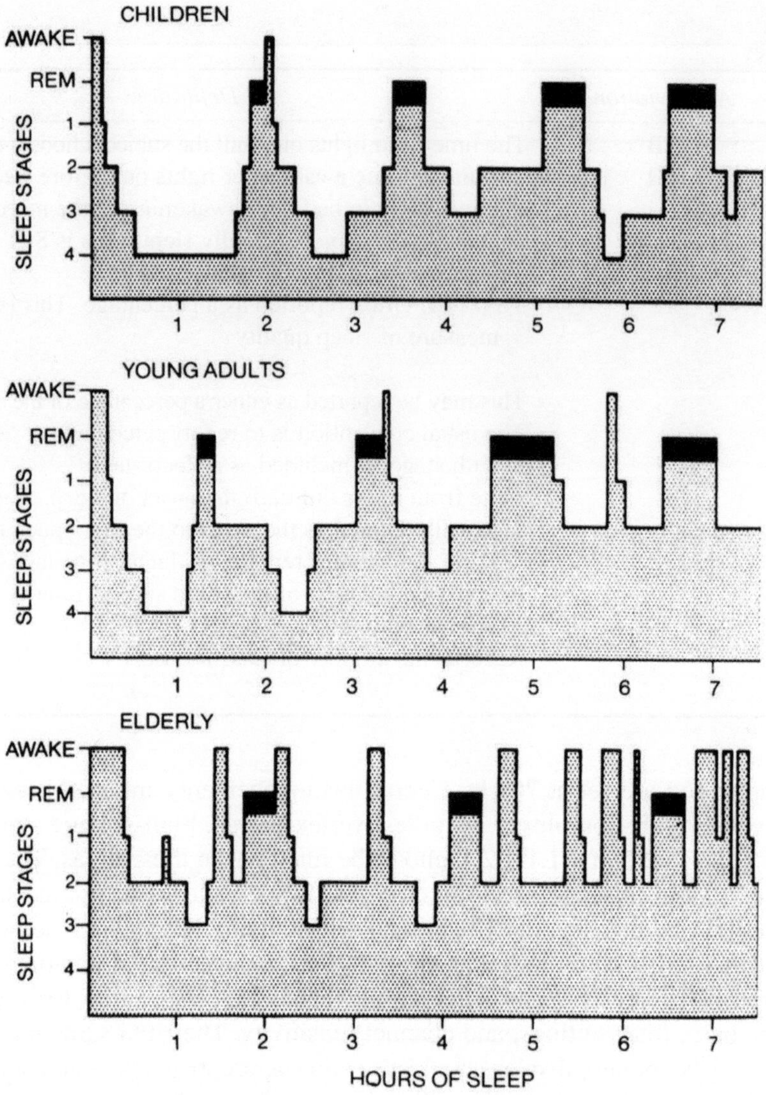

Fig. 1. Normal sleep cycles. Rapid eye movement (REM) sleep (darkened area) occurs cyclically throughout the night at intervals of approximately 90 min in all age groups. REM sleep shows little variation in the different age groups, whereas stage 4 sleep decreases with age. In addition, the elderly have frequent awakenings and a marked increase in total wake time. From ref. *17* with permission.

tone is necessary to identify normal and abnormal REM sleep. In the normal condition, REM sleep is characterized by a well-sustained, typically sudden, drop in tone with some intermittent phasic activity. The activity in the chin electrode is used in the formal staging of REM sleep discussed in Section 4.1.1.1.2.8. Three electrodes are placed, one on the chin and two on the muscles under the chin, one on each side. Only two electrodes are used, referred to each other, but if one fails because of movement, the third is available without waking the patient. Surface EMG electrodes are filtered to remove slow artifacts and allow the high-frequency muscle activity to be observed. A low-frequency setting of 10 Hz and a high-frequency setting of 70 Hz are commonly used.

Table 5
Normative Values for Sleep Stages in Adults

Sleep stages	*20–29 yr*		*30–39 yr*		*40–49 yr*		*50–59 yr*		*60–69 yr*	
%SPT	*Males*	*Females*	*Males*	*Females*	*Males*	*Females*	*Males*	*Females*	*Males*	*Females*
Wake	1.26 ± 1.08	0.53 ± 0.49	1.47 ± 1.94	1.84 ± 3.97	6.29 ± 5.56	1.63 ± 1.30	4.33 ± 2.33	4.95 ± 6.48	7.73 ± 6.02	8.93 ± 8.47
1	4.44 ± 1.62	4.18 ± 2.39	5.71 ± 3.43	4.17 ± 1.65	7.56 ± 3.03	5.64 ± 2.00	7.56 ± 3.94	4.85 ± 2.20	9.73 ± 3.97	7.69 ± 4.12
2	45.54 ± 5.17	52.37 ± 5.89	56.89 ± 7.36	53.77 ± 7.73	54.75 ± 11.14	54.01 ± 8.55	61.71 ± 10.30	57.80 ± 6.50	56.79 ± 8.76	54.78 ± 8.59
3 and 4	20.76 ± 4.78	17.69 ± 6.73	12.46 ± 5.58	14.00 ± 7.26	8.54 ± 6.84	12.05 ± 8.31	4.92 ± 7.70	10.63 ± 6.07	2.66 ± 5.05	7.17 ± 6.81
REM	28.00 ± 5.66	25.23 ± 3.63	23.47 ± 3.86	26.22 ± 5.27	22.85 ± 4.00	26.67 ± 4.10	21.48 ± 4.01	21.77 ± 3.26	23.09 ± 3.59	21.43 ± 4.04

REM, rapid eye movement sleep.
Abstracted from ref. *16*.

Table 6
Medication Effects on the Polysomnogram

Medication	*SWS*	*REM*	*Miscellaneous*
TCAs	⇔	⇓	
SSRIs/SNRIs	⇔	⇓	Venlafaxine is a 5HT and NE uptake inhibitor SSRIs and SNRIs ⇑ Non-REM slow eye movements
Trazodone	⇔	⇓	
Nefazadone	⇔	⇑	
Bupropion	⇔	⇑	NE and DA uptake inhibitor
Mirtazapine	⇔	⇔	
MAOIs	⇔	⇓	
Lithium	?	⇓	
BZDs	⇓	⇔	⇑ Spindle activity and stage 2
Zolpidem	⇔	⇔	
Dopaminergic drugs	?	?	Mixed results
Anticonvulsants			Minimal data
Phenytoin	⇑	?	
Barbiturates	⇓	⇓	⇑ Spindle activity
Carbamazepine	⇑	⇓	
Tiagabine	⇑	⇔	
Gabapentin	⇑	⇔	
Lipophilic beta-blockers	⇓	⇓	
Clonidine	?	⇓	
Opioids	⇓	⇓	Can ⇓ respiratory drive
Amphetamines	⇔	⇓	
Caffeine	⇓	?	Adenosine antagonist; adenosine ⇑ SWS
Alcohol (acute)	?	⇓	After ETOH metabolism there is a REM rebound
Alcohol (chronic)	⇓	⇓	
Sodium oxybate	⇑	⇑⇓	Approved for cataplexy and excessive sleepiness

4.1.1.1.2. Sleep Stage Scoring Rules

A summary of sleep stage scoring rules abstracted from the R and K manual is abstracted below and in Table 3.

4.1.1.1.2.1. Stage W Stage W (Fig. 2) refers to the waking state. Wakefulness is characterized by a low-voltage mixed-frequency EEG and/or the alpha rhythm. A posteriorly predominant rhythm in the alpha frequency (8–13 Hz), which attenuates with eye opening, is characteristic of the relaxed wakeful state with the eyes closed. This rhythm is generally referred to as the alpha rhythm, but is sometimes referred to as the Berger rhythm by electroencephalographers. In disease states (e.g., encephalopathy or hypothyroidism), the alpha rhythm may not reach the alpha frequency, an example of why it is important to not confuse the term "alpha rhythm" with the term "alpha frequency." When the eyes are open, low-voltage mixed-frequency rhythms are observed. Wake is usually also accompanied by the REMs of normal exploratory visual behavior, associated with high chin tone.

4.1.1.1.2.2. Movement Time Movement time is a scoring term applied to epochs that are obscured by movement artifact for more than 50% of the 30-s epoch, and occur before or

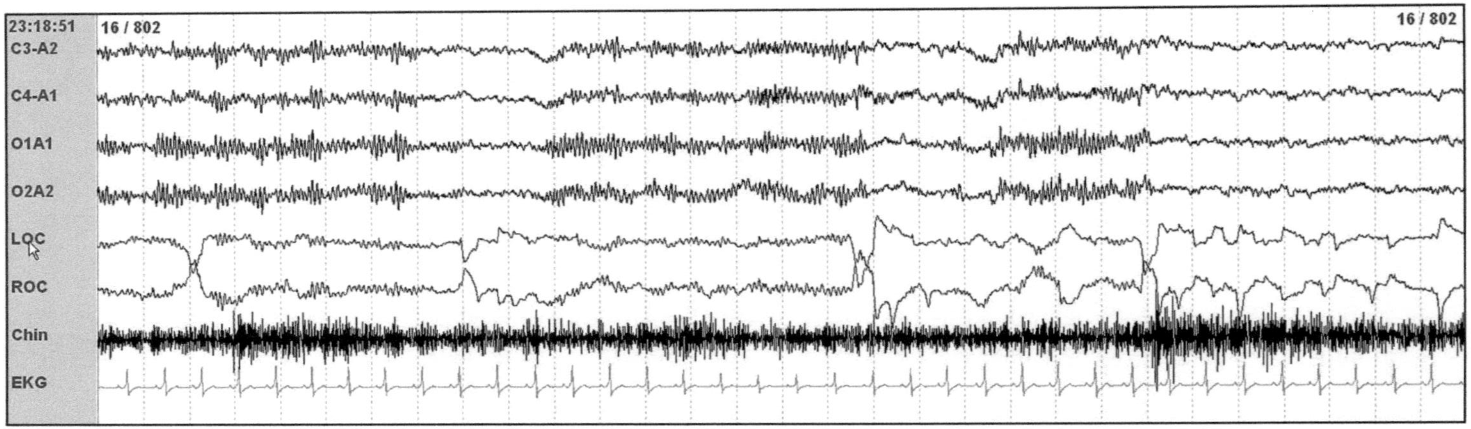

Fig. 2. Wake. The posterior dominant alpha rhythm block with eye opening. Fast eye movements typical of wakefulness are noted as rapid out-of-phase deflections. (30-s epoch)

after scorable sleep epochs. Movement time is not quantitated as either sleep or wake, but scored as a separate category.

4.1.1.1.2.3. Movement Arousal Movement arousal is a term used in the R and K manual that applies primarily to sleep stage scoring criteria. A movement arousal is defined as "any increase in EMG on any channel, which is accompanied by a change in pattern on any additional channel."

4.1.1.1.2.4. EEG Arousal Intrusions during sleep that seem to cause a disturbance with lightening of sleep, but without full awakening are termed "EEG arousals." A formal definition of EEG arousal has been generated by The Atlas Taskforce of the American Sleep Disorders Association *(8)* and is not in the R and K manual. An EEG arousal is defined as: "An abrupt shift in EEG frequency which may include theta, alpha, and/or frequencies greater than 16 Hz, but not spindles, subject to the following rules and conditions:

1. The subject must be asleep for at least 10 s before an arousal can be scored.
2. Minimum interval of sleep between two arousals is 10 s.
3. The EEG frequency shift must be 3 s or greater in duration.
4. Arousals do not require increases in chin EMG in NREM sleep.
5. Arousals require increased chin EMG in REM sleep.
6. EMG changes alone are not sufficient for scoring an arousal.
7. K-complexes and delta waves are not scored as arousals unless they occur within a frequency shift as described above.
8. Blocking artifact is not an arousal unless accompanied by the EEG frequency shift.
9. Three seconds of alpha frequency activity during sleep is not scored as an arousal unless there has been a 10-s period free of alpha
10. Transitions from one sleep stage to another are not in themselves arousals." *(8)*

4.1.1.1.2.5. Stage 1 Subjects typically descend from stage W to stage 1 sleep (Fig. 3). This stage is defined by the presence of a relatively low-voltage mixed-frequency EEG with a predominance of slower activity in the 3- to 7-Hz range. Sharp negative centrally predominant waves called vertex waves may be observed. Faster-frequency activity of 12 to 14 Hz may occur. At sleep onset, stage one is identified by the drop out of the alpha rhythm that predominates in the relaxed state before sleep. When greater than 50% of a 30-s epoch shows the slower activity of stage 1 sleep, the epoch is scored as stage 1. Slow, rolling eye movements that help the scorer identify the transition to sleep frequently accompany this stage. Stage 1 sleep may also appear intermittently throughout the night, often in response to a disturbance and is, thus, considered a "light" sleep stage. An increase in stage 1 sleep is, thus, one important measure of sleep disruption. This stage is frequently not perceived by the subject as sleep, particularly early in the night.

4.1.1.1.2.6. Stage 2 Stage 2 sleep (Fig. 4) is defined by the appearance of K-complexes and or sleep spindles, each lasting at least 0.5 s, in the absence of slow activity sufficient to meet the scoring requirements of stage 3 or 4 sleep (*see* Section 4.1.1.1.2.7.). A sleep spindle is a series of 12- to 14-Hz waves with a fusiform morphology. The K-complex in the R and K criteria is defined differently than that definition accepted by the EEG community. The K-complex in R and K is defined as "EEG wave form having a well-delineated negative sharp wave which is immediately followed by a positive component. Waves of 12 to 14 cps may or may not constitute a part of the complex" *(4)*. In the EEG literature, a K-complex refers to a vertex sharp wave accompanied by a sleep spindle. There is no amplitude criterion for a K-complex. As soon as either a spindle or K-complex of 0.5-s duration appears, stage 2 is

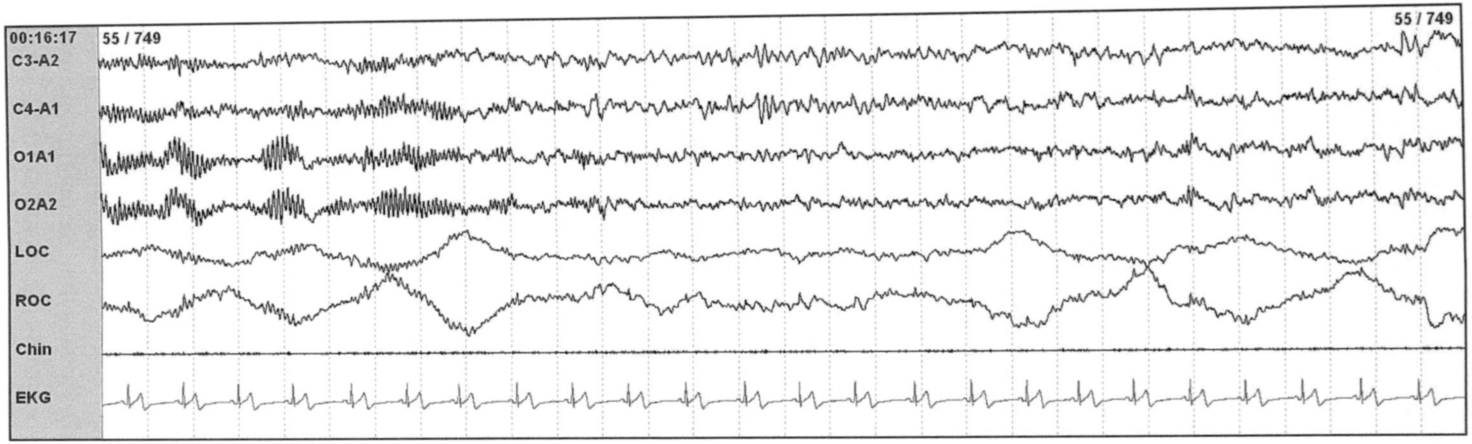

Fig. 3. Stage 1. Slow roving eye movements occur and the posterior dominant alpha rhythm comprises less than 50% of the epoch. (30-s epoch)

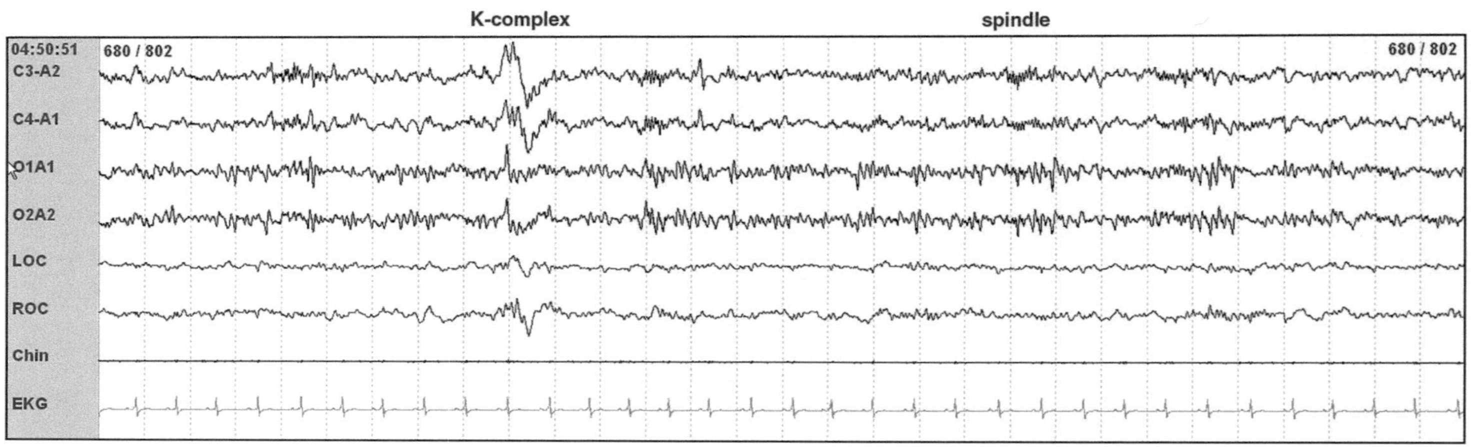

Fig. 4. Stage 2. This is a typical epoch of stage 2 demonstrating sleep spindles and a K-complex. This stage is defined by the presence of either K-complexes or spindles, each lasting at least 0.5 s in duration. (30-s epoch)

established. K-complexes and spindles appear intermittently. To score uneventful intervals between the appearance of these transients, the "3-min rule" is applied: "If less than 3 min of record which would ordinarily meet the requirements for stage 1 intervene between sleep spindles and/or K-complexes, these intervening epochs are to be scored stage 2, if there is no indication of movement arousal or pronounced increase in muscle tone during the interval in question" *(4)*. Stage 2 is the most abundant and easily identified sleep stage.

4.1.1.1.2.7. Stages 3 and 4 Stages 3 and 4 are characterized by the presence of high-voltage slow activity. The slow waves that define these stages must be 2 Hz or less and >75 μV in amplitude. If slow waves meeting these criteria are observed for at least 20% but not more than 50% of the epoch, that epoch is scored as stage 3 (Fig. 5). Stage 4 sleep (Fig. 6) is characterized by the same high-voltage slow waves for more than 50% of the epoch. Stages 3 and 4 sleep are commonly referred to as "delta sleep" or "slow-wave sleep." The terminology differences between the EEG and the sleep literature are again notable. A delta wave in sleep terminology is high-amplitude and 2 Hz or less, whereas a delta wave in EEG terminology is 3 Hz or less. It is worth pointing out, again, that slow-wave filters can distort the amplitude of the waveform. If a standard EEG filtering of 1 Hz is applied, the amplitude of waves of 1 Hz will be reduced by 20%.

4.1.1.1.2.8. Stage REM Scoring of REM sleep (Fig. 7) is more complicated than the other sleep stages. In REM, tonic and phasic physiological changes are reflected in EEG, EOG, and EMG; characteristic changes in all three leads are required to identify the stage. REM sleep is defined by a low-voltage mixed-frequency EEG, episodic REMs, and chin EMG activity that is at the lowest level recorded during the study. The drop in EMG activity is a hallmark of normal REM sleep and correlates with generalized skeletal muscle atonia (REM atonia). Sawtooth waves, a notched theta-frequency waveform evident in frontal and vertex areas, are often observed in association with eye movements. Although saw-tooth waves are unique to REM and aid identification of the stage, they are neither sufficient nor necessary for REM stage scoring.

Alpha activity in REM is usually faster than in stage 1 but slower than in wakefulness. Spindles occasionally appear in REM sleep, but any period less than 3 min between two spindles is scored as stage 2 sleep if no eye movements or movement arousals are observed in that interval (if movement arousals occur the stage reverts to stage 1, assuming other criteria for that stage are met). Difficulties in determining the precise beginning and end of the REM stage arise because:

1. EEG, EOG, and EMG do not typically change simultaneously.
2. REMs are episodic events.

Epochs are scored as REM from the point of EMG amplitude decrease, providing that REMs occur in subsequent epochs before the reappearance of spindles, K-complexes or a movement arousal. If one of these events occurs in the interval between the drop in tone and the first eye movement, REM is scored from that event forward, assuming that EMG is still at the REM stage level (i.e., the lowest level of the record). Epochs that are contiguous with stage REM epochs with low-voltage mixed-frequency EEG activity are scored as REM, if the EMG remains at the REM stage level and there is no movement arousal, regardless of whether REMs are present on the EOG in that epoch. A detailed explanation of REM rules is available in the R and K manual.

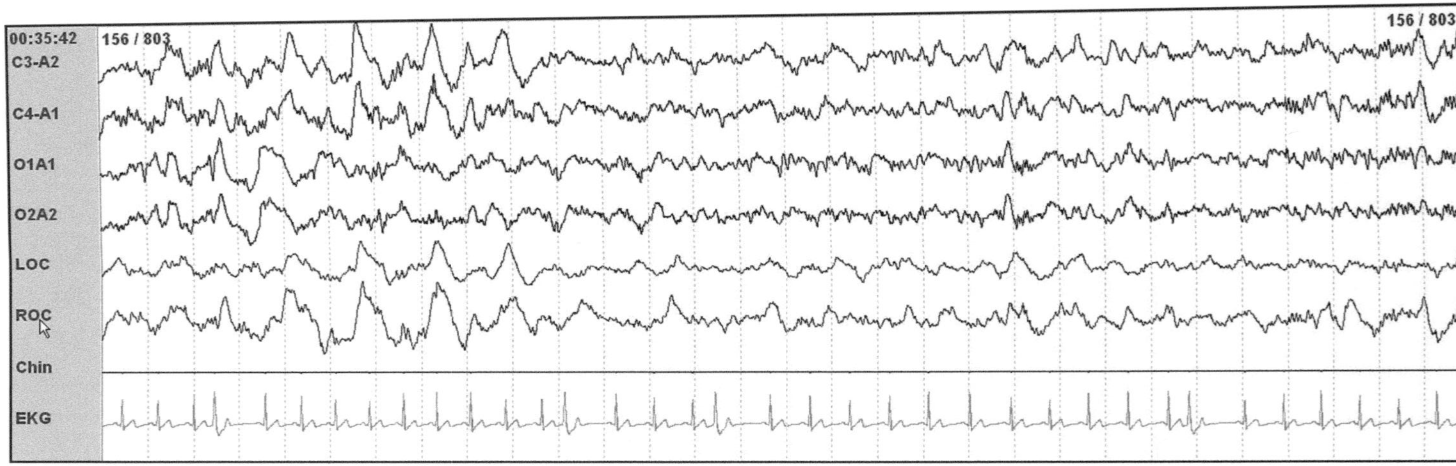

Fig. 5. Stage 3. This stage requires at least 20% and not more than 50% of the epoch to demonstrate activity ≤2 Hz with an amplitude >75 mV.

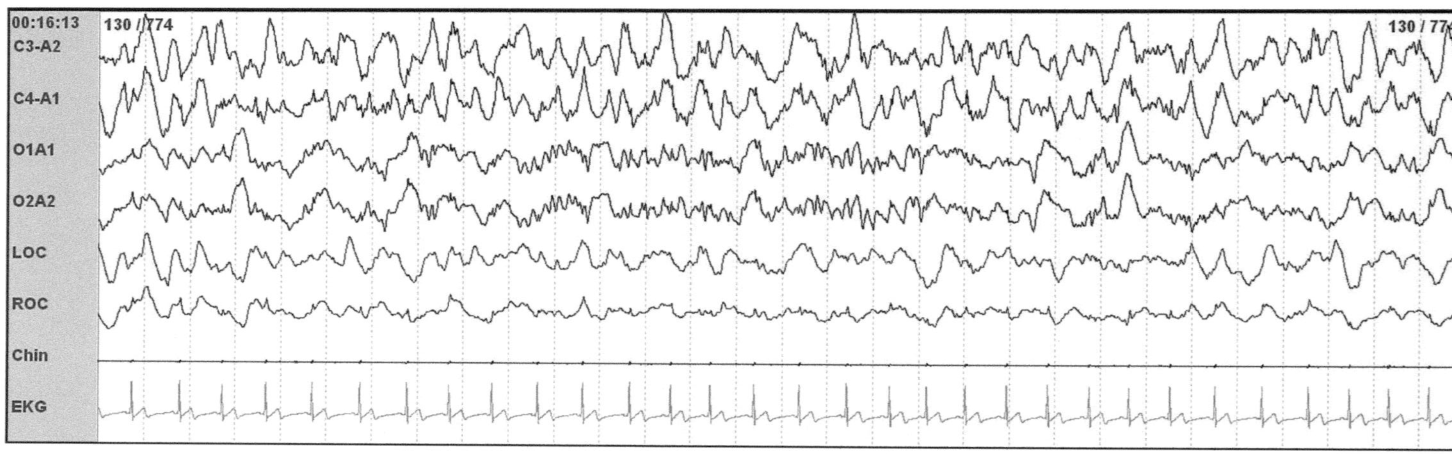

Fig. 6. Stage 4. This stage requires more than 50% of the epoch to demonstrate activity ≤2 Hz with an amplitude >75 mV.

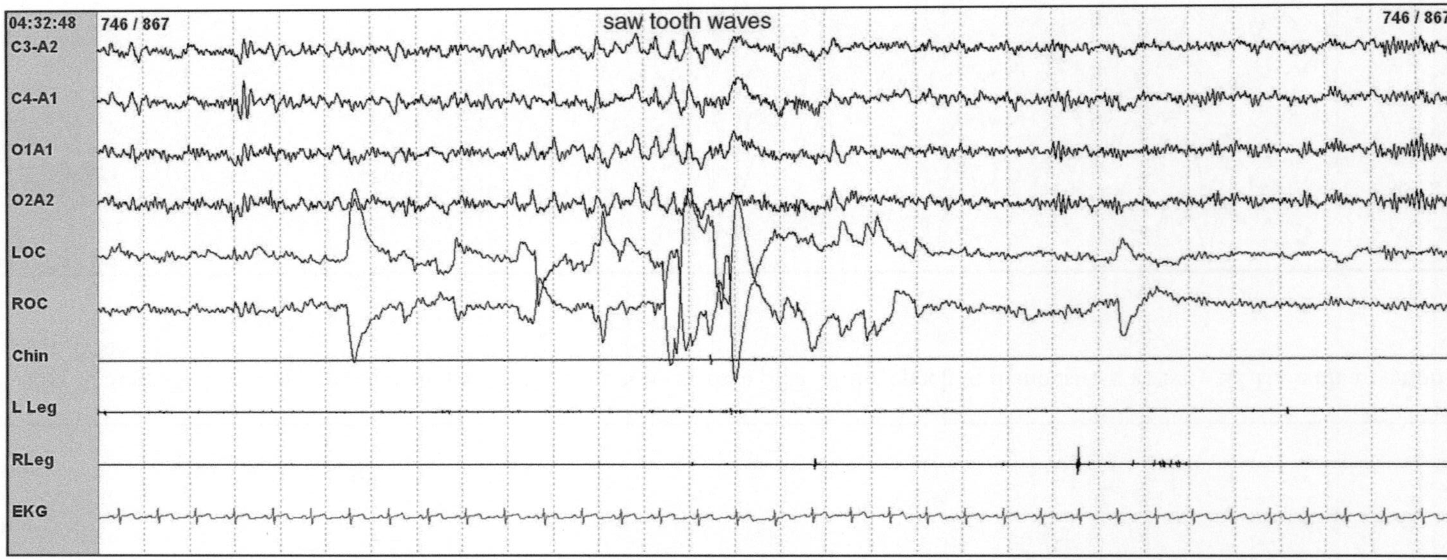

Fig. 7. Rapid eye movements. Rapid eye movements are observed with the tonic EMG level recorded at the lowest level during the study. Occasionally, sawtooth waves and phasic increases in EMG activity occur. (30-s epoch)

4.1.2. Respiratory Measures

The recording of airflow, respiratory effort, and oxygen saturation allows for the determination of abnormal breathing patterns during sleep. These abnormalities underlie the sleep-related breathing disorders. A variety of abnormal breathing events exist, including: apnea, a complete cessation of airflow; hypopnea, a reduction in airflow; and respiratory effort-related arousal (RERA), an EEG arousal induced by respiratory effort but not meeting criteria for an apnea or hypopnea. Surprisingly, there has been a lack of uniform definition of these events, based in large part on differing recording methodologies. Current definitions that have the most widespread clinical use are based on guidelines provided by the Medicare Administration and endorsed by position statements by the AASM; these will be detailed in Sections 4.1.2.2.2. and 4.1.2.2.1. The definition and interpretation of abnormal respiratory events varies widely in practice and in publications. When interpreting the summary report of a PSG, it is essential that the clinician is aware of the laboratory's working definitions.

4.1.2.1. Methodology of Recording Respiratory Measures

4.1.2.1.1. Airflow

Pneumotachographs are the gold-standard for the measure of airflow. Flow is measured across a resistor by a differential manometer. Pneumotachographs generally require a cumbersome mask that is leak-free and seals both mouth and nose. This technique is used in research laboratories but not in routine diagnostic clinical studies. Positive-pressure machines may have built in pneumotachographs, however, that assist in determining flow during the application of pressure in therapeutic studies.

Thermistors or thermocouples placed at both the nose and the mouth are the most commonly used devices for measuring airflow during PSGs. These devices identify the differences in temperature of inspired and expired air, resulting in an easily measured signal. Because change in temperature is not dependent on the volume of air moved, thermistors and thermocouples are sensitive to any airflow, but cannot measure airflow volumes. They have the advantage of measuring airflow at mouth and nose simultaneously, but have the disadvantage of overestimating airflow and underestimating abnormalities.

Nasal pressure transducers are used in many laboratories to assess airflow. These devices are more sensitive than thermistors or thermocouples to subtle changes in flow consistent with obstruction. The pressure transducer is inserted into a nasal cannula. The pressure tracing obtained is a direct function of flow. The shape of the signal reflects the airflow limitation that is observed in partial or complete obstruction of the airway and is an indirect measure of airway resistance. Flattening of the inspiratory signal and/or amplitude reduction indicates limitation of airflow (Fig. 8). The device overestimates decreased airflow when the patient breathes through the mouth. As a result, it is generally recommended that nasal pressure transducers are used with a thermistor or thermocouple.

Respiratory inductance plethysmography (RIP) is a technique used to measure changes in volume of the chest and abdomen. These volume changes can provide a measure of tidal volume. Bands around the chest and abdomen incorporate coils that expand and contract during breathing. Changes in length of the bands induce changes in the oscillating frequency of the circuit. An output signal can be calibrated for volumes in the abdominal and chest compartments. To accurately measure airflow, the system must be carefully calibrated in supine and upright positions. Calibration in obese patients is difficult. Most often, the system is not calibrated and is used as a qualitative measure of chest and abdominal effort.

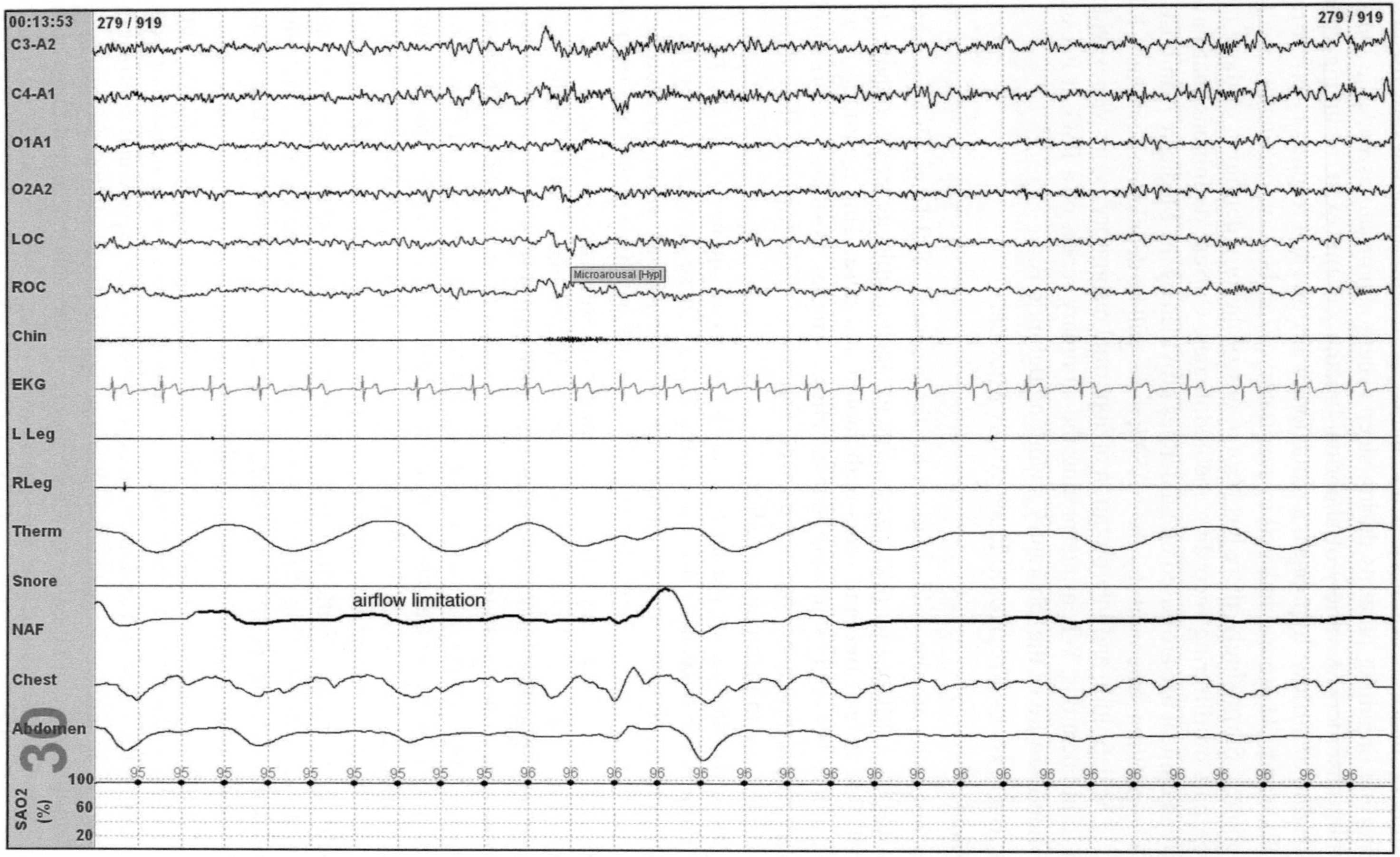

Fig. 8. RERA. Airflow limitation on the nasal air pressure transducer (NAF) is associated with an EEG arousal and steady oximetry readings. This epoch also demonstrates an arousal. (30-s epoch)

4.1.2.1.2. Respiratory Effort

Intra-esophageal pressure measurement, which reflects pressure changes in the intrathoracic compartment, is the gold-standard for measurement of chest effort. Increased airway resistance results in increased negative pressures. A manometer is inserted in the esophagus through a pediatric feeding tube that remains in place for the duration of the study. The discomfort and inconvenience of inserting the manometer make this a technique that is used in only a few centers.

RIP, when uncalibrated, provides a qualitative measurement of movement of the chest and abdomen with respiration.

Stain gauges and piezoelectric bands both also provide qualitative measure of the movement of chest and abdomen. One band is placed around the chest and the other around the abdomen. Piezoelectric bands are the most commonly used instruments for these measurements because of ease of use and low cost, but reliability is less than with RIP.

Intercostal EMG is a simple technique in which standard EEG electrodes are applied to an intercostal space close to the diaphragm. Respiratory effort is evident with a surface EMG recording, similar to the chin EMG. These recordings are a useful second measure of respiratory effort, but are prone to failure in obese patients.

4.1.2.1.3. Oxygen Saturation

Pulse oximetry is a noninvasive method of monitoring the percentage of hemoglobin that is saturated with oxygen. The pulse oximeter consists of a probe attached to the patient's finger or ear lobe. A source of light originates from the probe at two wavelengths. By calculating the absorption at the two wavelengths, the processor can estimate the proportion of hemoglobin that is oxygenated.

4.1.2.2. Types of Respiratory Events

4.1.2.2.1. Apnea

An apnea is defined as the absence of airflow for at least 10 s. There are three types:

1. Obstructive apnea (Fig. 9): absence of airflow for at least 10 s with evidence of persistent respiratory effort.
2. Central apnea (Fig. 10): absence of airflow for 10 s without evidence of any of respiratory effort.
3. Mixed apnea (Fig. 11): absence of airflow for 10 s with initial absence of effort followed by a return of respiratory effort before resumption of airflow.

4.1.2.2.2. Hypopnea

The term hypopnea refers to a decrease in airflow (Fig. 12). There have been many definitions of this event proposed. The current clinical definition that is recognized by Medicare and endorsed by the AASM is as follows: "Hypopnea in adult patients is defined as an abnormal respiratory event lasting at least 10 s with at least a 30% reduction in thoracoabdominal movement or airflow as compared to baseline, and with at least a 4% oxygen desaturation." Hypopneas can be associated with airway obstruction or a centrally mediated reduction in respiratory effort. In clinical practice, it is difficult to distinguish obstructive hypopneas from central hypopneas unless an esophageal pressure monitor is used. Because most effort and airflow monitors are not quantitative, there is debate regarding whether the criterion of 30% reduction in these measures is practical.

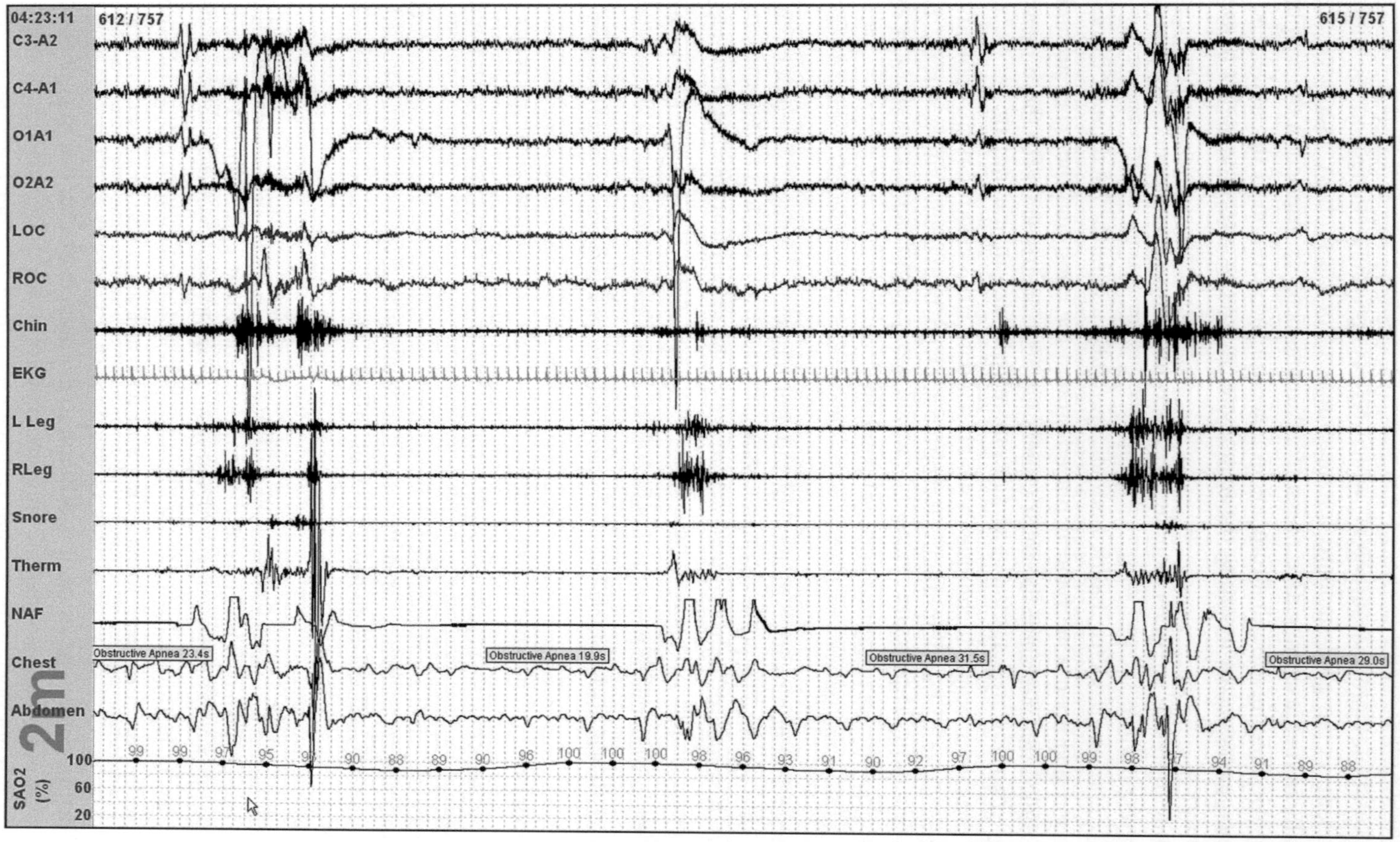

Fig. 9. Obstructive sleep apnea. Each apnea demonstrates persistent thoracoabdominal effort and is associated with airflow obstruction lasting at least 10 s. The epoch also reveals cyclic drops in oxygen saturation and respiratory-induced limb movements, which are not required for scoring apneas. (2-min window)

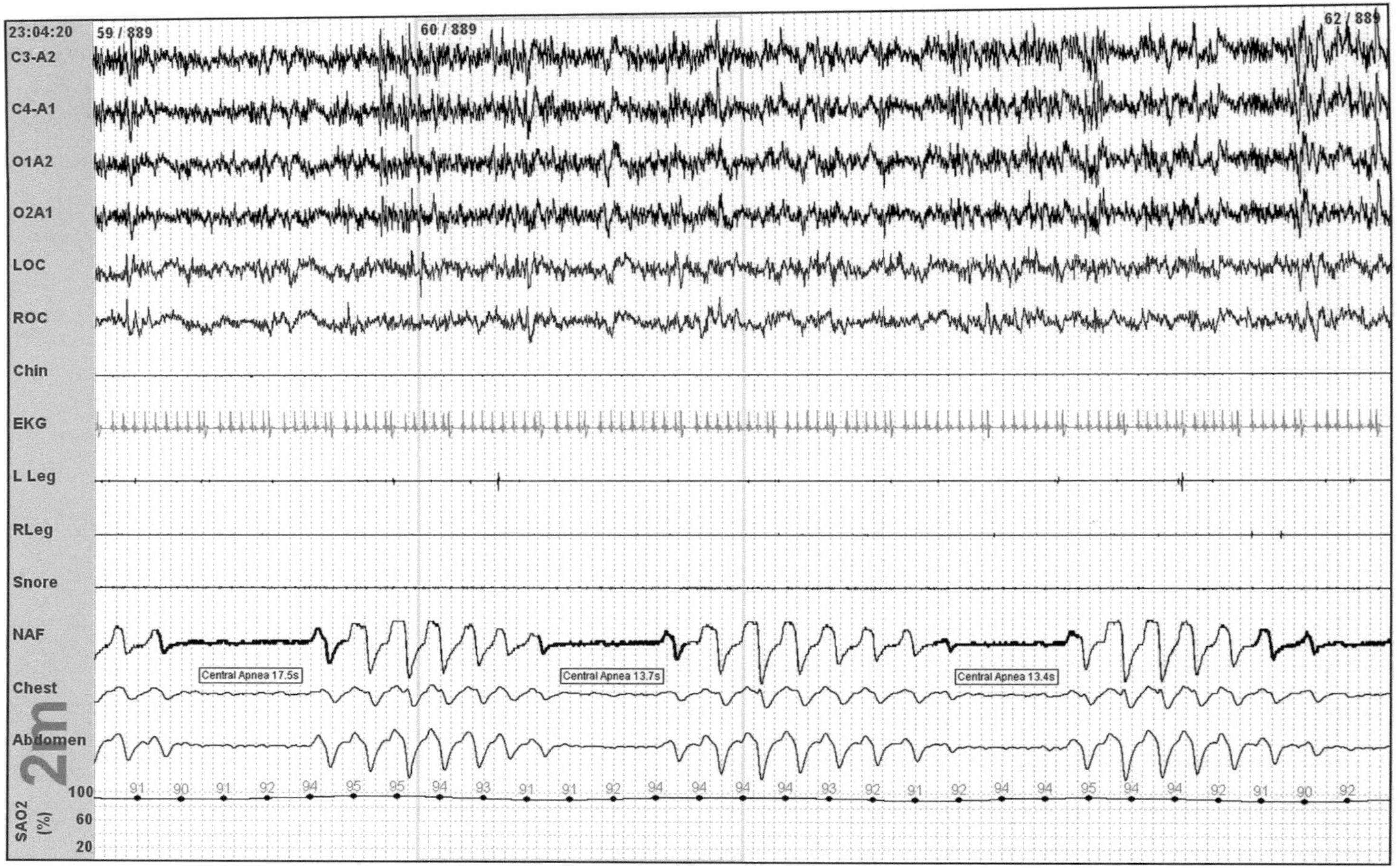

Fig. 10. Central sleep apnea. Each apnea is characterized by an absence of airflow and thoracoabdominal effort for at least 10 s. Desaturations are seen with the events but are not required. (2-min window)

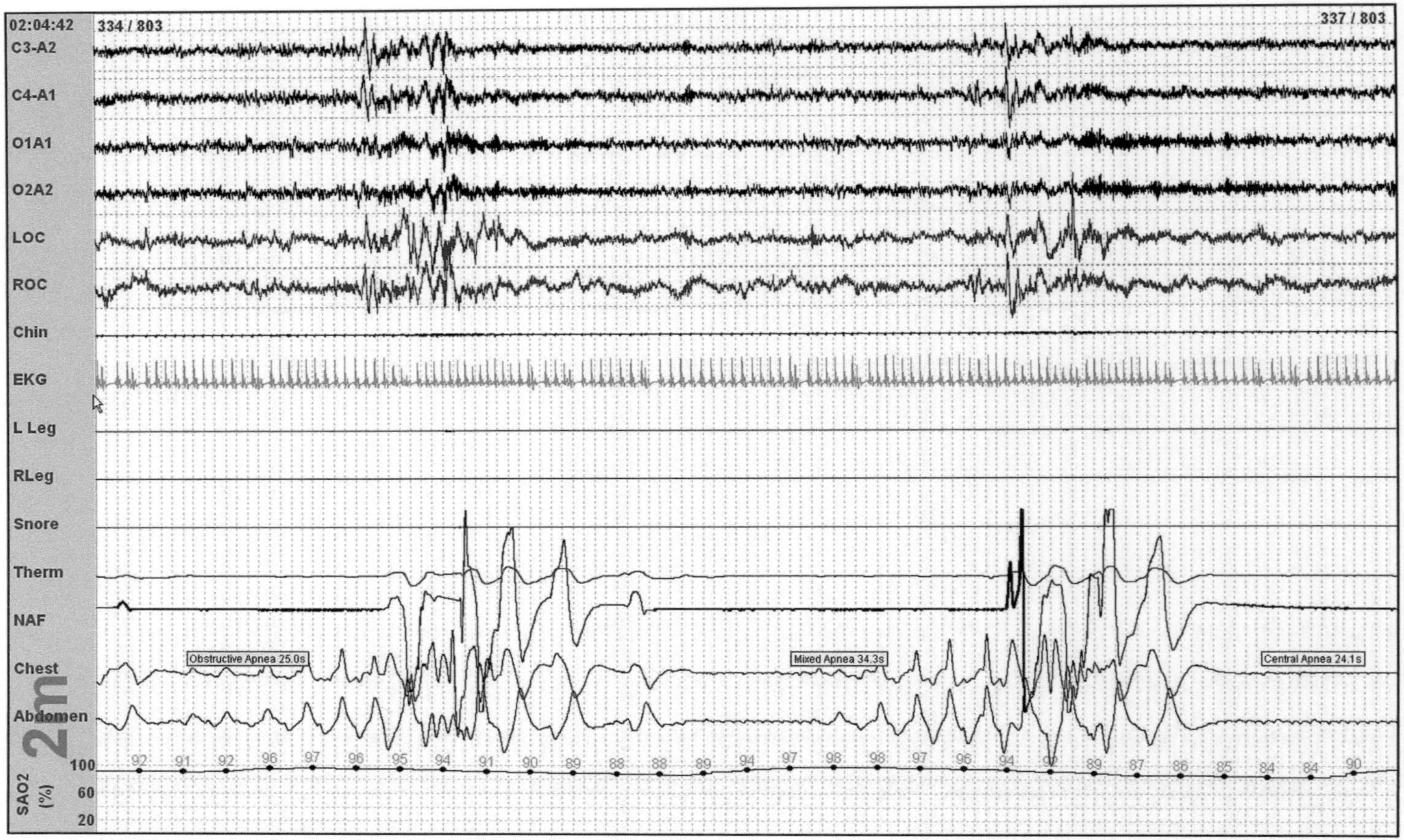

Fig. 11. Mixed apnea. This recording exhibits three apneas. The second is termed a mixed apnea because the event starts as a central event and evolves into an obstructive apnea. Mixed apneas often occur in association with obstructive and central apneas. In this case, an obstructive apnea precedes the event and a central apnea follows it. (2-min window)

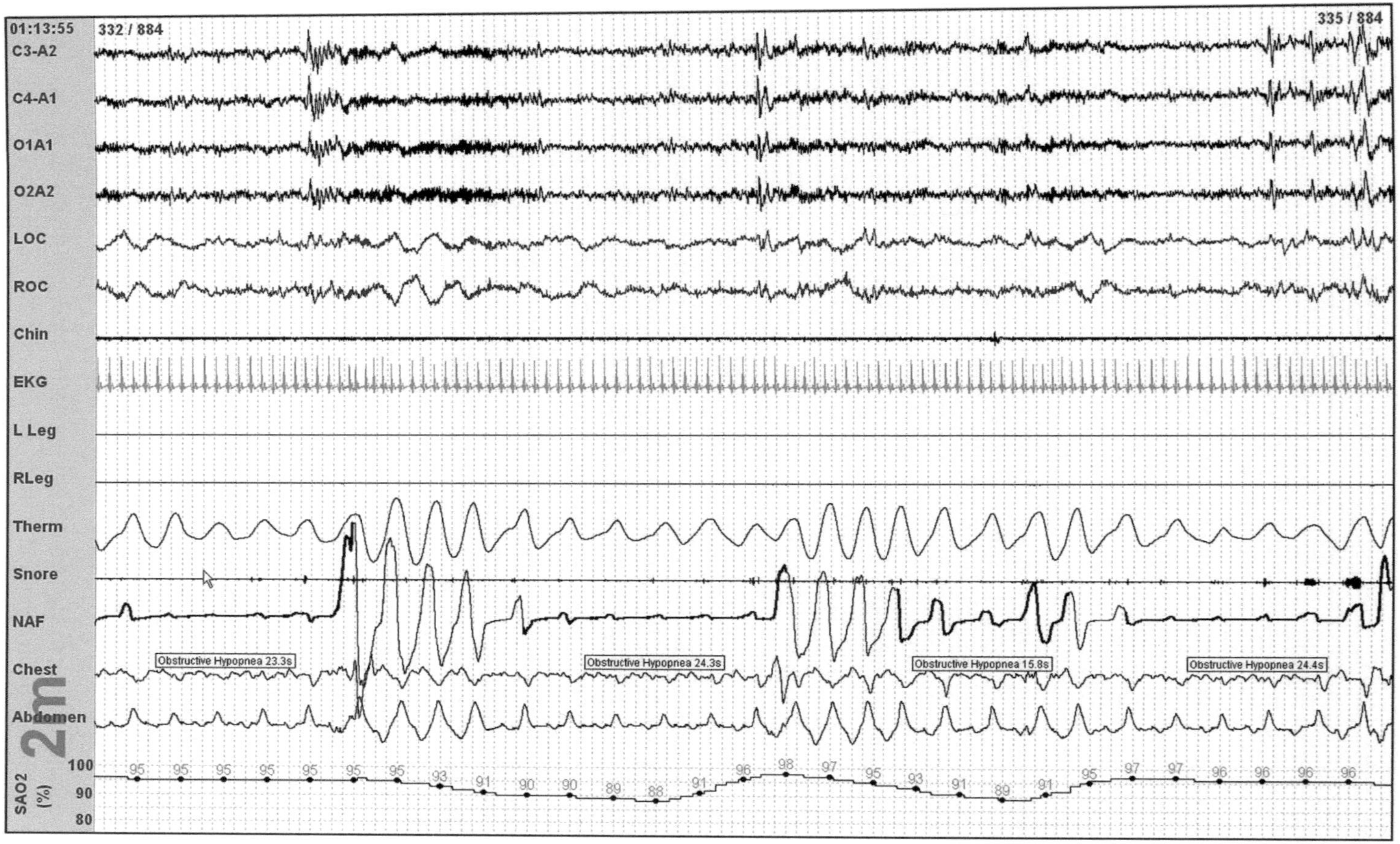

Fig. 12. Hypopnea. This is an example of recurrent "Medicare hypopneas." Medicare requires at least a 30% reduction in airflow or thoracoabdominal movement, lasting at least 10 s, with a drop in oxygen saturation of at least 4%. (2-min window)

4.1.2.2.3. Respiratory Effort-Related Arousal

When airway resistance increases, oxygen saturation and tidal volume may stay the same as respiratory effort increases to overcome the obstruction. The result of increased respiratory effort may be an arousal that disturbs sleep. These abnormalities are termed RERAs (Fig. 8) *(7)*. Evidence of increased respiratory effort in association with arousal is best evidenced by an esophageal manometer, but that technique is rarely available. Pressure transducers can show evidence of flow limitation before arousal. Changes in breathing patterns before arousal may also be correlated with increased airway resistance. The event should last 10 s. RERAs are not scored in all laboratories; some laboratories refer to these events as hypopneas or "hypopneas not meeting Medicare criteria."

4.1.2.3. Quantification of Respiratory Abnormality

1. Apnea hypopnea index (AHI): the apnea hypopnea index is calculated by adding the total number of apneas and hypopneas and dividing by the number of hours of sleep. When used with the definitions above, the index is useful as a standardized measure that reflects severity of sleep-disordered breathing.
2. Respiratory disturbance index (RDI): all respiratory events scored are added together and divided by the number of hours of sleep to generate the RDI. The RDI is not a standard measure because different laboratories measure respiratory events differently. If a laboratory measures RERAs, for example, they will be reflected in the RDI.
3. Oxygen saturation: laboratories typically report several measure of oxygen saturation, including lowest saturation, baseline saturation, and percent time at various levels of saturation.

4.1.2.3.1. EKG

A single channel EKG is recorded. There is no standard lead placement. Some laboratories use a modified Lead II (right shoulder, left leg), most use a precordial lead.

Recording EKG allows for continuous review of cardiac rhythm during the night. Cardiac abnormalities often correlate with respiratory abnormalities and sleep stages. Respiratory obstruction may result in periods of vagally mediated cardiac slowing followed by sympathetically induced increased rates with arousal. Hypoxia and increased sympathetic drive during arousal can exacerbate arrhythmia. REM sleep may be associated with prominent autonomic changes reflected in the rhythm strip.

4.1.3. Limb EMG

Surface EMG recording of the limbs allows for analysis of movement disorders and movement arousals during sleep. Usually only the anterior tibialis muscles of both legs are studied, but recordings may also be obtained from the arms (extensor digitorum muscles). The anterior tibialis muscles are studied because they have a good correlation with the movements observed in periodic limb movement disorder (PLMD; Subheading 6.4.1.), a disorder that exists with and without the complaint of restless legs (Subheading 6.4.2.). Leg movements may also accompany respiratory events or other sources of arousal. Periodic leg movements, by definition, occur in periodic sequences, whereas other leg movements may not be periodic.

4.1.3.1. Techniques for Recording Limb Movements

Two surface electrodes are placed on the bellies of the anterior tibialis muscles of both legs and each recorded in a bipolar fashion. A high-frequency filter of at least 128 Hz is recommended. The surface EMG obtained is calibrated while the patient is awake with a 30° dorsiflexion and plantar flexion of the great toe without resistance. Activity during sleep is compared with this biological calibration.

4.1.3.2. Techniques for Scoring and Reporting Limb Movements

The AASM has established the following scoring guidelines *(10)*:

1. Leg movement: a burst of anterior tibialis activity with a duration of 0.5 to 5 s and with an amplitude of at least 25% of the calibration movements.
2. Periodic leg movement sequence is defined as four or more leg movements separated by between 5 s and 90 s. These should be scored during both sleep and wake. Periodic leg movements during wakefulness are suggestive of restless legs syndrome (RLS) and correlate with periodic limb movements during sleep (PLMS) (Fig. 13).
3. Leg movement with arousal: to assess the impact of the events on sleep, leg movements that cause arousal are counted. The arousal must not follow leg movement onset by more than 3 s.

Leg movements associated with respiratory events are counted and classified as respiratory related.

Leg movements are reported in terms of both an absolute number as well as indices of events per hour of sleep with and without arousal. Leg movements within sequences (and therefore periodic) should be distinguished from leg movements not within sequences.

4.1.4. Body Position

Body position sensors allow correlation of respiratory abnormalities with position. Severity of sleep-disordered breathing may vary significantly with position.

4.1.5. Video EEG–PSG

Because only a limited EEG montage is used for the analysis of sleep stage in a standard PSG, extended EEG montages are necessary to attempt to identify epileptic activity during sleep. These studies are usually obtained in patients who complain of unusual nocturnal events that could represent seizure activity. Video monitoring helps characterize the behavioral components of the disorder in relationship to the neurophysiological findings. Respiratory monitoring may identify respiratory precipitants to seizure activity that would have been overlooked in routine or extended EEG analysis. The number of channels recorded is dependent on the capability of the recording equipment; a full EEG is optimal. A common eight-channel EEG recording added to the routine PSG includes includes: F7–T3, T3–T5, T5–O1, F8–T4, T4–T6, T6–O2, F3–C3, and F4–C4.

4.1.6. Interpretation of the PSG

Polysomnographic interpretation requires both an independent analysis of the component measurements and an overall synthesis of how these variables interact with each other.

Most reports include a quantitative report of sleep architecture, respiratory measures, EKG, limb movements, and a limited description of EEG. Diagnostic requirements for specific disorders using these measures are described in Tables 7 to 11.

Qualitative interpretation of the PSG is also important, and relies on the polysomnographer's ability to identify patterns of interaction between variables that vary from the norm and may be disease specific. A summary of statistics does not replace an epoch-by-epoch review of the record. Interpretation is dependent on the clinical context, making it important that the clinician approaches the process of reading a sleep study with an understanding of the patient's medical background and the specific questions posed. For example, an arousal from slow-wave sleep may be an insignificant polysomnographic finding, and subtle periods of decreased airflow are also common. In the context of a patient who presents with episodes of sudden fear and panic during sleep, however, the presence of a sudden arousal from slow-wave

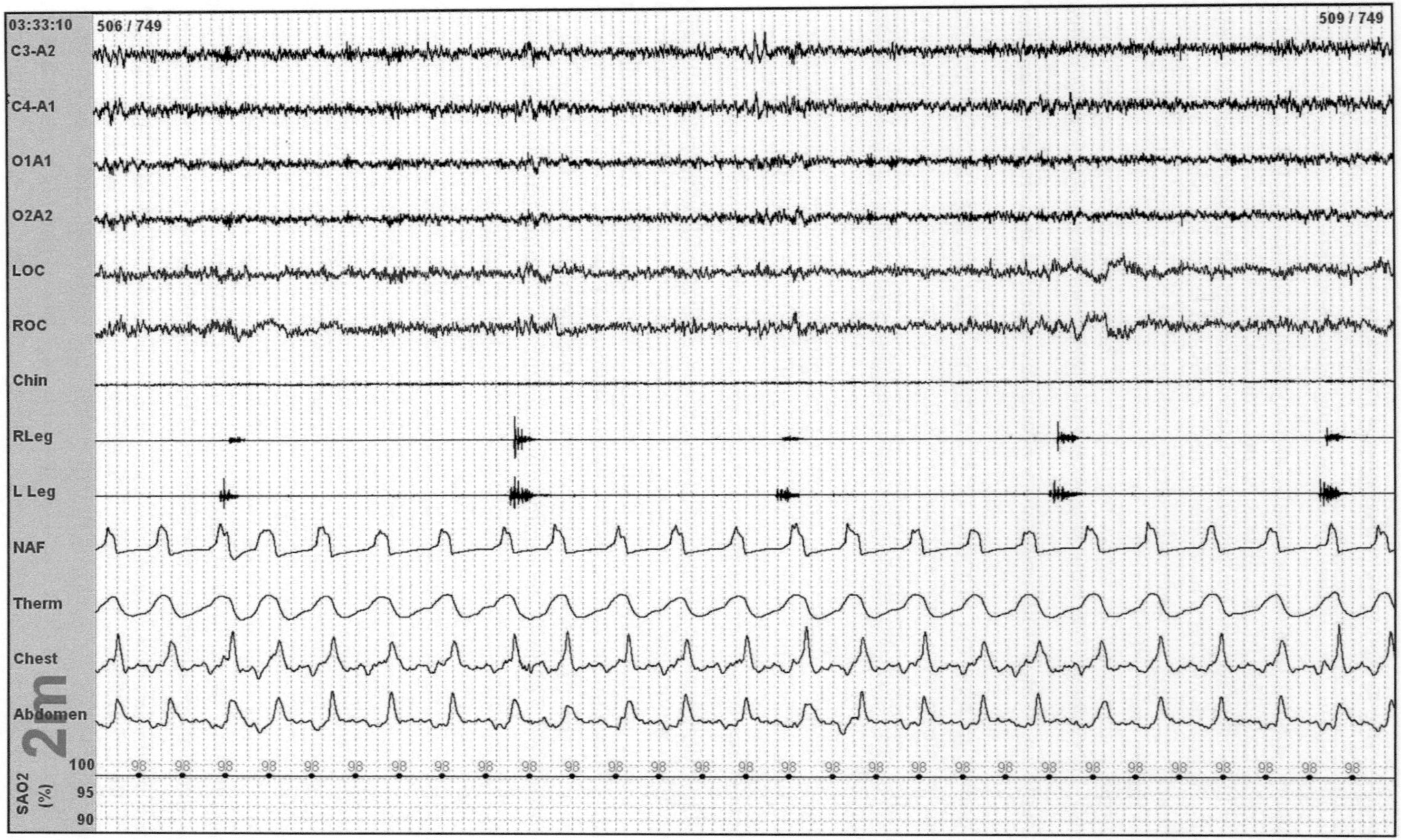

Fig. 13. Periodic limb movements during sleep. The limb movements in this case occur bilaterally and are independent of respiratory instability. The morphology is easily identified by its periodicity. (2-min window)

Table 7
Central Sleep Apnea

Disorder	*Clinical criteria*	*Polysomnographic criteria*	*Other requirements*	*Other features*
Primary central sleep apnea	At least one of the following: • Excessive daytime sleepiness • Frequent arousals and awakenings during sleep or insomnia complaints • Awakening short of breath	Five or more central apneas per hour of sleep	The disorder is not better explained by another current sleep disorder, another medical or neurological disorder, medication use, or substance use disorder.	$PaCO_2$ low normal <40 during wakefulness O_2 desaturation is usually mild Sleep is fragmented by respiratory events
Cheynes Stokes breathing pattern	No symptoms are required Patients often report symptoms of sleep disruption, insomnia, excessive daytime sleepiness, or awakening short of breath.	At least 10 central apneas and hypopneas per hour of sleep in which the hypopnea has a crescendo–decrescendo pattern of tidal volume accompanied by frequent arousals from sleep and derangement of sleep architecture	The disorder occurs in association with a serious medical illness, such as heart failure, stroke, or renal failure. The disorder is not better explained by another current sleep disorder, another medical or neurological disorder, medication use, or substance use disorder.	Oxygen desaturations usually not less than 80%. Distinguished from hypoventilation syndromes by $PaCO_2$ <45 Sleep is fragmented. Arousals are frequently associated with crescendo breathing. Attenuated by REM sleep

(Continued)

Table 7 (Continued)

Disorder	*Clinical criteria*	*Polysomnographic criteria*	*Other requirements*	*Other features*
High-altitude periodic breathing	Recent ascent to altitude of at least 4000 m	Recurrent central apneas, >5/hr, with cycle length 12–34 s, primarily in NREM	None	Sleep is fragmented. Attenuated by REM
Central sleep apnea due to drug or substance	The patient has been taking a long-acting opioid for at least 2 mo.	Five or more central apneas or periodic breathing (10 or more central apneas or hypopneas per hour of sleep with crescendo–descrescendo pattern and arousals)	The disorder is not better explained by another current sleep disorder or medical or neurological disorder.	May complicate other sleep-related breathing disorders
Primary sleep apnea of infancy (apnea of prematurity, <37 wk; apnea of infancy >37 wk)	No requirements	Prolonged central respiratory pauses of 20 s or more in duration, or shorter episodes with obstructive or mixed features with clinical compromise (e.g., bradycardia, hypoxemia, cyanosis, or hypotonia)	The disorder is not better explained by another current sleep disorder, medical or neurological disorder, or medication.	Exacerbated by REM sleep Findings may not occur nightly; polysomnographic study may be normal. Additional EEG and esophageal pH employed frequently.

Abstracted from ref. *5*, with permission.

Table 8
Obstructive Sleep Apnea

Disorder	*Clinical criteria*	*Polysomnographic criteria*	*Other requirements*	*Other features*
Obstructive sleep apnea	At least one of the following: • The patient complains of unintentional sleep episodes during wakefulness, daytime sleepiness, unrefreshing sleep, fatigue, or insomnia • The patient wakes with breath holding, gasping, or choking • The bed partner reports loud snoring, breathing interruptions, or both during the patient's sleep	With clinical criteria noted: Five or more apneas, hypopneas or RERAs per hour of sleep with evidence of respiratory effort during all or a portion of each respiratory event Without clinical criteria noted: Fifteen or more apneas, hypopneas, or RERAs per hour of sleep with evidence of respiratory effort during all or a portion of each respiratory event	The disorder is not better explained by another current sleep disorder, another medical or neurological medication use, or substance use disorder.	In this definition of sleep apnea there is no requirement that respiratory events are associated with oxygen desaturation. The Medicare-approved criteria for obstructive sleep apnea justifying treatment with positive pressure are different than the diagnostic criteria established by the International Classification of Sleep Disorders-2. Medicare does not recognize RERAs. Hypopneas by Medicare criteria require ≥4% desaturation and ≥30% decrease in respiratory effort or airflow. Using these criteria, positive pressure will be covered under Medicare in adult patients if (1) the AHI is ≥15 or (2) the AHI is ≥5 and ≤14 with documented symptoms of excessive daytime sleepiness, impaired cognition, mood disorders, or insomnia; or documented hypertension, ischemic heart disease or history of stroke.

Abstracted from ref. *5* with permission.

Table 9
Sleep-Related Hypoventilation/Hypoxemic Syndromes

Disorder	*Clinical criteria*	*Polysomnographic criteria*	*Other requirements*	*Other features*
Sleep-related nonobstructive alveolar hypoventilation syndrome	Symptoms not required Patients often report symptoms of sleep disruption, insomnia, or excessive daytime sleepiness	Episodes of shallow breathing longer than 10 s in duration associated with arterial oxygen desaturation and either frequent arousals from sleep or brady-tachycardia	No primary lung diseases, skeletal malformations, or peripheral neuromuscular disorders that affect ventilation are present The disorder is not better explained by another current sleep disorder, another medical or neurological disorder, medication use, or substance use disorder.	Carbon dioxide levels show an increase during episodes of hypoventilation. Daytime blood gases may be normal or abnormal.
Congenital central alveolar hypoventilation syndrome	Patient exhibits shallow breathing or cyanosis and apnea of perinatal onset during sleep	Polysomnographic monitoring during sleep demonstrates severe hypercapnia and hypoxia, predominantly without apnea	Hypoventilation is worse during sleep than during wakefulness. Rebreathing ventilatory response to hypoxia or hypercapnia is absent or diminished. The disorder is not better explained by another current sleep disorder, another medical or neurological medication use, or substance use disorder.	Rare disorder, recently associated with mutations of *PHOX2B* gene
Sleep-related hypoventilation/ hypoxemia due to pulmonary parenchymal or vascular pathology	Lung parenchymal disease or pulmonary vascular disease is present and believed to be the primary cause of hypoxemia	Polysomnography or sleeping arterial blood gas shows at least one of the following: • Sp O_2 during sleep <90% for more than 5 min with a nadir of at least 85% • More than 30% of total sleep time at an SpO_2 <90%	The disorder is not better explained by another current sleep disorder, another medical or neurological disorder, medication use, or substance use disorder.	$PaCO_2$ is not routinely measured during PSG. End tidal CO_2 is measured in some laboratories.

		• Sleeping $PaCO_2$ that is abnormally high or disproportionately increased relative to levels during wakefulness		
Sleep-related hypoventilation/ hypoxemia due to lower airways obstruction	Lower airways obstructive disease is present and is believed to be the primary cause of the hypoxemia.	Polysomnography or sleeping arterial blood gas shows at least one of the following: • SpO_2 during sleep of <90% for >5 min with a nadir of at least 85% • More than 30% of total sleep time at an SpO_2 <90% • Sleeping $PaCO_2$ that is abnormally high or disproportionately increased relative to levels during wakefulness.	The disorder is not better explained by another current sleep disorder, another medical or neurological disorder, medication use, or substance use disorder.	$PaCO_2$ is not routinely measured during PSG. End tidal CO_2 is sometimes measured in some laboratories.
Sleep-related hypoventilation/ hypoxemia due to neuromuscular and chest wall disorders	A neuromuscular or chest wall disorder is present and believed to be the primary cause of hypoxemia.	Polysomnography or sleeping arterial blood gas shows at least one of the following: • SpO_2 during sleep of <90% for more than 5 min with a nadir of at least 85% • More than 30% of total sleep time at an SpO_2 <90% • Sleeping $PaCO_2$ that is abnormally high or disproportionately increased relative to levels during wakefulness.	The disorder is not better explained by another current sleep disorder, another medical or neurological disorder, medication use, or substance use disorder.	Exacerbated by REM sleep because of muscle atonia Present in a wide variety of disorders that impair chest wall function including obesity, kyphoscoliosis, and any disorder causing muscular weakness, such as the muscular dystrophies, ALS, and myasthenia Exacerbated by disorders that also reduce chemosensitivity, such as myotonic dystrophy

SpO_2, saturation by pulse oximetry; PSG, polysomnogram; REM, rapid eye movement sleep; ALS, amyotrophic lateral sclerosis.
Abstracted from ref. *5*, with permission.

Table 10
Narcolepsy and Idiopathic Hypersomnia

Disorder	*Clinical criteria*	*Polysomnographic criteria*	*Other requirements*	*Other features*
Narcolepsy with cataplexy	Complaint of excessive daytime sleepiness occurring almost daily for at least 3 mo A definite history of cataplexy, defined as sudden and transient episodes of loss of muscle tone triggered by strong emotions, most reliably, laughing	If cataplexy is clinically unequivocal, or CSF hypocretin fulfills criteria noted, polysomnographic study with PSG and MSLT are recommended but not required as confirmatory evidence. Full night PSG with minimum of 6 h sleep on night preceding MSLT Confirmatory MSLT must demonstrate: mean sleep latency ≤8 min with minimum of 2 SOREMPs.	CSF hypocretin-1 levels less than 110 pg/mL or 1/3 normal mean control can replace PSG and MSLT. At the time of PSG/MSLT, patients must be free of medications that influence sleep ×5 half-lives of longest acting metabolite. Prior to PSG/MSLT sleep–wake cycle should be standardized for at least 7 d, confirmed by sleep logs or actigraphy. The disorder is not better explained by another current sleep disorder, another medical or neurological disorder, medication use, or substance use disorder.	CSF hypocretin may be considered as alternative to PSG/MSLT if clinical situation warrants. PSG often shows other abnormalities including some sleep-disordered breathing, leg movements, fragmented sleep and REM behavior disorder Clinical correlation is required to determine whether these abnormalities are primary or secondary. Many patients cannot undergo a valid MSLT because sleep-influencing medications, especially antidepressants, cannot be discontinued. MSLT in this situation may be misleading.

Narcolepsy without cataplexy	Complaint of excessive daytime sleepiness occurring almost daily for at least three months Typical cataplexy is not present, atypical cataplexy-like episodes may be reported	Full-night PSG with minimum of 6 h sleep on night preceding MSLT required MSLT must demonstrate: mean sleep latency ≤8 min with minimum of 2 SOREMPs.	CSF hypocretin-1 levels usually not abnormal At the time of PSG/MSLT, patients must be free of medications that influence sleep ×5 half-lives of longest acting metabolite. Prior to PSG/MSLT, sleep wake cycle should be standardized for at least 7 d, confirmed by sleep logs or actigraphy. The disorder is not better explained by another current sleep disorder, another medical or neurological disorder, mental disorder, medication use, or substance use disorder.	CSF hypocretin unlikely to be helpful as alternative to PSG/ MSLT PSG often shows other abnormalities, including some sleep-disordered breathing, leg movements, fragmented sleep, and REM behavior disorder. Clinical correlation is required to determine whether these abnormalities are primary or secondary. Many patients cannot undergo a valid MSLT because sleep-influencing medications, especially antidepressants, cannot be discontinued. MSLT in this situation may be misleading.
Narcolepsy due to medical condition	Complaint of excessive daytime sleepiness occurring almost daily for at least three months A definite history of cataplexy, defined as sudden and transient episodes of loss of muscle tone triggered by strong emotions, most reliably, laughing, OR	If cataplexy is unequivocal or CSF orexin fulfills criteria noted, polysomnographic study with PSG and MSLT are recommended but not required as confirmatory evidence. Full-night PSG with minimum of 6 h sleep on night preceding MSLT Confirmatory MSLT must demonstrate: mean sleep latency ≤8 min with minimum of 2 SOREMPs.	A significant underlying medical or neurological disorder accounts for the symptoms. The disorder is not better explained by another current sleep disorder, another medical or neurological disorder, medication use, or substance use disorder.	CSF hypocretin may be considered as alternative to PSG/MSLT if clinical situation warrants.

(Continued)

Table 10 (Continued)

Disorder	*Clinical criteria*	*Polysomnographic criteria*	*Other requirements*	*Other features*
	hypocretin-1 levels in CSF ≤110 pg/mL or 1/3 of normal mean control values			
Idiopathic hypersomnia with long sleep time	The patient has a complaint of excessive daytime sleepiness occurring almost daily for at least 3 mo. Prolonged nocturnal sleep, >10 h documented by interviews, actigraphy or, sleep logs Waking patient is difficult in the morning or at the end of naps.	PSG demonstrates no other sleep disorder. PSG sleep latency is short and sleep duration is ≥10 h. If MSLT performed, sleep latency is <8 min with fewer than 2 SOREMPs.	The disorder is not better explained by another current sleep disorder, another medical or neurological disorder, mental disorder, medication use, or substance use disorder.	Waking patient is typically difficult. MSLT not always performed because of difficulty both waking patient at the end of naps and keeping patient awake between naps PSG analysis of subtle respiratory events often identifies patients with sleep-disordered breathing as alternative diagnosis.
Idiopathic hypersomnia without long sleep time	The patient has a complaint of excessive daytime sleepiness occurring almost daily for at least 3 mo. The patient has normal nocturnal sleep (>6 h but <10 h), documented by interviews, actigraphy, or sleep logs.	Nocturnal PSG demonstrates a major sleep period that is normal in duration (>6 h but <10 h). PSG has excluded other causes of daytime sleepiness. MSLT after overnight PSG demonstrates a mean sleep latency of <8 min and fewer than 2 SOREMPs.	The disorder is not better explained by another current sleep disorder, another medical or neurological disorder, mental disorder, medication use, or substance use disorder.	PSG analysis of subtle respiratory events often identifies patients with sleep-disordered breathing as an alternative diagnosis.

CSF, cerebrospinal fluid, MSLT, multiple sleep latency test; PSG, polysomnogram; SREMPs, sleep-onset REM periods; REM, rapid eye movement sleep.
Abstracted from ref. *5*, with permission.

Table 11
REM Behavior Disorder and Disorders of Arousal From NREM

Disorder	*Clinical criteria*	*Polysomnographic criteria*	*Other requirements*	*Other features*
REM behavior disorder (RBD)	At least one of the following: • Sleep-related injurious, potentially injurious, or disruptive behaviors by history • Abnormal REM behaviors documented during polysomnographic monitoring	Presence of REM sleep without atonia: the EMG finding of excessive amounts of sustained intermittent elevation of submental EMG tone or excessive phasic submental or limb EMG twitching	Absence of EEG epileptiform activity during REM sleep unless RBD can be distinguished from concurrent REM sleep related seizure disorder. The disorder is not better explained by another current sleep disorder, another medical or neurological disorder, mental disorder, medication use, or substance use disorder.	Video recording and additional EMG limb leads (arms) are helpful. Often associated with and may precede parkinsonian syndromes Patients often report that behaviors were consistent with complex dream mentation.
Disorders of arousal from NREM	Characterized by confused behaviors emerging from NREM (typically slow-wave sleep) sleep including ambulation, routine, or inappropriate behaviors, or manifestations of terror, typically associated with partial or complete amnesia for the episode	None required Sudden spontaneous arousals from slow wave sleep may be noted with or without behavioral correlate. Polysomnogram useful to identify contributing causes of arousal	The disorder is not better explained by another current sleep disorder, another medical or neurological disorder, mental disorder, medication use, or substance use disorder.	Typically emerge in the first third of the night when slow-wave sleep is prominent Familial tendency, may be precipitated by another cause of arousal such as sleep-disordered breathing or leg movements Differential diagnosis includes seizure disorder and REM behavior disorder.

REM, rapid eye movement sleep; NREM, non-REM sleep.
Abstracted from ref. *5*, with permission.

sleep with evidence of preceding subtle airflow limitation, would suggest that sleep-disordered breathing was contributing to the manifestation of a disorder of arousal from NREM sleep. Ultimately, the interpretation of these qualitative features is dependent on the experience of the interpreting clinician.

4.2. The MSLT

The MSLT is a limited-montage polysomnographic study that evaluates subjects in a series of opportunities to nap during the day. It is a validated objective measure of the ability or tendency to fall asleep, and allows an opportunity to assess the presence of abnormal sleep-onset REM periods. The AASM published new practice guidelines for the clinical use of the MSLT in 2005 *(11)*.

4.2.1. Indications for the MSLT

As described in the current practice parameters, the MSLT is indicated for:

"1. The evaluation of patients with suspected narcolepsy to confirm the diagnosis.
2. The MSLT may be indicated as part of the evaluation of patients with suspected idiopathic hypersomnia to help differentiate idiopathic hypersomnia from narcolepsy."

The study is always preceded by a full-night PSG during the subject's usual sleep period. Total sleep time on this study must be at least 6 h.

4.2.2. Techniques of the MSLT

The montage for the MSLT includes scoring channels (EEG, chin EMG, and EOG) and EKG. Occipital leads (O1–A1, O2–A2) are added to better identify the drop out of alpha activity characteristic of sleep onset (Table 2). Occasionally, limited respiratory channels are added.

The standard MSLT provides five opportunities to nap at 2-h intervals. The study starts 1.5 to 3 h after termination of nocturnal PSG. The patient is studied in a dark quiet room; the instruction is to close the eyes and attempt to sleep.

Sleep logs or actigraphy should be obtained for at least 1 wk before the study to ascertain sleep–wake cycles to assure that naps are obtained during the patient's usual wake times.

For diagnostic purposes, the study is not valid if sleep-influencing drugs are either present or recently withdrawn (an exception would be to evaluate the effect of a drug on sleep latency, but this is typically a research application). A wide variety of medications are stimulants, depressants, or REM inhibitors. These medications need to be discontinued before the study by an interval of at least five half-lives of the longest active metabolite. When planning drug discontinuation before study, it is important to recognize that withdrawal effects on sleep architecture may persist after washout of the drug. Adequate time should be allowed for return of the patient's sleep to a baseline condition. Drug screening is usually obtained in the morning. Smoking is not allowed 30 min before the naps. Caffeine is also prohibited during the study.

4.2.3. Scoring of the MSLT

During each nap, the patient is given a 20-min opportunity to fall asleep. If no sleep occurs, the sleep latency is recorded as 20 min and the nap is ended.

Sleep onset is defined as the latency to the first epoch of any stage of sleep. An epoch of sleep is defined as greater than 15 s cumulative sleep in a 30-s epoch. If sleep occurs, the study is ended 15 min after the onset of sleep, by "clock time" not sleep time. Thus, if a patient fell asleep at minute 19, the study would end at minute 34, whether the patient were asleep or awake. If the patient fell asleep at minute 2, the study would end at minute 17.

REM latency is defined as the time from the first epoch of sleep to the first epoch of REM including any wake epochs that may occur. A REM epoch requires greater than 15 s of REM in a 30-s epoch.

Reporting should include start and end times of naps, mean sleep latency averaged over the five naps, number of sleep-onset REM periods, and latency to each REM period.

4.2.4. *Interpretation of the MSLT*

The primary measures of importance in the MSLT are:

1. The presence or absence of sleep-onset REM.
2. The mean sleep latency.

The occurrence of REM sleep at any time during a nap is defined as sleep-onset REM. Because REM sleep typically occurs 90 min after sleep onset, REM onset within 15 min of sleep onset during a nap is an abnormal finding; most narcoleptics have two sleep-onset REM periods in five naps, as will be discussed later, in Subheading 6. *(11)*. However, sleep-onset REM is not a specific finding. Previous sleep deprivation, medication (especially medication withdrawal), and circadian rhythm disturbances may result in sleep-onset REM. This fact underscores the fact that the study can only be interpreted when there is a careful analysis of the patient's clinical history and the conditions of the test. A recent extensive review of the MSLT by a Task Force of the Standards of Practice of the AASM concluded that the mean sleep latency on the MSLT does not discriminate well between populations of patients with sleep disorders and normal patients because of a large standard deviation of the mean *(12)*. The mean latency to sleep in normal subjects is approx 10 min, with a two standard deviation of 1.8 to 19 min. However, in the clinical context of the evaluation of a sleepy patient who, on clinical grounds, may have narcolepsy, the combination of assessment of sleep latency and the presence of sleep-onset REM is of use and is included in formal diagnostic criteria (Subheading 6.2.1.). In narcolepsy, the mean sleep latency is usually less than 8 min. In a meta-analysis of 255 patients with narcolepsy, the mean latency was reported as 3.1 min, with a standard deviation of 2.9 min.

4.3. *The MWT*

The MWT is another limited-montage laboratory polysomnographic study. It is a validated objective measure of the ability to stay awake for a defined time. Subjects are studied in a manner similar to the MSLT in daytime nap opportunities, but are instructed to stay awake rather than fall asleep. Protocols exist for both 20-min and 40-min trials, but the 20-min protocol is of limited use. New guidelines for the MWT were also recently published by the AASM *(8)*.

4.3.1. *Indications for the MWT per AASM Practice Parameters are as Follows:*

"1. The MWT 40-min protocol may be used to assess an individual's ability to remain awake when his or her inability to remain awake constitutes a public or personal safety issue.
2. The MWT may be indicated in patients with excessive sleepiness to assess response to treatment." *(11)*

4.3.2. *Techniques of the MWT*

The MWT uses the same montage as the MSLT. A previous-night PSG is not necessary but may be performed depending on the question to be answered by study. A four-trial protocol with each trial 40 min in duration is preferred. Four trials are performed at 2-h intervals, with the first starting 1.5 to 3 h after the patient's usual wake time. The room should be lit by low light, 0.1 to 0.13 lux, equivalent to a 7.5-watt night light behind the patient's head. The patient is seated in bed. The subject is instructed to sit still and try to stay awake. Alerting measures such as singing or clapping are not allowed. Prescription medications, including stimulants, caffeine, and tobacco are allowed if the study is designed to assess ability to stay awake with these substances in place; these should be documented.

4.3.3. Scoring of the MWT

Trials are ended at 40 min if no sleep occurs. In contrast, if the subject experiences the occurrence of three continuous epochs of stage 1 sleep or one epoch or any other sleep stage (termed "unequivocal sleep"), the patient is awoken and trial ended.

Sleep onset is defined as the first epoch of greater than 15 s of cumulative sleep in a 30-s epoch.

Reporting should include start and stop times for each trial, sleep latency, total sleep time, stages of sleep achieved, and mean sleep latency for the four trials. Absence of sleep is recorded as a latency of 40 min.

4.3.4. Interpretation of the MWT

Similar to the MSLT, the MWT has a wide range of normal values that makes it a poor discriminator between normal and abnormal populations. However, staying awake for four 40-min trials is considered strong objective evidence of the ability to stay awake in similar nonstimulating environments. It is understood that the test may not reliably predict sleepiness in another environment in which conditions, such as previous sleep, may be different. On the other hand, a mean latency of less than 8 min is abnormal. Latencies between 8 and 40 min are of unknown significance.

5. ACTIGRAPHY

Actigraphy is an ambulatory study that uses a small portable digital device, the actigraph, to record body movements of a subject during long periods of time, typically days, and, thus, enabling the assessment, by inference, of the subject's rest–activity cycle. The device is usually worn on the wrist, but may be worn on the trunk or ankle. The actigraph uses accelerometers to detect movement, which is sampled multiple times per second, averaged in epochs of 30 s or 1 min, and stored on the device for downloading at the end of the interval studied. A computer analysis of the data produces a histogram that demonstrates the activity level over successive 24-h periods. Other indices, such as sleep latency may be derived from this data. Actigraphy is best used with a sleep diary. Recording of sleep–wake behaviors with either a careful sleep log, or actigraphy is now required in the diagnostic criteria of the circadian rhythm disorders. Actigraphy is useful before the MSLT to document the preceding sleep–wake patterns. This avoids studying the patient under conditions of sleep deprivation or disturbed circadian phase. The technique is deemed a reliable method for detecting sleep in normal populations. However, using movement as a measure of wakefulness may not be reliable in patients with movement disorders. Nonetheless, the technique has been modified for use in detecting periodic leg movements in research applications.

6. CLINICAL APPLICATION OF TESTING

The sleep disorders described next are those that are particularly well-suited for polysomnographic analysis. The revised ICSD, Diagnostic and Coding Manual *(5)* establishes both clinical and polysomnographic criteria for the diagnosis of sleep disorders, which are abstracted in Tables 7 to 11. As noted in Subheading 4., the AASM approved indications for polysomnography include:

1. Suspicion of sleep-related breathing disorders.
2. Treatment and follow up of sleep-related breathing disorders.
3. In combination with the MSLT for suspected narcolepsy.
4. Evaluation of sleep-related behaviors that are violent, potentially injurious, or do not respond to conventional therapy.
5. To assist in the diagnosis of paroxysmal arousals that are suggestive of seizure disorder (with additional video and EEG).
6. Evaluation of sleep-related movement disorders.

Even if polysomnographic analysis is not specifically required in the assessment of a sleep complaint, clinicians often find that the judicious use of polysomnography may be remarkably revealing. For example, although the diagnoses of the various insomnias do not include the requirement of a PSG, polysomnographic study may reveal subtle sleep-disordered breathing, leg movements inducing arousal, or show that no objective impairment of sleep correlates with the patient's subjective report.

6.1. Sleep-Related Breathing Disorders

Sleep-related breathing disorders are disorders in which respiration is abnormal during sleep. The PSG is a necessary test in the diagnosis and management of these disorders.

Disorders in this classification include central sleep apnea, OSA, and sleep-related hypoventilation syndromes.

Patients with sleep-related breathing disorders may present with a wide variety of complaints, including daytime sleepiness, insomnia, inattentiveness, cognitive decline, loud snoring, nocturnal gasping, witnessed apneas, nocturnal chest pain, nonrestorative sleep, and morning headaches. Nocturnal hypoxia exacerbates ischemic heart disease and promotes the development of pulmonary hypertension. OSA is known to be associated with excessive daytime sleepiness and the development of hypertension. Associations with insulin resistance, nocturnal arrhythmia, stroke, myocardial infarction, insomnia, and mood disorders are likely.

Central sleep apnea syndromes (Table 7) are characterized by episodes of decreased respiratory effort that are either cyclic or intermittent (Fig. 10). OSA syndromes (Table 8) are characterized by airway obstruction with persistent respiratory effort (Fig. 9). Lastly, sleep-related hypoventilation syndromes (Table 9) include a variety of disorders, either primary idiopathic or secondary to other medical conditions, that result in sleep-induced or sleep-exacerbated hypercapnia and hypoxia. These abnormalities are quantified using the respiratory monitoring techniques described in Subheading 4.1.

The PSG in each of these disorders is distinctive (Tables 7–9), but some abnormalities are common to this group of disorders. Abnormal respiratory events of any type tend to fragment sleep. Polysomnographic evidence of sleep fragmentation includes increased stage 1 sleep, delayed REM latency, decreased REM and stages 3 and 4 sleep, increased arousals, increased awakenings, and decreased sleep efficiency. Sleep latency may be prolonged.

Polysomnography is also an important tool in treatment of sleep-related breathing disorders. Positive pressure applied as either constant positive airway pressure or bi-level positive airway pressure is the mainstay of treatment for obstructive and some central apneas. The appropriate at-home pressures are determined through the process of "titration" in the sleep laboratory, in which pressures are gradually adjusted during the course of the night to eliminate apneas, hypopneas, desaturations, and arousals. Supplemental oxygen is also frequently applied, and the effects monitored, during therapeutic PSGs with and without positive pressure, especially in patients with central sleep apnea or hypoventilation syndromes. Automatic titrating machines for at home use are available but have generally not replaced in-laboratory positive pressure titration.

6.2. Hypersomnias of Central Origin

Hypersomnias of central origin refer to a group of disorders that result in excessive daytime sleepiness but are not caused by disturbed nocturnal sleep or disorders of circadian rhythms. Excessive daytime sleepiness is defined as "the inability to stay awake and alert during the major waking episodes of the day, resulting in unintended lapses into drowsiness or sleep." Patients with these disorders may or may not demonstrate excessive sleep during a 24-h period. Diagnostic and Clinical Criteria of the ICSD for narcolepsy and idiopathic hypersomnia are outlined in Table 10. For the rarer disorders in this category, the reader is referred to the ICSD-2 *(2)*.

6.2.1. Narcolepsy

The most important and common disorder in this category is narcolepsy, with a prevalence of 0.02 to 0.18% in the United States. Narcolepsy is now classified as narcolepsy with cataplexy, narcolepsy without cataplexy, and narcolepsy secondary to medical condition. The last is extremely rare. Cataplexy refers to episodes of muscle weakness associated with strong emotion, often laughter. These episodes are typically brief, often involve the knees and/or face, and are unassociated with a change of consciousness, although sleep sometimes follows immediately. Cataplexy is pathognomonic of narcolepsy, but need not be present. Other common features of narcolepsy include hallucinations at sleep onset (hypnagogic hallucinations), sleep paralysis, inattentive "automatic behavior," and poorly maintained nocturnal sleep. Most of the characteristic symptoms of the disorder seem to reflect a disorder of the control mechanisms that regulate REM sleep. Episodes of REM occur at the wrong time, intruding on wakefulness, and the physiological components of REM sleep dissociate and appear independently. Cataplexy and sleep paralysis, for example, are manifestations of the muscle atonia of REM sleep appearing during wakefulness. Recent discoveries suggest that narcolepsy is caused by the loss of hypothalamic neurons containing the neuropeptide, hypocretin-1. In the past, neurophysiological testing with both the PSG and MSLT provided the only confirmatory evidence for the diagnosis of narcolepsy with and without cataplexy. New diagnostic guidelines include the option of obtaining cerebrospinal fluid levels of hypocretin-1, an assay that can be obtained by sending the sample to specialized centers. Approximately 90% of patients who demonstrate cataplexy have low hypocretin levels, whereas only 10 to 20% of patients classified as having narcolepsy without cataplexy show low hypocretin levels. Narcolepsy with cataplexy is highly associated with HLA subtype DQB1*0602, but because 12 to 38% of controls are positive for this antigen, HLA typing is not used as a diagnostic criterion. Narcolepsy without cataplexy has a less strong association with DQB1*0602. Many patients who carry the diagnosis of narcolepsy without cataplexy are likely to have been misdiagnosed and have another sleep disorder. The MSLT is a study that requires close attention to procedural detail and patient preparation, as described in Subheading 4.2. Improper administration of the MSLT results in both false-positive and false-negative results. The highest specificity (99.2%) and positive predictive value (87%) for MSLT findings are with the criteria of three or more sleep-onset REM periods combined with a mean sleep latency of less than 5 min *(13)*. Narcolepsy caused by medical condition, also referred to as symptomatic narcolepsy or secondary narcolepsy, is usually associated with pathology of the hypothalamus and may also result in low hypocretin levels. Although this presentation is rare, a variety of disease processes have been identified in these patients, including tumors, cerebral infarct, sarcoidosis, Niemann-Pick type C, multiple sclerosis, disseminated encephalomyelitis, and paraneoplastic syndromes.

6.2.2. Idiopathic Hypersomnia

If no other condition can be identified that explains excessive daytime sleepiness, the diagnosis "idiopathic hypersomnia" is applied. Polysomnographic studies as well as a careful medical neurological and psychiatric assessment are particularly important in these disorders to exclude subtle or occult abnormalities. Brain MRI is appropriate. The ICSD-2 now classifies these patients as idiopathic hypersomnia with prolonged sleep time, and idiopathic hypersomnia without prolonged sleep time. When subjected to detailed study using the more-sensitive measures of respiratory effort noted in Subheading 4.1.2., many patients who carry this diagnosis are identified as having RERAs, indicative of a sleep-related breathing disorder.

6.3. Parasomnias

The term parasomnia refers to undesirable events that occur during sleep, sleep onset, or on arousal from sleep. Parasomnias often include complex behaviors. Although the behaviors may seem to be goal directed, they are not under conscious control. These disorders typically are sleep-stage specific. The classification includes:

1. Disorders of arousal (from NREM sleep).
2. Parasomnias usually associated with REM sleep.
3. Other parasomnias.

By ICSD-2 standards, polysomnographic analysis is required only for the diagnosis of REM behavior disorder. In practice, however, polysomnographic analysis is extremely useful both to document the disorders and, more importantly, to identify other sleep disorders that may act as precipitants to arousal and subsequent behavioral manifestations.

6.3.1. REM Behavior Disorder

REM behavior disorder (Table 11) is characterized by abnormal behaviors that emerge during REM sleep. The typical complaints are violent thrashing, hitting, or yelling, accompanied by nightmares. The disorder is more common in men older than 50 yr. Polysomnographic study shows that the behavioral outbursts are associated with the intermittent loss of the muscle atonia that normally characterizes the REM sleep state (Fig. 14). If awoken spontaneously or by others during the episode, patients are often, but not always, able to correlate the actions to dream mentation. The disorder is highly correlated with parkinsonian states, including Parkinson disease, dementia with Lewy bodies, and multisystem atrophy. The disorder can emerge years before these diagnoses are clinically apparent, but idiopathic and drug-induced REM behavior disorder presentations are also recognized.

6.3.2. Disorders of Arousal (From NREM Sleep)

Disorders of arousal from NREM sleep (Table 11) include sleep terrors, sleepwalking, and confusional arousals. These disorders are characterized by confusion and automatic behavior after sudden arousal from NREM, usually stages 3 or 4 sleep (slow-wave sleep) (Fig. 15). As a result, symptoms tend to occur early in the night, when slow-wave sleep is prominent. Patients may sit bolt upright with a blood-curdling scream and autonomic activation (sleep terror),

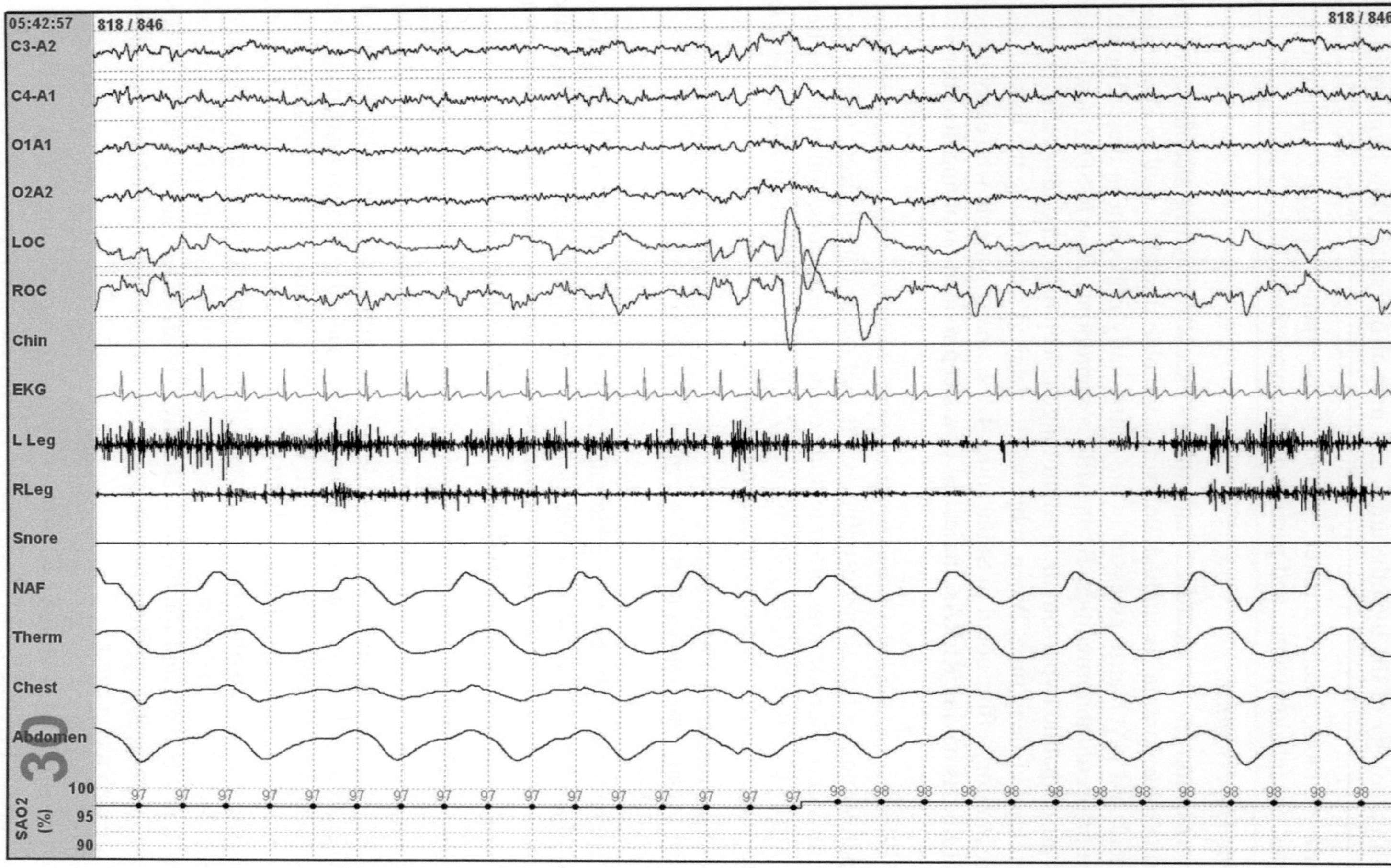

Fig. 14. Rapid eye movement (REM) behavior disorder. In this epoch, during REM there is a loss of the usual REM atonia. Excessive motor activity is captured in the leg EMG leads. The chin EMG tone remains atonic, as expected in REM. The elevation in motor tone associated with this disorder is rarely generalized. (30-s epoch)

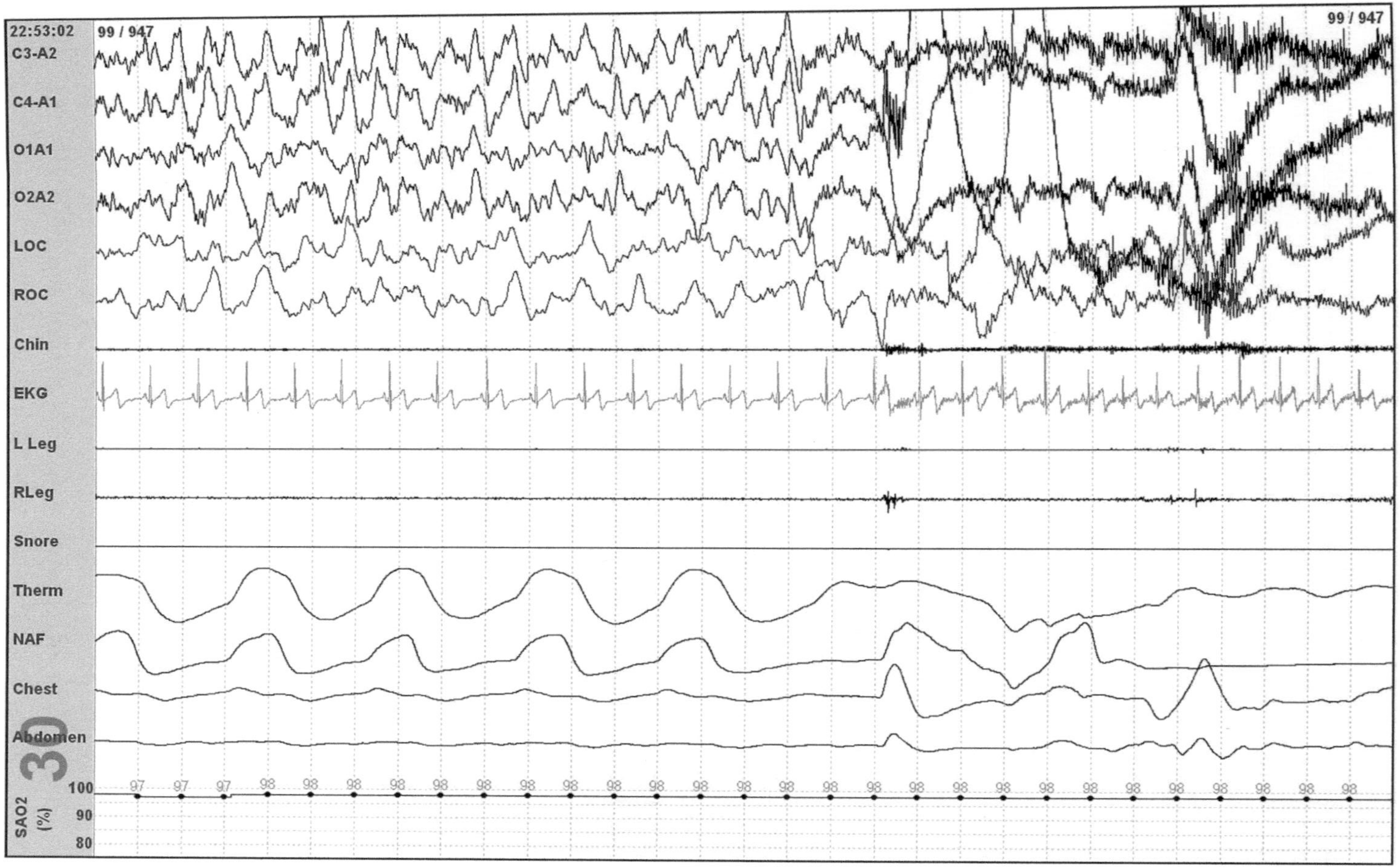

Fig. 15. Arousal from delta sleep. A sudden unprovoked arousal occurs from hypersynchronous delta sleep.

engage in complex acts (sleep walking), or arouse in a confused state (confusional arousal). Patients are difficult to fully awaken and are mostly amnestic for the episode. The tendency to arouse spontaneously from delta sleep tends to be a familial trait, with first presentation in childhood and resolution by adolescence. Stress, sleep deprivation, or any factors that contribute to sleep disruption, such as sleep-disordered breathing, are exacerbants that may result in the re-emergence of the behavioral syndrome in adulthood. Polysomnographic study may demonstrate the typical behaviors emerging from slow-wave sleep, sudden unexplained arousals from slow-wave sleep, or an underlying cause of arousal, such as sleep-disordered breathing or leg movements.

6.4. Sleep-Related Movement Disorders

Sleep-related movement disorders include conditions in which simple stereotyped movements are present during sleep and induce sleep disruption. Difficulty initiating and/or maintaining sleep are the typical complaints.

6.4.1. Periodic Limb Movement Disorder

The most prevalent of these disorders is PLMD. PLMD (Table 12) is characterized by periods of repetitive stereotyped leg movements that disturb sleep (PLMS) (Fig. 13). The leg movements are similar to the triple flexion response of the Babinski reflex, but arm movements may also occur. Other types of leg movements may be present without periodicity, or without sleep disruption, or may be secondary to other primary sleep disorders, in which case, the term PLMS is not appropriate. Leg movements, for example, frequently accompany arousals from sleep-disordered breathing. Many medications are implicated in the induction of periodic and aperiodic leg movements, most commonly selective serotonin reuptake inhibitors and tricyclic antidepressants.

6.4.2. Restless Legs Syndrome

RLS (Table 12) is a disorder characterized by an urge to move accompanied by uncomfortable sensations, predominantly in the legs, that are relieved by movement, occur when sedentary, and are worse in the evening *(14)*. This syndrome is closely associated with PLMD because 80 to 90% of patients with this disorder have PLMS. Some of these patients also demonstrate periodic limb movements while awake. RLS, however, is a syndrome based on clinical, not polysomnographic criteria. The disorder is familial in 50% of cases. Abnormalities of both dopamine and iron metabolism are implicated in the underlying pathophysiology. In patients with RLS, PET studies have shown decreased dopamine D2 binding and decreased dopamine storage in striatum. It has been postulated that brain iron deficiency underlies dopamine dysfunction in this disorder. Recent postmortem evidence suggests a deficiency of iron acquisition in the substantia nigra of patients with RLS. Iron deficiency is known to exacerbate or precipitate restless legs and periodic leg movements in familial and nonfamilial cases. Increasing iron stores by long-term iron supplementation is often therapeutic even in patients with low normal ferritin levels (less than 50). Dopaminergic agonists are highly successful treatments. Iron and dopamine may be functionally linked by the fact that tyrosine hydroxylase is a cofactor in dopamine metabolism. Secondary causes of RLS are subject to some debate because of limited data, and include uremia, neuropathy, medications (especially antidopaminergic drugs and selective serotonin reuptake inhibitors), and caffeine.

Table 12
Periodic Limb Movement Disorder and Restless Legs Syndrome

Disorder	*Clinical criteria*	*Polysomnographic criteria*	*Other requirements*	*Other features*
Periodic limb movement disorder	There is a clinical sleep disturbance or a complaint of daytime fatigue.	PSG: Stereotyped limb movements during sleep (PLMS) that are: • 0.5–5 s in duration • Of amplitude ≥25% of toe dorsiflexion during calibration • In a sequence of 4 or more movements • Separated by an interval of >5 s and <90 s PLMS index (events/hour of sleep) exceeds 15/h in adults, 5/h in children	PLMS are not better explained by another current sleep disorder, medical or neurological disorder, mental disorder, medication use, or substance use disorder.	If PLMS are present without sleep disturbance, the finding can be noted as a polysomnographic finding, but criteria are are not met for a diagnosis of PLMD PLMS are found in a variety of other sleep disorders, including narcolepsy, RBD and RLS. The term PLMD should not be used when diagnostic criteria for these disorders are fulfilled. Screen for low serum ferritin, neuropathy, and renal disease Leg movements decrease with low-dose dopaminergic agonists

(Continued)

Table 12 (Continued)

Disorder	*Clinical criteria*	*Polysomnographic criteria*	*Other requirements*	*Other features*
Restless legs syndrome	The patient reports an urge to move the legs, usually accompanied or caused by uncomfortable and unpleasant sensations in the legs. The urge to move or the unpleasant sensations begin or worsen during periods of rest or inactivity, such as lying or sitting. The urge to move or the unpleasant sensations are partially or totally relieved by movement, such as walking or stretching. The urge to move or the unpleasant sensations are worse, or only occur, in the evening or night.	None PLMS occur in 80–90% of patients PLMW may be present	The disorder is not better explained by another current sleep disorder, another medical or neurological disorder, mental disorder, medication use, or substance use disorder.	Often familial Screen for low serum ferritin, neuropathy, and renal disease Symptoms improve with low-dose dopaminergic agonists.

PLMD, periodic limb movement disorder; RBD, REM behavior disorder; REM, rapid eye movement sleep; RLS, restless legs syndrome; PLMW, periodic leg movements during wakefulness.

Abstracted from ref. *5*, with permission.

REVIEW QUESTIONS

1. Polysomnography:
   A. Is routinely performed in an ambulatory setting; only very sick, unstable patients require close monitoring in a laboratory.
   B. Is not required for the diagnosis of restless legs syndrome.
   C. Is essential for the diagnosis of insomnia.
   D. Is indicated only for the diagnosis of sleep apnea.
2. Stage 2 sleep is defined by the presence of:
   A. K-complexes or spindles of at least 0.5 s in duration in the absence of sufficient slow wave acitivity to define the presence of stages 3 and 4 sleep.
   B. Saw tooth waves.
   C. Absence of alpha activity for 50% of the epoch.
   D. K-complexes of at least 75 microvolts in amplitude.
3. Stage 1 sleep is characterized by all of the following except:
   A. Occasional vertex waves of less than 0.5 s.
   B. An epoch with less than 50% alpha activity.
   C. Slow rolling eye movements.
   D. Spindles lasting 0.5 s or more.
4. Stage 3 sleep is defined by epochs (30 s) with:
   A. Not more than 50%, but greater than or equal to 20% slow activity of 3 Hz or less with an amplitude of 75 μV.
   B. Not more than 50% but greater than or equal to 20% slow wave activity of 2 Hz or less with no amplitude criterion.
   C. Not more than 50% but greater than or equal to 20% slow wave activity of 2 Hz or less with an amplitude of >75 μV.
   D. At least 50% high-voltage slow activity of at least 2 Hz with an amplitude of 75 μV.
5. Scoring REM sleep requires:
   A. Saw tooth waves.
   B. EEG with an at least 8-channel montage.
   C. EMG of chin, eye movement recording (EOG), and EEG with central leads.
   D. EMG of both chin and anterior tibialis muscles.
   E. Phasic changes in respiration.
6. Current Center for Medicare and Medicaid Services (CMS) guidelines define hypopnea as:
   A. Any decrease in airflow.
   B. Any decrease in airflow or thoracoabdominal movement with an arousal or oxygen desaturation.
   C. 30% Decrease in airflow with 4% oxygen desaturation or an arousal.
   D. 30% Decrease in airflow or thoracoabdominal movement with at least 4% desaturation.
7. A mixed apnea is defined as:
   A. Absence of airflow for 10 s without any other criterion.
   B. Absence of airflow for 10 s with decreased but persistent respiratory effort on thoracoabdominal monitors throughout the event.
   C. Absence of airflow for less than 10 s with or without respiratory effort.
   D. Absence of airflow for 10 s with initial absence of respiratory effort followed by resumption of respiratory effort before airflow resumes.
8. Proper administration of a multiple sleep latency test (MSLT):
   A. Involves a patient sitting up in a minimally lit room to stay awake during the testing.
   B. Requires discontinuation of all drugs that affect sleep for a period of 15 days, or for a period of at least 5 times the half life of the drug and its longest acting metabolite.
   C. Consists of a neurophysiological recording during two 20-min long nap opportunities, typically scheduled on the afternoon preceding a whole-night PSG.
   D. Requires a minimum of 6 h of sleep deprivation on the night before testing.

9. A periodic leg movement sequence is scored when:
   A. There is any repetitive EMG activity in the anterior tibialis muscle of 0.5–5 s in duration, with an amplitude of at least 25% of the calibration EMG activity.
   B. There is a series of leg movements lasting at least 90 s, with an amplitude of 25% of the calibration EMG.
   C. Synchronous repetitive contractions of agonist and antagonist leg muscles are noted while awake or asleep.
   D. There is a series of 4 or more bursts of anterior tibialis EMG activity, each of 0.5–5 s in duration, with an amplitude of 25% of the calibration EMG, separated by more than 5 and less than 90 s.
10. Growth hormone secretion is associated with:
    A. The onset of REM.
    B. Stage one sleep.
    C. Arousal from slow-wave sleep.
    D. Slow wave sleep.
11. In a patient with excessive daytime sleepiness and equivocal cataplexy, a diagnosis of narcolepsy with cataplexy may be confirmed with which of the following:
    A. CSF fluid hypocretin measurements.
    B. MSLT alone.
    C. PSG showing no other sleep disorder accounting for sleepiness.
    D. PSG showing no other sleep disorder and MSLT showing sleep onset REM in any one of 5 naps and mean latency of at most 8 min.
12. Which of the following is not required for the diagnosis of restless legs syndrome:
    A. The patient reports the urge to move the legs accompanied by unpleasant sensations.
    B. Symptoms are precipitated by rest and relieved by activity.
    C. Symptoms are worse in the evening or night.
    D. Periodic limb movements are demonstrated on PSG.
13. By current guidelines, Medicare will approve treatment of obstructive sleep apnea with positive pressure in which of the following situations:
    A. Respiratory disturbance index of 15 or greater.
    B. Apnea hypopnea index of 15 or greater.
    C. Apnea hypopnea index of 15, only when there is an associated lowest oxygen desaturation to 88%.
    D. Apnea hypopnea index of 15, only if there are associated clinical symptoms.
14. The term "Disorder of Arousal" refers to:
    A. Any disorder that results in frequent arousals from sleep.
    B. A disorder characterized by abnormal tone during REM sleep.
    C. A disorder characterized by sudden arousal from NREM, usually slow wave sleep.
    D. A disorder characterized by marked difficulty waking after a normal sleep period.
15. REM behavior disorder is characterized by all of the following except:
    A. Often precedes the onset of parkinsonian disorders.
    B. Lack of the usual atonia during REM sleep.
    C. Violent and potentially self-injurious behaviors.
    D. Patients are typically amnestic for the behaviors.
16. Which of the following is not an effect of aging on sleep architecture:
    A. A reduction in slow wave sleep.
    B. A reduction in the percentage of REM sleep.
    C. A reduction in total sleep time.
    D. An increase in arousals.
17. What is the utility of the maintenance of wakefulness test (MWT)?
    A. It is used to diagnose narcolepsy.
    B. It measures both the ability to fall asleep and to stay awake in a controlled environment.
    C. It is a measure of the ability to stay awake in the provided testing conditions.
    D. It predicts the risk of an accident due to inappropriate episodes of sleep in the real world.

18. Restless leg syndrome responds to:
    A. Treatment with selective SSRIs.
    B. Treatment with dopaminergic antagonists.
    C. Treatment with dopaminergic agonists.
    D. No effective treatment is available.
19. The diagnosis of obstructive sleep apnea:
    A. Always requires the presence of obstructive apneas.
    B. Always is associated with significant oxygen desaturation.
    C. Requires that the patient demonstrate excessive daytime sleepiness.
    D. May be made in the absence of clinical criteria if the polysomnogram shows at least 15 obstructive events an hour including RERAs, hypopneas, and apneas.
20. A RERA (respiratory effort related arousal) is:
    A. Associated with increased negative intrathoracic pressure.
    B. Is obvious on chest and abdomen effort bands.
    C. Is typically associated with oxygen desaturation.
    D. Is never a major cause of sleep disruption.

REVIEW ANSWERS

1. B. Polysomnography is not required to establish a diagnosis of restless legs syndrome. The diagnosis of restless legs is made based on four essential criteria including: (1) an urge to move accompanied by uncomfortable sensations, predominantly in the legs that are (2) relieved by movement, (3) occur when sedentary and (4) are worse in the evening.
2. A. Stage 2 is defined by the presence of sleep spindles (≥0.5 sec) and/or K complexes (≥0.5sec) in the absence of sufficient slow activity to define stages 3 or 4 sleep.
3. B. The transition from drowsiness to stage 1 is characterized by slowing of the EEG. As this transition occurs, when less than 50% or a 30 second epoch demonstrates alpha activity, the epoch is scored stage 1. Slow rolling eye movements typically herald and occur during stage one sleep, but are not required for scoring.
4. C. The definition of Stage 3 is based on EEG frequency, amplitude, and epoch composition of delta activity. The delta activity required for stage 3 is <2 Hz, >75μV, and ≥20% but ≤50% of the epoch.
5. C. REM sleep is defined by low amplitude, asynchronous frequency EEG activity, REMs, in combination with the lowest tonic submental EMG level during sleep (usually atonia). Saw tooth waves are frequently present but are neither sufficient nor necessary for scoring REM.
6. D. Medicare (CMS) adopted the hypopnea definition used in the Sleep Heart Health Study an on-going large-scale epidemiologic study of sleep-disordered breathing. This definition does not include any measure of sleep disturbance.
7. D A mixed apnea is so called because of the presence of both central and obstructive features. The event begins with absence of respiratory effort and airflow. When respiratory effort resumes, airflow does not resume simultaneously because of obstruction of the airway. The total duration of absence of airflow is 10 s.
8. B. Medications that affect sleep architecture distort the findings on the MSLT by influencing both sleep latency and the presence of REM sleep. The current recommendation of the ICSD-2 is that prior to the administration of the MSLT the patient should be free of medications that influence sleep for 15 days, or at least five times the half- life of the drug and its longest metabolites. Rapid withdrawal of REM- influencing drugs may induce REM rebound and false positives. REM- inhibiting drugs produce false negative studies. Often, a meaningful MSLT cannot be obtained because the patient's underlying disorders do not allow for medication withdrawal.
9. D. Leg movements fulfilling the described requirements are necessary for the diagnosis of PLMD. Most adult patients demonstrate an index of greater than 15 leg movements per hour.
10. D. Maximal growth hormone release occurs within minutes of the onset of slow wave sleep.

11. A. Cerebrospinal hypocretin-1 <110 pg/ml or less than 1/3 normal control values is found in 90% of patients with narcolepsy with cataplexy and almost never in controls or other patients. The test may be useful when REM suppressant medications cannot be discontinued or when an MSLT is difficult to interpret. The test is not influenced by concurrent sleep disorders and psychotropic medications/substances.
12. D. Restless legs syndrome is a clinical diagnosis based on symptoms alone (*see* Question 1.) The disorder usually associated with PLMD, but periodic limb movements are not required to confirm the diagnosis of restless legs syndrome.
13. B. Medicare and other insurance carries adhering to Centers for Medicare and Medicaid Services (CMS) coverage criteria will approve treatment for obstructive sleep apnea if the AHI ≥15 in an asymptomatic patient or ≥5 with a history of any one or more of the following: excessive daytime sleepiness, impaired cognition, mood disorders or insomnia, or documented hypertension, ischemic heart disease, or history of stroke.
14. C. These non-REM parasomnias most often arise from slow wave sleep and are therefore occur more often during the first third of the night. A family history disorders of arousal is common and provocative factors include sleep deprivation, stress, medications, alcohol, caffeine and other sleep disorders that increase the probability of arousal such as sleep-related breathing disorders or PLMD.
15. D Patients with REM behavior disorder often report detailed dream activity consistent with observed dream-enacting behaviors. On the other hand, patients with non-REM parasomnias are typically amnestic for the demonstrated behaviors.
16. B. Under normal circumstances, the percentage of REM sleep in an older individual is similar to that of a younger adult. Slow wave sleep (stages 3 and 4 sleep) and total sleep time decrease with age. Arousals and sleep fragmentation increase with age.
17. C. The maintenance of wakefulness test (MWT) was designed to assess an individual's ability to resist sleep in a standardized, non-stimulating environment. The test has received criticism due to a lack of recreating realistic circumstances. Nonetheless, the MWT is often utilized to assess therapeutic response of a sleep disorder to treatment, and to evaluate an individual's ability to stay awake when impaired alertness poses a safety hazard.
18. C. Dopaminergic agonists are an effective treatment for RLS. Dopaminergic antagonists and SSRIs can exacerbate RLS.
19. D. Obstructive sleep apnea syndrome describes a spectrum of sleep related breathing disorders characterized by an absence or reduction in airflow despite continued respiratory effort. Some advocate that a response to CPAP aids in the diagnosis and others adhere to CMS guidelines as the definition. According to the AASM, OSA exists when 5 or more obstructive respiratory events are accompanied with a clinical presentation. Variation in recording methods of respiratory parameters exists and therefore a variation in diagnostic sensitivities continues to exist.
20. A. A RERA is characterized a sequence of increasing effort and/or breathing, not meeting the definition for an apnea or hypopnea, and terminating with an arousal.

REFERENCES

1. Loomis AL, Harvey EN, Hobart G. Cerebral states during sleep, as studied by human brain potentials. J Exp Psychol 1937;21:127–144.
2. Aserinsky E, Kleitman N. Regularly occurring periods of eye motility, and concomitant phenomena, during sleep. Science 1953;118:273–274.
3. Dement W, Kleitman N. Cyclic variations in EEG during sleep and their relation to eye movements, body motility, and dreaming. Electroencephalogr Clin Neurophysiol Suppl 1957;9:673–690.
4. Rechtschaffen AKA. A Manual Standardized Terminology: Techniques and Scoring System for Sleep Stages of Human Subjects. Los Angeles: UCLA Brain Information Service/Brain Research Institute, 1968.
5. The International Classification of Sleep Disorders: Diagnostic & Coding Manual, ICSD-2. Westchester, IL: American Academy of Sleep Medicine, 2005.

6. Espana RA, Scammell TE. Sleep neurobiology for the clinician. Sleep 2004;27:811–820.
7. Kushida CA, Littner MR, et al. Practice parameters for the indications for polysomnography and related procedures: an update for 2005. Sleep 28(4):499–521.
8. EEG arousals: scoring rules and examples: a preliminary report from the Sleep Disorders Atlas Task Force of the American Sleep Disorders Association. Sleep 1992;15:173–184.
9. Meoli AL, Casey KR, Clark RW, et al. Hypopnea in sleep-disordered breathing in adults. Sleep 2001;24:469–470.
10. Recording the scoring leg movements. The Atlas Task Force. Sleep 1993;16:748–759.
11. Littner MR, Kushida C, Wise M, et al. Practice parameters for clinical use of the multiple sleep latency test and the maintenance of wakefulness test. Sleep 2005;28:113–121.
12. Arand D, Bonnet M, Hurwitz T, Mitler M, Rosa R, Sangal RB. The clinical use of the MSLT and MWT. Sleep 2005;28:123–144.
13. Aldrich MS, Chervin, RD, Malow BA. Value of the multiple sleep latency test (MSLT) for the diagnosis of narcolepsy. Sleep 1997;20:620–629.
14. Allen RP, Picchietti D, Hening WA, Tenkwalder C, Walters AS, Montplaisi J. Restless legs syndrome: diagnostic criteria, special considerations, and epidemiology. A report from the restless legs syndrome diagnosis and epidemiology workshop at the National Institutes of Health. Sleep Med 2003;4:101–119.
15. Kryger MH, Roth T, Dement WC. Principles and practice of sleep medicine. Philadelphia, PA: Elsevier/Saunders, 2005:xxxiii.
16. Williams RL, Karacan I, Hursch CJ. Electroencephalography (EEG) of human sleep: clinical applications. New York, Wiley, NY, 1974:xiv.
17. Kales A, Kales JD. Sleep disorders. Recent findings in the diagnosis and treatment of disturbed sleep. NEJM 1974;290:487–499.

24

Autonomic Nervous System Testing

Frederick K. Nahm and Roy Freeman

Summary

Autonomic testing encompasses an array of procedures that can be used to assess a variety of symptoms ranging from lightheadedness and dizziness to anhydrosis to constipation and urinary incontinence. A number of procedures are available for testing of the many varied aspects of both the parasympathetic and sympathetic nervous systems. These tests include Valsalva maneuver testing, RR interval testing, tilt-table testing, microneurography, and the thermoregulatory sweat test. This chapter reviews the basic neuroanatomy and neurophysiology of the autonomic nervous system and the tests that are most effective in their evaluation.

Key Words: Anhydrosis; hypotension; lightheadedness; neuropathy; parasympathetic; sympathetic.

1. INTRODUCTION

Disorders of the autonomic nervous system (ANS) can be found in many conditions that span the disciplinary boundaries of medicine. The neurophysiological evaluation of the human autonomic nervous system function is challenging because most autonomic structures are located at a distance from the skin, and, thus, less amenable to direct observation or study. In addition, most neurophysiological measures of the ANS reflect end-organ function (smooth muscle, cardiac muscle, and glandular organs), rather than direct neural activity in the sympathetic and parasympathetic nerves. Despite these methodological difficulties, there are a number of clinically useful techniques that test the functional integrity of the sympathetic and parasympathetic nervous system. The aim of this chapter is to provide a basic understanding of neurophysiological testing methods that are used to evaluate autonomic disorders.

Neurophysiological tests of autonomic function are extensions of the clinical exam, and their appropriate use requires a thorough history and physical examination. In addition, a careful review of autonomic symptoms should be taken, with attention to systems commonly affected by autonomic dysfunction. As in all clinical tests, the pretest clinical probability affects the power of ANS testing, and a clear, well-formulated question should precede ANS testing. A brief list of the more common autonomic symptoms are listed in Table 1. A few of the more common indications for ANS testing are listed in Table 2.

2. NEUROANATOMIC AND NEUROPHYSIOLOGICAL PRINCIPLES

A systematic approach to ANS testing is important given its neuroanatomical complexity. A useful conceptual approach for understanding sympathetic and parasympathetic function is to think in terms of neuronal circuits and autonomic reflex arcs (Fig. 1). Afferent information arises

From: *The Clinical Neurophysiology Primer*
Edited by: A. S. Blum and S. B. Rutkove © Humana Press Inc., Totowa, NJ

Table 1
Common Autonomic Symptoms

Orthostatic intolerance
Dizziness
Lightheadedness
Fatigue
"Coathanger" headache
Nausea
Palpitations
Near-syncope and syncope
Genitourinary
Bladder urgency or frequency
Incontinence
Nocturia
Erectile dysfunction
Ejaculatory disturbances
Gastrointestinal
Diarrhea
Constipation
Fecal incontinence
Postprandial fullness, cramping, or bloating
Sudomotor
Hyperhidrosis
Hypohidrosis and anhidrosis

Table 2
Indications for Autonomic Nervous System Testing

Syncope
Central autonomic degenerations (e.g., multiple system atrophy with autonomic failure, and Parkinson disease)
Pure autonomic failure
Postural tachycardia syndrome
Autonomic and small-fiber peripheral neuropathies (e.g., diabetic peripheral neuropathy, amyloid peripheral neuropathy, and hereditary peripheral neuropathies)
Sympathetically mediated pain
Evaluating response to therapy
Differentiating benign symptoms (e.g., dizziness) from life-threatening autonomic disorders

from end-organ receptors, such as baroreceptors in vessel walls, and mechanoreceptors and metaboreceptors in skeletal muscle. Signals arising from these structures are directed to central autonomic structures via unmyelinated and myelinated autonomic nerves. The nucleus tractus solitarius, located in the dorsal medulla, receives much of the afferent peripheral autonomic input. This structure has reciprocal connections to central regions regulating both sympathetic and parasympathetic function and plays an important role in feedback regulatory circuits.

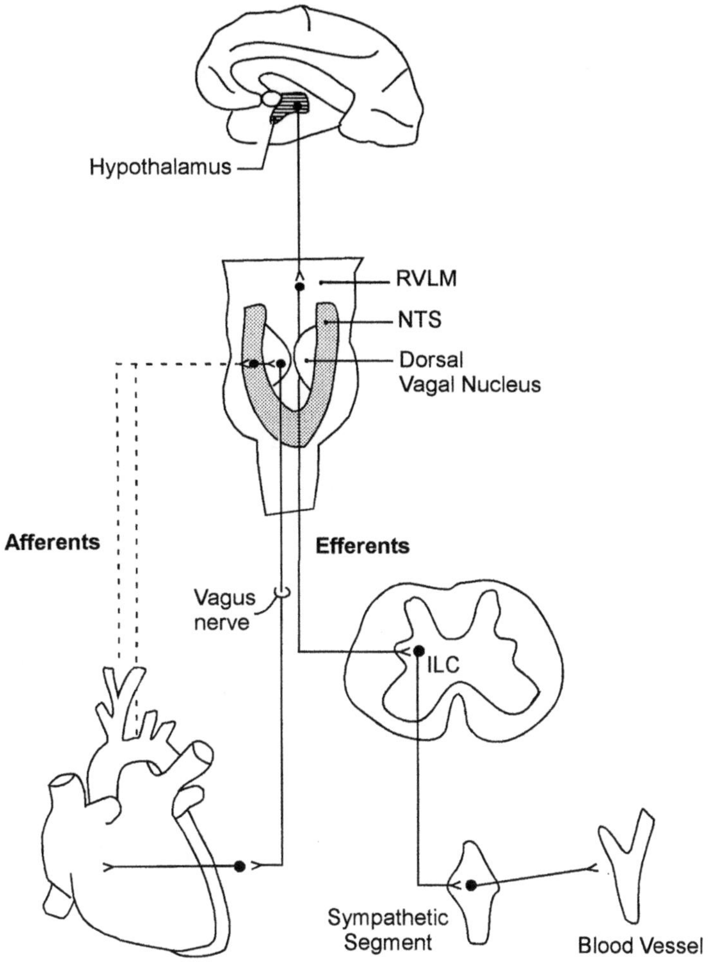

Fig. 1. Schematic representation of autonomic reflex arc, showing baroreceptor afferents in dashed lines, central processors, and efferents (solid lines) to end organs. ILC, intermedial lateral columns; RVLM, rostral ventral lateral medulla; NTS, nucleus tractus solitarius.

Central regions with input to sympathetic preganglionic neurons centers include the paraventricular nuclei of the hypothalamus, the amygdala, and the insular and medial prefrontal cortices. Preganglionic sympathetic neurons are also located in the rostral ventrolateral medulla, the caudal ventrolateral medulla, the caudal raphe nuclei, and the A5 region of the pons.

Efferent parasympathetic fibers originate in the Edinger–Westphal nucleus, ciliary ganglion, superior and inferior salivary nuclei, otic ganglion, and the nucleus ambiguous and the dorsal motor nucleus of the vagus. These fibers are carried in cranial nerves III, VII, IX, and X, and synapse on both end organs and parasympathetic ganglia located within or in close proximity to their target organs. Preganglionic parasympathetic fibers related to genitourinary function originate in the spinal cord at sacral cord levels 2, 3, and 4. Efferent preganglionic sympathetic fibers originate in the intermediolateral cell columns at levels T1–L2. These fibers synapse on postganglionic fibers in paravertebral, prevertebral, and previsceral ganglia.

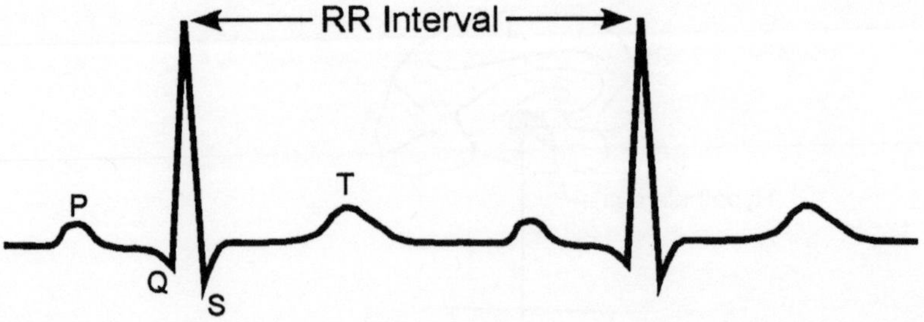

Fig. 2. The RR interval is the duration, measured in milliseconds, between successive R waves in the QRS complex.

3. PATIENT PREPARATION FOR AUTONOMIC TESTING

Before ANS testing, patients should refrain from heavy meals, coffee, and nicotine for at least 3 to 4 h before studies. Medications with either anticholinergic effects or those with α- and β-receptor agonist or antagonist activity should be withheld (if medically appropriate) 24 to 48 h before testing, because these can significantly affect the outcome of ANS tests.

Most autonomic testing sessions are 60 to 90 min. Patients should wear loose-fitting clothes for comfort and to facilitate the application of cardiac electrodes and blood pressure (BP) cuffs.

4. ELECTROPHYSIOLOGICAL METHODS

4.1. Heart Rate Recording and the RR Interval

Heart rate (HR) is routinely recorded with a three-lead electrocardiograph (ECG) machine. Tests evaluating HR variability can only be performed in the setting of sinus rhythm. The ECG leads are placed at anterior chest locations, which minimize movement artifact, and the reference electrode is usually placed at the mid-axillary line at approximately the T4 level. Autonomic tests of HR variability may be based on HR measure in beats per minute or its inverse, the RR interval, measured in milliseconds (Fig. 2).

4.2. BP Recording

BP during autonomic testing can be measured intermittently or continuously. Some methods rely on brachial artery cuffs, whereas others use devices that are applied to the wrist or the finger. The intermittent recording techniques rely on occlusive cuff methods. Techniques for detecting arterial lumen opening include auscultatory and oscillometric methods. Auscultation for Korotkoff sounds is a widely available, and invaluable, bedside tool. The oscillometric technique is based on the observation that during cuff deflation, the point of maximal oscillation between systolic and diastolic pressure corresponds to the mean arterial pressure. Although both the auscultatory and oscillometric techniques are relatively easy to perform, they provide only periodic sampling of BP.

Plethysmography and arterial tonometry are methods used for continuous, beat-to-beat, noninvasive BP recording. Plethysmography is an indirect measure of BP, whereby the pulse pressure is derived from the blood volume. The volume clamp method of Peñaz, also called arterial counterpulsation, uses a servo-loop circuit that continuously counterbalances arterial pressure, thereby clamping the volume of blood flow to maintain a transmural pressure of zero. The counterpulsation pressure provides a real-time measure of arterial pulse pressure. The volume clamp technique of Peñaz has been used with digital pressure cuffs and finger photoplethysmography. A potential drawback to this method is its sensitivity to decreases in distal limb temperature, which may significantly alter the measured BP.

Arterial tonometry is a method in which arrays of pressure sensors compress and partially flatten (applanation) an artery at a constant pressure. The intra-arterial forces measured by the sensor array are then translated into arterial pressure waveforms displayed in real time. The tonometric sensors, which are placed over the radial artery, may be affected by both improper sensor placement as well as tremors. Both methods may be inaccurate in subjects with peripheral vascular occlusive disease, and both are dependent on arm position. Recordings should be made with the arm abducted or elbow flexed so that the recording site is at the level of the heart throughout autonomic testing.

During formal tests of autonomic function, both intermittent and continuous BP recordings are usually obtained. Oscillometric BP recordings may be used to verify the accuracy of the continuous BP recordings, and may serve as a more reliable means of BP recording in the event of marked decreases in BP.

5. PARASYMPATHETIC FUNCTION TESTS

Neurophysiological tests of cardiovascular parasympathetic function evaluate the integrity of a reflex arc that includes the vagus and glossopharyngeal nerves; cardiopulmonary baroreceptors located in structures, such as the carotid sinus and aortic arch; pulmonary receptors; cardiac atria and ventricles; and pulmonary arteries and veins. Afferent signals from the receptors are carried in both myelinated and unmyelinated fibers, and travel in sympathetic nerves to the spinal cord and in the vagus and glossopharyngeal nerves to the nucleus tractus solitarius and other areas in the medulla that play a role in BP and HR control.

5.1. HR Response to Valsalva Maneuver

The Valsalva ratio (VR) is an index of HR changes that occur during a Valsalva maneuver (*see* Table 3). The Valsalva maneuver is performed by having the patient exhale for 15 s, while maintaining an expiratory pressure of 40 mmHg. Expiratory pressure can be measured by having the patient blow into a mouthpiece connected to a pressure transducer. Forced expiration to maintain an expiratory pressure of 40 mmHg produces the most consistent change in BP. Typically, a baseline HR and BP will be obtained during 3 min before the Valsalva maneuver. The patient is then instructed to inhale deeply, exhale completely, then take another deep breath in and blow into the mouth piece. Both HR and BP are recorded during the maneuver and up to 60 s after the Valsalva maneuver is terminated. A mild degree of lightheadedness is often reported during this procedure. Because of the need for prolonged expiratory effort, the elderly and those with any underlying pulmonary disorders may find the Valsalva maneuver difficult to perform. Typically, one to two trials are obtained for analysis, with an adequate time between trials (3–5 min) to allow subjects to return to a baseline state. Because intraocular pressure increases during Valsalva maneuver, recent retinal surgery or hemorrhage is a relative contraindication to performing this test.

The VR is a measure of the HR response to BP changes resulting from the mechanical and cardiovascular effects of the Valsalva maneuver. There are four phases of BP, designated I to IV, which delineate the cardiovascular changes during the Valsalva maneuver. Phase I consists of a transient rise in BP that results from the increased intrathoracic and intra-abdominal

Table 3
Valsalva Ratio

Stimulus	Expiration of 40 mmHg for 15 s
Afferent	Baroreceptors, glossopharyngeal, and vagus nerves
Central	Nucleus tractus solitarius
Efferent	Vagus and sympathetic nervous system
Response	1. Heart rate response to blood pressure changes
	2. Fall and rise in blood pressure (phases I–IV)

pressures. During phase II, there is a fall and recovery of BP. During phase III, there is a decrease in BP caused by the release of intrathoracic pressure. During phase IV, there is an "overshoot" caused by the increase cardiac output into a vasoconstricted peripheral circulation.

As the BP falls in phase II, there is an increase in HR caused by parasympathetic withdrawal that is then followed by sympathetic activation. This results in an increase in HR during and shortly after the Valsalva maneuver. In response to the BP overshoot during phase IV, approx 15 to 30 s after the end of the maneuver, the HR falls and produces a transient bradycardia that persists, in the normal state, until after the BP overshoot. This results in the minimum HR. The VR is calculated as the maximum HR, which occurs during or shortly after the Valsalva maneuver, divided by the minimum HR, which occurs after the cessation of the Valsalva maneuver (VR = max HR/min HR). A normal HR and BP response to Valsalva maneuver is depicted in Fig. 3.

A normal VR reflects an intact baroreceptor-mediated rise and fall in HR. The VR values are age related, and decline with increasing age. A reduced VR reflects baroreceptor and cardiovagal dysfunction. However, it is important that an adequate and steady expiratory pressure is maintained during the maneuver for these results to be meaningful.

5.2. HR Variability With Respiration (Respiratory Sinus Arrhythmia)

The expiratory/inspiratory (E/I) ratio, similar to the VR, is a measure of parasympathetic function (*see* Table 4). Timed breathing potentiates the normal sinus arrhythmia that occurs normally during respiratory cycles. The physiological mechanism of respiratory sinus arrhythmia is complex and includes pulmonary stretch and cardiac mechanoreceptor activation, baroreceptor activation, local cardiac reflexes and central factors. During inspiration, there is a reflex withdrawal of parasympathetic activity, increased excitability of cardiac pacemaker cells in the sinoatrial node, and an increase in HR. Respiratory sinus arrhythmia provides a measure of parasympathetic control of HR variation throughout the respiratory cycle.

The respiratory sinus arrhythmia is recorded with the patient supine and breathing at a fixed rate. A respiratory frequency of six breaths per minute, with slow inhalation and exhalation (5 s each) provides close to maximum HR variability (Fig. 4). The timed breathing is performed with the aid of either verbal coaching, or an electronic visual cue. This is a safe, easily performed test, without contraindications. Typically, six to eight cycles are recorded, and one to two trials are performed. The variation in HR across this cycle of six to eight breaths is often expressed as the mean difference between the maximum and minimum HR. Alternatively, one can calculate the E/I ratio, which is the sum of the longest RR intervals divided by sum of the shortest RR intervals (E/I = Σ longest RR/Σ shortest RR). Other

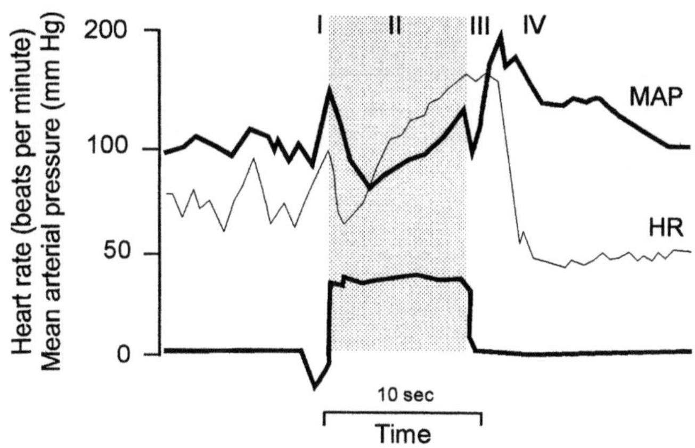

Fig. 3. Blood pressure and heart rate response to Valsalva maneuver. Top tracing represents mean arterial pressure (MAP) with phases denoted by Roman numerals. Middle tracing represents heart rate (HR). Bottom tracing shows expiratory pressure during Valsalva maneuver. The shaded area delimits both early and late phase II.

Table 4
Expiratory/Inspiratory Ratio

Stimulus	Deep breathing (6 cycles/second)
Afferent	Pulmonary receptors, cardiac mechanoreceptors, vagus and glossopharyngeal nerves, respiratory center
Central	Nucleus tractus solitarius
Efferent	Vagus
Response	1. Heart rate increase during inspiration 2. Heart rate decrease during expiration

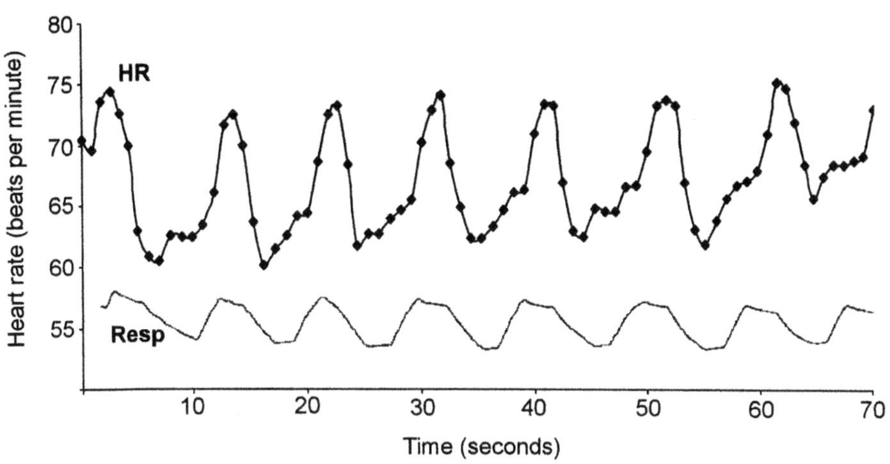

Fig. 4. Top tracing shows the variation in heart rate (HR) during metronomic breathing. Lower trace represents corresponding respiratory (resp) pattern.

analyses used to describe HR variability include measurements of HR standard deviation, mean consecutive difference, mean square consecutive difference, mean circular resultant, and power spectral analysis.

Table 5
Active Standing (30:15 Ratio)

Stimulus	Decreased central blood volume
Afferent	Baroreceptors, ergoreceptors, "central command," and vagus and glossopharyngeal nerves
Central	Nucleus tractus solitarius, rostral ventrolateral medulla
Efferent	Vagus
Response	1. Heart rate increase at beat ~15 2. Heart rate decrease at beat ~30

HR variability with respiration decreases with increasing age. Reduced HR variability with respiration can be observed in cases of autonomic failure caused by autonomic peripheral neuropathies and central autonomic degenerations. Other factors known to influence the HR response to deep breathing include poor respiratory effort, hypocapnia, salicylates, positioning, and obesity.

5.3. The 30:15 Ratio

The 30:15 ratio, similar to the VR and E/I ratio, is a measure of parasympathetic cardiovagal function (*see* Table 5). Immediately after standing, there is an increase in HR that reflects an exercise reflex and withdrawal of parasympathetic tone. There is a further HR increase at approx 15 s that reflects a compensatory response to decreased venous return and decreased cardiac output and a fall in BP. This is followed by a second period of relative bradycardia at approx 30 s. The 30:15 ratio is the RR interval at approx 30 s, divided by the RR interval at approx 15 s.

A baseline HR should be obtained in the supine position, and the patients are then asked to quickly stand upright onto their feet. A chair may be placed behind subjects to sit on should they become symptomatic suddenly. The HR variability is recorded for at least 1 min of active standing. A normal ratio is greater than unity, and reflects intact vagally mediated HR variation. An abnormal ratio suggests parasympathetic cardiovagal dysfunction. A number of general medical conditions, such as hypovolemia, medical deconditioning, and hypothyroidism can lead to misinterpretation of the 30:15 ratio and, thus, these factors should be taken into account when interpreting the test results.

6. SYMPATHETIC FUNCTION TESTS

6.1. Head-Up Tilt-Table Testing

During the tilt-table test, the BP and HR response to an orthostatic challenge is used to provide a measure of sympathetic function (*see* Table 6). Similar to the early response to active standing, the early cardiovascular response to head-up tilt is largely caused by blood volume redistribution to the lower extremities. However, tilt-table testing is more sensitive to such redistribution because there is minimal contraction of lower extremity muscles, thus, further reducing the amount of venous return. This test is used to assess orthostatic intolerance caused by sympathetic nervous system dysfunction and also to uncover a predisposition to neurally mediated (vasovagal syncope).

The patient lies supine on the tilt table. Beat-to-beat and oscillometric BP instruments are attached to each arm. EKG monitoring should take place throughout the test. A large waist

Table 6
Tilt-Table Testing

Stimulus	Decreased central blood volume
Afferent	Baroreceptors, vagus and glossopharyngeal nerves
Central	Nucleus tractus solitarius, rostral ventrolateral medulla, hypothalamus
Efferent	Sympathetic vasomotor
Response	1. Pattern, rate, and degree of blood pressure changes
	2. Heart rate rise or fall

belt should be placed around patients to secure them in case of syncope or unexpected falls. Other safety measures include the availability of cardiovascular medications, including atropine, and a resuscitation cart that can be used in the event of cardiac asystole and other unexpected cardiac arrhythmias.

Once the patient is comfortable, with feet resting on the footboard, a baseline BP is recorded for at least 3 min. The patient is then slowly tilted upright to an angle of 60° to 80°. During testing, the patient is asked to report any symptoms. Both BP and HR are recorded throughout tilt-table testing, after which the patient is returned to a horizontal supine position. BP and HR should be monitored in the supine position until patients return to baseline. Tilt-table test results are influenced by a number of conditions, such as hypovolemia, cardiovascular deconditioning, medications, hypothyroidism, infection and sepsis, and reduced myocardial function. These conditions must be accounted for during the interpretation of abnormal tilt-table findings.

Some laboratories use pharmacological measures to potentiate the orthostatic challenge of head-up tilt. Intravenous administration of isoproterenol, a β-adrenergic agonist that increases ventricular contractility and HR, is frequently used. This intervention may increase the sensitivity of the test but is associated with reduced specificity. In addition, the sight of needles and blood, as well as pain may further reduce the specificity of the test. The use of sublingual nitroglycerin may increase the sensitivity of the test while avoiding these precipitants.

A normal tilt-table test is one in which there are no symptoms, and only a modest fall in systolic BP. Orthostatic hypotension is defined as a decrease in systolic BP of greater than 20 mmHg, diastolic BP greater than 10 mmHg accompanied by symptoms of orthostatic intolerance. The pattern, the temporal characteristics, and the degree of changes that occur in both the BP and HR define the test abnormality. The use of the term "positive tilt-table test" is incomplete, uninformative, and should be avoided.

Three well-described patterns of neurally mediated syncope can occur during head-up tilt-table testing:

1. Vasodepression resulting in hypotension without bradycardia.
2. Cardioinhibition with a marked bradycardia (fewer than 40 beats per minute) with or without significant hypotension.
3. Mixed, with both bradycardia and hypotension (Fig. 5).

6.2. Valsalva Maneuver

As discussed in Subheading 5.1., there are four described phases of the Valsalva maneuver. Phase I is an initial rise in BP, largely caused by mechanical factors. Phase II is divided into early (II_e) and late (II_l) components. The fall in BP during phase II_e is caused by a decrease in preload and cardiac output. Phase II_l represents a rectification of the fall in BP

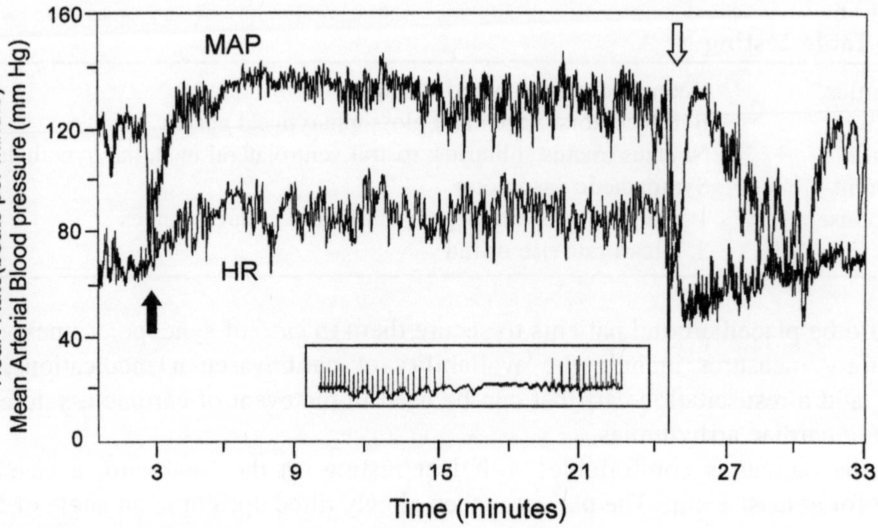

Fig. 5. Blood pressure and heart rate (HR) response to prolonged 45-min tilt-table test at 60° of upright position. Top tracing is mean arterial blood pressure (MAP) curve; bottom tracing is HR. The dark arrow depicts the point of head-up tilt. The hollow arrow depicts return to supine position. At 25 min into the test, there is a precipitous fall in both mean arterial pressure and heart rate. The inset is a sample of the ECG occurring at the 25 min mark, which shows heart rate slowing and asystole. *See* text for further details.

caused by an α-adrenergic sympathetically mediated increase in total peripheral vascular resistance and the increase in HR. The abrupt fall in BP during phase III represents a withdrawal of intrathoracic pressure at the termination of forced expiration. Phase IV, the last phase of the Valsalva maneuver, reflects a BP overshoot caused by the increase in α-adrenergic-mediated peripheral vascular resistance, and β-adrenoreceptor-mediated increase in cardiac output.

An abnormal response to the maneuver includes an excessive BP fall in the early phase II, an absent or incomplete recovery failure in late phase II, and/or the absence of a BP overshoot during phase IV. These abnormalities are observed in the presence of sympathetic dysfunction.

6.3. Sympathetic Cholinergic Sweat Function

The best-known neurophysiological test of sympathetic sudomotor function is the sympathetic skin response (SSR), or galvanic skin response. Increased sympathetic nerve fiber sudomotor activity causes release of sweat from glands in the skin, which can be detected as a change in surface conductivity. A standard EMG instrument can be used to measure this response. A small recording electrode is attached to the palmar and dorsal surface of both hands and feet, with a long sweep speed (10 s) setting, and a bandpass filter of 1 Hz to 2000 Hz. After a stable baseline is obtained, a stimulus is administered and the SSR is recorded. Any stimulus that activates the sympathetic system can be administered, such as an electrical stimulus, a loud noise, or a bright light. Simultaneous recording from both hand and foot can be performed, which demonstrates the length-dependent latency difference between these sites. Typically, three to four trials are obtained at each site. A normal SSR is a monophasic or biphasic deflection that habituates with time and repeated trials (Fig. 6). An abnormal

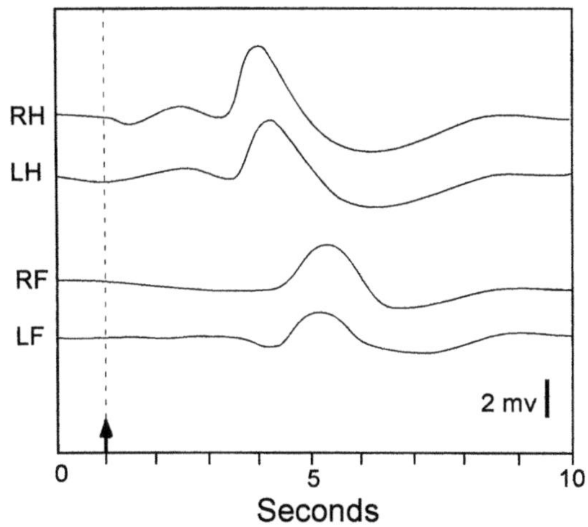

Fig. 6. Sympathetic skin response (SSR) measured from the left hand (LH) and right hand (RH), and left foot (LF) and right foot (RF) after a 30-mA stimulation to the proximal left upper extremity. The onset of the stimulus is denoted by the arrow and the dashed line. Note the delay in onset in the SSR recorded from the feet as compared with the arms.

response is an absent SSR, and reflects sympathetic sudomotor dysfunction. This test does not differentiate between preganglionic or postganglionic sympathetic dysfunction.

In the silastic imprint test, a plastic or silicone film is applied to the area of sweat gland activity provoked by the iontophoresis of a cholinergic agonist. The film is then removed, and the number of active sweat glands, their density, and the volume of sweat secreted can be determined by digital image analysis.

The thermoregulatory sweat test is performed by the application of an indicator, such as alizarin red, which changes color in the presence of moisture. The indicator is applied to the body of the patient, who is then placed in a thermally controlled environment, and the core temperature is raised above baseline. The pattern of color changes, which can be expressed as percentage of body surface, is used to quantify hypohypohidrosis and anhidrosis.

Iontophoresis of cholinergic agonists can also be used to evoke an axon reflex-mediated response from a more distal population of sweat glands. The response is recorded using a sudorometer, which measures changes in humidity resulting from sweat production. This quantitative sudomotor axon reflex test (Q-SART) can be used to measure both sweat volume and sweat pattern (Fig. 7). It is largely a measure of postganglionic sympathetic function, but may be reduced with severe preganglionic disorders.

7. MICRONEUROGRAPHY

Intraneural microneurography enables the direct measurement of muscle sympathetic nerve activity. A tungsten microelectrode of approx 5 μm in size is inserted into a fascicle of a distal sympathetic nerve to the skin or muscles. Using this technique, sympathetic outflow to skin and muscle can be measured at rest and in response to various physiological perturbations (Fig. 8). Muscle sympathetic nerve activity, consisting predominantly of efferent muscle blood vessel vasoconstrictor impulses, are grouped in relationship to the cardiac rhythm. Skin sympathetic activity consists of vasoconstrictor impulses to skin capillaries, arteriovenous shunts, and sudomotor and pilomotor impulses. They appear as irregular bursts

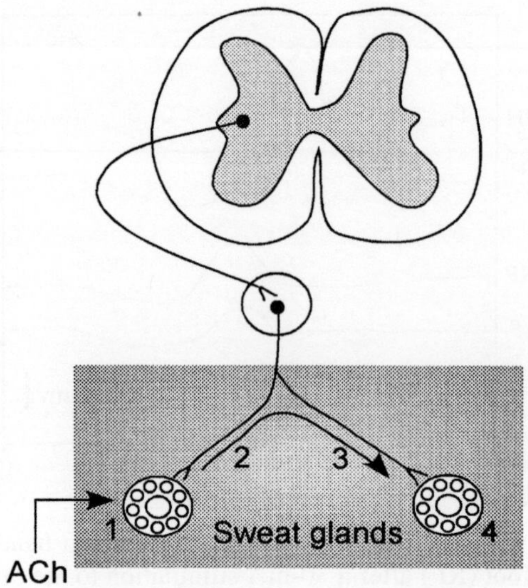

Fig. 7. Neurophysiology of the axon reflex test. Acetylcholine (ACh) is iontophoresed onto a sweat gland **(1)**, initiating an antidromic signal **(2)** in the distal postganglionic sympathetic fiber, which is then transmitted orthodromically **(3)** down an axon branch, stimulating glandular secretion at a distal site **(4)**. The hatched area represents structures within the skin.

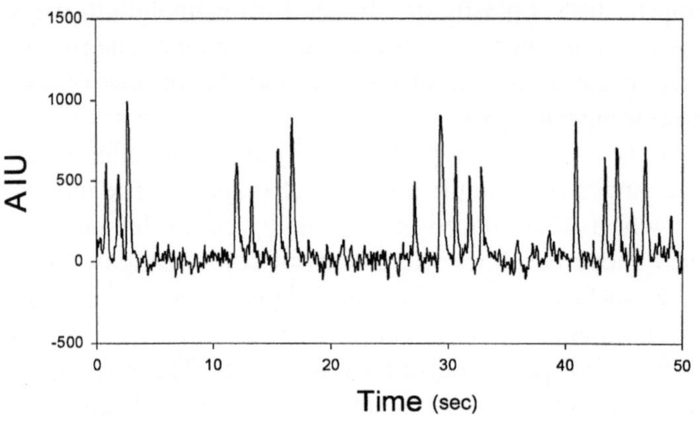

Fig. 8. Muscle sympathetic nerve activity displayed as integrated units (AIU) in response to sympathetic activation provoked by a nitroprusside bolus.

of varying duration, often occurring in relationship to the respiratory rhythm. This technique is primarily used in research, rather than in clinical laboratories.

SUGGESTED READING

Clinical autonomic testing report of the therapeutics and technology assessment subcommittee of the American Academy of Neurology. Neurology 1996;46:873–880.

Low PA. Laboratory evaluation of autonomic function. In: Clinical Autonomic Disorders, 2nd ed. (Low PA, ed.). Lippincott-Raven, Philadelphia, PA, 1997.

Primer on the Autonomic Nervous System (Robertson D, Low PA, Polinsky RJ, eds.). Academic Press, San Diego, CA, 1996.

Autonomic Failure—A Textbook of Clinical Disorders of the Autonomic Nervous System, 4th ed. (Mathias CJ, Bannister R, eds.). Oxford University Press, Oxford, England and New York, NY, 2001, p. 592.

Freeman R. Noninvasive evaluation of heart rate variability. In: Clinical Autonomic Disorders, 2nd ed. (Low PA, ed.). Lippincott-Raven, Philadelphia, PA, 1997, pp. 297–307.

The Autonomic Nervous System: An Introduction to Basic and Clinical Concepts, 5th ed. (Appenzeller O, Oribe E, eds.). Elsevier Science, New York, NY, 1997, p. 868.

REVIEW QUESTIONS

1. HR variability with respiration is a measure of:
   A. Parasympathetic nervous system function.
   B. Sinus arrhythmia.
   C. Vagal nerve activity.
   D. All of the above.
2. Head-up tilt-table testing may be used to measure:
   A. Parasympathetic nervous system function.
   B. Sympathetic nervous system function.
   C. A predisposition to vasovagal syncope.
   D. B and C.
3. A normal early phase II in the Valsalva maneuver refers to:
   A. HR increase.
   B. HR fall.
   C. BP overshoot.
   D. BP fall.
4. A decrease in the RR interval occurs when:
   A. The HR increases.
   B. The HR decreases.
   C. Vagal nerve activity increases.
   D. Vagal nerve activity decreases.
   E. E and D.
   F. B and C.
5. Which of the following agents can be used to induce axon reflex sweating:
   A. Isoproterenol.
   B. Propranolol.
   C. Nitroprusside.
   D. Acetylcholine.
6. The E/I ratio reflects:
   A. Sympathetic nervous system function.
   B. Parasympathetic function.
   C. BP fluctuations.
   D. HR variability.
   E. A and C.
   F. B and D.
7. The Valsalva maneuver is primarily used to evaluate:
   A. Sympathetic autonomic function.
   B. Parasympathetic baroreceptor function.
   C. A subject's expiratory capacity.
   D. A and B.
   E. A, B, and C.
8. The patterns of neurally mediate syncope include:
   A. Vasodepression.
   B. Mixed.

C. Cardioinhibitory.
D. All of the above.

9. Intravenous isoproterenol administered during a tilt-table test may:
A. Increase HR.
B. Reduce the specificity of the test.
C. Increase cardiac contractility.
D. All of the above.

10. The sudomotor axon reflex test is primarily a measure of these fibers:
A. Preganglionic sympathetic cholinergic.
B. Preganglionic sympathetic adrenergic.
C. Postganglionic sympathetic cholinergic.
D. Postganglionic sympathetic adrenergic.

REVIEW ANSWERS

1. The correct answer is D. The RR interval is the time between successive QRS complexes, and is a measure of cardiac vagal outflow. During normal breathing, a sinus arrhythmia can be observed in the RR interval, which is a normal physiological condition.
2. The correct answer is D. The head-up tilt-table test is used to assess orthostatic tolerance in patients with orthostatic hypotension caused by sympathetic nervous system dysfunction. It is also used to demonstrate a predisposition to neurally mediated syncope.
3. The correct answer is D. The early phase II section of Valsalva maneuver refers to the fall in BP that occurs during forced expiration. A BP overshoot can also be observed during phase IV of the maneuver; this overshoot occurs after the termination of forced expiration.
4. The correct answer is E. The RR interval is a measure of the time (in milliseconds) between successive QRS wave complexes. A decrease in the RR interval occurs when the HR increases. Because HR is under vagal control, a decrease in the RR interval, corresponding to an increase in the HR, occurs with the withdrawal of parasympathetic tone observed during a decrease in vagal nerve activity.
5. The correct answer is D. The cholinergic agonists are used to provoke axon reflex-mediated sweating. These agents are used to measure sudomotor function in the quantitative sudomotor axon reflex test and the silastic imprint method.
6. The correct answer is F. The E/I ratio is a measure of sinus arrhythmia. It is largely an index of cardiac vagal parasympathetic activity.
7. The correct answer is D. The HR and BP response to Valsalva maneuver is influenced by both the sympathetic and parasympathetic nervous system. Although an adequate forced expiration of 40 mmHg for 10 to 15 s is required to obtain reliable data during the Valsalva maneuver, it is not a formal test of pulmonary capacity or forced expiratory volume.
8. The correct answer is D. There are three patterns of cardiovascular responses that have been described during neurally mediated syncope. Vasodepression refers to a primary fall in BP without significant HR changes. The mixed form observed during head-up tilt-table testing is characterized by both hypotension and relative bradycardia, and cardioinhibitory neurally mediated syncope refers to a primary bradycardia below 40 beats per minute.
9. The correct answer is D. Intravenous isoproterenol increases cardiac inotropy and chronotropy. Although this intervention increases the sensitivity of the test, it reduces the specificity and can result in false positive studies.
10. The correct answer is C. The sudomotor axon reflex test is based on the fact that postganglionic nerve fibers can be antidromically stimulated by the application of a cholinergic agonist to the skin. This results in a signal traveling to a branch point, and then orthodromically down an adjoining postganglionic fiber, which results in the release of acetylcholine and the stimulation of eccrine glands in the skin. This test is used to evaluate the postganglionic sympathetic cholinergic system.

25
Visual Evoked Potentials

Frank W. Drislane

Summary

The visual evoked potential (VEP) is primarily a relatively large, positive polarity wave generated in the occipital cortex in response to visual stimulation. It measures the conduction time of neuronal activity from the retina to the occipital cortex and is used clinically as a measure of the integrity and function of that pathway. The optic nerve is the primary structure examined. The standard VEP averages many responses, time-locked to a photic stimulus. Of primary interest is the latency of the positive wave at a midline occipital EEG electrode, usually at approx 100 ms after stimulation, called the P100. This chapter summarizes the methodology for recording the VEP, provides an approach to its interpretation, and discusses its role in clinical practice.

Key Words: Evoked potential; visual; optic nerve; demyelination; latency; multiple sclerosis; P100; optic chiasm; occipital; blindness.

1. INTRODUCTION

The visual evoked potential (VEP) is primarily a relatively large, positive polarity wave generated in the occipital cortex in response to visual stimulation. It measures the conduction time of neuronal activity from the retina to the occipital cortex and is used clinically as a measure of the integrity and function of that pathway. The optic nerve is the primary structure examined.

The VEP is of large enough voltage that it can be seen occasionally on a routine EEG as an occipital waveform within the first 150 ms after a single photic stimulus. The standard VEP averages many such waveforms, time-locked to the stimulus. Of primary interest is the latency of the positive wave at a midline occipital EEG electrode, usually at approx 100 ms after stimulation, called the P100. This P100 peak is usually easy to recognize and measure.

2. RECORDING THE VEP

2.1. *Stimulus*

The standard stimulus for VEPs is a checkerboard pattern in which the squares alternate from black to white—the pattern reversal VEP (PRVEP). Dark squares become light and vice versa, without a change in the overall luminance of the display. Typically, the pattern is reversed 100 or 128 times at 1 to 2 Hz, and the results are then averaged. Usually a repeat trial of averaged stimuli is also recorded.

PRVEPs require maintaining visual fixation on the center of the pattern. The occipital cortex is particularly sensitive to the perception of edges, and a sharp-bordered checkerboard

From: *The Clinical Neurophysiology Primer*
Edited by: A. S. Blum and S. B. Rutkove © Humana Press Inc., Totowa, NJ

produces a strong and measurable response. PRVEPs are remarkably precise and constant for a given subject (who has no clinical change) and are very sensitive to dysfunction in the visual conducting system.

Alternatively, a flashing light in a strobe sequence (flash VEPs) or even alternating intensities or luminances of light can be the stimuli. They are used typically when a subject is unable to cooperate, for instance, neonates or patients with impaired mental status. Fixation is not required, and the eyes may be closed. Flash VEPs assess the integrity of the visual system at least through the lateral geniculate nucleus and can help to determine whether the optic nerve is intact. Flash potentials have greater latency variability and less sensitivity to visual conduction defects.

The PRVEP stimulus is generally presented on a television screen or video monitor positioned 1 m in front of the subject who is asked to focus on the center of the display. Inadequate fixation reduces the voltage (or amplitude) of the VEP—and, in the limiting case, will eliminate it. Eyes are tested one at a time, with the other eye covered by a patch. Simultaneous binocular testing could not localize an abnormality to one optic nerve or the other.

2.2. *Check Size*

Rather than describe the size of the stimulus, it is more appropriate to specify the visual angle it subtends. One degree (or 1/360th of a circle) is divided into 60 min (60′). An individual checkerboard square usually has a visual angle of 30′, with 8° for the entire stimulus or video screen. Smaller checks are more sensitive in detecting visual system defects, but visual acuity can be a problem. Peripheral vision is stimulated better by larger checks. Large checks produce more variable responses, perhaps because the subject focuses on areas of different luminance rather than on the edges. Checks larger than 50′ may help to compensate for poor visual acuity, but sensitivity to contrast in central vision is better with smaller checks.

2.3. *Contrast*

Contrast is the difference in luminance (or brightness) of the dark and light areas divided by the sum of their luminances. Low contrast decreases the P100 amplitude and increases the latency. Greater luminance decreases the P100 latency and increases the amplitude. The patient's pupils should not be dilated pharmacologically because this will alter luminance at the retina. Luminance is not one of the most important variables, but it must be kept constant over trials and among patients for standardization in the laboratory.

2.4. *Repetition Frequency*

The pattern reversal rate is usually approx two per second. Some components of the VEP last hundreds of milliseconds after the stimulus. Thus, repetition faster than four per second can produce overlap (interference of one potential with the next) and distort the waveform; it also may increase the P100 latency. Slower repetition rates prolong testing and might produce a varied response because of diminished attention. All stimuli and recording techniques must be kept constant in a laboratory.

2.5. *Averaging*

The time-locked voltage signals are averaged over 100 to 200 trials, usually with a duration of 500 ms each. A signal sampling rate of 1000 samples in 500 ms (2000 samples per second) is high enough to avoid distortion of the waveform. VEPs are usually amplified by a factor of 50,000. VEPs have a relatively high signal-to-noise ratio, and a larger number of trials is not required. This is in contrast to the subcortical or far-field auditory and somatosensory evoked potentials (EPs), which have much lower signal-to-noise ratios, requiring up to thousands of repetitions to produce reliable recordings. Averaging the waveforms eliminates the variation unrelated to the stimulus, distinguishing the VEP from the EEG background. The two separate trials (each following averaging) should be nearly superimposable. The recording computer usually includes artifact-rejecting programs; many provide an additional smoothing function.

2.6. Hardware

The recording EEG electrodes are placed 5 cm above the occipital midline (OZ) position and 5 cm to the right and left of this electrode—designated MO, RO, and LO. Usually, four channels are recorded. The first three channels consist of electrodes MO, RO, and LO, each referenced to a midfrontal electrode (MF). The fourth channel records MF referenced to an ear electrode and shows the active component of the MF electrode at approx 100 ms (with a negative polarity in contrast to that of the P100) and helps to explain some P100 distortions. MO–MF is the primary channel used for most readings. Noncephalic references generally contain too much artifact to be useful.

2.7. Filters

The low-frequency filter is usually set at 1 Hz and the high-frequency filter at 100 to 300 Hz (the shape of the standard P100 has a frequency of approx 15–20 Hz). A lower high-frequency filter frequency may cause an apparent increase in P100 latency.

2.8. Patient Factors

The patient should be alert and comfortable. No noise should accompany the stimulus; this could cause artifact. It is important to be sure that the stimulus can be seen clearly. Visual acuity must be tested. Usual eyeglasses are used to optimize visual acuity. There should be no pharmacological pupillary dilation. One eye is tested at a time. The technologist should ascertain that the patient is actually focusing on the center of the target throughout the test. This is particularly difficult for children and infants (note that fixation is not required with flash stimuli; they can be used even with comatose patients).

It should be determined before testing whether the patient has a significant visual field abnormality. If so, the distribution of the P100 field may be distorted and the latency prolonged somewhat. It may need to be recorded with more laterally placed electrodes.

If there are difficulties, it is important to check the patient's visual acuity and ability to cooperate. Low-amplitude VEPs may arise from inattention to the stimulus. The technologist should ask whether the stimulus is seen clearly. If visual acuity is a problem, increasing check size may help. It is also appropriate to check the integrity of the recording apparatus. Technical problems often eliminate all potentials, so the finding of some normal and some abnormal potentials suggests that there is a true clinical deficit.

3. NORMAL AND ABNORMAL VEPS

The VEP is a measure of physiologic function rather than primarily reflecting neuroanatomic lesions. EPs are particularly helpful clinically in determining a physiologic abnormality where the neurologic and ophthalmologic examinations are normal. VEPs are

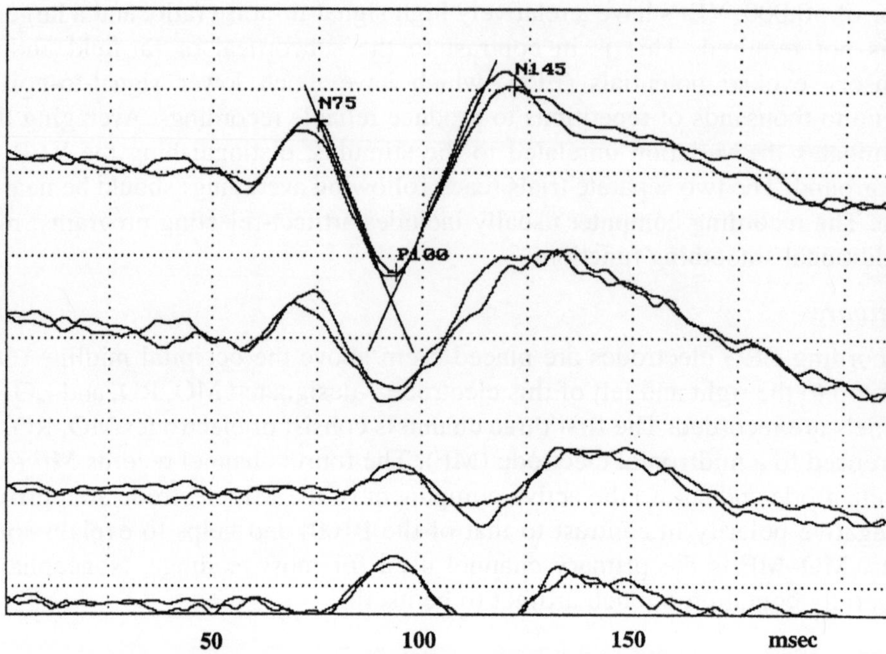

Fig. 1. Normal right eye pattern reversal visual evoked potential in a 39-yr-old woman. The N75 and N145 peaks help to identify the P100 (not necessary in this case). Triangulation lines help to determine the P100 latency precisely, at 93 ms in this case.

extremely sensitive and can detect dysfunction not discernible to neurologic, ophthalmologic, or other examinations. The primary measurement of interest clinically is the P100 latency after pattern reversal stimulation. Over time in the same healthy subject, P100 latencies may vary by as much as 10 ms, and interocular differences by as much as 9 ms. Most subjects have much less variability, and the results are often remarkably precise and reproducible.

3.1. Technical Factors

Latencies are determined from the stimulus signal to the peak of a deflection rather than at the onset of a deflection (as is common in peripheral nerve conduction testing). The clinical neurophysiologist should determine latencies and peak durations, with knowledge of the clinical neurophysiology and recording technique, often using paper printouts, rather than by computer cursors. Peaks may be determined simply by marking the nadir of the curve. Because many peaks are not so symmetric or regular, however, it is common to draw tangents to the slope of the primary downward and upward sides of the curve and assess the P100 latency at the intersection of those tangents (Fig. 1). Whatever the method of determination, it must be kept standardized for each laboratory.

As in EEG convention, the tracing has an upward deflection when the first electrode listed becomes more negative; the second electrode listed is considered the reference electrode. The full-field VEP displays a maximum voltage peak at the occipital midline. Thus, in most VEP recordings, with MO the first electrode in a channel, the P100 is a large downward deflection, appearing at approx 85 to 115 ms. Often, there is a smaller negative (or upward) peak at 60 to 80 ms, commonly designated the N75 peak. Another negative peak may follow the

P100. Latencies of the negative peaks are generally not of clinical interest, but their presence may assist in recognition of the P100.

There is a roughly normal distribution of P100 latencies in a healthy population; latencies 2.5 to 3 standard deviations beyond the mean are considered abnormal. Depending on the laboratory, this tends to be 114 to 117 ms. The upper limit of normal must be determined for a healthy population, with standard recording variables kept constant over time. VEPs are usually very precise for an individual over repeated trials, even months and years apart.

An excessive difference in the latencies on the two sides (usually 6–10 ms) is also considered abnormal. If the two peaks have normal latencies but an excessive difference between them, this raises concern for an abnormality in conduction in the visual pathway with the longer latency. This is not as reliable clinically as a prolonged latency itself.

Because the MF electrode is active and produces a negative potential at approximately the same time as the P100, a *W-shaped wave* can occur. The distorted P100 wave may also be caused by projections from different parts of the calcarine or occipital cortex, and stimulation of inferior visual fields may improve the waveform and demonstrate a normal (or abnormal) latency. This is not always recognized in the laboratory before the interpretation, and this realization may come too late to clarify an individual EP. Use of triangulation from the slopes on either side may overcome this problem. Use of a different reference electrode may do the same, but this also must be determined during the testing rather than at a subsequent reading. The actual shape or waveform is not constant enough to be considered normal or abnormal.

The amplitude of the P100 peak is variable, but often approx 10 μV, and, thus, would blend into the normal EEG background. Amplitude measurements are generally not of clear clinical significance or interest. The wave duration is determined from the peak of preceding N75 wave to the point at which the P100 wave returns to the earlier baseline. P100 duration, however, is not usually a clinically useful variable. Most markedly prolonged durations are associated with delayed latencies, as well.

VEPs are considered long latency and near-field potentials, coming after a significant delay (latency) after the stimulus. Near-field EPs appear to represent the excitatory and inhibitory postsynaptic potentials of cortical neurons. The visual (striate) cortex is generally considered the generator for N75 and P100. Large lesions in these areas can abolish the peaks.

3.2. Subject-Related Factors

3.2.1. Age

Under ideal conditions, PRVEPs may be recorded in infants and young children, but most infants cannot fixate on the pattern reversal target. Flash VEPs are generally abandoned once the subject can focus on that target.

P100 latencies decrease substantially during the first year of life. They assume typical adult values by age 1 yr, at least for larger check sizes, and certainly by age 5 yr for smaller checks. Each laboratory needs separately determined pediatric statistical norms. Latency may decrease slightly during adolescence but changes very little thereafter until it increases slightly after age 60 yr, possibly by approx 3 ms per decade.

3.2.2. Gender

Women have slightly shorter P100 latencies (by ~3 ms); this may be related to head size. Amplitudes are slightly higher in women. The P100 latency may increase slightly with increased body temperature. Most medications have little effect.

3.2.3. Acuity

If the image is sharp (not blurred), visual acuity is usually sufficient to get a reliable P100 latency. Extreme miosis or visual acuity less than 20/200 may increase the P100 latency somewhat, but the amount of light reaching the retina is usually not a major factor. Compensation for poor visual acuity can be made in part by increasing the check size. Smaller checks may be seen less well and increase the latency, particularly in children. Most differences in stimulation factors cause relatively small changes in latency and are not a problem, but these variables must be controlled and kept constant within a laboratory over different patients.

4. INTERPRETATION

4.1. Anterior Pathways

The primary role of VEPs is to assess the anterior visual pathway on each side. Monocular, full visual field stimulation restricts interpretation of an individual potential to the visual pathway anterior to the chiasm. A normal P100 latency indicates normal conduction from the retina to the occipital cortex. A delayed potential after stimulation of one eye (with a normal potential after stimulation of the other) implies a defect in conduction in the optic pathway anterior to the chiasm on that side.

If a P100 is absent, it must be determined whether this is caused by technical factors (the most common explanation) before it can be considered abnormal. Technical problems can include faulty electrodes, amplifiers, or averaging hardware, or failure of the subject to focus on the target. If not a technical artifact, complete absence of the VEP is generally caused by severe ocular or optic nerve disease, such as complete blindness in an eye or interruption of the optic nerve.

Pathways posterior to the optic chiasm share conduction of activity from both retinae. Bilaterally delayed latencies after stimulation of each eye separately could be caused by bilateral optic nerve lesions, but they could also be caused by chiasmatic or widespread posterior lesions, so the finding is not as useful clinically. Without hemifield stimulation, posterior visual pathways cannot be assessed.

Most abnormally prolonged P100 latencies are caused by disease affecting the optic nerve, particularly demyelination, compression, and other optic neuropathies. In clinical practice, VEPs are used primarily to detect lesions when they are not easily demonstrable by clinical examination or indicated by history. Often, the disease process (e.g., multiple sclerosis [MS]) is suspected, and the VEP is part of a search for lesions caused by that disease.

4.2. Posterior Pathways

Full-field stimulation of either eye activates both occipital cortices. The occipital cortex on each side generates a response large enough to be recorded over a wide area, including the opposite occipital region. Because cortical activity in either occipital lobe can produce a P100 wave with a normal latency at the occipital midline, and because the signals from each occipital lobe cannot be separated, full-field stimulation can produce a normal VEP even if there is a large lesion in posterior optic pathways.

Hemifield stimulation is required to assess posterior pathways. This is more difficult technically and is not performed in many laboratories; standardization is also difficult. Stimuli and patient considerations are the same. The patient focuses on the center of the screen. A randomization program may be used to present standardized stimuli alternately to either visual field at random times to maintain fixation on the central target. The recording is registered by computer to the appropriate side of stimulation. Stimulation is typically with 35′ checks, but half-field testing often uses larger check sizes to improve stimulation at the periphery.

With hemifield stimulation there may be no VEP produced or recordable in the midline. Therefore, the usual montage focuses on the laterally placed LO and RO electrodes ipsilateral to the stimulated visual field, referenced to MF. Paradoxically and interestingly, the VEP, although generated in the occipital cortex opposite the stimulated field, produces the largest scalp amplitude recording over the occipital lobe *ipsilateral* to the stimulated field—likely related to the orientation of the stimulated occipital neurons. This has been shown with well-recorded hemifield VEPs even after hemispherectomy!

Lesions in the chiasm itself can interfere with crossing of conduction from the retina to the opposite hemisphere (from the nasal retina, i.e., temporal visual field). In theory, chiasmal lesions may produce bilateral temporal hemifield abnormalities after stimulation of each eye. Stimulation of the nasal field (temporal retina) activates the occipital cortex ipsilateral to the eye stimulated (but the potential is seen best over the opposite occipital area). With chiasmal lesions, the nasal hemifield stimulation should remain normal.

Retrochiasmal lesions should produce VEPs of prolonged latency in the affected field after stimulation of *each* eye separately (normal and abnormal potentials should not change depending on the eye stimulated). An absent VEP or prolonged VEP latency after hemifield stimulation of one eye alone (with hemifield testing results of the other eye remaining normal) suggests a *partial* lesion in that eye or the optic nerve on the side affected.

In summary, it is only very large posterior pathway lesions that disrupt the VEP significantly. Many such lesions produce a dramatic visual field abnormality and can be detected by neurologic examination. With improved neuroimaging by MRI, hemifield VEP testing is seldom used today.

5. CLINICAL APPLICATIONS

The most common and important VEP abnormality is a delay in the P100 latency after full-field stimulation of a single eye. This indicates a defect in the optic conducting system anterior to the chiasm on that side—usually optic nerve disease. The abnormality is particularly clear if the P100 is normal after stimulation of the opposite eye (Fig. 2). Chiasmal and more posterior lesions, or widespread brain dysfunction can cause *bilateral* delay of the P100, usually with similar delays on testing of each eye.

Optic neuropathies can occur on both sides. Bilateral demyelinating lesions typically produce asymmetric and abnormal latencies on the two sides. Metabolic and degenerative disorders typically cause delays that are similar on the two sides. Also, P100 latency prolongations caused by retrochiasmal lesions are usually similar after stimulation of each eye separately.

5.1. Optic Neuritis

Optic neuritis and MS are the primary concerns in the assessment of the optic nerve. Approximately 90% of patients with a definite history of optic neuritis have delayed P100s. Acutely, a normal VEP makes the diagnosis of optic neuritis very unlikely, especially with severe clinical symptoms. Acute optic neuritis, whether part of MS or not, produces a marked prolongation of the P100 in the affected eye. Over weeks to months, visual acuity usually

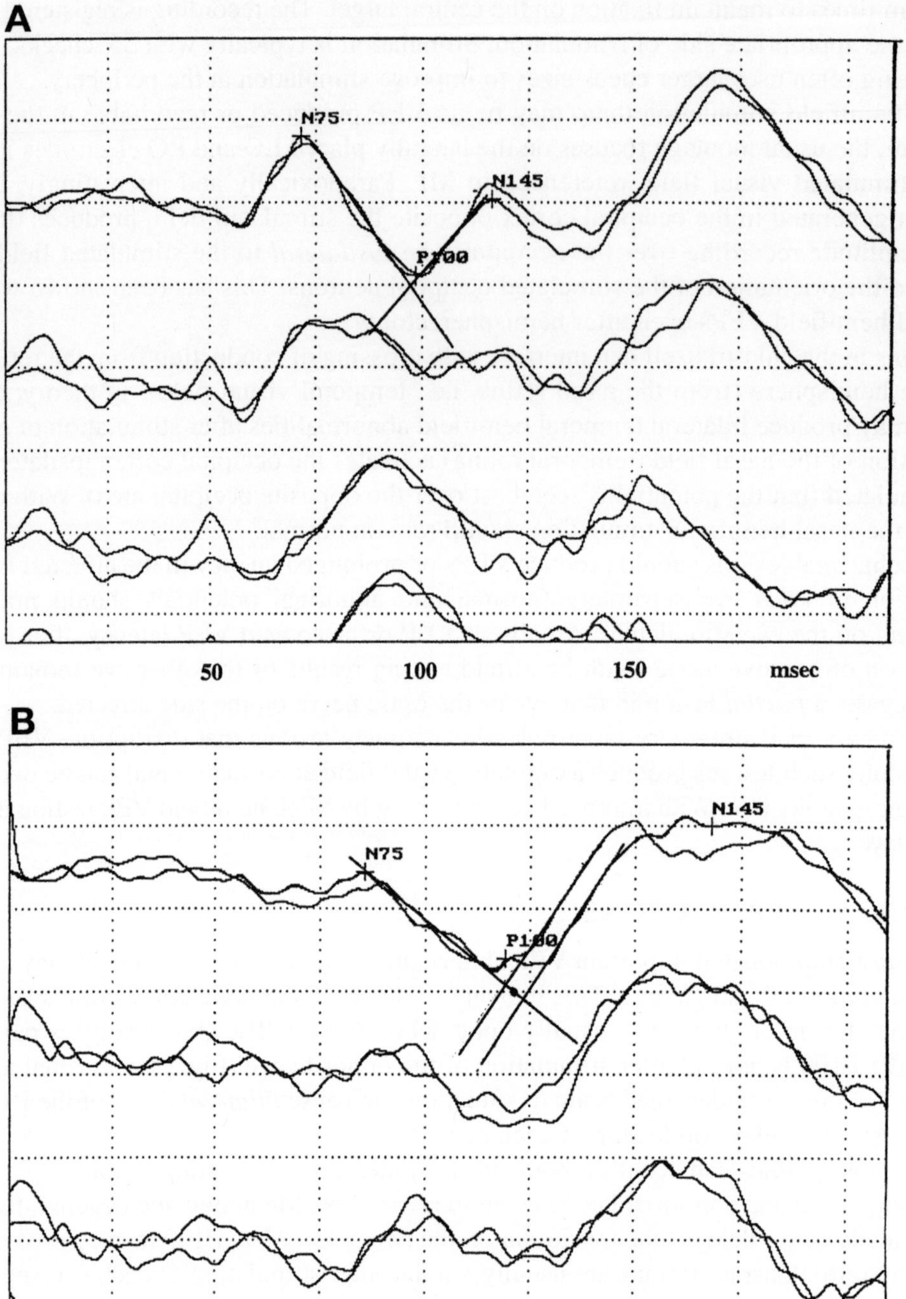

Fig. 2. (**A**) Normal pattern reversal visual evoked potential on the left (P100, 98 ms) and (**B**) significantly delayed P100 on the right (P100, 122 ms) in a 46-yr-old woman with right eye pain and abnormal T2-weighted spots on her MRI of the brain, considered clinically to have "possible MS."

improves, and approx 5% may return to normal after years. An abnormal VEP in an eye that appears normal clinically suggests an old optic neuritis with clinical recovery. Often, the VEP will improve also, though it seldom returns to normal. If an optic neuropathy is evident clinically before testing, the VEP will not add to the clinical impression.

5.2. *Multiple Sclerosis*

A clinically silent optic neuropathy found by VEP can be an aid in making the diagnosis of MS (although the diagnosis is often made now with greater reliance on MRI scans). Within the context of a diagnosis of MS, the VEP can be used to answer the question of whether the optic nerve is involved or not.

In MS with optic nerve involvement, there is typically a marked delay (e.g., 20–30 ms), but not complete absence of the P100. Very asymmetric bilateral abnormalities suggest bilateral optic neuropathies, possibly at different stages of development or recovery. P100s delayed to a similar degree on both sides may represent bilateral optic neuropathies but also raise the possibility of chiasmal, posterior, or more widespread metabolic disturbances.

VEPs are the single most useful EP in the evaluation of MS. They are abnormal in approx 90% of patients with definite MS and in up to 25% of patients with normal MRI results who are being evaluated for the possibility of MS. VEPs are more reproducible and reliable than somatosensory EPs and narrow the abnormality to a smaller region. Brainstem auditory potentials are far less sensitive in a diagnosis of MS. For study of the optic nerve alone, VEPs have sensitivity comparable to that of MRI. Neither, however, is completely specific for MS.

In patients with suspected "probable" or "possible" (rather than definite) MS but no history of optic neuritis, VEPs may show a clinically unsuspected lesion in 50% of patients and help to establish a diagnosis. If the diagnosis of optic neuritis or another cause of optic neuropathy is clear clinically, the VEP will add little additional information. Fewer than 10% of prolonged P100s return to normal in optic neuritis in the setting of MS, but this may reach one-third of patients with isolated optic neuritis (without MS).

In the case of suspected transverse myelitis, the VEP may offer evidence of demyelination far removed from the clinically obvious lesion. This may help to distinguish an isolated transverse myelitis from a multifocal disease process (such as MS) that includes a spinal cord lesion.

Finally, the VEP may also be used to monitor the progress of patients with optic neuritis or MS, but this may not be superior to clinical assessment alone. VEPs may be used as an assay for following progress in MS clinical treatment trials, but lesion burden on the MRI is generally considered a more reliable surrogate for the clinical status. VEPs may worsen even without a clinical recurrence of optic neuritis, so the correlation with disease burden is imprecise.

5.3. *Ischemic Optic Neuropathy*

Ischemic optic neuropathy may delay the P100. It often reduces the amplitude but typically does not affect the latency as much as demyelination does.

5.4. *Compressive Optic Nerve Lesions*

Compressive optic nerve lesions, such as those caused by tumors, may increase the latency of the P100 but they also tend to decrease the amplitude and may even abolish the

VEP altogether. MRI is a more useful test to look for compressive lesions, although the VEP can discern whether the optic nerve is physiologically involved by a process, if uncertain clinically. More posterior tumors causing papilledema may affect vision but not necessarily delay the P100.

Some compressive lesions involve the chiasm. These may produce visual field abnormalities, typically assessed better by neuro-ophthalmologic examination. Nevertheless, the VEP will often show an abnormality. Retrochiasmal mass lesions are unlikely to affect VEPs unless they are so large that they are obvious clinically and by MRI. Usually, the P100 latency delay is much less prominent than that caused by demyelination.

5.5. Traumatic Visual Loss

After blindness caused by trauma, a normal VEP offers a relatively good prognosis for recovery of vision, and an absent VEP is a very poor prognostic sign.

5.6. High-Pressure States

Pseudotumor cerebri usually does not affect the VEP substantially, although the high pressure can cause visual loss. In pseudotumor, VEP abnormalities may precede visual loss, but this is not reliable, and VEP changes may occur too late to be of clinical aid. Similarly, a mild latency change may be seen with glaucoma.

5.7. Neurodegenerative Diseases

Alzheimer's disease typically leaves the PRVEP unaffected. A minority of Parkinson disease patients has prolonged P100s, even without evident neuropathological changes in the optic conducting system; the significance of this is uncertain. Diseases involving white matter tracts are more likely to produce abnormalities. Leukodystrophies, such as adreno leukodystrophy, delay the P100. Most patients with Friedreich's ataxia have bilateral but relatively mild P100 delays. Abnormal VEPs are relatively common in spinocerebellar degenerations. Leber's hereditary optic neuropathy increases the P100 latency substantially, but also reduces the amplitude and eventually abolishes the VEP.

5.8. Various Medical Illnesses

Various medical illnesses, such as sarcoidosis and vitamin B12 deficiency, can lead to VEP abnormalities, but they are clearly very nonspecific. Hyperthyroidism may prolong the P100 latency; changes may resolve with treatment. Whether the VEP is abnormal depends on the nature of a disease and which structures it involves. The VEP is seldom helpful in the diagnosis of these many medical or degenerative conditions.

5.9. Operative Monitoring

Brainstem auditory and somatosensory EPs do not require a patient's attention and can be recorded even in coma or when a patient is under anesthesia. They can help establish whether the corresponding neuronal structures and pathways are intact. The intactness of pathways subtending brainstem auditory and somatosensory EPs can be very useful positive prognostic signs in coma.

PRVEPs require focus on a target and are, thus, not feasible during general anesthesia or coma. As noted earlier, flash VEPs may help to establish the integrity of the optic nerve or its severe disruption by trauma. They may indicate problems resulting from surgery (e.g., for an optic glioma) or other masses near the optic nerve or chiasm. Many patients with significant disease in these areas may have retained normal VEPs, however, so they cannot be used as screening tests.

5.10. Chiasmal Lesions

Chiasmal lesions often produce dramatic and unusual visual field defects, best detected clinically. They may produce bilaterally delayed P100 latencies, and, at times, hemifield stimulation may be abnormal. In most cases, however, the visual field examination and MRI scan are of greater use in detecting chiasm area lesions.

5.11. Retrochiasmal Lesions

Retrochiasmal lesions are rarely detectable by full-field stimulation. Bilateral posterior pathway lesions may delay the P100 bilaterally, typically in a symmetric fashion. Hemifield stimulation is necessary to detect unilateral posterior pathway lesions but, again, these lesions must generally be very large, clinically apparent to visual field or neurologic testing, and essentially always evident on MRI scans.

5.12. Ocular Problems

Ocular problems diminishing luminance at the retina tend to lower P100 amplitude, but any increase in VEP latency should be minimal. This can occur with eye closure, ptosis, corneal opacities, cataracts, severe miosis, or hemorrhages and other material in the cornea or liquid chambers of the eye. Retinopathy and glaucoma may increase the P100 latency slightly. Cataracts and other causes of loss of visual acuity do not typically affect the P100 wave latency until visual impairment becomes severe. Most ocular causes of abnormal VEPs are associated with markedly abnormal visual acuity and are often evident clinically.

Some ocular problems can be overcome by using a larger check size, particularly if there is diminished visual acuity. By varying check sizes and looking for responses, one can establish a rough estimate of visual acuity in very young children or others unable to report on their own vision.

5.13. Psychiatric Blindness

Psychiatric blindness can be evaluated to some degree with VEPs. Most activities leading to an intentional diminution of VEP recording (such as poor fixation or eye closure) can be seen by the technologist, and they typically will not affect latency anyway. As long as the subject looks at a target stimulus to some extent, a normal latency VEP may be obtained. This suggests normal conduction from the retina to the cortex, but it may not exonerate the entire occipital cortex from involvement with disease—cortical blindness can coexist with a normal VEP (flash VEPs test continuity of the visual pathway through the lateral geniculate alone and do not bear on the function of the posterior pathways or occipital cortex). Usually, however, VEPs are abnormal in cortical blindness. Thus, the normal VEP cannot be used to determine with certainty that blindness is psychiatric in origin, although a normal VEP would be unusual with symptoms of severe visual loss. Conversely, however, an abnormal VEP would be quite helpful in establishing biological disease in a patient suspected of psychiatric blindness.

SUGGESTED READING

Celesia GG. Visual evoked potentials and electroretinograms. In: Electroencephalography, Basic Principles, Clinical Applications, and Related Fields, 3rd ed. (Niedermeyer E, Lopes Da Silva F, eds.). Williams & Wilkins, Baltimore, MD, 1993, pp. 911–936.

Chiappa KH. Pattern-shift visual evoked potentials: methodology and interpretation. Chapters 2 and 3 in: Evoked Potentials in Clinical Medicine (Chiappa KH, ed.), Lippincott Raven Press, New York, NY, 1997, pp. 31–130.

Drislane FW. Use of evoked potentials in the diagnosis and follow-up of multiple sclerosis. Clin Neurosci 1994;2:196–201.

Misulis KE. Chapters 6 to 10: Visual evoked potentials. In: Spehlmann's Evoked Potential Primer, 2nd ed. (Misulis KE, ed.). Butterworth-Heinemann, Boston, MA, 1994, pp. 53–112.

Quality Standards Subcommittee of the American Academy of Neurology. Practice parameter: the usefulness of evoked potentials in identifying clinically silent lesions in patients with suspected multiple sclerosis (an evidence-based review). Neurology 2000;54(9):1720–1750.

REVIEW QUESTIONS

1. What is a VEP? What is the P100?
2. What does the PRVEP require of the patient? When is flash VEP used?
3. Which decrease the P100 amplitude: low or high contrast, and low or high luminance?
4. Which increase the P100 latency: low or high contrast, and low or high luminance?
5. What is the implication of a completely abolished VEP after stimulation of one eye?
6. What is the implication of a delayed VEP after stimulation of one eye?
7. What is the implication of bilaterally, but very asymmetrically delayed, VEPs after monocular stimulation of each eye?
8. What VEP technique is theoretically valuable to detect retrochiasmal lesions?
9. What is the value of the VEP in patients in whom MS is suspected?
10. Is there a role for VEP in the evaluation of patients with suspected psychogenic blindness?

REVIEW ANSWERS

1. The VEP is the waveform evoked from the occipital cortex after visual stimulation. The P100 is the name of the waveform measured using scalp occipital electrodes, time-locked to visual stimulation. It is usually a large positive deflection at approx 100 ms after stimulation, hence the name P100.
2. The PRVEP requires the patient to maintain visual fixation on the center of the target screen. The patient should be able to see the stimulus adequately. Whenever possible, optimization of the patient's visual acuity is desirable. Check size can be increased, if need be, to accommodate for poor visual acuity. Flash VEP can be used in subjects who are unable to fixate, such as neonates, or patients with impaired mental status.
3. Low contrast and low luminance lower the amplitude of the P100.
4. Low contrast and low luminance increase the P100 latency.
5. A completely absent VEP after monocular stimulation of one eye may suggest a lesion involving the anterior optic pathway on that side, but also may simply relate to technical factors, such as failure to fixate, or improperly functioning electrodes or hardware. In addition, severe ocular disease may be the explanation, rather than optic nerve disease.
6. A unilaterally delayed VEP response suggests a lesion of the optic pathway on that side, anterior to the chiasm.
7. Bilateral, but very asymmetrically, delayed VEPs usually suggest bilateral optic nerve lesions, possibly at different stages of recovery. Nevetheless, one cannot exclude chiasmatic or more broadly distributed postchiasmatic lesions.
8. Hemifield stimulation can theoretically be applied to sort out retrochiasmal pathways. This technique is not commonly performed, however, because MRI can often address such concerns much more easily.
9. VEP testing may be useful to identify clinically silent lesions in those with possible or probable MS based on other criteria. VEPs are abnormal in 90% of those with definite MS and up to 25%

of those with suspected MS and normal MRI results. Less than 10% of those with previous optic neuritis with MS will have normalization of the VEP over time.

10. VEPs may be helpful in the evaluation of suspected psychogenic blindness. In particular, an abnormal VEP speaks to an organic pathological process at work. A normal VEP can be supportive of psychogenic blindness, but cortical blindness can occasionally be associated with normal VEPs. Thus, a normal VEP does not exclude cortical blindness altogether.

26

Brainstem Auditory Evoked Potentials

Jacob R. Berger and Andrew S. Blum

Summary

Brainstem auditory evoked potentials (BAEPs) are electrical field potentials generated by stimulation of the auditory pathways. With repetitive auditory stimulation, reproducible electrical potentials can be elicited and recorded from scalp electrodes. These waves are generated by specific brain regions and occur at predictable intervals. Clinically, this neurophysiological property is useful to evaluate the integrity of auditory pathways (plus, by extension, neighboring CNS structures) and to localize defective transmission. This chapter summarizes the methodology and clinical application of BAEPs in the investigation of disorders affecting auditory pathways and the surrounding brainstem.

Key Words: Acoustic nerve; cerebellopontine angle; hearing; intraoperative monitoring; latency; midbrain; pons.

1. INTRODUCTION

Brainstem auditory evoked potentials (BAEPs) are electrical field potentials generated by stimulation of the auditory pathways. With repetitive auditory stimulation, reproducible electrical potentials can be elicited and recorded from scalp electrodes. Differences in electrical potential between active and reference electrodes are displayed graphically as a function of time after the auditory stimulus. These waves are generated by specific brain regions and occur at predictable intervals. Clinically, this neurophysiological property is useful to evaluate the integrity of auditory pathways (plus, by extension, neighboring CNS structures) and to localize defective transmission.

2. ANATOMY OF AUDITORY PATHWAYS AND CORRELATION WITH BAEP GENERATORS

To accurately interpret BAEP waveforms, a working knowledge of auditory pathway anatomy and physiology is important. The auditory pathway starts in the cochlea. Vibrations from sound waves cause hydrostatic pressure changes that are sensed by hair cells in the organ of Corti. Hair cells communicate with bipolar neurons that have cell bodies in the cochlea. These neurons project to the pontomedullary junction via the acoustic nerve. After synapsing in the ventral (low-frequency tones) or dorsal (high-frequency tones) cochlear nuclei in the rostral, lateral medulla, second-order neurons project to ipsilateral and contralateral superior olivary nuclei. Note that acoustic information is bilaterally represented at this point onward throughout the auditory system. Neurons in the superior olive then project

From: *The Clinical Neurophysiology Primer*
Edited by: A. S. Blum and S. B. Rutkove © Humana Press Inc., Totowa, NJ

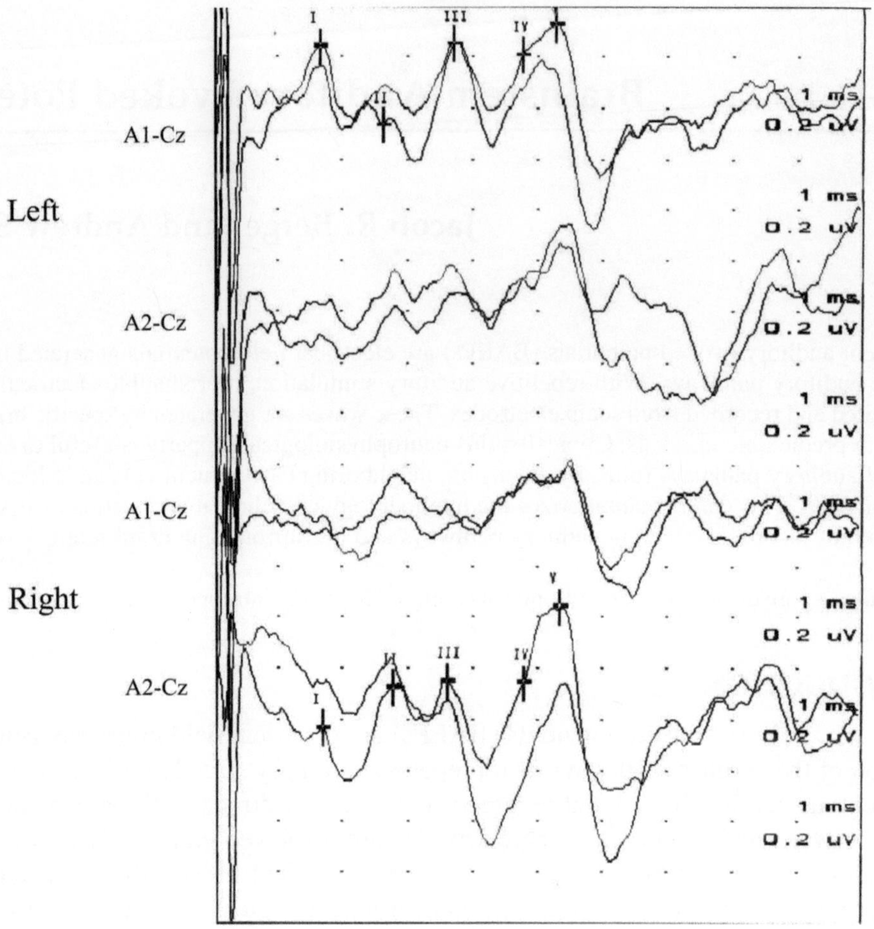

Fig. 1. Normal BAEP waveforms. These waveforms are from a 30-yr-old woman undergoing a work-up for possible multiple sclerosis. She had no auditory complaints. The top two channels represent results after left ear stimulation and the bottom two channels after right ear stimulation. Recording montages are indicated. Each study was performed in duplicate. Note the replicable BAEP waveforms, numbered I to V, as described in the text. Waveform latencies are measured for each and interwave latencies are derived including I to III, III to V, and I to V intervals. These are then compared with normative values. These BAEPs are normal and symmetric, side-to-side.

through the lateral lemniscus in the pontine tegmentum to the inferior colliculus. The inferior colliculus projects to the medial geniculate nucleus of thalamus, which projects forward to Heschl's gyrus (superior temporal gyrus), the primary auditory cortex.

BAEPs are made up of a sequence of stereotyped waveforms; five well-recognized waves are observed, numbered I to V (Fig. 1). Two later-appearing waves (VI and VII) are less reliable and of limited clinical value. These waveforms, elicited by auditory stimulation, are generated by specific structures in the auditory pathway described in the previous paragraph. All BAEP waveforms represent far-field potentials, caused by activity of deep gray and/or white matter structures, relatively far from the surface recording electrodes. Table 1 lists the known or likely generators for each of these waves.

Table 1
Brainstem Auditory Evoked Potential Waveforms and Their Associated Physiological Generators

Wave	*Generator*
I	Proximal acoustic nerve (segment near cochlea)
II	Distal acoustic nerve (segment near brainstem) or cochlear nuclei
III	Superior olive and projections to the lateral lemniscus; medial nucleus of trapezoid body might generate a part of wave III
IV	Most likely the lateral lemniscus, but data is not definitive
V	High pontine or lower midbrain structures: probably the lateral lemniscus, inferior colliculus, or a combination thereof
VI	Most likely the medial geniculate nucleus or projections from the inferior colliculus
VII	Most likely the auditory radiations to primary auditory cortex

3. METHODOLOGY

3.1. Stimulation and Recording Technique

Auditory evoked potentials (AEPs) are classified as short, middle, or long latency potentials depending on their latency from stimulus onset. The short latency AEPs, occurring within 10 to 15 ms of stimulus onset, consist of the electrocochleogram and BAEPs. The electrocochleogram, middle, and late-onset AEPs are of less clinical relevance and will not be included in this review.

To elicit and record BAEPs, an aural stimulus is delivered to the patient via headphones or indwelling earphones. Surface electrodes placed at A1 (left ear or mastoid), and A2 (right ear or mastoid) positions record voltage differences generated by stimulation of auditory pathways, using Cz (vertex) as a common reference. Waveforms are recorded from ipsilateral and contralateral pathways simultaneously, allowing easier recognition of individual peaks. Low- and high-frequency filters are commonly set to frequencies of 10 to 3000 Hz, respectively. Low-frequency filters can be raised to 100 to 200 Hz if muscle or mechanical artifact proves problematic. However, as the bandpass narrows, distortion of waveform morphology, amplitude, and latency can occur because of exclusion of low-frequency elements.

Although pure tones are routinely used to analyze the specific frequency spectra of hearing loss, they are generally not used to elicit BAEPs. The most common acoustic stimuli used in BAEP testing are broadband (wide frequency spectra) 2-ms-long clicks, which are generated by a 100 μs square-wave pulse delivered to the diaphragm of the speaker membrane. As one ear is stimulated with clicks, the other is masked with white noise, an equal mixture of all of the frequencies within the range of human hearing (typically 20 Hz–20 kHz). This method helps to prevent the undesired coactivation of the contralateral ear caused by bony conduction from the ipsilateral stimulated ear.

As with visual and somatosensory evoked potentials (SSEPs), and owing to the relatively small amplitude of the evoked auditory potentials, it is essential to average the BAEP waveforms produced by multiple stimulations to increase the signal-to-noise ratio. In general, 1000 to 4000 stimuli are given and averaged in BAEP testing. Eight to ten stimuli are given per second in most laboratories. Higher rates of frequency cause a rate-dependent reduction of amplitude, dispersion of waveforms, and longer interpeak latencies. Thus, if stimulation rates exceed 10 Hz, a different set of normative values must be used.

Click intensity can impact BAEP generation. With decreasing click intensity, absolute latencies are increased and amplitudes are diminished. However, interpeak latencies remain relatively constant. Click stimulus intensity is measured in units of decibel sensation level (dBSL), decibel hearing level (dBHL), or decibel peak equivalent sound pressure level (dBpeSPL) rather than the more familiar decibels. Units of dBSL are used when hearing thresholds are determined for the *individual ear* being examined. If hearing thresholds are established using a *group* of healthy people, dBHL units are used. To ensure maximal and consistent BAEP acquisition, the stimulus intensity is set to 65 dB above dBSL or dBHL. As an example, if a person or group studied had a mean perceptual threshold of 10 dB for pure tone audiometry in the testing environment, an appropriate click stimulus intensity for BAEP acquisition would be 75 dB. Both the dBSL and dBHL are subjective measurements. If more objective measures are required, the dBpeSPL can be used. To determine the dBpeSPL, the amplitude of the headphone speaker's membrane response to a brief click is recorded. This amplitude is matched to a longer duration sine-wave stimulus of defined dB intensity. In general, the dBpeSPL is approx 30 dB greater than the dBSL or dBHL.

Stimulus polarity can also affect BAEP waveforms. The movement of the speaker membrane towards the eardrum is called *condensation*; movement away from the eardrum is termed *rarefaction*. Most BAEPs are recorded during rarefaction, but it is often helpful to switch polarity (record during condensation) to distinguish artifact from true BAEP waveforms. Artifact should reverse in polarity whereas BAEPs should not be altered. Wave I is the BAEP most affected by changes in stimulus polarity, with rarefaction producing a shorter latency. It may be necessary to change stimulus polarity to adequately see individual waveforms in some patients. Wave V in particular may seem to be absent until stimulus polarity is reversed.

3.2. Patient-Related Variables Affecting BAEPs

A number of patient-related factors may also impact the BAEP recording. These include the individual's age, gender, level of arousal, body position, temperature, medications, and preexisting hearing loss. Of these, body position, level of arousal, medications, and patient temperature can be adjusted to optimize results.

Muscle artifact can interfere with the facile recording of BAEPs in 30 to 50% of waking patients. Sleep, most anesthetics, and coma of metabolic origin reduce this artifact. In the outpatient arena, it is often helpful to encourage the patient to try to sleep. Proper body positioning can also reduce the degree of muscle artifact caused by the cervical musculature as well as providing a situation more conducive to sleep.

Although minor differences in BAEP amplitudes and latencies occur as an individual matures to adulthood, the most pronounced changes occur in infancy and early childhood. Neonates and infants have thinner skulls and smaller head sizes than adults. As a result, their scalp recording electrodes are in closer proximity to the generators for these potentials. This results in higher amplitude BAEPs. Premature infants less than 30 wk of age may not have detectable BAEP waveforms. By age 3 to 6 mo, waveform morphology should approximate that observed in adulthood; by 1 to 2 yr, latencies shorten to adult values. Changes occur most rapidly in early infancy, reflecting the rate of myelination, nerve fiber growth, and development of synaptic efficiency occurring in the developing nervous system.

Female patients typically have shorter peak and interpeak latencies than male patients. Higher amplitudes are also frequently observed in female patients. Although effects of temperature on BAEPs are well-established, they are rarely clinically relevant, except for cases of severe hypothermia. Decreasing temperature prolongs both absolute and interpeak latencies. Waveforms can be observed at 27°C and above. If necessary, correction factors can be used. Hyperthermia has unclear effects on BAEP waveforms.

Numerous toxins and medications affect BAEPs. Ethanol intoxication prolongs both absolute and interpeak latencies without affecting amplitude; ethanol withdrawal has unclear effects. Barbiturates, benzodiazepines, chloral hydrate, and most anesthetics do not significantly alter BAEPs in all but extreme doses. Paralytic agents can reduce muscle artifact and facilitate waveform recognition. Aminoglycosides and salicylates cause prolonged wave I latency and reduced BAEP amplitudes; these changes are usually reversible with medication cessation. Platinum-containing chemotherapeutic medications prolong all BAEP latencies and raise the threshold needed to elicit BAEPs. Some anticonvulsants, including phenytoin, carbamazepine, and valproate, prolong absolute and interpeak latencies.

Sensorineural or conductive hearing loss can significantly impact the recording of BAEP waveforms. Mild hearing loss prolongs absolute latencies but will not affect interpeak latencies, as long as the stimulus intensity is strong enough. Severe hearing loss results in dispersed waveforms, phase cancellation, and difficult-to-interpret BAEP peaks. This may appear as absence of all waveforms. The I to V interpeak latency may be shorter with hearing loss, because of relative prolongation of the absolute latency of wave I. Cochlear dysfunction, as observed in Ménière's disease or in presbyacusis, is characterized by a hearing deficit, which is maximal at threshold. Increasing stimulus intensity can result in normalization of waveforms, presumably by augmenting recruitment of cochlear neurons.

4. INTERPRETATION

It is important to note that, in many normal individuals, all BAEP waves are not always present. Waves III and V should be detectable in all healthy individuals; wave I should also be observed but will only be present ipsilaterally. Wave II is often absent and wave IV is frequently buried and, therefore, indistinguishable from wave V. Wave III often has decreased amplitude on the side contralateral to that being stimulated. Waves VI and VII appear variably after wave V. Amplitudes of BAEP waveforms are extremely variable and, therefore, generally not of clinical significance. In most laboratories, latencies are considered abnormal if they fall beyond three standard deviations of the mean. Laboratories usually establish their own normative database. However, published tables of normal absolute and interpeak latencies are available (e.g., Chiappa, 1997).

BAEPs have been studied in humans since the 1950s, although routine clinical use did not occur for another 20 yr. Historically, BAEPs were largely used to diagnose suspected lesions of the brainstem or the VIIIth cranial nerve that were unable to be visualized by imaging techniques in the pre-MRI era. BAEPs have played a particularly important role in the diagnosis of multiple sclerosis and cerebellopontine angle (CPA) tumors and helped differentiate between metabolic and structural causes of coma. The advent of MRI and other advanced neuroimaging techniques have dramatically decreased the use of BAEPs because of the relative ease and sensitivity of these modern imaging modalities for posterior fossa processes.

Despite its fall from favor as the initial test of choice in suspected brainstem or VIIIth cranial nerve disease, important clinical roles for BAEPs still exist. It should be remembered

Table 2
Overall Patterns of Brainstem Auditory Evoked Potential Abnormalities[a]

Brainstem auditory evoked potential finding	*Lesion*
Prolonged wave I latency	Distal CN VIII
Prolonged I–III interpeak latency	Between proximal CN VIII and pons (CPA masses)
Prolonged III–V interpeak latency	Lesion between caudal pons and midbrain (stroke, tumor, MS, ICH, AVM, etc.)
Prolonged I–III and III–V latencies	Both rostral pons or midbrain and acoustic nerve or caudal pons
Absent wave I with normal III and V	Mild to moderate peripheral hearing loss
Absent wave III with normal I and V	Normal variant
Absent wave V with normal I and III	Above the caudal pons
Absence of all waves	Severe hearing loss
Absence of all waves except I (and possibly II)	Brain death

[a]CN, cranial nerve; CPA, cerebellopontine angle; MS, multiple sclerosis; ICH, intracranial hemorrhage; AVM, arteriovenous malformation.

that BAEPs assess *function* of auditory pathways whereas neuroimaging studies examine *structure.* BAEPs are still a test of choice if physical examination and imaging findings are unrevealing but suspicion remains high for a brainstem or VIIIth cranial nerve functional deficit. In individuals who cannot undergo MRI secondary to implanted pacemakers, defibrillators, claustrophobia, and other reasons, BAEPs may provide valuable information regarding brainstem integrity.

Because BAEPs are sensitive to auditory pathway dysfunction, a suspicion for clinically silent disease can be supported via evoked potential abnormalities. In demyelinating conditions, such as multiple sclerosis, BAEP testing can detect subclinical lesions, helping to define the extent and distribution of disease. However, BAEPs are less sensitive than VEPs and SSEPs in detecting abnormalities in patients with multiple sclerosis. The sensitivity for BAEPs in all those with suspected or confirmed multiple sclerosis is approx 46%. Those without overt brainstem symptoms have an even lower yield. In addition, BAEPs have a fairly low specificity for multiple sclerosis because numerous brainstem processes may result in abnormal responses. Nonetheless, testing enables the clinician to establish a baseline for comparison with future studies. In this way, improvement or progression of disease can be monitored objectively with electrophysiological data.

On the other hand, BAEPs exhibit much greater sensitivity for acoustic schwannomas and other tumors of the CPA. Sensitivities ranging from 75 to 100% have been reported in the detection of various tumors in this locale. BAEPs may even be higher in sensitivity than CT imaging, although inferior to MRI for tumors of the CPA. As in demyelinating diseases, the specificity of BAEPs for recognizing cerebellopontine tumors is less impressive than is its sensitivity.

BAEPs are also routinely used to assess hearing function in populations who cannot reliably cooperate with standard audiological testing (e.g., neonates and infants, and individuals with cognitive impairment). In addition, BAEPs are occasionally monitored in patents receiving potentially ototoxic chemotherapeutic agents.

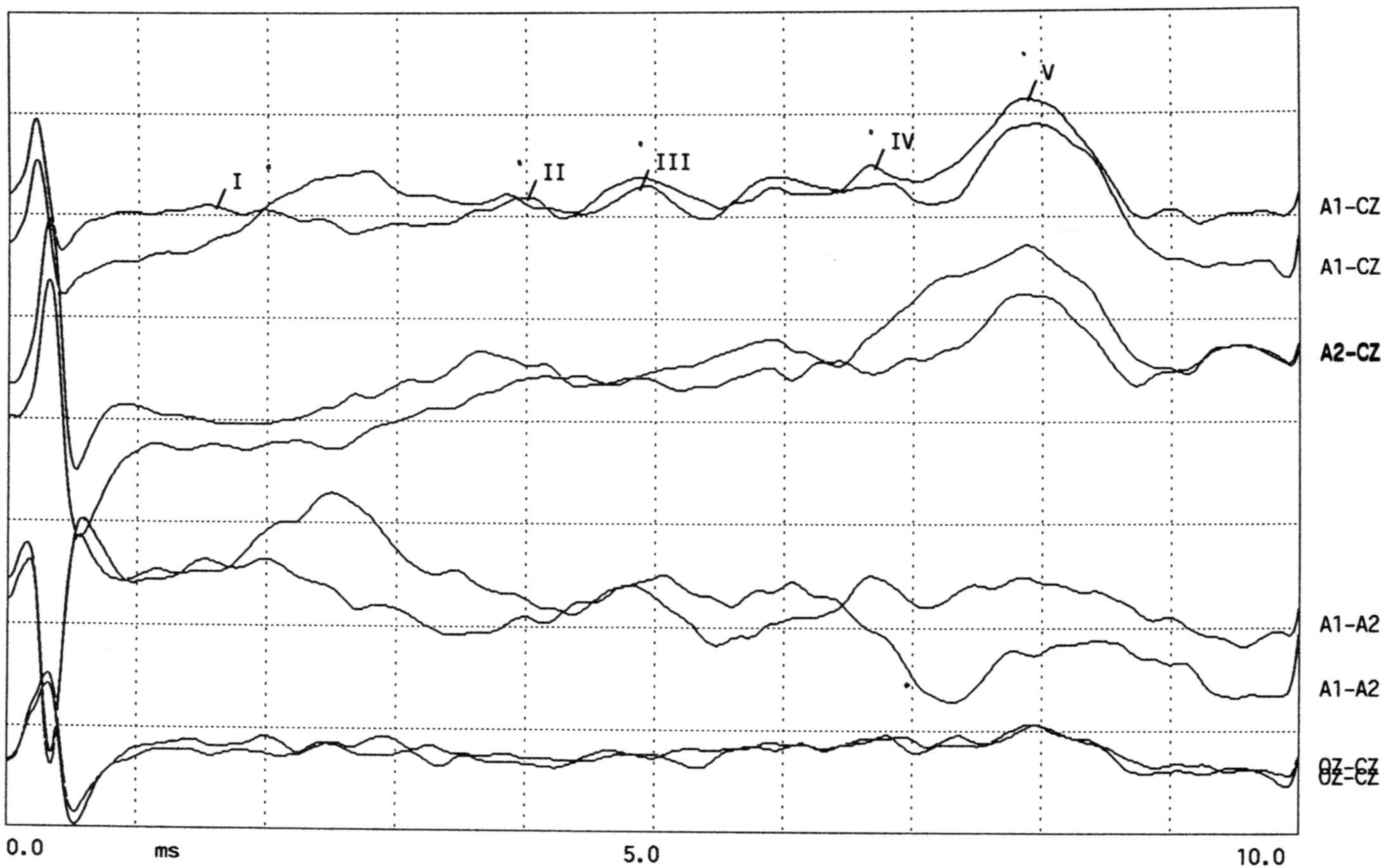

Fig. 2. Abnormal BAEPs. These waveforms are from a 59-yr-old man with bilateral intranuclear ophthalmoplegia. There were no reported auditory symptoms. These recordings were performed with left ear stimulation. Montages are listed. Wave I is poorly formed but evident and has a normal latency. All other latencies are delayed. Furthermore, the I to III, III to V, and I to V interwave latencies are delayed. Right-sided results are very similar (not shown). This suggests bilateral defects in the auditory pathways both between the medulla and the lower midbrain, and between the auditory nerves and medulla.

Today, intraoperative monitoring has emerged as one of the most important uses of BAEPs. Often, BAEPs are requested during posterior fossa surgery to monitor brainstem and acoustic nerve integrity in "real time." BAEPs are not affected by most anesthetic agents and remain reliable measures of brainstem function under general anesthesia. If BAEP waveforms become abnormally prolonged or lost during surgery, the procedure can be adjusted in an attempt to avoid long-lasting neurological injury. Examples of surgeries in which BAEPs are often used include resection of acoustic neuromas, microvascular cranial nerve decompression, and clipping/repair of posterior circulation aneurysms.

Table 2 outlines the common patterns of BAEP abnormalities and identifies their most commonly associated pathological causes. The most useful and informative measures derived from BAEP testing are the following:

1. Absolute latency of wave I: prolongation implies a lesion of the VIIIth cranial nerve. This can be observed in sensorineural or conductive hearing loss. Most schwannomas spare wave I.
2. Wave I to III interpeak latency: prolongation implies a lesion between the proximal segment of the eighth cranial nerve and the superior olivary nucleus. Often this reflects damage to structures at the CPA. Meningiomas and schwannomas at the CPA are typical examples of neoplasms that can cause prolonged I to III interpeak latency.
3. Wave III to V interpeak latency: prolongation suggests a lesion in pathways traveling from the caudal pons to the midbrain. Demyelinating plaques, infarcts, and neoplasms in the brainstem are often associated with increased III to V latency.

Figure 2 illustrates an example of an abnormal BAEP evaluation with accompanying interpretation. In addition to hearing loss, demyelinating lesions, CPA tumors, brainstem tumors, and infarcts affecting the auditory pathways, BAEP abnormalities have been described in numerous other disease states. Texts such as that of Chiappa (1997) have summarized these findings, which are beyond the scope of this review.

SUGGESTED READING

Celesia GG, Brigell M. Auditory evoked potentials. In: Electroencephalography, Basic Principles, Clinical Applications, and Related Fields, 3rd ed. (Neidermeyer E, Lopes Da Silva F, eds.). Williams & Wilkins, Baltimore, MD, 1993, pp. 937–956.

Chiappa KH. Evoked potentials in clinical medicine, 3rd ed. Lippincott-Raven, Philadelphia, PA, 1997.

Erwin CW, Husain AM. Brainstem auditory evoked potentials. In: Current practice of clinical electroencephalography (Ebersole JS, Pedley TA, eds.). Lippincott Williams & Wilkins, Philadelphia, PA, 2003, pp. 864–891.

Huszar L. Clinical utility of evoked potentials. eMedicine [serial online] 2006. Available at www.emedicine.com/neuro/topic69.htm.

Misulis KE. Spehlmann's Evoked Potential Primer, 2nd ed. Butterworth-Heinemann, Boston, MA, 1994.

Misulis KE, Head TC. Brainstem auditory-evoked potentials. In: Essentials of Clinical Neurophysiology, 3rd ed. Butterworth-Heinemann, Boston, MA, 2003, pp. 211–220.

REVIEW QUESTIONS

1. A 66-yr-old man with bilateral watershed infarcts from complications of cardiac bypass surgery seems to have pure word deafness. BAEP testing would most likely reveal which of the following:
   A. Absence of all waveforms.
   B. Bilateral absence of wave V, VI, and VII.
   C. No abnormalities.
   D. Prolonged III to V interpeak latency.

2. A 22-yr-old woman, deaf in both ears because of complications of meningitis in infancy, develops symptoms suggestive of multiple sclerosis. BAEPs will most likely reveal which of the following:
   A. Prolonged absolute latency of wave I.
   B. Prolonged III to V interpeak latency.
   C. Prolonged I to III interpeak latency.
   D. Absence of all waveforms.
   E. No abnormalities.
3. A prolonged I to III interwave latency would most likely be caused by which of the following:
   A. Coma.
   B. Meningioma.
   C. Demyelinating disease.
   D. Brainstem hemorrhage.
4. BAEPs in a patient meeting criteria for brain death would show which of the following:
   A. Absence of all waves.
   B. Intact waveforms I and II.
   C. No abnormalities.
   D. Absence of wave I.
5. When recording BAEPs, improving the signal-to-noise ratio is largely accomplished by which of the following:
   A. Increasing stimulus intensity.
   B. Increasing rate of repetitions.
   C. Increasing amplification of elicited waveforms.
   D. Giving multiple stimulus repetitions.
6. A patient with sensorineural hearing loss may have which of the following changes in BAEP waveforms:
   A. Global increase in waveform amplitude.
   B. Global increase in absolute latencies.
   C. Increased I to V interpeak latency.
   D. Rapid correction of prolonged latency with increased stimulus intensity.
   E. B and D.
   F. B and C.
7. Changes in stimulus polarity affect which BAEP waveform latency the most consistently:
   A. Wave I.
   B. Wave III.
   C. Wave IV.
   D. Wave V.
8. A BAEP study notable for a prolonged I to V interpeak latency might indicate the presence of which of the following:
   A. Pontine glioma.
   B. Neurofibromatosis type II.
   C. Multiple sclerosis.
   D. All of the above.
9. The inferior colliculus likely plays a role in generating which of the following waveforms:
   A. Wave I.
   B. Wave II.
   C. Wave III.
   D. Wave IV.
   E. Wave V.
10. The least-sensitive diagnostic assay for multiple sclerosis is which of the following:
    A. MRI of the brain with contrast enhancement.
    B. BAEPs.
    C. VEPs.
    D. SSEPs.

REVIEW ANSWERS

1. The correct answer is C. Cortical lesions should not affect BAEP waveforms. It is possible that wave VII might be affected by a lesion involving the auditory radiations, but waves VI and VII are not used routinely in clinical evaluation.
2. The correct answer is D. Severe hearing loss can cause BAEP amplitudes to be reduced so much that they are not detectable.
3. The correct answer is B. Masses at the CPA often cause this pattern of abnormalities on BAEP testing. Common lesions include vestibular schwannomas and meningiomas.
4. The correct answer is B. Patients who meet criteria for brain death typically have intact wave I responses. In approx 10% of individuals, wave II may also be present.
5. The correct answer is D. Giving multiple stimulus repetitions is essential to achieving an adequate signal-to-noise ratio. Hundreds to thousands of stimuli are given to differentiate small AEPs from EEG potentials.
6. The correct answer is E. Mild-to-moderate hearing loss globally prolongs absolute latencies. Interpeak latencies are not appreciably changed. Increasing stimulus intensity can result in normalization of waveforms.
7. The correct answer is A. Rarefaction causes shorter absolute wave I latency than condensation. Although changing polarity can affect the amplitudes and latencies of other waveforms, there is more variation between individuals. Use of alternating polarity clicks is not routinely used because of the potential of phase cancellation. In most cases, rarefaction clicks produce the most reliable waveforms. If artifact is suspected or certain waveforms seem absent, a trial with condensation clicks is warranted.
8. The correct answer is D. All of the above are correct. A pontine glioma, depending on its location, could impact the I to III and/or III to V interpeak latencies. Neurofibromatosis type II is often characterized by bilateral acoustic schwannomas, resulting in prolonged I to III latencies. Demyelinating lesions in the brainstem can cause prolonged III to V latency. All of these would, in turn, cause I to V interpeak latency prolongation.
9. The correct answer is E. Wave V is probably, at least in part, generated by activity in the inferior colliculus.
10. The correct answer is B. BAEPs have the least sensitivity for detecting abnormalities in multiple sclerosis.

27
Somatosensory Evoked Potentials

Jacob R. Berger and Andrew S. Blum

Summary

Somatosensory evoked potentials (SSEPs) are electrical potentials generated by various portions of the ascending sensory pathways in response to stimulation of peripheral sensory nerves. SSEPs can be easily elicited and recorded and can be used to examine the functional integrity of somatosensory pathways. This chapter summarizes the methodology for the recording of SSEPs, as well as their role in the evaluation of processes that may affect ascending sensory pathways (e.g., demyelination), and highlights their particular usefulness as an intraoperative tool during spinal cord surgery.

Key Words: Demyelination; dorsal columns; intraoperative monitoring; latency; multiple sclerosis; signal averaging.

1. INTRODUCTION

Somatosensory evoked potentials (SSEPs) are electrical potentials generated by various portions of the ascending sensory pathways in response to stimulation of peripheral sensory nerves. SSEPs can be easily elicited and recorded and can be used to examine the functional integrity of somatosensory pathways. SSEPs conceptually resemble brainstem auditory evoked potentials (BAEPs) and visual evoked potentials (VEPs); both of these are produced after stimulation of auditory and visual pathways, respectively, and have been discussed in previous chapters (Chapters 25 and 26). Motor evoked potentials (MEPs), recorded from peripheral muscles in response to cortical stimulation, also have important and emerging clinical uses, but, historically, have been used with less frequency than SSEPs. MEPs will not be discussed in this chapter.

2. PHYSIOLOGY AND ANATOMY

As with electroencephalography, evoked potential surface electrodes record changes in extracellular voltage at the skin or scalp surface. SSEP waveforms reflect the activity of summated dendritic potentials of underlying gray matter as well as the propagating electrical activity of nerve fiber conduction along white matter pathways. Most elicited SSEP waveforms probably have components of both gray and white matter activity, but generally one or the other predominates. When gray matter activity in the cortex or spinal cord is responsible for the main electrical generators of the recorded SSEP waveform, the evoked potential is termed a "near-field" potential. "Far-field" potentials reflect electrical activity distant to the recording electrode, typically operant in white matter tracts or subcortical structures. Unlike

From: *The Clinical Neurophysiology Primer*
Edited by: A. S. Blum and S. B. Rutkove © Humana Press Inc., Totowa, NJ

near-field potentials that may be missed by small adjustments in electrode placement, far-field potentials are diffusely detectable over the scalp.

In general, SSEPs are recorded from a number of sites along the neuraxis, proximal to the point of peripheral nerve stimulation. These sites permit recording the activity of various physiological generators along the somatosensory pathway. These include a portion of the peripheral nerve or plexus, the lumbar and/or cervical spinal cord, the subcortical relay nuclei, and the somatosensory cortex. For most clinical situations, the most important potentials are those derived from the activity of the cervical spinal cord, relay nuclei, and cortex. Lumbar potentials are very variable and difficult to obtain in certain individuals, and their absence should be interpreted with caution. Nerve conduction and EMG studies are often more effective in evaluating instances of peripheral nerve, root, or plexus dysfunction.

It is possible to stimulate virtually any sensory nerve in the body and record associated SSEPs, but because of their ease of stimulation, reliability of recording sites, and ample existent normative data, median nerve and posterior tibial nerve responses are most commonly used. Ulnar and peroneal nerves are also occasionally used, especially if preexisting factors specific to the individual preclude proper examination of median or posterior tibial nerve responses.

The median nerve receives contributions from both the medial and lateral cords of the brachial plexus and contains fibers spanning from the C5 to T1 roots. Stimulation of sensory receptors in the skin initiates activation of peripheral sensory nerves, which extend through the brachial plexus to the dorsal root ganglia. These bipolar neurons transmit this physiological activation centrally through the appropriate spinal root and into the spinal cord. Fine touch, proprioception, and vibration sense are conducted rostrally through the ipsilateral cuneate tract in the dorsal columns of the spinal cord before synapsing in the nucleus cuneatus in the lower brainstem. From here, fibers project through the brainstem tegmentum to the contralateral ventral posterior lateral nucleus of thalamus (VPL). The contralateral VPL has widespread connections to the contralateral (to site of stimulation) somatosensory cortex in the parietal lobe.

Posterior tibial nerve SSEPs assess the integrity of somatosensory pathways originating in the posterior tibial nerve. These pathways ascend in a similar manner as described for the median nerve, through analogous structures, including the lumbar plexus, lumbosacral roots, ipsilateral gracile tract in the dorsal columns, ipsilateral gracile nucleus of the lower brainstem, and contralateral VPL nucleus of thalamus, before reaching the contralateral medial parietal cortex.

3. METHODOLOGY

3.1. Stimulation

To elicit SSEPs, a brief electrical pulse is administered to the distal portion of a peripheral nerve. In nerves with both motor and sensory components, the stimulus intensity is raised to just above motor threshold, until slight muscle twitching is reliably observed. This level of intensity is used to assure proper stimulation of low-threshold myelinated sensory nerve fibers. For pure sensory nerves, the stimulus intensity is set to 2.5 to 3 times the sensory threshold. Pulses are given at a rate of 4 to 7 per second. This allows enough time to prevent distortion of waveforms from cancellation or synergy of potentials elicited from sequential stimulations. Late latency peaks can be examined by reducing the rate of stimulus, but are generally not of clinical import. A stimulus duration of 100 to 300 μs is typically used.

3.2. Recording

As with VEPs and BAEPs, multiple (500–2000 for SSEPs) stimulus trials are given and averaged to maximize the signal to noise ratio. Standard surface recording electrodes are placed at key points along each sensory pathway. Recording parameters should follow guidelines proposed by the American Electroencephalographic Society. Reasonable settings include a wide bandpass with a low-frequency filter set at 5 to 30 Hz and high-frequency filter set at 2500 to 4000 Hz.

Evoked potential waveforms are named for the characteristic polarity of their voltage peak and the characteristic time to maximal amplitude, as measured in milliseconds after stimulation. For example, an upward (negative by convention) deflection maximal at 9 ms after stimulation would be termed the N9 response. Note that different texts use slightly different names to refer to the same significant peaks (e.g., N9 as opposed to N10), because of slight variation in normative values established by individual laboratories. Ideally, each laboratory should establish its own set of normal values from the population from which it draws. In addition, sets of normal values may be found in publication.

3.3. Median Nerve SSEPs

SSEP recording from the median nerve is the most common evoked potential test to assess the integrity of somatosensory pathways involving the upper limbs. To elicit median nerve SSEPs, stimulation is given 2 cm proximal to the wrist crease, over the median nerve. Standard recording sites for median SSEPs are at Erb's point, over the cervical spine, and on the scalp. Erb's point is located above the clavicle, just lateral to the edge of the sternocleidomastoid muscle. Proper placement can be assured if stimulation at Erb's point induces abduction of a flexed arm. A second electrode is placed over the C2 or C5 spinous process. This site is identified in relation to the prominent spinous process of the C7 vertebra. Scalp electrodes are placed 2 cm posterior to the C3 and C4 electrode placements, as defined by the International 10-20 system. This region overlies the primary somatosensory cortices. Reference (Ref) electrodes are often placed at the contralateral Erb's point (EPc) or at Fz, per the 10–20 system, although other sites, such as the elbow or distal arm may be used.

The recording sites are labeled as EPi, Erb's point (ipsilateral); C2s, C2 spinous process; C5s, C5 spinous process; CPc, centroparietal cortex (contralateral); CPi, centroparietal cortex (ipsilateral); and Ref (EPc, Fz).

A common montage used to view median nerve SSEPs involves channel 1: CPc–CPi, detects near-field cortical potentials; channel 2: CPi–Ref (Fz), detects subcortical far-field potentials; channel 3: C5s–Ref (Fz), records cervical cord activation; and channel 4: EPi–Ref (EPc), records activity of afferents under Erb's point.

This montage allows for comparison of evoked potentials to a common reference and emphasizes the comparison of evoked responses between the ipsilateral and contralateral sensory cortex. In certain situations, other derivations and montages can be used to optimize localization, but these are used infrequently.

Although it is prudent to wait 60 to 80 ms to exclude the possibility of prolonged potentials, the most clinically relevant median nerve SSEP waveforms typically occur within 4 to 40 ms after stimulation. These include the N9 (displayed best by channel 4), the P13/14 (displayed best in channel 2), and the N20 (observed best in channels 1 and 2). Figure 1 shows an example of a normal median nerve SSEP recording.

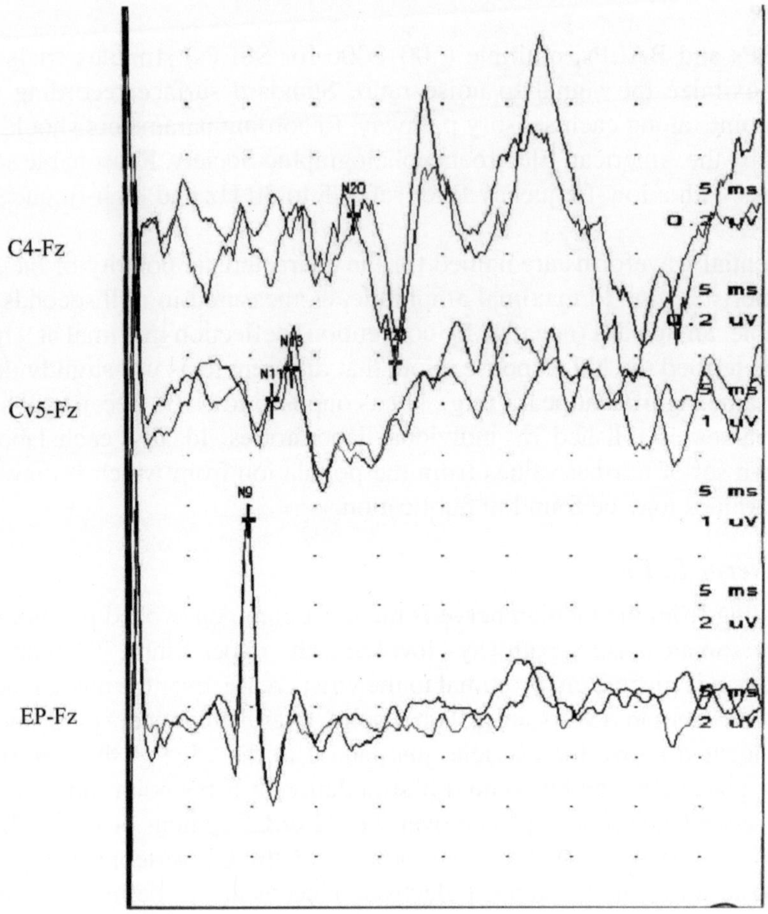

Fig. 1. Normal median nerve SSEPs from a 30-yr-old woman undergoing a work-up for possible multiple sclerosis. She reported tingling in the arms. Recording montages are indicated. The studies were performed in duplicate. This represents stimulation of the left median nerve. Recognizable waveforms are seen and labeled accordingly. The N9 potential is seen clearly in the bottom channel. The N/P13 is evident in the middle channel. The N20 response is recognized in the upper channel. Waveform latencies are measured; interwave latencies are similarly derived from these measurements. These values are then compared with normal values. In this instance, all such values were normal. Results after right median nerve stimulation were symmetric.

Depending on montage and electrode placement, waves generated by different CNS structures can have the same polarity and latency. For example, both subcortical and cortical potentials having maximum positivity 24 ms after stimulation could be equally termed the P24. It is essential to realize that proper interpretation of a named waveform is not possible unless its generator is understood.

Despite the fact that SSEPs have been used clinically for decades, the areas of the nervous system responsible for generating many SSEP waveforms have not been definitively elucidated. Table 1 identifies the most probable generators associated with their respective SSEP waveforms, based on the available data.

Table 1
Median Nerve Somatosensory Evoked Potential Waveforms and Corresponding Generators

Waveform	*Generator*
N9	This clavicular potential, recorded by channel 4, represents a near-field potential, generated by afferent action potentials in the brachial plexus, near EPi. This potential can be strongly influenced by changes in shoulder position.
N13	This near-field cervical potential, recorded by channel 3, is likely generated by dorsal horn neurons and from ascending afferents in the cuneate tract (dorsal columns).
P/N 13/14	This far-field potential, recorded by channel 2, is likely generated by the caudal medial lemniscus in the brainstem. As with most subcortical (far-field) potentials, it is observed in all scalp electrodes. Latency and morphology varies widely between individuals but is stable in an individual. The P13/14 can be monophasic, biphasic, or triphasic, possibly reflecting input from multiple generators.
N18	This long-duration potential recorded by channel 2 likely stems from postsynaptic activity of tectal and pretectal nuclei, which receive input from the medial lemniscus. Thalamic lesions do not affect the N18, implying a neuronal generator that is more caudal. The N18 probably represents a composite waveform with multiple generators.
N20	This is a near-field potential recorded by channels 1 and 2, over the contralateral parietal cortex. It is likely generated from thalamocortical radiations projecting from the ventral posterior lateral thalamus. The N20 may also represent a composite waveform.

3.4. Posterior Tibial SSEPs

Posterior tibial nerve SSEPs are elicited by stimulating in between the Achilles tendon and medial malleolus, using a stimulus intensity just strong enough to cause plantar flexion of the toes. Most of the relevant waveforms appear within 60 ms of stimulation. If none appear, an interval of 100 ms should be used to exclude the possibility of extremely prolonged waveforms. Recording electrodes are placed as CPi, the same locus as for median nerve SSEP recording, except ipsilateral; CPz, midline, between Cz and Pz; C5s, same location as for median nerve SSEP recording; T12s over the T12 spinous process; and Ref, common sites include the iliac crest, elbow, or Fpz. (frontopolar midline, per 10–20 system).

A common montage for posterior tibial nerve SSEPs using these electrodes is channel 1: Cpi–Ref (Fpz), detects near-field potentials over cortex; channel 2: CPz–Ref (Fpz), detects near-field potentials over cortex; channel 3: Fpz–Ref (C5s), detects subcortical far-field potentials; and channel 4: T12–Ref (iliac crest), detects near-field potentials from lumbar cord.

Figure 2 illustrates a set of normal posterior tibial nerve SSEP responses using the above montage. As illustrated, there are three main identifiable waveforms that appear. As before, they are named according to their polarity and latency. These are the LP/N22 (or lumbar potential), the N34, and the P37/38. As mentioned, the LP/N22 response is highly variable because of technical factors (e.g., interfering EMG-derived artifact). The responses that are

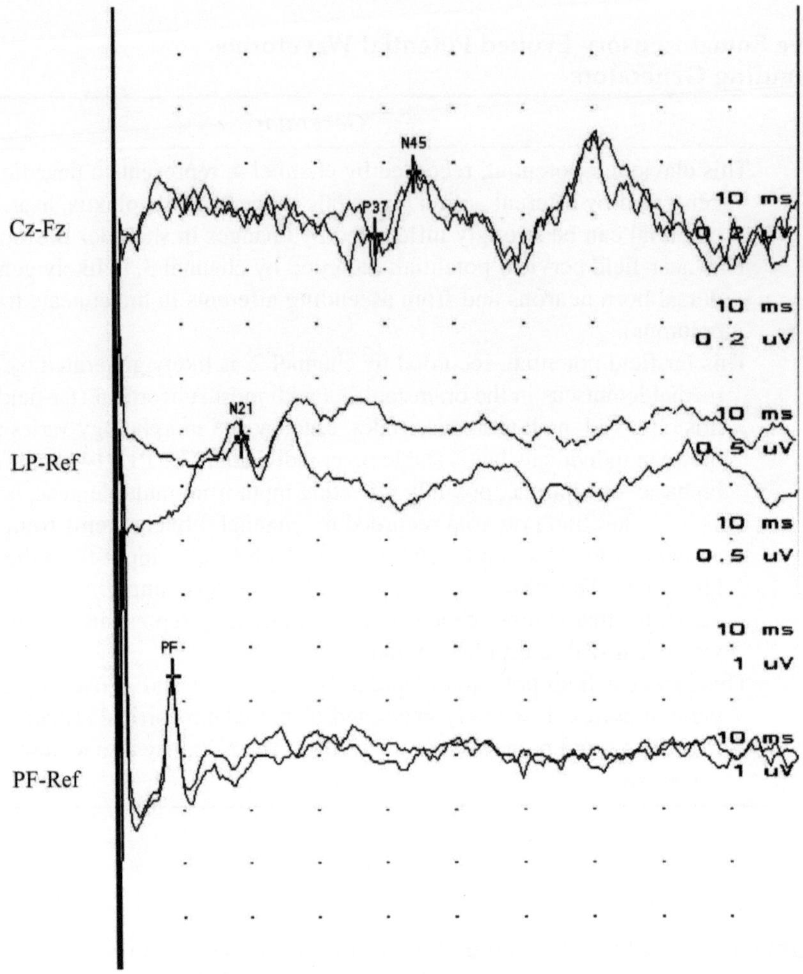

Fig. 2. Normal posterior tibial nerve somatosensory evoked potentials from a 27-yr-old man with ataxia, dysarthria, and white matter changes on MRI. Recording montages are listed. The studies were performed in duplicate. This is from stimulation of the left leg. Recognized waveforms are labeled. The potential from the popliteal fossa is seen clearly in the bottom channel. The N21 (lumbar potential) response is observed in the middle channel. The uppermost channel displays the cortical potential (P37). Latencies are measured and interwave latencies are derived. Values are compared with a normative sample. Values for this study were found to be within normal limits.

more rostral are more consistently and cleanly recorded (with higher signal-to-noise ratio) and, therefore, of more clinical value. These posterior tibial nerve SSEP responses and their (likely) corresponding physiological generators are outlined in Table 2.

Remember that the sensory cortex for the lower leg is located on the medial aspect of the hemisphere. As a result, it is thought that evoked tangential dipoles involving bilateral sensory cortices result in the predominant waveform appearing ipsilateral to the stimulation. By contrast, stimulation of lower extremity sensory nerves that are more proximal, with relatively more lateral cortical representation evoke waveforms that are observed best in the contralateral hemisphere.

Table 2
Posterior Tibial Nerve Somatosensory Evoked Potential Waveforms and Corresponding Generators

Waveform	*Generator*
LP/N22	This lumbar potential is likely generated by activity in the dorsal roots, dorsal root entry zone, and by postsynaptic activity in the lumbar cord enlargement. It is maximally detected by electrodes overlying the lower thoracic or upper lumbar spine. In the described montage, it appears best in channel 4. It is analogous to the N13 wave observed with median nerve somatosensory evoked potentials.
N34	This potential likely reflects subcortical activity from multiple brainstem nuclei along the somatosensory pathway. This appears best in channel 3.
P37/38	This near-field response is a large waveform with a field restricted to the *ipsilateral* parasagittal area, near Cz or Pz. A simultaneously occurring negative waveform can often be observed with frontocentral leads (the N38).

4. INTERPRETATION

The use of SSEPs, similar to BAEPs and VEPs, has decreased dramatically as novel neuroimaging techniques have become better at identifying structural lesions in the spinal cord and brainstem. Still, in situations in which imaging cannot be performed or when it is unrevealing, evoked potential testing can provide useful information regarding the functional integrity of sensory pathways.

SSEPs are sensitive for sensory pathway dysfunction and can detect abnormalities in patients with normal or inconclusive imaging studies, especially in demyelinating disease, such as multiple sclerosis or neuromyelitis optica (Devic's). Abnormal SSEPs are found in 91% of patients with definite multiple sclerosis. SSEP testing in individuals with probable and possible multiple sclerosis reveals a sensitivity of 79% and 42%, respectively. Posterior tibial nerve SSEPs are more sensitive than median nerve SSEPs; however, the yield is highest when both are combined. Sensitivity is increased when sensory symptoms and/or signs are present. Still, 40 to 50% of multiple sclerosis patients without clinical sensory involvement will have abnormal SSEPs. Figure 3 illustrates an abnormal posterior tibial nerve SSEP study in a patient with suspected multiple sclerosis.

In addition to their role in diagnosis, SSEPs can help establish a baseline to better monitor disease progression or effect of therapy. The use of SSEP monitoring in spine surgery has also become commonplace. It is difficult to imagine advances in imaging techniques that would supplant the usefulness of real-time functional monitoring of spinal cord integrity in the operative setting.

As with other evoked potential studies, the most important information is gleaned from the presence or absence of specific waveforms, as well as latencies of such waveforms, and interpeak latencies. Amplitudes in evoked potential studies are highly variable and rarely of clinical use. Waveforms with isolated abnormalities of amplitude but with normal latencies should be interpreted to be within the broad spectrum of normal. Most laboratories consider latencies within three standard deviations of the mean to be within normal limits.

Among the various SSEP waveforms, those of cortical origin (e.g., the N20 and P37) are the most tenuous. They are often prolonged or abolished by sleep, sedation, and many

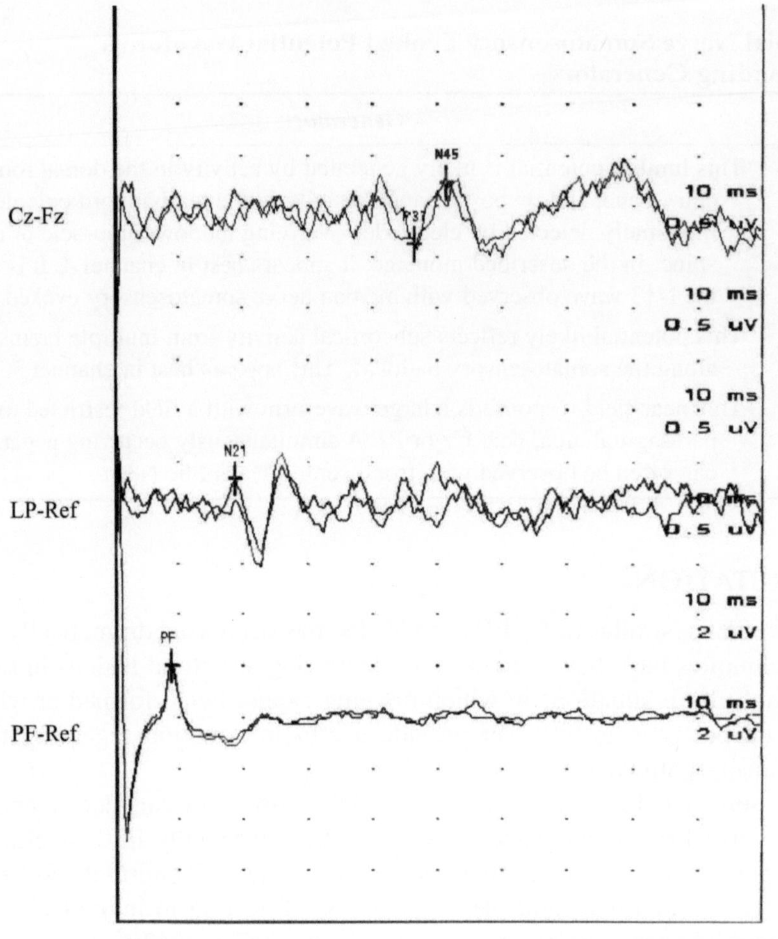

Fig. 3. Abnormal posterior tibial nerve somatosensory evoked potentials SSEPs. Same patient as in Fig. 1. She also reported tingling sensations in the legs. Montages are listed. The studies are in duplicate. This shows results of right leg stimulation. Waveforms are identified; latencies have been measured, and interwave latencies derived accordingly. Her popliteal fossa (PF) and lumbar (N21) potentials exhibit normal latencies (lower two channels). However, her cortical potentials (P37/N45) are delayed (46 and 51 ms, respectively) and interpeak latencies involving these cortical potentials are prolonged. Cortical responses were not observed with left leg stimulation (not shown). This suggests an abnormality affecting the central somatosensory pathway between the lumbar cord and the sensory cortex. When this study is considered together with her normal median nerve SSEP results (Fig. 1), a lesion is suggested more specifically between the lumbar and cervical cord.

anesthetic agents. Barbiturates, benzodiazepines, and nitrous oxide can also cause attenuation or absence of cortical waveforms, although typically this is less pronounced than with the halogenated anesthetic agents (e.g., halothane and isoflurane). Downregulated activity of thalamocortical projections likely contributes to this effect. Subcortical potentials, however, are generally not affected by anesthesia. This property makes the more resilient subcortical SSEPs particularly valuable in the intraoperative arena. Etomidate, an intravenous anesthetic agent, increases the amplitude of cortical waveforms. Thus, etomidate can be administered to patients to enhance the visibility of cortical SSEPs when they are deemed especially vital.

However, some authors have expressed concern that its usage might artificially potentiate cortical SSEPs and, thereby, mask important changes in somatosensory pathway function.

Most data on normative values for SSEPs have been obtained from adults. Median and posterior tibial nerve SSEPs can be recorded in infancy, although an incompletely developed nervous system renders markedly different recordings than those of adults. Although subcortical components should be detectable in full term neonates, the N20 may not be observed until 2 mo of age. Reproducible P37/38 responses should be observed by 1 yr of age. The delay of reliable cortical SSEPs is likely multifactorial, including incomplete myelination and increased time spent in sleep states, which are known to attenuate cortical SSEP waveforms. As CNS myelination progresses with maturation, conduction time decreases. This maturational change occurs most rapidly in early infancy and more slowly in early childhood. Normal median nerve SSEP latencies are observed by age 6 to 8 yr; normal posterior tibial nerve latencies are observed by age 5 to 7 yr.

5. APPLICATIONS: CORRELATION WITH VARIOUS PATHOLOGIES

Numerous pathological mechanisms may impact somatosensory conduction, and these may all evoke abnormalities in SSEP testing. These include, but are not limited to, ischemia, tumor, spinal cord compression, and demyelination. Individuals with vitamin B12 deficiency, vitamin E deficiency, HIV, amyotrophic lateral sclerosis, myotonic dystrophy, diabetes, and some hereditary neurodegenerative disorders often exhibit abnormal SSEP results. Even asymptomatic subjects heterozygotic for mutations associated with adrenoleukodystrophy and adrenomyeloneuropathy demonstrate SSEP abnormalities. Certain progressive myoclonic syndromes are associated with abnormally high-voltage ("giant") SSEP responses (e.g., Lafora Body disease or Unverricht–Lundborg disease). The absence of certain SSEP waveforms can be associated with brain death.

Median nerve SSEPs are frequently used in conjunction with lower-extremity SSEP studies to assist clinical localization. Assuming intact peripheral nerves, a normal median nerve SSEP response paired with abnormal lower extremity SSEPs implies a lesion between the cauda equina and the cervical spinal cord. Alternatively, normal posterior tibial nerve studies with abnormal median nerve responses suggest a peripheral lesion along the median nerve pathway. Tables 3 and 4 illustrate various patterns of SSEP (median and posterior tibial nerve, respectively) abnormalities, with their associated localizing values.

6. APPLICATIONS: INTRAOPERATIVE MONITORING

Intraoperative monitoring has emerged as one of the major uses for SSEPs, providing a real-time measure of central nervous system integrity. Many surgeons use SSEP monitoring routinely for certain spine procedures, such as scoliosis corrections and spinal fusions. Loss of scalp responses to stimulation alerts the surgeon to potential CNS injury and allows for adjustment in the operative approach. Lack of corrective intervention in response to SSEP signal changes during spinal corrective surgeries has been associated with poorer outcomes. SSEPs can also help to define essential neuroanatomic structures that may not be easily distinguishable with visual inspection. For instance, SSEPs have been used to define the primary somatosensory cortex and central sulcus in patients undergoing epilepsy surgery in the frontoparietal region.

To monitor SSEPs in the intraoperative setting, a preoperative baseline test is performed to ensure that reliable potentials can be obtained. Intraoperative measurements are continuously

Table 3
Median Nerve Somatosensory Evoked Potentials: Response Patterns and Localization

Somatosensory evoked potentials findings	*Localization of lesion*
Absent N9 Normal P14, N20 absolute latencies	Normal study
Prolonged or absent N9 Equally prolonged P14 and N20 absolute latencies with normal interpeak latencies	Distal to brachial plexus
Prolonged N9–P14 interval Normal P14–N20 interval	Between brachial plexus and lower medulla
Prolonged P14–N20 interval Otherwise normal	Between lower medulla and cerebral cortex (often observed with multiple sclerosis plaques)
Bilaterally absent N20s	Associated with poor prognosis in coma patients if not medication effect Observed in brain death
Intact cervical N13 Abnormal scalp P14	Between cervical cord and medial lemniscus (e.g., cervical cord compression)

Table 4
Posterior Tibial Nerve Somatosensory Evoked Potentials: Response Patterns and Localization

Somatosensory evoked potentials findings	*Localization*
Prolonged or absent N22 Normal interpeak latencies Nonrecordable potentials	Severe peripheral neuropathy Cauda equina lesion (less likely)
Prolonged N22–P37 interpeak latency Median nerve SSEPs either not performed or abnormal	Between cauda equina and cortex Even if median nerve SSEP findings suggest a lesion rostral to the mid-cervical cord, other more caudal lesions are possible
Prolonged N22–P37 interpeak latency Normal median nerve SSEPs	Between cauda equina and mid cervical cord
Inferior displacement of the N22 recording site	Tethered cord

updated throughout the surgery, with special attention given to the periods before instrumentation, during instrumentation, after instrumentation is complete, and before closing. In most cases, posterior tibial nerve SSEPs are used, with median nerve responses serving as a control. As mentioned in Section 4, subcortical potentials are much more resilient to the effects of anesthesia. They provide a key backup potential in the operative environment to the cortical responses, which are much more labile.

Although SSEPs have been used to monitor CNS integrity during carotid endarterectomy, monitoring with electroencephalography is used more frequently. This is primarily because of the ease of monitoring, the ability to examine broader areas of cortical function, and the rapidity with which ischemic changes in the EEG can be identified.

SUGGESTED READING

American Electroencephalographic Society. Guidelines on evoked potentials. J Clin Neurophysiol 1994;11:40–73.

Chiappa KH. Evoked Potentials in Clinical Medicine, 3rd ed. Lippincott–Raven, Philadelphia, PA, 1997.

Emerson RG, Adams DC. Intraoperative monitoring. In: Current Practice of Clinical Electroencephalography (Ebersole JS, Pedley TA, eds.). Lippincott, Williams & Wilkins, Philadelphia, PA, 2003, pp. 936–954.

Emerson RG, Pedley TA. Somatosensory evoked potentials. In: Current Practice of Clinical Electroencephalography (Ebersole JS, Pedley TA, eds.). Lippincott, Williams & Wilkins, Philadelphia, PA, 2003, pp. 892–922.

Erwin CW, Rozear MP, Radtke RA, Erwin AC. Somatosensory evoked potentials and surgical monitoring. In: Electroencephalography, Basic Principles, Clinical Applications, and Related Fields, 3rd ed. (Neidermeyer E, Lopes Da Silva F, eds.). Williams & Wilkins, Baltimore, MD, 1993, pp. 957–974.

Misulis KE. Spehlmann's Evoked Potential Primer, 2nd ed. Butterworth-Heinemann, Boston, MA, 1994.

REVIEW QUESTIONS

1. All of the following medications may diminish cortical SSEP waveform amplitude except which of the following:
   A. Isoflurane.
   B. Halothane.
   C. Midazolam.
   D. Nitrous oxide.
   E. Phenobarbital.
   F. None of the above.
2. SSEP testing will show abnormalities if which of the following tracts is damaged:
   A. Gracile.
   B. Corticospinal.
   C. Spinothalamic.
   D. Lateral lemniscus.
3. Increased amplitude of cortical SSEP waveforms can be associated with which of the following:
   A. Nitrous oxide use.
   B. Myoclonic epileptic syndromes.
   C. Anxiety.
   D. Etomidate.
   E. More than one of the above.
4. SSEPs in children typically demonstrate which of the following characteristics:
   A. Normal median SSEP latencies by age 5 yr.
   B. Most marked change in latency after age 4 yr.
   C. The presence of N20 in all term neonates.
   D. Subcortical potentials in all term neonates.
5. Intraoperative posterior tibial SSEP testing identifies loss of P37. Concurrent median SSEPs show normal N9 but loss of subcortical and cortical potentials recorded from scalp electrodes. The site of damage is most likely at which of the following sites:
   A. Peripheral nerves.
   B. Cauda equina.

C. Between the cauda equina and mid-cervical spinal cord.
D. Above the level of the mid-cervical spinal cord.
E. Two of the above.
F. Definitely one, and possibly two of the above.

6. A waveform, which, when averaged over multiple trials, has a maximum downward deflection 34 ms and upward deflection at 36 ms after stimulation would be termed:
   A. N34.
   B. P34.
   C. P36.
7. The most important methodological factor in maximizing signal-to-noise ration in SSEP testing is:
   A. Relaxation of the patient.
   B. Averaging of multiple trials.
   C. Intensity of the stimulus.
   D. Strength of signal amplification.
8. A term neonate with a brachial plexus injury could have all of the following on median SSEP testing except:
   A. Absent N20.
   B. Absent P14.
   C. Absent N9.
   D. Prolonged P14.
9. Absent N22 waveform from a recording electrode at T12 could indicate all of the following except:
   A. Inferior displaced lumbar enlargement/tethered cord.
   B. Peripheral neuropathy.
   C. Cauda equina lesion.
   D. Thoracic cord lesion.
10. An isolated demyelinating lesion in the mid thoracic cord would be expected to have which of the following patterns:
    A. Normal N22 latency, normal N22–P37 latency, abnormal N9–P14 latency.
    B. Normal N22 latency, abnormal N22–P37 latency, abnormal N9–N20 latency.
    C. Normal N22 latency, abnormal N22–P37 latency, normal N9–N20 latency.
    D. Abnormal N22 latency, abnormal N22–P37 latency, normal N9–N20 latency.

REVIEW ANSWERS

1. The correct answer is F. All of the above agents may diminish the cortical SSEP waveform. Benzodiazepines, phenobarbital, and low-concentration nitrous oxide typically have a lesser effect than halogenated anesthetics, such as isoflurane and halothane.
2. The correct answer is A. SSEP testing primarily tests the dorsal columns of the spinal cord, where the cuneate and gracile tracts lie. The corticospinal tract can be assessed with MEP testing. The spinothalamic tract runs lateral to the dorsal columns and contains input from less myelinated and nonmyelinated fibers carrying information regarding pain and temperature. The lateral lemniscus can be assessed with BAEPs.
3. The correct answer is E. Myoclonic epileptic syndromes and the use of etomidate, a neuromuscular blocker, have been associated with heightened amplitude of SSEP waveforms. Anxiety might mask cortical waveforms because of increased muscle artifact. Nitrous oxide use may attenuate cortical waveforms.
4. The correct answer is D. Subcortical potentials should be observed in all term neonates. Median SSEPs reach normal values after tibial SSEPs, typically at age 6 to 8 yr. The most rapid change in conduction speed occurs in early infancy, not after age 4 yr. N20 potentials may not be present until 2 mo of age.

5. The correct answer is F. The abnormalities on median SSEPs implies a lesion above the level of the mid-cervical cord. This, alone, could cause the abnormalities observed in the tibial studies. However, a second lesion, inferior to the cervical cord may also be present.
6. The correct answer is B. Waveforms are named by their polarity and the latency at which they have maximal amplitude. Keep in mind that the waveform represents the averaged amplitude and latency of hundreds of stimulations. It is essential to know the derivation and context from which the waveform is derived. For example, N22 means two very different things if recorded from T12 after peroneal stimulation or from the scalp after median nerve stimulation.
7. The correct answer is B. Averaging of multiple trials is essential in maximizing signal-to-noise ratio. It is necessary to help factor out EEG potentials that often mask the relatively low-voltage SSEP potentials. Although relaxing the patient will improve the recording, averaging remains the single most important factor.
8. The correct answer is B. Brachial plexus injury in neonates and adults may cause absent N9 with equally delayed, but present subcortical potentials. Adults should also have equally delayed cortical potentials. Neonates may not have N20 responses present until 2 mo of age.
9. The correct answer is D. A thoracic cord lesion would result in prolonged N22–P37 conduction time but N22 latency would remain normal. Electrodes that are placed more inferiorly would help to identify a tethered cord. Peripheral nerve conduction studies may help to identify a peripheral neuropathy if abnormal N22 is found on SSEP testing.
10. The correct answer is C. Only conduction between the lumbosacral cord and thoracic lesion should be affected. Median potentials should be normal, as should conduction from peripheral nerves to the lumbosacral cord.

28
Transcranial Magnetic Stimulation

Masahito Kobayashi and Alvaro Pascual-Leone

Summary

During the past two decades, transcranial magnetic stimulation (TMS) has emerged as an important modality for the exploration of cerebral function and assessing the integrity of human motor pathways. In TMS, a strong magnetic pulse activates neural elements oriented predominantly horizontally to the brain surface, and a motor evoked potential can be recorded in the activated muscles. In single-pulse TMS, stimulation can be applied to different levels of the nervous system, including the spinal cord, to assist in localizing a lesion to a specific level and helping to characterize it as demyelinating or axonal in nature. A central motor conduction time can also be calculated; this is defined as the latency difference between the motor evoked potentials induced by stimulation of the motor cortex and those evoked by spinal (motor root) stimulation. A variety of additional testing paradigms have been created over the years, including the use of paired-pulse techniques and repetitive stimulation, the latter potentially assisting in treating a variety of disorders, including depression and Parkinson disease.

Key Words: Central conduction time; excitability; motor evoked potential; repetitive stimulation; silent period; transcranial magnetic stimulation.

1. INTRODUCTION

The purpose of this chapter is to introduce transcranial magnetic stimulation (TMS) and review its applications in clinical neurophysiology to date and its future potential. TMS, as currently used, was introduced by Anthony Barker from the University of Sheffield in 1985. Since its introduction, the use of TMS in clinical neurophysiology, neurology, neuroscience, and psychiatry has spread widely, mostly in research applications, but also increasingly with clinical aims in mind. TMS provides a noninvasive and safe method of activating the human motor cortex, and assessing the integrity of the central motor pathways. It has the advantage over transcranial electric stimulation (TES) of being practically painless.

2. BASIC PRINCIPLES OF MAGNETIC STIMULATION

TMS is based on the principle of electromagnetic induction, discovered by Michael Faraday in 1838. As a pulse of current passes through a coil of wire, a magnetic field is generated perpendicular to it. The rate of change of that magnetic field determines the induction of a secondary current in any nearby conductor. If a current pulse passing through a coil placed over a subject's head has sufficient strength and short enough duration, a rapidly changing magnetic pulse is generated that penetrates scalp and skull, reaching the brain with negligible attenuation, inducing a secondary ionic current in the brain. Properly induced currents can

From: *The Clinical Neurophysiology Primer*
Edited by: A. S. Blum and S. B. Rutkove © Humana Press Inc., Totowa, NJ

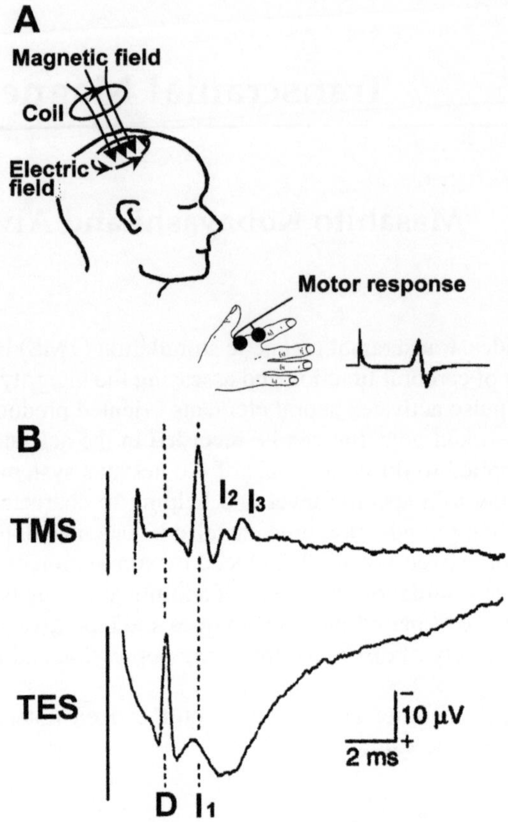

Fig. 1. (**A**) Principle of transcranial magnetic stimulation (TMS). The current flowing briefly in the coil generates a changing magnetic field that induces an electric current in the tissue, in the opposite direction. (**B**) Direct recording from the surface of the spinal cord after transcranial magnetic stimulation (TMS) and transcranial electric stimulation (TES). Both types of stimuli can evoke and early spike termed a direct (D) wave and up to four further spikes, termed indirect (I) waves. The current evoked by TMS flows in the posterior to anterior direction. TMS preferentially evokes I-waves, whereas TES evokes D-waves.

depolarize neurons that are appropriately oriented in relationship to the magnetic fields and the secondarily induced current, and generate action potentials (Fig. 1A).

During the past 15 yr, the term "TMS" has become established, even though it is a misnomer. It is true that the neurons are stimulated and depolarized transcranially by the TMS coil applied to the scalp, even though the behavioral outcome may be disruptive, rather than activating (i.e., stimulating). It is, however, not the case that magnetic fields exert the main effect. Indeed, TMS is better conceptualized as transcranial *electrodeless electric* stimulation through electromagnetic induction, because electrical charges flow into an excitable cell membrane, causing a change in transmembrane potential. This can result in the initiation of an action potential, which spreads along the neural tissue.

Although TMS depolarizes neurons by virtue of the induced electric current, there are some differences between electromagnetic induction and direct electric stimulation. The onset latency of motor evoked potentials (MEPs) induced by TMS of the motor cortex in contralateral hand muscles is often longer than that induced by TES. Direct recording from the surface of the spinal cord has shown that both types of stimulation, TMS and TES, can evoke an early spike, termed a direct (D) wave and up to four further spikes, termed indirect (I) waves (Fig. 1B). TMS will evoke I-waves preferentially activating the pyramidal cells indirectly, that is, transsynaptically, whereas TES activates pyramidal tract neurons directly at their axon hillock, eliciting D-waves. For TMS, fast-conducting axons (>75 m/s) have a lower threshold for D-waves, whereas slowly conducting axons (<55 m/s) have a lower threshold for I-waves. Conversely, for TES, the majority of axons have a lower threshold for D-waves or a similar threshold for D- and I-waves. In addition, with the stronger stimuli, the site of activation by TES will shift below the cortex, whereas TMS will still excite axons mostly within the cortex, even with much stronger intensity of stimuli. The different sites of activation by TMS and TES as well as the different distribution of D- and I-waves that they evoke may cause the differential timing of MEPs and, in general, the differential effects of these noninvasive stimulation effects. However, further work on the mechanisms of action of TMS and TES on the brain is needed to fully explore these issues.

The precise mechanisms underlying the brain effects of TMS remain filled with questions and unknowns. Currents induced in the brain by TMS flow parallel to the plane of the stimulation coil (approximately parallel to the brain's cortical surface when the stimulation coil is held tangentially to the scalp). Therefore, in contrast to electrical cortical stimulation, TMS preferentially activates neural elements oriented horizontally to the brain surface. Exactly what neural elements are activated by TMS remains unclear and may, in fact, be variable across different brain areas and different subjects. The combination of TMS with other neuroimaging and neurophysiological techniques provides an enhanced understanding of the mechanisms of action of TMS and a novel approach to the study functional connectivity between different areas in the human brain.

3. EQUIPMENT

The design of magnetic stimulators is relatively simple. They consist of a main unit and a stimulating coil (Fig. 2). The main unit is composed of a charging system, one or more energy storage capacitors, a discharge switch, and circuits for pulse shaping, energy recovery, and control functions. Different charging systems are possible; the simplest design uses step-up transformers operating at a line frequency of 50 to 60 Hz. Energy storage capacitors can also be of different types. The essential factors in the effectiveness of a magnetic stimulator are the speed of the magnetic field rise time and the maximization of the peak coil energy. Therefore, large energy storage capacitors and very efficient energy transfer from the capacitor to the coil are important. Typically, energy storage capacity is approx 2000 J, and 500 J are transferred from the capacitors into the stimulating coil in less than 100 μs via a thyristor, an electronic device that is capable of switching large currents in a few microseconds. The peak discharge current needs to be several thousand amperes to induce currents in the brain of sufficient magnitude to depolarize neural elements (~10 mA/cm^2).

During *transcranial* brain stimulation only, the stimulating coil needs to come in close contact with the subject. Stimulating coils consist of one or more well-insulated coils of copper wire frequently housed in a molded plastic cover. Stimulating coils are available in a variety of shapes and sizes. The focality of the magnetic field and, hence, of the current induced by TMS varies depending on the geometry of the stimulation coil used. Two different shapes of coils are most commonly used, an "8"-shaped coil and a circular coil (Fig. 2). The former provides a more focal means of stimulation, making fairly detailed mapping of cortical outputs

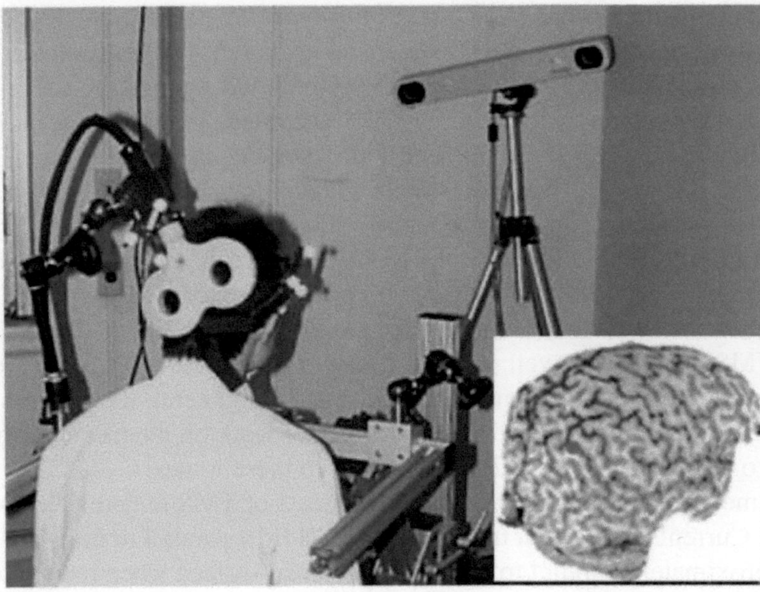

Fig. 2. Stereotactic frameless navigation system. Under the navigation according to anatomic or functional imaging, the coil can be placed precisely on the subject's scalp just above a targeted brain region.

in cortical functions possible. Current knowledge suggests, largely based on mathematical modeling, that the most focal forms of TMS currently available affect an area of 0.5×0.5 cm at the level of the brain cortex. Stimulation is restricted to rather superficial layers in the convexity of the brain (cortex or gray–white matter junction) and direct effect onto deep brain structures is not possible. Circular coils, on the other hand, can induce a more widely distributed electric field, possibly with greater depth penetration, and, thus, evoke MEPs more readily or stimulate a larger brain area, both hemispheres simultaneously, which is often desirable, particularly in the study of central motor conduction times (CMCTs).

It is obviously desirable to localize the TMS coil precisely at the targeted area of the brain, a task that can be challenging if the targeted area is outside the primary motor or visual cortex. In such instances, TMS may not evoke a readily, immediately measurable effect, and scalp-to-brain relations can be quite variable from subject to subject. Digitization of the subject's head and registration of the TMS stimulation sites onto the MRI of the subject's head addresses such issues of anatomic specificity of the TMS effects by identifying the actual brain target in each experimental subject. The use of optical digitization and frameless stereotactic systems represents a further improvement. Such systems were originally developed for neurosurgical interventions. With minimal adaptations to the specific needs of TMS, frameless stereotactic devices can be used to guide the placement of the stimulation coil on the subject's scalp to precisely target a particular brain region defined anatomically or by functional neuroimaging (Fig. 2).

4. SAFETY CONSIDERATIONS

TMS, especially repetitive TMS (rTMS, *see* Section 5.3), remains an experimental technique, side effects are possible, and strict safety and ethical guidelines need to be followed. In 1996, a consensus conference on the safety of TMS/rTMS was held under the auspices of

Table 1
Safe Duration/Number of Pulses for Single Trains of Repetitive Transcranial Magnetic Stimulation in Normal Volunteers

	Repetitive transcranial magnetic stimulation intensity (% of motor threshold)						
Frequency (Hz)	100	110	120	130	140	150	160
1	1800 s/1800	1800 s/1800	360 s/360	50 s/50	50 s/50	50 s/50	50 s/50
5	10 s/50	10 s/50	10 s/50	10 s/50	7.6 s/38	5.2 s/16	3.6 s/8
10	5 s/50	5 s/50	4.2 s/42	2.9 s/29	1.3 s/13	0.8 s/8	0.7 s/7
20	2.05 s/41	1.6 s/32	1.0 s/20	0.55 s/11	0.35 s/7	0.25 s/5	0.25 s/5
25	1.28 s/32	0.84 s/21	0.4 s/10	0.24 s/6	0.2 s/5	0.2 s/5	0.2 s/5

[a]To add an additional margin of safety, reduce the allowable duration of a train by 25%.

the National Institutes of Health and specific guidelines and precautions recommended that should be strictly adhered to. There are relative and absolute contraindications to TMS. Examples of contraindications include metal anywhere in the head (excluding the mouth), cardiac pacemakers and implanted medication pumps, intracranial or intracardiac electrodes, raised intracranial pressure, pregnancy, a history of seizures, a family history of epilepsy, and medications that might increase the risk of seizures.

The main safety concern in using TMS is its potential to induce a seizure, even in subjects without any predisposing illness. This risk is low (on the order of 1 in 1000 studies or fewer) and essentially limited to the application of rTMS. However, it is important to note that even when safety guidelines are followed, induction of a seizure is possible and, hence, laboratories should be set up to recognize and promptly treat a possible seizure. This is particularly the case if rTMS is applied at higher stimulation frequencies. Table 1 shows that the maximum safe durations of single trains of rTMS at various frequencies and intensities determined from experience in the National Institutes of Neurological Disorders and Stroke (NINDS). It is important to note that the NINDS researchers did reduce the allowable duration of a train by 25% in the studies, if the potential clinical benefits would be speculative or if no clinical benefit would be expected. If repeated trains of rTMS are used, the inter-train interval should be considered because of a cumulative increase in cortical excitability. According to Chen et al. (1997), an inter-train interval of 5 s seems to be safe if 10 trains of rTMS are used with the inte nsity of 110% of motor threshold and at the frequency of 20 Hz. Combination of any of the stimulating parameters should also be safe with an inter-train interval of 5 s. Even rTMS with fewer than 10 trains should follow these guidelines for inter-train intervals because of possible spread of excitation. For stimulus intensities of 120% of motor threshold or higher, the inter-train interval should be longer than 1 min, because no potentiation was observed at the inter-train interval of 1 min.

Approximately 10 to 20% of subjects studied with TMS develop a muscle tension headache or a neck ache. Generally, these mild discomforts respond promptly to aspirin, acetaminophen, or other common analgesics. If the subjects do not wear earplugs during the rTMS studies, rTMS can also cause ringing in the ears or even transient hearing loss. Therefore, adequate ear protection is strongly recommended. Finally, it is important to realize that TMS has only been studied for approx 15 yr and the data in humans are still few. Although animal studies have not shown any risks of brain damage or long-term injury to the brain or its functions after TMS, caution is still imperative. This is particularly the case regarding cognitive deficits that might be induced by TMS. Neuropsychological assessment, appropriately tailored to the brain area stimulated, should be considered after rTMS to exclude undesirable side effects.

5. METHODS AND MEASUREMENTS OF TMS

5.1. Single-Pulse TMS

Single-pulse TMS delivered to different levels of the neuraxis (motor system) will provide information regarding the excitability of the motor cortex and the conduction along corticospinal, corticonuclear, and callosal fibers. The excitability of nerve roots and the conduction along the peripheral motor pathway to the muscles will also influence the measures obtained by TMS. In the clinical point of view, the patterns of findings will help us to localize the level of a lesion within the nervous system, to distinguish between a predominantly demyelinating or axonal lesion in the motor tracts, or to predict the functional motor outcome (Table 2). Similar to other neurophysiological tests, however, the abnormalities revealed by TMS are not always disease specific. It remains to be fully evaluated whether such neurophysiological findings provide earlier prognostic indicators as those that might be obtained from careful, longitudinal neurological exams.

5.1.1. Motor Threshold

When TMS is applied to the motor cortex at an appropriate stimulation intensity, MEPs can be recorded from contralateral extremity muscles. Motor threshold refers to the lowest TMS intensity necessary to evoke MEPs in the target muscle if single-pulse TMS is applied to the representation of the targeted muscle on the motor cortex. In the majority of recent TMS studies, motor threshold is defined as the lowest intensity required to elicit MEPs of more than 50-μV peak-to-peak amplitude in at least 50% of successive trials, in resting or activated (slightly contracted) target muscles. Motor threshold is thought to reflect excitability of the membrane of mainly corticospinal neurons and interneurons projecting onto these neurons in the motor cortex, as well as the excitability of motor neurons in the spinal cord, neuromuscular junctions, and muscle. In addition to the membrane excitability, the motor threshold should reflect the activities of neural inputs that may affect membrane excitability, that is, tonic inhibitory and excitatory drives onto the cortical output neurons, and also the efficacy of a chain of synapses from presynaptic cortical neurons to muscles. The change of the motor threshold has been studied in various diseases. Motor threshold is often increased in diseases that may affect the corticospinal tract, such as multiple sclerosis, stroke, and brain or spinal cord injury.

When single-pulse TMS is applied over the occipital lobe, phosphenes can be induced in a large number of subjects. A similar methodology as the motor threshold can determine the threshold for the phosphene, which may reflect the excitability of occipital cortex. Studies on phosphene threshold can provide novel insights into the neurophysiology of the human visual cortex. A decrease of the phosphene threshold, that is, increased excitability of visual cortex, has been demonstrated in healthy subjects during mental visual imagery. From a more clinical viewpoint, previous studies have investigated phosphene thresholds in patients suffering from migraines (both with and without visual aura) for the study of the pathophysiology of migraine aura. According to their studies, phosphene thresholds are significantly lower in

Table 2
Diagnostic Application of Transcranial Magnetic Stimulation

TMS measure	*Abnormal findings*	*Diseases and symptoms*
CMCT	Long	MS, ALS, stroke, secodary parkinsonism, secondary dystonia, brain injury, SCI/CS
MEP	Dispersed	MS, ALS, stroke
	Small or absent	MS, ALS, stroke, brain injury, SCI/CS (19), hydrocephalus, Bell's palsy
	Large	PD, dystonia
MEP with triple-stimulation technique	Central conduction failure[b]	MS, ALS (with upper neuron damage), stroke, secondary parkinsonism, brain injury, SCI/CS, hydrocephalus
Silent period	Long	MS, stroke[c], brain injury, SCI/CS, polyradiculitis, demyelinating polyneuropathy, epilepsy (63)
	Short	ALS, PD, dystonia agenesis of corpus callosum
	Absent	SCI/CS
Interhemispheric conduction	Long latency[d]	MS, stroke, brain injury (with transcallosal lesion), dysgenesis of corpus callosum, hydrocephalus
	Reduced interhemispheric inhibition	MS, ALS
	Interhemispheric inhibition absent	Stroke (with transcallosal lesion), dysgenesis of the corpus callosum, hydrocephalus
Motor cortex excitability	High motor threshold[e]	MS, stroke, agenesis of corpus callosum, brain injury, SCI/CS
	Low motor threshold[e]	ALS, hydrocephalus, epilepsy (116)
	Increased intracortical inhibition	Early-stage ALS
	Decreased intracortical inhibition	PD, SCI/CS, epilepsy (116), stroke
	Enlarged cortical representation	Dystonia

[a]TMS, transcranial magnetic stimulation; CMCT, central motor conduction time; MS, multiple sclerosis; ALS, amyotrophic lateral sclerosis; SCI, spinal cord injury; CS, cervical spondylosis; MEP, motor evoked potential; PD, Parkinson disease.

[b]Central conduction failure indicates smaller size of the test MEP than that of control examined by triple-stimulation technique.

[c]Prolonged duration with normal MEP and CMCT may be observed in the motor syndrome, with exaggerated inhibition within the motor cortex, resembling motor neglect. *See* text also.

[d]The latency for transcallosal inhibition (ipsilateral silent period) after single-pulse TMS (Fig. 5B).

[e]High or low value of the motor threshold indicates that they are higher or lower compared with intact hemisphere or healthy subjects.

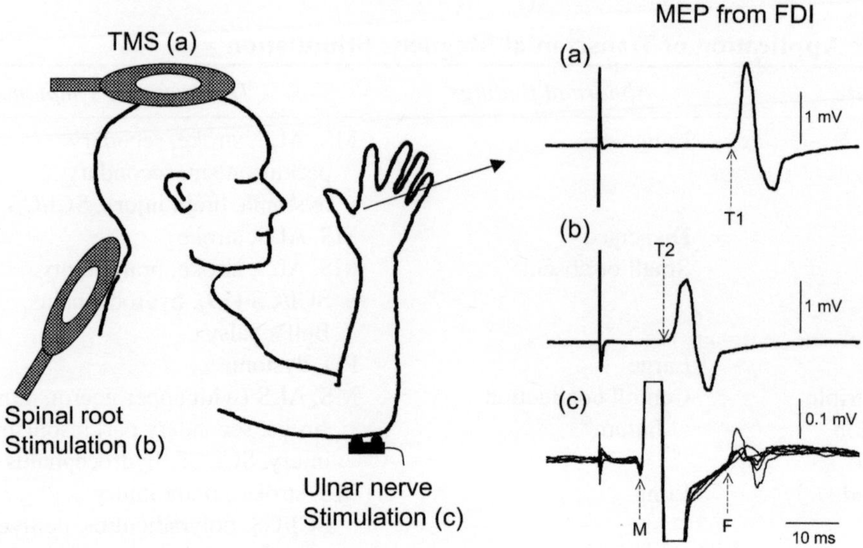

Fig. 3. Schematic representation of the calculation of central motor conduction time (CMCT). (**a**) Motor evoked potential (MEP) induced by transcranial magnetic stimulation (TMS). (**b**) MEP after cervical spinal root stimulation. (**c**) F-waves after ulnar nerve electric stimulation. CMCT is estimated by onset latency of T1 minus onset latency of T2. Using F-wave latency, it can be evaluated more precisely as T1 – (1/2)(F + M – 1), where T1 is the onset latency of MEP elicited by TMS, T2 is the onset latency of the MEP elicited by the coil placed on the back of cervical spine, M is the onset latency of the M-wave by electrical ulnar nerve stimulation, and F is the onset latency of the F-wave by electrical ulnar nerve stimulation. FDI, first dorsal interosseous.

migraineurs (i.e., greater visual cortical excitability) as compared with sex- and age-matched healthy controls, even in asymptomatic intervals. However, these thresholds may vary depending on other factors, such as thickness of skull and scalp or medication and so on, and that, thus, at times, it can be difficult to compare them among individual subjects and between hemispheres.

5.1.2. Central Motor Conduction Time

CMCT is defined as the latency difference between the MEPs induced by stimulation of the motor cortex and those evoked by spinal (motor root) stimulation. It is calculated by the subtraction of the latency of peripheral segment of the motor pathway (spinal motor root to muscle) from that of the entire motor pathway (motor cortex to muscle) (Fig. 3). If a TMS coil is placed over the back of the neck or spine, the magnetic pulse will stimulate spinal nerve roots but not the descending spinal tracts themselves. Bone is the major governor of induced current in the human body because ofits extremely low conductivity. The induced electric field and its first spatial derivative increase remarkably at the neuroforamen of the spine, whereas those induced in the spinal canal are relatively small. Indeed, it is extremely difficult to stimulate the trunk of spinal cord with magnetic stimulation of the neck, and the latency of elicited MEPs will not shift even when moving the coil along the rostrocaudal axis of the spine. The CMCT calculated from the data of magnetic stimulation, therefore, includes both the true time for central motor conduction plus the time taken for at least one synaptic delay at the spinal level and the time from the proximal root to the intervertebral foramen.

Table 3
Normal Values for Central Conduction[a]

Muscle	*Age (yr)*	*Total conduction time (ms)*	*CMCT-F (ms)*	*CMCT-S (ms)*
APB	18–83	20.4 ± 1.5 (n = 150)	4.3 ± 0.8 (n = 19)	6.5 ± 1.2 (n = 222)
ADM	17–75	19.3 ± 1.2 (n = 52)	5.0 ± 1.2 (n = 100)	6.2 ± 1.2 (n = 152)
FDI	17–84	21.1 ± 1.6 (n = 51)	5.9 ± 1.1 (n = 51)	
Pollicis	25–62	20.1 ± 1.0 (n = 11)	5.8 ± 1.1 (n = 11)	
EDC	20–83	15.2 ± 1.5 (n = 150)		6.4 ± 1.2 (n = 150)
Biceps	18–87	11.8 ± 1.2 (n = 49)		5.5 ± 1.2 (n = 122)
Pectoralis	47–87			5.6 ± 1.2 (n = 53)
Triceps	47–87			5.3 ± 0.7 (n = 53)
Tibialis anterior	17–76	27.4 ± 2.3 (n = 202)	13.1 ± 3.8 (n = 150)	13.6 ± 1.8 (n = 106)
Abductor hallucis	19–74	39.3 ± 2.4[b] (n = 48)	12.7 ± 1.6 (n = 50)	17.3 ± 1.8[b] (n = 48)

[a]All data except "[b]" were calculated from the MEPs obtained with facilitation. CMCT-F, calculated using F-wave latencies; CMCT-S, calculated using data obtained by the transcranial magnetic stimulation of spinal roots; APB, abductor pollicis brevis; n, number of subjects in each study; ADM, abductor digiti minimi; FDI, first dorsal interosseous; EDC, extensor digitorum communis.

[b]MEPs were recorded from the relaxed muscles.

More precise peripheral central conduction time can be calculated using F-wave latency instead of spinal root TMS (Fig. 3). Table 3 shows normal values of the latency of the MEPs and CMCT for the muscles used frequently for central conduction studies. Most previous studies obtained MEPs under the facilitation of muscles, that is, slight voluntary contraction, to obtain stable MEPs with constant latencies.

Abnormalities of CMCT have been demonstrated in a variety of neurological disorders. Significant prolongation of CMCT suggests demyelination of central motor pathways, whereas low-amplitude MEPs with little delay or absence of responses are more suggestive of loss of neurons or axons. The CMCT also correlates with the grade of motor deficits after stroke. Measurement of CMCT can provide supporting evidence for the diagnosis of various diseases, and serial measures of CMCT can be used as objective markers of disease progression and prognosis. However, changes in CMCT are not specific for any one particular disease.

5.1.3. Motor Evoked Potentials

The amplitude of the MEP reflects not only the integrity of the corticospinal tract but also the excitability of motor cortex and nerve roots and the conduction along the peripheral motor pathway to the muscles, as the motor threshold. With some lesions or diseases that affect corticospinal tract, abnormal findings will be observed in the MEPs. After stroke lesions, emergence of the MEPs from paretic extremities induced by TMS of the motor cortex of the affected side correlate with a favorable recovery, whereas the absence of MEPs suggests a poor outcome.

Some previous clinical studies with TMS, however, have been addressing mainly the slowing CMCT when evaluating corticospinal tract function. The reduced size of MEPs should reflect the central motor conduction failure, but, at times, their evaluation has been less successful. Even in healthy subjects, the sizes of MEPs can vary among trials for one individual subject as well as among subjects, leading to a broad range of normal values. In addition,

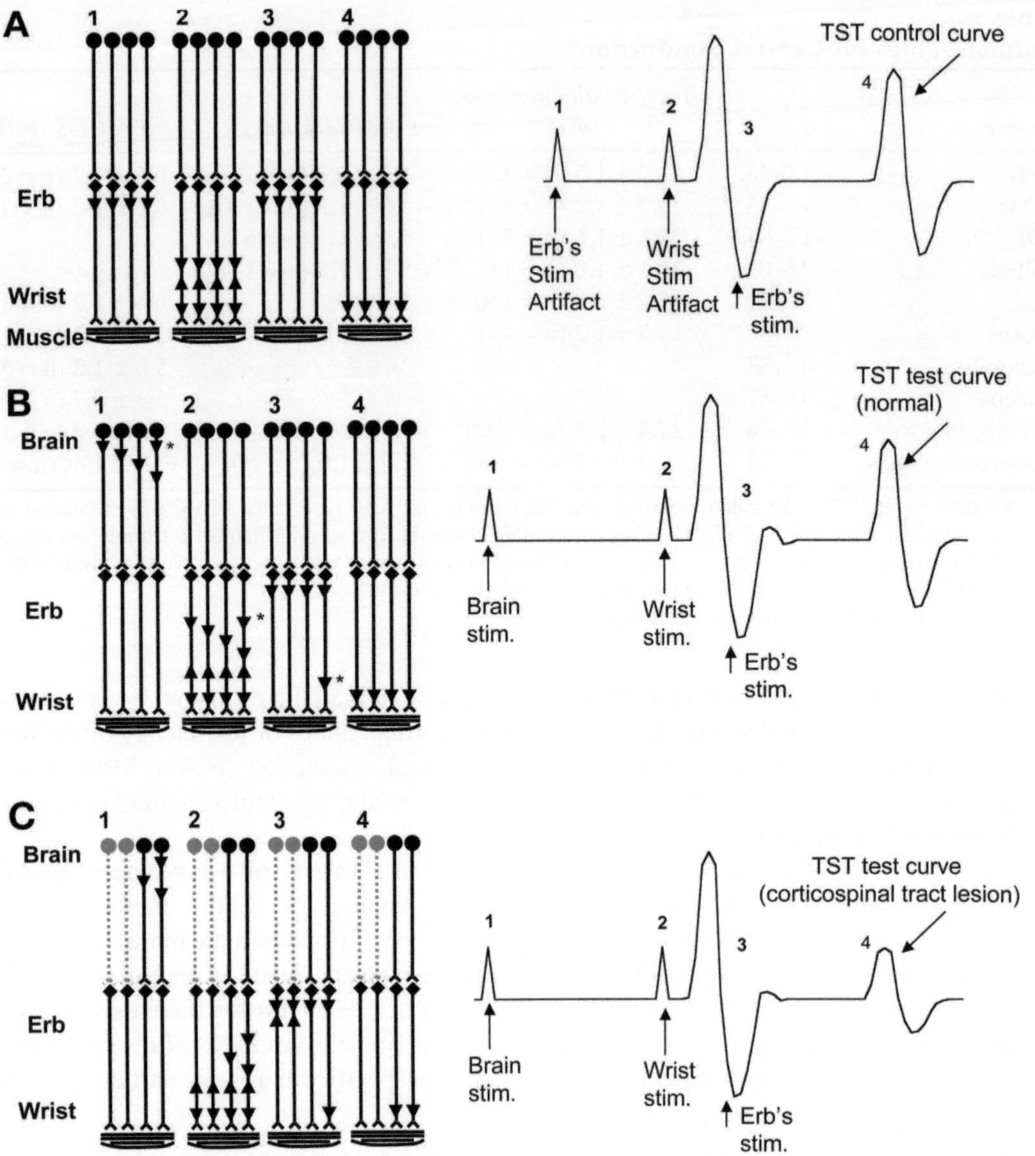

Fig. 4. The principle of the triple-stimulation technique (TST), consisting of a single-pulse transcranial magnetic stimulation (TMS) and peripheral nerve stimuli at Erb's point and the wrist. The size of response to the third stimuli (one at Erb's point) is studied (TST curve). (**A**) The control condition. (**A1**) A maximal stimulus is applied to Erb's point. (**A2**) After a delay, a maximal stimulus is given at the wrist that evokes the first large response (orthodromic action potentials) and collides with descending action potentials and cancels them (antidromic action potentials). (**A3**) After a delay, a maximal stimulus is applied again to Erb's point and (**A4**) a synchronized response is recorded as a control response. (**B**) TST on a healthy subject. In this example, maximal TMS excites all axons. and desynchronization is assumed to occur. (**B1**) Maximal TMS is applied to the motor area and evokes descending volleys with various latencies, including multiple volleys (*). (**B2**) After a delay, a maximum stimulus is given at the wrist and cancels the descending action potentials with various latencies. (**B3**) The second discharge (*) on the axon is not cancelled and causes a small response. (**B4**) After a delay, a maximal stimulus is applied at Erb's point, and the synchronized response

MEPs by TMS are usually much smaller than compound muscle potentials evoked by peripheral nerve stimulation, and TMS does not seem to accomplish depolarization of all spinal motor neurons.

Solving these problems, Magistris et al. (1999) developed a triple-stimulation technique, which has provided a quantitative electrophysiological measurement of the central motor conduction failure (Fig. 4). With their method, applying peripheral stimuli at Erb's point and wrist stimulation as well as TMS, they have induced two collisions in the peripheral motor neurons and have successfully suppressed the desynchronization of MEPs caused by multiple descending volleys evoked by TMS. Their methods demonstrate that TMS can depolarize almost all spinal motor neurons supplying target muscle in healthy subjects. According to their data, the sizes of MEPs by TMS get smaller than those of compound muscle potentials because of "phase cancellation," in which the negative phases of individual motor unit potentials are cancelled by the positive phases of others, because of their various latencies. Their new technique provides new insights into corticospinal tract conduction of healthy subjects and, when applied to patients with corticospinal dysfunction, is 2.75 times more sensitive than conventional MEPs in detecting corticospinal conduction failures. This technique, thus, improves detection and also gives quantitative information on deficits of the corticospinal conduction. Further studies are required, but the triple-stimulation technique will be useful in following clinical courses of patients and be beneficial for assessment and following treatment of disorders affecting central motor conduction.

5.1.4. Silent Period

If a subject is instructed to maintain muscle contraction of a targeted muscle and a single TMS pulse with suprathreshold intensity is applied over the motor cortex representing the targeted muscle, the EMG activity of the muscle will be arrested for a few hundred milliseconds after the MEP (Fig. 5A). This period of EMG suppression is referred to as a "silent period," often defined as the time from the end of the MEP to the return of voluntary EMG activity. It is, at times, difficult to define the end of the MEPs, and some reports define the duration of the silent period as the time interval from stimulus delivery to the return of the voluntary activity. Most of this period is thought to be caused by cortical inhibitory mechanisms of the motor cortex, whereas spinal inhibitory mechanisms, such as Renshaw inhibition, are considered to contribute only to the first 50 or 60 ms of this suppression. The neuronal elements responsible for the silent period are most likely mediated by γ-aminobutyric acid-B receptors. Silent periods of abnormally short or long duration are observed in patients with various movement disorders.

5.1.5. Transcallosal Conduction

TMS of a single pulse delivered to one motor cortex can suppress ongoing voluntary EMG activity in small hand muscles ipsilateral to the side of the TMS. This can be demonstrated by having the subjects maintain the intrinsic hand muscles contracted with more than 50% of

---

Fig. 4. *(Continued)* from all axons excited initially by TMS is recorded as a second response. **(C)** TST on a subject with a partial failure of central conduction. **(C1)** Even maximal TMS could not excite all axons. **(C2–C4)** Because of the conduction failure, only a part of the peripheral nerves is synchronized, and the TST test curve is smaller than that of the TST control curve. Stim, stimulus.

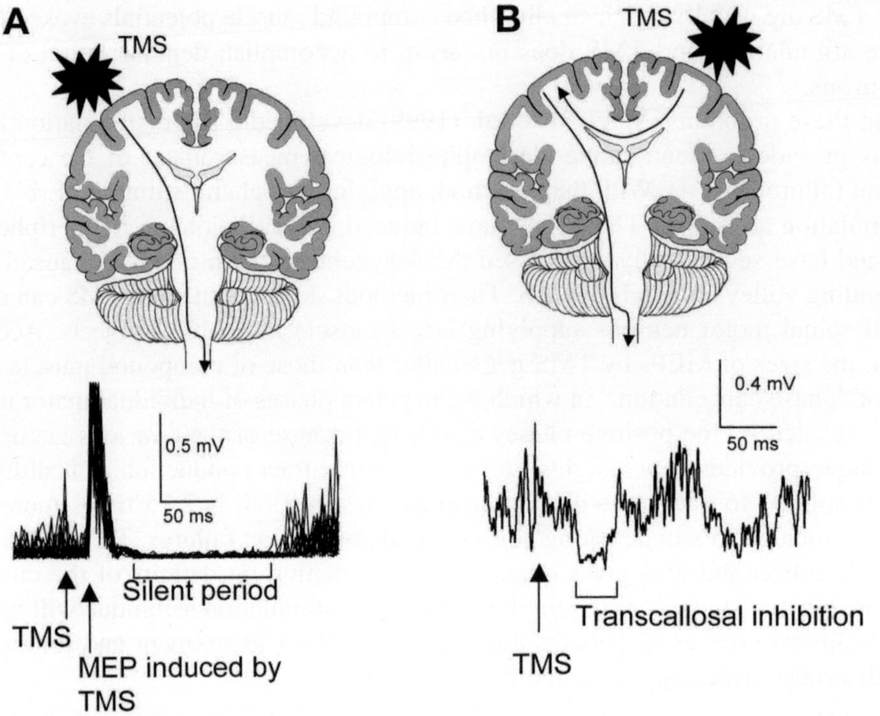

Fig. 5. The effect of transcranial magnetic stimulation (TMS) of the hand motor area of each hemisphere on a voluntarily contracted muscle. Responses are recorded from the first dorsal interosseus muscle (FDI), which is kept contracted with 20% of maximal voluntary force pressing a force transducer. The intensity of TMS is set at 130% of the minimum intensity to elicit motor evoked potentials (MEPs) of 50-µV peak-to-peak amplitude (motor threshold). (**A**) Example of silent period to TMS. Stimulation of the hand motor cortex of one hemisphere elicits MEPs in the contralateral FDI, which is followed by a suppression of tonic voluntary EMG activity. Fifteen consecutive rectified EMG responses are superimposed. (**B**) Transcallosal inhibition. Stimulation of the ipsilateral motor cortex inhibits tonically firing corticospinal neurons in the motor cortex of the unstimulated hemisphere and, thus, produces a transcallosal inhibition of tonic EMG activity in the ipsilateral FDI. Twenty consecutive rectified EMG responses are averaged.

maximum force and applying a single TMS pulse strong enough (80–100% of maximum power of the magnetic stimulator) to the motor cortex ipsilateral to the contracting muscles. This period suppression is thought to be mediated via transcallosal pathways reflecting a function of transcallosal conduction, and is named "transcallosal inhibition" (Fig. 5B). This period of inhibition begins 10 to 15 ms after the minimum corticospinal conduction time to the recorded hand muscle, that is, 30 to 40 ms after TMS, with a duration of approx 30 ms in healthy subjects. In the patients with lesions in the corpus callosum, this transcallosal inhibition is either delayed or absent. For example, multiple sclerosis often involves lesions in the corpus callosum, which can be clinically silent, but is thought to be associated with poor prognostic value regarding cognitive functions. Thus, this transcallosal technique may be worthwhile to study the function of transcallosal conduction in patients with suspected multiple sclerosis, with a support of the neuroanatomic studies with MRI. This TMS method can be associated with the paired-pulse TMS technique to investigate interhemispheric interactions further (*see* Section 5.2.2.).

5.2. *Paired-Pulse TMS*

5.2.1. *Paired-Pulse TMS to Examine Intracortical Inhibitory and Facilitatory Mechanisms*

Combining a subthreshold conditioning stimulus with a suprathreshold test stimulus at different short (1–20 ms) interstimulus intervals through the same TMS coil (Fig. 6A), it is possible to study inhibitory and facilitory interactions in the cortex and examine the functional integrity of intracortical neuronal structures. The effects of the conditioning TMS on the size of test MEP depend on the stimulus intensity and the interstimulus interval. In a number of previous studies using this paired-pulse TMS paradigm, the conditioning TMS is most commonly set at 60 to 80% of the resting motor threshold because the conditioning TMS with this intensity can induce stable effects on the test MEP and will not cause any descending volleys, avoiding any modification of the excitability at the spinal level. If a conditioning stimulus is set at 60 to 80% of the resting motor threshold, maximum inhibitory effects are found at short interstimulus intervals of 1 to 4 ms. The maximum amount of this inhibition is usually 20 to 40% of the test MEP. Facilitory effects of the conditioning TMS pulse with the same subthreshold intensity onto the test MEP can be observed at intervals of 7 to 20 ms. The magnitude of this facilitation is very variable across subjects, from 120 to 300% of the test MEPs. The amount of these effects can vary depending on the size of test MEPs, and, thus, for this paired-pulse TMS study, it is critical to maintain the relaxation of patients and adjust the size of test MEPs across trials and subjects. Previous studies, applying this method on the motor representation of various muscles, showed that these phenomena of intracortical inhibition and facilitation are very similar for intrinsic hand muscles, lower face, leg, or proximal arm muscles, indicating that these intracortical mechanisms are similar across different motor representations. The intracortical inhibition and facilitation measured by this technique are induced by separate mechanisms and their effects seem to originate at the motor cortical level, but not at subcortical or spinal levels.

This paired-pulse technique has also been used to examine the effects of CNS-active drugs on the human motor cortex. In this context, paired-pulse TMS might be useful in selecting the best-suited medication for a given patient by matching the identified abnormality in a given disorder with the effects of different pharmaceutical agents. Such a neurophysiology-based approach to medication selection in epilepsy or psychosis would certainly be desirable, although systematic studies are needed.

Paired-pulse TMS can also reveal changes of the excitability in the motor cortex induced by various neurological and psychiatric diseases (Table 2). These results, however, appear to be nonspecific for each disease. For example, essentially the same abnormalities in the paired-pulse curve can be seen in dystonia and idiopathic Parkinson disease (PD). Furthermore, disorders without clear motor cortex pathology, such as schizophrenia, depression, or obsessive–compulsive disorder have been found to cause alterations of the intracortical inhibition and facilitation, hence, raising further questions regarding the specificity of these findings. Nevertheless, longitudinal studies of the clinical neurophysiology using this paired-pulse TMS technique may well have prognostic significance for neurological and psychiatric diseases.

5.2.2. *Paired-Pulse Paradigm to Examine Interhemispheric Interactions*

The term paired-pulse TMS can also be used to refer to the application of single stimuli to two different brain regions. For instance, interhemispheric interactions and transcallosal conduction

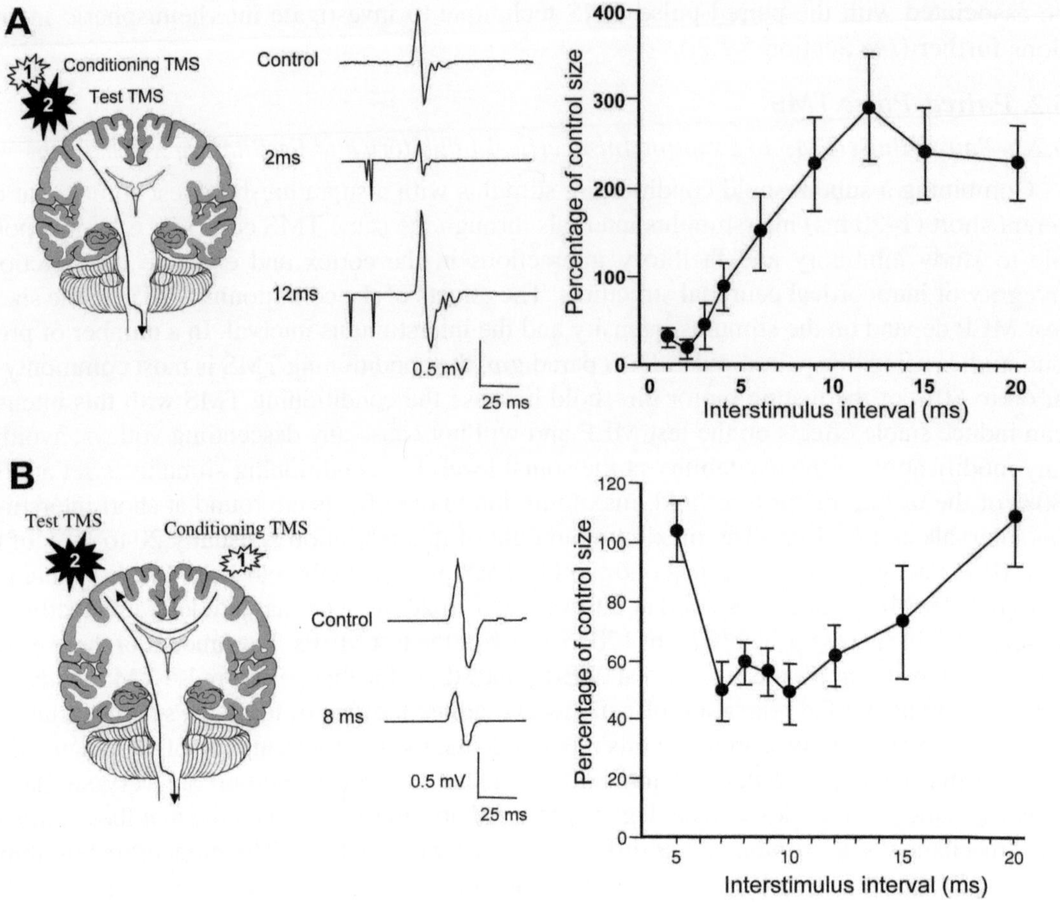

Fig. 6. (**A**) The change of motor evoked potential (MEP) sizes obtained by the paired conditioning–test–stimulus paradigm from the first dorsal interosseus (FDI) muscle. The intensity of conditioning transcranial magnetic stimulation (TMS) was set to 80% of resting motor threshold. The intensity of test TMS was adjusted to produce a control MEP of an average peak-to-peak amplitude of 1 mV when given alone. In the graph, the size of the MEPs is expressed as a percentage of the control, unconditioned MEPs, and plotted against the interstimulus interval. Data are means of eight healthy subjects (mean age, 31.6 ± 4.9 yr). Error bars indicate standard errors. Note that the conditioning stimulus inhibited the test MEP at short interstimulus intervals (1–4 ms), but facilitated the test MEP at longer intervals (6–20 ms). (**B**) Modulation of the amplitude of the MEPs induced in the FDI by TMS of the contralateral motor cortex as a consequence of a conditioning TMS pulse applied to the motor cortex of the opposite hemisphere (ipsilateral to the target FDI). The intensity of the conditioning TMS was set to 110% of resting motor threshold (for induction of MEPs in the contralateral FDI), and the intensity of test TMS was adjusted to produce a control MEP of an average peak-to-peak amplitude of 1 mV when given alone. In the graph, the size of MEPs is expressed as a percentage of the control, unconditioned TMS, and plotted against the interstimulus interval. Data are means of eight healthy subjects (mean age, 31.6 ± 4.9 yr). Error bars indicate standard errors. Note the significant inhibition at the interstimulus interval of 7 to 12 ms.

times between motor cortices in both sides can be examined, applying a conditioning suprathreshold stimulus to one motor cortex and the following test TMS pulse after a short interval (4–30 ms) to the other motor cortex. This paradigm was first introduced by Ferbert et al. (1992), who showed that the cortical excitability of one motor cortex is decreased

7 to 15 ms after suprathreshold TMS of the opposite motor cortex (Fig. 6B). This interhemispheric interaction is influenced by the intensity of the conditioning TMS; the stronger the conditioning TMS, the greater and longer the induced interhemispheric inhibition. In addition to interhemispheric inhibition, Ugawa et al. (1993) have reported early interhemispheric facilitation, which can be observed at interstimulus intervals of 4 to 5 ms, using conditioning TMS of relatively low intensity.

This methodology allows the investigation of interhemispheric interactions in a variety of circumstances. For instance, musical instrumentalists have been shown to have less interhemispheric interactions than nonmusicians. Right-handed people have more significant interhemispheric influence of the right, nondominant side by the dominant side, than in the opposite direction. Patients with cortical myoclonus show no such interhemispheric interactions, suggesting affected transcallosal or cortical inhibitory interneurons. Patients with mirror movements or those recovering from a stroke are likely to show changes in these interhemispheric influences.

5.3. *Repetitive TMS*

Generally, rTMS refers to the application of a train of TMS pulses of the same intensity to a single brain area at a given frequency. The frequency can range from 1 stimulus per second to 20 or more. The higher the stimulation frequency and intensity, the greater is the disruption of cortical function *during* the applied train of stimulation. However, after such immediate effects, a train of rTMS can also induce a modulation of cortical excitability, which may last beyond the duration of the rTMS itself. This effect may range from inhibition to facilitation, depending on the stimulation parameters (particularly frequency of stimulation). Lower frequencies of rTMS with the 1-Hz stimulation applied over the motor cortex can suppress excitability of the targeted motor cortex, whereas 20-Hz stimulation trains seem to lead to a temporary increase in cortical excitability. Although these effects vary among individuals, the effect of low-frequency rTMS is relatively robust and long lasting and can be applied to the motor cortex and to other cortical region to study brain–behavior relations.

The mechanisms of this modulation of cortical excitability beyond the duration of the rTMS train are still unclear. Long-term potentiation and depression of cortical synapses or closely related neuronal mechanisms have been drawn as possible mechanisms to explain the effect of high- and low-frequency rTMS, respectively. Animal studies suggest that modulation of neurotransmitters and gene induction may contribute to these long lasting modulatory effects of rTMS.

The lasting modulation of cortical activity by rTMS is not limited to motor cortical areas. There is also evidence that these long-lasting effects of rTMS can be induced in areas outside the motor cortex and be associated with measurable behavioral effects, including visual, prefrontal, and parietal cortex as well as the cerebellum. This approach, using long-lasting modulation of cortical activity, so-called "off-line effect of rTMS," both locally and along functional neural networks, can be extremely useful to study brain–behavior relations.

5.3.1. *Clinical Use of rTMS*

The lasting, modulatory effects of rTMS on the excitability of the cerebral cortex have also been observed in patients with neurological and psychiatric disorders, leading to behavioral and possibly therapeutic effects. A number of studies on a variety of neurological disorders are providing tantalizing results on applications of rTMS. To establish a clinical therapeutic indication, however, well-controlled, multicenter randomized clinical trials with sufficiently high numbers of patients are required, and it may be worth pursuing trials for some neurological disorders.

The potential clinical application of rTMS for major depression is the most thoroughly studied. Lasting beneficial effects have been seen in approx 40% of patients with medication-resistant depression in recent publications. It is reported that both high-frequency rTMS of the left dorsolateral prefrontal cortex and low-frequency rTMS of the right side can improve depression. It seems that patients with decreased cerebral metabolism may respond better to high-frequency stimulation and those with hypermetabolism may respond better to low-frequency stimulation, which is in line with the frequency-dependant effects of rTMS on the motor cortical excitability.

In PD, sub-motor threshold rTMS at high frequency (5 Hz) to the motor cortex has been shown to improve contralateral hand performance in some studies. However, other studies have not been able to confirm this effect. There are two rationales for rTMS trials in PD:

1. Increasing cortical excitability to thalamocortical drive, which is believed to be lacking in this disease.
2. Modifying catecholamine metabolism subcortically through cortical stimulation.

However, at this point, any beneficial effect of rTMS in PD is unproven and speculative.

Because the physiological studies of task-specific dystonia have suggested hyperexcitability of the motor cortex or a failure of intracortical inhibition, rTMS of the motor cortex at 1 Hz has resulted in a clinical improvement in patients with writers' cramp. In tic disorders, a similarly abnormal increase of cortical excitability has been reported, and 1-Hz rTMS of the motor cortex may reduce the frequency of tics. These effects are transient and not meaningful for all patients, but the data suggest the impaired inhibitory mechanisms in the motor cortex of these patients.

Using low-frequency rTMS, several other studies have shown successful reduction in the frequency of seizures or abnormal movements caused by intractable seizures or cortical myoclonus, although in very small numbers of patients. Similar logic might be applicable to spasticity, intractable neurogenic pain, or schizophrenia, in which suppression of abnormally increased cortical excitability might achieve desirable symptomatic relief.

Outcome after stroke may be favorably influenced by rTMS, suppressing abnormal, maladaptive cortical plasticity and enhancing adaptive cortical activity to promote neurorehabilitation. Some activities in the uninjured brain could reflect beneficial cortical reorganization promoting functional recovery, but some may be maladaptive, generating the behaviors that, if suppressed, would improve functional outcome. Indeed, after stroke, some patients may suffer from some additional symptoms, involving abnormal hyperactivity of the residual brain. Contralesional spatial neglect after stroke is not only caused by the lesion but primarily caused by the hyperactivity of the intact hemisphere and 1-Hz rTMS of the unaffected parietal lobe; suppressing excitability of the intact hemisphere can improve contralesional visuospatial neglect after stroke. In addition, it patients with Broca's aphasia caused by left frontal stroke may improve their naming after 1-Hz rTMS of the right Brodmann's area 45, which is supposedly overactivated in patients with unrecovered, nonfluent aphasia. It is premature to propose these observations as realistic therapeutic applications, but rTMS can show us the property of plastic changes of the cortical circuitry after brain damages and may hint at novel clinical interventions.

SUGGESTED READING

Berardelli A. Transcranial magnetic stimulation in movement disorders. Electroencephalogr Clin Neurophysiol 1999;51(Suppl):276–280.

Chen R, Gerloff C, Classen J, et al. Safety of different inter-train intervals for repetitive transcranial magnetic stimulation and recommendations for safe ranges of stimulation parameters. Electroencephalogr Clin Neurophys 1997;105:415–421.

Ferbert A, Priori A, Rothwell JC, J, et al. Interhemispheric inhibition of the human motor cortex J Physiol 1992;453:525–546

Fitzgerald PB, Brown TL, Daskalakis ZJ. The application of transcranial magnetic stimulation in psychiatry and neurosciences research. Acta Psychiatr Scand 2002;105:324–340.

Gugino LD, Romero JR, Aglio L, et al. Transcranial magnetic stimulation coregistered with MRI: a comparison of a guided versus blind stimulation technique and its effect on evoked compound muscle action potentials. Clin Neurophysiol 2001;112:1781–1792.

Kobayashi M, Pascual-Leone. Transcranial magnetic stimulation in neurology. Lancet Neurology 2003;3:145–156.

Kujirai T, Caramia MD, Rothwell JC, et al. Corticocortical inhibition in human motor cortex. J Physiol 1993;471:501–519.

Magistris MR, Rosler KM, Truffert A, et al. A clinical study of motor evoked potentials using a triple stimulation technique. Brain 1999;122:265–279.

Mills KR. Magnetic Stimulation of the Human Nervous System. Oxford University Press, Oxford, England, 1999.

Pascual-Leone A, Davey N, Wassermann EM, Rothwell J, Puri BK, eds. Handbook of Transcranial Magnetic Stimulation. Arnold, London, UK, 2001.

Pascual-Leone A, Valls-Sole J, Wassermann EM, Hallett M. Responses to rapid-rate transcranial magnetic stimulation of the human motor cortex. Brain 1994;117:847–858.

Rossini PM. Is transcranial magnetic stimulation of the motor cortex a prognostic tool for motor recovery after stroke? Stroke 2000;31:1463–1464.

Rossini PM, Barker AT, Berardelli A, et al. Non-invasive electrical and magnetic stimulation of the brain, spinal cord and roots: basic principles and procedures for routine clinical application. Report of an IFCN committee. Electroencephalogr Clin Neurophysiol. 1994;91:79–92.

Rossini PM, Rossi S. Clinical applications of motor evoked potentials. Electroencephalogr Clin Neurophysiol 1998;106:180–194.

Sparing R, Mottaghy FM, Ganis G, et al. Visual cortex excitability increases during visual mental imagery: a TMS study in healthy human subjects. Brain Res 2002;938:92–97.

Ugawa Y, Hanajima R, Kanazawa I. Interhemispheric facilitation of the hand area of the human motor cortex. Neurosci Lett 1993;160:153–155

Walsh V, Pascual-Leone A. Neurochronometrics of Mind: TMS in Cognitive Science. MIT Press, Cambridge, MA, 2003.

Wassermann EM, Lisanby SH. Therapeutic application of repetitive transcranial magnetic stimulation: a review. Clin Neurophysiol 2001;112:1367–1377.

Weber M, Eisen AA. Magnetic stimulation of the central and peripheral nervous systems. Muscle Nerve 2002;25:160–175.

Index

MIX
Papier aus verantwortungsvollen Quellen
Paper from responsible sources
FSC® C105338

If you have any concerns about our products,
you can contact us on
ProductSafety@springernature.com

In case Publisher is established outside the EU,
the EU authorized representative is:
Springer Nature Customer Service Center GmbH
Europaplatz 3, 69115 Heidelberg, Germany

Printed by Libri Plureos GmbH
in Hamburg, Germany